Fish & Fisheries Series

Volume 45

Series Editor

Kai Lorenzen, University of Florida, Gainesville, USA

The Fish & Fisheries series aims to publish key books on all aspects of fish and fisheries science, conservation and management: biology and ecology of fishes and invertebrates of fisheries interest, fisheries assessment and management, aquaculture, and the social and economic dimensions of aquatic resource use and conservation. For over thirty years, the Fish & Fisheries series has been at the forefront of providing in-depth syntheses of knowledge in this vibrant and fast-moving field. The series is directed at researchers in the fish and fisheries sciences, senior undergraduate and postgraduate students, and at professionals in fisheries, aquaculture and marine and freshwater conservation.

Prospective authors and/or editors are encouraged to consult the Publishing Editor for more details. Comments or suggestions for future volumes are welcome:

Éva Lőrinczi
Publishing Editor
eva.loerinczi@springer.com

Kevin L. Pope • Robert Arlinghaus
Len M. Hunt • Abigail J. Lynch
Brett T. van Poorten
Editors

Understanding Recreational Fishers

Disciplinary and Interdisciplinary Approaches for Fisheries Management

Springer

Editors
Kevin L. Pope
Cooperative Research Units Program
United States Geological Survey
Lincoln, NE, USA

Robert Arlinghaus
Department of Fish Biology, Fisheries
and Aquaculture
Leibniz Institute of Freshwater Ecology
and Inland Fisheries
Berlin, Germany

Division of Integrative Fisheries Management
Humboldt-Universität zu Berlin
Berlin, Germany

Len M. Hunt
Ontario Ministry of Natural Resources
Thunder Bay, ON, Canada

Brett T. van Poorten
School of Resource and Environmental
Management
Simon Fraser University
Burnaby, BC, Canada

Abigail J. Lynch
National Climate Adaptation Science Center
United States Geological Survey
Reston, VA, USA

Brett T. van Poorten (Simon Fraser University)
The open-access fee for this book was primarily provided by Brett van Poorten (Simon Fraser University), with additional support from all co-editors and their respective institutions.

ISSN 2542-890X ISSN 2542-8926 (electronic)
Fish & Fisheries Series
ISBN 978-3-031-99738-9 ISBN 978-3-031-99739-6 (eBook)
https://doi.org/10.1007/978-3-031-99739-6

This Springer imprint is published by the registered company Springer Nature Switzerland AG
The registered company address is: Gewerbestrasse 11, 6330 Cham, Switzerland

Managing fisheries involves the interaction of fish, fishers, and management institutions. Each component is often studied in isolation; true understanding comes from an inter- or trans-disciplinary perspective. Picture credit goes to Florian Möllers/AVN

Preface

Objectives

Despite more than 50 years of research into the human dimensions of recreational fisheries, there is no textbook to present the theoretical grounding, operationalisation, and interpretation of the most elemental social components involved in fisheries management – namely, outcomes and trade-offs, behaviours (and antecedents or predictors of it), and the relationships among actors (fishers, fisher groups, managers, and management organisations). The objective of this edited volume is to reduce this knowledge gap. The book comprises four parts: (I) recreational fisheries as social-ecological systems, (II) disciplinary views of fishers' behaviours and outcomes, (III) engagement of fishers to support effective social science, and (IV) interdisciplinary integration for recreational fisheries management. *Part I* provides the general motivation for an interdisciplinary approach to understand recreational fisheries as coupled systems. *Part II* provides the foundations of various social science disciplines and highlights linkages and differences in the underlying theories and concepts used. *Part III* describes various qualitative and quantitative methods to study the social aspects of recreational fisheries in more detail, and *Part IV* suggests approaches to achieve standardisation and integration of different data streams, including how to link social and ecological data. The book chapters have been written by leading scholars of recreational fisheries science, involving social and economic scientists, political ecologists, and fisheries ecologists with a tradition in either applying human dimensions or working towards integration.

Framing

Recreational Fisheries as Coupled Social-Ecological Systems

Part I describes the foundation of our thinking of recreational fisheries as complex social-ecological systems. In fact, the inextricable linkages of fishers and fish stocks provide the opportunity for the social-ecological system of recreational fishing to serve as an empirically tractable model for understanding fundamental general drivers inherent in human-environment relationships (Carpenter and Brock 2004; Brock and Carpenter 2007; Roth et al. 2007; Biggs et al. 2009; Horan et al. 2011; Hunt et al. 2011, 2013; Ward et al. 2016; Arlinghaus et al. 2017; Solomon et al. 2020). Social dimensions exist within all key structural elements of the social-ecological systems as well as in the formation and evaluation of goals and objectives for fisheries; *Part I* is intended to address these key structural elements. Note that although this edited volume focuses on the behaviour of recreational fishers, policymakers and managers are also relevant actors that can be studied with the same methods, as well as the interactions between the social and ecological sub-systems through organisations and the rule systems (institutions) invented by humans to guide behaviours and achieve desired outcomes.

Disciplinary Views

It is trite to say that fisheries management is mainly about managing fishers and their expectations and behaviours (Hilborn 2007). Even so, this truism, which extends to recreational fisheries (Hunt et al. 2013), helps to explain the many calls by fisheries professionals and researchers to better understand these fishers, their behaviours, and more generally the human dimensions of recreational fisheries (Moeller and Engelken 1972; Knopf et al. 1973; Diver and Knopf 1976; Hampton and Lackey 1976; Voiland and Duttweiler 1984; Aas and Ditton 1998; Ward et al. 2016; Arlinghaus et al. 2017; Weber et al. 2019). Though understanding fishers and their behaviours is important, specific guidance on exactly what to understand is often unclear and tends to be shaped by one's disciplinary lens (e.g., Ditton 1996). Interdisciplinary inquiry proceeds through disciplines, which often are the theoretical and methodological engine of scientific progress. Yet, individuals who are not trained in the social sciences may struggle to understand the research paradigms and approaches of different social science disciplines. *Part II* is meant to address this gap by providing foundational content to help readers understand the key disciplines that focus on recreational fishers and best practices for conducting disciplinary-specific research. The chapters in *Part II* look at facets of behaviour when we present the various social science disciplines that are involved in studying aspects of the human dimensions of recreational fisheries, and we attempt to outline the foundations of each discipline, with a view of helping the user (e.g., student or professional) to move towards interdisciplinary inquiry of recreational fisheries or to use specific disciplines to address specific social questions. These chapters do not present specific analytical approaches within each discipline. Rather, the structure highlights the foundations of the social science disciplines that are applied to recreational fisheries.

Methods of Social Sciences in Recreational Fisheries

Researchers from each discipline hold different assumptions about how knowledge is generated and what methods are accepted to produce knowledge. Consequently, recreational fisheries researchers use a variety of methods: (i) to collect data from fishers and fisheries; (ii) to design and analyse research; and (iii) to synthesise data and information. In *Part III*, the various methods used to gather information on fisher behaviour are compiled. *Part III* also attempts to outline the linkages in the different methodological approaches that are used by social science researchers in recreational fisheries.

Interdisciplinary Integration for Recreational Fisheries Management

Evaluating recreational fisheries as complex social-ecological systems can improve understanding of local, regional, and global fisheries dynamics, and will inform a much broader scientific literature on general human-environment interactions. *Part IV* of this book aims to further interdisciplinary work on recreational fisheries specifically and complex social-ecological systems generally.

Outlook

The ultimate intention of this book is that readers will fully embrace recreational fisheries as complex, adaptive social-ecological systems that most often need to be studied through interdisciplinary approaches for better management of these important resources. This is not intended to conclude that disciplinary approaches to recreational fisheries are fruitless. Quite to the contrary—disciplinary approaches have a key role and will continue to do so. Yet, what this book attempts to drill down is that various disciplines of the social sciences exist next to each other, that each of the disciplines and their methods are of value, and that, depending on the question or practical problem that is to be solved, reliance on one or several approaches within interdisciplinary projects is advisable.

The book is particularly directed at students, many of whom have a biological or ecological background, to appreciate what social sciences can offer to understand recreational fisheries dynamics and how to link social and ecological sciences within interdisciplinary inquiry. The book also makes an important claim that stakeholders and rights holders are important knowledge holders whose integration in research can provide much needed context for solving the pressing management issues faced by our community.

The following epitaph may be read as a call to embrace interdisciplinary inquiry and to appreciate what different disciplinary approaches from economics, psychological, and political ecology have to offer to understand the dynamics of recreational fisheries (inspired by Larkin 1977).

> Here lies the single disciplinary focus,
> It advocated for a specific way of doings,
> And didn't spell out alternative workings,
> We challenge it with the best of wishes,
> Especially on behalf of recreational fishers and fishes.
> We suggest what will take its place,
> Multi- and interdisciplinarity will be that space.

Lincoln, NE, USA	Kevin L. Pope
Berlin, Germany	Robert Arlinghaus
Thunder Bay, ON, Canada	Len M. Hunt
Reston, VA, USA	Abigail J. Lynch
Burnaby, BC, Canada	Brett T. van Poorten

References

Aas Ø, Ditton RB (1998) Human dimensions perspective on recreational fisheries management: implications for Europe. In: Hickley P, Tompkins H (eds) Recreational fisheries: social, economic, and management aspects. Fishing News Books, London, pp 153–164

Arlinghaus R, Alós J, Beardmore B, Daedlow K, Dorow M, Fujitani M, Hühn D, Haider W, Hunt LM, Johnson BM, Johnston F, Klefoth T, Matsumura S, Monk C, Pagel T, Post JR, Rapp T, Riepe C, Ward H, Wolter C (2017) Understanding and managing freshwater recreational fisheries as complex adaptive social-ecological systems. Rev Fish Sci Aquac 25(1):1–41

Biggs R, Carpenter SR, Brock WA (2009) Turning back from the brink: detecting an impending regime shift in time to avert it. Proc Natl Acad Sci 106(3):826–831

Brock WA, Carpenter SR (2007) Panaceas and diversification of environmental policy. Proc Natl Acad Sci 104(39):15206–15211

Carpenter SR, Brock WA (2004) Spatial complexity, resilience, and policy diversity: fishing on lake-rich landscapes. Ecol Soc 9(1):8

Ditton R (1996) Human dimensions in fisheries. In: Ewert AW (ed) Natural resource management: the human dimension. Westview Press, Inc. and Routledge, New York

Driver BL, Knopf RC (1976) Temporary escape: one product of sport fisheries management. Fisheries 1(2):21–29

Hampton EL, Lackey RT (1976) Analysis of angler preferences and fisheries management objectives with implications for management. Proc Ann Conf Southeastern Association Game Fish Comm 29:310–316

Hilborn R (2007) Managing fisheries is managing people: what has been learned? Fish Fish 8:285–296

Horan RD, Fenichel EP, Drury KL, Lodge DM (2011) Managing ecological thresholds in coupled environmental-human systems. Proc Natl Acad Sci 108(18):7333–7338

Hunt LM, Arlinghaus R, Lester N, Kushneriuk R (2011) The effects of regional angling effort, angler behavior, and harvesting efficiency on landscape patterns of overfishing. Ecol Appl 21: 2555–2575

Hunt LM, Sutton SG, Arlinghaus R (2013) Illustrating the critical role of human dimensions research for understanding and managing recreational fisheries within a social-ecological system framework. Fish Manag Ecol 20:111–124

Knopf RC, Driver BL, Bassett JR (1973) Motivations for fishing. Trans North Am Wildlife Natl Resourc Conf 38:191–204

Larkin PA (1977) An epitaph for the concept of maximum sustained yield. Trans Am Fish Soc 106(1):1–11

Moeller GH, Engelken JH (1972) What fishermen look for in a fishing experience. J Wildl Manag 36:1253–1257

Roth BM, Kaplan IC, Sass GG, Johnson PT, Marburg AE, Yannarell AC, Havlicek TD, Willis TV, Turner MG, Carpenter SR (2007) Linking terrestrial and aquatic ecosystems: the role of woody habitat in lake food webs. Ecol Model 203:439–452

Solomon CT, Dassow CJ, Iwicki CM, Jensen OP, Jones SE, Sass GG, Trudeau A, van Poorten BT, Whittaker D (2020) Frontiers in modelling social-ecological dynamics of recreational fisheries: a review and synthesis. Fish Fish 21:973–991

Voiland MP, Duttweiler MW (1984) Where's the humanity? A challenge and opportunity for the fisheries community. Fisheries 9(4):10–12

Ward HG, Allen MS, Camp EV, Cole N, Hunt LM, Matthias B, Post JR, Wilson K, Arlinghaus R (2016) Understanding and managing social-ecological feedbacks in spatially structured recreational fisheries: the overlooked behavioral dimension. Fisheries 41(9):524–535

Weber CT, Borit M, Aschan M (2019) An interdisciplinary insight into the human dimension in fisheries models. A systematic literature review in a European union context. Front Mar Sci 6: 369

Acknowledgments

Dedication is the hallmark of an edited book such as ours. We are appreciative of the investments and efforts expended by our distinguished teams of authors. We asked the lead authors to assemble author teams composed of more than three individuals from multiple organisations to encompass as much diversity—discipline, expertise, geographic location of work, gender, race, age—as possible in authorship, and they all embraced and exceeded our expectations.

Peer review is the heart of scientific publication. We engaged an illustrious and insightful set of reviewers to provide the scientific rigour and peer review desired in this volume, and this book is better for their efforts. In general, we sought to obtain two reviews for each chapter, one from a co-author of a different chapter in hopes of generating greater cohesion among chapters and one from an individual external to this book. Reviewers, listed in alphabetical order, were: Marcy Anderson, Nathan Bennett, Brock Bergseth, Matt Burgess, Ed Camp, Steve Carpenter, Christopher Chizinski, Elizabeth Golebie, Alan Graefe, Jon Helge Volstad, Chris Henderson, Jorge Holzer, Brian Irwin, Olaf Jensen, Brett Johnson, Fiona Johnston, Michelle Jones, Mark Kaemingk, Sophia Kochalski, Rebecca Krogman, Gerard Kyle, Michael Lacy, Richard Ladle, Adam Landon, Kirsten Leong, Philip Loring, Loren McClenachan, Julia McQuillan, Norman Mercado-Silva, Matthew Navarro, Erika Palmer, Andrew Paul, Daniel Phaneuf, Bill Pine, Carsten Riepe, Kelly Robinson, Jens Rommel, Karina Ryan, Valerio Sbragaglia, Michael Sorice, Susan Steffen, Stian Stensland, Harry Strehlow, Patrick J. Sullivan, Stephen Sutton, Ruth Thurstan, Jerry Vaske, John Whitehead. We thank them immensely for their help.

Robert Arlinghaus and Len Hunt are particularly grateful for our key academic mentor in the human dimensions of recreational fisheries, specifically Wolfang Haider (RIP), and acknowledge the collective work with a range of colleagues across the world that has shaped how we think about the social side of recreational fisheries, specifically Robert B. Ditton, Steve Sutton, Gerard Kyle, Øystein Aas, Eli Fenichel, Josh Abbott and our "HD" PhD students and postdocs Malte Dorow, Ben Beardmore, Philipp Freudenberg, Marie Fujitani, Carsten Riepe, and Max Birdsong.

We are grateful to Éva Lőrinczi and Bibhuti Sharma (Springer Nature) for their patience and guidance through the publication process as well as the Fish and Fisheries Series' Editor, Kai Lorenzen. We thank Susan Pope for countless hours of assistance formatting chapter manuscripts. The Nebraska Cooperative Fish and Wildlife Research Unit is jointly supported by a cooperative agreement among the U.S. Geological Survey, the Nebraska Game and Parks Commission, the University of Nebraska, the U.S. Fish and Wildlife Service, and the Wildlife Management Institute. Any use of trade, firm, or product names within this volume is for descriptive purposes only and does not imply endorsement by the U.S. Government. The open-access fee for this book was paid for by the individual co-editors and their respective institutions. RA acknowledges the financial support to the open-access fees by the Leibniz Institute of Freshwater Ecology and Inland Fisheries and by the Humboldt-Universität zu Berlin.

Contents

Part I
Recreational Fisheries

Complexity and Integration of Recreational Fisheries

Abigail J. Lynch, Len M. Hunt, A. Ben Beardmore, Brett T. van Poorten, Kevin L. Pope, and Robert Arlinghaus

Abstract Recreational fisheries are interconnected, complex, adaptive systems characterized by multiple direct and indirect interactions among ecological and human subsystems. This is important for many reasons, including that feedbacks between the social and ecological dimensions lead to difficult-to-predict, often entirely unexpected, outcomes and because many management and governance systems have multiple objectives that can involve social (e.g. fisher satisfaction), economic (e.g. license revenue), and ecological (e.g. fish conservation) dimensions. Embracing a social-ecological-system perspective can usher in an improved era of recreational fisheries science and management. Interdisciplinary approaches that unite experts across disciplines (e.g. social and ecological sciences) to create a unique theoretical, conceptual, and methodological identity are needed to gather crucial information from recreational fishers, quantify and predict fisher behaviours and outcomes from these behaviours, and integrate these findings into fisheries

A. J. Lynch (✉)
National Climate Adaptation Science Center, U.S. Geological Survey, Reston, VA, USA
e-mail: ajlynch@usgs.gov

L. M. Hunt
Ontario Ministry of Natural Resources, Thunder Bay, ON, Canada

A. B. Beardmore
Environmental Analysis and Sustainability Program, Wisconsin Department of Natural Resources, Madison, WI, USA

B. T. van Poorten
School of Resource and Environmental Management, Simon Fraser University, Burnaby, BC, Canada

K. L. Pope
Cooperative Research Units Program, United States Geological Survey, Lincoln, NE, USA

R. Arlinghaus
Department of Fish Biology, Fisheries and Aquaculture, Leibniz Institute of Freshwater Ecology and Inland Fisheries, Berlin, Germany

Division of Integrative Fisheries Management, Humboldt-Universität zu Berlin, Berlin, Germany

K. L. Pope et al. (eds.), *Understanding Recreational Fishers*, Fish & Fisheries Series 45, https://doi.org/10.1007/978-3-031-99739-6_1

management. In this chapter, we lay the conceptual foundation for recreational fisheries as coupled social-ecological systems that are also complex and adaptive, and discuss the interdisciplinary approach to operationalize this book's vision.

Keywords Adaptive systems · Complex systems · Coupled social-ecological systems · Fisher behaviours · Fisher outcomes · Fishery science

1 Recreational Fisheries

Recreational fishing is the act of attempting to catch fish or other aquatic animals with leisure as the primary motivation as opposed to meeting basic nutritional needs or sustaining livelihoods (FAO 2012). *Recreational fisheries* are the systems in which recreational fishing occurs. This fishing occurs with many gear types (e.g. gill nets, cast nets, spears, bows), but hook-and-line (i.e. angling) is most prevalent across the globe. It is also important to note that these definitions are somewhat opaque and that recreational fisheries can surely make substantial contributions to both food and livelihoods (e.g. Embke et al. 2022; Lynch et al. 2024; Nyboer et al. 2022).

Today, recreational fisheries are a major use of wild fish in freshwaters and many coastal areas, particularly in high-income countries (Arlinghaus et al. 2021). Recreational fisheries are socially and economically important (Weithman 1999; FAO 2012; Tufts et al. 2015; Arlinghaus et al. 2019), and bestow benefits (i.e. ecosystem services, Pope et al. 2016) to more than 220 million people globally (Arlinghaus et al. 2019). These benefits include nutrition and food, jobs dependent on expenditures by fishers, physiological and physical health benefits from nature contact (Griffiths et al. 2017), development of aquatic stewardship (Shephard et al. 2022), investments and motivation for conservation (Granek et al. 2008), and other social-psychological benefits enjoyed by participating in recreational fishing (Weithman 1999). Some of these benefits can be quantifiably measured using metrics such as utility and welfare (Perman et al. 2003), jobs generated (Tufts et al. 2015), and satisfaction of recreational fishers (Birdsong et al. 2021).

Management of recreational fisheries is focused on habitat, biota, and human users (Hubert and Quist 2010) and the interactions of the three elements (FAO 2012; Arlinghaus et al. 2016, 2017). Unfortunately, management of recreational fisheries often occurs in isolation from other important uses of aquatic systems (e.g. other water-based recreation, commercial fisheries, coastal development, hydroelectricity generation, drinking water, navigation, agriculture, flood control, and other ecosystem services, MEA 2005). Sometimes, the management objectives and associated activities (see Barber and Taylor 1990; Krogman et al.2026) tailored to recreational fisheries can even be in opposition with other uses and users (Lynch et al. 2017). To manage recreational fisheries sustainably, managers can acknowledge existing externalities and operate within their sphere of influence (Carpenter et al. 2017) through iterative improvement in their understanding of habitat, biota, and recreational fishers' behaviours and outcomes across disciplinary and sectorial divides (Arlinghaus et al. 2017).

The application of social-ecological-system thinking in recreational fisheries (Hunt et al. 2013; Arlinghaus et al. 2017) is now widely embraced and continues to grow (see review in Solomon et al. 2020). Yet, even still, independent theories and approaches of the natural and the social sciences applied to fisheries have hindered adoption of a social-ecological-system approach within fisheries science and management (Fenichel et al. 2013a; Arlinghaus et al. 2017). Historically, fisheries science has focused on understanding biota and habitat (principally via fisheries ecology, see Post et al. (2026)) with much less attention paid to understanding people and their behaviours (via social psychology, resource economics, political ecology, and other human-dimension studies, Arlinghaus et al. (2008b), Hunt et al. (2013)). It is important to note that these social sciences are all distinct disciplines where siloed thinking can also constrain cross-referencing and integration of knowledge of recreational fisheries (Fenichel et al. 2013a).

To advance this social-ecological-system approach, biologically and ecologically trained professionals—the group of professionals who dominate the science and management cadres of recreational fisheries (Barber and Taylor 1990)—need to understand and value the strength, diversity, and limits of various social science approaches to understand people's attitudes (i.e. mindset, outlook, and feelings toward someone or something that are typically reflected in behaviours; degrees to which people have favourable or unfavourable evaluations of specific behaviours), norms (i.e. shared standards of acceptable behaviours; perceived social pressures to perform—or to not perform—specific behaviours), and behaviours (i.e. range of actions and mannerisms made by individuals or organizations; ways in which people act or respond to a given situation) as well as the social and economic outcomes of recreational fishing. Likewise, those trained in social science disciplines need to recognize that siloed thinking constrains cross-referencing and integration of knowledge within the social sciences of recreational fisheries (Fenichel et al. 2013a). This book explicitly focuses on the social side of recreational fisheries and the utility of interdisciplinary approaches: common blind spots in fisheries professional training. It explains the backbone, methodological approach, and future of the social sciences of recreational fisheries, with the intention of providing a common understanding of social-scientific approaches to improve the integration of natural and social sciences within that framework. Ultimately, the aim is to support the continued integration of the ecological and social sciences for more sustainable management of recreational fisheries.

2 Social-Ecological-Systems Framework

We define a social-ecological system as one or more ecological subsystems interacting with one or more mutually dependent social, socio-economic, or socio-technical subsystems (Berkes et al. 2003; Folke et al. 2005; Ostrom 2007;

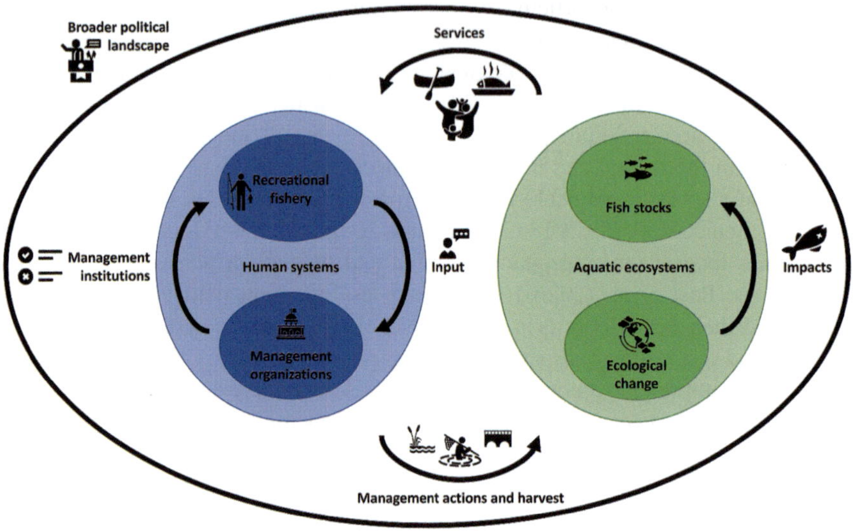

Fig. 1 A basic depiction of a recreational fishery as a coupled social-ecological-system

Arlinghaus et al. 2017; Colding and Barthel 2019). Note that "social" is used here exclusively with reference to humans as opposed to the social dynamics in ecological systems (e.g. collective behaviour of fish). The multiple direct and indirect feedbacks among the social and ecological systems are key to understanding difficult-to-predict system outcomes in recreational fisheries (e.g. sustainability, resilience), and they shape system trajectories and responses to interventions (Ward et al. 2016). Very often, the bidirectional interactions of social and ecological systems in recreational fisheries are directly impacted by the institutions (i.e. rules, norms, and strategies and tactics; Ostrom 1990) that regulate, and respond to, the social and ecological systems or by higher-order related ecosystems or political systems (Fig. 1).

A unifying framework for the study of social-ecological systems (also referred to as human-environmental systems or coupled human and natural systems in the literature) was first proposed by Berkes and Folke (1998) to improve resilience and sustainability of natural systems that are interdependent and co-evolving with human systems. The social-ecological-system framework that explicitly refers to human behavioural feedbacks builds on the foundation of earlier work in bioeconomics (e.g. Gordon 1954; McConnell and Sutinen 1979) and human dimensions of fish and wildlife (e.g. Hendee 1974; Lee and Driver 1998) recognizes interdependencies between social and ecological dimensions of exploited natural resource systems. Ostrom (2007, 2009) helped interpret recurrent patterns within these systems (see Arlinghaus et al. (2017) for an application of the Ostrom social-ecological-systems framework to recreational fisheries). Many studies and frameworks have followed from different research traditions to enhance or sharpen the integrated framework for particular contexts (Anderson 1993; Ditton 2004; Liu et al. 2007; Schlüter et al. 2012; Colding and Barthel 2019; Abbott et al. 2022).

3 Recreational Fisheries as Complex, Adaptive Social-Ecological Systems

Recreational fisheries constitute a prime example of a coupled social-ecological system (Carpenter and Brock 2004; Hinkel et al. 2014, 2015; Schlüter et al. 2014) where fish are directly (e.g. harvest, stocking) or indirectly (e.g. catch-and-release, habitat management) impacted by one or more highly mobile user types (Johnston et al. 2010; Post 2013) and associated management agency (Fig. 1). The economic welfare (i.e. the economists' approach to wellbeing) of heterogenous recreational fishers are impacted by fish through their utility contributions for catch rates, species, sizes, and other catch-related attributes of the fishery (Hunt et al. 2019). Managers aim to achieve objectives for the fishery and are impacted by changes in the habitats, the fish, and the fishers as well as the broader socio-political climate and broad ecological changes (Carpenter et al. 2017). Beyond how recreational fishers and managers perceive and behave, an important aspect of a social-ecological-system framework is also how organizations of actors devise rule-based institutions that affect both fish (through stocking and habitat changes, van Poorten et al. (2011), Ziegler et al. (2017), Radinger et al. (2023)) and fishers (through harvest regulations and restrictions, Aas et al. (2000), Cox et al. (2003)).

Social-ecological systems are one example of complex adaptive systems that are characterized by multiple nonlinear feedbacks acting across heterogeneous actors functioning over multiple spatial- and temporal scales (Levin et al. 2013; Arlinghaus et al. 2017). Complex, adaptive systems constantly evolve and are very difficult to understand, partly because micro-level interactions (e.g. among recreational fishers and local lakes) can create emergent macro-level outcomes (e.g. overfishing at the landscape) that are impossible to be understood at the local level alone. This is because local-level interactions create (emergent) outcomes at broader levels that then change the selection environment for local-level interactions. Complex, adaptive systems may be hierarchically organized at multiple spatial, temporal, or organizational scales (e.g. individual fish within populations or stocks, nested in lakes nested within a landscape of lakes). Interactions and feedback among hierarchies within a recreational fishery can be adaptive to actions by actors at all scales (e.g. agencies or recreational fishing organizations) as well as to changes in fish populations or the landscape (Gunderson and Holling 2001; Allen et al. 2014).

A fundamental question that demands thorough scientific understanding is how to steer complex, adaptive social-ecological systems towards a politically desired state given the multitude of complex feedbacks and outcomes (Arlinghaus et al. 2017). A central tenet proposed here and elsewhere (Biggs et al. 2012; Arlinghaus et al. 2019) is to explicitly manage relevant feedbacks to keep a recreational fishery in a desired state (i.e. increase resilience) and prevent it from flipping to an undesired one (i.e. new stable state where target fish(es) are no longer available). This is a fundamentally new perspective and challenge because managers are accustomed to managing site or fishery-specific outputs generated by localized interactions (e.g. yield, catch rate, fisher satisfaction) or simply fish population states (e.g. spawning stock biomass) in selected ecosystems (Post et al. 2026) rather than focusing on managing feedbacks generating outcomes and states. A focus on managing feedback, however, demands a mechanistic understanding of how recreational fishers (i.e. people) respond to changes—a core tenet of this book.

4 Understanding Recreational Fishers' and Managers' Behaviours Through Various Social Science Disciplines

To understand a social-ecological system as complex and adaptive, variability in behaviours (for both fish and human subsystems), heterogeneity (e.g. among actors and ecological systems), relationships and considerations of power and privilege among fishers and non-fishers, and the broader implications of micro-level localized interactions should be included. Properly representing human behaviour (e.g. of recreational fishers and managers) or other behaviour-related outputs (e.g. support of management policies by recreational fishers) is the focus of this book. Different disciplines approach this need from different foundations with different focal points (Hunt et al. 2026). They vary in how recreational fisher behaviour and outcomes are conceptualized and measured (Fig. 2), what research questions are asked, which

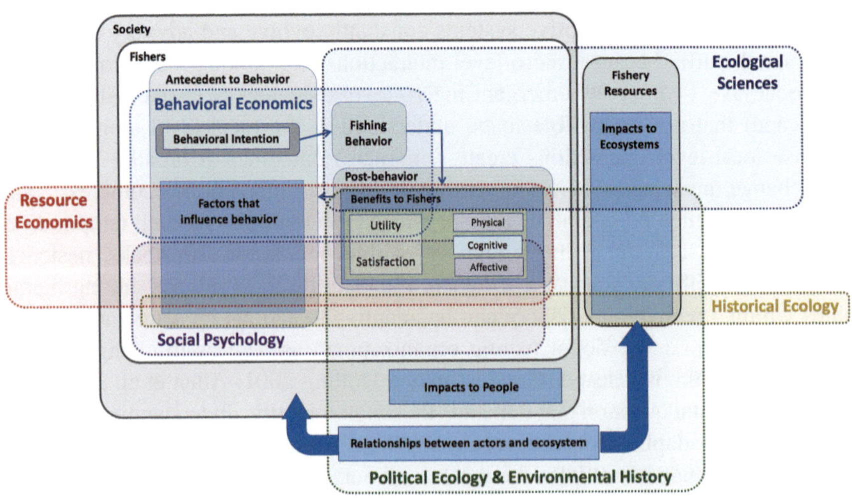

Fig. 2 The relationship of different disciplinary approaches covered in this book. (Modified from Beardmore (2013))

concepts are used to address questions, and how questions and concepts are related. Disciplines also vary if they focus on the individual fisher or on patterns and processes leading to social outcomes (e.g. power structures). When examined in detail, the various social science disciplines that have dominated recreational fisheries science to date complement each other well (Fig. 2) and will in turn be presented in greater detail in other chapters in this book. A brief summary of the similarities and differences of various key social science disciplines follows to motivate the structure and organization of the various chapters in this book. Of course, our coverage of disciplines here is incomplete, and some important disciplines such as sociology or institutional economics, are not mentioned below. These and other disciplines have contributed and continue to contribute towards understanding recreational fishers and managing recreational fisheries.

Historically, the dominant research paradigm for recreational fisheries has been a rational actor model that assumes people who engage in fishing make optimal choices to maximize personal benefits (Hendee 1974; Driver 1985; Anderson 1993). In social psychology (Landon et al. 2026), personal benefits are expected psychological outcomes from fishing (Driver 1985; Manfredo et al. 1996) and may be measured as satisfaction, which has physical, cognitive, and affective components (Fig. 2; Birdsong et al. 2021). Social psychologists focus on how antecedents (e.g. values, beliefs, attitudes, personal norms) additively or synergistically influence intentions and behaviours (e.g. the theory of reasoned action, Ajzen and Fishbein (1980); the cognitive hierarchy model, Fulton et al. (1996)). This compositional approach to understand behaviours (termed this because the individual cognitions are composed to affect behaviour, Parkkila et al. (2010)) contrasts with the decompositional approach preferred by resource economists (sensu Lancaster 1966). Economists also favour a rational actor model, which assumes people make decisions that maximize personal utility (Fig. 2). However, they typically associate observed or stated behavioural choices (e.g. selecting a fishing site) with the attributes describing those choice options (e.g. different lakes with different catch rates or harvest regulations). Using statistical techniques, they in turn decompose the behavioural choices to learn about the preferences of people for attributes of fishing and to quantify the economic benefits from changes to an activity such as recreational fishing (Melstrom et al. 2026). Economics thus typically directly studies the behavioural choices of fishers, whereas social psychologists study the antecedents that form behavioural intentions and ultimately shape behaviours. Despite over 40 years of research on recreational fisheries and following the same rational actor idea, social psychologists and resource economists have used different units for measuring similar constructs (e.g. satisfaction or utility to measure wellbeing), have published in different journals, have shown little cross-referencing, and have had different research motivations (Fenichel et al. 2013a, see also Hunt et al. 2026). Furthermore, although behavioural economics, with its efforts to link economic and psychological theories, offers much promise to understand the behaviours of recreational fishers, few applications exist (Mackay et al. 2026), leading to little awareness or overlap with research conducted by resource economists or social psychologists (Fenichel et al. 2013a).

Other social science disciplines in fisheries science, including political ecology (Boucquey et al. 2026) and environmental history (Thurstan et al. 2026), have focused less on understanding individual behaviours, but rather study social units (e.g. communities of recreational fishers) and how these units are shaped by incentives, opportunities, constraints, and power structures (Fig. 2). Perhaps differences in scales of analysis (i.e. individual vs. social group) and time frames (i.e. past vs. present), in addition to ontological and epistemological differences among disciplines, further inhibits integration with other social science disciplines. Yet, social-psychologists and behavioural economists may also focus on outputs other than behaviour (e.g. norms, conflicts, procedural justice) because they affect the acceptability of rules, perceptions of the procedural fairness of management decisions, and support for management policies. As norms, conflicts, and perceptions of justice affect agency and power dynamics, we see ample opportunities for more crosstalk among the various social science disciplines involved in recreational fisheries.

Limited cross-fertilization and cross-referencing exist among social science disciplines in fisheries science. The lack of clarity about specific needs combined with the fact that research on recreational fishers emerged in isolation among communities of researchers situated in different departments and faculties has resulted in a body of largely disciplinary-specific research (Arlinghaus et al. 2014) that lacks integration, let alone awareness of other disciplines (e.g. Fenichel et al. 2013a). This lack of understanding and appreciation for different disciplinary-specific approaches represents a key barrier for conducting interdisciplinary research (Hunt et al. 2026) that is directed at fundamental understanding of recreational fisheries dynamics with a view toward supporting management decisions.

5 Interdisciplinary Integration

Despite the importance of interdisciplinary work to understand and solve recreational fisheries issues from a social-ecological-systems perspective, many constraints exist in the contemporary academic system (reviewed in Heberlein (1988), Campbell (2005), Arlinghaus et al. (2014), Hunt et al. (2026)). Academic "silos" have prevented the effective integration of social and ecological sciences in recreational fisheries (Heck et al. 2015), thereby constraining our understanding of the dynamics and effects of feedbacks within and among the social and ecological components of recreational fisheries (Arlinghaus et al. 2008a, 2014; Ward et al. 2016). Better appreciating the various feedbacks and associated thresholds among humans and nature in a structured way and learning how these feedbacks, along with all involved structures (e.g. organizations), give rise to specific outcomes—such as conflicts or biologically unsustainable exploitation—are essential for building sustainability in recreational fisheries and in other coupled social-ecological systems (Arlinghaus et al. 2017; Camp et al. 2026; Jensen et al. 2026; Wilberg et al. 2026).

In recent years, there has been an increasing emphasis on interdisciplinary studies linking social and ecological dynamics in recreational fisheries (Carpenter and Brock 2004; Biggs et al. 2009; Johnston et al. 2010, 2013, 2015; Horan et al. 2011; Hunt et al. 2011; Matsumura et al. 2019). Many of the key feedbacks, between fishers and managers, and between fishers and aquatic ecosystems, involve behavioural dynamics, which is why studying interactions in recreational fisheries from the perspective of the behavioural sciences has become prominent both in the social and in the ecological domains (Cooke et al. 2026; de Kerckhove et al. 2026; Arlinghaus et al. 2026; Golebie et al. 2026; Kochalski et al. 2026; Kyle et al. 2026; van Putten et al. 2026; Venturelli et al. 2026; Vaske et al. 2026a,b). Evaluating recreational fisheries as complex social-ecological systems will help us better understand local, regional, and global fisheries dynamics (Sievert et al. 2026), will inform a much broader scientific literature on general human-environment interactions, and, perhaps most importantly, will support better management of these complex social-ecological systems (Arlinghaus et al. 2017; Robinson et al. 2026; van Poorten et al. 2026).

We believe that it is less helpful to approach the science of recreational fisheries from a particular discipline than to tackle the problem with the disciplines that are necessary to understand the problem (e.g. eliminating overharvest requires ecological understanding of fish population dynamics and social psychological understanding of fisher behaviour). In many cases, the knowledge may already have been produced in disciplinary ways, in other cases, new disciplinary and often interdisciplinary inquiry may be needed to integrate across scales from the individual to the collective and from the local lake to the landscape of fisheries. One key factor in these dynamics is the behaviour of recreational fishers and managers (Fulton et al. 2011). This book describes: (1) the foundations of each discipline (i.e. ecological sciences, resource economics, social psychology, political ecology, history, and behavioural economics), (2) the methods to be used (i.e. surveys, experiments, digital data mining, content analysis, mixed methods, evidence synthesis), and, (3) ways to integrate concepts across the social and ecological domains (i.e. integrated models, participatory modelling, adaptive management, standardization).

6 Evaluating Outcomes

Though the goals of recreational fisheries often include statements such as “satisfying anglers,” “conserving the resource for future generations,” and “ecological integrity and sustainability,” quantitative objectives are rarely stated explicitly in management plans (Lackey 1974; FAO 2012). This lack of clarity on quantitative objectives has made it difficult to evaluate the effectiveness of recreational fisheries management because we cannot compare where we are with where we would like to be (McMullin and Pert 2010). The importance of explicitly stating objectives was demonstrated in Johnston et al. (2010), who showed that different quantitative objectives lead to different fishery outcomes. Within the context of fisheries as social-ecological systems, objectives should normally include quantitative measures of social, ecological, and economic values, as proposed in the context of Optimal Social Yield (Malvestuto and Hudgins 1996; Johnston et al. 2010).

The concept of Optimal Social Yield is equivalent to the economic concept of welfare. One key assumption is that the economic welfare of recreational fishers integrates ecological objectives (e.g. when the ecosystem and its fish stocks diminish, so also does the quality of fishing). Johnston et al. (2010) showed in modelling that optimizing a fishery to Optimal Social Yield thus would also be ecologically sustainable. However, a few aspects are problematic here. First, there might be other objectives that a management tailored to optimize fishing quality might compromise (e.g. the recreational quality of other non-fishing lake users). Second, optimization might be easily achieved in models, but not in reality. Third, for Optimal Social Yield to be operationalized, full knowledge of what determines the value of fishing to people must be known. The key point is that most recreational fisheries are multi-objective fisheries from a management perspective and at least some objectives will have a social dimension. Therefore, evaluation outcomes of social-ecological interactions demand interdisciplinary inquiry and knowledge.

The difficulty in stating objectives is especially important when considering fisheries as social-ecological systems because fisheries managers are typically trained in biology or ecology (Brooks et al. 2015). Although it may be that ecologically minded managers can at least qualitatively determine whether a fishery is at a conservation risk (i.e. not meeting sustainability objectives), it is much more difficult and contentious to determine the social value of a fishery and whether improvements can be made. Moreover, it is relatively easy to determine if catch rates or fish size are satisfactory (a possible biological determinant of social utility and satisfaction), but if recreational fishers are heterogenous in their relative importance of these attributes, managers may inadvertently reduce the social value of their fisheries. There is also the question of who is privileged with deciding what is meant by desirable outcomes (Fenichel et al. 2013b). By carefully considering what constitutes the social objectives of the fishery and how to measure those objectives, it will be possible to understand social measures of fishery value and to apply those measures to improve fisheries for the sake of recreational fishers.

Finally, management and governance are, by design, full of trade-offs (Fenichel et al. 2013a). Power structure and normative questions of how to allocate usually scarce goods are key considerations that demand social and ethical sciences to have a seat at the table. This book will hopefully provide some methodological basis to address these concerns or be informed about them.

7 Book Outlook

Recreational fisheries are complex, adaptive social-ecological systems. Yet historically, management of recreational fisheries did not apply interdisciplinary practices or provide foundational training for fisheries professionals in interdisciplinary approaches. Holistic thinking can enhance disciplinary strategies and training, and it is needed to understand the complexity of recreational fisheries, specifically how fishers behave and the outcomes that recreational fisheries provide to people, ecosystems, and society. With the chapters to follow, we believe this book can help instil new fundamentals in how we conceptualize and study complex, adaptive social-ecological systems and can help transition how we manage recreational fisheries by integrating appropriate elements from multiple social science disciplines to ensure sustainable use of recreationally exploited fish species. The chapters that follow showcase examples of disciplinary background and methods to study the social aspects of recreational fisheries and ways to integrate different knowledge domains with a view to improve fisheries management.

Acknowledgements We thank Christopher Chizinski and Mark Kaemingk for conducting constructive reviews of prior versions of this chapter. Any use of trade, firm, or product names is for descriptive purposes only and does not imply endorsement by the U.S. Government. The Nebraska Cooperative Fish and Wildlife Research Unit is jointly supported by a cooperative agreement among the U.S. Geological Survey, the Nebraska Game and Parks Commission, the University of Nebraska, the U.S. Fish and Wildlife Service, and the Wildlife Management Institute.

References

Aas Ø, Haider W, Hunt L (2000) Angler responses to potential harvest regulations in a Norwegian sport fishery: a conjoint-based choice modeling approach. N Am J Fish Manag 20:940–950

Abbott JK, Lew DK, Whitehead JC, Woodward RT (2022) The future of fishing for fun: the economics and sustainable management of recreational fisheries. Rev Environ Econ Policy 16: 262–281

Ajzen I, Fishbein M (1980) Understanding attitudes and predicting social behavior. Prentice-Hall, Englewood Cliffs

Allen CR, Angeler DG, Garmestani AS, Gunderson LH, Holling CS (2014) Panarchy: theory and application. Ecosystems 17:578–589

Anderson LG (1993) Toward a complete economic theory of the utilization and management of recreational fisheries. J Environ Econ Manag 24:272–295

Arlinghaus R, Bork M, Fladung E (2008a) Understanding the heterogeneity of recreational anglers across an urban-rural gradient in a metropolitan area (Berlin, Germany), with implications for fisheries management. Fish Res 92:53–62

Arlinghaus R, Johnson BM, Wolter C (2008b) The past, present and future role of limnology in freshwater fisheries science. Int Rev Hydrobiol 93:541–549

Arlinghaus R, Hunt LM, Post JR, Allen MS (2014) Not fish, not meat: some guidance on how to study fisheries from an interdisciplinary perspective. In: Taylor WW, Lynch AJ, Léonard NJ (eds) Future of fisheries: perspectives for emerging professionals. American Fisheries Society, Bethesda, pp 223–230

Arlinghaus R, Lorenzen K, Johnson BM, Cooke SJ, Cowx IG (2016) Management of freshwater fisheries: addressing habitat, people and fishes. In: Craig JF (ed) Freshwater fisheries ecology. Wiley, Oxford, pp 557–579

Arlinghaus R, Alós J, Beardmore B, Daedlow K, Dorow M, Fujitani M, Hühn D, Haider W, Hunt LM, Johnson BM, Johnston F (2017) Understanding and managing freshwater recreational fisheries as complex adaptive social-ecological systems. Rev Fish Sci Aquac 25:1–41

Arlinghaus R, Abbott JK, Fenichel EP, Carpenter SR, Hunt LM, Alós J, Klefoth T, Cooke SJ, Hilborn R, Jensen OP, Wilberg MJ, Post JR, Manfredo MJ (2019) Governing the recreational dimension of global fisheries. Proc Natl Acad Sci USA 116:5209–5213

Arlinghaus R, Aas Ø, Alós J, Arismendi I, Bower S, Carle S, Czarkowski T, Freire KMF, Hu J, Hunt LM, Lyach R, Kapusta A, Salmi P, Schwab A, Tsuboi J-i, Trella M, McPhee D, Potts W, Wołos A, Yang Z-J (2021) Global participation in and public attitudes toward recreational fishing: international perspectives and developments. Rev Fish Sci Aquac 29:58–95

Arlinghaus R, Fujitani M, Ehrlich E, Engel M, Cooke SJ (2026) Knowledge co-production in recreational fisheries science and management. In: Pope KL, Arlinghaus R, Hunt LM, Lynch AJ, van Poorten BT (eds) Understanding recreational fishers: disciplinary and interdisciplinary approaches for fisheries management. Springer Nature, Cham, p 533–572

Barber WE, Taylor JN (1990) The importance of goals, objectives, and values in the fisheries management process and organization: a review. N Am J Fish Manag 10:365–373

Beardmore AB (2013) The importance of understanding angler heterogeneity for managing recreational fisheries. Thesis, Simon Fraser University

Berkes F, Folke C (eds) (1998) Linking social and ecological systems: management practices and social mechanisms for building resilience, first. Cambridge University Press, Cambridge

Berkes F, Colding J, Folke C (eds) (2003) Navigating social-ecological systems: building resilience for complexity and change. Cambridge University Press, Cambridge

Biggs R, Carpenter SR, Brock WA (2009) Turning back from the brink: detecting an impending regime shift in time to avert it. Proc Natl Acad Sci USA 106:826–831

Biggs R, Schlüter M, Biggs D, Bohensky EL, BurnSilver S, Cundill G, Dakos V, Daw TM, Evans LS, Kotschy K, Leitch AM, Meek C, Quinlan A, Raudsepp-Hearne C, Robards MD, Schoon ML, Schultz L, West PC (2012) Toward principles for enhancing the resilience of ecosystem services. Annu Rev Environ Resour 37:421–448

Birdsong M, Hunt LM, Arlinghaus R (2021) Recreational angler satisfaction: what drives it? Fish Fish 22:682–706

Boucquey N, Fly J, Boonstra WJ, Drakopulos L, Leong KM, Marjadi MN, Wise S, Fairbanks L (2026) Political ecology for understanding recreational fishers and fisheries. In: Pope KL, Arlinghaus R, Hunt LM, Lynch AJ, and van Poorten BT (eds) Understanding recreational fishers: disciplinary and interdisciplinary approaches for fisheries management. Springer Nature, Cham, Switzerland, p 197–232

Brooks K, Schirmer J, Pascoe S, Triantafillos L, Jebreen E, Cannard T, Dichmont CM (2015) Selecting and assessing social objectives for Australian fisheries management. Mar Policy 53: 111–122

Campbell LM (2005) Overcoming obstacles to interdisciplinary research. Conserv Biol 19:574–577

Camp EV, van Poorten BT, Robinson KF, Harrison HL, NguyenV , Crandall CA, Lorenzen K, Casola W, Dedual M, Klizentyte K (2026) Integrating diverse stakeholder knowledge in fisheries decision making. In: Pope KL, Arlinghaus R, Hunt LM, Lynch AJ, and van Poorten BT (eds) Understanding recreational fishers: disciplinary and interdisciplinary approaches for fisheries management. Springer Nature, Cham, Switzerland, p 607–660

Carpenter SR, Brock WA (2004) Spatial complexity, resilience and policy diversity: fishing on lake-rich landscapes. Ecol Soc 9:8

Carpenter SR, Brock WA, Hansen GJA, Hansen JF, Hennessy JM, Isermann DA, Pedersen EJ, Perales KM, Rypel AL, Sass GG, Tunney TD, Vander Zanden MJ (2017) Defining a safe operating space for inland recreational fisheries. Fish Fish 18:1150–1160

Colding J, Barthel S (2019) Exploring the social-ecological systems discourse 20 years later. Ecol Soc 24(1):2

Cooke SJ, Rytwinski T, Harper M, Howarth A, Nakagawa S, Hunt LM (2026) Evidence synthesis in recreational fisheries science and management. In: Pope KL, Arlinghaus R, Hunt LM, Lynch AJ, and van Poorten BT (eds) Understanding recreational fishers: disciplinary and interdisciplinary approaches for fisheries management. Springer Nature, Cham, Switzerland, p 573–604

Cox SP, Walters CJ, Post JR (2003) A model-based evaluation of active management of recreational fishing effort. N Am J Fish Manag 23:1294–1302

de Kerckhove DT, Tucker CM, Gutowsky L, Morgan G, Guzzo M, Barajas VL (2026) Optimizing creel surveys. In: Pope KL, Arlinghaus R, Hunt LM, Lynch AJ, and van Poorten BT (eds) Understanding recreational fishers: disciplinary and interdisciplinary approaches for fisheries management. Springer Nature, Cham, Switzerland, p 387–409

Ditton RB (2004) Human dimensions of fisheries. In: Manfredo MJ, Vaske JJ, Field D, Brown PJ, Bruyere B (eds) Society and natural resources: a summary of knowledge prepared for the 10th International Symposium on Society and Resource Management. Modern Litho, Jefferson, Missouri. University Press of Colorado, Denver, pp 199–208

Driver BL (1985) Specifying what is produced by management of wildlife by public agencies. Leis Sci 7:281–295

Embke HS, Nyboer EA, Robertson AM, Arlinghaus R, Akintola SL, Atessahin T, Badr LM, Baigun C, Basher Z, Beard TD Jr, Boros G, Bower SD, Cooke SJ, Cowx IG, Franco A, Gaspar-Dillanes MT, Granada VP, Hart RJ, Heinsohn CR, Jalabert V, Kapusta A, Krajč T, Koehn JD, Lopes G, Lyach R, Magqina T, Milardi M, Nattabi J, Nyaboke H, Phang S, Potts WM, Ribeiro F, Mercado-Silva N, Sreenivasan N, Thorpe A, Treer T, Ustups D, Weyl OLF, Wood LE, Zengin M, Lynch AJ (2022) Global dataset of species-specific inland recreational fisheries harvest for consumption. Sci Data 9:488

FAO (2012) Recreational fisheries. FAO technical guidelines for responsible fisheries, no. 13. FAO, Rome

Fenichel EP, Abbott JK, Huang B (2013a) Modelling angler behaviour as a part of the management system: synthesizing a multi-disciplinary literature. Fish Fish 14:137–157

Fenichel EP, Gentner B, Arlinghaus R (2013b) Normative considerations for recreational fishery management: a bioeconomic framework for linking positive science and normative fisheries policy decisions. Fish Manag Ecol 20:223–233

Folke C, Hahn T, Olsson P, Norberg J (2005) Adaptive governance of social-ecological systems. Annu Rev Environ Resour 30:441–473

Fulton DC, Manfredo MJ, Lipscomb J (1996) Wildlife value orientations: a conceptual and measurement approach. Hum Dimens Wildl 1:24–47

Fulton EA, Smith ADM, Smith DC, Van Putten IE (2011) Human behaviour: the key source of uncertainty in fisheries management. Fish Fish 12:2–17

Gordon HS (1954) The economic theory of a common-property resource: the fishery. J Polit Econ 62:124–142

Golebie E, van Riper CJ, Kochalski S, Nunan F (2026) Mixed methods research: applications and considerations for the human dimensions of recreational fisheries. In: Pope KL, Arlinghaus R, Hunt LM, Lynch AJ, and van Poorten BT (eds) Understanding recreational fishers: disciplinary and interdisciplinary approaches for fisheries management. Springer Nature, Cham, Switzerland, p 481–506

Granek EF, Madin EM, Brown MA, Figueira W, Cameron DS, Hogan Z, Kristianson G, de Villiers P, Williams JE, Post J, Zahn S, Arlinghaus R (2008) Engaging recreational fishers in management and conservation: global case studies. Conserv Biol 22:1125–1134

Griffiths SP, Bryant J, Raymond HF, Newcombe PA (2017) Quantifying subjective human dimensions of recreational fishing: does good health come to those who bait? Fish Fish 18: 171184

Gunderson LH, Holling CS (eds) (2001) Panarchy: understanding transformations in human and natural systems. Island Press, Washington, DC

Heberlein TA (1988) Improving interdisciplinary research: integrating the social and natural sciences. Soc Nat Resour 1:5–16

Heck N, Stedman RC, Gaden M (2015) The integration of social science information into Great Lakes fishery management: opportunities and challenges. Fish Res 167:30–37

Hendee JC (1974) A multiple-satisfaction approach to game management. Wildl Soc Bull 2:104–113

Hinkel J, Bots PWG, Schlüter M (2014) Enhancing the Ostrom social-ecological system framework through formalization. Ecol Soc 19(3):51

Hinkel J, Cox ME, Schlüter M, Binder CR, Falk T (2015) A diagnostic procedure for applying the social-ecological systems framework in diverse cases. Ecol Soc 20(1):32

Horan RD, Fenichel EP, Drury KLS, Lodge DM (2011) Managing ecological thresholds in coupled environmental-human systems. Proc Natl Acad Sci USA 108:7333–7338

Hubert WA, Quist MC (eds) (2010) Inland fisheries management in North America, 3rd edn. American Fisheries Society, Bethesda

Hunt LM, Arlinghaus R, Lester N, Kushneriuk R (2011) The effects of regional angling effort, angler behavior, and harvesting efficiency on landscape patterns of overfishing. Ecol Appl 21: 2555–2575

Hunt LM, Sutton SG, Arlinghaus R (2013) Illustrating the critical role of human dimensions research for understanding and managing recreational fisheries within a social-ecological system framework. Fish Manag Ecol 20:111–124

Hunt LM, Camp E, van Poorten B, Arlinghaus R (2019) Catch and non-catch-related determinants of where anglers fish: a review of three decades of site choice research in recreational fisheries. Rev Fish Sci Aquac 27:261–286

Hunt LM, Boucquey N, Beardmore B, Christensen J, Fulton DC, Mackay M, Melstrom RT, Post JR, Schroeder SA, Thurstan RH, van Putten EI, Arlinghaus R (2026) Introducing disciplinary perspectives and interdisciplinary possibilities for understanding recreational fishers and fisheries. In: Pope KL, Arlinghaus R, Hunt LM, Lynch AJ, van Poorten BT (eds) Understanding recreational fishers: disciplinary and interdisciplinary approaches for fisheries management. Springer Nature, Cham, pp 57–94

Jensen OP, Arlinghaus R, Golden AS, Janssen MA, Solomon CT, van Poorten BT (2026) Integrated models of the social-ecological dynamics of recreational fisheries. In: Pope KL, Arlinghaus R, Hunt LM, Lynch AJ, and van Poorten BT (eds) Understanding recreational fishers: disciplinary and interdisciplinary approaches for fisheries management. Springer Nature, Cham, Switzerland, p 661–692

Johnston FD, Arlinghaus R, Dieckmann U (2010) Diversity and complexity of angler behaviour drive socially optimal input and output regulations in a bioeconomic recreational-fisheries model. Can J Fish Aquat Sci 67:1507–1531

Johnston FD, Arlinghaus R, Dieckmann U (2013) Fish life history, angler behaviour and optimal management of recreational fisheries. Fish Fish 14:554–579

Johnston FD, Beardmore B, Arlinghaus R (2015) Optimal management of recreational fisheries in the presence of hooking mortality and noncompliance—predictions from a bioeconomic model incorporating a mechanistic model of angler behavior. Can J Fish Aquat Sci 72:37–53

Kochalski S, Barz F, Pita P, Harrison HL (2026) Quantitative and qualitative content analysis of text and images. In: Pope KL, Arlinghaus R, Hunt LM, Lynch AJ, and van Poorten BT (eds) Understanding recreational fishers: disciplinary and interdisciplinary approaches for fisheries management. Springer Nature, Cham, Switzerland, p 507–532

Krogman RM, Taylor SM, Adams AJ, Boucek RE, Lombardo SM, Crandall CA, Steffen SF, Acosta A, Boné Morón E, Haukebo S, Hyder K, Radford ZM, Potts WM (2026) Role of recreational fisher information in fisheries management. In: Pope KL, Arlinghaus R, Hunt LM, Lynch AJ, and van Poorten BT (eds) Understanding recreational fishers: disciplinary and interdisciplinary approaches for fisheries management. Springer Nature, Cham, Switzerland, p 21–54

Kyle G, Landon AC, Pilgreen D, Golebie E (2026) Measurement instruments. In: Pope KL, Arlinghaus R, Hunt LM, Lynch AJ, and van Poorten BT (eds) Understanding recreational fishers: disciplinary and interdisciplinary approaches for fisheries management. Springer Nature, Cham, Switzerland, p 355–385

Lackey RT (1974) Priority research in fisheries management. Wildl Soc Bull 2:63–66

Lancaster KJ (1966) A new approach to consumer theory. J Polit Econ 74:132–157

Landon AC, Schroeder SA, Stefen SF, Arlinghaus R (2026) Social psychology for understanding recreational fishers and fisheries. In: Pope KL, Arlinghaus R, Hunt LM, Lynch AJ, and van Poorten BT (eds) Understanding recreational fishers: disciplinary and interdisciplinary approaches for fisheries management. Springer Nature, Cham, Switzerland, p 157–196

Lee ME, Driver BL (1998) Benefits-based management: a new paradigm for managing amenity resources. In: Burch W (ed) Ecosystem management. CRC Press, Boca Raton, pp 143–154

Levin S, Xepapadeas T, Crépin A-S, Norberg J, de Zeeuw A, Folke C, Hughes T, Arrow K, Barrett S, Daily G, Ehrlich P, Kautsky N, Mäler K-G, Polasky S, Troell M, Vincent JR, Walker B (2013) Social-ecological systems as complex adaptive systems: modeling and policy implications. Environ Dev Econ 18:111–132

Liu J, Dietz T, Carpenter SR, Alberti M, Folke C, Moran E, Pell AN, Deadman P, Kratz T, Lubchenco J, Ostrom E, Ouyang Z, Provencher W, Redman CL, Schneider SH, Taylor WW (2007) Complexity of coupled human and natural systems. Science 317:1513–1516

Lynch AJ, Cooke SJ, Beard TD Jr, Kao YC, Lorenzen K, Song AM, Allen MS, Basher Z, Bunnell DB, Camp EV, Cowx IG, Freedman JA, Nguyen VM, Nohner JK, Rogers WM, Siders ZA, Taylor WW, Youn S-J (2017) Grand challenges in the management and conservation of North American inland fishes and fisheries. Fisheries 42:115–124

Lynch AJ, Embke HS, Nyboer EA, Wood LE, Thorpe A, Phang AC, Viana DF, Golden CD, Milardi M, Arlinghaus R, Baigun C, T Beard Jr TD, Cooke SJ, Cowx IG, Koehn JD, Lyach R, Potts W, Robertson AM, Schmidhuber J, Weyl OLF (2024) Inland recreational fisheries contribute nutritional benefits and economic value but are vulnerable to climate change Nature Food 5(5):433-443. https://doi.org/10.1038/s43016-024-00961-8

Mackay M, van Putten EI, Richter A, Abatayo AL, Chuah S-H, Dankel DJ, Schadeberg A, Snyder H (2026) Behavioural economics for understanding recreational fishers and fisheries. In: Pope KL, Arlinghaus R, Hunt LM, Lynch AJ, and van Poorten BT (eds) Understanding recreational fishers: disciplinary and interdisciplinary approaches for fisheries management. Springer Nature, Cham, Switzerland, p 261–284

Malvestuto SP, Hudgins MD (1996) Optimum yield for recreational fisheries management. Fisheries 21:6–17

Manfredo MJ, Driver BL, Tarrant MA (1996) Measuring leisure motivation: a meta-analysis of the recreation experience preference scales. J Leis Res 28:188–213

Matsumura S, Beardmore B, Haider W, Dieckmann U, Arlinghaus R (2019) Ecological, angler, and spatial heterogeneity drive social and ecological outcomes in an integrated landscape model of freshwater recreational fisheries. Rev Fish Sci Aquac 27:170–197

McConnell KE, Sutinen JG (1979) Bioeconomic models of marine recreational fishing. J Environ Econ Manag 6:127–139

McMullin SL, Pert E (2010) The process of fisheries management. In: Hubert WA, Quist MC (eds) Fisheries management in North America, 3rd edn. American Fisheries Society, Bethesda, pp 133–155

Melstrom RT, Jungers B, Bi X, Abbott JK (2026) Resource economics for understanding recreational fishers and fisheries. In: Pope KL, Arlinghaus R, Hunt LM, Lynch AJ, and van Poorten BT (eds) Understanding recreational fishers: disciplinary and interdisciplinary approaches for fisheries management. Springer Nature, Cham, Switzerland, p 123–156

Millennium Ecosystem Assessment (MEA) (2005) Millennium Ecosystem Assessment reports

Nyboer EA, Embke HS, Robertson AM, Arlinghaus R, Bower S, Baigun C, Beard D, Cooke SJ, Cowx IG, Koehn JD, Lyach R, Milardi M, Potts W, Lynch AJ (2022) Overturning stereotypes: the fuzzy boundary between recreational and subsistence inland fisheries. Fish Fish 23:1282–1298

Ostrom E (1990) Governing the commons. Cambridge University Press, New York

Ostrom E (2007) A diagnostic approach for going beyond panaceas. Proc Natl Acad Sci USA 104: 15181–15187

Ostrom E (2009) A general framework for analyzing sustainability of social-ecological systems. Science 325(5939):419–422

Parkkila K, Arlinghaus R, Artell J, Gentner B, Haider W, Aas Ø, Barton D, Roth E, Sipponen M, Hickley P (2010) Methodologies for assessing socio-economic benefits of European inland recreational fisheries, EIFAAC occasional paper no 46. FAO, Ankara

Perman R, Ma Y, McGilvray J, Common M (2003) Natural resource and environmental economics, 3rd edn. Pearson Addison Wesley, Harlow

Pope KL, Pegg MA, Cole NW, Siddons SF, Fedele AD, Harmon BS, Ruskamp RL, Turner DR, Uerling CC (2016) Fishing for ecosystem services. J Environ Manag 183:408–417

Post JR (2013) Resilient recreational fisheries or prone to collapse? A decade of research on the science and management of recreational fisheries. Fish Manag Ecol 20:99–110

Post JR, Wilson KL, Johnston FD, Sass GG, Askey PJ, Ward HGM, Allen MS (2026) Ecology for understanding recreational fishers and fisheries. In: Pope KL, Arlinghaus R, Hunt LM, Lynch AJ, and van Poorten BT (eds) Understanding recreational fishers: disciplinary and interdisciplinary approaches for fisheries management. Springer Nature, Cham, Switzerland, p 95–121

Radinger J, Matern S, Klefoth T, Wolter C, Feldhege F, Monk CT, Arlinghaus R (2023) Ecosystem-based management outperforms species-focused stocking for enhancing fish populations. Science 379:946–951

Robinson KF, Arlinghaus R, Camp EV, Post JR, Stedman R, Sutton SG (2026) Adaptive management for recreational fisheries decisions in the face of uncertainty. In: Pope KL, Arlinghaus R, Hunt LM, Lynch AJ, and van Poorten BT (eds) Understanding recreational fishers: disciplinary and interdisciplinary approaches for fisheries management. Springer Nature, Cham, Switzerland, p 719–746

Schlüter M, McAllister RR, Arlinghaus R, Bunnefeld N, Eisenack K, Hölker F, Milner-Gulland EJ, Müller B, Nicholson E, Quaas M, Stöven M (2012) New horizons for managing the environment: a review of coupled social-ecological systems modeling. Nat Resour Model 25:219–272

Schlüter M, Hinkel J, Bots PWG, Arlinghaus R (2014) Application of the SES framework for model-based analysis of the dynamics of social-ecological systems. Ecol Soc 19(1):36

Shephard S, List CJ, Arlinghaus R (2022) Reviving the unique potential of recreational fishers as environmental stewards of aquatic ecosystems. Fish Fish 24:1–13

Sievert N, Krogman RM, Embke H (2026) Data standardisation and management to facilitate large-scale and interdisciplinary approaches. In: Pope KL, Arlinghaus R, Hunt LM, Lynch AJ, and van Poorten BT (eds) Understanding recreational fishers: disciplinary and interdisciplinary approaches for fisheries management. Springer Nature, Cham, Switzerland, p 747–782

Solomon CT, Dassow CJ, Iwicki CM, Jensen OP, Jones SE, Sass GG, Trudeau A, van Poorten BT, Whittaker D (2020) Frontiers in modelling social–ecological dynamics of recreational fisheries: a review and synthesis. Fish Fish 21:973–991

Thurstan RH, Chong-Montenegro C, Christensen J, Clark A, Gaynor A (2026) Environmental history and historical ecology for understanding recreational fishers and fisheries. In: Pope KL, Arlinghaus R, Hunt LM, Lynch AJ, and van Poorten BT (eds) Understanding recreational fishers: disciplinary and interdisciplinary approaches for fisheries management. Springer Nature, Cham, Switzerland, p 233–259

Tufts BL, Holden J, DeMille M (2015) Benefits arising from sustainable use of North America's fishery resources: economic and conservation impacts of recreational angling. Int J Environ Stud 72:850–868

van Poorten BT, Arlinghaus R, Daedlow K, Haertel-Borer SS (2011) Social-ecological interactions, management panaceas, and the future of wild fish populations. Proc Natl Acad Sci USA 108: 12554–12559

van Poorten BT, Hunt LM, Arlinghaus R, Lynch AJ, Pope KL (2026) Best practices for understanding recreational fishers. In: Pope KL, Arlinghaus R, Hunt LM, Lynch AJ, and van Poorten BT (eds) Understanding recreational fishers: disciplinary and interdisciplinary approaches for fisheries management. Springer Nature, Cham, Switzerland, p 783–810

van Putten EI, Trudeau A, Mackay M, Obregón C, Post JR, Richter A, Schuch E, Solomon CT (2026) Economic experiments in recreational fishing studies. In: Pope KL, Arlinghaus R, Hunt LM, Lynch AJ, and van Poorten BT (eds) Understanding recreational fishers: disciplinary and interdisciplinary approaches for fisheries management Springer Nature, Cham, Switzerland, p 411–442

Vaske JJ, Needham MD, Miller CA (2026a) Fundamentals of sampling and implementing human dimensions of fishing surveys. In: Pope KL, Arlinghaus R, Hunt LM, Lynch AJ, and van Poorten BT (eds) Understanding recreational fishers: disciplinary and interdisciplinary approaches for fisheries management. Springer Nature, Cham, Switzerland, p 337–353

Vaske JJ, Needham MD, Miller CM, Sponarski C (2026b) Guidelines for writing and constructing recreational fisher surveys. In: Pope KL, Arlinghaus R, Hunt LM, Lynch AJ, and van Poorten BT (eds) Understanding recreational fishers: disciplinary and interdisciplinary approaches for fisheries management. Springer Nature, Cham, Switzerland, p 287–325

Venturelli P, Skov C, Audzijonyte A, Belmaker J, Ana Cardoso AS, Gordon L, Gundelund C, Hyder K, Jarić I, Lennox RJ, Potts WM, Vacura K, Vaz AS, Sbragaglia V (2026) Digital data mining. In: Pope KL, Arlinghaus R, Hunt LM, Lynch AJ, and van Poorten BT (eds) Understanding recreational fishers: disciplinary and interdisciplinary approaches for fisheries management. Springer Nature, Cham, Switzerland, p 443–479

Ward HG, Allen MS, Camp EV, Cole N, Hunt LM, Matthias B, Post JR, Wilson K, Arlinghaus R (2016) Understanding and managing social–ecological feedbacks in spatially structured recreational fisheries: the overlooked behavioral dimension. Fisheries 41:524–535

Weithman S (1999) Socioeconomic benefits of fisheries. In: Kohler CC, Hubert WA (eds) Inland fisheries management in North America, 2nd edn. American Fisheries Society Press, Bethesda, pp 193–213

Wilberg MJ, Irwin BJ, Jones ML (2026) Participatory modelling to support evaluation of management actions for recreational fisheries. In: Pope KL, Arlinghaus R, Hunt LM, Lynch AJ, and van Poorten BT (eds) Understanding recreational fishers: disciplinary and interdisciplinary approaches for fisheries management. Springer Nature, Cham, Switzerland, p 693–718

Ziegler JP, Golebie EJ, Jones SE, Weidel BC, Solomon CT (2017) Social-ecological outcomes in recreational fisheries: the interaction of lakeshore development and stocking. Ecol Appl 27: 56–65

Role of Recreational Fisher Information in Fisheries Management

Rebecca M. Krogman, Stephen M. Taylor, Aaron J. Adams, Ross E. Boucek, Steven M. Lombardo, Chelsey A. Crandall, Susan F. Steffen, Alejandro Acosta, Eduardo Boné Morón, Sepp Haukebo, Kieran Hyder, Zachary M. Radford, and Warren M. Potts

Abstract Fisheries managers work with the three components of fisheries—biota, environment, and people—to achieve recreational, stewardship, and socioeconomic goals. Although manipulation of biota and the aquatic environment remains popular (e.g. harvest regulation, protecting or enhancing habitats, and guarding water quality), the human dimension of fisheries must also be considered throughout the fisheries management process. Managers can and should engage fishers in a myriad of ways, from identification of problems to guidance on alternatives, and from measurement of change to evaluation of success. In this chapter, we review common

R. M. Krogman (✉)
Fisheries Research Section, Iowa Department of Natural Resources, Chariton, IA, USA

S. M. Taylor
Department of Primary Industries and Regional Development, Western Australian Fisheries and Research Laboratories, Western Australia, Australia
e-mail: Steve.Taylor@dpird.wa.gov.au

A. J. Adams · R. E. Boucek · S. M. Lombardo
Bonefish & Tarpon Trust, Miami, FL, USA
e-mail: aaron@bonefishtarpontrust.org; ross@bonefishtarpontrust.org; steven@bonefishtarpontrust.org

C. A. Crandall
Fish and Wildlife Research Institute, Florida Fish and Wildlife Conservation Commission, Gainesville, FL, USA
e-mail: chelsey.crandall@myfwc.com

S. F. Steffen
Kansas Department of Wildlife and Parks, Emporia, KS, USA
e-mail: susan.steffen@ks.gov

A. Acosta
Fish and Wildlife Research Institute, Florida Fish and Wildlife Conservation Commission, Marathon, FL, USA

E. B. Morón · S. Haukebo
Environmental Defense Fund, Austin, TX, USA
e-mail: ebone@edf.org; shaukebo@edf.org

K. L. Pope et al. (eds.), *Understanding Recreational Fishers*, Fish & Fisheries Series 45, https://doi.org/10.1007/978-3-031-99739-6_2

fisheries management goals and objectives; detail information needs stemming from those objectives that can be answered by recreational fishers; suggest areas where increased use or application would more proactively address management objectives; and provide examples of integration of recreational fisher information into the decision-making process. Relevant examples of fisheries management frameworks that require recreational fisher information include the multinational Gulf of Mexico and Caribbean flats fishery; collaborative data collection in United Kingdom sea angling fisheries and Florida Everglades largemouth bass fisheries; co-management of mixed commercial-recreational fisheries in Western Australia; co-operative management of marine fisheries in New Zealand; and co-design of fisheries management plans for UK fisheries

Keywords Adaptive fishery management · Fisheries management process · Human dimensions · Stakeholders

1 Typical Goals and Objectives of Fisheries Management

Recreational fisher information constitutes an essential component of what characterizes and drives a recreational fishery, and therefore should be integrated into fisheries management. Fisheries managers work with the three components of fisheries—biota, environment, and people—to achieve desirable recreational outcomes, as well as aquatic resource stewardship and broader societal benefits (McMullin and Pert 2010; Elmer et al. 2017). These broader and more qualitative outcomes may be considered fisheries management *goals*. Commonly cited goals may include:

- "To maintain a high-quality fishery for largemouth bass, including the opportunity to catch trophy sized fish" (McMullin and Pert 2010).
- "To provide fishing opportunities to every Iowan" (Crane and Olcott 1933).
- "To improve crappie and yellow perch fisheries" (Mosel et al. 2015).
- "To get more Victorians fishing, more often" (VFA 2022).

K. Hyder
Centre for Environment, Fisheries and Aquaculture Science, Suffolk, UK

School of Environmental Sciences, University of East Anglia, Norfolk, UK
e-mail: kieran.hyder@cefas.gov.uk

Z. M. Radford
Centre for Environment, Fisheries and Aquaculture Science, Suffolk, UK

School of Life Sciences, University of Bristol, Norfolk, UK
e-mail: zachary.radford@cefas.gov.uk

W. M. Potts
Department of Ichthyology and Fisheries Science, Rhodes University, Eastern Cape, South Africa
e-mail: w.potts@ru.ac.za

Tangible fisheries management goals translate directly to more detailed operational *objectives*. Objectives should be specific, measurable, achievable, realistic, and time bound (McMullin and Pert 2010). In other words, they are quantitative in such a way that failure or success in meeting them can be determined.

For instance, the goal "To provide fishing opportunities to every Iowan", an early goal of the Iowa Conservation Commission, translated to an objective to construct State-operated recreational fishing lakes such that the maximum distance an Iowan would have to drive to a lake was 40 km, within 25 years (Crane and Olcott 1933). Specifications for said artificial lakes defined morphometry and fish species composition. The broad goal to improve crappie and yellow perch fisheries is made tangible by specific objectives of "transitioning from an aggregate statewide daily creel limit of 25 panfish to species-specific daily creel limits <25 fish per fisher or implementing state-wide minimum length limits would reduce harvest of black crappie and yellow perch in Wisconsin by 25%" (Mosel et al. 2015). The broad goal of getting more people fishing more often in Victoria, Australia, translated to an objective to increase fishing participation to "target one million" fishers by 2020 (VFA 2022). Objectives frequently orient around fishers themselves, their fishing experience, and their desires, making recreational fisher data essential not only for developing effective fisheries management objectives but also for measuring success.

Recreational fishery management goals and objectives may need to be balanced with other goals and objectives for an aquatic resource. For instance, in marine environments, recreational fishing often co-occurs with commercial fishing, which can be a real challenge where recreational fisheries are not embedded in governance systems (Potts et al. 2020) and have different data governance and regulatory requirements (Arlinghaus et al. 2019). This can bring the sectors into conflict (Hyder et al. 2018, 2020; Obregón et al. 2022). The US's National Oceanic and Atmospheric Administration identifies several goals of its recreational fisheries policy for marine systems, including maintenance of sustainable fishery resources, promotion of recreational fishing as a source of social, cultural, and economic value, and facilitation of fishing participation through science-based management (NOAA 2015). Simultaneously, it recognizes the "need for stewardship approaches able to best accommodate [both recreational and commercial fisheries] while achieving fishery conservation and management". To establish and operationalize this balance between potentially conflicting interests, recreational fishers and other stakeholders should be integrated into the fishery management process, either directly or indirectly as sources of information and feedback (Hyder et al. 2018, 2020; Potts et al. 2020). This is especially relevant given the greater number of recreational fishers relative to commercial fishers and their substantial role in fish landings (Arlinghaus et al. 2019).

2 Recreational Fishers' Role in the Fishery Management Process

The role of recreational fisher information has long been recognized (e.g. Tunison et al. 1949) and gained some research attention beginning in the 1950s (Wales and Lane 1958). Early work focused on fisher opinions and behaviour toward fish (Smith and Gross 1955; Wales and Lane 1958), which continues to be relevant today and is frequently studied using creel surveys and other forms of catch reporting (Pollock et al. 1994; Hyder et al. 2018). In response to the information gained, fisheries managers often took actions that supplemented the fishery or created new fisheries (e.g. stocking fish), protected desirable fish or limited harvest (e.g. establishing protective regulations or shifting quotas), or enhanced the angling environment (e.g. adding access or manipulating fish habitat).

This "piscicentric" approach (Arlinghaus et al. 2021) fails to recognize the complexity of recreational fisheries as social–ecological systems, affecting and being affected by the broader evolution of the social and ecological landscape. The nature of a recreational fishery may be affected by shifting societal norms (e.g. fish welfare concerns), attitudes and expectations (e.g. activity-general, non-catch related motives vs activity-specific, catch-related), and participation by new people (e.g. new migrants). The approach also fails to identify the numerous touchpoints at which recreational fisher information could be used. For example, fishing regulations used to be developed with minimal public input, relying on the public's trust in professional experience and knowledge (Smith and Gross 1955; McMullin and Pert 2010).

As a somewhat extreme demonstration of the ability of recreational fishers to participate in fishery management, fishing clubs in West Germany govern the entire management process independently of a formal overseeing government body. Instead, those local clubs manage specific areas, focusing on access, stocking, and compliance with regulations. Incorporation of fishers into the fishery management process can shift constituents' perspectives to be more proactive and forward-looking, thereby being more protective of the fishery resource (Arlinghaus et al. 2019). In the UK, marine recreational fisheries are now fully integrated in the management process alongside commercial fisheries in the Fisheries Act (UK Government 2020), with management approaches being co-designed by all stakeholders through the development of Fisheries Management Plans. In addition, co-management approaches have been highlighted as an important research question for recreational fisheries science (Holder et al. 2020a), and fishers need to be central to decision-making processes to ensure responsible and sustainable recreational fisheries (Cooke et al. 2019).

Fishery management constitutes a cyclic process, sometimes called adaptive fishery management (McMullin and Pert 2010). This process could include: (i) identification of problems or opportunities; (ii) acquisition of relevant information; (iii) evaluation of management alternatives; (iv) selection and implementation of a management action; and (v) assessment of outcomes. Recreational fisher-derived information may be applied at any step of the fisheries management process through both direct and indirect pathways. The likelihood of active or direct engagement with recreational fishers depends on the scale of the fishery itself, the management issue faced, and fisher stewardship orientation (Granek et al. 2008). The overall benefit of co-design processes with contributions from the fisher community is to improve management outcomes and compliance with measures (Fidler et al. 2022). Throughout this chapter, we discuss ways recreational fisher information may be used in each of the 5 steps, with tangible examples from both marine and freshwater recreational fisheries around the world.

The central tenet of this adaptive fishery management concept is that problem identification, data collection, and management design are done in a collaborative process (Camp et al. 2026) that includes a combination of recreational fishers, resource management agencies, and researchers, the latter from academia or non-governmental organizations. The collaborative process builds trust among the participants and increases information flow in all directions (Fig. 1). Indeed, numerous fisheries-focussed conservation organizations engaged in and even leading this type of collaboration have been classified as Habitat-Dependent Outdoor Recreation and Conservation Organizations (Raynal et al. 2020). Fisheries-focussed Habitat-Dependent Outdoor Recreation and Conservation Organizations enhance engagement of recreational fishers in the research, management, and advocacy process, making them more likely to engage in large-scale conservation efforts (Raynal et al. 2020). This process greatly reduces the likelihood that fishers will advocate for short-sighted management changes that benefit their portion of the constituency to

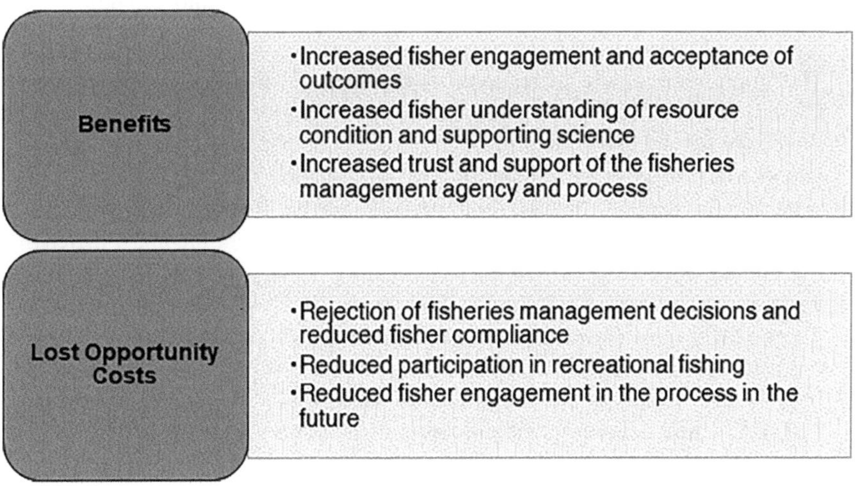

Fig. 1 Benefits of incorporation of recreational fisher information throughout the fishery management process, and potential costs of failing to do so

the longer-term detriment of the resource—indeed, in some cases, fishers engaged with Habitat-Dependent Outdoor Recreation and Conservation Organizations will counterbalance sectors of the fishery that oppose conservation measures. In short, investment in relationships and communication with fishers increases positive engagement and, importantly, compliance with regulations. Importantly, Bonefish & Tarpon Trust (BTT) is a Habitat-Dependent Outdoor Recreation and Conservation Organization strategically placed at the centre of a conceptual triangle—the points represented by (1) the fishing community, (2) resource management agencies, (3) academia—to maximize information exchange, building relationships, and engaging fishers in science, advocacy, and management. The integration with fishers is structured as either an informed nested or parallel mixed-methods approach (Kinnebrew et al. 2021). This means that fishers are queried about perceptions of the fishery to help guide research efforts, are consulted about projects as the projects are designed, participate in data collection, and advocate for application of research findings to management and conservation. For example, fishers worked with BTT to collect information on the permit (*Trachinotus falcatus*) fishery and advocate for additional regulations focussed on perceived fishery declines and a long-term management outlook (Box 1). Similarly, fishing guides in the Bahamas took part in research to identify important bonefish habitats, advocated (with BTT) for habitat protections, and in some areas were active in the community to ensure compliance. Thus, collaborative engagement does not eliminate the risk of special-interest fisher groups negatively influencing fisheries management, but chances are greatly reduced, and a growing portion of the fishing community are becoming conservation advocates.

Box 1 Fisher-Led Data Collection on Bonefish, Permit and Tarpon

Challenge

The importance of the recreational flats fishery in the U.S. Florida Keys, Eastern Gulf of Mexico, Bahamas, Cuba, and Caribbean is recognized with protective regulations now in place in certain countries (see 2.1). Top conservation concerns now are habitat loss and degradation from development and water quality declines, making both juvenile and adult habitat high priorities for identification and protection. Coastal habitats worldwide are under increasing threat from coastal development, including nursery habitats for economically important species (Schmitter-Soto et al. 2017). For example, Atlantic Tarpon (*Megalops atlanticus*) is an economically important species throughout its range, including in the Florida Keys where it is part of the recreational flats fishery with an economic impact exceeding $465 million (Smith et al. 2023). Juvenile tarpon are obligate users of mangrove and marsh habitats as nurseries (Adams and Murchie 2015) and thus impacted by development and

(continued)

Box 1 (continued)
land use changes (Sklar and Browder 1998; Beach 2002). Similarly, bonefish and permit spawning sites lack protections, so are susceptible to fishing impacts on spawning aggregations (Sadovy and Domeier 2005), and development impacts to spawning sites (Adams et al. 2019).

However, habitat evaluations can be challenging for species like bonefish, permit, and tarpon due to their spatially expansive and seasonal habitat use patterns and generally low catch rates. A major challenge in habitat conservation for juvenile tarpon habitats, for example, is that most juvenile habitat locations are unknown; moreover, the conditions of these nursery habitats are only now being characterized (Wilson et al. 2019).

Approach

Bonefish & Tarpon Trust (BTT) collaborates with recreational fishing guides and fishers on identifying conservation threats and research needs, study design, data collection, and advocacy for application of findings to management and conservation. BTT engaged fishers in data acquisition as an integral part of fishery management for bonefish habitat protection in the Bahamas, spatial management of permit in Florida, and identification of juvenile tarpon habitat. Importantly, in each case, a conservation threat and research need were identified, and collaborative research conducted to obtain appropriate data. Only after research findings were finalized were management and conservation measures proposed. Each of the cases below reflect a process and outcomes based on long-developed and maintained relationships between BTT, recreational fishers, and resource management agencies, which enabled the research to be designed and conducted (BTT-fishers collaboration) and findings applied to management (BTT-fishers-management agency collaboration).

Outcomes

Bonefish Habitat Protection

Recreational fishing guides helped document bonefish habitat use patterns via mark-recapture (Boucek et al. 2019), including home ranges and pre-spawning aggregation locations (Adams et al. 2019). They also identified possible pre-spawning aggregation sites for bonefish via interviews (i.e. Traditional Ecological Knowledge); BTT incorporated this information into a parallel mixed-methods approach that identified seven pre-spawning aggregation sites on five islands (Kinnebrew et al. 2021, Fig. 2) and contributed to the creation of five new national parks and expansion of one park to provide habitat protections (Adams et al. 2019).

(continued)

Box 1 (continued)

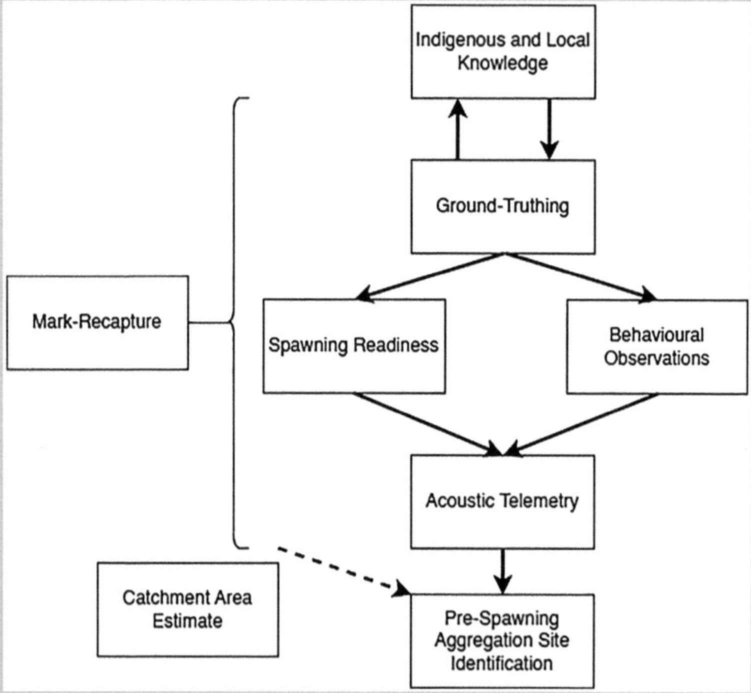

Fig. 2 Acquisition of relevant information regarding pre-spawning aggregation sites for bonefish (adapted from Adams et al. 2019)

Permit Spatial Management.

In 2009, recreational fishers in the Florida Keys requested a re-evaluation of the permit fishery due to an apparent decline in their fishing catch rates on the flats. In response, FWC conducted statewide stakeholder workshops, which resulted in the creation of the Special Permit Zone (SPZ) encompassing the Florida Keys, where the flats fishery for permit occurs. The SPZ has regulations and seasonal regulations that are more restrictive than the remainder of Florida, including a harvest closure for spawning season from May through July (summarized in Brownscombe et al. 2019). BTT conducted acoustic telemetry and received mark-recapture assistance from fishing guides to determine that fewer than 10% of permit crossed SPZ boundaries, making the

(continued)

Box 1 (continued)
protected area highly effective (Boucek et al. 2022). Subsequent collaborative research between BTT and fishing guides found that a spawning area called Western Dry Rocks (WDR) was an important spawning location for Lower Florida Keys flats permit, and that depredation rates were unacceptably high. In cooperation with other non-governmental organizations, BTT and the Florida Keys guides advocated for and achieved a total spawning season fishing closure at WDR (https://myfwc.com/fishing/saltwater/recreational/wdr/).

Juvenile Tarpon Habitat Identification and Restoration

In 2016, BTT began working with fishers and fishing guides to identify juvenile tarpon habitats in Florida (Wilson et al. 2019). BTT scientists used social media, print media, personal connections, and presentations at fishing club meetings to ask fishers and fishing guides to report locations where juvenile tarpon (<30 cm fork length) had been observed within the previous 2 years. More than 200 sites were reported. These sites are being incorporated into ongoing collaborations with FWC and other management agencies to prioritize habitat protection and restoration. For example, two previously identified sites are now in the process of a large-scale habitat restoration project being conducted in collaboration with the Florida Department of Environmental Protection.

Trust and engagement built on a long-term relationship between BTT and the fishing community were essential for this effort. Fishers characterized the reported sites as "natural" or "altered" based on their experience, and ground truthing of habitat characteristics by BTT scientists revealed high accuracy in fisher-reported habitat condition. Confidentiality of the reported information was promised and has been maintained; site data are only shared outside of BTT when specific habitat protection, restoration efforts, and plans are being formulated. Importantly, the engagement of fishers and fishing guides in research also increased awareness of the importance of habitat to fisheries. In a survey conducted at the 2016 Snook Symposium—an FWC-sponsored fisher outreach effort focused on popular recreational species with juvenile habitat requirements similar to tarpon—46% of >350 fishers in attendance voted for juvenile habitats as the top concern (the next highest topic garnered 15% of votes) (Wilson et al. 2023). This shift in fisher conservation perspective enabled BTT to work with FWC colleagues to propose snook as a Flagship Umbrella species for protection of coastal mangrove wetland habitats (Wilson et al. 2023), further engaging recreational fishers in resource management.

2.1 Identification of Problems and Opportunities

Recreational fishers can provide early warning of impending issues in a fishery and contribute to early stages of planning by identifying potential improvements. Because they use the fishery, they may be aware of changes over space and time, which may be difficult to detect in ecological sampling, and they may have preferences for fishery development or needs that are not currently met.

As an example, fishers and constituents directly involved in recreational fishing have been integral in advocating for management changes for flats-inhabiting fishes in the US Florida Keys, Eastern Gulf of Mexico, The Bahamas, Cuba, and the Caribbean Sea (hereafter referred to collectively as the flats fishery). Recreational fisheries across the region are extremely diverse, resulting in a broad range in the degree of fishery regulation by each country. Few countries have national policies and frameworks specific to recreational fishery management, whereas others have none or very limited policies (Potts et al. 2020). Before 2017, recreational fishing guides in the region were concerned about the management of the flats fishery for bonefish (*Albula vulpes*), tarpon (*Megalops atlanticus*), and permit, specifically regarding the lack of government policies in place to ensure fishery sustainability.

Bahamian fishing guides were also concerned about the lack of secure fishing rights and the lack of training programs available to develop future guides (Haukebo and Boné Morón 2020). Individual guides and larger recreational guiding organizations worked collectively to advocate for fisheries reforms, and in 2017, the Bahamian government passed the Flats Fishing Regulations. The law established a requisite flats fishing license for individuals and guides, a license marketplace and affiliated conservation fund, a ban on commercial harvest of certain flats species and the use of nets for capturing certain species, and several other provisions. A national guide training program is also now available through the Bahamas Agriculture and Marine Science Institute. Similarly, in Cuba, fishers pushed recommendations forward for developing a data collection program, improving communication between fishers and managers, and integrating fishers through a participatory decision-making process as essential components of new regulations in Cuba's first national fisheries law (Angulo-Valdes et al. 2022).

Concomitant with this recognition of recreational flats fishery issues in South and Central American countries, the US Florida, Eastern Gulf of Mexico, and Atlantic flats fishery faced similar challenges. Again, fishing guides provided some of the earliest warning of a potential decline in the abundance of permit in the flats fishery, working in partnership with Bonefish and Tarpon Trust (BTT). In 2009 and 2010, guides in the Florida Keys communicated concerns to the Florida Fish and Wildlife Conservation Commission (FWC) with a request to revise management of permit and pompano (*Trachinotus carolinus*), which were previously managed in Florida as a single fishery. Their call to action led to data acquisition in partnership with fishers and to revised management of the Florida Keys' permit fishery (see more in 2.2). Then in 2016, recreational fishing guides reported observing permits at offshore spawning sites beginning in April, one month before the spawning season harvest closure. Acoustic telemetry confirmed these observations, and in February 2018, FWC extended the spawning season harvest closure to include April (Brownscombe et al. 2019). Telemetry research found that 70% of permit tagged on the flats spawned at a single location Western Dry Rocks, and although the harvest season was closed, catch and release fishing for permit at spawning locations continued. Fishers then reported high rates of depredation of permits by sharks. Depredation rates reached 35% at spawning sites, and zero on the flats (Holder et al. 2020b). A campaign from February 2020 to February 2021 headed by recreational fishing guides and supported by BTT and other fishing-focused organizations (Lower Keys Guides Association, American Sportfishing Association, International Game Fish Association, Wild Oceans, Coastal Conservation Association, and Florida Keys Fishing Guides Association) advocated to close Western Dry Rocks to all fishing for the permit spawning season (April–July). In April 2021, FWC enacted a 4-month fishing closure of Western Dry Rocks to protect the spawning permit. Recreational fishers were essential in not only identifying potential problems in the flats fishery, but also in further steps of the management process.

In addition to identifying problems, recreational fishers can provide guidance toward new opportunities, such as new fisheries or outdoor experiences that were not previously provided or considered. This is especially true for fishers who may not meet the typical description and may have unique preferences and motives. For example, in Iowa, USA, urban fishery development and diversification entailed engaging potential fishers, not only active participants. Human dimensions research conducted by the Iowa Department of Natural Resources (IDNR) targeted all residents of urban locations to assess the preferences of potential fishers who had not yet been recruited to the activity (Box 2). By doing so at the advent of the IDNR's Community Fishing Program, the IDNR was able to quantify potential for growth in the urban-residing recreational fishing population (up to 148,000 people: Krogman and Stubbs 2021a), confirming the need for the program. It also clarified the need for specific features at fishing locations, especially for those facing constraints to participation. These features, including restrooms, parking areas, pedestrian access, shade, and picnic areas, are now standard recommendations for any urban pond and park renovation or acquisition (T. Stubbs, IDNR Community Fishing Program Biologist, personal communication, 25 April 2023). Potential growth and expansion in urban fisheries depended on a strong foundation of knowledge about fishers, including those who had not previously been recruited to fishing.

Box 2 Characterizing Urban and Suburban Recreational Fishers in Iowa

Challenge

The Iowa Department of Natural Resources (IDNR) manages hundreds of public-access recreational fisheries across the state of Iowa, USA. Faced with rapid urbanization and declines in fisher participation, the IDNR initiated the

(continued)

Box 2 (continued)

Community Fishing Program in 2017. The program was focused entirely on improving fishing opportunities for urban Iowans but had limited information to drive program priorities. Previous surveys targeted licensed fishers' behaviour and preferences, leaving a knowledge gap regarding unlicensed and unrecruited fishers in general. Although some studies had investigated factors constraining participation in fishing in Iowa (e.g. IDNR and RBFF 2013), none had determined common, measurable, or considered constraints that might prevent a person from being initiated into fishing in the first place. Finally, rapid demographic change during the previous two decades produced an urban population in Iowa that was not reflected in the licensed fisher population (USCB 2018), and little was known regarding whether these factors would correlate with a unique fisher profile in terms of interests, preferences, and constraints.

Approach

To address these knowledge gaps about potential, lapsed, and active recreational fishers, a general population survey was designed which included questions regarding previous fishing behaviour and preferences, factors reducing fishing participation, desired amenities and outreach programs, and demographics (Krogman and Stubbs 2021a). The invitation was mailed with a US$2 up-front incentive to a representative sample from every community in the Community Fishing Program. Responses were sought for potential fishers who had not yet been recruited to license purchasing, so a public mailing list was used in lieu of the existing fishing license database of addresses. The survey instrument was provided both online and via mailed paper copy with return postage paid. The survey yielded a 28% response rate with responses from every community targeted.

Outcomes

Important findings included the potential for growth in urban fisheries, as indicated by high numbers of potential and lapsed fishers; the need for expanded amenities at urban fishing locations; the need for not only provision of viable fishing opportunities close to home, but also better communication of where opportunities were; and the spatial variation in desire for certain fish species, amenities, and programs. Several examples reveal a positive impact of Community Fishing Program actions:

- IDNR created a mobile-friendly urban fishing map documenting all urban fishing locations with public access; this page is located at https://www.iowadnr.gov/Fishing/Fish-Local and receives more website visits than any other IDNR webpage.

(continued)

Box 2 (continued)

- As many fishing locations are not owned by IDNR, IDNR communicated the importance of certain amenities and park features to partners, as well as providing letters of support and education. Partners have increasingly incorporated these improvements in renovations and new acquisitions since 2017.
- Spatial prediction of interest in channel catfish (*Ictalurus punctatus*) and flathead catfish (*Pylodictis olivaris*) facilitated identification of neighbourhoods where catfish stocking could be introduced or reduced, making stocking more efficient for IDNR (Krogman and Stubbs 2021b, Fig. 3).
- IDNR tailors outreach and education topics to more specific target audiences, knowing that potential fishers' interests differed from those already fishing. Marketing of those topics and course availability is also conducted using neighbourhood-based apps with spatially targeted ads.

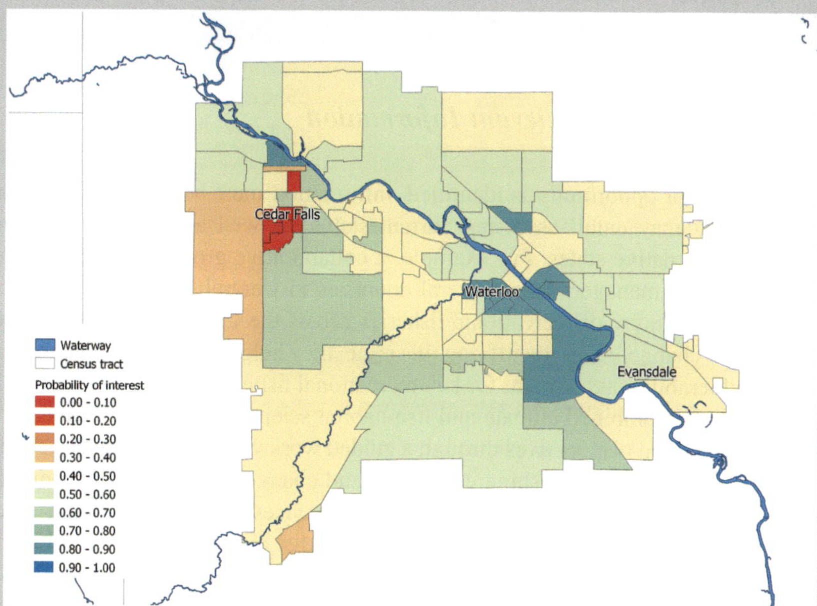

Fig. 3 Predicted interest for recreational fishing targeting channel catfish in a northern Iowa city and its suburbs, based on survey of the public living in urban areas

Indirect sources of information, such as data mining (Venturelli et al. 2026), may also reveal fisher preferences and concerns before they escalate, allowing fishery managers to proactively address issues. Scraping text and imagery from social media, including fishing-oriented apps (e.g. MyCatch) as well as forums and other broader networks, can reflect fisher sentiments, behaviour, and more (Cooke et al. 2021; Kochalski et al. 2026; Lennox et al. 2022). Likewise, content analysis of open-ended responses to surveys or dictation from interviews may reveal common concerns not included in formal survey questions. For instance, the nature and extent of conflicts among small boat fishers in Hawaii were assessed using both manual and automated content analysis of open-ended responses from a fisher survey (Suan et al. 2021). Although the survey was conducted for another broader purpose, data from an open-ended question characterized conflicts in the fishery, and the relationships and processes involved in that conflict. This indirect approach to using fisher data yielded identification of one potential issue which needed management attention (a rise in tangible disputes nested in frayed relationships among fishers), as well as the recommendation that strengthening relationships in the Hawaii small boat fishery was the appropriate, more long-term approach to reducing emergent conflicts (Suan et al. 2021).

2.2 *Acquisition of Relevant Information*

After an issue or opportunity is identified, information must be acquired to characterize it. This may entail assessing current status, as well as potential risks and benefits to alternative states. For example, a collaborative group of fishing guides, researchers, and managers from several countries in Central America shared concerns about the future of recreational fisheries across the Caribbean due to impacts from climate change. After identifying this issue, they helped organize and conduct a climate vulnerability assessment for key recreational fisheries (Braynen et al. 2022). The assessment combined climate and fish habitat scientific data with local ecological knowledge and perspectives through a guided workshop. Workshop participants provided insight into local fishing conditions and changes, socioeconomic implications, and uncertainty, as well as revealing differences in their perception of the severity of climate change impacts. Their knowledge and expertise lent a broader perspective on the potential impacts of climate change, communication needs, and management alternatives than would have been possible with ecological data alone.

Recreational fishers may be directly involved in the acquisition of relevant information by providing their opinions, experiences, knowledge, or even participating in biological data collection. Engaging directly in relevant data collection can be a major motivation for recreational fishers and other stakeholders, contributing to their long-term dedication, self-identification as a conservation-minded person, and positive outlook on fishery management, as demonstrated for the California groundfish fishery (Mason et al. 2020), the Bahamian billfish fishery (Lauretta and Goodyear 2018), and the flats fishery of the Gulf of Mexico and Caribbean (Box 1).

Informally sharing catch data through social media or mobile applications like MyCatch can yield data like that collected in a creel survey (Venturelli et al. 2017; Johnston et al. 2021). Metrics include fish species targeted, catch, harvest, fish size, and fishing effort (Johnston et al. 2021) and may be comparable to biological samples collected by managers (Gundelund et al. 2021). Active participation of fishers through citizen science has been shown to increase confidence in ultimate management outcomes (Hartley and Robertson 2006) and will continue to become a more important part of fishery management in the future (Skov et al. 2021).

Voluntary data collection may be part of a formalized structure, as demonstrated in the UK and South African marine recreational fisheries. Although creel surveys are a common tool used to directly obtain fishing trip data from recreational fishers (Pollock et al. 1994), they may not be as efficient in fisheries that cover a large area with many dispersed fishers, such as in marine fisheries (Cooke et al. 2000, Venturelli et al. 2017, Weir et al. 2022, Box 1). In the UK, sea angling is not licensed, meaning there is no list of fishers available nor mandatory catch reporting requirements; therefore, surveys of catch and effort are required to fully quantify the activity. In addition, response rates to telephone and mail surveys are very low, such that obtaining a random sample of fishers to contribute to off-site surveys like catch diaries is very difficult (Hyder et al. 2020, 2021). Instead, fishers voluntarily complete a logbook survey as part of the "Sea Angling Diary Project" (https://seaangling.org/), a program with clear conservation messaging but also incentivized by prizes for participation (Hyder et al. 2020, 2021). This is a large citizen science project where volunteer sea fishers report all fish kept or released and trip characteristics (e.g. location, date, duration, method) for all fishing trips throughout the year using an online system or mobile app (Hyder et al. 2020, 2021). Then, researchers generate multilevel Bayesian models of catches of all species by individual fishers, considering various demographic and skill characteristics of the fishers. The predictions from these models are subsequently used to post-stratify catch rates to the total number of fishers via a technique called "multilevel regression and poststratification", enabling the total number and tonnage of fish caught by marine recreational fishers to be quantified. In South Africa's shore-based marine recreational fishery, a national competitive angling league (i.e. Rock and Surf Super Pro League) and the South African Fisheries Ecology Research Lab from Rhodes University have developed a partnership over the last decade. The research lab developed a database to collect species, length, and catch and effort information from the league, and this database was used by the league to calculate points and prize winners for their competitions. After 10 years, this database now represents the largest georeferenced dataset of species composition, size distribution, and catch and effort information for the recreational fishery in South Africa. These data will be used to inform fisheries management (e.g. stock assessments, Fisheries Management Plans) and marine spatial planning (e.g. protected areas, planning).

As another example, the Coastal Angler Science Team (CAST) was developed as a collaboration between fishers and researchers to study fisheries in the Florida Everglades, USA, a massive ecosystem with thousands of square kilometres of water. The CAST program is a product of long-term monitoring efforts before and after the world's largest coastal restoration, enhancing freshwater flow to Everglades National Park, and its long duration (ongoing for 18 years: Rehage and Loftus 2007) allowed the development of consistent and long-term relationships with fishers and fishing guides in the area. Trust built in CAST facilitated the sharing of confidential logbooks that provided detailed catch records and specific locations of each fish caught dating back to the early 1970s, which were used to explore relationships between fisher catch rate for bass and snook to seasonal hydrologic factors directly linked to Everglades restoration targets (Sklar and Browder 1998). For instance, fisher catch rates for Florida largemouth bass tracked fishery-independent catch records. Furthermore, 56% of the variation in fisher catch rate was explained by the amount of time the coastal Everglades did not receive freshwater the previous year, a focal performance measure of Everglades restoration (Boucek and Rehage 2015). Long-term relations built through CAST also inspired a cooperative mark-recapture study conducted by Florida International University, in which fishers scanned their captured fish with PIT tag readers and recorded measurements. These fishers were trained at the beginning of the study on how to properly use the readers and record data, facilitating evaluation of the survival of Florida largemouth bass in coastal refuge habitat across years of varying drying severity (Pierce et al. 2021). Including CAST fisher data improved largemouth bass recapture probabilities and increased sample size, yielding a mechanistic linkage between variations in fisher catch rate with relevant hydrologic factors like drying.

Alternatively, requisite reporting can also be used to capture fishery metrics and is a commonly used approach in marine fisheries (especially commercial but sometimes recreational; Hyder et al. 2018). Co-management and allocation of fishing rights frequently demand the contribution of data to track stock abundance and fishing pressure. In Western Australia, fisher engagement from both recreational and commercial sectors is needed to guide allocations of Western rock lobster (*Panulirus cygnus*), Roe's abalone (*Haliotis roei*), West Coast demersal scalefish, and other species (Box 3). Stakeholder collaboration is considered an essential component to the conduct of cost-effective fishery-independent surveys, which accompany the shared responsibility of fishing rights in Western Australia (Caputi et al. 2021).

Box 3 Co-management of Recreational and Commercial Fisheries in Western Australia

Challenge

Western Australia occupies a third of the Australian continent and provides a vast variety of marine, estuarine, and freshwater recreational fishing opportunities. With an estimated 753,000 recreational fishers and a participation rate of 28% (DPIRD 2024), understanding the magnitude of recreational catches is requisite for the management of more than 40 commercial fisheries managed within 21 resources under the principles of Ecosystem Based Fisheries

(continued)

Box 3 (continued)

Management (EBFM, Newman et al. 2021). For some sought-after species, such as snapper (*Pagrus auratus*) and the West Australian dhufish (*Glaucosoma hebraicum*), the estimated recreational catch exceeds that of the commercial catch. The recognized importance of supporting recreational fishing-based tourism in certain areas has also resulted in some fisheries being essentially "recreational-only". Western Australia is also home to the world's first commercial fishery, Western rock lobster (*Panulirus cygnus*) to meet the Marine Stewardship Council Standard, and a substantial part of the State's fishery value comes from independently certified sustainable fisheries. Achieving and maintaining this accreditation requires an in-depth understanding of how and where recreational fishing occurs.

Approach

The Western Australian Department of Primary Industries and Regional Development (DPIRD) is the state government agency responsible for managing sustainability of the community's aquatic ecological resources to generate economic and social outcomes. The peak sector body for recreational fishers, Recfishwest, is responsible for representing the interests of recreational fishers. Six separate species and recreational fishery license categories have been implemented, providing useful sampling frames for probabilistic surveys.

Outcomes.

Recreational fisher-contributed data in Western Australia have facilitated management action execution in numerous ways.

- Estimated catches for key species are important data inputs for stock assessment and implementation of Harvest Strategies which provide a pre-agreed framework for making fisheries management decisions. Historically, DPIRD adopted an Integrated Fisheries Management Policy whereby an allowable sustainable harvest was set for a resource, allocations were then determined between user groups, and each sector's catch was then managed within their allocation (Crowe et al. 2013). These catch shares were allocated for three recreational fisheries: Western rock lobster, Roe's abalone (*Haliotis roei*), and West Coast demersal scalefish (Ryan et al. 2016).

Fishery management is fully coordinated with input from stakeholders to deliver resource-level objectives established by the Minister on behalf of the community (Fig. 4). Ongoing recreational fishing surveys provide reliable and regular information to meet current and future management needs. The need for robust information on recreational fishing has motivated the completion of

(continued)

Box 3 (continued)
more than 34 separate on-site and off-site surveys since 1976 (Henry and Lyle 2003; Tate et al. 2020) and will continue to be relevant for Western Australia fishery management in the future.

Recreational fishing provides economic and social benefits for the Western Australian community, with a recent survey indicating that annual expenditure by recreational fishers is approximately AUD$2.4 billion (McLeod and Lindner 2018) and economic contribution is approximately AUD$1.1 billion (Moore et al. 2018). Participation in recreational fishing and percentage of users satisfied with DPIRD's management are both measured annually from a telephone survey and used as key performance indicators of the EBFM process. The most recent satisfaction rating (for 2023/24) was 89%, consistent with values between 86% and 90% since 2011/12 (DPIRD 2024).

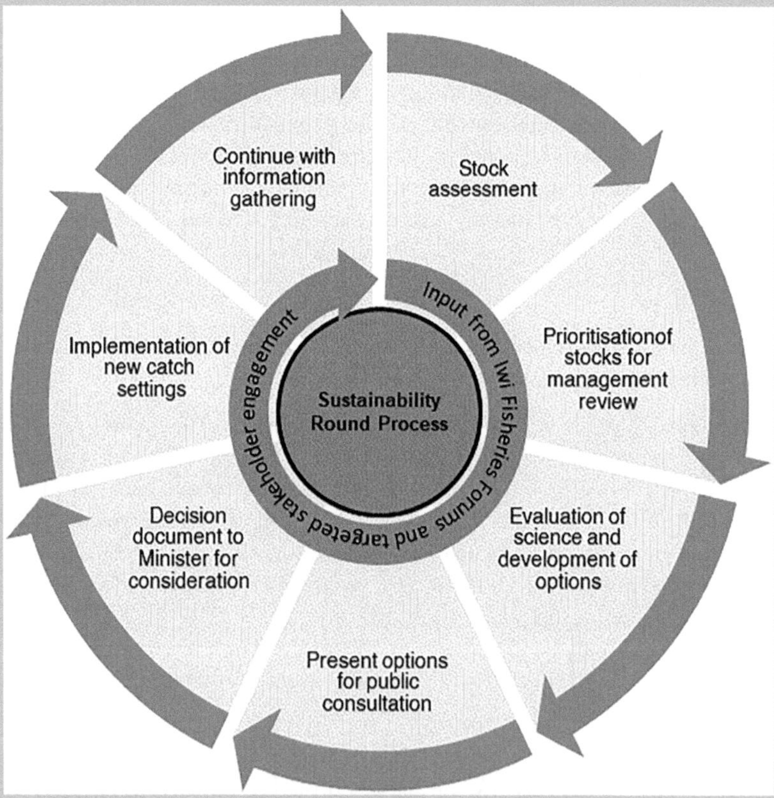

Fig. 4 Cyclical decision-making process used by fisheries New Zealand for rock lobster fisheries management (adapted from Fisheries New Zealand 2020)

Recreational fishers may provide data indirectly through remote monitoring methods or by contributing information to large datasets that can be mined (Lennox et al. 2022). For example, analysis of the mobile application and accompanying website MyCatch yielded similar areas of fisher travel origin, fishing trip behaviours, and catch rates to creel survey data collected by the management agency in Alberta, Canada (Johnston et al. 2021). App-derived data were not comparable to biological surveys' metrics of fish abundance and community composition. More recently, the same app's data limited to highly specialized fishers (walleye tournament fishers in Iowa) yielded higher catch rates, longer fishing trips, and larger fish than indicated in concurrent creel surveys at the same locations (R. Krogman, unpublished data), but aligned with certain biological samples by gear and season. App-based data can provide reasonable estimates of both effort and catches in specialized fisheries (Gundelund et al. 2021). Anonymous location data, or the use of anonymous data regarding mobile phone users' travel behaviour, may also be mined to determine how far people may travel to get to a fishing location, how long they stay, and how frequently they visit. Other remote monitoring may include the use of drones, counters, or cameras to capture water usage and fishing pressure (e.g. Taylor et al. 2018).

Finally, recreational fisher information may be obtained indirectly through meta-analysis or identification of general trends from a compilation of studies on the same. Meta-analysis and large-scale reviews are frequently used to leverage relevant information, such as recreational fisher site choice, which may derive from many data-poor or variable sources. For instance, site choice was examined across 114 manuscripts by Hunt et al. (2019), helping identify general and site-specific attributes that may affect fisher selection. Their review indicated that cost had a negative effect, whereas destination waterbody size, quality of catch, facilities or amenities, and environmental quality had a positive effect on site choice, a finding that could guide management actions such as investing in site amenities or stocking fish. Not only did site choice valuation identify areas where management intervention may be worthwhile but it also provided willingness-to-pay estimates, which further guide management alternative selection.

2.3 Evaluation of Management Alternatives

Fisher information can also be used to help evaluate management alternatives. Recreational fishers may share their preferences for a suite of management options and provide information regarding the potential or anticipated impacts of management alternatives. This is a common component of Ecosystem Based Fisheries Management (EBFM; Box 3). Two examples where management is informed by a collaborative group containing representatives of both recreational and customary interests include the National Rock Lobster Management Group (NRLMG) and the Southern Scallop Working Group (SSWG) in New Zealand. Customary fishing rights are guaranteed to tangata whenua (people of the land with authority in a particular place) under the Treaty of Waitangi, and therefore Māori participation in weighing management alternatives is essential. Almost all of New Zealand's fisheries are managed by a single agency, the Ministry for Primary Industries, which focuses on ensuring the sustainability of New Zealand's wild fish stocks, aquaculture, and the wider aquatic environment.

New Zealand's 13,020-km coastline provides a wealth of recreational fishing and tends to be more productive than in nearby Australia, although fewer species are caught by recreational fishers (Hartill et al. 2012, New Zealand Initiative 2016). An estimated 9% of New Zealand residents are considered recreational fishers, collectively harvesting around 11 million marine fish across numerous species a year (Wynne-Jones et al. 2019). National panel surveys provide periodic updates of recreational catches on a stock basis (Wynne-Jones et al. 2019), and numerous on-site surveys provide catch estimates for specific areas of the North Island (e.g. Hartill et al. 2012).

The NRLMG was set up in 1992 and is the longest-standing multi-stakeholder fisheries group in New Zealand. Group members represent the various customary, recreational, and commercial fishing sectors, environmental interests, fisheries compliance, and Fisheries New Zealand. The goal of the NRLMG is to ensure that rock lobster fisheries will be managed and maintained at or above the assessed and agreed biological reference points, using a comprehensive approach that recognizes a range of commercial, customary, non-commercial, recreational, and environmental concerns and values. At present, each year the NRLMG reviews catch settings for rock lobster stocks and provides advice on any regulatory changes needed to support Total Allowable Catch decisions (NRLMG 2021). Fisheries New Zealand is currently reviewing the NRLMG's role and function for improvement of the process, as indicated by the cyclic and constantly improving nature of the fishery management cycle.

In the SSWG, a multi-sectoral working group was set up in 2018 to address sustainability concerns for scallops at the top of the South Island. The SSWG brings together fisheries managers, scientists, iwi (a Māori community or people), and representatives of the commercial and recreational fishing sectors, with the overall aim of ensuring that the scallop stocks rebuild to a healthy level (SSWG 2021). The SSWG provides a platform for stakeholders to provide input on what an appropriate framework might be, including areas such as research, harvest, and restoration (SSWG 2021). Customary rights are protected by law in the Treaty of Waitangi (Fisheries Claims) Settlement Act of 1992 and the 1992 Deed of Settlement. Under the Deed of Settlement, obligations to Māori exist to provide for customary fishing practices and the traditional gathering of fish. This legislation also includes the need to make better provision for Māori participation in the management and conservation of New Zealand's fisheries. The SSWG provides a similar function to the NRLMG as a multisector working group, but does not implement the regulatory framework.

In a more recent example under the 2020 Fisheries Act, the UK is developing bespoke Fisheries Management Plans (FMPs) for managing sustainable fisheries, supporting a vibrant fishing industry, and supporting a healthy and productive environment. Although 43 FMPs have been identified for work, the first five FMPs have been initiated. For example, stocks of European sea bass (Moronidae: *Dicentrarchus labrax*) have recently declined, leading to tension between recreational and commercial sectors. Its FMP is one of the top priorities, and government decision makers have taken the options identified through co-design forward, with the aim of finalizing the sea bass FMP during 2023. Co-design, a process led by social scientists, includes shared experience, collective intelligence debates, and co-design and co-refine workshops. The goal is to seek views from all sectors with an interest in the management of sea bass and identify recommendations that are palatable to all stakeholders.

A collection of fisher information through surveys or other indirect methods can also clarify management alternatives without direct stakeholder engagement. For example, in rainbow trout fisheries in British Columbia, Canada, Ward et al. (2013) were able to identify four fisher groups who differed in fishing behaviour, catch and harvest, and avidity and would likely react differently to varying management scenarios. Differences among regions in fisher group composition indicated that region-specific trout management (e.g. restrictive regulations, fish stocking densities) was appropriate. Likewise, different fisher groups in a northern pike fishery had variable responses to and preferences for restrictive fishing regulations, and the optimal regulation depended on both biology and fisher composition in the fishery (Johnston et al. 2010). Use of recreational fisher information can proactively avoid the pursuit of undesirable management alternatives and highlight options that may be most acceptable to fishers. This facilitates adoption and compliance, as well as supporting fisher satisfaction.

2.4 Selection and Implementation of a Management Action

After management alternatives have been considered, an action must be chosen and implemented. Although involving recreational fishers in this step is somewhat less common (e.g. Crandall et al. 2019), there are several ways fishers can contribute meaningfully. They may conduct fishery enhancements as leaders or collaborators in line with the manager's plan (e.g. stocking fish or establishing fish habitat); encourage compliance and regulation enforcement; share information with others regarding the action; and enforce institutions that guide others' behaviour (Granek et al. 2008; Hunt et al. 2013).

As mentioned in Box 1, trust and meaningful engagement were essential for building a strong relationship between recreational fishers and a fishery-focused non-governmental organization, such that fishers and fishing guides willingly initiate communication of fishery issues. Their involvement in habitat identification translated directly to widespread self-reported understanding of the importance of juvenile fish habitat to fishery sustainability (Box 1). Although this example comes from Florida and the surrounding region, inclusion in marine recreational fisheries management decision-making in general was still considered lacking by stakeholders in Florida; only 15% of survey respondents felt they had opportunities to do so (Crandall et al. 2019). Importantly, perception of their ability to influence decisions correlated positively with satisfaction in fisheries management decisions.

Participatory processes should promote transparency in decision-making, as well as recognizing trade-offs and uncertainties, and may follow several tenets to be as successful as possible in establishing consensus (Helvey 2004; Irwin et al. 2011; Gleason et al. 2013). As demonstrated in Box 3 in Sect. 2.3, co-management scenarios based on EBFM principles include fishers in this part of the fishery management decision-making process.

2.5 *Assessment of Outcomes*

Recreational fisher information can also help assess the outcomes of a management action and evaluate its effectiveness. Fishers are affected by management actions and may respond with direct feedback, such as that provided by surveys, or may alter their behaviour in some way (Ward et al. 2016). An obvious fisher-contributed metric assessing outcomes is fisher satisfaction, a question common in fisher surveys (Box 4) and frequently used as a key metric of success (Box 3). Changes in behaviour, such as fishing effort, travel distance or frequency, and fish targeting and harvesting may also reflect tangible changes in a recreational fishery because of management intervention (Box 4). Recreational fishers' behaviour in response to an action may be affected by their awareness and perceptions of the legitimacy of the action, their own preferences and experience, and their understanding of the norms and legalities around the action (Ward et al. 2016). Unfortunately, empirical research of fisher reactions and behavioural feedback to management actions is somewhat lacking.

Box 4 Using Creel Surveys to Measure Change in Fisher Use and Behaviour Following a Lake Renovation

Challenge

Creel intercept surveys have been instrumental in Iowa, USA as part of the lake restoration process, particularly in gathering pre-renovation information from recreational fishers to guide improvements and to measure changes in fisher behaviour after renovation is complete. Lake-based recreation in general draws US$1.1 billion in annual expenditures, with recreational fishing being one of the most important activities in which people participate. IDNR invests US$9.6 million annually to improving water quality in Iowa's public lakes, often as part of a broader watershed- and park-level renovations.

(continued)

Box 4 (continued)

Before renovation, Three Mile Lake attracted 0.08 fisher-hours/acre/day (Krogman 2022). Recreational fishers caught 1.6 fish/fisher-hour and harvested 0.9 fish/fisher-hour, with 43% of harvest being crappies (*Pomoxis* spp). Most fishers released fish because they were too small (83%) or because the fisher was not interested (12%). Fishers primarily targeted crappies, followed by walleye (*Sander vitreus)* and largemouth bass (*Micropterus salmoides).* Yellow bass (*Morone mississippiensis)* were considered a major hindrance in the recreational fishery and were a substantial part of the reason for fishery renovation, in addition to common carp (*Cyprinus carpio).* Recreational fisher satisfaction averaged 3.7/5.0, where 5.0 was extremely satisfied and 1.0 was extremely dissatisfied; they also reported concerns about algae blooms and high turbidity, both issues echoed by ambient lake monitoring (AQUIA 2022). Recreational fishers travelled an average of 132 km and spent US$151/trip to visit Three Mile Lake.

Approach

Surveys at Three Mile Lake, an 880-acre reservoir in southwest Iowa, were conducted before and after a large-scale renovation in 2016, which involved reservoir drawdown, piscicide application, habitat additions, sedimentation and shoreline erosion prevention, and amenity additions including a fish cleaning station. Surveys included questions regarding fisher effort, catch and harvest, reasons for release behaviour, satisfaction, and trip-related expenditures.

Outcomes

After renovation, visitation and fishing pressure more than tripled to 0.28 fisher-hours/acre/day (Krogman 2022). Fishing catch and harvest rates decreased somewhat, which was partially attributed to a shift in focal target species from crappies to walleye and partially to reduction of yellow bass. Highly desirable walleye accounted for most of the harvest, and fisher satisfaction averaged 4.6/5.0, indicating a substantial improvement in fisher perception of the recreational fishery (Fig. 5). Recreational fishers travelled less than before renovation (92 km), which could have related to the 2020 pandemic which drastically reduced travel distances for outdoor recreation in general, but also to a shift in which cities fishers most often travelled from (i.e. from the Des Moines metropolitan area instead of the Omaha and Council Bluffs area). Spending per trip increased to US$286/trip. Although environmental challenges remained consistent over time at Three Mile Lake, their severity was reduced by renovation. Fishers reported environmental concerns half as frequently as before, and ambient lake monitoring indicated that turbidity, total dissolved solids, and microcystins had been reduced (AQUIA 2022).

(continued)

Box 4 (continued)

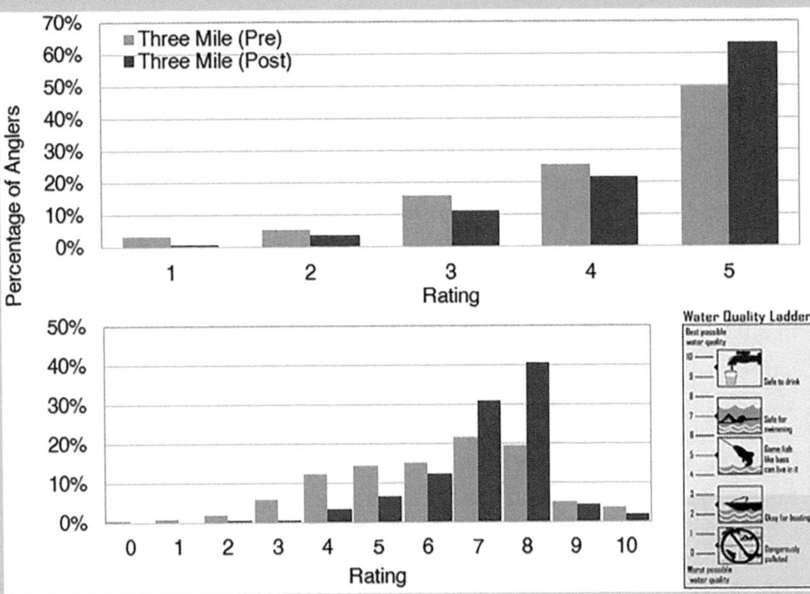

Fig. 5 Fishing satisfaction (upper panel) and water quality (lower panel) rating for Three Mile Lake before and after lake renovation (fishing satisfaction ranges from 1 to 5, with 1 = very dissatisfied and 5 = very satisfied; water quality ratings are based on the US Environmental Protection Agency Water Quality Ladder, which ranges from 0 to 10)

In general, the renovation at Three Mile Lake was considered successful based on these creel results. In addition to traditionally included metrics on fishing effort, catch, and satisfaction, human dimensions questions such as spending and travel behaviour were considered essential components measuring success. Creel intercept surveys were also conducted concurrently at two sister lakes nearby, with the initial hypothesis that fishing effort would shift from Three Mile Lake to the others during renovation. Instead, it was found that the three-lake complex acted as a single destination for recreational fishers, and that the renovation of Three Mile Lake had a long-term positive effect on the other two locations as well.

Sentiment and other reactions to management-induced changes can also be tracked and quantified by mining online forums, social media, and other websites to which users either passively or actively provide data (Lennox et al. 2022). Again, active contribution of data may include commenting on a fishing forum or posting fish catch photos. For example, Allison et al. (2023) categorized the accompanying comments to posts on the Facebook group "SaltfishSA" as pro-environmental, neutral, or anti-environmental. They found that the pro-environmental comments in response to posts of "illegal catches" (such as retaining more fish than specified in the bag limit) increased over time, suggesting that anglers were becoming increasingly supportive of the management regulations. Similarly, Allison et al. (2023) used pictures posted by "SaltfishSA" members to examine changes in the catch and release behaviour of shore-based marine anglers in South Africa over time. Using standardized metrics from the photographs, they showed that even without interventions from scientists or managers, the catch-and-release behaviour of shore-based anglers had improved since the group's inception. Although social media contributors may not be representative of all fishers (e.g. due to lower participation by certain demographic groups) or of all fishers using social media (e.g. due to the tendency to post either positive or negative content), these approaches can provide large volumes of text- or image-based feedback in a timely manner with very little data acquisition effort. Passive data generation may also include anonymous location or website visitation behaviours. In terms of assessing outcomes, content or sentiment analysis of recreational fishers' comments on their fishing trip or response to official information from the fisheries management entity can reveal positive or negative feelings toward fishers' experiences.

3 Importance of Applied Research and Engagement of Recreational Fishers to Guide Fisheries Management

As demonstrated above, recreational fisher information has numerous applications in fisheries management. It demands greater incorporation and study during the management process rather than as an afterthought or tangential to biological fisheries data (Ditton 1996; Granek et al. 2008). The information needed from recreational fishers depends on the management application and should derive directly from a specific portion of the management process and a specific objective. First of all, are fishers being involved in the development of management goals and objectives for a particular recreational fishery? In a new or newly managed fishery, recreational fishers may contribute to broader goal identification, as in the development of FMPs in the UK (see Sect. 2.3). Next, consider which steps in the management process achieving that objective can best be informed by recreational fisher data, and how those data may be obtained. Whether direct or indirect, engagement of recreational fishers should occur early and proactively to incorporate the information gained into subsequent steps. Therefore, applied human dimensions research and engagement should be a concurrent and ongoing component of fishery management.

Applied human dimensions research and stakeholder engagement benefit fisheries management by increasing understanding and prioritization of constituent needs, desires, and constraints; deriving alternative management actions that may be palatable to constituents; predicting fisher behaviour regarding the effects of management actions; and explaining reasons for failure of a management action. Stakeholder engagement—especially of participating recreational fishers but also related stakeholders—can illuminate issues in the fishery of which the manager may not have been aware; produce fisher-derived data; produce diverse and creative alternatives and solutions; and provide meaningful feedback regarding a management action's effects. Engaging recreational fishers in the management process in a tangible way not only reduces costs from taking the entire workload in terms of time and resources, but also reduces lost opportunity costs from taking inappropriate or inefficient actions. To be successful, it must be meaningful and feasible (Box 1, Arlinghaus et al. 2019, Reed et al. 2020).

Alternately, failing to utilize recreational fishers as a resource of information and energy can have a significant cost. These failings may include dissatisfied fishers (e.g. UK recreational fishers' role in participatory management: Reed et al. 2020); alienated stakeholders (e.g. failure of consensus in California marine protected area design and subsequent top-down decision-making: Helvey 2004); poaching and other forms of noncompliance (e.g. suspected poaching in no-take zones in the Great Barrier Reef, McCook et al. 2010); conflict and feelings of marginalization (e.g. "salmon wars" in Alaska: Harrison and Loring 2014, Reed et al. 2020); distrust for management or governance (e.g. fisher distrust of managers due to a failed regulation that was rescinded: Boxrucker 2002, Reed et al. 2020); ballot initiatives and other legal action to circumvent manager autonomy (e.g. recreational fisher-led ban of commercial gill-nets in Florida, Loring 2017); economic loss (e.g. $41.8 million revenue loss due to the unregulated black market recreational fisheries in Algeria: Babali et al. 2018); and ecological damage (e.g. recreational fishers' role in declining marine finfish stocks in the USA: Coleman et al. 2004; scallop fishery collapse in Lyme Bay: Fleming and Jones 2012).

As one example, in 1999, the California Fish and Game Commission and a group of concerned citizens initiated a process of evaluating a marine protected area in the Channel Islands National Marine Sanctuary (Helvey 2004). A Marine Reserves Working Group composed of recreational fishers, resource managers, commercial interests, and others was established, which proceeded to identify ground rules, problems and mission statements, and potential management alternatives. However, when a final design of the marine protected area's areal coverage was needed, the group could not reach consensus, and eventually, resource management staff established a design independently with their own authority. Helvey (2004) suggested a number of primary causes of the failure in consensus: (i) weighting certain goals disproportionate to their overall importance, often without intending to do so; (ii) not acknowledging the scientific uncertainty in predicted benefits or costs of alternatives; (iii) not having all relevant information available in time for fair consideration; (iv) not utilizing available scientific expertise, and (v) less developed conflict negotiation skills and willingness for flexibility. The last two points highlight the necessity of not only the engagement of key recreational fishers and stakeholders but also of fisheries scientists in effective co-management processes. Despite the controversy around the Channel Islands Marine Protected Areas, they were successfully integrated into California's Marine Life Protection Act plan, and lessons learned from the early process were applied in subsequent statewide marine protected area planning (Gleason et al. 2013). Specifically, roles and rules were established early, deliverables and timelines were identified, and more resources, including scientists were engaged in the process.

Integrating recreational fishers into the fisheries management process may face substantial resistance despite its demonstrated benefits. Governance systems for recreational fisheries can adapt to increased fisher information integration but may have to cycle through a period of evaluation and re-organization after overcoming high inertia and lack of trust. This was clearly demonstrated for German recreational fisheries by Daedlow et al. (2011). In West Germany, recreational fisheries were managed at a local level by small, self-organized fishing clubs which governed fishery management and access to waterbodies in their respective districts. Entrusting these fishing clubs with management responsibilities was challenging for East Germany, which had been governed under a system of larger umbrella associations (still fisher-led) before 1989, and ultimately led to high resistance to West Germany's governance adoption. In that case, the pre-existing system was retained after a critical evaluation of each framework. In contrast, a voluntary, bottom-up decision-making process in scallop fishery management ultimately led to fishery endangerment and substantial loss of social capital before the passage of the protective Marine Act (Fleming and Jones 2012). In this case, broader, more structured stakeholder involvement and greater top-down interventions were determined to be necessary to provide adequate protections for the scallop fishery and were therefore embraced as part of a new decision-making framework.

4 Conclusions

Recreational fisher information helps managers identify problems and concerns in a fishery, acquire needed data (through direct and indirect methods), evaluate management options, implement actions, and assess outcomes. Every step of the fisheries management process involves goals and specific objectives, which may be guided by recreational fishers from their initial formulation to final assessment of their success or failure.

Engaging recreational fishers in the fishery management process can be challenging, but the "real world" examples in this chapter demonstrate its salience for fisheries management. Furthermore, managers are increasingly propelled to understand more about the fishers themselves beyond simply trying to quantify catches (Fig. 1). The benefits of incorporating fisher information into the management process may include the following:

- Improved or positive outcomes.
- Satisfied and engaged fishers.
- Increased fisher compliance.
- Reduced feelings of marginalization.

- Greater knowledge of the research and management process, and a better understanding of conservation measures, which leads to.
- Increased political support for governing agency actions and processes.
- Greater trust and confidence in the agency and the fisheries management process overall.

Failure to incorporate fisher information into the management process may cause:

- Thwarted or unforeseen outcomes.
- Dissatisfied fishers.
- Decreased fisher compliance.
- Political or legislative action to alter management and its processes.
- Declining fisher participation.
- Distrust and reduced confidence toward the governing agency.

We encourage managers to proactively collect fisher information and enthusiastically seek fisher participation and feedback throughout the fisheries management process. The techniques presented in the remainder of this book collectively represent a toolbox, with each tool designed to engage fishers or utilize fisher information in a particular way. Some tools may be more accessible to you than others, but the intent of this chapter is to convince you that they are worth learning to use for better fisheries management.

Acknowledgements We thank C. Baird for their useful insights regarding New Zealand fishery management and three peer reviewers whose comments improved a previous version of this manuscript

References

Adams AJ, Murchie KJ (2015) Recreational fisheries as conservation tools for mangrove habitats. Am Fish Soc Symp 83:43–56

Adams AJ, Shenker JM, Jud ZR, Lewis JP, Carey C, Danylchuk AJ (2019) Identifying pre-spawning aggregation sites for bonefish (*Albula vulpes*) in the Bahamas to inform habitat protection and species conservation. Environ Biol Fish 102(2):159–173

Allison C, Winkler AC, Childs A-R, Muller C, Potts WM (2023) Can social media platforms be used to foster improved environmental behaviour in recreational fisheries? Fish Res 258:106544

Angulo-Valdes J, Pina-Amargos F, Figueredo-Martin T, Fujita R, Haukebo S, Miller V, Boné-Morón E, Whittle D (2022) Managing marine recreational fisheries in Cuba for sustainability and economic development with emphasis on the tourism sector. Mar Pol 145:105254

AQUIA (2022) Water quality monitoring and assessment database version 4.2.0.11557. Iowa Department of Natural Resources, Des Moines. https://programs.iowadnr.gov/aquia/. Accessed 19 Oct 2022

Arlinghaus R, Abbott JK, Fenichel EP, Carpenter SR, Hunt LM, Alós J, Klefoth T, Cooke SJ, Hilborn R, Jensen OP, Wilberg MJ, Post JR, Manfredo MJ (2019) Governing the recreational dimension of global fisheries. Proc Natl Acad Sci 116(12):5209–5213

Arlinghaus R, Aas Ø, Alós J, Arismendi I, Bower S, Carle S, Czarkowski T, Freire KM, Hu J, Hunt LM, Lyach R, Kapusta A, Salmi P, Schwab A, Tsuboi J-i, Trella M, McPhee D, Potts W, Wołos A, Yang Z-J (2021) Global participation in and public attitudes toward recreational fishing: international perspectives and developments. Rev. Fish Sci Aquac 29(1):58–95

Babali M, Kacher M, Belhabib D, Louanchi F, Pauly D (2018) Recreational fisheries economics between illusion and reality: the case of Algeria. PloS One 13(8):e0201602

Beach D (2002) Coastal sprawl: the effects of urban design on aquatic ecosystems in the United States. Pew Oceans Commission. https://www.pewtrusts.org/-/media/legacy/uploadedfiles/wwwpewtrustsorg/reports/protecting_ocean_life/envpewoceanssprawlpdf.pdf. Accessed 25 June 2015

Boucek RE, Rehage JS (2015) A tale of two fishes: using recreational angler records to examine the link between fish catches and floodplain connections in a subtropical coastal river. Estuar Coast 38(1):124–135

Boucek RE, Lewis JP, Stewart BD, Jud ZR, Carey E, Adam AJ (2019) Measuring site fidelity and homesite-to-pre-spawning site connectivity of bonefish (*Albula vulpes*): using mark-recapture to inform habitat conservation. Environ Biol Fishes 102(2):185–195

Boucek RE, Ellis RE, Forauer AR, Adams AJ (2022) A decade-long connectivity study of Permit (*Trachinotus falcatus*) in Florida supports a spatial management approach. Environ Biol Fish 106:181–192

Boxrucker J (2002) Rescinding a 254-mm minimum length limit on white crappies at Ft. Supply Reservoir, Oklahoma: the influence of variable recruitment, compensatory mortality, and angler dissatisfaction. N Am J Fish Manag 22:1340–1348

Braynen M, Perez VB, Callwood K, Carroll G, Castañeda A, Dahlgren C, Garbutt D, Gittens L, Glazer B, Green F, Haukebo S, Moore R, Poon S, Requena N, Sherman K, Sho V, Tate S (2022) A climate vulnerability assessment for Caribbean recreational fisheries. Environmental Defense Fund. https://fisherysolutionscenter.edf.org/sites/default/files/2022-09/A%20Climate%20Vulnerability%20Assessment%20for%20Caribbean%20Recreational%20Fisheries.pdf. Accessed 30 Jan 2023

Brownscombe JW, Adams AJ, Young N, Griffin LP, Holder PE, Hunt J, Acosta A, Morley D, Boucek R, Cooke SJ, Danylchuk AJ (2019) Bridging the knowledge-action gap: a case of research rapidly impacting recreational fisheries policy. Mar Policy 104:210–215

Camp EV, van Poorten BT, Robinson KF, Harrison HL, NguyenV CCA, Lorenzen K, Casola W, Dedual M, Klizentyte K (2026) Integrating diverse stakeholder knowledge in fisheries decision making. In: Pope KL, Arlinghaus R, Hunt LM, Lynch AJ, van Poorten BT (eds) Understanding recreational fishers: disciplinary and interdisciplinary approaches for fisheries management. Springer Nature, Cham, p 607–660

Caputi N, de Lestang S, Newman SJ, Jackson G, Smith K (2021) Stakeholder-government collaboration in developing cost-effective fishery-independent surveys in rights-based and co-managed fisheries. Mar Policy 128:104510

Coleman FC, Figueira WF, Ueland JS, Crowder LB (2004) The impact of United States recreational fisheries on marine fish populations. Science 305(5692):1958–1960

Cooke SJ, Dunlop WI, Macclennan D, Power G (2000) Applications and characteristics of angler diary programmes in Ontario, Canada. Fish Manag Ecol 7:473–487

Cooke SJ, Twardek WM, Reid AJ, Lennox RJ, Danylchuk SC, Brownscombe JW, Bower SD, Arlinghaus R, Hyder K, Danylchuk AJ (2019) Searching for responsible and sustainable recreational fisheries in the Anthropocene. J Fish Biol 94(6):845–856

Cooke SJ, Venturelli P, Twardek WM, Lennox RJ, Brownscombe JW, Skov C, Hyder K, Suski CD, Diggles BK, Arlinghaus R, Danylchuk AJ (2021) Technological innovations in the recreational fishing sector: implications for fisheries management and policy. Rev. Fish Biol Fish 31:253–288

Crandall CA, Monroe M, Dutka-Gianelli J, Lorenzen K (2019) Meaningful action gives satisfaction: stakeholder perspectives on participation in the management of marine recreational fisheries. Ocean Coast Manag 179(1):104872

Crane JL, Olcott GW (1933) Report on the Iowa twenty-five year conservation plan. Iowa State Board of Conservation, Iowa Fish and Game Commission, Des Moines

Crowe FM, Longson IG, Joll LM (2013) Development and implementation of allocation arrangements for recreational and commercial fishing sectors in Western Australia. Fish Manag Ecol 20:201–210

Daedlow K, Beckmann V, Arlinghaus R (2011) Assessing an adaptive cycle in a social system under external pressure to change: the importance of intergroup relations in recreational fisheries governance. Ecol Soc 16(2):3

Department of Primary Industries and Regional Development (DPIRD) (2024) Annual report 2023/2024. Department of Primary Industries and Regional Development, Western Australia. https://www.wa.gov.au/system/files/2024-10/dpird_annual_report_2023-24.pdf. Accessed 27 June 2025.

Ditton RB (1996) Human dimensions in fisheries. In: Ewert AW (ed) Natural resource management, first edn. Routledge, New York, pp 73–90

Elmer LK, Kelly LA, Rivest S, Steell SC, Twardek WM, Danylchuk AJ, Arlinghaus R, Bennett JR, Cooke SJ (2017) Angling into the future: ten commandments for recreational fisheries science, management, and stewardship in a good Anthropocene. Environ Manage 60(2):165–175

Fidler RY, Ahmadia GN, Amkieltiela A, Cox C, Estradivari GL, Handayani C, Mahajan SL, Mascia MB, Pakiding F, Andradi-Brown DA, Campbell SJ, Claborn K, De Nardo M, Fox HE, Gill D, Hidayat NI, Jakub R, Le DT, Purwanto VA, Harborne AR (2022) Participation, not penalties: community involvement and equitable governance contribute to more effective multiuse protected areas. Sci Adv 8(18):eabl8929

Fisheries New Zealand (2020) Fisheries New Zealand Sustainability Round Process. https://www.mpi.govt.nz/dmsdocument/45697-Fisheries-New-Zealand-Sustainability-round-proces. Accessed 6 July 2023.

Fleming DM, Jones PJS (2012) Challenges to achieving greater and fairer stakeholder involvement in marine spatial planning as illustrated by the Lyme Bay scallop dredging closure. Mar Policy 36(2):370–377

Gleason M, Fox E, Ashcraft S, Vasques J, Whiteman E, Serpa P, Saarman E, Caldwell M, Frimodig A, Miller-Henson M, Kirlin J, Ota B, Pope E, Weber M, Wiseman K (2013) Designing a network of marine protected areas in California: achievements, costs, lessons learned, and challenges ahead. Ocean Coast Manag 74(2013):90–101

Granek EF, Madin EM, Brown MA, Figueira W, Cameron DS, Hogan Z, Kristianson G, de Villiers P, Williams JE, Post J, Zahn S, Arlinghaus R (2008) Engaging recreational fishers in management and conservation: global case studies. Conserv Biol 22(5):1125–1134

Gundelund C, Venturelli P, Hartill BW, Hyder K, Olesen HJ, Skov C (2021) Evaluation of a citizen science platform for collecting fisheries data from coastal sea trout anglers. Can J Fish Aquat Sci 78(11):1576–1585

Harrison HL, Loring PA (2014) Larger than life: the emergent nature of conflict in Alaska's upper Cook Inlet salmon fisheries. SAGE Open 4(4):1–14

Hartill BW, Cryer M, Lyle JM, Rees EB, Ryan KL, Steffe AS, Taylor SM, West L, Wise BS (2012) Scale- and context-dependent selection of recreational harvest estimation methods: the Australasian experience. N Am J Fish Manag 32(1):109–123

Hartley TW, Robertson RA (2006) Stakeholder engagement, cooperative fisheries research and democratic science: the case of the Northeast Consortium. Hum Ecol Rev. 13(2):161–171

Haukebo S, Boné Morón E (2020) Recreational fisheries management in the Gulf of Mexico and the Caribbean: recent discoveries, current challenges, and future opportunities. GCFI73 Special Session Report. English edition: https://www.gcfi.org/download/75/edf-workshop-report-gcfi73/9995/gcfi73-recreational-fisheries-special-session-report_update-1.pdf. Accessed 1 Dec 2020.

Helvey M (2004) Seeking consensus on designing marine protected areas: keeping the fishing community engaged. Coast Manag 32(2):173–190

Henry GW, Lyle JM (2003) The national recreational and indigenous fishing survey. FRDC Project No. 99/158. Australian Government Department of Agriculture, Fisheries and Forestry

Holder PE, Jeanson AL, Lennox RJ, Brownscombe JW, Arlinghaus R, Danylchuk AJ, Bower SD, Hyder K, Hunt LM, Fenichel EP, Venturelli PA, Thorstad EB, Allen MS, Potts WM, Clark-Danylchuk S, Claussen JE, Lyle JM, Tsuboi J-i, Brummett R, Freire KMF, Tracey SR, Skov C, Cooke SJ (2020a) Preparing for a changing future in recreational fisheries: 100 research questions for global consideration emerging from a horizon scan. Rev. Fish Biol Fish 30:137–151

Holder PE, Griffin LP, Adams AJ, Danylchuk AJ, Cooke SJ, Brownscombe JW (2020b) Stress, predators, and survival: exploring Permit (*Trachinotus falcatus*) catch-and-release fishing mortality in the Florida Keys. J Exp Mar Biol Ecol 524:151289

Hunt LM, Sutton SG, Arlinghaus R (2013) Illustrating the critical role of human dimensions research for understanding and managing recreational fisheries within a social-ecological system framework. Fish Manag Ecol 20(2–3):111–124

Hunt LM, Camp E, van Poorten B, Arlinghaus R (2019) Catch and non-catch-related determinants of where anglers fish: a review of three decades of site choice research in recreational fisheries. Rev. Fish Sci Aquac 27(3):261–286

Hyder K, Weltersbach MS, Armstrong M, Ferter K, Townhill B, Ahvonen A, Arlinghaus R, Baikov A, Bellanger M, Birzaks J, Borch T, Cambie G, de Graaf M, Diogo HMC, Dziemian Ł, Gordoa A, Grzebielec R, Hartill B, Kagervall A, Kapiris K, Karlsson M, Kleiven AR, Lejk AM, Levrel H, Lovell S, Lyle J, Moilanen P, Monkman G, Morales-Nin B, Mugerza E, Martinez R, O'Reilly P, Olesen HJ, Papadopoulos A, Pita P, Radford Z, Radtke K, Roche W, Rocklin D, Ruiz J, Scougal C, Silvestri R, Skov C, Steinback S, Sundelöf A, Svagzdys A, Turnbull D, van der Hammen T, van Voorhees D, van Winsen F, Verleye T, Veiga P, Vølstad J-H, Zarauz L, Zolubas T, Strehlow HV (2018) Recreational sea fishing in Europe in a global context—participation rates, fishing effort, expenditure, and implications for monitoring and assessment. Fish Fish 19(2):225–243

Hyder K, Maravelias CD, Kraan M, Radford Z, Prellezo P (2020) Marine recreational fisheries—current state and future opportunities. ICES J Mar Sci 77(6):2171–2180

Hyder K, Brown A, Armstrong M, Bell B, Hook SA, Kroese J, Radford Z (2021) Participation, effort, and catches of sea anglers resident in the UK in 2018 & 2019. Centre for Environment Fisheries & Aquaculture Science. https://www.gov.uk/government/publications/sea-angling-in-the-uk-report-2018-and-2019. Accessed 30 Jan 2023.

Iowa Department of Natural Resources & Recreational Boating and Fishing Foundation (IDNR and RBFF) (2013) First-time and repeat angler analysis, Des Moines

Irwin BJ, Wilberg MJ, Jones ML, Bence JR (2011) Applying structured decision making to recreational fisheries management. Fish 36(3):113–122

Johnston FD, Arlinghaus R, Dieckmann U (2010) Diversity and complexity of angler behaviour drive socially optimal input and output regulations in a bioeconomic recreational-fisheries model. Can J Fish Aqua Sci 67(9):1507–1531

Johnston F, Simmons S, van Poorten B, Venturelli P (2021) Comparative analyses with conventional surveys reveal the potential for an angler app to contribute to recreational fisheries monitoring. Can J of Fish Aqua Sci 79(1):31–46

Kinnebrew E, Shoffner E, Farah-Pérez M-NM, Siegel K (2021) Approaches to interdisciplinary mixed methods research in land-change science and environmental management. Conserv Biol 35(1):130–141

Kochalski S, Barz F, Pita P, Harrison HL (2026) Quantitative and qualitative content analysis of text and images. In: Pope KL, Arlinghaus R, Hunt LM, Lynch AJ, van Poorten BT (eds) Understanding recreational fishers: disciplinary and interdisciplinary approaches for fisheries management. Springer Nature, Cham, p 507–532

Krogman RM (2022) Fisheries and habitat response to lake restoration efforts. Annual Report to the U.S. Fish and Wildlife Service Sport Fish Restoration Program, Iowa Department of Natural Resources, Des Moines

Krogman RM, Stubbs TJ (2021a) Factors affecting interest in fishing, program preferences, and constraints by potential, lapsed, and active anglers in urban areas of Iowa. Hum Dim Wild 28(1): 1–17

Krogman RM, Stubbs TJ (2021b) Strategic planning for catfish angling in the future: an example from Iowa. N Am J Fish Manag 41(S1):S298–S313

Lauretta M, Goodyear P (2018) Blue marlin (*Makaira nigricans*) standardized indices of abundance from the U.S. pelagic longline and recreational tournament fisheries. Collect Vol Sci Pap ICCAT 75(5):948–964

Lennox RJ, Sbragaglia V, Vollset KW, Sortland LK, McClenachan L, Jarić I, Guckian ML, FerterK DAJ, Cooke SJ, Arlinghaus R, Twardek WM (2022) Digital fisheries data in the Internet age: emerging tools for research and monitoring using online data in recreational fisheries. Fish Fish 23:926–940

Loring PA (2017) The political ecology of gear bans in two fisheries: Florida's net ban and Alaska's salmon wars. Fish Fish 18(1):94–104

Mason ET, Kellum AN, Chiu JA, Waltz GT, Murray S, Wendt DE, Starr RM, Semmens BX (2020) Long-term participation in collaborative fisheries research improves angler opinions on marine protected areas. PeerJ 8:e10146

McCook LJ, Ayling T, Cappo M, Choat JH, Evans RD, De Freitas DM, Heupel M, Hughes TP, Jones GP, Mapstone B, Marsh H, Mills M, Molloy FJ, Pitcher CR, Pressey RL, Russ GR, Sutton S, Sweatman H, Tobin R, Wachenfeld DR, Williamson DH (2010) Adaptive management of the Great Barrier Reef: a globally significant demonstration of the benefits of networks of marine reserves. P Natl Acad Sci 107(43):18278–18285

McLeod P, Lindner R (2018) Economic dimension of recreational fishing in Western Australia. https://recfishwest.org.au/wp-content/uploads/2019/03/Economic-Dimensions-of-Recreational-Fishing-in-Western-Australia-Report-2.pdf. Accessed 30 Sept 2022.

McMullin SL, Pert E (2010) The process of fisheries management. In: Hubert WA, Quist MC (eds) Inland Fisheries Management in North America, third edn. American Fisheries Society, Bethesda, pp 133–156

Moore A, Schirmer J, Magnusson A, Keller K, Hinten G, Galeano D, Woodhams J, Wright D, Maloney L, Dix AF (2018) National social and economic survey of recreational fishers 2018–2021. Fisheries Research & Development Corporation, Deakin. https://www.frdc.com.au/sites/default/files/products/2018-161-NRFS_main%20report_FINAL_19Feb2023.pdf. Accessed 17 Jul 2023.

Mosel KJ, Isermann DA, Hansen JF (2015) Evaluation of daily creel and minimum length limits for black crappie and yellow perch in Wisconsin. N Am J Fish Manag 35:1–13

National Oceanic and Atmospheric Administration (NOAA) (2015) National saltwater recreational fisheries policy. Office of Sustainable Fisheries, NOAA Fisheries Directorate, Silver Spring. https://www.fisheries.noaa.gov/national/recreational-fishing/national-saltwater-recreational-fisheries-policy. Accessed 13 Sep 2022.

National Rock Lobster Management Group (NRLMG) (2021) National rock lobster management group annual report 2020/21. https://www.mpi.govt.nz/dmsdocument/48577-National-Rock-Lobster-Management-Group-Annual-Report 2020/21. https://www.mpi.govt.nz/dmsdocument/48577-National-Rock-Lobster-Management-Group-Annual-Report-202021. Accessed 31 Jan 2023.

Newman SJ, Wise BS, Santoro KG, Gaughan DJ (2021) Status reports of the fisheries and aquatic resources of Western Australia 2020/21: the state of the fisheries. Department of Primary Industries and Regional Development, Western Australia. https://www.fish.wa.gov.au/Documents/sofar/status_reports_of_the_fisheries_and_aquatic_resources_2020-21.pdf. Accessed 13 Sept 2022.

New Zealand Initiative (2016) What's the catch? The state of recreational fisheries management in New Zealand. Edited by Randall Bees. The New Zealand Initiative, Wellington, New Zealand. https://www.nzinitiative.org.nz/reports-andmedia/reports/whats-the-catch/document/81

Obregón C, Hughes M, Tweedley JR, Loneragan NR (2022) Feeling the pinch: perceived marginalization of small-scale commercial crab fishers by an expanding recreational sector. In: Jentoft S, Chuenpagdee R, Said AB, Isaacs M (eds) Blue justice: small-scale fisheries in a sustainable ocean economy. Springer, Heidelberg, pp 295–314

Pierce JL, Lauretta MV, Rezek RJ, Rehage JS (2021) Survival of Florida largemouth bass in a coastal refuge habitat across years of varying drying severity. Trans Am Fish Soc 150(4): 435–451

Pollock KH, Jones CM, Brown TL (1994) Angler survey methods and their applications in fisheries management. American Fisheries Society., Special Publication 25, Bethesda

Potts WM, Downey-Breedt N, Obregon P, Hyder K, Bealey R, Sauer WHH (2020) What constitutes effective governance of recreational fisheries?—a global review. Fish Fish 21(1):91–103

Raynal JM, Weeks R, Pressey RL, Adams AJ, Barnett A, Cooke SJ, Sheaves M (2020) Habitat-dependent outdoor recreation and conservation organizations provide potential for recreational fishers to contribute to conservation of coastal marine ecosystems. Global Ecol Conserv 24: e01342

Reed M, Courtney P, Lewis N, Freeman R, Chiswell H, Black J, Urquhart J, Phillipson J (2020) Assessing participation of commercial fishers and recreational anglers in fisheries science and management in England. A report prepared for DEFRA. The Countryside and Community Research Institute, University of Gloucestershire, Gloucestershire. https://eprints.glos.ac.uk/10042/1/10042-Reed-%282020%29-Assessing-participation-of-commercial-fishers-and-recreational.pdf. Accessed 31 January 2023

Rehage JS, Loftus WF (2007) Seasonal fish community variation in headwater mangrove creeks in the southwestern Everglades: an examination of their role as dry-down refuges. Bull Mar Sci 80(3):625–645

Ryan KL, Trinnie FI, Jones R, Hart AM, Wise BS (2016) Recreational fisheries data requirements for monitoring catch shares. Fish Manag Ecol 23(3–4):218–233

Sadovy Y, Domeier ML (2005) Are aggregation-fisheries sustainable? Reef fish fisheries as a case study. Coral Reefs 24:254–262

Schmitter-Soto JJ, Aguilar-Perera A, Cruz-Martínez A, Herrera-Pavón RL, Morales-Aranda AA, Cobián-Rojas D (2017) Interdecadal trends in composition, density, size, and mean trophic levels of fish species and guilds before and after coastal development in the Mexican Caribbean. Biodivers Conserv 27:459–474

Sklar FH, Browder JA (1998) Coastal environmental impacts brought about by alterations in freshwater flow in the Gulf of Mexico. Environ Manag 22:547–562

Skov C, Hyder K, Gundelund C, Ahvonen A, Baudrier J, Borch T, deCarvalho S, Erzini K, Ferter K, Grati F, van derHammen T, Hinriksson J, Houtman R, Kagervall A, Kapiris K, Karlsson M, Lejk AM, Lyle JM, Martinez-Escauriaza R, Moilanen P, Mugerza E, Olesen HJ, Papadopoulos A, Pita P, Pontes J, Radford Z, Radtke K, Rangel M, Sagué O, Sande HA, Strehlow HV, Tutiņš R, Veiga P, Verleye T, Vølstad JH, Watson JW, Weltersbach MS, Ustups D, Venturelli PA (2021) Expert opinion on using angler Smartphone apps to inform marine fisheries management: status, prospects, and needs. ICES J Mar Sci 78(3):967–978

Smith RF, Gross RW (1955) An evaluation of the fifteen-inch minimum size limit on pickerel in three New Jersey lakes. New Jersey Fisheries Laboratory, Department of Conservation and Economic Development, Division of Fish and Game

Smith M, Fedler A, Adams AJ (2023) Economic assessments of recreational flats fisheries provide leverage for conservation. Environ Biol Fishes 106:131–145

Southern Scallop Working Group (SSWG) (2021) Implementation plan. Southern scallop strategy: Marlborough Sounds. https://www.mpi.govt.nz/dmsdocument/44902-Implementation-plan-Southern-Scallop-Strategy-Marlborough-Sounds. Accessed 31 Jan 2023.

Suan A, Leong KM, Oleson KLL (2021) Automated content analysis of the Hawai'i small boat fishery survey reveals nuanced, evolving conflicts. Ecol Society 26(4):9

Tate A, Ryan KL, Smallwood CB, Desfosses CJ, Taylor MS, Lai EK, Blight SJ (2020) Review of recreational fishing surveys in Western Australia. Fisheries Research Report No. 301. Department of Primary Industries and Regional Development, Western Australia

Taylor SM, Blight SJ, Desfosses CJ, Steffe AS, Ryan KL, Denham AM, Wise BS (2018) Thermographic cameras reveal high levels of crepuscular and nocturnal shore-based recreational fishing effort in an Australian estuary. ICES J Mar Sci 75(6):2107–2116

Tunison AV, Mullin SM, Meehean OL (1949) Survey of fish culture in the United States. Prog Fish-Cult 11(1):31–69

U.S. Census Bureau (USCB) (2018) American community survey, total population, 1-year estimates, Table B01003. http://data.census.gov. Accessed 8 Jan 2018.

UK Government (2020) Fisheries Act 2020. https://www.legislation.gov.uk/ukpga/2020/22/enacted. Accessed 30 Jan 2023.

Venturelli PA, Hyder K, Skov C (2017) Angler apps as a source of recreational fisheries data: opportunities, challenges and proposed standards. Fish Fish 18(3):578–595

Venturelli P, Skov C, Audzijonyte A, Belmaker J, Ana Cardoso AS, Gordon L, Gundelund C, Hyder K, Jarić I, Lennox RJ, Potts WM, Vacura K, Vaz AS, Sbragaglia V (2026) Digital data mining. In: Pope KL, Arlinghaus R, Hunt LM, Lynch AJ, van Poorten BT (eds) Understanding recreational fishers: disciplinary and interdisciplinary approaches for fisheries management. Springer Nature, Cham, p 443–479

Victorian Fisheries Authority (VFA) (2022) Target one million—phase one. https://vfa.vic.gov.au/recreational-fishing/go-fishing-victoria/target-one-million. Accessed 30 Jan 2023.

Wales JH, Lane JP (1958) Public opinion analysis in trout fishery management. Trans Am Fish Soc 87(1):163–171

Ward HGM, Quinn MS, Post JR (2013) Angler characteristics and management implications in a large, multistock spatially structured recreational fishery. N Am J Fish Manag 33(3):576–584

Ward HGM, Allen MS, Camp EV, Cole N, Hunt LM, Matthias B, Post JR, Wilson K, Arlinghaus R (2016) Understanding and managing social-ecological feedbacks in spatially structured recreational fisheries: the overlooked behavioral dimension. Fish 41(9):524–535

Weir JL, Vacura K, Bagga J, Berland A, Hyder K, Skov C, Attby J, Venturelli PA (2022) Big data from a popular app reveals that fishing creates superhighways for aquatic invaders. PNAS Nexus 1(3):pgac075

Wilson JK, Adams AJ, Ahrens RN (2019) Atlantic tarpon (*Megalops atlanticus*) nursery habitats: evaluation of habitat quality and broad-scale habitat identification. Environ Biol Fishes 102(2): 383–402

Wilson JK, Stevens PW, Blewett DA, Boucek R, Adams AJ (2023) A new approach to define an economically important fish as an umbrella flagship species to enhance collaborative stakeholder-management agency habitat conservation. Environ Biol Fishes 106(2):237–254

Wynne-Jones J, Gray A, Heinemann A, Hill L, Walton L (2019) National panel survey of marine recreational fishers 2017–18. New Zealand Fisheries Assessment Report 2019/24. https://rescuefish.co.nz/wp-content/uploads/2020/07/Recreational-harvest-estimates-2108.pdf. Accessed 30 Jan 2023.

Part II
Disciplinary Views of Fishers' Behaviours and Outcomes

Introducing Disciplinary Perspectives and Interdisciplinary Possibilities for Understanding Recreational Fishers and Fisheries

Len M. Hunt, Noëlle Boucquey, Ben Beardmore, Joseph Christensen, David C. Fulton, Mary Mackay, Richard T. Melstrom, John R. Post, Susan A. Schroeder, Ruth H. Thurstan, E. Ingrid van Putten, and Robert Arlinghaus

Abstract Individuals from many disciplines conduct research to understand the social dimension of recreational fisheries. This diverse inquiry has produced a comprehensive understanding of the behaviours of recreational fishers, the outcomes from fishing, and the relationships among fishers, others, and the natural and human environment. The associated body of research, however, is largely disconnected across disciplines. Our goals here are to help readers understand the similarities and differences among disciplines and to identify opportunities for interdisciplinary-

L. M. Hunt (✉)
Ontario Ministry of Natural Resources, Thunder Bay, ON, Canada
e-mail: len.hunt@ontario.ca

N. Boucquey
Department of Environmental Studies, Eckerd College, St. Petersburg, FL, USA

B. Beardmore
Environmental Analysis and Sustainability Program, Wisconsin Department of Natural Resources, Madison, WI, USA

J. Christensen
School of Humanities, The University of Western Australia, Crawley, WA, Australia

D. C. Fulton
U.S. Geological Survey, Minnesota Cooperative Fish and Wildlife Research Unit, University of Minnesota, St. Paul, MN, USA

M. Mackay · E. I. van Putten
Commonwealth Scientific and Industrial Research Organisation Environment, Hobart, TAS, Australia

Centre for Marine Socioecology, University of Tasmania, Hobart, TAS, Australia

R. T. Melstrom
School of Environmental Sustainability, Loyola University Chicago, Chicago, IL, USA

J. R. Post
Department of Biological Sciences, University of Calgary, Calgary, AB, Canada

K. L. Pope et al. (eds.), *Understanding Recreational Fishers*, Fish & Fisheries Series 45, https://doi.org/10.1007/978-3-031-99739-6_3

based research on recreational fishers. Before addressing these goals, we begin by answering basic questions: What is a discipline? What are the primary disciplines used to study recreational fishers? And what is the genealogy of research on recreational fishers? Seven key disciplines are then classified by multi-criteria related to both focus and epistemological and methodological approaches. This classification reveals clear connections among: (i) ecological science and resource economics, and to a lesser extent historical ecology; (ii) environmental history and political ecology; and (iii) behavioural economics and social psychology. We next describe and provide possible remedies for barriers that inhibit interdisciplinary research, including different epistemologies, nomenclature, and reward bias. Finally, we highlight opportunities to conduct interdisciplinary research by describing the types of benefits that each discipline can provide when conducting interdisciplinary research on recreational fisheries.

Keywords Behaviours · Context · Epistemologies · Institutions · Outcomes · Relationships

1 What Is a Discipline and Interdisciplinary Research?

The idea of an academic discipline is surprisingly unclear. Whereas we might initially think of disciplines as the traditional "subjects" we encountered in school (e.g. mathematics, physics, history, or chemistry), in reality, a discipline is multidimensional and fluid and quickly separates into sub-disciplines (sometimes called fields). Although many definitions exist, we follow Schummer (2004, p. 436), who noted that a discipline is "a body of knowledge and a social body that generates, evaluates, communicates, and teaches the corresponding knowledge".

Disciplines foster the development of unique professional identities amongst their group members (e.g. scientists within a department in a faculty), using "technical terms that facilitate precise communication", and having shared sets of key questions along with typical methods of data analysis (Buker 2003, p. 74). For example, the scientific discipline of "chemistry" offers many different fields, such as organic and

S. A. Schroeder
Department of Fisheries, Wildlife, and Conservation Biology, Minnesota Cooperative Fish & Wildlife Research Unit, University of Minnesota, St. Paul, MN, USA

R. H. Thurstan
Centre for Ecology and Conservation, University of Exeter, Cornwall, UK

R. Arlinghaus
Department of Fish Biology, Fisheries and Aquaculture, Leibniz Institute of Freshwater Ecology and Inland Fisheries, Berlin, Germany

Division of Integrative Fisheries Management, Humboldt-Universität zu Berlin, Berlin, Germany inorganic chemistry, and "history" includes ancient history, medieval history, women's history, and so forth. What ultimately makes a discipline go beyond a group of people with a shared interest in a topic. Disciplines have unique cognitive, social, and communicative dimensions, with shared traditions, widely accepted concepts (e.g. ecosystem or population in ecological science or utility in economics), vocabularies, and jargon (Sugimoto and Weingart 2015). In practical terms, disciplines are usually considered mature once they are institutionalized to some degree with their own journals, textbooks, conferences, and academic departments (Sugimoto and Weingart 2015; Hammarfelt 2018). Indeed, it is this social and financial institutionalization of disciplines, along with the organized training of new students, which differentiates disciplines from other descriptors like sub-disciplines, domains, topics, fields, or subjects (Hammarfelt 2018), though these descriptors are often used interchangeably. It is also worth keeping in mind that though the history of knowledge production is long, the modern structure of disciplines is only a little over a century old (Klein 1996; Wellmon 2015).

The production of new knowledge (and the revision and reinterpretation of what we thought we knew before) remains a core goal of every academic discipline. Questions about what constitutes legitimate knowledge and what methods of inquiry produce such knowledge are thus often debated within and between disciplines. Indeed, differing epistemologies (i.e. theories of valid knowledge production) are a key challenge to overcome when working with people from different disciplines. Epistemologies are influenced by the ontological beliefs of a discipline (i.e. its theories on the nature of reality), and in turn affect a discipline's methodology, which is based on beliefs about how researchers can best obtain knowledge in the world (Lincoln et al. 2017). Epistemologies thus matter because they influence how disciplines structure data collection and analysis (e.g. what methods of collecting information about the world would a particular discipline accept as producing legitimate and useful knowledge?). Further, in management agencies filled with people trained in particular disciplines, people's acknowledged or taken-for-granted epistemologies affect what information is deemed legitimate for policymaking decisions.

In modern science, a range of paradigms (worldviews) exist that share key sets of ontological, epistemological, and methodological beliefs (Guba and Lincoln 1994; Slevitch 2011; Lincoln et al. 2017). Though many variations exist, here we briefly discuss three basic paradigms: positivism, critical realism, and constructivism (Table 1). Positivists believe in a single true reality, which humans can come to know via objective empirical study. The positivist methodology is that of deduction (e.g. moving from general toward more specific knowledge) using experimentation and hypothesis-testing (Lincoln et al. 2017). Critical realists also maintain that an independent material reality exists; however, they argue that our knowledge of this reality must always be partial because it is filtered through human thought, which is inevitably situated in historical, cultural, and social contexts (Proctor 1998). Critical realists are methodological pragmatists, often using diverse methods with the goal to

Table 1 Three social science paradigms and their associated characteristics

Paradigm	Positivism	Critical realism	Constructivism
Ontology (theory of the nature of reality)	One reality independent of human thought	One reality independent of human thought	Multiple realities; reality co-created via our interactions with the world
Epistemology (beliefs about how we obtain valid knowledge)	Objective empirical research; humans can ultimately obtain truths about reality	Situated empirical research; humans can only ever develop a partial understanding of reality given all knowledge is filtered through specific human contexts	Knowledge created via dialectical relationships and subjective interpretation; consensus sought
Methodology (beliefs about the best methods of investigation)	Hypothesis testing; experimentation	Pragmatic choice based on nature of the research questions	Dialectical methods; often qualitative
Typical disciplines	Physical sciences, natural sciences, some social sciences	Social sciences	Some social sciences, some humanities
Potential approach to recreational fishing research	Testing hypotheses about fish population dynamics	Mixed-methods research aimed at understanding key patterns of social-ecological relationships and their underlying causes	Oral histories, open-ended interviews aimed at understanding fishers' experiences

Partially adapted from Lincoln et al. (2017)

triangulate findings from different sources to get as close to a material or social truth as possible (Archer et al. 1999; Fletcher 2017). Constructivists' ontological beliefs differ from positivists and critical realists in that they posit that multiple realities may exist simultaneously (Lincoln et al. 2017; Slevitch 2011). For constructivists, knowledge emerges via a dialectical (back-and-forth) relationship between individuals and their world, and it is always tied to specific places and contexts (Obermeister 2017). Because for constructivists, meaning arises via relationships, qualitative methods are usually employed. Besides these predominantly Western social science paradigms, it is important to recognize the many other paradigms around the world, sometimes termed indigenous, local, or traditional (Berkes 2008; Obermeister 2017; Reid et al. 2021).

The variety of research paradigms available suggests that researchers can benefit from using more than one discipline when answering applied research questions in fisheries sciences. Consequently, much interest exists in developing teams to conduct research using non-disciplinary approaches. Although researchers frequently use descriptors like multidisciplinary and interdisciplinary research to describe non-disciplinary research, confusion reigns as these terms are not often carefully defined or understood. Here, we follow the definitions provided by van den Besselaar and Heimeriks (2001, p. 706):

> In multidisciplinary research, the subject under study is approached from different angles, using different disciplinary perspectives. However, neither the theoretical perspectives, nor the findings of the various disciplines are integrated in the end. An interdisciplinary approach, on the other hand, creates its own theoretical, conceptual, and methodological identity. Consequently, the results of an interdisciplinary study of a certain problem are more coherent, and integrated.

Implicit in these definitions is that interdisciplinary research also requires individuals to learn from other team members to co-develop knowledge. This co-developed knowledge can provide better insights for answering applied research questions and helping to develop research competencies and personally rewarding experiences for team members (Campbell 2003; Arlinghaus et al. 2017). Though disciplinary-specific research is often sufficient to answer some research questions (e.g. what is the economic value of catching a fish species), when warranted, we encourage interdisciplinary over multidisciplinary research. This encouragement is based both on the benefits of conducting interdisciplinary research and the view that recreational fisheries are a coupled social-ecological system (e.g. Carpenter and Brock 2004; Massey et al. 2006) or even a complex adaptive system (Arlinghaus et al. 2017). Whenever social and ecological systems interact, interdisciplinary knowledge is critical to understand how the system responds to actions or changes (Arlinghaus et al. 2017). These "system" views of fisheries provide the shared conceptual framework that is a necessary condition for conducting interdisciplinary research.

2 What Is the Genealogy of Research on Recreational Fishers?

We describe seven different disciplines and sub-disciplines that researchers use to study recreational fishers and their behaviours. For convenience, we hereafter describe these seven areas as disciplines, though some may represent sub-disciplines or even interdisciplinary social science approaches. Our descriptions of the disciplines include a very brief genealogy of their application to understand recreational fishers. The disciplines are presented in chronological order of their appearance in the primary literature beginning in the 1920s (Fig. 1). We focus on these seven disciplines because we believe they comprise the bulk of research on recreational fishers and our authorship team includes subject matter experts, including lead or key authors of discipline-specific chapters in this book (Boucquey et al. 2026; Landon et al. 2026; Mackay et al. 2026; Melstrom et al. 2026; Post et al. 2026; Thurstan et al. 2026). Readers might, nevertheless, consider how other disciplines such as ecological economics, institutional economics (e.g. Daedlow et al. 2013), human geography, and environmental psychology can and have contributed to understanding fishers.

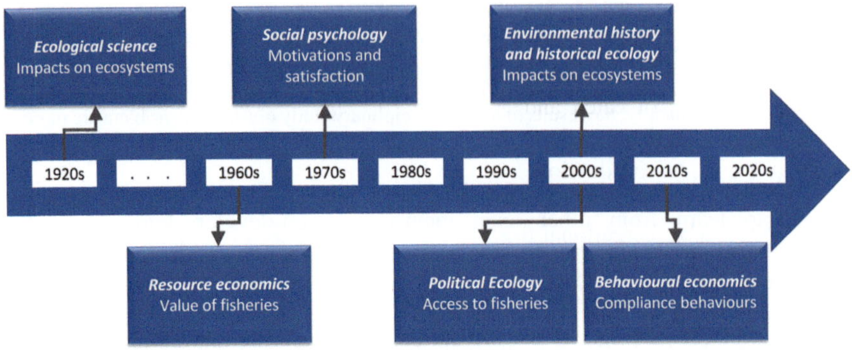

Fig. 1 Disciplines characterized by decade (1920s–2020s) of first primary publication on recreational fishers and fisheries (based on review by authors)

2.1 Ecological Science (1920s to Present)

Ecological scientists have seldom considered the behaviours of recreational fishers as specific targets for their research. Instead, ecological science focuses primarily on fish population dynamics with a resulting literature on how fish populations change in response to changes in age- or size-dependent reproduction and mortality rates, whether through natural processes or harvest (e.g. Ricker 1945; Larkin 1978; Hilborn and Walters 1992). When ecologists focus on responses of exploited (harvested) fish populations, the science is described as fisheries biology or fisheries ecology. Applied ecological scientists were the first to publish "social" research about recreational fishers (e.g. Westerman 1929, see Fig. 1) by using them as informants of the resource or what some term surrogate biology (Brown 1987). By inspecting the contents of the wicker "creel" baskets that fishers historically used to keep harvested fish, researchers began to estimate the impact of fishing on fish populations through mortality. Researchers using these creel surveys continue to collect information about the size of fish, the fate of caught fish (i.e. harvested or released), and the effort, which is the time that individuals spent fishing (see de Kerckhove et al. (2026) and Guthrie et al. (1991) for more information). Fisheries professionals did not examine other aspects of recreational fishers for almost 50 years when they included additional questions in creel surveys or conducted separate social surveys to describe fishers' attitudes, management preferences, and satisfaction (e.g. Hampton and Lackey 1976; Alcorn 1981; Ditton and Hunt 2001). These efforts, however, often lacked a formal connection between theories of human behaviours and the questions posed to fishers.

Although Larkin (1978) noted that understanding behaviours of recreational fishers is critically important for fisheries management, ecologists were slow to adopt methods tailored to study these behaviours. One key innovation in this discipline was a seminal paper by Johnson and Carpenter (1994), which argued that fishers are dynamic players in systems (i.e. fish population characteristics shape fishers' behaviours and, in turn, recreational harvest shapes characteristics of fish populations). Thus, ecologists studying recreational fisheries theorized that fishers are a different type of "predator" who seek fishing sites that balance the expected benefits in terms of catch and non-catch site amenities versus the cost of travel distance and time (Post et al. 2008). Fishing effort, rather than being a fixed pressure on the biological production system, responds to changes in the system. The importance of this perspective was amplified in an analysis of fishing effort and the propensity of recreational fisheries to collapse, especially when combined with fishers' behavioural characteristics, fish community structure, and management policy (Post et al. 2002; Carpenter and Brock 2004). This perspective led to explicit analyses of the dynamic behaviour of recreational fishers in time and space, where recreational fishers' behaviours responded to variations in catch and non-catch-related attributes (e.g. catch rates for fish and boat launch availability) among different fishing sites (e.g. Post et al. 2008; Johnston et al. 2010; Hunt et al. 2011; Wilson et al. 2019).

A second key innovation was recognition of the heterogeneity (variability) in fishers' behaviours, largely because fishers place differing levels of importance on catch- and non-catch-related factors of the experience. Therefore, adopting an aggregate measure of fishing effort can hide important outcomes to fisheries that arise from the heterogeneity present among fishers (Johnston et al. 2010, 2013; Matsumura et al. 2019). Although many researchers have examined heterogeneity among fishers (e.g. Fisher 1997; Wilde and Ditton 1994; Hunt 2005; Ward et al. 2013; Dabrowska et al. 2014), only more recently was heterogeneity included to analyse dynamic interactions between fishers and fish, both in single fisheries (Johnston et al. 2010, 2013, 2015) and across landscapes of fisheries (Hunt et al. 2011; Carruthers et al. 2018; Matsumura et al. 2019). Over the last several decades, therefore, we witnessed a fundamental shift among ecological scientists from viewing fishers as a fixed source of stress to fisheries (e.g. mortality) to a critical component of a fishery that both affects and is affected by feedbacks among fish, the larger aquatic ecosystem, and fisheries managers (Ward et al. 2016).

Box 1 What Is Ecological Science? (for more information, see Post et al. 2026)

Ecological science provides one lens through which one can study and attempt to understand the behaviour of fishers when faced with single or landscapes of fisheries for their recreational benefit. This approach is grounded in well-established theories of predator and prey dynamics in which the behaviour of both interactors jointly governs the outcomes. The analogy to recreational fisheries is that fisher behaviour and fish production are key processes resulting in variability in fish abundance and angler success over space and time. This variation results in density-dependent processes with feedbacks to recreational fisher behaviour and fishery quality and sustainability. As fishers often prefer larger sizes in their catch, the predator-prey dynamic involves

(continued)

Box 1 (continued)
ecological concepts related to both biological life history theory and size-structured population interactions. The primary goal is often to understand the dynamic outcomes of fisher behaviour and ecological fish production, measured by ecologists as catch rates and other characteristics of the catch, such as size, species, and propensity to release catch. Secondarily, catch-related performance is often related to decisions by fishers on where and how much to fish, so feedbacks exist between fisher behaviour, harvest decisions, and harvest impacts on fish populations and communities. It is this dynamic system of feedback and its impacts on both fisher behaviour and sustainability of the fish resource that is the focus of application of the ecological sciences to recreational fishery systems.

Many ecological applications to recreational fisheries have a strong empirical focus and use catch and effort data to infer proximate characteristics of fishers' behaviours, rather than attempting to understand human behaviour directly. As such, ecologists typically count anglers, as they would predators, and lump them into a measure of fishing effort (e.g. measures such as total hours spent fishing). Because researchers often contrast fisheries of assorted sizes, a common unit would be fishers' effort density (e.g. hours fished per hectare of lake or per km of stream). The emergent pattern of fishing effort density across multiple fisheries that vary in catch-related outcomes is then used to infer how fishers make decisions with their feet on where to fish based on their expected catch. In harvested fisheries, this sets up a dynamic interaction between fishers and their prey fish.

2.2 Resource Economics (1960s to Present)

Resource economics developed in the 1960s in response to policymakers queries about the benefits of government infrastructure projects, including the recreational benefits of public reservoirs. Shortly after Trice and Wood (1958) described ways economists could value recreation benefits, including fishing. Brown et al. (1965) assessed the economic value of Pacific salmon (*Oncorhynchus*) fisheries in Oregon. Early applications of economic valuation methods, such as Brown et al. (1965) used information on the number of recreation trips and travel distances to *reveal* the value of individual sites (e.g. accounting for the costs of travel and sometimes the time spent travelling from different origins). In the 1980s (e.g. Milon 1988a, b), multiple-site models (also known as random utility or even choice models) replaced single-site models as they were better suited to model fishers' behaviours within a system of fishing sites. Researchers continue to use these random utility models to estimate fishers' preferences for catch- and non-catch-related attributes, to predict choices of fishing sites, and to estimate changes to the economic value of fishing (see Hunt et al. 2019).

The 1980s also saw the onset of stated preference methods that focus on recreational fishers' choices, with the most prominent being the contingent valuation method (Daubert and Young 1981). This method is used to value resources through individuals' *stated* willingness to pay or preference for hypothetical improvements to environmental conditions, such as increases in fish abundance. Unlike random utility models, the contingent valuation method is suitable to study fishers' choices for general policy options (e.g. willingness to pay to avoid a reduced season length for fishing).

In the 1990s, resource economists further advanced economic valuation methods by estimating choice models from data on actual and stated (hypothetical) behaviours (Adamowicz et al. 1994). These combined stated and revealed preference choice models address the weaknesses from each constituent model type such as assuming that stated choices represent actual choices or avoiding collinearity problems among attributes that plagues models based on actual choices (Louviere et al. 2000). This combined method, however, was applied to recreational fishing much later (Morey and Breffle 2006) as even stated preference choice models were late to appear in this literature (Aas et al. 2000). The 2000s also saw the realization of McConnell and Sutinen's (1979) vision of applying bioeconomic (analogous to a coupled social-ecological system) models to recreational fishing (e.g. Carpenter and Brock 2004; Massey et al. 2006).

Box 2 What Is Resource Economics? (for more information, see Melstrom et al. 2026)

Resource economics examines the economic value and socioeconomic consequences of decisions that affect natural resources. For recreational fishing, these decisions can range from fishing at a favorite spot, to constructing public boats ramps, to putting restrictions on harvest. Resource economics focuses on the social consequences of these decisions because individuals, communities, and natural resources are interdependent. Failure to account for the social consequences (e.g. by focusing on individual profits rather than long-run social benefits) can lead to policy decisions that produce suboptimal outcomes. The field is, thus, designed to answer questions in recreational fisheries about management and policy such as: How important is catch in determining the number of fishing days? How will a change in catch limits affect the demand for fishing in the short and long run? What is the effect of increasing licence prices on fishing-associated business income or agency revenue? The outcomes of interest can relate to fisheries production (e.g. fish harvested or income generated) or consumption (e.g. the avidity of a fisher). The way people and governments spend their money in recreational fishing is relevant to resource economics, but these researchers study more than money flows. A key focus of resource economics is showing the value of natural resources not usually expressed in units of money.

(continued)

Box 2 (continued)

Like the other disciplines in this book, resource economics emphasizes empirical research to understand decision-making and its consequences. This does not mean that theory has no role, because it certainly does, particularly in model building and hypothesis formation. Theory can also have a role in identifying the parameters that link conditions in the fishery to the behaviour and outcomes of interest. In resource economics, as in many other modern economic fields, though, empirical research is viewed as crucial for understanding behaviour. It is also crucial in probing the sensitivity of parameter estimates and predictions to assumptions that analysts make in their model or research design, such as how fishers compare alternatives in their decision making, whether comparisons are made using static thinking, and if the behaviour of other fishers factor into individual decisions. Most resource economists apply empirical methods to models that describe choices of individuals or managers as the outcome of a set of physical, economic, and institutional conditions. Fundamentally, resource economics seeks to link choice or environmental conditions to welfare outcomes (i.e. the effect on wellbeing).

2.3 *Social Psychology (1970s to Present)*

The 1970s saw the first publication of social psychological research in recreational fisheries (Fig. 1) across three primary research domains: (i) motivations and satisfactions; (ii) specialization and involvement; and (iii) attitudes and other cognitions (e.g. values and beliefs) within a cognitive hierarchy. Driver and colleagues (e.g. Driver 1975; Driver and Brown 1975) began developing a conceptual framework and psychometric measurement instruments for understanding motivations, or recreational experience preferences, for outdoor recreation, including fishing in the early 1970's. Similar research on motivations and satisfactions was initiated for fishing and other forms of outdoor recreation during the same time (e.g. Moeller and Engelken 1972; Hendee 1974). This focus on understanding the diversity of motivations and satisfactions continues to the present (see Birdsong et al. 2021 for a review). Key developments during this time include a focus on better understanding activity-general (e.g. nature experience) and activity-specific and situation-specific motivations, such as catch-related motivations (e.g. Graefe 1980; Fedler and Ditton 1994) and the recognition of an apparent disconnect between motivations and satisfaction (Arlinghaus 2006).

The sociologist Hobson Bryan (1977, p. 175) introduced a framework to formally describe heterogeneity of recreationists, calling his concept specialization, which is defined as "a continuum of behaviour from the general to the particular, reflected by equipment and skills used in the sport and activity setting preferences". Despite that seminal publication, researchers could not agree on how to operationalize the concept in subsequent studies (Scott and Shafer 2001). The study of recreation specialization evolved from examining experience, which typically incorporated behavioural measures of participation (e.g. days or years of participation), to include cognitive and psychological measures (Scott and Shafer 2001). Modern researchers describe recreation specialization as a comprehensive, multidimensional measure (Bricker and Kerste 2000; Beardmore et al. 2011) that incorporates (i) psychological aspects such as attraction, self-expression, and centrality (sometimes operationalized as involvement, Kyle et al. 2007); (ii) cognitive components such as skills and knowledge related to an activity; and (iii) behavioural components such as experience measures, equipment ownership, and group memberships. Trying to understand how heterogeneity in specialization, commitment, involvement, and centrality affects fishers' behaviour and support or opposition to management actions remains an important area of inquiry (e.g. Kyle et al. 2007; Schroeder et al. 2018; Slaton et al. 2023).

Researchers have used several concepts linked to the cognitive hierarchy in studies of fisheries, conceptualized by higher-order cognitions such as values or beliefs that affect higher-level cognitions such as attitudes and ultimately behavioural intentions and behaviours (Fulton et al. 1996). Specifically, the social psychological concept of attitude (defined as an evaluation of an object) has been applied within the larger human dimensions of fisheries and wildlife fields since the early 1970's to understand or summarize preferences for management options (Riepe and Arlinghaus 2014) and policies as well as to try to predict fishers' participation behaviours. Many attitude studies conducted in fisheries management during the past 50 years have drawn on the social psychological construct of attitude in case-specific descriptions of fishers' attitudes toward existing or proposed management strategies and programs without being theoretically oriented. Some studies have applied theoretical frameworks such as the Theory of Reasoned Action (Fishbein and Ajzen 1975) or Theory of Planned Behaviour (Ajzen 1991) to try to predict and change fisher participation behaviour or regulatory compliance with modest success. The role of value orientations in influencing attitudes and behaviour has also been examined (Fulton et al. 1996; Riepe and Arlinghaus 2014). As in other areas of outdoor recreation research and management, research has examined the role of norms in understanding perceptions of crowding and fisher behaviour and management preferences (Needham et al. 2014). Research on affect and identity has received limited attention in social psychology research on fishers.

Box 3 What Is Social Psychology? (for more information, see Landon et al. 2026)

One broadly cited definition of social psychology is the study of how an individual's thoughts, feelings, and actions are affected by the actual, imagined, or symbolically represented presence of other people (Mitchell 2009). Since its emergence as a field of theoretical and research inquiry in the early

(continued)

Box 3 (continued)

twentieth century, social psychology has been an intellectual space shared between psychologists and sociologists. Though a shared field, psychological social psychology tends to put greater emphasis on internal psychological processes such as cognitions and emotions of individuals, whereas sociological social psychology focuses on factors that affect social life, such as status, role, and class. The theoretical focus of social psychology has waxed and waned over the decades, yet three central topics have tended to remain dominant: (i) the study of attitudes, opinions, and beliefs including their formation, their function, their dimensions, how they might be changed, and their influence on behaviour; (ii) the study of social perceptions and self-perception; and (iii) intra-group and intergroup processes and their influences on individual and group behaviour (Ross et al. 2010).

Social psychology is an extremely broad field with diverse foundational theories only some of which researchers have used to study recreational fishers. In general, three research paradigms in the human dimensions of recreational fisheries have been pursued during the past five decades. The first is a "cognitive approach" grounded in social psychological theory and methods for understanding attitudes and beliefs and their antecedents and consequences including behaviour. This approach assumes that an individual's behaviour (e.g. fishing, following catch limits, choosing a fishing location) is a function of an interconnected hierarchy of cognitions including values, attitudes, and beliefs with a strong correspondence between these cognitions and actual behaviour. A second long-standing paradigm has focused on the motivations that lead to participation in fishing. This approach applies social psychological theory and methods to describe the outcomes or "satisfactions" that fishers think they will achieve from fishing. A third approach has focused on self-perception and identity formation to identify types of fishers based on their specialization or level of involvement in fishing. These approaches build on social perception and self-perception theories in social psychology. Of course, researchers have conducted other applications of social psychological theories to study recreational fishers.

2.4 Political Ecology (2000s to Present)

Political ecologists are relative newcomers to the study of recreational fishing. Though political ecologists have more widely studied commercial fisheries (since the 1990s), the earliest engagements with recreational fishing came in the early 2000s, often as part of research on the effects of tourism, or in relation to questions about competition between fishery sectors. A key component of this early work centred on who benefits from or loses access to fishery resources. For example, Doyon (2002) examined the pressures a recreational fishing industry placed on artisanal fishing practices in a Mexican coastal community in conjunction with a push to attract more tourists. Similarly, Holen (2004) documented an influx of urban recreational fishers into the subsistence fishing grounds of the Ahtna people along Alaska's Copper River in parallel with the development of the oil industry. In later work, political ecologists have focused more explicitly on the recreational sector (though always in relation to other resource users), and they have expanded the scope of their analyses. Besides questions of power and resource distribution, which are always central to political-ecological work, researchers have investigated questions about the role of different knowledges in recreational fisheries management (Palmer 2004), the politically influenced history of interactions between recreational fishers and fished resources (McClenachan 2013; Engie and Quiroga 2014), the social-ecological entanglements of urban fishing (Kalina et al. 2019; Marjadi et al. 2021), further questions around recreational industries and tourism (Karrow and Thompson 2016; Hall 2021), and continuing political-economic tensions and associated policy performances between commercial and recreational sectors (May 2016; Boucquey 2020). We remain in the early days of political ecology attention to recreational fisheries, and extensive opportunities exist for political ecology to engage with recreational fishing practices.

Box 4 What Is Political Ecology? (for more information, see Boucquey et al. 2026)

Political ecology emerged as a new field in the early 1980s as scholars in a variety of social and natural sciences began to question explanations for environmental change that failed to consider broader socioeconomic and political conditions. Most famously, political ecology is described as a field that "combines the concerns of ecology and a broadly defined political economy" (Blaikie and Brookfield 1987, p. 17). In other words, political ecology is interested in how political-economic decisions affect "on the ground" human-environment relationships and wellbeing. Researchers in political ecology share several key commitments, including a theoretical commitment to understand the role of power in organizing social-ecological relationships, a methodological commitment to "in-depth, direct observation", and a normative commitment to research in the service of social-ecological justice (Perreault et al. 2015, p. 7). In its focus on power relationships and their effects on human-environment interactions, political ecology is interested in human (and non-human) behaviours. However, political ecology does not promote any single approach to understanding behaviours. Rather, political ecology conceptualizes human (and non-human) agency as always conditioned by particular constellations of historical and contemporary cultural, political, environmental, and economic processes. These features shift in space and time; thus, political ecologists generally reject standardized models of

(continued)

Box 4 (continued)
behaviour and often very explicitly reject rational-actor models that fail to account for the importance of context (Peet and Watts 2004).

Applied to recreational fishing, political ecologists are interested in assessing several key features of a fishery. First, political ecologists are interested in understanding the power dynamics at work (e.g. what social groups have or lack access to the fishery? What groups have influence in policymaking concerning the fishery? What groups are benefitting economically from the fishery? Who creates knowledge about this fishery, and how is it disseminated?). Second, political ecologists are interested in a detailed understanding of the particular social-ecological relationships in the fishery (e.g. what is the ecological condition of the fishery and how does it relate to people's attitudes toward and use of the fishery? What cultural norms or political-economic decisions have historically or are now affecting how the fishery is utilized?). Third, political ecologists often want to know how discourses (collections of texts, narratives, and images) surrounding a fishery may be affecting the policies and practices associated with the fishery (e.g. what are the dominant discourses around the fishery? How do these discourses shape people's beliefs, scientific and political decision-making, and ultimately people's material interactions with the fishery?). Finally, in keeping with the field's normative commitments, following their research political ecologists will want to make recommendations for improving the social-ecological wellbeing of the fishery.

2.5 Environmental History and Historical Ecology (2000s to Present)

Environmental and historical ecology are combined here as they are covered in one discipline-specific chapter in this text (Thurstan et al. 2026). Both disciplines developed their aquatic focus through an emphasis on commercial fishing. Studying commercial fishing was facilitated by more extensive and accessible historical records by virtue of past scale, economic importance, and preferential monitoring over recreational fishing (Bogue 2000; Bolster 2006). In contrast, recreational fishing was long viewed as economically and ecologically insignificant (McPhee et al. 2002), whereas the individualized and non-commercial nature of recreational fishing activities precluded the ready generation of formal records. Today, this emphasis is changing as research from multiple disciplines demonstrates the ecological, social, and cultural significance of recreational fishing (Arlinghaus et al. 2002, 2021; Holder et al. 2020), and as technological and scholarly advances facilitate the study of sources that record recreational fishing activities, such as newspapers, special interest magazines, and oral histories (Daedlow et al. 2013). Early historical studies on recreational fisheries examined changes in the species caught (McClenachan 2009a) and their environmental impacts (McClenachan 2009b, 2013), before expanding to include conflicts between commercial and recreational fishers, technological change, and the development of management regimes in recreational fisheries (e.g. Christensen and Jackson 2014), and social and cultural changes in the recreational or amateur sector (e.g. Frawley 2015). Despite this recent growth in research, studies on the history and historical ecology of recreational fishing remain limited compared with studies on commercial fisheries, and much remains to be discovered about recreational fishery development, past behaviours and practices, and how these changes have influenced aquatic environments through time.

Box 5 What Is Environmental History? (for more information, see Thurstan et al. 2026)

History seeks to interpret and understand and the past through critical analysis of historical evidence. Whereas attempts to illuminate the past using documentary and oral accounts have occurred for centuries, historical interest in natural environments and human-environment interactions—today termed environmental history—began developing only in the late-1960s, as the historical profession responded to increasing public awareness of environmental problems and the rise of modern environmental movements. Initially, this new discipline focused on terrestrial human-environment relations. Aquatic environments remained largely overlooked until the late 1990s, when heightened consciousness of human-induced changes in marine and freshwater ecosystems fostered greater interest in past interactions with aquatic environments. Marine environmental history is now recognized as a distinct field of environmental history (Bolster 2006). Developing in conversation with marine historical ecology, it has adopted innovative and multi-disciplinary approaches to overcome the epistemological challenge of historicising aquatic ecosystems through the diverse and fragmentary evidence that survives in the present (Christensen and Tull 2014; Schwerdtner Máñez and Poulsen 2016). Nonetheless, practitioners of marine environmental history generally adhere by the core methodology of their larger discipline: the rigorous evaluation of evidence, or "sources", according to relevant historical and cultural contexts; a focus on evaluating human actions, behaviours, institutions and values, in the context of human-environment interactions; and the communication of research findings in a narrative form accessible to both specialized and general audiences (Bolster 2006; Keiner 2013).

Environmental history today is studied across multiple contexts, varying widely across spatial and temporal scales and in terms of the societies and cultures being examined. Considering this, and because of the variety of source materials constituting historical evidence, no overarching theoretical framework is applied to defining or explaining human behaviour as it relates to

(continued)

Box 5 (continued)
natural and human-altered environments in the past. However, environmental historians frequently focus upon the dynamics and outcomes of group behaviour. This focus can involve examining how social or technological changes influence human-environment relationships and environmental outcomes (Christensen and Jackson 2014; McClenachan 2013), or the effects of cultural, institutional, or regulatory change upon group behaviours and how this shapes participation and practices in recreational fishing (Frawley 2015). Beyond this focus, marine environmental history can pursue any aspect of interactions between fishers and aquatic environments across time. This pursuit includes studies of conflicts between the recreational and commercial sectors, and studies of fisher values or behaviours that extend across timeframes that cover both open-access and different regulatory frameworks or management regimes. Marine environmental historians can also work collaboratively with other disciplines to uncover useful sources for extending temporal frameworks for understanding ecosystem change, placing past fisher values and behaviour in historical context, and assisting social and biological scientists to recognize the cultural and political dimensions of their unique disciplinary practices.

2.6 Behavioural Economics (2010s to Present)

Behavioural economics is a very new addition to the disciplinary toolbox available to study recreational fisher behaviour. The theoretical underpinning of the approach has, however, had a long- and well-established grounding. Economics and psychology came together as behavioural economics as a "proper" discipline in the 1980s, with a growth spurt after Daniel Kahneman and Amos Tversky in 1979 developed prospect theory (Kahneman 2013), which explains an asymmetry in how people value gains versus losses (e.g. losing a $20 bill has larger effect on an individual than does finding a $20 bill). Their Nobel Memorial Prize in Economics in 2002 and that of Richard Thaler in 2017 for nudge research cemented the role of behavioural economics. Several popularized books (e.g. Thaler and Sunstein 2008; Kahneman 2013) have further increased academic and public awareness of behavioural economics and its approaches. Behavioural economics even had a fast introduction into the governance domain as behavioural or nudge units were implemented by governments around the world to test, for example, new health and energy policies and to use simple nudges to encourage people to pay outstanding fees and fines to governments.

The introduction and application of behavioural economics into fisheries research lagged for several years after its implementation in the related fields of environmental management and conservation. Applications of behavioural economics in fisheries research initially mainly focused on compliance (Battista et al. 2018). Traditional perspectives on compliance assumed monetary incentives were the main driver in compliance decisions, but over time, it is increasingly accepted that social norms, legitimacy, and fairness play a big role in compliance decisions. Fisheries compliance models have thus extended the rational choice model to use concepts from psychology and sociology (Sutinen and Kuperan 1999). However, only in the last decade have researchers applied these compliance models and concepts to recreational fishing (Bergseth and Roscher 2018). To build or test theory, behavioural economics often relies on behavioural economic experiments (i.e. conducted in a lab, field, or lab-in-field, Stoop et al. 2012). In an experimental setting, the relative importance of deterrence and behavioural instruments such as nudges based on social norms can test recreational fisher decision making (Mackay et al. 2019). A growing appetite exists for applications of behavioural economics in recreational fisheries beyond compliance, which will likely be an important focus over the next decades.

Box 6 What Is Historical Ecology? (for more information, see Thurstan et al. 2026)

As environmental history was emerging in the late 1960s, other academic disciplines were also engaging with the past in attempts to situate our contemporary understanding of ecosystems, and human impacts upon these, within wider temporal frameworks. Historical ecology as a term has been used in multiple contexts, with origins in the disciplines of history, ecology, geography, and anthropology (Szabó 2015). Despite these diverse roots, it is today universally acknowledged as a research framework or community of practice that integrates multi-disciplinary perspectives and methods to understand the transformation of ecosystems and landscapes (Armstrong et al. 2017; Crumley 2021). Interest in aquatic historical ecology awakened during the late 1990s in response to growing alarm at the scale of ecological degradation in oceans and seas, and a new appreciation of the tendency for overlooking ecosystem changes on an intergenerational level (Pauly 1995). Marine historical ecology is now recognized as a distinct field of research that embraces studies of past species occurrence and distribution, long-term shifts in population dynamics and ecosystem functioning, the human drivers of, and responses to, environmental change, and how cultural values and social norms affect interactions with marine ecosystems across time (Kittinger et al. 2015; Schwerdtner Máñez and Poulsen 2016).

Historical ecology relies upon diverse forms of evidence, or proxy data, to reveal patterns of ecosystem change and the outcomes of human-environment interaction across extended periods. This evidence can include archaeological and anthropological artefacts, paleo-environmental archives, statistical and proto-statistical datasets, and many types of documentary records, oral testimony, and visual sources including photographs, illustrations and art. Similar to marine environmental history, marine historical ecology has developed

(continued)

Box 6 (continued)

through multi-disciplinary collaborations, and as a result, is highly diverse in terms of analytical techniques and the temporal and spatial frameworks employs, and without a dominant theory or conceptual model. Applied to recreational fishing, it can uncover long-term transitions in fisher effort and behaviour (e.g. Thurstan et al. 2018), provide baseline information to complement recent surveys and monitoring data, and reveal patterns to the past distribution and abundance of key recreationally targeted species. Historical ecology also contributes to improving understanding of recreational fishers as long-term components of dynamic social-ecological system, in challenging present-day assumptions of what constitutes "normal" in marine and freshwater environments, and in building support for enhanced conservation strategies and restoration targets in modified or degraded marine ecosystems (Thurstan 2022).

Box 7 What Is Behavioural Economics? (for more information, see Mackay et al. 2026)

To understand behavioural economics, one first needs to understand the principles of economics. (Neoclassical) Economics assumes individuals have full information and always make the best decisions based on a careful, rational analysis of the costs and benefits of the available options, motivated by self- interest. This is not the case, in reality. To better understand the drivers of individual behaviour and decision-making, it is important to go beyond assumptions of selfish rationality and to consider psychological elements in individual behaviour and decision-making. This is what behavioural economics does; it extends rather than abandons traditional economics theory (Ho et al. 2006). The discipline of behavioural economics is a relatively young field of economics expanding since the 1980s and especially since Daniel Kahneman and Amos Tversky won the Nobel Memorial Prize in Economics in 2002 and Richard Thaler in 2017 for behavioural economics research. Although the study of behavioural mechanisms and cognitive biases is as old as human existence.

Fisher behaviour is a key consideration and source of uncertainty for recreational fisheries management (Fulton et al. 2011). All human behaviour is complex and thus difficult to understand, predict, and manage. Understanding fisher behaviour in the commercial context where financial gains apply is more aligned with profit maximization and rational choice theory but deviates

(continued)

Box 7 (continued)
from this on occasion. The overall approach to recreational fisheries management has tended to mirror that of the commercial fisheries with a heavy emphasis on the use of regulatory tools, such as effort regulation and catch limits (Cooke and Cowx 2006). Even so, managing recreational fishers, who do not have a profit motive, is particularly difficult. For compliance management in both sectors instrumental factors, such as economic incentives and deterrence, have mainly been used (Nielsen 2003). However, the absence of formal management infrastructure in recreational fisheries (i.e. landing obligations, logbooks, electronic monitoring, or on-board observers) renders this approach less effective and suggests the need for an alternative approach. The drivers of recreational fishing and the socio-psychological benefits derived from the activity fit neatly under the behavioural economic umbrella. The topic of behavioural economics enriches our understanding of fisher behaviour by studying how psychological mechanisms such as cognitive biases, social norms and preferences, and emotions affect behaviour.

3 How Do We Classify Disciplines?

We classified the seven disciplines used to understand recreational fishers using criteria related to research focus and epistemological and methodological approach. The resulting taxonomy identifies some disciplines that share a high degree of overlap. The classification also helps to reveal how the different disciplines can each contribute efforts towards a comprehensive understanding of recreational fishers and fisheries, including outcomes from fishing.

3.1 Taxonomic Criteria

The primary research purpose is the first criterion associated with the focus theme (Table 2). Purpose consists of three broad categories: (i) estimating outcomes from the fishery, (ii) understanding factors that influence fishers' behaviours; and (iii) understanding relationships among fishers, other actors, and the ecosystem. We further separate the category of outcomes into: (i) impacts on the ecosystem, such as over-harvesting fish; (ii) benefits that accrue to people, such as the economic value of fishing or satisfaction; and (iii) negative impacts on people such as experiencing crowding or congestion at fishing sites.

The second and third criteria for focus assess the degree of institutional coverage. Institutions are "… the prescriptions that humans use to organize all forms of repetitive and structured interactions … at all scales" (Ostrom 2005, p. 3). Types of institutions consist of rules, norms, and strategies (or tactics). Strategies involve an aim (e.g. catching fish), attributes (e.g. the gear and equipment used to catch fish),

Table 2 Evaluating disciplines by classification criteria

Criteria	Behavioural economics	Ecological sciences	Environmental history	Historical ecology	Political ecology	resource economics	social psychology
Purpose							
Outcome: impacts on ecosystems		1	2	1	2		
Outcome: benefits to people	2	2	2		2	1	2
Outcome: impacts on people			2		2		2
Factors that influence behaviours	1	2		2		2	1
Relationships among actors and ecosystem			1		1		
Institutional type							
Rules	•	•	•	•	•	•	•
Norms	•		•	•	•		•
Strategies and tactics					•		•
Institutional scale							
Operational	•	•	•	•	•	•	•
Collective choice					•		
Constitutional					•		
Research paradigm							
Positivism	2	1		2		1	2
Critical realism	1		1	1	1	2	1
Constructivism			2		2		
Assumed behaviours							
Cost-benefit type calculus by fishers	•	•			•	•	•
Social context influence on fishers	•		•	•	•		•
Institutional context influence on fishers		•	•	•	•	•	
Analytical approach							
Quantitative	•	•				•	•
Qualitative			•		•		•
Mixed				•	•		•
Subject scale							
Individual	1	2		2	2	1	1
Aggregate		1	1	1	1		

1 primary, 2 secondary, • employed

and conditions (e.g. where in a river to use specific equipment and styles of fishing) whereas norms also include deontic logic that describes permissions or restrictions (e.g. a fisher shall not cast over other fishers' lines while fishing) and rules that include formal sanctions for non-compliance such as penalties for fishing with too many lines (Crawford and Ostrom 1995). Institutions are also aligned at different scales that consist of operational (e.g. the rules that a fisher must follow), collective choice (e.g. the formal processes that managers must follow to change an operational rule), and constitutional levels such as the recognition of who is privileged with the right to manage recreational fisheries (Ostrom 2005).

Four criteria comprise the epistemological and methodological approaches (Table 2). The first criterion is the research paradigm that we previously explained (Sect. 1 and Table 1). The research paradigm heavily influences the three remaining criteria of assumed behaviours of recreational fishers, analytical approach, and subject scale. The assumed behaviours of fishers are characterized by three potentially overlapping options: (i) behaviours arise from an individual type of cost-benefit calculation (e.g. maximizing utility or some other fitness measures such as catch rate); (ii) behaviours are shaped by their social context (e.g. norms within a community influence behaviours); and (iii) behaviours are influenced by the context they face (e.g. imbalances in power within a community of fishers can result in inequity in accessing fishing resources). For the analytical approach, we characterize the research as quantitative, qualitative, or mixed methods. Mixed methods involve using both qualitative and quantitative methods, such as a quantitative survey to understand broad patterns in thought or behaviour, followed by in-depth interviews to investigate why those patterns exist (see Golebie et al. 2026). Finally, the subject scale describes whether researchers examine individual or groups of fishers to understand behaviours and outcomes.

3.2 Contrasting and Classifying Disciplines

We begin by contrasting key differences between social psychology and resource economics, given that researchers from these disciplines have conducted most of the research on the behaviours of recreational fishers. Researchers from these disciplines differ both in epistemological and methodological approaches and focus resulting in a fundamentally different way that each come to understand, model, and predict behaviours of fishers (Table 2 and Fig. 2). Resource economists adopt a positivist paradigm to model fishers' behaviours to understand how people benefit from the fishery. These rational actor models are developed by beginning with behaviours (e.g. choices of fishing sites) and then associating these choices with the attributes (e.g. expected catch rates, travel costs, environmental quality) that describe the choice options. These economists use this decompositional approach to model behaviours to estimate preferences for attributes that in turn are used to predict fishers' behaviours and benefits and the value of recreational fishing (see Melstrom et al. 2026).

Social psychologists typically adopt a critical realism perspective to understand aspects of the decision-making process of fishers. These researchers begin from the antecedents of behaviours such as values, attitudes, behavioural control, and norms to understand how these antecedents shape intended and actual behaviours (see Landon et al. 2026). This compositional approach to modelling behaviours focuses

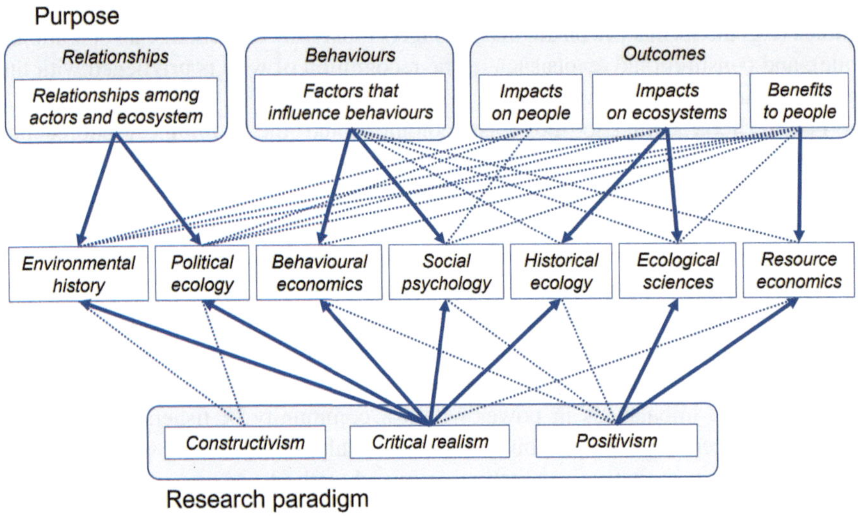

Fig. 2 Classifying disciplines by purpose and research paradigm (solid lines and arrows represent a primary emphasis, whereas dotted lines represent a secondary emphasis)

much more heavily on the decision-making process than outcomes from these decisions. Therefore, although resource economists and social psychologists both conduct research to understand the behaviours of fishers, differences in their approaches and focus yield much different methods and models, and result in little overlap in research (Fenichel et al. 2013).

We identify three pairs of disciplines that share similarities in focus and epistemological and methodological approach. Though these disciplinary pairings highlight some broad similarities and potentially complementary goals, we emphasize that projects can also often benefit from interdisciplinary work by participants from very different disciplines.

The one discipline without a strong pairing is historical ecology. On the one hand, historical ecology shares the same focus as ecological science to understand impacts on ecosystems, to examine operational and often rules-based institutions, and to rely on aggregated or collective observations from fishers (see Table 2 and Fig. 2). On the other hand, unlike ecological scientists, historical ecologists primarily rely on critical realism over positivism, focus on social and institutional factors over the individual cost-benefit calculus to understand behaviours, and use mixed methods over quantitative methods.

3.2.1 Ecological Science and Resource Economics

The first strong pairing of the seven disciplines involves ecological science and resource economics. Both disciplines share a similar focus with a primary emphasis on assessing outcomes from recreational fisheries and on considering how operational rules might influence those outcomes. Whereas resource economists almost exclusively focus on understanding the benefits to people (and especially recreational fishers), ecological scientists focus primarily on assessing impacts on the ecosystem.

Researchers from ecological science and resource economics primarily adopt a positivist paradigm, suggesting that a single reality exists and that researchers can uncover this reality by using objective methods. Consequently, these researchers use quantitative methods and develop statistical models from theories that posit a type of individual cost-benefit (and rational actor) approach to decision making (e.g. Post et al. 2008; Hunt et al. 2019). They differ, however, in their analyses of fishers with resource economists examining individual behaviours, whereas ecological scientists typically examine aggregated behaviours such as fishing effort.

3.2.2 Environmental History and Political Ecology

A second strong pairing involves environmental history and political ecology (Table 2 and Fig. 2). Environmental historians and political ecologists share considerable focus with an emphasis on understanding relationships among the human and nonhuman actors in the system. Although they are interested in understanding how particular rules and norms affect these relationships, they are also concerned with how larger social, political, and economic institutions shape these norms. Political ecologists are also interested in how power operates among individual actors, and they focus on and between various institutional scales.

Environmental historians and political ecologists primarily adopt a critical realism paradigm, though some researchers adopt constructivism. These individuals embrace the importance of social and institutional contexts in influencing behaviours of fishers and typically employ qualitative data. Some differences, however, exist in the approaches used by these researchers. Political ecologists differ from environmental historians through their: (i) use of mixed-methods approaches; (ii) belief that while people may make choices influenced by their perceived costs and benefits, a person's available choices are always heavily circumscribed by the political-economic context and the person's intersectional identity; and (iii) application on largely contemporary studies (albeit often containing historical components) that focus on understanding how social-ecological relationships are affected by larger political-economic processes.

3.2.3 Behavioural Economics and Social Psychology

The final strong pairing involves behavioural economics and social psychology (Fig. 2 and Table 2). Researchers from these disciplines are primarily interested in understanding the factors that influence behaviours of fishers at the operational scale. The roles of rules and norms on behaviours are explicitly examined by many of these researchers. Social psychologists sometimes focus on impacts on people as measured through crowding (e.g. Shelby and Vaske 2007) and researchers from both disciplines also examine the benefits to fishers as measured through satisfaction for social psychologists (e.g. Birdsong et al. 2021) and through the same measures that resource economists use for behavioural economists (e.g. economic valuation).

Researchers from these disciplines primarily employ a critical realism lens, though positivist-based research is also common (Table 2). These researchers often consider that individuals use a cost-benefit calculus (rational actor model) to make decisions, although the role of social context remains paramount (e.g. social norms influence behaviours). Behavioural economists, however, also explore how biases infiltrate decision making and thus question a rational actor model (Bazerman 1994) and how researchers and others can use these biases to nudge people to more desirable behaviours (Thaler and Sunstein 2008), though the question of who defines desirable remains unanswered. Although behavioural economists almost exclusively use quantitative methods and statistical models, social psychologists also employ qualitative and mixed-methods approaches.

4 How Do We Overcome Barriers That Inhibit Interdisciplinary Research?

Several barriers exist that limit the ability of individuals to conduct effective interdisciplinary research, even within the social sciences (Fenichel et al. 2013), let alone the integration of ecological and social sciences. Here, we offer some broad remedies to overcome three common barriers: (i) epistemological differences, (ii) nomenclature, and (iii) reward structure. Other barriers, including academic training (Turner et al. 2015), lack of integration (Eigenbrode et al. 2007), and imbalance of disciplinary perspectives on research teams (Campbell 2003), are covered elsewhere. For more specific ideas on conducting successful interdisciplinary projects, see books devoted to the topic (e.g. Repko and Szostak 2021; Lury et al. 2018).

4.1 Bridging Epistemological Differences

As discussed in Sect. 1, researchers are trained in different paradigms across the social and natural sciences. These paradigms have unique epistemologies (theories of how legitimate knowledge is produced), which may be in direct conflict with each other. These differences can create challenges for interdisciplinary collaboration, particularly if they go unacknowledged (Miller et al. 2008; Turner et al. 2015). Some key problems that stem from failing to address epistemological differences can include: (i) a tendency to privilege a single epistemological viewpoint, thus missing potentially relevant insights from others; (ii) a failure to adequately conceptualize the objects or systems of inquiry (particularly salient when attending to coupled social-ecological systems); (iii) a potential to exacerbate existing power differences among the project team (e.g. based on career stage, gender, race), especially if one epistemology is in the minority; and (iv) a failure to adequately communicate among members of the project team, potentially affecting the quality of results that might emerge from the interdisciplinary work (Campbell 2005; Miller et al. 2008).

So, how can we best address differing epistemologies and promote successful interdisciplinary collaborations in recreational fishing research where warranted? At the heart of successful efforts lies a fundamental sense of humility and curiosity on the part of the individuals involved. Indeed, "by recognising that (groups of) people create or discover and assess knowledge in a variety of ways, we can begin to open our minds to multiple ways of knowing" (Moon et al. 2021, p. 5). Beyond entering the collaboration with an open mind, some practical suggestions include: (i) holding explicit conversations at the outset of a project that discuss researchers' guiding paradigms (worldviews) and epistemologies, along with any associated values or advocacy people might be engaged in; (ii) acknowledging potential power differences amongst team members due to career stage, gender, race, funding or institutional ties, and strategizing ways to prevent the marginalization of team members (iii) holding these conversations in conjunction with the development of the practical aspects of the project (e.g. research questions, overall framework for investigation, methods, integrative models, anticipated publications and authorship), so that the effort to address them is more than superficial; (iv) trying to balance the research team among those with different epistemologies so that one is not dominant; and (v) encouraging frequent communications among the project team, including "fun" elements to help develop friendships that can overcome epistemological tensions (Campbell 2005; Glaeser et al. 2009; Moon et al. 2021).

It is also helpful to keep in mind that epistemological challenges on a project team can be reframed as opportunities. Indeed, "challenges to deeply held, sometimes unquestioned beliefs can be opportunities for growth. This can be one of the most rewarding aspects of interdisciplinary research", though addressing such challenges at the outset of a project is more likely to lead to growth than trying to come to terms with them in the results and analysis phase (Campbell 2005, p. 575). In recreational fishing research, interdisciplinary teams can already benefit from their shared interest in this fascinating social-ecological system and from a desire to solve an applied research question. Keeping this shared interest in mind while working through any epistemological challenges will ideally lead to productive, useful, and professionally rewarding research.

4.2 *Overcoming Nomenclature Barriers*

Another way to facilitate interdisciplinary research is to overcome the barrier of disciplinary-specific language and nomenclature (Eigenbrode et al. 2007; Morse et al. 2007). This is important when disciplines use different jargon to describe the same phenomenon. For example, consider the terms "place attachment" and "state dependence" from social psychology and resource economics, respectively. Place attachment represents a bond between people and their environments (Moore and Graefe 1994) that develops through meanings that individuals attach to places (Stedman 2002). Kyle et al. (2004) note that the two main dimensions of place attachment are place dependence (e.g. a good place to catch a lot of fish) and place identity (e.g. a place that one feels connected to because of, for example, fond memories of visits to the site as a child). The identity form of place attachment results in increased fidelity for an individual to fish the same site. This fidelity, however, cannot be explained by the quality of attributes associated with the site, such as travel cost or catch rate. Economists use the term (substantive) state dependence to describe the exact same phenomenon of fidelity to a fishing site that is unrelated to the attributes of the site (Heckman 1981). Consequently, social psychologists and resource economists have both observed this same phenomenon, but they use different jargon to describe it. The jargon, therefore, creates a barrier that camouflages existing bridges between disciplines.

A second and potentially more troubling problem with nomenclature occurs when researchers from different disciplines use and define the same term in different ways. An example using the concept of value will help illustrate this problem. For a social psychologist, values are: "underlying thoughts that people use to examine themselves, other people, and occurrences for which actions are upheld. They are few, strongly held, and enduring" (Landon et al. 2026). For a resource economist, a value is "… essentially a trade-off, [of] one item for another" (Melstrom et al. 2026). Thus, social psychologists focus on the behavioural process as values influence higher-level antecedents such as attitudes and ultimately behaviours (Fulton et al. 1996). Resource economists focus on the behavioural outcome (economic well-being) as they believe value can only be measured in relation to a baseline state, such as valuing an increase of catching three compared to two fish per hour at a fishing site (Phaneuf and Requate 2017, p. 391). Owing to their different epistemology, behavioural economists differ from resource economists in their view and definition of value. Behavioural economists believe that value may differ when it is assessed as an outcome of the experience versus as a factor that influences the decision-making process of people (Kahneman 2013, p. 446).

Possible remedies to the nomenclature barrier (beyond those listed in Sect. 4.1) focus on learning and communication (Alexander et al. 2019). Arlinghaus et al. (2014) also note the importance of having project leaders who act as information brokers and using common frameworks. If these information brokers are well-versed in multiple disciplines, they can help team members to learn about different disciplines and their nomenclature. Likewise, information brokers can publish articles that synthesize and illustrate connections (network analysis) among literature from different disciplines to help break down language barriers. For example, Fenichel et al. (2013) reviewed research by ecological scientists, resource economists, and others (primarily social psychologists) to contrast research methods, explain key disciplinary-specific concepts, and to illustrate the lack of awareness that exists among researchers of the other discipline. Here, we contribute to this latter effort by compiling a glossary of key disciplinary-specific terms related to recreational fishers and fishing outcomes (see the glossary).

4.3 Overcoming Existing Reward Structures

A final barrier for conducting interdisciplinary research is the bias of rewarding researchers who conduct disciplinary versus interdisciplinary research. Interdisciplinary research usually takes longer to conduct than does disciplinary research (Campbell 2005; Arlinghaus et al. 2014). This increased time can reduce research productivity when measured as the number of publications (Alexander et al. 2019), and thus, inhibit the achievement of academic rewards and tenure (Turner et al. 2015). Although interdisciplinary research contributes both to academic and applied goals (see Lynch et al. 2026), disciplinary-specific researchers often focus on the former. The emphasis on academic goals can result in a pathology that discourages applied and interdisciplinary research and thus can limit important insights for managing recreational fisheries.

One possible remedy to overcome the barrier of the rewards bias is for researchers to begin their careers conducting disciplinary-specific research before they venture into conducting interdisciplinary research (Arlinghaus et al. 2014). Another remedy is to foster research networks that generate additional opportunities for publication and communicating research findings.

5 What Opportunities Exist for Conducting Interdisciplinary Research?

Many researchers use a coupled social-ecological system framework to facilitate interdisciplinary research on recreational fisheries (e.g. Carpenter and Brock 2004; Massey et al. 2006; Arlinghaus et al. 2017). The resulting applications, however, are largely confined to ecological science and resource economics disciplines. It is no coincidence that these disciplines share strong similarities in ontological, epistemological, and methodological beliefs and approaches, and that they both focus on outcomes from fisheries. We believe that these common approaches instil familiarity among all researchers, and thus, encourage acceptance of the approaches that are used to develop knowledge about recreational fishers and fisheries.

One might be tempted to conclude that disciplines that share similar approaches and focus might represent the best opportunities for conducting new interdisciplinary research. We caution readers that this conclusion overlooks a critical factor that helps to drive interdisciplinary research: researchers from each discipline should expect a tangible benefit from including researchers from other disciplines. For example, interdisciplinary research among resource economists and ecological scientists is often successful not because it provides a competing (and duplicated) way to model the behaviours of recreational fishers. Instead, the success may arise because ecological scientists have largely adopted models from resource economists to predict fishers' behaviours, given that these economic models represent a theoretically sound way to model behaviours and to estimate benefits to people. The relationship can also benefit resource economists who rely on ecological scientists to model fish population dynamics. Thus, a symbiotic relationship likely exists among these researchers in their efforts to model recreational fisheries as part of a social-ecological system. Further evidence supporting the importance of tangible benefits in facilitating interdisciplinary research comes from a blending of ecological science, resource economics, and social psychology disciplines in interdisciplinary research (Johnston et al. 2010, 2013). Here, social psychology (and particularly recreational specialization) is included to address the limited understanding that exists in resource economic models of why diversity (heterogeneity) exists within a community of recreational fishers. By addressing this limitation, researchers can do more than simply examine how preference heterogeneity among fishers yields different expected outcomes of impacts on ecosystems and benefits to people. The researchers can better theorize why heterogeneity exists, how heterogeneity might change in the future, and how heterogeneity influences the effectiveness of different management interventions. Therefore, researchers may be more likely to conduct research with individuals from different disciplines if they expect a benefit from their inclusion. Although this expectation is facilitated by having a similar focus and approach, addressing epistemological barriers within teams (see Sect. 4.1) can also help with expectations.

We feel that opportunities exist for conducting interdisciplinary research on recreational fishers and fisheries using all seven disciplines described in this chapter. Consider a simple case where an interest lies in understanding how a recreational fishery impacts fish (exploitation), yet benefits people. The natural tendency is to engage ecological scientists and resource economists to develop and integrate methods that account for the feedbacks between fish populations and fishers and to borrow methods to measure economic benefits and exploitation outcomes. We already noted that these researchers can also benefit from social psychology by adopting approaches to identify and understand the heterogeneity that exists within populations of fishers who differ in their avidity, setting preferences, and decisions to retain caught fish. Historical ecologists can help contextualize the fishery and assess the past impacts on the aquatic ecosystem, and especially fish populations. These researchers can also provide information to assess whether generational shifts among fisheries managers, ecological scientists, and even fishers have occurred over time, leading to new norms of what constitutes a quality fishery (analogous to Pauly's (1995) shifting baseline syndrome).

Social psychologists and behavioural economists offer the potential to better understand the antecedents of the behaviours of fishers. For example, researchers could integrate social psychology theories such as the cognitive hierarchy (Fulton et al. 1996) with classic models from resource economics to predict fishing site choices. These integrated models would not only more thoroughly capture the decision-making process of fishers but also they could be used to understand why different fishers exhibit different preferences for attributes of fishing sites (e.g. catch-related fishing quality) and how these preferences may change and evolve over time. Social psychologists and behavioural economists can also help to understand how social norms influence fishers' decisions to release or harvest fish and how fisheries managers can use this understanding to nudge fishers towards releasing more of the fish that they catch.

Environmental historians and political ecologists can help to understand why these preferences and norms exist among fishers, how these preferences and norms evolved over time, and how historical and political interactions between fishers, fisheries managers, and enforcement officers have affected the relationships and levels of trust between fishers and managers. For example, recreational fisheries in Germany are managed to limit impacts on individual fish, and thus, releasing caught fish that are legal is socially problematic and sometimes is even prohibited (Arlinghaus et al. 2021). In North America (and most English-speaking countries), recreational fisheries are managed to limit impacts on the population, and thus, releasing caught fish is highly encouraged and is becoming a social norm among some fishing communities (Arlinghaus et al. 2021). Both political ecologists and environmental historians can help to understand why these recreational fisheries have evolved in much different ways, and thus, provide insights into any existing social conflicts as well as how recreational fishers and fisheries might continue to evolve. For example, these researchers can help illuminate when, where, and why fishers might resist some fishing regulations.

Even with our example that narrowly focuses on understanding impacts and benefits from recreational fishing, we hope that the reader sees the opportunities that are available to conduct more fulsome interdisciplinary research. Of course, our example is poor because we began by privileging ecological science and resource economics over the other disciplines. In reality, all researchers can equally help set the scope for any interdisciplinary study and identify and provide solutions to overcome barriers to interdisciplinary research, particularly epistemological differences.

Acknowledgements We thank two anonymous reviewers and Kevin Pope for providing very helpful suggestions for reorganizing materials and adding content.

References

Aas Ø, Haider W, Hunt LM (2000) Angler responses to potential harvest regulations in a Norwegian sport fishery: a conjoint-based choice modeling approach. N Am J Fish Manage 20(4): 940–950

Adamowicz W, Louviere J, Williams M (1994) Combining revealed and stated preference methods for valuing environmental amenities. J Environ Econ Manag 26(3):271–292

Ajzen I (1991) The theory of planned behavior. Organ Behav Hum Dec 50(2):179–211

Alcorn SR (1981) Fishing quality in two urban fishing lakes, St. Louis, Missouri. N Am J Fish Manage 1(1):80–84

Alexander KA, Hobday AJ, Cvitanovic C, Ogier E, Nash KL, Cottrell RS, Fleming A, Fudge M, Fulton EA, Frusher S, Kelly R, MacLeod CK, Pecl GT, van Putten I, Vince J, Watson RA (2019) Progress in integrating natural and social science in marine ecosystem-based management research. Mar Freshw Res 70(1):71

Archer M, Sharp R, Stones R, Woodiwiss T (1999) Critical realism and research methodology. Alethia 2(1):12–16

Arlinghaus R (2006) On the apparently striking disconnect between motivation and satisfaction in recreational fishing: the case of catch orientation of German anglers. N Am J Fish Manag 26(3): 592–605

Arlinghaus R, Mehner T, Cowx IG (2002) Reconciling traditional inland fisheries management and sustainability in industrialized countries, with emphasis on Europe. Fish Fish 3(4):261–316

Arlinghaus R, Hunt LM, Post JR, Allen MS (2014) Not fish, not meat: some guidance on how to study fisheries from an interdisciplinary perspective. In: Taylor WW, Lynch AJ, Léonard NJ (eds) Future of fisheries: perspectives for emerging professionals. American Fisheries Society, Bethesda, pp 223–230

Arlinghaus R, Alós J, Beardmore B, Daedlow K, Dorow M, Fujitani M, Hühn D, Haider W, Hunt LM, Johnson BM, Johnston F, Klefoth T, Matsumura S, Monk C, Pagel T, Post JR, Rapp T, Riepe C, Ward H, Wolter C (2017) Understanding and managing freshwater recreational fisheries as complex adaptive social-ecological systems. Rev Fish Sci Aquac 25(1):1–41

Arlinghaus R, Aas Ø, Alós J, Arismendi I, Bower S, Carle S, Czarkowski T, Freire KMF, Hu J, Hunt LM, Lyach R, Kapusta A, Salmi P, Schwab A, Tsuboi J, Trella M, McPhee D, Potts W, Wołos A, Yang Z-J (2021) Global participation in and public attitudes toward recreational fishing: international perspectives and developments. Rev Fish Sci Aquac 29(1):58–95

Armstrong CG, Shoemaker AC, McKechnie I, Ekblom A, Szabó P, Lane PJ, McAlvay AC, Boles OJ, Walshaw S, Petek N, Gibbons KS, Quintana Morales E, Anderson EN, Ibragimow A, Podruczny G, Vamosi JC, Marks-Block T, LeCompte JK, Awâsis S, Nabess C, Sinclair P, Crumley CL (2017) Anthropological contributions to historical ecology: 50 questions, infinite prospects. PLoS One 12(2):e0171883

Battista W, Romero-Canyas R, Smith SL, Fraire J, Effron M, Larson-Konar D, Fujita R (2018) Behavior change interventions to reduce illegal fishing. Front Mar Sci 5:403

Bazerman MH (1994) Judgment in managerial decision making. Wiley, New York

Beardmore B, Haider W, Hunt LM, Arlinghaus R (2011) The importance of trip context for determining primary angler motivations: are more specialized anglers more catch-oriented than previously believed? N Am J Fish Manage 31(5):861–879

Bergseth BJ, Roscher M (2018) Discerning the culture of compliance through recreational fisher's perceptions of poaching. Mar Policy 89:132–141

Berkes F (2008) Sacred ecology, 2nd edn. Routledge, New York

Birdsong M, Hunt LM, Arlinghaus R (2021) Recreational angler satisfaction: what drives it? Fish Fish 22:682–706

Blaikie PM, Brookfield HC (1987) Land degradation and society. Methuen, London

Bogue MB (2000) Fishing the Great Lakes: an environmental history, 1783–1933. University of Wisconsin Press, Madison

Bolster JW (2006) Opportunities in marine environmental history. Environ Hist 11(3):567–597

Boucquey N (2020) The "nature" of fisheries governance: narratives of environment, politics, and power and their implications for changing seascapes. J Political Ecol 27(1):169–189

Boucquey N, Fly J, Boonstra WJ, Drakopulos L, Leong KM, Marjadi MN, Wise S, Fairbanks L (2026) Political ecology for understanding recreational fishers and fisheries. In: Pope KL, Arlinghaus R, Hunt LM, Lynch AJ, van Poorten BT (eds) Understanding recreational fishers: disciplinary and interdisciplinary approaches for fisheries management. Springer Nature, Cham, p 197–232

Bricker KS, Kerste DL (2000) Level of specialization and place attachment: an exploratory study of whitewater recreationists. Leis Sci 22(4):233–257
Brown TL (1987) Typology of human dimensions information needed for Great Lakes sport-fisheries management. Trans Am Fish Soc 116(3):320–324
Brown WG, Singh A, Castle EN (1965) Net economic value of the Oregon salmon-steelhead sport fishery. J Wildl Manage 29(2):266–279
Bryan H (1977) Leisure value systems and recreational specialization: the case of trout fishermen. J Leis Res 9(3):174–187
Buker E (2003) Is women's studies a disciplinary or an interdisciplinary field of inquiry? NWSA J 15(1):73–93
Campbell LM (2003) Challenges for interdisciplinary sea turtle research: perspectives of a social scientist. Mar Turt Newsl 100:28–32
Campbell LM (2005) Overcoming obstacles to interdisciplinary research. Conserv Biol 19(2): 574–577
Carpenter SR, Brock WA (2004) Spatial complexity, resilience, and policy diversity: fishing on lake-rich landscapes. Ecol Soc 9(1):8
Carruthers TR, Dabrowska K, Haider W, Parkinson EA, Varkey DA, Ward HGM, McAllister MK, Godin T, van Poorten BT, Askey PJ, Wilson KL, Hunt LM, Clarke AD, Newton E, Walters C, Post JR (2018) Landscape scale social and ecological outcomes of dynamic angler and fish behaviours: processes, data, and patterns. Can J Fish Aquat Sci 76(6):970–988
Christensen J, Jackson G (2014) Shark Bay snapper: science, policy, and the decline and recovery of a marine recreational fishery. In: Christensen J, Tull M (eds) Historical perspectives of fisheries exploitation in the Indo-Pacific, MARE publication series, vol 12. Springer, Dordrecht
Christensen J, Tull M (2014) Historical perspectives of fisheries exploitation in the Indo-Pacific. Springer, New York
Cooke SJ, Cowx IG (2006) Contrasting recreational and commercial fishing: searching for common issues to promote unified conservation of fisheries resources and aquatic environments. Biol Conserv 128(1):93–108
Crawford SES, Ostrom E (1995) A grammar of institutions. Am Polit Sci Rev 89(3):582–600
Crumley CL (2021) Historical ecology: a robust bridge between archaeology and ecology. Sustainability 13(15):8210
Dabrowska K, Haider W, Hunt LM (2014) Examining the impact of fisheries resources and quality on licence sales. J Outdoor Recreat Tour 5–6:58–67
Daedlow K, Beckmann V, Schlüter M, Arlinghaus R (2013) Explaining institutional persistence, adaptation, and transformation in East German recreational-fisheries governance after the German reunification in 1990. Ecol Econ 96:36–50
Daubert JT, Young RA (1981) Recreational demands for maintaining instream flows: a contingent valuation approach. Am J Agric Econ 63(4):666–676
de Kerckhove DT, Tucker CM, Gutowsky L, Morgan G, Guzzo M, Barajas VL (2026) Optimizing creel surveys. In: Pope KL, Arlinghaus R, Hunt LM, Lynch AJ, van Poorten BT (eds) Understanding recreational fishers: disciplinary and interdisciplinary approaches for fisheries management. Springer Nature, Cham, p 387–409
Ditton RB, Hunt KM (2001) Combining creel intercept and mail survey methods to understand the human dimensions of local freshwater fisheries: combining creel intercept and mail survey methods. Fish Manage Ecol 8(4–5):295–301
Doyon S (2002) The structural marginalization of artisanal fishing communities: the case of La Boquita. Anthropologica 44(1):83–98
Driver BL (1975) Quantification of outdoor recreationists' preferences. In: Smissen B, Myers J (eds) Research: camping and environmental education, HPEP series 11. Pennsylvania State University, University Park, pp 165–187
Driver BL, Brown PJ (1975) A social psychological definition of recreation demand, with implications for recreation resource planning. In: Assessing demand for outdoor recreation, vol 12. The National Academies Press, pp 63–88

Eigenbrode SD, O'Rourke M, Wulfhorst JD, Althoff DM, Goldberg CS, Merrill K, Morse W, Nielsen-Pincus M, Stephens J, Winowiecki L, Bosque-Pérez NA (2007) Employing philosophical dialogue in collaborative science. Bioscience 57(1):55–64

Engie K, Quiroga D (2014) The emergence of recreational fishing in the Galápagos marine reserve: adaptation and complexities. In: Denkinger J, Vinueza L (eds) The Galapagos marine reserve. Springer, New York, pp 203–226

Fedler AJ, Ditton RB (1994) Understanding angler motivations in fisheries management. Fisheries 19(4):6–13

Fenichel EP, Abbott JK, Huang B (2013) Modelling angler behaviour as a part of the management system: synthesizing a multi-disciplinary literature. Fish Fish 14(2):137–157

Fishbein M, Ajzen I (1975) Belief, attitude, intention and behavior: an introduction to theory and research. Addison-Wesley, Reading

Fisher MR (1997) Segmentation of the angler population by catch preference, participation, and experience: a management-oriented application of recreation specialization. N Am J Fish Manag 17(1):1–10

Fletcher AJ (2017) Applying critical realism in qualitative research: methodology meets method. Int J Soc Res Methodol 20(2):181–194

Frawley J (2015) Kissing fish: rex Hunt, popular culture, sustainability and fishing practices. J Aust Stud 39(3):307–325

Fulton DC, Manfredo MJ, Lipscomb J (1996) Wildlife value orientations: a conceptual and measurement approach. Hum Dimens Wildl 1(2):24–47

Fulton EA, Smith ADM, Smith DC, van Putten IE (2011) Human behaviour: the key source of uncertainty in fisheries management: human behaviour and fisheries management. Fish Fish 12(1):2–17

Glaeser B, Bruckmeier K, Glaser M, Krause G (2009) Social-ecological systems analysis in coastal and marine areas: a path toward integration of interdisciplinary knowledge. In: Lopes P, Begossi A (eds) Current trends in human ecology. Cambridge Scholars Publishing, Newcastle upon Tyne, pp 183–203

Golebie E, van Riper CJ, Kochalski S, Nunan F (2026) Mixed methods research: applications and considerations for the human dimensions of recreational fisheries. In: Pope KL, Arlinghaus R, Hunt LM, Lynch AJ, van Poorten BT (eds) Understanding recreational fishers: disciplinary and interdisciplinary approaches for fisheries management. Springer Nature, Cham, p 481–506

Graefe AR (1980) The relationship between level of participation and selected aspects of specialization in recreational fishing. Dissertation, Texas A&M University, College Station

Guba EG, Lincoln YS (1994) Competing paradigms in qualitative research. In: Denzin N, Lincoln Y (eds) Handbook of qualitative research. Sage, Thousand Oaks, pp 163–194

Guthrie D, Hoenig JM, Holliday M, Jones CM, Mills MJ, Moberly SA, Pollock KH, Talhelm DR (1991) Creel and angler surveys in fisheries management. American Fisheries Society, Bethesda

Hall CM (2021) Tourism and fishing. Scand J Hosp Tour 21(4):361–373

Hammarfelt B (2018) What is a discipline? The conceptualization of research areas and their operationalization in bibliometric research. In: Costas R, Franssen T, Yegros-Yegros A (eds) STI2018. Centre for Science and Technology Studies (CWTS), Leiden, pp 197–203

Hampton EL, Lackey RT (1976) Analysis of angler preferences and fisheries management objectives with implications for management. Proc SE Assoc Game Fish Comm 29:310–316

Heckman JJ (1981) Statistical models for discrete panel data. In: Manski CF, McFadden D (eds) Structural analysis of discrete data with econometric applications. MIT Press, Cambridge, pp 114–178

Hendee JC (1974) A multiple-satisfaction approach to game management. Wildl Soc Bull 2:104–113

Hilborn R, Walters CJ (1992) Quantitative fisheries stock assessment. Springer, Boston

Ho TH, Lim N, Camerer CF (2006) Modeling the psychology of consumer and firm behavior with behavioral economics. J Mark Res 43(3):307–331

Holder PE, Jeanson AL, Lennox RJ, Brownscombe JW, Arlinghaus R, Danylchuk AJ, Bower SD, Hyder K, Hunt LM, Fenichel EP, Venturelli PA, Thorstad EB, Allen MS, Potts WM, Clark-Danylchuk S, Claussen JE, Lyle JM, Tsuboi J, Brummett R, Freire KMF, Tracey SR, Skov C, Cooke SJ (2020) Preparing for a changing future in recreational fisheries: 100 research questions for global consideration emerging from a horizon scan. Rev Fish Biol Fish 30(1):137–151

Holen DL (2004) The Atna' and the political ecology of the Copper River fishery, Alaska. Arctic Anthropol 41(1):58–70

Hunt LM (2005) Recreational fishing site choice models: insights and future opportunities. Hum Dimens Wildl 10(3):153–172

Hunt LM, Arlinghaus R, Lester N, Kushneriuk R (2011) The effects of regional angling effort, angler behavior, and harvesting efficiency on landscape patterns of overfishing. Ecol Appl 21(7):2555–2575

Hunt LM, Camp E, van Poorten B, Arlinghaus R (2019) Catch and non-catch-related determinants of where anglers fish: a review of three decades of site choice research in recreational fisheries. Rev Fish Sci Aquac 27(3):261–286

Johnson BM, Carpenter SR (1994) Functional and numerical responses: a framework for fish-angler interactions. Ecol Appl 4:808–821

Johnston FD, Arlinghaus R, Dieckmann U (2010) Diversity and complexity of angler behaviour drive socially optimal input and output regulations in a bioeconomic recreational-fisheries model. Can J Fish Aquat Sci 67(9):1507–1531

Johnston FD, Arlinghaus R, Dieckmann U (2013) Fish life history, angler behaviour and optimal management of recreational fisheries. Fish Fish 14(4):554–579

Johnston FD, Beardmore B, Arlinghaus R (2015) Optimal management of recreational fisheries in the presence of hooking mortality and noncompliance — predictions from a bioeconomic model incorporating a mechanistic model of angler behavior. Can J Fish Aquat Sci 72(1):37–53

Kahneman D (2013) Thinking, fast and slow. Farrar, Straus and Giroux, New York

Kalina MR, Mbereko A, Botes A, Maharaj B (2019) Subsistence marine fishing in a neoliberal city: a political ecology analysis of securitization and exclusion in Durban. S Afr J Polit Ecol 26(1): 363–380

Karrow T, Thompson T (2016) Political ecology of the flats fishing industry in The Bahamas. In: Political ecology and tourism. Routledge, London, pp 63–77

Keiner C (2013) How scientific does marine environmental history need to be? Environ Hist 18(1): 111–120

Kittinger JN, McClenachan L, Gedan KB, Blight LK (eds) (2015) Marine historical ecology in conservation: applying the past to manage for the future. University of California Press, Oakland

Klein JT (1996) Crossing boundaries: knowledge, disciplinarities, and interdisciplinarities. University Press of Virginia, Charlottesville

Kyle G, Graefe A, Manning R, Bacon J (2004) Effects of place attachment on users' perceptions of social and environmental conditions in a natural setting. J Environ Psychol 24(2):213–225

Kyle G, Absher J, Norman W, Hammitt W, Jodice L (2007) A modified involvement scale. Leis Stud 26(4):399–427

Landon AC, Schroeder SA, Stefen SF, Arlinghaus R (2026) Social psychology for understanding recreational fishers and fisheries. In: Pope KL, Arlinghaus R, Hunt LM, Lynch AJ, van Poorten BT (eds) Understanding recreational fishers: disciplinary and interdisciplinary approaches for fisheries management. Springer Nature, Cham, p 157–196

Larkin PA (1978) Fisheries management – an essay for ecologists. Annu Rev Ecol Syst 9:57–73

Lincoln YS, Lynham SA, Guba EG (2017) Paradigmatic controversies, contradictions, and emerging confluences, revisited. In: Denzin NK, Lincoln YS (eds) The SAGE handbook of qualitative research, 5th edn. SAGE Publications, pp 163–188

Louviere JJ, Hensher DA, Swait JF (eds) (2000) Stated choice methods: analysis and applications. Cambridge University Press, New York

Lury C, Fensham R, Heller-Nicholas A, Lammes S, Last A, Michael M, Uprichard E (eds) (2018) Routledge handbook of interdisciplinary research methods. Routledge, London, New York

Lynch AJ, Hunt LM, Beardmore B, van Poorten BT, Pope KL, Arlinghaus R (2026) Complexity and integration of recreational fisheries. In: Pope KL, Arlinghaus R, Hunt LM, Lynch AJ, van Poorten BT (eds) Understanding recreational fishers: disciplinary and interdisciplinary approaches for fisheries management. Springer Nature, Cham, p 3–19

Mackay M, Yamazaki S, Jennings S, Sibly H, van Putten IE, Emery TJ (2019) The influence of nudges on compliance behaviour in recreational fisheries: a laboratory experiment. ICES J Mar Sci 77(6):2319–2332

Mackay M, van Putten EI, Richter A, Abatayo AL, Chuah S-H, Dankel DJ, Schadeberg A, Snyder H (2026) Behavioural economics for understanding recreational fishers and fisheries. In: Pope KL, Arlinghaus R, Hunt LM, Lynch AJ, van Poorten BT (eds) Understanding recreational fishers: disciplinary and interdisciplinary approaches for fisheries management. Springer Nature, Cham, p 261–284

Marjadi MN, Drakopulos L, Guo LW, Koehn JZ, Panchang SV, Robertson D (2021) Negative socio-environmental feedback loop may foster inequality for urban marine subsistence fishers. Environ Sci Pol 121:68–77

Massey DM, Newbold SC, Gentner B (2006) Valuing water quality changes using a bioeconomic model of a coastal recreational fishery. J Environ Econ Manag 52(1):482–500

Matsumura S, Beardmore B, Haider W, Dieckmann U, Arlinghaus R (2019) Ecological, angler, and spatial heterogeneity drive social and ecological outcomes in an integrated landscape model of freshwater recreational fisheries. Rev Fish Sci Aquac 27(2):170–197

May CK (2016) Visibility and invisibility: structural, differential, and embedded power in collaborative governance of fisheries. Soc Nat Resour 29(7):759–774

McClenachan L (2009a) Historical declines of goliath grouper populations in South Florida, USA. Endanger Species Res 7:175–181

McClenachan L (2009b) Documenting loss of large trophy fish from the Florida keys with historical photographs. Conserv Biol 23(3):636–643

McClenachan L (2013) Recreation and the "right to fish" movement: anglers and ecological degradation in the Florida keys. Environ Hist 18(1):76–87

McConnell KE, Sutinen JG (1979) Bioeconomic models of marine recreational fishing. J Environ Econ Manag 6:127–139

McPhee DP, Leadbitter DA, Skilleter G (2002) Swallowing the bait: is recreational fishing in Australia ecologically sustainable? Pac Conserv Biol 8(1):40–51

Melstrom RT, Jungers B, Bi X, Abbott JK (2026) Resource economics for understanding recreational fishers and fisheries. In: Pope KL, Arlinghaus R, Hunt LM, Lynch AJ, van Poorten BT (eds) Understanding recreational fishers: disciplinary and interdisciplinary approaches for fisheries management. Springer Nature, Cham, p 123–156

Miller TR, Baird TD, Littlefield CM, Kofinas G, Chapin FS III, Redman CL (2008) Epistemological pluralism: reorganizing interdisciplinary research. Ecol Soc 13(2):46

Milon JW (1988a) A nested demand shares model of artificial marine habitat choice by sport anglers. Mar Resour Econ 5(3):191–213

Milon JW (1988b) Travel cost methods for estimating the recreational use benefits of artificial marine habitat. J Agric Appl Econ 20(1):87–101

Mitchell JP (2009) Social psychology as a natural kind. Trends Cogn Sci 13(6):246–251

Moeller GH, Engelken JH (1972) What fishermen look for in a fishing experience. J Wildl Manag 36(4):1253–1257

Moon K, Cvitanovic C, Blackman DA, Scales IR, Browne NK (2021) Five questions to understand epistemology and its influence on integrative marine research. Front Mar Sci 8:574158

Moore RL, Graefe AR (1994) Attachments to recreation settings: the case of rail-trail users. Leis Sci 16(1):17–31

Morey ER, Breffle WS (2006) Valuing a change in a fishing site without collecting characteristics data on all fishing sites: a complete but minimal model. Am J Agric Econ 88(1):150–161

Morse WC, Nielsen-Pincus M, Force JE, Wulhorst JD (2007) Bridges and barriers to developing and conducting interdisciplinary graduate-student team research. Ecol Soc 12(2):8

Needham MD, Vaske JJ, Whittaker D, Donnelly MP (2014) Extending the encounter-norm-crowding generalization to angler evaluations of other social and resource indicators. Hum Dimens Wildl 19(3):288–299

Nielsen JR (2003) An analytical framework for studying: compliance and legitimacy in fisheries management. Mar Policy 27(5):425–432

Obermeister N (2017) From dichotomy to duality: addressing interdisciplinary epistemological barriers to inclusive knowledge governance in global environmental assessments. Environ Sci Pol 68:80–86

Ostrom E (2005) Understanding institutional diversity. Princeton University Press, Princeton

Palmer L (2004) Fishing lifestyles: 'territorians', traditional owners and the management of recreational fishing in Kakadu National Park. Aust Geogr Stud 42(1):60–76

Pauly D (1995) Anecdotes and the shifting baseline syndrome of fisheries. Trends Ecol Evol 10(10): 430

Peet R, Watts M (eds) (2004) Liberation ecologies: environment, development, and social movements, 2nd edn. Routledge, Taylor & Francis Group, London

Perreault TA, Bridge G, McCarthy J (eds) (2015) The Routledge handbook of political ecology. Routledge, Taylor & Francis Group, London

Phaneuf DJ, Requate T (2017) A course in environmental economics: theory, policy and practice. Cambridge University Press, Cambridge

Post JR, Sullivan M, Cox S, Lester NP, Walters CJ, Parkinson EA, Paul AJ, Jackson L, Shuter BJ (2002) Canada's recreational fisheries: the invisible collapse? Fisheries 27(1):6–17

Post JR, Persson L, Parkinson EA, Kooten TV (2008) Angler numerical response across landscapes and the collapse of freshwater fisheries. Ecol Appl 18(4):1038–1049

Post JR, Wilson KL, Johnston FD, Sass GG, Askey PJ, Ward HGM, Allen MS (2026) Ecology for understanding recreational fishers and fisheries. In: Pope KL, Arlinghaus R, Hunt LM, Lynch AJ, van Poorten BT (eds) Understanding recreational fishers: disciplinary and interdisciplinary approaches for fisheries management. Springer Nature, Cham, p 95–121

Proctor JD (1998) The social construction of nature: relativist accusations, pragmatist and critical realist responses. Ann Assoc Am Geogr 88(3):352–376

Reid AJ, Eckert LE, Lane J, Young N, Hinch SG, Darimont CT, Cooke SJ, Ban NC, Marshall A (2021) "Two-eyed seeing": an indigenous framework to transform fisheries research and management. Fish Fish 22(2):243–261

Repko AF, Szostak R (2021) Interdisciplinary research: process and theory, 4th edn. SAGE, Los Angeles

Ricker WE (1945) A method of estimating minimum size limits for obtaining maximum yield. Copeia 1945(2):84–94

Riepe C, Arlinghaus R (2014) Explaining anti-angling sentiments in the general population of Germany: an application of the cognitive hierarchy model. Hum Dimens Wildl 19(4):371–390

Ross L, Lepper M, Ward A (2010) History of social psychology: insights, challenges, and contributions to theory and application. In: Fiske ST, Gilbert DT, Lindzey G (eds) Handbook of social psychology, 5th edn. John Wiley & Sons, New York, pp 3–50

Schroeder SA, Fulton DC, Altena E, Baird H, Dieterman D, Jennings M (2018) The influence of angler values, involvement, catch orientation, satisfaction, agency trust, and demographics on support for habitat protection and restoration versus stocking in publicly managed waters. Environ Manag 62(4):665–677

Schummer J (2004) Multidisciplinarity, interdisciplinarity, and patterns of research collaboration in nanoscience and nanotechnology. Scientometrics 59(3):425–465

Schwerdtner Máñez K, Poulsen B (2016) Of seascapes and people: multiple perspectives on oceans past. In: Schwerdtner Máñez K, Poulsen B (eds) Perspectives on oceans past. Springer, Dordrecht, pp 1–10

Scott D, Shafer CS (2001) Recreational specialization: a critical look at the construct. J Leis Res 33(3):319–343

Shelby LB, Vaske JJ (2007) Perceived crowding among hunters and anglers: a meta-analysis. Hum Dimens Wildl 12(4):241–261

Slaton C, Koemle D, Birdsong M, Arlinghaus R (2023) Explaining attitudes to management actions and beliefs about other user groups and conservation with angler characteristics: a case study in a coastal pike (*Esox lucius*) fishery in the southern Baltic Sea, Germany. Fish Res 263:106669

Slevitch L (2011) Qualitative and quantitative methodologies compared: ontological and epistemological perspectives. J Qual Assur Hosp Tour 12(1):73–81

Stedman RC (2002) Toward a social psychology of place: predicting behavior from place-based cognitions, attitude, and identity. Environ Behav 34(5):561–581

Stoop J, Noussair CN, van Soest D (2012) From the lab to the field: cooperation among fishermen. J Polit Econ 120(6):1027–1056

Sugimoto CR, Weingart S (2015) The kaleidoscope of disciplinarity. J Doc 71(4):775–794

Sutinen JG, Kuperan K (1999) A socio-economic theory of regulatory compliance. Int J Soc Econ 26(1/2/3):174–193

Szabó P (2015) Historical ecology: past, present and future: historical ecology. Biol Rev 90(4): 997–1014

Thaler RH, Sunstein CR (2008) Nudge: improving decisions about health, wealth, and happiness. Yale University Press, New Haven

Thurstan RH (2022) The potential of historical ecology to aid understanding of human-ocean interactions throughout the Anthropocene. J Fish Biol 101(2):351–364

Thurstan RH, Buckley SM, Pandolfi JM (2018) Trends and transitions observed in an iconic recreational fishery across 140 years. Glob Environ Chang 52:22–36

Thurstan RH, Chong-Montenegro C, Christensen J, Clark A, Gaynor A (2026) Environmental history and historical ecology for understanding recreational fishers and fisheries. In: Pope KL, Arlinghaus R, Hunt LM, Lynch AJ, van Poorten BT (eds) Understanding recreational fishers: disciplinary and interdisciplinary approaches for fisheries management. Springer Nature, Cham, p 233–259

Trice AH, Wood SE (1958) Measurement of recreation benefits. Land Econ 34(3):195–207

Turner VK, Benessaiah K, Warren S, Iwaniec D (2015) Essential tensions in interdisciplinary scholarship: navigating challenges in affect, epistemologies, and structure in environment-society research centers. High Educ 70(4):649–665

van den Besselaar P, Heimeriks G (2001) Disciplinary, multidisciplinary, interdisciplinary: concepts and indicators. In: Davis M, Wilson CS (eds) Proceedings of the 8th International Conference on Scientometrics & Informetrics (ISSI). University of New South Wales, Sydney, pp 705–716

Ward HGM, Quinn MS, Post JR (2013) Angler characteristics and management implications in a large, multistock, spatially structured recreational fishery. N Am J Fish Manag 33(3):576–584

Ward HGM, Allen MS, Camp EV, Cole N, Hunt LM, Matthias B, Post JR, Wilson K, Arlinghaus R (2016) Understanding and managing social-ecological feedbacks in spatially structured recreational fisheries: the overlooked behavioral dimension. Fisheries 41(9):524–535

Wellmon C (2015) Organizing enlightenment: information overload and the invention of the modern research university. Johns Hopkins University Press, Baltimore

Westerman FA (1929) Progress in trout propagation in Michigan. Trans Am Fish Soc 59(1): 234–237

Wilde GR, Ditton RB (1994) A management-oriented approach to understanding diversity among largemouth bass anglers. N Am J Fish Manag 14(1):34–40

Wilson KL, De Gisi J, Cahill CL, Barker OE, Post JR (2019) Life-history variation along environmental and harvest clines of a northern freshwater fish: plasticity and adaptation. J Anim Ecol 88(5):717–733

Ecology for Understanding Recreational Fishers and Fisheries

John R. Post, Kyle L. Wilson, Fiona D. Johnston, Greg G. Sass, Paul J. Askey, Hillary G. M. Ward, and Micheal S. Allen

Abstract We examine recreational fisher behaviour and recreational fishery systems through the lens of an ecologist to understand the dynamical properties of these social-ecological systems. From the perspective of an ecologist, recreational fishers and the fish they capture can be viewed as analogous to predator-prey systems. Our understanding of predator-prey interactions is supported by a richness of empirical and conceptual research, primarily developed within sub-fields of behavioural, population, and community ecology. We develop this analogy between predator-prey ecology and fisher behaviour by examining the key underlying processes within a conceptual framework based on simple models. We then characterize several processes inherent to recreational fisheries that can, at least in part, decouple these simple predator-prey interactions. We examine the impacts on fisher behaviour and fishery outcomes of non-random spatial distributions of fishers and fish, heterogeneity of fisher behaviour, and multi-species fisheries, and develop an enhanced

J. R. Post (✉)
Department of Biological Sciences, University of Calgary, Calgary, AB, Canada
e-mail: jrpost@ucalgary.ca

K. L. Wilson
School of Resource and Environmental Management, Simon Fraser University, Burnaby, BC, Canada

F. D. Johnston
School of Resource and Environmental Management, Simon Fraser University, Burnaby, BC, Canada

G. G. Sass
Office of Applied Science, Wisconsin Department of Natural Resources, Bolder Junction, WI, USA

P. J. Askey
Freshwater Fisheries Society of British Columbia, Summerland, BC, Canada

H. G. M. Ward
Ministry of Water, Land and Resource Stewardship, Penticton, BC, Canada

M. S. Allen
School of Forest, Fisheries and Geomatics Science, University of Florida, Gainesville, FL, USA

K. L. Pope et al. (eds.), *Understanding Recreational Fishers*, Fish & Fisheries Series 45, https://doi.org/10.1007/978-3-031-99739-6_4

framework to understand these dynamic interactions. Population ecology and density-dependent feedbacks are important concepts underlying fish population dynamics, and also set limits to the sustainability of fisher harvest. Predator-prey theory is helpful in understanding fisher behaviour and its feedback with fish production, fishery quality, and sustainability. As fishers often prefer larger sizes in their catch, the predator-prey dynamic involves ecological concepts related to life-history theory and size-structured interactions. Although some recreational fisheries target a single species, many involve multi-species fisheries, such that food web theory is also important in understanding ecological feedbacks between fisher behaviour and fishery outcomes. In addition, individual recreational fisheries are typically embedded within landscapes of alternative fisheries, so the spatial configuration of fishing opportunities and the spatial behaviour of fishers is central in understanding fishery outcomes across landscapes. We discuss field methods used to measure fisher behaviour and provide empirical examples from well-studied recreational fisheries. Field methods to measure catchability, catch-per-unit effort, and landscape distribution of fisher effort are informed by the ecological theory of functional and numerical responses. Methods and implications of catch-and-release behaviour, multi-species fisheries, and non-catch related fisher site choice are discussed as related to our understanding of fisher behaviour and fisheries outcomes.

Keywords Catch-and-release · Fisher effort · Fishery management · Food webs · Functional response · Ideal free distribution · Landscapes · Numerical response · Predator-prey theory

1 The Discipline of Ecology

Ecology is the scientific study of the processes and interactions that determine the distribution and abundance of organisms (Krebs 2009). Distribution refers to where organisms are found, and abundance refers to how many are found within an area. In ecology, we seek to understand: why this distribution, and why this abundance? Three approaches to answering these questions are descriptive, functional, or evolutionary. Descriptive approaches involve the natural history of organisms across space and time. Functional ecology identifies the processes of interaction among organisms and their interactions with their environment, structured within a cause-and-effect framework. Evolutionary approaches focus on organisms and their interrelationships, considering their evolutionary history. These three approaches provide the core of the science of ecology and are prerequisites to understanding the distribution and abundance of organisms in nature.

Sub-fields of ecology include several levels of integration: individuals, populations, communities, and ecosystems. Ecology of individuals encompasses aspects of behaviour and physiology that shape their interactions and success. Population ecology focusses on demographic processes such as birth, death, immigration, and emigration rates and how these interact in controlling abundance over time and space. Community ecology focusses on interactions among species, such as competition and predation, and how these processes result in species abundance and composition. And ecosystem ecology focusses on the roles that organisms play in larger global biogeochemical processes within terrestrial, aquatic, and atmospheric systems.

Our approach in this chapter is to examine the behaviour of recreational fishers through an ecological lens based on a combination of field observations, experimentation, simulation models, and theory (Carpenter 1998). Intuitively, recreational fishers and the fish they capture can be viewed as analogous to predator-prey systems. In ecology, our understanding of predator -prey interactions is supported by a richness of empirical and conceptual research on predator-prey interactions, primarily developed within sub-fields of behavioural, population, and community ecology. Ecologically, predator behaviour is presumed to reflect evolutionary selection of behavioural traits that maximize fitness. The ultimate expression of fitness in the natural world is the production of offspring who perpetuate the most successful genotypes. Proximate measures of fitness of the predator are often measured as prey consumption, translating into their growth, survival, and reproductive success. Of course, this analogy is not directly relevant to recreational fisheries where a successful fisher does not necessarily have greater reproductive success! Instead, success is typically measured as high catch rates or the capture of large fish. But, in addition, fishing success (as measured by sociologists and economists as satisfaction and utility; Landon et al. 2026; Melstrom et al. 2026) includes a host of non-catch related characteristics of the fishing experience (e.g. fishers' values for place and culture).

We develop this analogy between predator-prey ecology and fisher behaviour by examining the key underlying processes within a conceptual framework based on simple models. We then characterize several processes that can partially decouple these simple predator-prey interactions, complicating the simple models, and which are likely important within recreational fisheries. The final component of the conceptual framework is the incorporation of predator and prey heterogeneity, that is, several types of fishers and multiple fish species, typical of many recreational fisheries. And finally, we describe the types of data that can inform these ideas and ways in which they are used by ecologists to provide the information needed to understand fisher behaviour and manage recreational fisheries.

1.1 Ecological Predator-Prey Interactions and Fisher Behaviour

Predator-prey interactions can be characterized by two important processes. First is the combination of predator and prey behaviour that results in searching, handling, and consumption of prey by predators, as modified by the avoidance behaviour of prey. In ecology, this aggregate of behaviours results in what is called a "functional response". The second process involves aggregative behaviour of predators across a landscape in response to heterogeneous prey abundance and is termed a "numerical response".

Predator functional responses describe the relationship between prey abundance and the resulting prey consumption rates by predators. Functional responses are determined by predator behaviours such as search time, search efficiency, attack rate, handling time, digestion time, and physiological constraints such as energy content and satiation. In recreational fisheries, functional responses are characterized by the rate of fish capture by an average fisher (expressed as catch-per-unit-effort where effort is in units of time—CPUE) as a function of the abundance of catchable fish. These responses can be proportional, asymptotic, or sigmoid in shape (referred to as Type I, II, or III functional responses) (Fig. 1). In a recreational fishery, the CPUE functional response can be shaped by characteristics like fishing gear selection, use of technology, fisher experience, and fish behaviour (e.g. schools of young fish or adults aggregating on spawning grounds).

Predator numerical responses represent the aggregation (or reproduction) of predators in prey patches with greater abundance or production of prey. Similarly, fishers are expected to choose to participate in fishing or move among fishing sites based on their perception of the quality fishing opportunities available to them. This behaviour leads to a feedback function that characterizes the degree to which fishers as a group are responsive to the assessed quality of the fishing opportunity (i.e. if

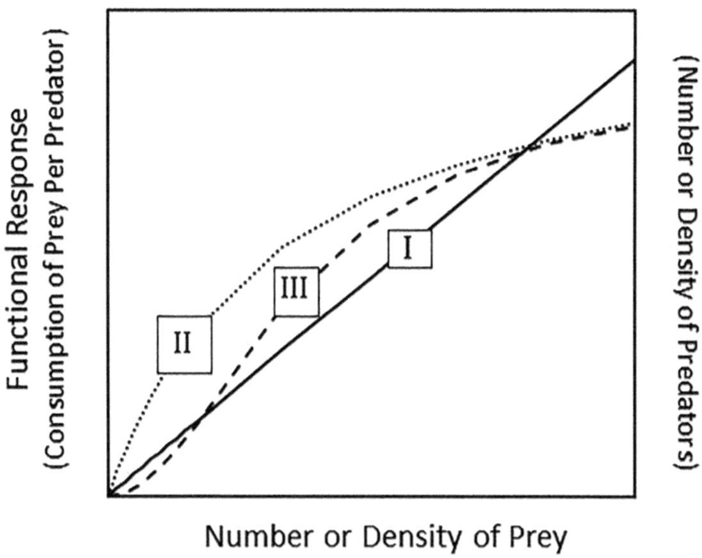

Fig. 1 Functional responses describe the relationship between the number or density of prey and the consumption rate by predators (they can be classified into three types, a linearly increasing consumption rate [Type I], a curvilinear and asymptotic relationship [Type II] and a sigmoid relationship [Type III]), whereas numerical responses describe how the density of predators are distributed across a gradient or prey density (and can be described with the same three types of responses as for functional responses); in recreational fisheries a functional response is measured by catch rates of fish and numerical response by the number of fishers, both as functions of the number or density of fish within targeted fisheries

fishing quality is good, then more effort is attracted, and if poor, then effort is repelled). Therefore, a numerical response is the aggregate outcome of the participation rate and site choice behaviour of a population of fishers.

Predator numerical responses represent behaviours presumed to maximize the fitness of individuals across prey patches when in competition for prey resources with other predators. Under certain ecological assumptions, the spatial pattern that results from these predator behaviours has been characterized conceptually by ecologists as the Ideal Free Distribution theory (Fretwell and Lucas 1969; Gillis et al. 1993). The Ideal Free Distribution concept posits that individual predators will choose prey patches that minimize competition and maximize their fitness (i.e. patches that provide the highest consumption rates). It also assumes that predators have perfect knowledge of the quality of all patches, that predators are ideal—selecting patches that maximize their fitness, and free—implying no costs or constraints on patch choice. Because predators consume prey, increasing predator abundance, and therefore increasing intra-specific competition, will act to disperse predators among patches in such a way that the marginal fitness is equivalent across prey patches. An equal distribution of prey quality across patches will also result in an equal distribution of predator abundances and fitness among patches. But if prey production varies across patches, then predator abundance will be positively related to prey productivity in such a way as to equalize marginal fitness across patches. In recreational fisheries comprised of multiple fishing sites or opportunities (i.e. a set of lakes or streams in a region), the prediction from ideal free distribution theory is that fishers will distribute among these opportunities such that their measures of success become equivalent among all possible fishing sites. The outcome of this fisher behaviour is homogeneity of fisher success across opportunities, but spatial variation in the fisher abundance that is proportional to fish production at individual sites.

In recreational fisheries, the base assumptions underlying the ideal free distribution are often not strictly met, requiring new concepts that better approximate functional and numerical responses. In recreational fisheries, we know there are constraints on getting to and among fishing opportunities based on travel time and costs. Therefore, ideal free distribution "with costs" alters the prediction of homogeneity in satisfaction and fishing quality due to the trade-offs between the benefits of fishing and costs associated with travel. This leads to predictions of "optimal" fisher behaviour where fishers distribute among opportunities in relation to the relative productivity of the various patches, and inversely related to distance as the travel time and costs increase (i.e. fitness in a particular patch is devalued by the costs to acquire the benefits of that patch). In a simple landscape where the productivity of patches is equivalent, the Ideal Free Distribution expectation is a negative exponential decline in the spatial distribution of effort as the distance from the source of the fishers (i.e. population centres) increases across a landscape of fishing opportunities (i.e. fishing lakes).

Application of ideal free distribution theory has provided a simple and robust hypothesis for the behaviour of recreational fishers when faced with a landscape of multiple fishing opportunities (Post et al. 2002; Cox et al. 2003; Parkinson et al. 2004) and provided a deductive engine for modelling studies (Post et al. 2008;

Carruthers et al. 2019; Matsumura et al. 2019). As would be predicted from Ideal Free Distribution, coarse spatial scale field data on fisher catch rates from the British Columbia rainbow trout fishery showed relative homogeneity within regions of similar access costs and inter-regional differences with total fishing effort inversely related to catch rates and positively related to distance from home (Post et al. 2002; Post 2013). This is interpreted as an emergent property of fisher site choice behaviour driven by catch-related fishing quality and access costs (Parkinson et al. 2004). Clearly, fisher site choice is more complex (Hunt 2005; Hunt et al. 2011), but fishing quality and access costs are often among the most important drivers of utility in recreational fisheries (Hunt et al. 2021). In lieu of more detailed descriptions of site-specific attributes, it appears as if catch-related and travel distance-dependent attributes can approximate spatial patterns of effort density and fishing quality at larger spatial scales (Hunt et al. 2011; Wilson et al. 2020) as would be predicted by the ecological foraging processes implied by Ideal Free Distribution theory.

1.2 Predator-Prey Models

Ecologists often build and exercise models designed to ask and answer quantitative questions (Krebs 2009; Hutchings 2021). A modelling approach can increase the rate at which we generate understanding of complex systems. Models are also useful tools to examine dynamic properties of systems based on rates of processes and variability in those rates. Useful models should reflect the true nature of one or more aspects of the system of interest to the researcher. Structurally, a model is a formal expression of the relationship between defined entities of the system in mathematical terms. Predator-prey systems, and fisher behaviour in response to prey characteristics, are complex systems that have been widely studied through the application of models. Functional and numerical responses within predator-prey system models are inherently quantitative and dynamic and provide an effective tool to study their system properties. Here, we develop simple models of predator and prey behaviour and their interactions and examine situations where there are important deviations from simple model representations of recreational fisheries.

At the core of our understanding of fisher behaviour within and among fisheries, measured as aggregate fishing effort, is the dynamic interaction between fishers and their fish resources. Fishing can alter abundance, size structure, behaviour, demographic rates, and species composition of fish communities. These changes to the fish community alter fishing quality as perceived by fishers, which in turn creates feedback among fisher behaviours, site choice, and total fishing effort. Therefore, we need to acknowledge this feedback to understand both fisher behaviours and the fishing outcomes.

To understand feedback between fisher behaviour and fish populations, we again borrow from ecological predator-prey theory and models (Clark 2006, 2018). Fisher harvest is analogous to predators consuming prey and prey (fish) populations are controlled by ecological production (the aggregate outcome of birth and death processes). In the simplest structure with a fixed fishing effort, we use a Gordon-Schaefer model:

$$dN/dt = \text{Production} - \text{Harvest} \tag{1}$$

$$\text{Production} = r \cdot N \cdot (1 - N/K) \tag{2}$$

$$\text{Harvest} = \text{Catch} = C = q \cdot E \cdot N \tag{3}$$

where dN/dt is the rate of change in the fish population, which is the difference between the production rate of the population, represented here by logistic population growth, and harvest by fishers. Harvest is the product of the number of fish in the population, N, total fishing effort, E, and the efficiency of that fishing effort, q, if it is assumed that all fish captured are harvested. This simple formulation describes a Type I functional response where the harvest is proportional to fish abundance based on a constant fisher efficiency (analogous to predator attack rate). The efficiency of the fisher-predator (q) is typically referred to as "catchability" in fisheries science and is a fundamental parameter representing the linkage between fishers and fish. Fisher perceptions of fish abundance and fishing quality are based on the number of fish captured compared to the time spent fishing, known as catch per unit effort (CPUE). Rearranging Eq. 3, we see that fisher CPUE (C/E) in this simple formulation is proportional to the fish abundance as $q{\cdot}N$. Therefore, catchability is just as important as fish abundance for determining fishing outcomes. Numerically, q represents the proportion of the fish population that is caught per unit of fisher effort. Many components of fisher and fishery manager behaviours are buried in this seemingly simple term (e.g. choice of gear, knowledge and skill of the fishers, catch regulations, restrictions), plus fish behaviours (e.g. feeding aggressiveness, degree of aggregation). Many of these fish and fisher behaviours can shift q from a constant to a fish density-dependent parameter, which leads to a non-linear functional response (e.g. Type II).

The next step in understanding how fisher behaviour and fishing quality interact is to build into the general model the idea that there is a numerical response to fishing quality. The relationship between CPUE and fish abundance is a key component of predicting harvest rates, and importantly, will guide the numerical (effort) response. Unlike the simple model above (Eq. 3) with a constant E, fisher effort is expected to respond dynamically to changes in fishing quality.

$$E = f(\text{quality}). \tag{4}$$

The CPUE is often a key indicator of fishing quality. Therefore, the relationship between CPUE and $q{\cdot}N$ has an enormous influence on how fisher effort (which drives harvest) responds to fish abundance (as with Type I, II, or III numerical responses in Ecology).

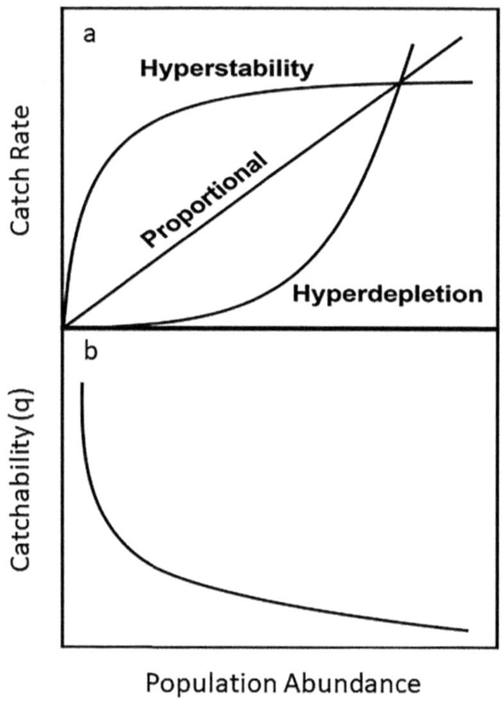

Fig. 2 (**a**) Catch rate (fish caught per time or effort) can be related to population abundance in three ways (many simple models presume that catch rate will increase proportionately with abundance fish in the population presuming random encounter of homogeneous fishers and fish, but other mechanisms can lead to either an asymptotically increasing or exponentially increasing catch with increasing fish abundance; these alternate mechanisms can lead to either hyperstability or hyperdepletion of catch rates as populations decline in response to harvest) (**b**) hyperstable catch rates imply that catchability (q) is low at high fish population abundance and increases as fish abundance declines (which has the effect of accelerating rates of harvest as fish populations decline due to harvest)

Traditional fisheries management tenets have often assumed that the relationship between fisher CPUE and fish population size is proportional, like the Type I functional response in ecological predator-prey interactions (Figs. 1 and 2). This assumes that fishers will leave a waterbody when catch rates become unsatisfactory and seek alternative, higher catch rate opportunities, consistent with predictions from the ideal free distribution theory. Self-regulating fisher behaviour could potentially lead to a sustainable equilibrium if fishers become dissatisfied and leave before the fishery becomes overharvested (Askey and Johnston 2013). However, non-linear functional responses lead to hyperstability or hyperdepletion and therefore can impede a self-regulating equilibrium.

1.3 Departures from Simple Predator-Prey Models

In its simplest formulation, predator-prey theory postulates that the local abundance of predators and their consumption are tightly linked to the abundance of local prey. In this case, if catch rates decline, then fisher effort will decline proportionately as fishers leave the system (Fig. 2a). However, there are many factors that may cause a partial disconnect between predator and prey populations, particularly in recreational fisheries. Such disconnects can have deleterious results for the prey population.

If fish aggregate, either in schools or in association with habitat types, and if fishers can find these aggregations, then catch rates can remain high despite reductions in density due to harvest (Post et al. 2002). Likewise, catch rates may saturate at higher fish densities due to fisher handling time (e.g. hooking, playing, harvesting, or releasing a fish) similar to a Type II functional response between predator consumption and prey density (Fig. 1). This density-dependent catchability, where q increases as populations decline, leads to an undesirable and pathological result in which CPUE is hyperstable, declining little with declining fish populations (Fig. 2b). If this is the case, then there is little or no signal to fishers that the fish population is declining, and therefore no resulting fisher numerical response to abandon a declining fish population. This depensatory process is an important outcome of the interaction between fish and fisher behaviours, and one that can lead to overfishing and collapse (Post et al. 2002; Sass et al. 2021). The opposite outcome, in which fishing in relatively pristine conditions selectively removes the most spatially or behaviourally vulnerable fish, resulting in a rapid decline in catch rate, is termed hyperdepletion (Fig. 2a). Such fisheries would likely lead to reduced effort as CPUE would not be attractive (Biro and Post 2008).

An early exploration of functional responses in recreational fisheries was for a Wisconsin walleye (*Sander vitreus*) fishery, where a linearly increasing (Type I) relationship between catch rate and fish abundance was identified (Johnson and Carpenter 1994). Yet, other analyses identified curvilinear relationships for both Ontario (Shuter et al. 1998; Post et al. 2002) and Yukon Territory lake trout (*Salvelinus namaycush*) (Wilson et al. 2020) and California marine bass (*Paralabrax* spp.) (Erisman et al. 2011). Hyperdepletion has been observed in recreational fisheries where, after an initial period of high catch rates of the most vulnerable fish in a population, CPUE stabilizes with reductions in stock size due to harvest, fish learning behaviour, and refractory (invulnerable) periods following capture and release (Cox and Walters 2002; Askey et al. 2006; Cline et al. 2012; Sass et al. 2018; Alós et al. 2019). In contrast, recreational fishers and fish behaviour can also lead to hyperstability through several mechanisms: effort sorting (i.e. more skilled fishers maintaining high catch rates as stock size declines), use of technology, fish spawning aggregations, and selective habitat use by fish (Hansen et al. 2005; Ward et al. 2013; van Poorten et al. 2016; Mrnak et al. 2018; Dassow et al. 2020; Feiner et al. 2020). Observed strengths of hyperstability and hyperdepletion patterns vary among geographic locations and among species in recreational fisheries (Mogensen et al. 2014; Mosley et al. 2022), and it needs to be recognized in these situations that catch rates provide biased measures of population status.

The behaviour of recreational fishers is not always driven by harvest, but instead by catch. Catch can include harvested and released fish. Moreover, the numerical response of fishers to catch can be weak if the primary site choice characteristics are not based on fish abundance. For example, fishing sites that are close to fishers, sites with improved access and facilities (paved roads, restrooms, boat ramps, lodging, or campsites), and aesthetics (e.g. pristine lake shorelines, water quality) can all influence fisher effort relatively independently of fish abundance. These factors can decouple the predator-prey dynamic as represented in Eq. 3, but in separate ways and with contrasting outcomes. Catch-and-release is interpreted by some as an absence of fishing mortality (although some level of discard mortality inevitably occurs) and would therefore eliminate the fisher numerical response presumed earlier to be driven by depletion. Revisiting the Gordon-Schaefer model:

$$\text{Harvest} = q \cdot E \cdot N \cdot (\alpha + \beta) \tag{5}$$

α is the proportion of the catch that is retained (i.e. harvested) and β is the proportion of released fish that die from capture, handling, and release. Both α and β are features of fisher behaviour—related to individual or legislated decisions to release fish and individual skills to effectively capture, handle, and release fish to minimize subsequent mortality. The other key component impacting harvest is fishing effort, E. If the proportion of the catch retained (α) and release mortality (β) are zero, then the impact of fishing on fish populations is benign. However, if effort is sufficiently high and harvest exceeds production, then overfishing can result.

Furthermore, fisher behaviour may not be driven by either harvest or catch, but instead by a variety of non-catch related attributes (Lewin et al. 2006). The influence of non-catch related attributes on where fishers fish can result in a disconnect within the predator-prey system. This can lead to fisher behaviour that is not tightly tied to fish abundance and therefore no longer self-regulating. Numerically, fishing effort is neither constant as depicted in Eq. 3 nor related to fishing quality (i.e. CPUE) as in Eq. 4. These departures from a strong linkage between predator and prey can have significant impacts on fisheries. For example, fisheries that are close to urban centres may maintain elevated levels of fishing pressure despite low fishing quality; in these instances, fishing pressure could be sufficient to collapse fish populations (Post et al. 2002, 2008; Hunt et al. 2011). Regulations and stocking practices may also influence where fishers choose to fish, potentially leading to exploitation rates that are not correlated with fish abundance (e.g. Beard et al. 2003; Johnston et al. 2007; Mee et al. 2016).

1.4 Multiple Predator "Species"

Now, let us move away from a single predator-prey system. Imagine a system where the prey population is targeted by multiple predator types. This implies that a single functional or numerical response will be inappropriate in characterizing predator and prey behaviours. There is widespread evidence in recreational fisheries supporting heterogeneity of fisher behaviours (Beardmore et al. 2013; Dabrowska et al. 2017). Therefore, an approach adopted by ecologists is to assess evidence for clusters of traits and behaviours of fishers and segmentation into different "species" or types of fishers. Traits often considered in this segmentation include the importance of catch rate, fish size (e.g. trophy fish), harvest tendency, fishing frequency (avidity), willingness to travel (costs), and skill (q). Pioneering work on specialization identified four to seven distinct groups of fishers that differentiate along a specialization continuum (Bryan 1977; Fisher 1997). More recent studies of fisher heterogeneity have found that three to four groups often adequately describe the variance and association among traits and behaviours of recreational fishers (e.g. Oh et al. 2005; Oh and Ditton 2006; Dorow et al. 2010; Beardmore et al. 2013; Ward et al. 2013). For example, some fishers may be more consumptively oriented and particularly interested in harvest, whereas others may be more trophy-oriented and place more emphasis on the size of fish caught rather than the number of fish caught (Johnston et al. 2010; Hunt et al. 2021). Furthermore, individual fisher's preferences may not be consistent temporally or among species. Harvest orientation may also decline over time or with age as fishers become more specialized (Bryan 1977, but see Dorow et al. 2010; Embke et al. 2020; Hunt et al. 2021). The impact on the fish resource of multiple fisher segments, and their relative abundance within a fishery, can differ due to differences in annual fishing effort, catchability, and thus catch rates, and propensity to harvest fish (e.g. Ward et al. 2013). Simulation models can be useful to assess the impacts of this heterogeneity in fisher behaviour within fisheries (e.g. Johnston et al. 2010; van Poorten and Camp 2019) where the ecological analogy is multiple predator species competing and behaving differentially in the predator-prey interaction.

Understanding the source of behavioural variability among fishers is crucial to effectively manage recreational fisheries. Recognizing that fishers do not behave homogeneously in response to catch, in willingness to travel and accepting additional costs, or in efficiency and skill level, indicates that a single functional or numerical response relationship is an inappropriate description of fisher behaviour (van Poorten and Camp 2019). However, functional and numerical responses of an entire population of fishers are simply phenomenological descriptions of behavioural traits of many individual fishers. In other words, there is a richness of individual behaviours that, in aggregate, result in fishery-wide patterns. Despite these limitations, simple phenomenological descriptors of population-level processes have been useful in understanding patterns seen in real-world fisheries (Post et al. 2002; Askey et al. 2013).

1.5 Multiple Prey Populations

In addition to multiple types of predators, there is also often heterogeneity in key attributes of fish resources distributed across landscapes (e.g. production, abundance, and size-structure, fisher demand, distance from fishers' homes, and a host

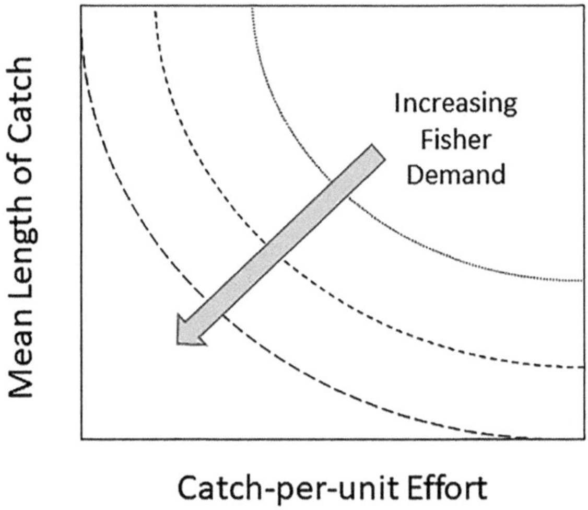

Fig. 3 Size-number trade-offs where the combination of density-dependent fish ecology and fisher behaviour constrain the combination of fish density, fishing catch per unit effort, and mean length of the catch (increasing fisher demand reduces the trade-offs between realized catch-per-unit effort and mean length of catch)

of other site-specific characteristics) (Cantin et al. 2021). Although fishers would likely prefer fisheries with high catch rates of large fish, density-dependent ecological processes typically result in fish populations with either high density of small fish or low density of large fish (Post and Parkinson 2012; Persson et al. 2014). The equilibrium outcome of the combination of these ecological trade-offs and heterogeneity among fisher preference in catch rate and fish size is an equal fishing quality isopleth within fisheries sharing other non-fish traits (Parkinson et al. 2004; Wilson et al. 2016) (Fig. 3). The greater the fisher demand, driven by non-catch related fishery traits such as travel time or distance, the closer the quality isopleth is to the axes with reduced catch rate and reduced fish size (Fig. 3). This outcome is an emergent property of heterogeneity of fisher behaviours and ecological processes distributed across landscapes of recreational fisheries.

Feedbacks between fisher behaviour and fish population characteristics create a dynamic where fisher behaviour is responsive to both local and landscape conditions and is therefore context specific. Even so, most empirical studies of fisher behaviour reveal only behavioural responses to the local conditions. Understanding processes that drive decision making in fisher behaviour is best explored with experimental manipulation of key drivers with controls (as developed in van Putten et al. 2026). Such experimental data are scarce, so models of site choice behaviour by multiple types of fishers can be simulated across complex landscapes of travel distances, fish abundances, and other site characteristics to predict equilibrium landscape patterns in the aggregate of fisher behaviour and fishery outcomes (Carruthers et al. 2019; Matsumura et al. 2019).

A further consideration is that although some recreational fisheries have a single species focus, others are multi-species. Fishers may target a focal species through their choice of gear and fishing locations, but influence more than one species (or sub-population) because of their fishing tactics. Now we see that a recreational fishery can have multiple fisher types that are fishing over multiple single or multi-species fishing sites that can vary in travel time and access costs across a diverse landscape. Further, recreational fishers are often influenced by regulations, where closed seasons or restrictive harvest limits can cause fisher behaviour to target alternate species in alternate places (Chagaris et al. 2019) or co-occurring species with more liberal regulations (e.g. open seasons, length limits; Murphy et al. 2019). Density-dependent prey switching behaviour by fishers may alter a fishery if fishers attempt to maximize their catch independent of species. Catchability also varies among species; thus, it may be that regardless of fisher stated preference, lower preference but higher q species will have higher catch rates and can lead to overharvest. Although not all recreational fisheries are this complex, these sources of variability in fisher behaviour highlight the diverse axes over which fisher behaviour can be expressed. Differential species preferences and harvest regulations interact with fisher behaviour types to modify site, gear, and harvest choices, which can have deleterious consequences for some fish populations (Johnston et al. 2013). The ecological approach to understanding fisher behaviours and resulting fishery dynamics in systems with multiple fisher types, multiple fish species, and management attempts to manipulate fishery outcomes can be characterized using multi-factorial predator-prey behavioural relationships, but these social-ecological systems are often complex (Ward et al. 2016). These systems are often best assessed through simulation of integrated social-ecological models (Carruthers et al. 2019; Matsumura et al. 2019; and see Jensen et al. 2026) and empirically using adaptive and experimental management experiments (van Putten et al. 2026).

An additional challenge in multi-species fisheries is the presence of a less preferred, but abundant, alternate species which might provide sufficient attraction to a fishery to maintain fisher effort on a waterbody despite a depleted preferred species, leading to concerns of collapse of the preferred species (Marshall and MacCrimmon 1970; Nguyen et al. 2012; Chagaris et al. 2019). But these observations are not universal—imposition of restrictive regulations can also repel effort, as was observed after implementing catch and release in a former harvest-oriented fishery for bull trout (*Salvelinus confluentus*) in Alberta (Johnston et al. 2007). This decoupling of fisher numerical response behaviour and regulations necessitates ecosystem-based fisheries management and explicit consideration of fisher behaviour as integral to these dynamic social-ecological systems (see van Putten et al. 2026; Jensen et al. 2026).

1.6 Abiotic Habitat Considerations

Not all decisions on where and when predators feed are related to prey availability. Predators also make behavioural decisions based on abiotic factors, such as habitat preferences, temperature, and competition with other predators. These variables may not be the primary motivation leading a predator to one site over another but play a role in their decision. Fishers also make choices on where and when to fish based on factors beyond the species, size, and number of fish they may catch. These are referred to as non-catch related attributes of fisher decision-making.

Both catch and non-catch related attributes are determinants of fisher satisfaction (Birdsong et al. 2021). Catch-related fishing quality includes attributes such as species, catch rate and harvest rate, and size of fish caught. Spatial behaviour is influenced by costs to access fish resources (e.g. travel distance, road access quality, licence fees, gear costs). Hunt (2005) and Hunt et al. (2019) have suggested five non-catch related attributes that, in addition to travel cost, play a significant role in fishers' motivations to choose a site: environmental quality, facility quality, regulations, congestion, and destination size. Other non-catch attributes include amenities (e.g. campgrounds, toilets, boat launch), lake characteristics (e.g. destination size, environmental quality), regulations and management (e.g. length restrictions, daily bag limits, gear and motor restrictions, stocking policies; Dabrowska et al. 2017). Each fisher will weigh the expected benefits of fishing a system (e.g. high catch rate, large fish, location on the landscape) with conditions that may be less favourable, resulting in implicit trade-offs among attributes. For example, fishers may choose to fish a system with a boat launch that has lower catch rates and smaller fish that is close to home, rather than a system with hike-in only access with higher catch rates and larger fish, that is farther away, and less crowded. Non-catch related attributes tend to be more stable over time (e.g. regulations change slowly, amenities are not regularly upgraded, and lake size and location do not change) than do catch-related attributes. Thus, non-catch attributes may be hard for managers to influence. However, satisfaction from non-catch related motives for going fishing, such as relaxation, getting out in nature, and socialization, may be more easily achieved or at least controlled by fishers when they choose a site (Arlinghaus et al. 2017; Beardmore et al. 2015), regardless of the fishing quality. In contrast, catch-related attributes are where most management actions are targeted, but they may be harder to predict, and catch-related motives for fishing (e.g. to catch lots of big fish) are often harder to satisfy. Thus, catch-related attributes may ultimately have a greater influence on fisher satisfaction and future fishing decisions (Arlinghaus et al. 2017; Beardmore et al. 2015; Birdsong et al. 2021).

Although the ecological approaches of functional and numerical responses and Ideal Free Distribution (dependent on catch-related attributes) have been useful in understanding fisheries dynamics, it is clear that non-catch-related attributes can also be important drivers of fisher behaviour. As the importance of non-catch-related attributes grows, the disconnect between the numerical response of the fisher to the fish prey increases, leading to a move away from a typical Type I response. For example, if distance is critical for fishers in determining where they fish, fisheries close to larger population centres will experience greater fishing pressure even when catch rates are reduced simply due to proximity. Some fish populations may not be able to support such pressure and could collapse. At a minimum, the increased pressure will lead to changes in fishing quality and the isopleth describing the trade-off between fish size and catch (Fig. 3). The importance on non-catch-related attributes in driving fisher behaviour is the reason that approaches that integrate socio-ecological dynamics are necessary to fully characterize the complex feedbacks within and dynamics of recreational fisheries.

1.7 Non-lethal Impacts

A key attribute of fisher behaviour is their propensity to harvest their catch—traditionally, harvest was a primary motivation for fishing. This behaviour aligned fishers, as predators, well with ecological predator-prey models. Although many recreational fishers remain harvest-oriented (Gaeta et al. 2013; Chagaris et al. 2019; Embke et al. 2020), social norms of nearly exclusive catch-and-release have emerged for some fish species, particularly in North America (e.g. black bass *Micropterus* spp., muskellunge *Esox masquinongy,* trout *Oncorhynchus* spp., and *Salvelinus* spp.) (Sass and Shaw 2020). Predominantly catch-and-release fisheries decouple fishers as predators in ecological models, especially in situations where the product of release mortality and fishing effort is low (i.e. β and E in Eq. 5) and compliance with mandatory release is high (Sullivan 2002; Post et al. 2003). Regulation-mandated fisher behaviour requiring release generally derives from one of two management goals. First is a conservation goal designed to ensure that overfishing neither compromises the sustainability of the targeted populations nor the social and economic values of the recreational fishery (Allen et al. 2008; Rypel et al. 2016). Second is a social goal designed to maintain or enhance trophy fish potential despite enhanced fishing effort (e.g. complete catch-and-release, restrictive minimum length limits, low bag limits). Voluntary release, on the other hand, is a fisher's personal behavioural choice often in support of a conservation ethos (Sass and Shaw 2020). Counter to a catch-and-release approach is a growing ethical argument against release behaviour grounded in animal welfare concerns. This is particularly predominant in some Western European countries and Indigenous cultures in North America (Arlinghaus et al. 2007, 2012).

The regulation of fisher behaviour via partial catch-and-release (e.g. length limits, bag limits) or complete catch-and-release is often used to attempt to structure fish populations to meet fisher and management goals for fishing quality. Although catch-and-release fishing has been promoted for fish conservation (and has been successful, and maybe overly successful in some cases) (Allen et al. 2008; Rypel et al. 2016), the practice can also result in species-specific over abundances, shifts in fish community composition, and regulation ineffectiveness to structure fish populations (Sinnatamby et al. 2023). Satisfying goals for catch rate and size structure (functional responses), and the provision of fishing opportunities (numerical responses) across gradients of fish abundance can be used to characterize the performance of fisheries. Liberal harvest regulations may result in reduced catch rates, higher individual fish growth rates, and greater trophy fish potential. In contrast, conservative regulations often result in the opposite results in fish populations. Therefore, managers are tasked with implementing harvest regulations that attempt to control fisher behaviour to achieve a balance among these three key ecological measures—catch rates, growth rates, and trophy-fish abundance. As such, predominant voluntary release can lead to regulation ineffectiveness in structuring fish populations. For example, a 28-year program of length-based harvest regulations on a Mississippi reservoir failed to restructure the largemouth bass population due to predominant voluntary release (α in Eq. 5) (Hessenauer et al. 2018; Miranda et al. 2017). Voluntary release, in the absence of discard mortality and sublethal effects, eliminates the role that fisher behaviour plays in structuring fish populations. Therefore, fisheries management must consider fisher behaviour where high release rates have the potential to decouple harvest from regulation effectiveness in structuring fish populations.

2 Measuring Fisher Behaviour and Fishery Outcomes

Assessing fisher behaviour, and fish population and community outcomes in recreational fisheries requires data on both fisher and fish. Key variables for assessing fisher behaviour include estimates of fisher effort (i.e. fish species-specific and total effort) and species-specific catch, harvest, and release rates. Data on the size of fish caught, catch and harvest rates, fishing effort, and species information are often collected using intercept surveys (roving or access point surveys) of fishers at a waterbody, often called creel surveys (Pollock et al. 1994; de Kerckhove et al. 2026). Species-specific release rates are critical for assessing fisher behaviour in fisheries managed for multiple species but have often been excluded from traditional creel surveys (but see Sass et al. 2022a). These surveys and their data provide an understanding of the impact of fishers on the fish populations (e.g. catch and harvest), their revealed preference in the conditions offered at that place and time (e.g. fishing effort), as well as an opportunity to informally talk with fishers about what attracted them to the fishery, things they dislike about the fishery and broader questions about regulations and fishery management. Methods for conducting these surveys and analysing the data to provide estimates of fisher effort, catch, and harvest aggregated over a season are available in many sources (e.g. Pollock et al. 1994; Jones and Pollock 2013). Fisher diary surveys (e.g. Gaeta et al. 2013), and mail or phone, or email surveys (e.g. Gundelund et al. 2020) have also been used to estimate effort, catch, harvest, and release rates. Stated preference surveys have also been used to predict fisher behaviour where fishers are offered a diversity of hypothetical site characteristics among which to choose (e.g. Beardmore et al. 2015; Dabrowska et al. 2017).

Technology is increasingly playing a role in supplementing information on fishers and their impact on fisheries. Citizen science projects and the expansion of digital platforms such as online forums or mobile phone apps are increasingly being used in fisheries data collection (e.g. Jiorle et al. 2016; Gundelund et al. 2020; Johnston et al. 2022; Venturelli et al. 2026). Traditionally, fisher effort has been estimated through creel surveys, although technology (e.g. car counters, drones, parking lot counts, satellite imagery, trail cameras, aerial flights, video surveillance) has improved our predictions of fisher effort dynamics over time (e.g. van Poorten et al. 2015; Hartill et al. 2016; van Poorten and Brydle 2018; Dutterer et al. 2020). Importantly, several of these high-frequency automated methods improve the precision of fishing effort estimates, thereby improving the ability of managers to detect fishing responses to changes in fishing conditions or regulations. New analytical techniques are also being developed that help borrow information across fishing sites, allowing for landscape patterns in fishing effort to be estimated (Trudeau et al. 2021). These technological advances hold great promise for quantifying fisher effort dynamics across fishing sites, and potentially across multi-species fisheries. For example, quantifying changes in fishing effort in response to changes in length limits and seasonal closures is a key need, and automated surveys can save substantial time and effort. Including fisher primary and secondary residence zip code information in creel surveys has proved invaluable for monitoring fisher effort across fisheries landscapes (Kaemingk et al. 2021; Lant et al. 2022). To monitor outcomes in fisheries managed for multiple species, common fisheries-independent surveys are used to estimate species-specific abundances (e.g. mark-recapture population estimates) or relative abundances (e.g. catch-per-unit-effort) and fish demographics (e.g. length, weight, age).

2.1 Functional Responses: Catchability and Catch-Per-Unit Effort

Functional responses in the context of recreational fisheries link a key metric, mean catch rate (in units of catch per unit effort; CPUE), to the availability of catchable fish (usually measured in units of abundance or density) (Johnson and Carpenter 1994). Catch rate is a key measure of success and satisfaction in recreational fisheries and is estimated from creel surveys of fishers, either as they complete their fishing trip or when they are in the process of fishing.

Establishing a functional response for a fishery is a data-intensive endeavour that requires experimental or empirical contrasts in fish densities and fishery-independent measures of abundance. For many fish species, estimating the total fish population size can be exceedingly difficult, and, in those cases, catchability can be estimated using intensive mark and recapture programs. In these programs, large numbers of fish are marked using a tag or other identifiable mark, and the number of harvested fish with a mark provides an estimate of the fishing mortality rate. Catchability is then estimated as the fishing mortality rate divided by the fishing effort that generated it. Given the substantial data requirements, functional responses and associated density-dependent catchability estimates are not often measured and reported in the literature.

2.2 Numerical Responses: Fisher Effort Across Landscapes

Fisher behaviour, through site choice, is often referred to as an effort response at the population level—equivalent to numerical response in ecological terms (Post 2013). An effort response is the result of the aggregate decision-making by a population of fishers to the quality of the fishery—where quality is measured by fish abundances or catch rates. Because fishing quality is often one of the primary determinants of site choice, effort is attracted by high and repelled by low fishing quality (Post et al. 2008; Post and Parkinson 2012). The numerical effort response behaviour by fishers on individual fisheries is typically measured through creel surveys designed to estimate total effort on a water body in relation to the abundance of catchable fish. This approach has been used in fisheries like Pacific salmon (*Oncorhynchus* spp.) and kokanee (*O. nerka*) troll fisheries in British Columbia (Cox et al. 2002; Askey 2016) and in an Ontario lake trout fishery (Shuter et al. 1998; Post et al. 2002). In other cases, these relationships are established by collecting data over many individual fishers within a lake district, as in Wisconsin walleye (Johnson and Carpenter 1994), British Columbia rainbow trout (*O. mykiss*) (Post et al. 2008; Carruthers et al. 2019; Cantin et al. 2021), and Yukon Territory lake trout fisheries (Wilson et al. 2020). In contrast, Camp et al. (2016) evaluated fishing effort and fish abundance in multiple marine fisheries and documented little evidence of a numerical response, suggesting multiple social factors may overwhelm the traditional link between these two variables.

In contrast to these empirical observations of effort and fish abundance across landscapes that are used to reveal a pattern, other studies use stated choice surveys to predict the probability of fishing given catch rates (Beardmore et al. 2015; Dabrowska et al. 2017). A review of shapes of effort response curves for multiple species and ecological groupings of recreational fisheries revealed Type I, II, and III numerical responses using this alternate approach (Golden et al. 2022). The functional responses differed among fisheries in both the range of effort and the responsiveness of effort to fish catch rates (Golden et al. 2022).

An important means of understanding numerical responses is through simulation models that take insights, theories, and methods described above to make predictions on how diverse fishers will distribute across the landscape and interact with fish populations. Latent class models, principal component analyses, or hierarchical cluster analyses are some of the methods used to test for the presence of groups of fisher types within datasets of fisher behaviours (Fisher 1997; Dorow et al. 2010; Beardmore et al. 2013; Ward et al. 2013; also see Melstrom et al. 2026; Mackay et al. 2026). Data for these types of analyses may come from choice experiments that state intended behaviour (e.g. Beardmore et al. 2013), or from sources such as creel surveys, mail and web surveys, and fisher diaries that reveal actual behaviour (e.g. Ward et al. 2013). Social-ecological modelling frameworks integrate feedback among fishers, managers, and fish populations to explore the impacts of various management scenarios (e.g. Johnston et al. 2018; Matsumura et al. 2019; Carruthers et al. 2019; Wilson et al. 2020). Understanding the diversity of fisher behaviour is useful in contrasting the effectiveness of regulations designed to sustain fisheries. Given the diversity of fishers in populations, studies suggest that managers should seek to provide a variety of fishing opportunities that appeal to heterogeneous fisher populations, rather than using one-size-fits-all policies (Ward et al. 2013; van Poorten and Camp 2019). These insights could not have been gained without careful integration of multiple lines of evidence through models.

2.3 Catch-and-Release Fishing Behaviour

Estimating species-specific release rates generally requires the collection of catch-and-release information during creel surveys or in fisher diary surveys (Gaeta et al. 2013; Gilbert and Sass 2016; Shaw et al. 2019; Sass et al. 2022b). Unfortunately, traditional creel surveys often only record information from harvested fish, not released fish. To estimate release rates, creel surveys and fisher diary surveys should be designed to specifically capture this aspect of recreational fisheries and be coupled with fisheries-independent surveys to assess population responses (Sass et al. 2022a, b). This can be accomplished by estimating and modelling the components of the Gordon-Schaefer model with additional components to incorporate fisher release mortality (as in Eqs. 1, 2, 3, 4, and 5), derived from ecological predator-prey theory. Understanding the dynamics of fisher behaviour in catch-and-release fisheries is crucial in managing these recreational fisheries.

2.4 Fisher Behaviour in Multi-species Fisheries

Although the methods of assessing fisher behaviour in multi-species fisheries do not differ from single-species fisheries, fisher behaviour in these multi-species systems can result in complex and often undesirable fishery outcomes. In fisheries where some species are predominantly released voluntarily and others are harvested, density-dependent effects on growth and alterations to the food web can occur. High voluntary release rates of largemouth bass (*Micropterus salmoides*) and muskellunge (*Esox masquinongy*) have resulted in reduced individual growth rates and a decrease in the catch of large fish (Hansen et al. 2015; Gilbert and Sass 2016; Shaw et al. 2019; Sass and Shaw 2020). High largemouth bass abundances and a lack of bullhead *Ameiurus* spp. harvest has also negatively influenced walleye recruitment (Hansen et al. 2018; Sikora et al. 2021). Even following fisheries closure for the harvested species, the alteration of food web structure in these situations can remain persistent and cause the fishery to collapse or stock recovery to be delayed (Walters and Kitchell 2001; Post et al. 2002; but see Cahill et al. 2022). In some multispecies Salmonid fisheries, fishers have been incentivized with rewards to encourage targeting and to discourage voluntary release of predatory rainbow trout and lake trout to recover a collapsed prey species, kokanee, which is also a popular fishery (Hansen et al. 2010; Warnock et al. 2022). These shifts in fish species composition favouring one species over another due to voluntary release have the potential to disrupt the balance between catch rates and trophy fish potential, not meet diverse fisher desires, lead to single-species regulation ineffectiveness, and highlight the need for ecosystem-based fisheries management and the recognition of recreational fisheries as social-ecological systems (Solomon et al. 2020).

2.5 *Non-catch-Related Fisher Site Choice*

Fisher preferences for various attributes of a fishery are often determined from revealed preference models (actual behaviour) or stated preference models (intended behaviour) (Hunt et al. 2019). These two approaches are disciplinarily rooted. The ecological approach of assessing numerical response of fishers to temporal and spatial variation in fishing quality typically involves observations of fishing effort, analogous to assessments of predator abundance in relation to prey abundance distribution (i.e. the revealed preference of the predators). Whereas stated preference models derive from projections of future behaviour, from statements of preference or intent, as developed in survey design in the social sciences. Linear and additive relationships describing the effects of various attributes of fisher utility (i.e. satisfaction, benefit) are generally assumed, but these assumptions are often not assessed (Hunt et al. 2019). Non-linearities in the relationships between realized utility (i.e. satisfaction, benefit) and the various attributes could alter the relative influence of different attributes on fishing decisions (Johnston et al. 2010; Hunt et al. 2019). For example, the change in utility from catching zero fish versus one fish may not be the same as the difference in utility between catching one fish versus two fish, or ten fish versus eleven fish. One might expect a diminishing benefit as catch rates increase (Johnston et al. 2010; Arlinghaus et al. 2014). For example, Patterson and Sullivan (2013) found that fishing effort on put-and-take rainbow trout fisheries in Alberta was strongly related to the distance from population centres but not to catch rates, and that satisfaction was logarithmically related to catch rates. Thus, catch rates may be an important predictor for fisher site choices when they are low, whereas other attributes may become critical in determining fishers' fishing decisions when catch rates are moderate or high.

3 Future Research

Our review of the ecological science perspectives and approaches to understanding fisher behaviour and the dynamic outcomes of their interaction with fishes has identified a number of gaps in understanding, providing a useful research agenda. These gaps can be characterized as a need for (1) enhanced understanding of processes, (2) understanding methodological biases and technological innovation, and (3) enhanced integration across the fields and methods of social, ecological, and data sciences.

Temporal trends in social and ecological processes have the potential to mischaracterize recreational fisheries if viewed as status quo or static. For example, trends away from a consumptive paradigm towards increasing conservation approaches have been observed broadly, but there is substantial variation among recreational fisheries in this trend. In addition, increasing concerns for animal welfare involving regulations preventing catch-and-release are growing, particularly within some jurisdictions (Arlinghaus et al. 2012). Temporal trends in demography involving deficits in recruitment of young fishers and aging of populations can shift fishing effort, species preferences, and site selection by fishers. And in the longer term, it is possible that changes in the proportion of people who live in cities, and changes in fishers' travel habits, as concerns about the climate impact of travel increase, may lead to broad-scale changes in how fishers distribute over the landscape. And lastly, climate change and its impact on aquatic ecosystems will likely alter production and species composition of recreational fishes and modify feedback to angler behaviour through functional and numerical responses. Spatial heterogeneity in the rates and directions of these trends will challenge our understanding of fisher behaviour, fishery outcomes, and assessment of sustainable recreational fisheries.

The predator-prey analogy applied to recreational fishers has been useful in identifying dynamic processes leading to either self-sustaining or overexploited fisheries. The complexity of fisher behaviour, when coupled to fish population and community dynamics, characterized by functional and numerical responses, is useful in identifying management challenges. For example, what are the characteristics of fisher and fish behaviour that result in pathological nonlinearities and risk of overharvest and collapse?

Methodological issues can lead to bias in data and functions derived from creel and stated preference surveys. Although not necessarily without bias, technological innovations in data collection, such as cameras, drones, fishing apps, and GPS trackers, can be used to collect copious amounts of data over substantial temporal and spatial scales that would be extremely expensive using more common methods. Rigorous ground truthing, bias assessment, and correction are important to characterize fisher behaviour for use in dynamic models. These assessments are lacking today.

Though the field of ecology and ecologists may have done much of the heavy lifting early in our exploration of fisher behaviour in recreational fisheries (Johnson and Carpenter 1994; Shuter et al. 1998; Post et al. 2002, 2008; Cox and Walters 2002; Beard et al. 2003; Cox et al. 2003; Parkinson et al. 2004; Askey et al. 2006), the future must involve the development and testing of process-oriented social-ecological systems models. Model validation must effectively integrate temporal and spatial data from fisheries stock assessments and fisher creel and stated preference surveys. Such an approach would provide the tools necessary to more fully understand the role that fisher behaviour plays in fishery outcomes and develop rigorous management strategy evaluations. It would also provide the predictive capacity for the development of adaptive management and structured decision approaches for reducing uncertainty and refining policy decisions—all with the end goal of optimizing recreational fisheries and sustaining their substantial ecological services for fishers and for society.

References

Allen MS, Walters CJ, Myers R (2008) Temporal trends in largemouth bass mortality, with fishery implications. N Am J Fish Manag 28(2):418–427

Alós J, Campos-Candela A, Arlinghaus R (2019) A modelling approach to evaluate the impact of fish spatial behavioural types on fisheries stock assessment. ICES J Mar Sci 76(2):489–500

Arlinghaus R, Cooke SJ, Lyman J, Policansky D, Schwab A, Suski C, Sutton SG, Thorstad EB (2007) Understanding the complexity of catch-and-release in recreational fishing: an integrative synthesis of global knowledge from historical, ethical, social, and biological perspectives. Rev Fish Sci 15(1–2):75–167

Arlinghaus R, Schwab A, Riepe C, Teel T (2012) A primer on anti-angling philosophy and its relevance for recreational fisheries in urbanized societies. Fisheries 37(4):153–164

Arlinghaus R, Beardmore B, Riepe C, Meyerhoff J, Pagel T (2014) Species-specific preferences of German recreational anglers for freshwater fishing experiences, with emphasis on the intrinsic utilities of fish stocking and wild fishes. J Fish Bio 85(6):1843–1867

Arlinghaus R, Alós J, Beardmore B, Daedlow K, Dorow M, Fujitani M, Hühn D, Haider W, Hunt LM, Johnson BM, Johnston F, Klefoth T, Matsumura S, Monk C, Pagel T, Post JR, Rapp T, Riepe C, Ward H, Wolter C (2017) Understanding and managing freshwater recreational fisheries as complex adaptive social-ecological systems. Rev Fish Sci Aquac 25(1):1–41

Askey PJ (2016) Managing dynamic fisheries with static regulations: an assessment of size-graded bag limits for recreational kokanee fisheries. N Am J Fish Manag 36(2):241–253

Askey PJ, Johnston NT (2013) Self-regulation of the Okanagan Lake kokanee recreational fishery: dynamic angler effort response to varying fish abundance and productivity. N Am J Fish Manag 33(5):926–939

Askey PJ, Richards SA, Post JR, Parkinson EA (2006) Linking angling catch rates and fish learning under catch-and-release regulations. N Am J Fish Manag 26(4):1020–1029

Askey PJ, Parkinson EA, Post JR (2013) Linking fish and fisher dynamics to assess stocking strategies for hatchery-dependent, open access recreational fisheries. N Am J Fish Manag 33: 557–568

Beard TD Jr, Cox SP, Carpenter SR (2003) Impacts of daily bag limit reductions on angler effort in Wisconsin walleye lakes. N Am J Fish Manag 23(4):1283–1293

Beardmore B, Haider W, Hunt LM, Arlinghaus R (2013) Evaluating the ability of specialization indicators to explain fishing preferences. Leis Sci 35(3):273–292

Beardmore B, Hunt LM, Haider W, Dorow M, Arlinghaus R (2015) Effectively managing angler satisfaction in recreational fisheries requires understanding the fish species and the anglers. Can J Fish Aquat Sci 72(4):500–513

Birdsong M, Hunt LM, Arlinghaus R (2021) Recreational angler satisfaction: what drives it? Fish Fish 22(4):682–706

Biro PA, Post JR (2008) Rapid depletion of genotypes with fast growth and bold personality traits from harvested fish populations. Proc Natl Acad Sci 105(8):2919–2922

Bryan H (1977) Leisure value systems and recreational specialization: the case of trout fishermen. J Leis Res 9(3):174–187

Cahill CL, Walters CJ, Paul AJ, Sullivan MG, Post JR (2022) Unveiling the recovery dynamics of walleye after the invisible collapse. Can J Fish Aquat Sci 79:708–723

Camp EV, Ahrens RN, Allen MS and Lorenzen K (2016) Relationships between angling effort and fish abundance in recreational marine fisheries. Fish Manag Ecol 23:264–275

Cantin A, Farineau A, Bender D, Post JR (2021) Predicting distribution, production, recreational demand and conservation concern of indigenous rainbow trout populations based on landscape characteristics. Can J Fish Aquat Sci 78:1096–1108

Carpenter SR (1998) The need for large-scale experiments to assess and predict the response of ecosystems to perturbation. In: Pace ML, Groffman PM (eds) Successes, limitations, and frontiers in ecosystem science. Springer, New York, pp 287–312

Carruthers TR, Dabrowska K, Haider W, Parkinson EA, Varkey DA, Ward H, McAllister MK, Godin T, van Poorten B, Askey PJ, Wilson KL, Hunt LM, Clarke A, Newton E, Walters CJ, Post JR (2019) Landscape scale social and ecological outcomes of dynamic angler and fish behaviours: processes, data, and patterns. Can J Fish Aquat Sci 76:970–988

Chagaris D, Allen M, Camp E (2019) Modeling temporal closures in a multispecies recreational fishery reveals tradeoffs associated with species seasonality and angler effort dynamics. Fish Res 210:106–120

Clark CW (2006) The worldwide crisis in fisheries: economic models and human behavior. Cambridge University Press, Cambridge

Clark CW (2018) Modelling the behaviour of fishers and fishes. ICES J Mar Sci 75(3):932–940

Cline TJ, Weidel BC, Kitchell JF, Hodgson JR (2012) Growth response of largemouth bass (*Micropterus salmoides*) to catch-and-release angling: a 27-year mark-recapture study. Can J Fish Aquat Sci 69(2):224–230

Cox SP, Walters C (2002) Modeling exploitation in recreational fisheries and implications for effort management on British Columbia rainbow trout lakes. N Am J Fish Manag 22(1):21–34

Cox SP, Beard TD, Walters C (2002) Harvest control in open-access sport fisheries: hot rod or asleep at the reel? Bull Mar Sci 70(2):749–761

Cox SP, Beard TD, Walters CJ, Post JR (2003) Active management of recreational fishing effort: a model based evaluation of policies for British Columbia rainbow trout lakes. N Am J Fish Manag 23:1294–1303

Dabrowska K, Hunt LM, Haider W (2017) Understanding how angler characteristics and context influence angler preferences for fishing sites. N Am J Fish Manag 37(6):1350–1361

Dassow CJ, Ross AJ, Jensen OP, Sass GG, van Poorten BT, Solomon CT, Jones SE (2020) Experimental demonstration of catch hyperstability from habitat aggregation, not effort sorting, in a recreational fishery. Can J Fish Aquat Sci 77(4):762–769

de Kerckhove D, Tucker C, Gutowsky L, Morgan G, Guzzo M, and Barajas VL (2026) Optimizing Creel Surveys. In: Pope KL, Arlinghaus R, Hunt LM, Lynch AJ, van Poorten BT (eds) Understanding recreational fishers: disciplinary and interdisciplinary approaches for fisheries management. Springer Nature, Cham, p 387–409

Dorow M, Beardmore B, Haider W, Arlinghaus R (2010) Winners and losers of conservation policies for European eel, *Anguilla anguilla*: an economic welfare analysis for differently specialised eel anglers. Fish Manag Ecol 17(2):106–125

Dutterer AC, Dotson JR, Thompson BC, Paxton CJ, Pouder WF (2020) Estimating recreational fishing effort using autonomous cameras at boat ramps versus creel surveys. N Am J Fish Manag 40(6):1367–1378

Embke HS, Douglas Beard Jr T, Lynch AJ, Vander Zanden MJ (2020) Fishing for food: quantifying recreational fisheries harvest in Wisconsin lakes. Fisheries 45(12):647–655

Erisman BE, Allen LG, Claisse JT, Pondella DJ, Miller EF, Murray JH (2011) The illusion of plenty: hyperstability masks collapses in two recreational fisheries that target fish spawning aggregations. Can J Fish Aquat Sci 68(10):1705–1716

Feiner ZS, Wolter MH, Latzka AW (2020) "I will look for you, I will find you, and I will [harvest] you": persistent hyperstability in Wisconsin's recreational fishery. Fish Res 230:105679

Fisher MR (1997) Segmentation of the angler population by catch preference, participation, and experience: a management-oriented application of recreation specialization. N Am J Fish Manag 17(1):1–10

Fretwell SD, Lucas HL (1969) On territorial behaviour and other factors influencing habitat distribution in birds. Acta Biotheor 19:16–36

Gaeta JW, Beardmore B, Latzka AW, Provencher B, Carpenter SR (2013) Catch-and-release rates of sport fishes in northern Wisconsin from an angler diary survey. N Am J Fish Manag 33(3): 606–614

Gilbert SJ, Sass GG (2016) Trends in a northern Wisconsin muskellunge fishery: results from a countywide angling contest, 1964–2010. Fish Manag Ecol 23(2):172–176

Gillis DM, Peterman RM, Tyler AV (1993) Movement dynamics in a fishery: application of the ideal free distribution to spatial allocation of effort. Can J Fish Aquat Sci 50(2):323–333

Golden AS, van Poorten B, Jensen OP (2022) Focusing on what matters most: evaluating multiple challenges to stability in recreational fisheries. Fish Fish 23(6):1418–1438

Gundelund C, Arlinghaus R, Baktoft H, Hyder K, Venturelli P, Skov C (2020) Insights into the users of a citizen science platform for collecting recreational fisheries data. Fish Res 229:105597

Hansen MJ, Beard TD Jr, Hewett SW (2005) Effect of measurement error on tests of density dependence of catchability for walleyes in northern Wisconsin angling and spearing fisheries. N Am J Fish Manag 25(3):1010–1015

Hansen MJ, Schill D, Fredericks J, Dux A (2010) Salmonid predator-prey dynamics in Lake Pend Oreille, Idaho, USA. Hydrobiologia 650(1):85–100

Hansen JF, Sass GG, Gaeta JW, Hansen GA, Isermann DA, Lyons J, Vander Zanden MJ (2015) Largemouth bass management in Wisconsin: intraspecific and interspecific implications of abundance increases. In: Tringali MD, Long JM, Birdsong TW, Allen MS (eds) Black bass diversity: multidisciplinary science for conservation. American Fisheries Society, Symp 82,, Bethesda, pp 193–206

Hansen GJ, Midway SR, Wagner T (2018) Walleye recruitment success is less resilient to warming water temperatures in lakes with abundant largemouth bass populations. Can J Fish Aquat Sci 75(1):106–115

Hartill BW, Payne GW, Rush N, Bian R (2016) Bridging the temporal gap: continuous and cost-effective monitoring of dynamic recreational fisheries by web cameras and creel surveys. Fish Res 183:488–497

Hessenauer JM, Vokoun J, Davis J, Jacobs R, O'Donnell E (2018) Size structure suppression and obsolete length regulations in recreational fisheries dominated by catch-and-release. Fish Res 200:33–42

Hunt LM (2005) Recreational fishing site choice models: insights and future opportunities. Hum Dimens Wildl 10(3):153–172

Hunt LM, Arlinghaus R, Lester N, Kushneriuk R (2011) The effects of regional angler effort, angler behaviour, and harvest efficiency on landscape patterns of overfishing. Ecol Appl 21:2555–2575

Hunt LM, Camp E, van Poorten B, Arlinghaus R (2019) Catch and non-catch-related determinants of where anglers fish: a review of three decades of site choice research in recreational fisheries. Rev Fish Sci Aquac 27:261–286

Hunt LM, Phaneuf DJ, Abbott JK, Fenichel EP, Rodgers JA, Buckley, Drake DA, Johnson TB (2021) The influence of human population change and aquatic invasive species establishment on future recreational fishing activities to the Canadian portion of the Laurentian Great Lakes. Can J Fish Aquat Sci 78(3):232–244

Hutchings JA (2021) A primer of life histories: ecology, evolution, and application. Oxford University Press, Oxford

Jensen OP, Arlinghaus R, Golden AS, Janssen MA, Solomon CT, van Poorten BT (2026) Integrated models of the social-ecological dynamics of recreational fisheries. In: Pope KL, Arlinghaus R, Hunt LM, Lynch AJ, van Poorten BT (eds) Understanding recreational fishers: disciplinary and interdisciplinary approaches for fisheries management. Springer Nature, Cham, p 661-692

Jiorle RP, Ahrens RNM, Allen MS (2016) Assessing the utility of a smartphone app for recreational fishery catch data. Fisheries 41(12):758–766

Johnson BM, Carpenter SR (1994) Functional and numerical responses: a framework for fish-angler interactions? Ecol Appl 4(4):808–821

Johnston FD, Post JR, Mushens CJ, Stelfox JD, Paul AJ, Lajeunesse B (2007) The demography of recovery of an overexploited bull trout, *Salvelinus confluentus*, population. Can J Fish Aquat Sci 64(1):113–126

Johnston FD, Arlinghaus R, Dieckmann U (2010) Diversity and complexity of angler behaviour drive socially optimal input and output regulations in a bioeconomic recreational-fisheries model. Can J Fish Aquat Sci 67(9):1507–1531

Johnston FD, Arlinghaus R, Dieckmann U (2013) Fish life history, angler behaviour and optimal management of recreational fisheries. Fish Fish 14(4):554–579

Johnston FD, Allen MS, Beardmore B, Riepe C, Pagel T, Hühn D, Arlinghaus R (2018) How ecological processes shape the outcomes of stock enhancement and harvest regulations in recreational fisheries. Ecol Appl 28(8):2033–2054

Johnston FD, Simmons S, van Poorten B, Venturelli P (2022) Comparative analyses with conventional surveys reveal the potential for an angler app to contribute to recreational fisheries monitoring. Can J Fish Aquat Sci 79:31–46

Jones CM, Pollock KH (2013) Recreational angler survey methods: estimation of effort, harvest, and released catch. In: Zale AV, Parrish DL, Sutton TM (eds) Fisheries techniques, 3rd edn. American Fisheries Society, Bethesda

Kaemingk MA, Bender CN, Chizinski CJ, Bunch AJ, Pope KL (2021) Temporal invariance of social-ecological catchments. Ecol Appl 31(2):e02272

Krebs CJ (2009) Ecology: the experimental analysis of distribution and abundance, 6th edn. Benjamin Cummings, San Francisco

Landon A, Schroeder SA, Steffen SF, and Arlinghaus R (2026) Social Psychology for Understanding Recreational Fishers and Fisheries. In: Pope KL, Arlinghaus R, Hunt LM, Lynch AJ, van Poorten BT (eds) Understanding recreational fishers: disciplinary and interdisciplinary approaches for fisheries management. Springer Nature, Cham, p 157–196

Lant MJ, Ogle DH, Feine ZS, Sass GG (2022) A comparison of fish catch rates among local, non-local, and non-residents anglers of three northern Wisconsin lakes. Fish Res 250:106286

Lewin WC, Arlinghaus R and Mehner T (2006) Documented and potential biological impacts of recreational fishing: insights for management and conservation. Rev Fish Sci 14(4):305–367

Mackay M, van Putten EI, Richter A, Abatayo AL, Chuah S-H, Dankel DJ, Schadeberg A, Snyder H (2026) Behavioural economics for understanding recreational fishers and fisheries. In: Pope KL, Arlinghaus R, Hunt LM, Lynch AJ, van Poorten BT (eds) Understanding recreational fishers: disciplinary and interdisciplinary approaches for fisheries management. Springer Nature, Cham, p 261–284

Marshall TL, MacCrimmon HR (1970) Exploitation of self-sustaining Ontario stream populations of brown trout (*Salmo trutta*) and brook trout (*Salvelinus fontinalis*). J Fish Res Bd Can 27(6): 1087–1102

Matsumura S, Beardmore B, Haider W, Dieckmann U, Arlinghaus R (2019) Ecological, angler, and spatial heterogeneity drive social and ecological outcomes in an integrated landscape model of freshwater recreational fisheries. Rev Fish Sci Aquac 27(2):170–197

Mee JA, Post JR, Ward H, Wilson KL, Newton ER, Cantin A (2016) Interaction of ecological and angler processes: experimental stocking in an open access, spatially structured fishery. Ecol Appl 26:1693–1707

Melstrom RT, Jungers B, Bi X, Abbott JK (2026) Resource economics for understanding recreational fishers and fisheries. In: Pope KL, Arlinghaus R, Hunt LM, Lynch AJ, van Poorten BT (eds) Understanding recreational fishers: disciplinary and interdisciplinary approaches for fisheries management. Springer Nature, Cham, p 123–156

Miranda LE, Colvin ME, Shamaskin AC, Bull LA, Holman T, Jones R (2017) Length limits fail to restructure a largemouth bass population: a 28-year case history. N Am J Fish Manag 37(3): 624–632

Mogensen S, Post JR, Sullivan MG (2014) Vulnerability to harvest by anglers differs across climate, productivity, and diversity clines. Can J Fish Aquat Sci 71(3):416–426

Mosley CL, Dassow CJ, Caffarelli J, Ross AJ, Sass GG, Shaw SL, Solomon CT, Jones SE (2022) Species differences, but not habitat, influence catch rate hyperstability across a recreational fishery landscape. Fish Res 255:106438

Mrnak JT, Shaw SL, Eslinger LD, Cichosz TA, Sass GG (2018) Characterizing the angling and tribal spearing walleye fisheries in the Ceded Territory of Wisconsin, 1990–2015. N Am J Fish Manag 38(6):1381–1393

Murphy R Jr, Scyphers S, Gray S, Grabowski J (2019) Angler attitudes explain disparate behavioral reactions to fishery regulations. Fisheries 44(10):475–487

Nguyen VM, Rudd MA, Hinch SG, Cooke SJ (2012) Differences in information use and preferences among recreational salmon anglers: implications for management initiatives to promote responsible fishing. Hum Dimens Wildl 17:248–256

Oh CO, Ditton RB (2006) Using recreation specialization to understand multi-attribute management preferences. Leis Sci 28:369–384

Oh CO, Ditton RB, Anderson DK, Scott D, Stoll JR (2005) Understanding differences in nonmarket valuation by angler specialization level. Leis Sci 27:263–277

Parkinson EA, Post JR, Cox SP (2004) Linking the dynamics of harvest effort to recruitment dynamics in a multi-stock, spatially structured fishery. Can J Fish Aquat Sci 61:1658–1670

Patterson WF, Sullivan MG (2013) Testing and refining the assumptions of put-and-take rainbow trout fisheries in Alberta. Hum Dimens Wildl 18(5):340–354

Persson L, Van Leeuwen A, de Roos AM (2014) The ecological foundation for ecosystem-based management of fisheries: mechanistic linkages between the individual-, population-, and community-level dynamics. ICES J Mar Sci 71(8):2268–2280

Pollock KH, Jones CM, Brown TL (1994) Angler survey methods and their applications in fisheries management. American Fisheries Society, Special Publication 25, Bethesda

Post JR (2013) Resilient recreational fisheries or prone to collapse? A decade of research on the science and management of recreational fisheries. Fish Manag Ecol 20:99–110

Post JR, Parkinson EA (2012) Temporal and spatial patterns of angler effort across lake districts and policy options to sustain recreational fisheries. Can J Fish Aquat Sci 69:321–329

Post JR, Sullivan M, Cox S, Lester NP, Walters CJ, Parkinson EA, Paul AJ, Jackson L, Shuter BJ (2002) Canada's recreational fisheries: the invisible collapse? Fisheries 27(1):6–17

Post JR, Mushens CJ, Paul AJ, Sullivan M (2003) Assessment of alternative management strategies for sustaining recreational fisheries: model development and application to bull trout, *Salvelinus confluentus*. N Am J Fish Manag 23:22–34

Post JR, Persson L, Parkinson EA, van Kooten T (2008) Angler numerical response across landscapes and the collapse of freshwater fisheries. Ecol Appl 18:1038–1049

Rypel AL, Lyons J, Griffin JDT, Simonson TD (2016) Seventy-year retrospective on size-structure changes in the recreational fisheries of Wisconsin. Fisheries 41(5):230–243

Sass GG, Shaw SL (2020) Catch-and-release influences on inland recreational fisheries. Rev Fish Sci Aquac 28(2):211–227

Sass GG, Gaeta JW, Allen MS, Suski CD, Shaw SL (2018) Effects of catch-and-release angling on a largemouth bass (*Micropterus salmoides*) population in a north temperate lake, 2001–2005. Fish Res 204:95–102

Sass GG, Feiner ZS, Shaw SL (2021) Empirical evidence for depensation in freshwater fisheries. Fisheries 46(6):266–276

Sass GG, Shaw SL, Renik KM (2022a) Celebrating 75 years of Wisconsin's Northern Highland fishery research area: the past, present, and future. Fisheries 47(2):55–67

Sass GG, Shaw SL, Sikora LW, Lorenzoni M, Luehring M (2022b) Plasticity in abundance and demographic responses of walleye to elevated exploitation in a north temperate lake. N Am J Fish Manag 42(3):535–548

Shaw SL, Sass GG, Eslinger LD (2019) Effects of angler harvest on adult muskellunge growth and survival in Escanaba Lake, Wisconsin, 1956–2016. N Am J Fish Manag 39(1):124–134

Shuter BJ, Jones ML, Korver RM, Lester NP (1998) A general, life history based model for regional management of fish stocks: the inland lake trout (*Salvelinus namaycush*) fisheries of Ontario. Can J Fish Aquat Sci 55(9):2161–2177

Sikora LW, VanDeHey JA, Sass GG, Matzke G, Preul M (2021) Fish community changes associated with bullhead removals in four northern Wisconsin lakes. N Am J Fish Manag 41: S71–S81

Sinnatamby RN, Cantin A, Paul AJ, Earle J, Post JR (2023) No evidence of recovery of native trout in response to sustained angling suppression of invasive brook trout. N Am J Fish Manag 43: 1294–1309

Solomon CT, Dassow CJ, Iwicki CM, Jensen OP, Jones SE, Sass GG, Trudeau A, van Poorten B, Whittaker D (2020) Frontiers in modelling social-ecological dynamics of recreational fisheries: a review and synthesis. Fish Fish 21(5):973–991

Sullivan MG (2002) Illegal angling harvest of walleyes protected by length limits in Alberta. N Am J Fish Manag 22(3):1053–1063

Trudeau A, Dassow CJ, Iwicki CM, Jones SE, Sass GG, Solomon CT, van Poorten B, Jensen OP (2021) Estimating fishing effort across the landscape: a spatially extensive approach using models to integrate multiple data sources. Fish Res 233:105768

van Poorten BT, Camp EV (2019) Addressing challenges common to modern recreational fisheries with a buffet-style landscape management approach. Rev Fish Sci Aquac 27(4):393–416

van Poorten BT, Carruthers TR, Ward HG, Varkey DA (2015) Imputing recreational angling effort from time-lapse cameras using a hierarchical Bayesian model. Fish Res 172:265–273

van Poorten BT, Walters CJ, Ward HG (2016) Predicting changes in the catchability coefficient through effort sorting as less skilled fishers exit the fishery during stock declines. Fish Res 183: 379–384

van Poorten BT and Brydle S (2018) Estimating fishing effort from remote traffic counters: opportunities and challenges. Fish Res 204:231–238

van Putten EI, Trudeau A, Mackay M, Obregón C, Post JR, Richter A, Schuch E, Solomon CT (2026) Economic experiments in recreational fishing studies. In: Pope KL, Arlinghaus R, Hunt LM, Lynch AJ, van Poorten BT (eds) Understanding recreational fishers: disciplinary and interdisciplinary approaches for fisheries management. Springer Nature, Cham, p 411–442

Venturelli P, Skov C, Audzijonyte A, Belmaker J, Ana Cardoso AS, Gordon L, Gundelund C, Hyder K, Jarić I, Lennox RJ, Potts WM, Vacura K, Vaz AS, Sbragaglia V (2026) Digital data mining. In: Pope KL, Arlinghaus R, Hunt LM, Lynch AJ, van Poorten BT (eds) Understanding recreational fishers: disciplinary and interdisciplinary approaches for fisheries management. Springer Nature, Cham, p 443–479

Walters C, Kitchell JF (2001) Cultivation/depensation effects on juvenile survival and recruitment: implications for the theory of fishing. Can J Fish Aquat Sci 58(1):39–50

Ward HGM, Quinn MS, Post JR (2013) Angler characteristics and management implications in a large multi-stock spatially structured recreational fishery. N Am J Fish Manag 33:576–584

Ward HGM, Allen MS, Camp E, Cole N, Hunt LM, Matthias B, Post JR, Wilson K, Arlinghaus R (2016) Understanding and managing social-ecological feedbacks in spatially structured recreational fisheries: the overlooked behavioral dimension. Fisheries 41:524–535

Warnock WG, Thorley JL, Arndt SK, Weir TJ, Neufeld MD, Burrows JA, Andrusak GF (2022) Kootenay Lake kokanee (*Oncorhynchus nerka*) collapse into a predator pit. Can J Fish Aquat Sci 79:234–248

Wilson KL, Cantin A, Ward HGM, Newton ER, Mee JA, Varkey D, Parkinson EA, Post JR (2016) Supply-demand equilibria and the size-number tradeoff in spatially structured recreational fisheries. Ecol Appl 26:1086–1097

Wilson K, Foos A, Barker O, Farineau A, De Gisi J, Post JR (2020) Social-ecological feedbacks drive spatial exploitation in a northern freshwater fishery: a halo of depletion. J Appl Ecol 57: 206–218

Resource Economics for Understanding Recreational Fishers and Fisheries

Richard T. Melstrom, Brenna Jungers, Xiang Bi, and Joshua K. Abbott

Abstract We describe how resource economics measures the economic benefit of recreational fishing. Information about economic benefits provides one way to evaluate the social and economic consequences of a management or policy action. We divide this chapter into two parts. First, we describe random utility maximization theory, which explains fishing choices as a function of fishing site attributes, recreational fishers' characteristics, and individual preferences. Models developed from random utility maximization theory and individual data have become the dominant approach in fisheries valuation. We emphasize that this value, though nearly always denominated in money units, is not equivalent to the actual amount of money spent on fishing. Second, we provide a pair of empirical demonstrations, one using actual fishing trip data and another using hypothetical data, based on a mail survey of license holders. We end this chapter with a discussion of using economic models to predict behaviours, frontier areas in recreation demand modelling, and integration with interdisciplinary research.

Keywords Demand model · Multiple site model · Nonmarket valuation · Welfare economics

R. T. Melstrom (✉)
School of Environmental Sustainability, Loyola University Chicago, Chicago, IL, USA
e-mail: rmelstrom@luc.edu

B. Jungers
Department of Agricultural Economics, Mississippi State University, Starkville, MS, USA

X. Bi
U.S. Government Accountability Office, Seattle, WA, USA

J. K. Abbott
School of Sustainability, Arizona State University, Tempe, AZ, USA

K. L. Pope et al. (eds.), *Understanding Recreational Fishers*, Fish & Fisheries Series 45, https://doi.org/10.1007/978-3-031-99739-6_5

1 Introduction to Resource Economics

Resource economics developed as a distinct field following the scholarship of several economists in the 1950s and 1960s on natural resource conservation. Their work provided answers to questions about protecting common property, natural resources, and valuing public investments in projects that produced "intangible benefits" (Brown et al. 2016). Before the twenthieth century, scholars considered land one of the three essential "factors of production," along with physical capital and labour (Hubacek and van den Bergh 2006). However, at the time, this essentialness was understood, or at least articulated, narrowly. Natural resources were thought to be valuable because they could be extracted from the environment and transformed into marketable goods like food, housing, and machines. This purely consumptive view of natural resources changed in the 1950s and 1960s. Resource economics developed to rigorously show that a resource could be an asset even if it did not factor into market production. It also showed that failure to fully account for this economic value in private use and social policy decisions left natural resources vulnerable to overexploitation.

This development occurred shortly after the US federal government began to emphasize cost–benefit analysis to evaluate major infrastructure projects (Hanemann 2006). This emphasis, codified in laws like the Flood Control Act of 1936, led to calls for economists to measure the value of outdoor recreation (Banzhaf 2010). Some federal decision makers suspected that proposed but rejected reservoir projects would have been accepted had the economic value of water-based recreation been factored into the benefit–cost analysis, instead of being treated as an intangible benefit (Trice and Wood 1958). Before the 1950s, however, there were no methods to measure values other than through market prices. Economic valuation was not applied to non-extractive water uses like recreational fishing and boating, partly due to criticism of the assumptions required for measurement and partly due to reluctance among economists to measure what they personally considered intangible and hence, unsuitable for a money measure (Banzhaf 2010). By the 1960s, however, theoretical and empirical advances in the profession made a more complete accounting of economic value possible.

Economic value measures the trade-offs people are willing and able to make, in terms of one item for another. Note that the trade-off a person is willing to make is not necessarily the same as what they actually have to trade. For example, a person might exchange 10 units of one item to get one unit of another, but perhaps they would have been willing to exchange up to 15 or 20. This distinction is important because a measure of a good's value is what a person or group would be willing to give up for it, and not what they actually gave up. One way to put this in dollar terms is to determine the maximum amount of money a person will exchange for an item, which is a measure resource economists refer to as willingness to pay. Another way is to measure the minimum amount of money a person will accept to forego receiving an item, which resource economists refer to as willingness to accept. Note that willingness to pay and willingness to accept are not mirror images of each other. Research suggests that, in general, willingness to accept significantly exceeds willingness to pay (see Mackay et al. 2026). Dissecting the difference goes beyond the scope of this chapter, but suffice it to say that whether a person believes they own (or are owed) a good matters in their value of the good.

It is crucial to recognize that willingness to pay and willingness to accept are not the same as market price. In fact, nothing limits the items in the example above to market goods with any price at all. The existence of a price does mean, though, that economic value can be decomposed into two parts: expenditure value and surplus value. Expenditure value is simply the actual amount paid. Surplus value is the remaining amount of value that the consumer has left over. Measuring this surplus is a central challenge in resource economics because it cannot be directly observed. In social policy decisions, decision makers often emphasize expenditure values, which are easier to measure and may appear more tangible. Nonetheless, the concept of surplus value is well established, widespread, and conceptually sound; though the term has not entered mainstream vernacular, it can be found in any modern economics textbook (Ekelund and Hébert 1989). Measuring economic value only in terms of expenditures also runs against our own personal intuitive sense of what is valuable, because we all know that goods provided for free or at minimal cost can have immense value.

Another reason why decision makers may desire information about expenditures is that they view spending by one household as a source of income for another. This idea has two defects, however. First, a dollar spent in one industry is not equivalent to a dollar of new income. In principle, money spent on, say, recreational fishing is money that could have been spent on some other activity, without producing any more income in an economy overall. To be sure, spending on recreational fishing can be an important component of income generation and economic value in some areas, but one cannot use expenditures alone to compare the value of fishing in developed local economies to fishing in natural areas where expenditures are zero by design, for example wildlife refuges and wilderness areas. Second, the idea ignores the importance of consumer (e.g. recreational fisher) values, some of whom may spend little actual money on the good (e.g. fishing). Other things equal, a person who spends fewer dollars on the same good gets more enjoyment, not less, out of their consumption, as a focus on expenditures would incorrectly imply.

Consider a simple example to demonstrate the logic about prices versus values. A recreational fisher (from here forth termed fisher) who happens to live next to a small neighbourhood pond enjoys fishing at the pond every day but spends essentially nothing to do so because they do not have to buy gas, food, lodging, or a boat. The fact that they fish rather than do some other activity, though, reveals that fishing is valuable to them. Now, suppose someone fences off the pond so the fisher must travel many miles to a public lake to fish. The fisher originally preferred the pond to the lake, so they must be worse off now that the pond is closed, which could be demonstrated by comparing their (consumer) surplus values at the pond and lake. Of course, while the fisher may be personally worse off, is closing the pond good for the local economy because the fisher now spends money on fishing at the lake? Not necessarily, because the money now spent at the lake means a reduction in the fisher's spending on other goods.

Below, we describe one way that resource economists measure surplus value. Sometimes we will refer to surplus value as consumer surplus, economic benefit, or, simply value. Keep in mind that this is the money measure most economists would recommend in decision-making. Best practice in cost–benefit analysis is that benefits should be measured in terms of surplus values (Griffin 1998).

2 Using Resource Economics to Understand the Value of Recreational Fisheries

Economic value can be separated into parts, such as expenditure and surplus value, and classified by type, such as market and nonmarket value. Nonmarket value includes the value of items not traded in markets, for example, fishing for sport or subsistence on a public lake, as well as intangibles such as existence value. Other types of value include use and non-use values. Use values are the benefits that individuals derive actively from interacting with an amenity, whereas non-use values are the benefits that individuals passively experience from a resource's existence. Use values can be either market or nonmarket (as in the case of fishing at a public location), but non-use values tend to be nonmarket in nature. A resource can be a source of many distinct types of economic value. For example, coastal waters can provide market value by supporting a commercial fishery, nonmarket use value through a recreational fishery, and non-use value through the spiritual connections people may feel with the water, all three of which might occur simultaneously. Resource economists also distinguish between marginal and total values. A marginal value measures the change in value from an incremental change in the quality of a resource. For example, economic valuation might be used to estimate the marginal value fishers place on the construction of a new boat launch. The total value is the overall value of a resource.

The two families of methods that economic researchers employ to collect data on values are known as revealed preference and stated preference methods. Revealed preference methods rely on the fact that a resource is a bundle of quality attributes that influence the demand for the resource. For example, trips demanded (and therefore, taken) are an observable behaviour through which individuals reveal their preferences for a recreational fishing site. Because revealed preference methods involve observing actual behaviour, they can only be used to estimate use values.

Rather than hinging on observed behaviour, stated preference methods elicit data by asking or surveying individuals about their willingness to pay in hypothetical choice scenarios. Stated preference methods do not require individuals to actually access or use a resource to reveal their preferences, and so they can be used to collect data on both use and non-use values. For example, a person who never intends to visit or fish at a particular site, nonetheless, may express a positive willingness to pay to protect it because they place existence value on that ecosystem. One additional benefit of stated preference data is that an analyst can design the hypothetical scenarios to shine light on the effects of policies that have not yet been implemented.

Besides the type of data, resource economics also classifies valuation methods by the type of trade-off. The travel cost method is arguably the most popular valuation method used to value fishing and site quality. This method works by constructing demand curves from data on site visits and prices based on visitors with different residential locations. This price is the aggregate cost of taking a trip (i.e. the travel cost, which can include items such as fuel or airline costs and the value of travel time), which tends to be higher for people who must travel farther to access the site. By relating the inclination for trips demanded at different prices, an analyst can construct demand curves for fishing sites, which can then be used to measure economic value.

The travel cost method can be linked to measures of economic value (Fig. 1). Panel (a) shows a possible demand curve for a recreational fishing site. This demand curve shows the amount that fishers are willing to pay for a given quantity on the margin. If quantity is measured by the number of trips, then for fishers whose travel cost is p_0, the number of trips is X_0, the expenditure is the rectangle between p_0 and X_0, and the surplus value is the triangle above p_0. Economic value is the sum of the rectangle and triangle. Panel (b) shows the increase in trips and surplus value when the travel cost decreases from p_0 to p_1. Panel (c) illustrates the increase in demand (i.e. willingness to pay), trips, and surplus when site quality improves. When resource economists use the travel cost method, typically they focus on measuring the surplus value areas, like in panels (a), (b), or (c). Although demand curves are not always linear as in the figure, the idea that economic and surplus values can be measured as areas underneath a demand curve always holds true.

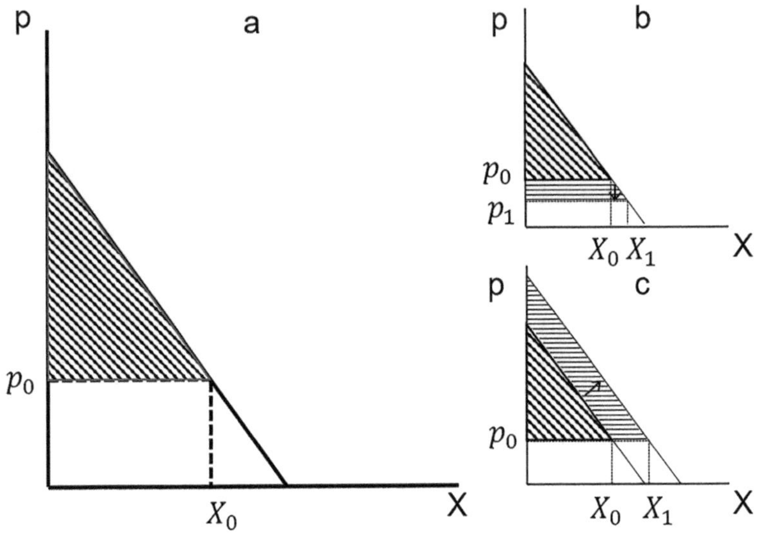

Fig. 1 The demand for fishing at a given site (panel (**a**) illustrates the demand curve for different combinations of quantity (X) and price (p), with consumer surplus illustrated by the shaded triangle; panel (**b**) shows the change in quantity demanded and consumer surplus when price decreases; panel (**c**) illustrates a shift in demand caused by an improvement in site quality)

The contingent valuation method is a common stated preference method. Simply put, contingent valuation asks whether an individual would be willing to pay a certain amount of money for an item or a quality change, contingent on a well-defined scenario for its provision and a mechanism for payment. For example, a person could be asked how much additional income tax they would be willing to pay to fund a conservation program to protect fish habitat. Another common stated preference method uses choice experiments that compare two or more items or "choice alternatives" with different attribute levels specified by the researcher; individuals state their preference by selecting the alternative they most prefer. The alternatives could be fishing sites with distinct species presence, catch rates, and travel distances as attributes. With travel distance included as an attribute, the researcher can apply the travel cost method to the experimental data to value changes in fishing quality and site access. Other attributes, valuation scenarios, and payment mechanisms are possible; for example, choice experiments have been used in resource economics to examine how fishing site choices are influenced by gear and harvest restrictions (Cha and Melstrom 2018; Jungers et al. 2023; Knoche and Lupi 2016).

3 Recreation Demand Modelling and Valuation

This section describes how economists use demand models and the travel cost method to value recreational fishing. The typical model in resource economics takes the form of a discrete choice model based on random utility maximization (RUM) theory. Recreation demand modelling has more nuance than we have the space to describe here fairly. The RUM model, as it is known, is a type of multiple-site model in which the quantity demanded is a discrete variable indicating if an individual visited a particular site or not in a given choice occasion. Alternatives to the standard RUM include the repeated RUM and corner solution models (Herriges et al. 1999). There are also single-site models in which the quantity demanded is the number of trips an individual takes to one location in a season (Boyer et al. 2017). Single-site models can provide an expedient way to measure the total value of unique recreational sites, but RUM models are more appealing when there are alternative sites with varying degrees of site quality. These RUM models are well suited for answering questions about the behaviours and values of recreational fishers because they are designed to analyse the trade-offs people are willing to make given many alternatives, such as fishing location and target species. Resource managers can use RUM models to assess how their initiatives affect fishing demand and economic value, and even predict how their proposed policies could impact demand and value prior to implementation.

The logic is that a fisher chooses some alternative j out of a set of J_i alternatives on a particular choice occasion i, and their choice of j is influenced by the attributes of j. In practice, these attributes are usually some combination of site-specific and individual-specific attributes such as facility availability, entry fees, travel cost,

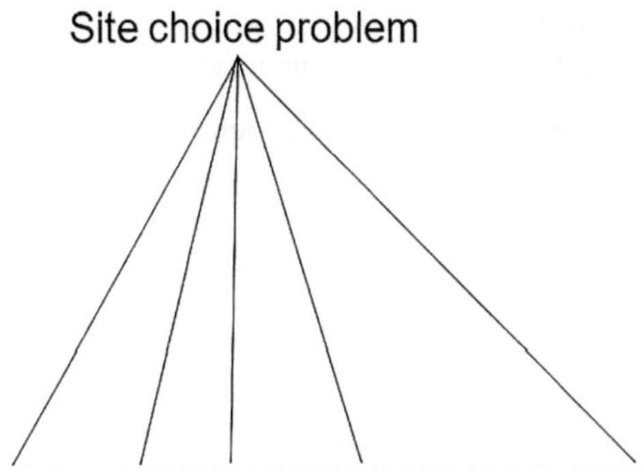

Fishing site:	*1*	*2*	*3*	*4*	…	*100*
Catch rate:	0.5	0.7	0.0	0.2	…	1.1
Dock:	0	1	1	0	…	1
Travel cost:	23	31	20	75	…	19

Fig. 2 Stylized example of a random utility maximization (RUM) model with fishing sites as the alternatives and three attributes

size distribution of the fish stock, target species availability, and expected catch rates (Hunt et al. 2019). In making their choice, each fisher must decide how much income they are willing to give up to access alternatives with more or less desirable combinations of these attributes. By identifying trade-offs between travel cost and the non-cost related attributes, analysts can use RUM models to measure economic values for fishing alternatives, groups of locations, and location attributes.

The structure of a RUM model for a choice with 100 alternatives is illustrated (Fig. 2). Each alternative is represented by a line coming from the decision-making node. At the end of the line is the number assigned to the alternative, which represents a fishing site. The figure also provides examples of the attributes describing these sites, including catch rate (in fish caught per hour), the presence of a dock (a value of 1 if present and 0 otherwise), and travel cost (in US dollars). An analyst could use this type of RUM model to value the 100 sites individually or in small groups, changes in catch rates, and expanding or restricting the availability of docks. An analyst could also use it to forecast changes in site demand and cross-site substitution because of opening or closing sites, increasing or decreasing catch rates, or changes in the availability of access points (e.g. docks).

The remainder of this section is organized around seven steps that an analyst must follow to value recreational fishing using an RUM model. These steps are as follows:

1. Identify the part of the fishery to be valued.
2. Define the population.
3. Define the choice set.
4. Define the structure of the model.
5. Collect data.

6. Estimate the model.
7. Measure economic value using the model.

Our descriptions are necessarily condensed due to space limitations, but there are several texts readers can turn to for more information, including Haab and McConnell (2002) and Parsons (2017).

The order of the steps is not set in stone. Resource economists will often conduct step 5 before step 3. Sometimes an analyst can only define the choice set after having collected data from fishers. Data limitations can also affect the structure of the model. An analyst should check what information might be available from fish and wildlife agencies before implementing a survey. Nearly all RUM models of recreational fishing are estimated using individual data collected via household surveys (which we implicitly assume herein) that can be a useful source of information about preferences. However, self-reported site information can be subjective (e.g. perceptions of crowding, water quality) and suffer from recall bias (e.g. timing of visit, number of fish caught), so it should be evaluated carefully before being used to describe site attributes (Lupi et al. 2020).

3.1 Identify the Part of the Fishery to Be Valued

The first step is to identify what needs to be valued. Interest in the economic value of recreational fishing is almost always motivated by questions about the value of one or more fishing areas or the potential impacts of fisheries management. Examples of impacts include opening a new fishing site, reducing bag limits, and prohibiting the movement of bait to prevent disease spread. These impacts should help guide the structure of the RUM model, the population of interest, and the data needs. For example, if the driving question is "what is the value of recreational fishing in Lake Michigan?" the analyst will need to design a RUM model of fishing for introduced salmon (e.g. *Oncorhynchus tshawytscha*), lake trout (*Salvelinus namaycush*), and walleye (*Sander vitreus*) (Melstrom and Lupi 2013). Or, if the question is "how much value could the paddlefish (*Polyodon spathula*) fishery in northeast Oklahoma lose if we implemented catch-and-release?" the analyst will need to design a RUM model of paddlefish fishing that includes catch-and-release as a site attribute. If catch-and-release has not yet been implemented, this will require collecting stated preference data based on a hypothetical catch-and-release attribute (Cha and Melstrom 2018).

3.2 Define the Population of Interest

The second step is to define the population of interest, which could be active fishers, both active and potential fishers, or a specific type of fisher, such as those targeting bass (e.g. *Micropterus salmoides*) or who fish aboard charter vessels. In principle, the population should be defined as broadly as possible to include non-users, so that the RUM model can flexibly account for reactivation and recruitment, which is often a crucial management consideration. Note that RUM models estimated from active fisher data can still be used to make predictions about recruitment and reactivation, though the predictions may have less validity than models that incorporate data from lapsed or potential fishers.

3.3 Define the Choice Set

The third step determines the extent of the choice set J_i. In principle, individual choice sets should only include alternatives that a fisher would consider choosing, excluding any alternatives they do not consider (though one should be careful to not exclude alternatives that a fisher may consciously filter out before making their final decision; see Haab and Hicks (1999)). For example, if a fisher has gear-specific fishing preferences, an analyst may want to exclude sites that exclude the use of these gear types. In practice, though, the distinction between relevant and irrelevant alternatives is not always clear. If the analyst is unsure whether an alternative is irrelevant, it is better to include it. Changes in site quality and access can cause fishers to substitute toward smaller, farther, less popular sites, and you want a model that can accommodate such substitution patterns.

Often, analysts simply construct the choice set from a list of alternatives collected via a survey and apply the same choice set to all decision makers (e.g. Haab et al. 2012). Sometimes rules-of-thumb are used to construct heterogeneous, individual-specific choice sets, for example by excluding or aggregating alternatives far from a fisher's home (e.g. Parsons et al. 2000). There is also research that has explored heterogeneous choices sets constructed empirically, as part of the estimation process (Haab and Hicks 1997).

An analyst also must define the nature of an alternative. Alternatives can be location, target species, method, mode, and time-specific. Many RUM model applications, though, simply pool fishing trips across targeting preferences, method, mode, and timing, and only differentiate the alternatives by location (i.e. fishing sites). Sometimes defining a fishing site is straightforward; for example, in the case of most lakes and ponds, which have definite boundaries and just one or two access points. However, large lakes, reservoirs, and streams can have hundreds of miles of shoreline that need to be separated into distinct alternatives (e.g. Melstrom et al. 2015). The analyst may also have to decide how to deal with large numbers of alternatives that matter to some fishers but are irrelevant to most. For example, there may be several thousand ponds and small lakes in a state, which no fisher has complete knowledge of individually, but may attract a non-negligible share of fishing trips in aggregate. The convention is to use an aggregate for alternatives in the same region that are similar albeit less popular while leaving the most popular sites as individual alternatives, or to randomly draw sites into the choice set during model estimation (Parsons et al. 2000). The policy questions motivating the research should also play a role in defining alternatives. An analyst should consider distinguishing alternatives by species, method, or mode if the policy could have a significant effect on substitution between species, method, or mode preferences. For example, managers could be interested in the cross-method substitution effects of a rule to limit trolling; in this scenario, distinguishing choice alternatives among casting, spinning, and trolling alternatives, in addition to fishing sites themselves, may be crucial.

Best practice is to include an alternative to not fish. Some fishers are more avid than others are, and there will be occasions when individuals decide against fishing even though an opportunity exists. Modelling this no-participation decision is important because it can be influenced by perceptions of site quality and individual constraints, just as the decision about where and what to fish. It will also allow the analyst to evaluate the potential effect of fisheries management on participation and avidity, allowing them to construct estimates of site-specific and system-wide fishing effort, and nest the RUM model within the larger social–ecological system. To include a no-participation alternative in the model, the analyst will need to know, for a given choice occasion or set of occasions (e.g. over a season), when an individual did and did not fish.

3.4 Define the Structure of the Model

Random utility maximization theory links recreational fishing choices and economic value to the attributes of alternative j by deriving part of the fisher's utility function U_{ij}. Utility measures an individual's well-being in that a higher utility measure indicates that an individual is happier. For example, if a fisher enjoys fishing for salmon (*Oncorhynchus* spp.) more than bass (*Micropterus* spp.), then their utility from salmon will be larger than from bass. Actual utility is not known to the analyst, but when a fisher chooses j over all other alternatives k in their choice set on a particular choice occasion, we learn that $U_{ij} > U_{ik}$ for all $j \neq k$. It is possible to relate this choice to income y_i, travel cost p_{ij}, and attributes describing the quality q_{ij} of j. Research in resource economics has shown that quality can be determined through some combination of fishing (e.g. expected catch rate, target species availability, and size distribution of the fish stock) and non-fishing attributes (e.g. water quality, shoreline amenities, and facility quality and availability) (Hunt et al. 2019). The form of the utility function is unknown, but we can use a linear approximation. A linear utility function is not necessary, but most applications of the RUM model in resource economics adopt this approach partly because calculating willingness to pay is more complicated with a nonlinear-in-income function (there is no closed-form solution) and partly because the relevant income measure for y_i is not clear (e.g. after-tax income or income after deducting food and housing costs) (Haab and McConnell 2002). This is because y_i disappears from the probability functions used to estimate the linear-in-utility RUM model. The linear approximation we use is:

$$U_{ij} = V_{ij} + \varepsilon_{ij}$$

where $V_{ij} = \alpha(y_i - p_{ij}) + \beta q_{ij}$ is the part of utility that analysts can observe (note that travel cost subtracts from the fisher's income y_i) and ε_{ij} is the part that is unobservable and idiosyncratic to each fisher. The parameters α and β measure the effect of income-less-travel cost and quality on utility. There will always be a part of utility that is unobservable, so the analyst must acknowledge that utility from their perspective is partially random—hence the name of the RUM model—and to gain any further traction on linking choices to the attributes in a utility-theoretic way, the analyst needs to assume something about ε_{ij}. One approach is to assume that ε_{ij} is distributed type I extreme value with scale parameter μ so that the probability that a fisher chooses j can be written as

$$P_{ij} = \frac{e^{\mu V_{ij}}}{\sum_k e^{\mu V_{ik}}}$$

which is a probability function that economists refer to as a conditional logit, which differs from a multinomial logit as the V_i terms vary among the alternatives; for a detailed explanation, see Haab and McConnell (2002). This function says that the probability of choosing is higher when an alternative has a greater observable utility. The scale parameter μ is not identified, so typically it is omitted from the formula. Thus, inserting the observable form of utility, we can write the probability as

$$P_{ij} = \frac{e^{\alpha(y_i - p_{ij}) + \beta q_{ij}}}{\sum_k e^{\alpha(y_i - p_{ik}) + \beta q_{ik}}}$$

which reduces to

$$P_{ij} = \frac{e^{-\alpha p_{ij} + \beta q_{ij}}}{\sum_k e^{-\alpha p_{ik} + \beta q_{ik}}}$$

With a linear specification, variables that vary across decision makers but not across alternatives, such as income, fall out of the probability function. In this way, the RUM model allows an analyst to link individuals' choice of a particular recreational alternative (behaviour) to their valuation of these alternatives as a whole and their quality attributes. The utility parameters can be estimated from data on fishers' choices alone by selecting the parameters that make these choices most likely in the sample data.

One weakness of the conditional logit is that it always forecasts that a change in one alternative will redistribute choices to the other alternatives in equal proportion (Hunt 2005). For example, closing a site that receives 10% of the effort when the total number of alternative sites is 10 will increase trips to the other sites by 11%. Though this may be true on average, it is unlikely in reality that every other site would experience an identical 11% effort increase. Alternative assumptions about the distribution of ε_{ij}, however, can produce probability functions that relax this assumption, such as the nested and mixed logits (Haab and McConnell 2002; Train 1998). Other alternatives to the conditional logit include the cross-nested logit and scale heterogeneity logit (Hunt et al. 2007; Melstrom et al. 2017). We use the conditional logit in the rest of this chapter because it has the simplest probability function.

3.5 Collect Data

The next step is to collect information on the population of interest and the alternatives in the choice set. Information about the population typically comes via an off-site survey about when, where, and what type of fishing trips occurred. The survey can be conducted over the internet, mail, or telephone (see Vaske et al. (2026a, b) for additional details). At a minimum, the survey needs to elicit the trip destination, date, and whether it was taken primarily for the purpose of fishing, if collecting revealed preference data. Typically, analysts focus on modelling primary purpose trips, as it may not be clear how to distribute a visitor's travel cost between activities and locations in multi-purpose trips (Lupi et al. 2020). Typically, a survey will also ask about targeted species, catch, method and mode, party size, and type of travel vehicle. Asking fishers only about their most recent trip will limit recall bias, but it constrains the number of choice occasions to one per fisher, and it may introduce a seasonal bias depending on the timing of the survey and the level of heterogeneity in fishing preferences and fishing site diversity (Kaemingk et al. 2018). For example, a survey with recall bias could overstate the overall demand for fishing if respondents tend to overestimate trips in the distant past (Connelly et al. 2000). Alternatively, one can ask about the last two or three trips, or ask fishers to indicate, from among a list of location alternatives, how many times they visited locations in a season.

Surveys can be the most resource-intensive part of developing a recreation demand model and analysts should follow best practices. Johnston et al. (2017) and Lupi et al. (2020) provided guidance on survey issues in valuation research that, in our opinion, are essential reading for analysts committed to RUM models (see also Vaske et al. (2026a, b) for general details about survey methods). Failure to follow universal standards and valuation best practices will likely lead to results that are less reliable and valid (i.e. more prone to bias).

Ideally, the analyst could collect their data from a survey of the general population. General population samples have the advantage that every individual could be a participant, which can yield insights into recruitment and willingness to engage in recreation, including fishing. Care should be taken to evaluate the representativeness of the sample, though, because avid recreators may be more likely to participate in the survey (Lupi et al. 2020). In practice, analysts tend to rely on offsite (e.g. mail or internet) surveys of license holders. On-site surveys (e.g. creel surveys) are rarely used because it is usually only possible to survey a few sites in a season, limiting the geographic scope of the analysis, and more avid fishers are more likely to be intercepted, creating a potential bias.

The analyst also needs to collect data on the attributes that link fishers' behaviours to the conditions at fishing sites. Site attributes come in two varieties: site-specific and individual-specific attributes. Site-specific attributes vary across sites but not fishers, such that all fishers perceive the same attribute level for any particular alternative. Site-specific attribute data can be collected from external sources, for example, the presence of docks based on travel guides, measures of water quality from satellite imagery, or catch rates from creel data (see de Kerckhove et al. 2026). Individual-specific attributes vary across sites and fishers and are usually collected from surveys or generated by combining site-specific attributes with fishers' characteristics. The most important of these is travel cost, which should include both the money cost (including fuel and miles-related depreciation, though not vehicle age-related depreciation) and the opportunity cost of travel time to and from each site. The basic travel cost formula is

$$\mathrm{tc}_{ij} = \mathrm{distance}_{ij} \times (\mathrm{fuel}_i + \mathrm{depreciation}_i) + \mathrm{traveltime}_{ij} \times \mathrm{timevalue}_i$$

where $\mathrm{distance}_{ij}$ is the round-trip distance in miles or kilometers from fisher i's home to site j, fuel_i is the per mile or kilometer fuel cost, $\mathrm{depreciation}_i$ is the per mile or kilometer maintenance and depreciation cost, $\mathrm{traveltime}_{ij}$ is round-trip travel time in hours, and $\mathrm{timevalue}_i$ is the per hour value of the fisher's time. Analysts may either use generic or fisher-specific fuel and vehicle depreciation costs, but the latter will require analysts to collect information about each fisher's vehicle type. The time value is usually some function of the fisher's wage or income, such as one-third of the fisher's personal income divided by 2000, which is the typical working hours in a year (Phaneuf and Smith 2005). The usual practice is not necessarily the best practice, though, and calculating the value of travel time is an ongoing topic of research (Lloyd-Smith et al. 2019). Analysts should recognize that their choice of the value of travel time would affect the resulting valuation measures; one recent assessment of recreation demand modelling practices has called for analysts to conduct sensitivity analyses of their choice of time value (Lupi et al. 2020).

It may also be necessary to transform the attribute data into measures, as fishers perceive them. Accurate estimation of the attribute coefficients β requires distinguishing between the perceptions of different fishers. Some of these differences can be inferred from fisher characteristics, including targeting preferences and fishing mode. For example, catch rates for bass may matter a great deal to those targeting bass but probably not at all to those targeting salmon, at least conditional on the decision to fish for salmon. It would, therefore, make little sense to allow catch rates for bass to enter equivalently into the catch perceptions of both bass and salmon fishers; the analyst could account for this by setting bass catch rates to zero across sites in salmon fishers' choice sets (in effect, forcing the effect of bass on salmon fisher utility to be zero). In general, this can be accomplished by inserting interaction variables between site-specific attributes and fisher characteristics in a standard RUM model.

Sometimes, resource economists combine revealed and stated preference methods to estimate a RUM model. Combining methods can help overcome weaknesses inherent in each alone; revealed preference methods can measure actual behaviours but cannot be used to measure how users will respond to choice attributes that do not exist, co-vary with unobserved attributes, or vary little across alternatives (Louviere et al. 2000). Stated preference methods can remedy problems with missing or co-varying data but may not be able to predict accurately choices in actual choice sets. One method of combining revealed and stated preference methods is to simply pool the associated data and jointly estimate a single RUM model. Another approach is to estimate the parameters with revealed preference data, then estimate parameters with stated preference data, and finally combine the parameters into a single utility function. However, parameters estimated in different RUM models are not directly comparable due to differences in the scale parameter μ, so some sort of adjustment for scale is necessary when combining parameters generated by different data. A simple solution is to multiply the stated preference parameters by the ratio of the revealed to stated preference travel cost parameters. However, the assumption that the travel cost parameters are different only up to the scale factor will not hold in all contexts. Alternatively, an analyst could estimate the stated preference and revealed preference parameters sequentially by estimating a model with the stated preference data, then fixing the shared revealed preference parameters equal to their stated preference counterparts, and finally estimating the remaining parameters along with a scale factor using the revealed preference data. Another, and our recommended, approach is to jointly estimate the RUM model including the estimation of scale differences. See von Haefen and Phaneuf (2008) and Whitehead and Lew (2020) for examples that use the last approach. Though a full treatment goes beyond the scope of this chapter, it suffices to say that combining stated and revealed preference data can be a complicated but worthwhile approach to learn about fisher behaviours.

3.6 *Estimate the Model*

The sixth step is to estimate the utility function coefficients in the RUM model, which is typically done using built-in commands in statistical software. The analyst will need to set up a spreadsheet that organizes the data into decision makers (fishers), choice occasions, the choice alternatives, and the dependent and independent variables. Data are typically arranged in long format, in which there are as many rows as decision makers × choice occasions × choice alternatives, as illustrated (Fig. 3). A fisher can be allowed to have more than one choice occasion. For example, an analyst may decide that the fishers they are modelling should have 52 choice occasions (i.e. that fishers decide or choose between the choice alternatives once per week for a year). The data set will need to include identifiers for fishers, their choice occasions, and the choice alternatives, so that they can keep track of which alternatives belong in which choice set on a given choice occasion. As

Fig. 3 Stylized example of a spreadsheet with two decision makers (id), two choice occasions (occasion), four choice alternatives (alternative), and the dependent variable indicating their choice (choice)

id	occasion	alternative	choice
1	1	1	0
1	1	2	1
1	1	3	0
1	1	4	0
1	2	1	0
1	2	2	1
1	2	3	0
2	2	4	0
2	1	1	0
2	1	2	0
2	1	3	1
2	1	4	0
2	2	1	0
2	2	2	0
2	2	3	0
2	2	4	1

is standard for logit-based models, the dependent variable should equal one for the alternative that a fisher chooses and zero otherwise.

3.7 *Measure Economic Values Using the Model*

To calculate economic value, one uses the utility function in the RUM model to compute the change in income that leaves utility the same after a site quality change. This income payment (irrespective of whether they actually have to pay it) measures the value of a beneficial change to a fisher. When quality changes leave a fisher worse off, they must be compensated to be just as well off without the change, and the payment amount will be negative.

If the quality change is the same at every alternative, the willingness to pay formula is

$$\mathrm{WTP} = \frac{\hat{\beta}\Delta q}{\hat{\alpha}}$$

where Δq is the quality change (Haab and McConnell 2002). Note that this formula does not apply to changes that occur at some alternatives and not others, nor does it apply to changes that vary across sites (see Appendix 5.1 for longhand formulas that can be used for changes that are more general).

An analyst can also use a RUM model to measure the economic value of one or more alternatives by calculating the loss associated with closing the alternative or group of alternatives in question. In these scenarios, the willingness to pay formula is

$$\text{WTP} = \frac{\ln\left(1 - P_j\right)}{\hat{\alpha}}$$

where P_j is the sample mean probability of visiting j (Haab and McConnell 2002). This expression implies that an analyst can estimate the value of an alternative simply by taking the natural logarithm of one minus the proportion of times the alternative is chosen in the sample, and then dividing by the negative of the travel cost parameter.

Note that the two WTP expressions above measure economic value per choice occasion for the average fisher. To calculate the total economic benefit or loss in the population, the analyst then multiplies the WTP amount by the number of trip occasions, T, as in

$$\text{TWTP} = \text{WTP} \times T.$$

4 An Application

We now provide a concrete demonstration of how to use a RUM model to measure the value of recreational fishing. We base the demonstration on a study by one of the authors of fishing in Oklahoma City. This work took place in 2017 in response to a request from the H.B. Parson Fish Hatchery, part of the Oklahoma City Parks & Recreation Department, for information about local fisher preferences. For the purposes of this chapter, we will address two of the study objectives. First, we will measure the economic value of adding non-fish amenities, such as playgrounds and shade trees, around city lakes and ponds. Second, we will measure local residents' fishing effort around the state, to gauge their propensity to substitute between fishing sites in and outside the city. These objectives were motivated by questions about the benefits of projects not directly related to fishing, and whether such projects could boost local demand for fishing.

To address the first objective, we will measure the economic value of adding playgrounds, restrooms, and bank trees at city lakes. Note that this cannot be done by simply comparing the value of fishing at city lakes with and without these amenities. With only about a dozen public fishing lakes and ponds in the city, there is too little variation in the presence of each amenity to identify their value. We therefore turn to stated preference data from a choice experiment that asked Oklahoma City fishers to choose between two city lakes and then asked whether the respondent would prefer the lake they chose or to stay home. See Melstrom and Kaefer (2020) for more information about the experimental design. We can use these data to construct a

Table 1 Site attributes in the RUM models of Oklahoma City fishing

	Stated preference data		Revealed preference data	
	Mean	Individual (fisher)-specific?	Mean	Individual (fisher)-specific?
Travel cost	15.28	Yes	43.16	Yes
Catfish catch	0.65	No	0.25	Yes
Sunfish catch	2.25	No	0.27	Yes
Bass catch	–	–	0.23	Yes
Restroom	0.33	No	–	–
Playground	0.09	Yes	–	–
Bank trees	0.34	No	–	–
Log(acres)	–	–	5.86	No
Oklahoma City	–	–	0.10	No
Pond	–	–	0.19	No
Other	–	–	0.02	No
Opt-out	0.33	No	–	–
Observations	2769		27,222	

choice set composed of two lakes and a stay-home option for a total of three choice alternatives.

The utility function in the model includes five site-specific and two individual-specific attributes (Table 1). Catfish (*Ictalurus punctatus*) and sunfish (*Lepomis* spp.) catch rates are denominated in fish per hour. All other attributes are indicator variables that equal 1 if the attribute is present and zero otherwise. We transformed playground presence into an individual-specific attribute by multiplying it by a dummy variable equal to one if the fisher's household included children. The idea here is that playgrounds are only salient to fishers with children. Alternatively, we could test this assumption by including a separate interaction between playgrounds and fishers without children. If, for example, fishers without children view playgrounds as a disamenity because they want to avoid children, then the effect of the interaction would be negative rather than null. The opt-out attribute is an alternative specific constant (i.e. a dummy variable for the stay-home option in the experiment).

We estimated the model in Stata, which has documentation for built-in commands available on its website (e.g. https://www.stata.com/manuals13/rclogit.pdf). Alternatives to the conditional logit version of the RUM model, such as the nested and mixed logits, are accessible through built-in or user-written commands in Stata and other software, including R, LIMDEP, and Python. Example code for R, MATLAB, and GAUSS can be found on Kenneth Train's website (https://eml.berkeley.edu/~train/software.html). The code and data used here are available at https://github.com/melstrom/okc-fishing (Appendix 5.2 contains a more detailed walk-through of the data and procedures).

The RUM model coefficients are presented (first and second columns of Table 2). All coefficients are statistically significant at the 0.05 level except for playground. If we use the formula $\text{WTP}_i = \hat{\beta}\Delta q/\hat{\alpha}$ to estimate attribute values, we find that adding

Table 2 Parameter estimates in the RUM models of Oklahoma City fishing

	Stated preference data		Revealed preference data	
	Coef.	St. error	Coef.	St. error
Travel cost	−0.019	0.003	−0.012	0.002
Catfish catch	0.263	0.074	0.419	0.182
Sunfish catch	0.093	0.025	0.366	0.077
Bass catch	–	–	0.295	0.250
Playground	0.177	0.190	–	–
Restroom	0.554	0.094	–	–
Bank trees	0.455	0.098	–	–
Log(acres)	–	–	0.471	0.054
Oklahoma City	–	–	0.296	0.173
Pond	–	–	2.745	0.437
All-else	–	–	5.991	0.431
Opt-out	0.488	0.161	–	–
Observations	2769		27,236	

restrooms, bank trees, and playgrounds to all alternatives is worth $29.16, $23.95, and $9.32 per trip, respectively, in 2017 U.S. dollars. However, this assumes the attributes are being added to all choice alternatives, which includes a stay-home alternative, for all decision makers. When we calculate the value of these amenities using the longhand formula instead (see Appendix 5.1), we find that adding restrooms, bank trees, and playgrounds is worth $19.43, $15.70, and $1.58 per trip to the affected sites, in 2017 US dollars.

To address the second objective, we will measure the effect of closing a city lake on trips to lakes in and outside Oklahoma City. In the same survey that administered the choice experiment, we asked respondents to provide details on their two most recent fishing trips in Oklahoma. We used this revealed preference data to construct a choice set of 51 alternatives, composed of the 40 individual lakes and ponds with the most visits in the sample, 10 aggregated alternatives for ponds grouped by city (e.g. Edmond ponds, Norman ponds), and an "all-else" alternative for all other locations—including lakes, ponds, and rivers—not represented by the other 50 alternatives. The alternatives include three Oklahoma City lakes (Lake Hefner, Lake Overholser, and Lake Stanley Draper), eight public ponds in the city (Crystal Lake, Dolese Youth Park Pond, Edwards Fishing Lake, Kids Lake, Kitchen Lake, Route 66 Park Ponds, South Lakes Park Ponds, Zoo Lake), and several dozen fishing sites outside the city. The revealed preference choice set does not include an opt-out or stay-home alternative, in contrast to the stated preference data. We admit that this oversimplifies the fishing alternatives available to Oklahoma City fishers, but it is sufficient for the purposes of demonstrating the model.

The utility function in our revealed-preference RUM model is a function of four site-specific and four individual-specific attributes (Table 1). Catch rates are individual-specific because we set the effect of catch rates to zero for ponds more than five miles from a fisher's home, as fishers are unlikely to know or easily find catch information for ponds outside their community (unlike larger lakes, which often have public fishing reports). The site-specific attributes include the natural logarithm of acres (to account for the number of fishing spots at a lake), a dummy variable for sites in Oklahoma City, a dummy variable for ponds, and an alternative specific constant for the all-else alternative.

The RUM model coefficients based on the revealed preference data are presented (third and fourth columns of Table 2). All parameter estimates for the attributes except bass catch rate are statistically significant at the 0.05 level. The bass catch rate is only significant at the 0.25 level. The estimates provide evidence that fishers prefer sites with lower travel costs, more catfish, more sunfish, larger surface area, and waterbodies located in Oklahoma City. The effect of ponds indicates that a significant share of fishers prefer to fish at smaller ponds, and the effect of the all-else variable indicates that a significant share prefer to fish at locations not included as individual or grouped alternatives in the choice set.

The predicted trip shares for Oklahoma City lakes are presented (Table 3). The model predicts that 21.8% of trips by fishers in the metro area are taken at city lakes and ponds. This percentage is remarkably close to the 20.7% of trips in the sample that take trips to city lakes and ponds.

We can use the model to predict the effect on economic value and behaviours of closing a location. The predictions from closing Lake Hefner, Lake Stanley Draper, and all city ponds (including Crystal Lake) are included (Table 3). Closing Lake Hefner lowers the percentage of visits to fishing spots in the city to 15.2%. Note that the percentage point reduction (6.6) is close to the percentage of trips (7.9) to Lake Hefner before the closure. In other words, the RUM model predicts that a large majority of trips will substitute for locations outside Oklahoma City. The same conclusion follows in the other two closure scenarios. This should not be surprising given the current distribution of trips and the conditional logit assumption; a large majority of fishing trips go to sites outside the city, and the conditional logit always predicts that if one alternative is lost, choices will redistribute to the other alternatives in equal proportion.

Table 3 Predicted share of trips to locations before and after hypothetical closures

	Current (%)	Closing Lake Hefner (%)	Closing Lake Stanley Draper (%)	Closing Crystal Lake and city ponds (%)
City ponds	1.95	2.13	1.95	–
Crystal Lake	1.23	1.34	1.30	–
Lake Hefner	7.92	–	8.36	8.20
Lake Overholser	5.56	6.09	5.88	5.76
Lake Stanley Draper	5.17	5.64	–	5.35
Oklahoma City total	21.83	15.20	17.49	19.31

We can use the formula for the per-choice occasion value of an alternative to measure the benefits of keeping one or more of the Oklahoma City lakes open for fishing, WTP $= \ln(1 - P_j)/\hat{\alpha}$. Given 7.9% of trips go to Lake Hefner, the RUM estimates indicate that the lake is worth \$6.86 per trip in 2017 U.S. dollars. We can measure the annual benefit of Lake Hefner by multiplying this value by the number of trips in the population. The average number of trips per fisher in Oklahoma is 20, and with 375,000 fishers in the metro region, the number of trip occasions in the population of interest is 7.5 million. Multiplying the annual per-trip value by the total number of trip occasions, we estimate that the annual benefit of Lake Hefner is \$51.45 million in 2017 US dollars.

5 Discussion

Resource economists use RUM models to understand and predict recreational fishing behaviours and measure economic value. These models are based on the idea that people assess trade-offs in a consistent, structured manner when making a choice between discrete alternatives. In the context of recreational fishing, fishers' actual or stated choices of where, when, and how much to fish under scenarios can be used to calibrate economic models that predict how changes in fishing quality and other potentially relevant site conditions affect behaviour. Furthermore, by observing how fishers trade off money (i.e. travel costs, for non-monetary attributes) resource economists can connect fishers' choices to estimates of their monetized loss or gain. Therefore, RUM models provide a theoretically grounded mathematical structure that, when combined with fishing site choice data, allows economists to generate both behavioural and welfare predictions for changes to policies or biophysical conditions that influence the quality or cost of the recreational experience. These capabilities have made RUM-based models go-to tools for policy analysis.

Of course, RUM models, like any behavioural model, are simplifications of actual decision processes. The deliberative structure of utility maximization at the heart of RUM models may be untenable in cases where fishers either rely on force of habit in routine decision contexts (i.e. fishing the same site day after day under a wide range of quality variation) or where they utilize heuristics to cope with novel contexts with high uncertainty. Furthermore, the predictions of RUM models depend to a significant extent on the size and the degree of aggregation of fishers' choice sets, which are generally unobserved by the analyst. Given their simple structure, RUM models can also prove awkward in their ability to capture repeated linked decision-making through time, particularly if a fisher prefers variety (i.e. diminishing marginal utility of trips to a given site) or if decisions are interconnected in the sense that whether to fish a site today may depend on its expected quality or cost at a later date. Finally, though a RUM model can estimate the direct economic value and behavioural effects of a quality or policy change, it is unable to directly evaluate the effects of behaviourally mediated feedback from a policy. For example, improving quality at a site may increase participation from competing fishers, creating congestion that reduces fishers' welfare. Alternatively, reducing bag limits likely decreases economic values in the near term but may ultimately increase values if stocks recover and support higher catch rates. RUM models do not directly lend themselves to evaluating the final effects of policies that produce such endogenous feedbacks.

Though these and other limitations of RUM models may compromise their descriptive realism, the ultimate proof of their value rests in their accuracy in predicting critical management statistics (e.g. effort or harvest) and their utility in helping decision makers evaluate management trade-offs. Unrealistic or simplified models can still provide useful and reasonably accurate predictions in many contexts, though perhaps requiring less extensive data, fewer person-hours in model development, and less specialized statistical skills from the analyst. Nevertheless, there are several frontier areas of research in recreation demand modelling that may be important for researchers who model behaviours of recreational fishers and recreational fisheries management practitioners.

First, multiple ways exist in which RUM models (and other recreation demand models) can be adjusted to improve their predictive value for managers. These RUM models are ultimately fit to data on fisher choices; therefore, they often perform well for within-sample predictions. Yet they may underperform for predictions out of the estimation sample, especially under novel quality or policy conditions. One approach that can improve out-of-sample predictions is the strategic combination of revealed and stated preference data. As noted above, stated preference questions can be structured to extend the range of scenarios in the data beyond those historically available. Though combining revealed and stated preference data is common, it has primarily been used where valuation is the primary concern. Future efforts could instead focus on how augmenting revealed preference data with choices from constructed scenarios improves the predictive functionality of RUM models. Another area with promise for improving prediction is the application of machine learning approaches to recreation demand models. Machine learning techniques have improved predictions elsewhere in economics and data science by developing algorithms to avoid the biases that come from overfitting to the estimation sample (Varian 2014). Though these techniques have seen little use in recreation demand modelling, their application to RUM models (e.g. Horowitz and Nesheim 2021) could improve the predictive accuracy of these models for forward-looking scenarios. Finally, even though our chapter has emphasized the modelling of a representative or "average" fisher, accounting for heterogeneity in fishers' preferences, skills, or expectations can be important for improving predictive accuracy, as well as for considering how policy changes may differentially affect sub-populations. Well-established extensions to the RUM model, such as the random parameters logit or latent class logit, have been extensively used to improve estimates of economic value. However, they may also improve predictions of fishers' behaviours and important management outcomes such as catch or fishing mortality. More research is needed to define the most important dimensions of heterogeneity to capture in recreation demand models.

Second, there is a greater need to apply insights from behavioural economics (see Mackay et al. 2026) and other cognitive sciences into recreation demand modelling for fishery management. These insights can be particularly useful to better structure choice experiments surveys, which potentially suffer from various biases such as anchoring or ordering effects (e.g. Day and Prades 2010; Scheufele and Bennett 2013). For example, researchers have utilized eye-tracking software to understand better survey respondents' attention to various aspects of the survey under alternative designs (Van Loo et al. 2018). Statistical approaches, combined with clever survey design, can also be used to identify agents in the data that appear to have utilized decision rules or heuristics that ignore certain choice attributes (Lew and Whitehead 2020). Many of these approaches are designed to identify and estimate the preferences of agents who are sufficiently "rational" in the sense that they pay attention to the choice context as described by the analyst so that their resulting welfare estimates are valid. However, if real-world fishers utilize similar heuristics in actual recreational decision-making, then it may not be sensible to ignore these "non-rational" (or rationally inattentive) individuals if the focus is to predict behaviour. A greater focus on understanding how hypothetical versus real-world fishing choices differ is needed, particularly if stated preference data are increasingly utilized to expand the predictive scope of RUM models. Finally, alternative statistical models exist bearing a family resemblance to RUM models that posit alternative psychological decision processes, such as random regret minimization models (Chorus et al. 2008; van Cranenburgh et al. 2015). These models, to our knowledge, have not been applied in fisheries management contexts. However, evidence from the transportation literature, which has seen a wider array of choice model applications, suggests that these alternatives can outperform RUM models, if sometimes only marginally (Hensher et al. 2013).

Third, a growing need exists for RUM models to model more flexibly and accurately recreation decisions across time and space. The most common approach for modelling repeated trips across one or more sites over a period, the repeated RUM (often the repeated nested logit, Morey et al. 1993), typically assumes perfect substitution across trips over time, except to the extent that quality or cost attributes change. This assumption ignores constraints in fishers' lives (e.g. vacation time, school breaks) that may make trips at certain times more desirable than others may and ignores the potential for current behaviours to be affected by anticipated conditions in the future. For example, a fisher may forego an otherwise desirable trip today in anticipation of better fishing conditions in a month. Dynamic discrete choice models (e.g. Provencher and Bishop 1997; Baerenklau and Provencher 2005) offer a rigorous but computationally expensive and coding-intensive alternative to the inherently static RUM model. Alternatively, "discrete-continuous" models that conceptualize choices as occurring in both a discrete and continuous fashion (i.e. deciding which alternatives to consume and then choosing each in a continuous quantity) can be refashioned to allow for forward-looking dynamics and more realistic patterns of temporal substitution (e.g. Lloyd-Smith et al. 2020). Such "Kuhn-Tucker" models (Phaneuf et al. 2000) have been deployed to show the welfare losses to marine fishers from compressed open access seasons (Abbott et al. 2018), and these models have many useful properties for integration with biological models of stock dynamics (Abbott and Fenichel 2013). Nevertheless, they are not as "plug and play" in terms of their integration with popular software packages.

Fourth, a need exists to better integrate recreation demand models, including RUM models, with models of natural and physical processes that affect fishing quality and fishers' behaviours, and which may in turn be influenced by fishing. Recreational fisheries are social-ecological systems that are subject to stressors that are both exogenous to fishing activities (e.g. climate change, population change, invasive species, nutrient loading) and endogenous to it (e.g. target and bycatch stock dynamics, congestion) (Abbott et al. 2022). Yet, traditional recreation demand models are static and do not directly evaluate the behavioural or welfare effects of exogenous stressors or endogenous feedbacks. Given projections of external stressors, it is relatively straightforward to forward-simulate their effects (e.g. Hunt et al. 2021). However, if fishing itself influences state variables, an integrated bioeconomic model is required. For example, Lee et al. (2017) integrated an age-structured model of groundfish stock dynamics with a RUM-based model of fishers' behaviours to evaluate recreational fishery policies (e.g. bag limits) based on their immediate effect on economic value and their ultimate effect as mediated through future stock dynamics. Massey et al. (2006) developed a bioeconomic model for summer flounder in which the effects of an external stressor (nutrient loading) affected fish stock dynamics, which are themselves influenced by fishing pressure. These models have significant potential value for decision makers because they lend themselves to robust scenario-based consideration of uncertainty as deployed in management strategy evaluation. However, few examples of empirically calibrated recreational bioeconomic models exist. This is perhaps due to greater data needs and the significant interdisciplinary effort required to develop an integrated model that passes economic and biological muster.

Fifth, recreation demand models may benefit from incorporating new data sources. Traditionally, recreational fishing models have drawn on creel data and recall-based surveys. Though these instruments share the benefit of rigorous sampling protocols, they have become increasingly cost-ineffective and therefore, prone to non-response bias as overall response rates to traditional mail and phone surveys decline (Stedman et al. 2019). The growing use of smartphones and recreational fishing apps may present ways to access otherwise inaccessible fisher sub-populations (Venturelli et al. 2017, 2026). Furthermore, these apps may allow for more accurate real-time reporting of catch rates and other outcomes than recall-based surveys while also permitting the study of granular behavioural patterns (e.g. high grading and regulatory discards) that have been inaccessible via more conventional survey methods. Nevertheless, research suggests that the populations utilizing such "citizen science" apps may not be representative in either preferences or demographics, and therefore, they should be adopted with caution (Gundelund et al. 2020). Increased integration of digital licensing within the official apps of natural resource management agencies could create a vehicle for more representative license frame-based catch reporting and preference elicitation (Abbott 2015).

Another new area of technology and supplemental fisher demand data, which can serve as a check on creel and recall surveys, is the use of drones (Provost et al. 2020).

As we argue, many areas exist in which new data, improvements to "off the shelf" RUM models, and better integration with biological models can increase the scope of use and predictive validity of recreational fishing demand models. Nonetheless, multiple decades of application have demonstrated the usefulness of RUM-based models in understanding fishing preferences and in predicting the effects of environmental conditions and policies on economic value and behaviours. Practitioners of human dimensions studies in recreational fishery policy are likely to rely heavily on RUM models for many decades to come.

Acknowledgements Data and code for the applications described in this chapter can be found at https://github.com/melstrom/okc-fishing.

Appendixes

Appendix 5.1

After estimating the RUM model, an analyst can predict the behavioural and welfare consequences of changes in the attributes as well as adding or removing alternatives. In terms of behaviour, one can compute the effect on the choice of alternative *j* using the average of the predicted choice probabilities

$$s_j = \frac{1}{N}\sum\nolimits_{i=1}^{N} \frac{e^{-\hat{\alpha}p_{ij}+\hat{\beta}q_{ij}}}{\sum\nolimits_k e^{-\hat{\alpha}p_{ik}+\hat{\beta}q_{ik}}}$$

where the ^ indicates an estimated parameter, *i* indicates a particular choice occasion, and *N* is the number of choice occasions in the data. s_j is the predicted share of choice occasions choosing *j*. The attribute levels determining p_{ij} and q_{ij} can be varied to simulate the impact of interest. Suppose the analyst wants to learn about the impact of changing q_{ij} to q_{ij}^* at *j*. The model predicts the share of fishers who choose *j* after the change as

$$s_j^* = \frac{1}{N}\sum\nolimits_{i=1}^{N} \frac{e^{-\hat{\alpha}p_{ij}+\hat{\beta}q_{ij}^*}}{\sum\nolimits_{k\neq j} e^{-\hat{\alpha}p_{ik}+\hat{\beta}q_{ik}} + e^{-\hat{\alpha}p_{ij}+\hat{\beta}q_{ij}^*}}$$

The analyst can then estimate the change in, for example, total trips to a particular site using the formula

$$\Delta T_j = \left(s_j^* - s_j\right) \times T$$

where T is the total number of trip occasions in the population, including the occasions that fishers choose not to fish, if the model includes a no-fish alternative. If the model does not include a no-fish alternative, then T is the observed or estimated number of fishing trips in the population.

To calculate the economic value (i.e. the benefit or loss to fishers) of a change, measured in terms of willingness to pay, the formula for a single choice occasion is

$$\mathrm{WTP}_i = \frac{\ln\left(\sum_{k\neq j} e^{-\hat{\alpha}p_{ik}+\hat{\beta}q_{ik}} + e^{-\hat{\alpha}p_{ij}+\hat{\beta}q_{ij}^*}\right) - \ln\left(\sum_k e^{-\hat{\alpha}p_{ik}+\hat{\beta}q_{ik}}\right)}{\hat{\alpha}}$$

which is the difference in utility caused by the change in conditions divided by the marginal utility of income. The average willingness to pay across all choice occasions in the sample is $\overline{\mathrm{WTP}} = \sum_{i=1}^{N} \mathrm{WTP}_i/N$. One can then calculate the total welfare change in the population as

$$\mathrm{WTP} = \overline{\mathrm{WTP}} \times T.$$

An analyst can use a RUM model to measure the economic value of one or more alternatives by calculating the loss associated with closing the alternative or group of alternatives in question. In RUM models, the loss from removing an alternative equals the economic value of keeping the alternative open. It is not necessary for removal or closure to have occurred. The analyst only needs to simulate the removal using the model. To do this, the analyst drops the alternative from the per choice occasion WTP_i formula:

$$\mathrm{WTP}_i = \frac{\ln\left(\sum_{k\neq j} e^{-\hat{\alpha}p_{ik}+\hat{\beta}q_{ik}}\right) - \ln\left(\sum_k e^{-\hat{\alpha}p_{ik}+\hat{\beta}q_{ik}}\right)}{\hat{\alpha}}$$

which simulates the effect of removing access to j. Actually, if the analyst wants to value existing sites, and all quality measures are site-specific, $q_{ik} = q_k$, there is no need for information about site attributes besides travel cost (e.g. English et al. 2018). One can just estimate the model with a full set of site dummy variables or alternative specific constants, δ_k, and a variable for travel cost, that is,

$$\mathrm{WTP}_i = \frac{\ln\left(\sum_{k\neq j} e^{-\hat{\alpha}p_{ik}+\hat{\delta}_k}\right) - \ln\left(\sum_k e^{-\hat{\alpha}p_{ik}+\hat{\delta}_k}\right)}{\hat{\alpha}}$$

The parameter estimates for the alternative specific constants account for all site-specific influences. In effect, they measure the unique desirability of each site conditional on travel cost, which is all one needs to model overall site demand. Furthermore, we can simplify the expression to

$$\mathrm{WTP} = \frac{\ln\left(1 - P_j\right)}{\hat{\alpha}}$$

where P_j is the sample mean probability of visiting j (Haab and McConnell 2002). This expression implies that an analyst can estimate the per-choice occasion value of an alternative simply by taking the natural logarithm of one minus the sample proportion of times the alternative is chosen in the sample, and then dividing by the negative of the travel cost parameter. However, this approach cannot be used to value a particular attribute.

Appendix 5.2

This appendix describes the application in Sect. 5.4 in more detail. We used the application to demonstrate how to use a RUM model to first measure the economic value of adding non-fish amenities and second to predict the economic consequences of removing fishing sites. The application used data collected from a survey of Oklahoma City fishers that was implemented by one of the authors (RTM) in consultation with personnel at the H.B. Parsons Fish Hatchery. The survey collected stated preference data from a choice experiment between two city lakes as well as a follow-up question asking whether the respondent would prefer the lake they chose or to stay home. The attributes of these hypothetical lakes included catfish catch rate, sunfish catch rate, presence of a playground, presence of restrooms, a tree-lined bank edge, and driving distance. One of the experimental choice occasions is illustrated (Fig. 4).

12. Suppose you are trying to decide whether to go fishing in the Oklahoma City area and there are two lakes to choose from. The lakes have different characteristics as shown in the table, but otherwise they are the same. Compare the lakes and check the box to indicate which you prefer.

Attributes	**Lake A**	**Lake B**
Lake size (circumference)	1 mile	1 mile
Typical catfish (all species) catch rate	1 every 2 hours	1 every hour
Typical sunfish (all species) catch rate	1 every hour	2 every hour
Playground	No	No
Restroom	Yes	No
Bank landscape	Grass	Tree-lined
Distance from your home (one-way)	1 mile	9 miles
Where would you prefer to fish?	☐	☐

13. If you had to choose between fishing at the lake you preferred above in Question 12, OR staying home (and doing some activity other than fishing), what would you do?

☐ Go fishing at the lake ☐ Stay home/different activity

Fig. 4 One of the choice problems used in the choice experiment in the Oklahoma City fisher survey

ATTENTION: The next few questions ask about your **MOST RECENT** trip to fish **ANYWHERE** in Oklahoma.

8a. When was this trip/outing?
______________Month/Year ☐ Weekday ☐ Weekend

8b. Was fishing the main purpose?
☐ Yes ☐ No

8c. How many hours did this trip last? If this was an overnight trip, please write the total number of hours you fished and the number of days the trip lasted.
______________Hours ______________Days

8d. Where did you fish most of the time on this trip?
______________Name of lake/river/pond ______________City you were in or nearest city

8e. About how much money did you spend on gear/bait and food/lodging and transportation?
$________Gear/bait $________Food/lodging $________Transportation (ex: gas, parking)

8f. Which fish types did you target and catch on this trip? *Please mark all that apply:*

Type of fish	Targeted	Number caught	Type of fish	Targeted	Number caught
Catfish (all species)	☐	___ fish	White/sand bass	☐	___ fish
Crappie	☐	___ fish	Trout	☐	___ fish
Sunfish	☐	___ fish	Walleye	☐	___ fish
Black bass (all species)	☐	___ fish	Other:__________	☐	___ fish

Fig. 5 One of the trip questions in the Oklahoma City fisher survey

The survey also collected revealed preference data by asking respondents about their two most recent fishing trips in Oklahoma. Respondents could provide information about the location, purpose, timing, species targeted, and number of fish caught by species. The portion of the survey instrument that collected data on the most recent trip is presented (Fig. 5).

The survey was administered through the mail to licensed fishers with addresses in the Oklahoma City metropolitan area, including Canadian, Oklahoma, and Cleveland counties. A random sample of 2000 fishers received a four-page questionnaire, an introductory letter, and a postage-paid reply envelope. The questionnaire presented subjects with two hypothetical choice scenarios; therefore, the RUM model included up to two choice occasions per fisher. Returned questionnaires yielded 923 responses to the choice experiment from 479 unique respondents for a response rate of 24%. Although this response rate appears low, response rates have trended downward, and mail surveys that contain stated preference questions and that target urban residents tend to achieve the lowest response rates (Whitehead 1993; Stedman et al. 2019).

The stated and revealed preference data are arranged such that the information about each choice alternative is placed in a row. The first 12 rows and 7 columns in the stated preference data are shown (Fig. 4). The first, second, and third columns contain identifiers for the fishers, choice occasions, and choice alternatives, respectively. We can see that there are two fishers, two choice occasions per fisher, and three alternatives per choice occasion. The fourth column shows the order of the choice occasions in the questionnaire. The fifth column indicates the alternative specific constant for the opt-out alternative. The sixth column shows the fisher's choice, and the final column contains the distance attribute (columns with the other attribute levels are not shown for brevity).

To calculate travel cost for the stated preference data, we multiplied the round-trip driving distance to the hypothetical site by a per mile driving and time cost. To calculate it for the revealed preference data, we multiplied the distances from the centroid of a fisher's home zip code to the coordinates of each fishing site in the choice set, except for the distance to the all-else alternative, which we fixed at 50 miles. We used a per-mile driving cost of $0.28 (including fuel and depreciation) and a travel time cost assuming a time value of one-half of the household income divided by 2000 and a driving speed of 40 miles per hour. The AAA Your Driving Costs 2017 Edition indicates, on average, the cost of driving per mile is $0.1026 in fuel, $0.0791 in maintenance and $0.102 in depreciation, for a total per mile vehicle cost of $0.2837. We imputed per-mile depreciation by dividing the added depreca-tion cost ($510) of the additional miles (5000) in the report. Additionally, before estimating the RUM model with the revealed preference data, we excluded locations more than two standard deviations past the average driving distance (about 170 miles) as these locations are less likely to be reasonable substitutes for Okla-homa City lakes given that they would require overnight trips to visit (with costs not captured in our travel cost formula).

Our Stata code and data can be found at https://github.com/melstrom/okc-fishing. To estimate the RUM model with the stated preference data, we used the command

```
clogit choice tcost catfish sunfish playground restrooms banktrees
optout, group(choice_id) cluster(angler_id)
```

which estimates the model as a conditional logit. The command includes the cluster option to adjust the standard errors for non-independence across choices made by the same fisher. We then used the estimates to measure the economic value of adding playgrounds, restrooms, and bank trees to city lakes. To do this, we first generated a new variable to indicate where the amenity will be placed, by typing

```
gen newamenity = (optout!=1)
```

which in plain language says that the new amenity is placed at every site except the stay-home alternative. Next, we calculated utility under "status quo" conditions:

```
gen exp_v = exp(_b[tcost]*tcost + _b[catfish]*catfish + _b[sunfish]
*sunfish
 + _b[optout]*optout)
by choice_id, sort: egen sumv = sum(exp_v)
```

where `_b[variable]` is the coefficient for *variable* and `egen sumv = sum (exp_v)` calculates the sum in the WTP function. Then, we calculated utility with the impact of interest, which happens to be the addition of bank trees:

```
gen exp_vstar = exp(_b[tcost]*tcost + _b[catfish]*catfish +
 _b[sunfish]*sunfish + _b[banktrees]*newamenity +
 _b[optout]*optout)
by choice_id, sort: egen sumvstar = sum(exp_vstar)
```

Finally, we calculated mean willingness to pay:

```
gen wtp = (ln(sumvstar) - ln(sumv))/(-1*_b[tcost])
sum wtp
```

One technique to verify that the willingness to pay formula was programmed correctly is to compute the value of a $1 increase in travel cost to each site. The WTP_i should be −$1 (i.e. fishers must receive $1 to be willing to accept the change) for every choice occasion.

To estimate the RUM model with the revealed preference data, we used the command

```
clogit choice tcost catfish sunfish bass logacres oklahomacity pond
allelse, group(choice_id) cluster(angler_id)
```

We then predicted the share of trips to lakes and ponds in Oklahoma City using the code

```
gen exp_v = exp(_b[tcost]*tcost + _b[catfish]*catfish + _b[sunfish]
*sunfish +
 _b[bass]*bass + _b[logacres]*logacres + _b[oklahomacity]
*oklahomacity
 + _b[pond]*pond + _b[allelse]*allelse)
by choice_id, sort: egen sumv = sum(exp_v)
gen s = exp_v/sumv
```

or simply

```
predict s, pc1
```

if we use Stata's built-in post-estimation command.

We can also estimate the value of keeping a lake, such as Lake Hefner, open, by estimating fishers' willingness to pay to avoid closing the lake. We did this by calculating the status quo utility

```
gen exp_v = exp(_b[tcost]*tcost + _b[catfish]*catfish + _b[sunfish]
*sunfish +
_b[bass]*bass + _b[logacres]*logacres + _b[oklahomacity]
*oklahomacity
+ _b[pond]*pond + _b[allelse]*allelse)
by choice_id, sort: egen sumv = sum(exp_v)
```

and then utility without the lake

```
gen exp_vstar = exp(_b[tcost]*tcost + _b[catfish]*catfish + _b[sunfish]
*sunfish +
_b[bass]*bass + _b[logacres]*logacres + _b[oklahomacity]
*oklahomacity
+ _b[pond]*pond + _b[allelse]*allelse) if waterbody!="Hefner"
by choice_id, sort: egen sumvstar = sum(exp_vstar)
```

and, finally, mean willingness to pay:

```
gen wtp = (ln(sumvstar) - ln(sumv))/(-1*_b[tcost])
sum wtp
```

The average value produced using this approach is almost the same as the value produced from the formula $\mathrm{WTP}_{\mathrm{Hefner}} = \ln(1 - P_{\mathrm{Hefner}})/\hat{\alpha}$, where P_{Hefner} is the sample share of trips to Lake Hefner, in Sect. 5.4.

As mentioned several times in this chapter, it is possible to combine revealed and stated preference data into a single model. For example, using our RUM models of Oklahoma City fishers, we could estimate the effect of adding bank trees in a scenario with revealed preference data by inserting the parameter on bank trees into the RUM model estimated from revealed preference data, after multiplying the parameter (i.e. 0.455) by 0.632 (i.e. 0.012/0.019), and collecting site attribute data on bank trees. In general, the parameters in different RUM models are not directly comparable due to underlying differences in scale (Louviere et al. 2000). However, if we assume that the travel cost coefficient identified in both datasets are identical up to the scale factor, which is probably a safe assumption in our application given that revealed and stated preference choice contexts are similar, we can then input the stated preference parameters into the RUM model estimated from revealed preference data after adjusting for the scale difference.

References

Abbott JK (2015) Fighting over a red herring: the role of economics in recreational-commercial allocation disputes. Mar Resour Econ 30(1):1–20

Abbott JK, Fenichel EP (2013) Anticipating adaptation: a mechanistic approach for linking policy and stock status to recreational angler behaviour. Can J Fish Aquat Sci 70(8):1190–1208

Abbott JK, Lloyd-Smith P, Willard D, Adamowicz W (2018) Status-quo management of marine recreational fisheries undermines angler welfare. Proc Natl Acad Sci USA 115(36):8948–8953

Abbott JK, Lew DK, Whitehead JC, Woodward RT (2022) The future of fishing for fun: the economics and sustainable management of recreational fisheries. Rev Environ Econ Policy 16(2):262–281

Baerenklau KA, Provencher B (2005) Static modeling of dynamic recreation behavior: implications for prediction and welfare estimation. J Environ Econ Manag 50(3):617–636

Banzhaf HS (2010) Consumer surplus with apology: a historical perspective on non-market valuation and recreation demand. Annu Rev Resour Econ 2:183–207

Boyer TA, Melstrom RT, Sanders LD (2017) Effects of climate variation and water levels on reservoir recreation. Lake and Reservoir Management, 33(3):223–233

Brown GM, Smith VK, Munro GR, Bishop R (2016) Early pioneers in natural resource economics. Annu Rev Resour Econ 8:25–42

Cha W, Melstrom RT (2018) Catch-and-release regulations and paddlefish angler preferences. J Environ Manag 214:1–8

Chorus CG, Arentze TA, Timmermans HJ (2008) A random regret-minimization model of travel choice. Transp Res B Methodol 42(1):1–18

Connelly NA, Brown TL, Knuth DA (2000) Assessing the relative importance of recall bias and nonresponse bias and adjusting for those biases in statewide angler surveys. Hum Dimens Wildl 5(4):19–29

Day B, Prades JLP (2010) Ordering anomalies in choice experiments. J Environ Econ Manag 59(3): 271–285

de Kerckhove DT, Tucker CM, Gutowsky L, Morgan G, Guzzo M, Barajas VL (2026) Optimizing creel surveys. In: Pope KL, Arlinghaus R, Hunt LM, Lynch AJ, van Poorten BT (eds) Understanding recreational fishers: disciplinary and interdisciplinary approaches for fisheries management. Springer Nature, Cham, p 387–409

Ekelund RB Jr, Hébert RF (1989) Consumer surplus: the first hundred years. In: Wood JC, Woods DN (eds) Sir John R. Hicks: critical assessments. Routledge, New York, pp 198–230

English E, von Haefen RH, Herriges J, Leggett C, Lupi F, McConnell K, Welsh M, Domanski A, Meade N (2018) Estimating the value of lost recreation days from the Deepwater Horizon oil spill. J Environ Econ Manag 91:26–45

Griffin RC (1998) The fundamental principles of cost-benefit analysis. Water Resour Res 34(8): 2063–2071

Gundelund C, Arlinghaus R, Baktoft H, Hyder K, Venturelli P, Skov C (2020) Insights into the users of a citizen science platform for collecting recreational fisheries data. Fish Res 229:10559

Haab TC, Hicks RL (1997) Accounting for choice set endogeneity in random utility models of recreation demand. J Environ Econ Manag 34(2):127–147

Haab TC, Hicks RL (1999) Choice set considerations in models of recreation demand: history and current state of the art. Mar Resour Econ 14(4):271–281

Haab TC, McConnell KE (2002) Valuing environmental and natural resources: the econometrics of non-market valuation. In Valuing environmental and natural resources. Edward Elgar Publishing

Haab T, Hicks R, Schnier K, Whitehead JC (2012) Angler heterogeneity and the species-specific demand for marine recreational fishing. Mar Resour Econ 27(3):229–251

Hanemann WM (2006) The economic conception of water. In: Rogers PP, Llamas MR, Cortina LM (eds) Water crisis: myth or reality. Taylor and Frances, London, pp 61–92

Hensher DA, Greene WH, Chorus CG (2013) Random regret minimization or random utility maximization: an exploratory analysis in the context of automobile fuel choice. J Adv Transp 47(7):667–678

Herriges JA, Kling CL, Phaneuf DJ (1999) Corner solution models of recreation demand: a comparison of competing frameworks. In: Herriges JA, Kling CL (eds) Valuing recreation and the environment: revealed preference methods in theory and practice. Edward Elgar, Cheltenham, pp 163–197

Horowitz JL, Nesheim L (2021) Using penalized likelihood to select parameters in a random coefficients multinomial logit model. J Econ 222(1):44–55

Hubacek K, van den Bergh JC (2006) Changing concepts of "land" in economic theory: from single to multi-disciplinary approaches. Ecol Econ 56(1):5–27

Hunt LM (2005) Recreational fishing site choice models: insights and future opportunities. Hum Dimens Wildl 10(3):153–172

Hunt LM, Boxall PC, Boots B (2007) Accommodating complex substitution patterns in a random utility model of recreational fishing. Mar Resour Econ 22(2):155–172

Hunt LM, Camp E, van Poorten B, Arlinghaus R (2019) Catch and non-catch-related determinants of where anglers fish: a review of three decades of site choice research in recreational fisheries. Rev Fish Sci Aquac 27(3):261–286

Hunt LM, Phaneuf DJ, Abbott JK, Fenichel EP, Rodgers JA, Buckley JD, Drake DAR, Johnson TB (2021) The influence of human population change and aquatic invasive species establishment on future recreational fishing activities to the Canadian portion of the Laurentian Great Lakes. Can J Fish Aquat Sci 78(3):232–244

Johnston RJ, Boyle KJ, Adamowicz W, Bennett J, Brouwer R, Cameron TA, Hanemann WM, Hanley N, Ryan M, Scarpa R, Tourangeau R, Vossler CA (2017) Contemporary guidance for stated preference studies. J Assoc Environ Resour Econ 4(2):319–405

Jungers B, Abbott JK, Lloyd-Smith P, Adamowicz W, Willard D (2023) "A la carte" management of recreational resources: evidence from the US Gulf of Mexico. Land Econ 99(2):161–181

Kaemingk MA, Chizinski CJ, Hurley KL, Pope KL (2018) Synchrony—an emergent property of recreational fisheries. J Appl Ecol 55(6):2986–2996

Knoche S, Lupi F (2016) Demand for fishery regulations: effects of angler heterogeneity and catch improvements on preferences for gear and harvest restrictions. Fish Res 181:163–171

Lee MY, Steinback S, Wallmo K (2017) Applying a bioeconomic model to recreational fisheries management: groundfish in the northeast United States. Mar Resour Econ 32(2):191–216

Lew DK, Whitehead JC (2020) Attribute non-attendance as an information processing strategy in stated preference choice experiments: origins, current practices, and future directions. Mar Resour Econ 35(3):285–317

Lloyd-Smith P, Abbott JK, Adamowicz W, Willard D (2019) Decoupling the value of leisure time from labor market returns in travel cost models. J Assoc Environ Resour Econ 6(2):215–242

Lloyd-Smith P, Abbott JK, Adamowicz W, Willard D (2020) Intertemporal substitution in travel cost models with seasonal time constraints. Land Econ 96(3):399–417

Louviere JJ, Hensher DA, Swait JD (2000) Stated choice methods: analysis and applications. Cambridge University Press, Cambridge

Lupi F, Phaneuf DJ, von Haefen RH (2020) Best practices for implementing recreation demand models. Rev Environ Econ Policy 14(2):302–323

Mackay M, van Putten EI, Richter A, Abatayo AL, Chuah S-H, Dankel DJ, Schadeberg A, Snyder H (2026) Behavioural economics for understanding recreational fishers and fisheries. In: Pope KL, Arlinghaus R, Hunt LM, Lynch AJ, van Poorten BT (eds) Understanding recreational fishers: disciplinary and interdisciplinary approaches for fisheries management. Springer Nature, Cham, p 261–284

Massey DM, Newbold SC, Gentner B (2006) Valuing water quality changes using a bioeconomic model of a coastal recreational fishery. J Environ Econ Manag 52(1):482–500

Melstrom RT, Kaefer M (2020) How diverse are preferences for urban fisheries? Evidence from a choice experiment. Hum Dimens Wildl 25(6):503–516

Melstrom RT, Lupi F (2013) Valuing recreational fishing in the Great Lakes. N Am J Fish Manag 33(6):1184–1193

Melstrom RT, Lupi F, Esselman PC, Stevenson RJ (2015) Valuing recreational fishing quality at rivers and streams. Water Resour Res 51(1):140–150

Melstrom RT, Jayasekera DH, Boyer TA, Jager C (2017) Scale heterogeneity in recreationists' decision making: evidence from a site choice model of sport fishing. J Outdoor Recreat Tour 18: 81–87

Morey ER, Rowe RD, Watson M (1993) A repeated nested-logit model of Atlantic salmon fishing. Am J Agric Econ 75:578–592
Parsons GR (2017) Travel cost models. In: Champ PA, Boyle KJ, Brown TC (eds) A primer on nonmarket valuation. Springer, Dordrecht, pp 187–233
Parsons GR, Plantinga AJ, Boyle KJ (2000) Narrow choice sets in a random utility model of recreation demand. Land Econ 76(1):86–99
Phaneuf DJ, Smith VK (2005) Recreation demand models. In: Maler KG, Vincent JR (eds) Handbook of environmental economics, 2nd edn. Elsevier, Amsterdam, pp 671–761
Phaneuf DJ, Kling CL, Herriges JA (2000) Estimation and welfare calculations in a generalized corner solution model with an application to recreation demand. Rev Econ Stat 82(1):83–92
Provencher B, Bishop RC (1997) An estimable dynamic model of recreation behavior with an application to Great Lakes angling. J Environ Econ Manag 33(2):107–127
Provost EJ, Butcher PA, Coleman MA, Kelaher BP (2020) Assessing the viability of small aerial drones to quantify recreational fishers. Fish Manag Ecol 27(6):615–621
Scheufele G, Bennett J (2013) Effects of alternative elicitation formats in discrete choice experiments. Aust J Agric Resour Econ 57(2):214–233
Stedman RC, Connelly NA, Heberlein TA, Decker DJ, Allred SB (2019) The end of the (research) world as we know it? Understanding and coping with declining response rates to mail surveys. Soc Nat Resour 32(10):1139–1154
Train KE (1998) Recreation demand models with taste differences over people. Land Econ 74(2): 230–239
Trice AH, Wood SE (1958) Measurement of recreation benefits. Land Econ 34(3):195–207
van Cranenburgh S, Guevara CA, Chorus CG (2015) New insights on random regret minimization models. Transp Res A Policy Pract 74:91–109
Van Loo EJ, Grebitus C, Nayga RM Jr, Verbeke W, Roosen J (2018) On the measurement of consumer preferences and food choice behaviour: the relation between visual attention and choices. Appl Econ Perspect Policy 40(4):538–562
Varian HR (2014) Big data: new tricks for econometrics. J Econ Perspect 28(2):3–28
Vaske JJ, Needham MD, Miller CA (2026a) Fundamentals of sampling and implementing human dimensions of fishing surveys. In: Pope KL, Arlinghaus R, Hunt LM, Lynch AJ, van Poorten BT (eds) Understanding recreational fishers: disciplinary and interdisciplinary approaches for fisheries management. Springer Nature, Cham, p 327–353
Vaske JJ, Needham MD, Miller CM, Sponarski C (2026b) Guidelines for writing and constructing recreational fisher surveys. In: Pope KL, Arlinghaus R, Hunt LM, Lynch AJ, van Poorten BT (eds) Understanding recreational fishers: disciplinary and interdisciplinary approaches for fisheries management. Springer Nature, Cham, p 287–325
Venturelli PA, Hyder K, Skov C (2017) Angler apps as a source of recreational fisheries data: opportunities, challenges and proposed standards. Fish Fish 18(3):578–595
Venturelli P, Skov C, Audzijonyte A, Belmaker J, Ana Cardoso AS, Gordon L, Gundelund C, Hyder K, Jarić I, Lennox RJ, Potts WM, Vacura K, Vaz AS, Sbragaglia V (2026) Digital data mining. In: Pope KL, Arlinghaus R, Hunt LM, Lynch AJ, van Poorten BT (eds) Understanding recreational fishers: disciplinary and interdisciplinary approaches for fisheries management. Springer Nature, Cham, p 443–479
Von Haefen RH, Phaneuf DJ (2008) Identifying demand parameters in the presence of unobservables: a combined revealed and stated preference approach. J Environ Econ Manag 56(1):19–32
Whitehead JC (1993) Total economic values for coastal and marine wildlife: specification, validity, and valuation issues. Mar Resour Econ 8(2):119–132
Whitehead JC, Lew DK (2020) Estimating recreation benefits through joint estimation of revealed and stated preference discrete choice data. Empir Econ 58(4):2009–2029

Social Psychology for Understanding Recreational Fishers and Fisheries

Adam C. Landon, Susan A. Schroeder, Susan F. Steffen, and Robert Arlinghaus

Abstract Social psychology is a diverse sub-discipline in psychology and sociology concerned with the effects of the social and physical environment on individuals' cognitions and behaviours. The human dimensions literature has a long history of adopting and extending concepts from social psychology to understand the cognitions and behaviours of recreational fishers, and the well-being outcomes of participation. In this chapter, we present a novel integrated conceptual model of the fishing experience from the perspective of social psychology. We use this model to review and synthesize the relevant literature on social psychology as applied to understand recreational fishers. Key elements of the model, and our review, include the cognitive antecedents of fishers' preferences for fishing experiences, the relation between preference, intention, and behaviour, and the post-behavioural well-being outcomes of fishing participation.

Keywords Cognition · Fisher behaviour · Preference · Social psychology · Well-being

A. C. Landon (✉)
Minnesota Department of Natural Resources, University of Minnesota, St. Paul, MN, USA

Department of Fisheries, Wildlife and Conservation Biology, University of Minnesota, St. Paul, MN, USA
e-mail: Adam.Landon@state.mn.us

S. A. Schroeder
University of Minnesota, Minneapolis, MN, USA

S. F. Steffen
Kansas Department of Wildlife and Parks, Emporia, KS, USA

R. Arlinghaus
Department of Fish Biology, Fisheries and Aquaculture, Leibniz Institute of Freshwater Ecology and Inland Fisheries, Berlin, Germany

Division of Integrative Fisheries Management, Humboldt-Universität zu Berlin, Berlin, Germany

K. L. Pope et al. (eds.), *Understanding Recreational Fishers*, Fish & Fisheries Series 45, https://doi.org/10.1007/978-3-031-99739-6_6

1 Introduction

Participation in recreational fishing involves choices about where to fish, when to fish, how to fish, what to fish for, and with whom, among many others (Hunt et al. 2019). These choices are conditioned on recreational fishers' preferences (e.g., Oh et al. 2005; Beardmore et al. 2013), the broader context in which the activity exists (e.g., Hunt and Ditton 1997; Hunt et al. 2013), and a plethora of other fishers' attributes (e.g., Aas and Ditton 1998; Arlinghaus and Mehner 2005). Diverse social science fields offer explanations of the processes that underpin the formation of fishers' preferences, the relationship between fishers and their context, their choices, and the outcomes that participation may generate (Aas and Ditton 1998). In this chapter, we review the social psychological approach for understanding recreational fisher behaviour and the implications of participation in recreational fisheries for well-being.

Social psychology is a sub-discipline within both psychology and sociology concerned with the influence of the social and physical environment on individuals' cognitions (i.e., values, beliefs, attitudes, and norms), emotions, and behaviours (Eagly and Chaiken 1993; Mitchell 2009). Like microeconomists and microsociologists, social psychologists are interested in understanding the collective action of individual people. Social psychologists focus on understanding how diverse cognitions (e.g., values, beliefs, attitudes, and norms) work together to affect a behavioural choice (Homer and Kahle 1988), and the relationship between those choices and their psychological underpinnings with the environment. Fishers exhibit great diversity in their fisheries-related cognitions (Fedler and Ditton 1986, 1994; Salz and Loomis 2005; Anderson et al. 2007) and behaviours (Bryan 1977; Sutton and Ditton 2001; Murphy Jr et al. 2019), and here the theories and methods of social psychology offer plausible accounts for that diversity (Hunt et al. 2023).

Many of the key human dimension's concepts of interest to recreational fisheries research and management fall under the umbrella of social psychology (Aas and Ditton 1998; Parkkila et al. 2010). This is, in part, because many of the pioneering scholars that studied recreational fisheries from a social perspective (e.g., Robert B. Ditton at Texas A&M, USA) had training in psychology or applied disciplines relying on such frameworks (e.g., outdoor recreation studies, Ditton 1996). It is also, in part, a function of the utility of social psychological concepts to management. For example, fishers' satisfaction with fishing experiences (Birdsong et al. 2021), motives related to recreational fishing (Fedler and Ditton 1994) and fishers' attitudes toward fisheries management (Arlinghaus and Mehner 2005) are of great interest to fishery managers, with all three concepts (satisfaction, motives, and attitudes) being psychological constructs (Heberlein 2012; Manning 2022). Research rooted in social psychology explains what contributes to fishers' satisfaction with fishing experiences (Hendee 1974; Dorfman 1979; Gundelund et al. 2022), and their beliefs, norms, and attitudes toward fisheries management (Manfredo et al. 1995; Schroeder et al. 2018), among numerous other critical topics (Aas and Ditton 1998).

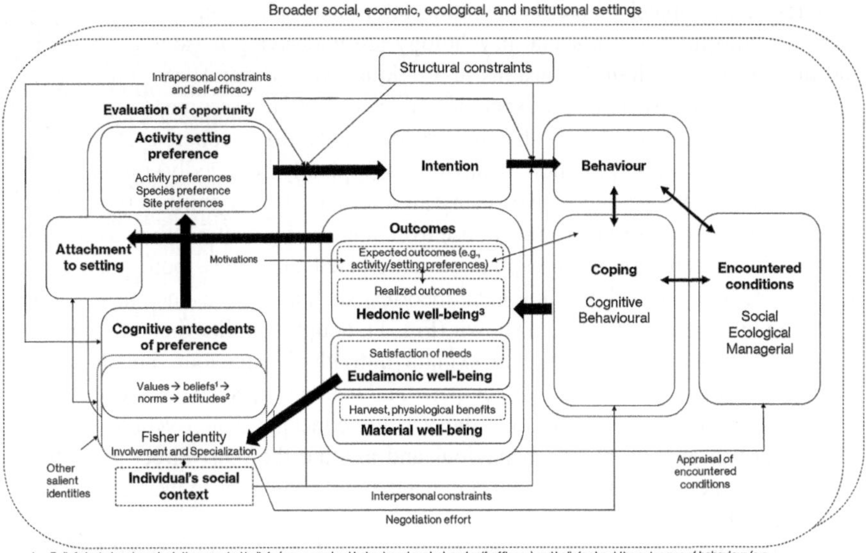

1. Beliefs include value orientations, control beliefs (e.g., perceived behavioural control, and self-efficacy) and beliefs about the outcomes of behaviour (e.g., motivations)
2. Attitudes include catch orientation
3. Hedonic well-being includes the satisfaction derived from the activity. Satisfaction is conceptualized to arise from the realization of expected outcomes of participation including those general to all outdoor recreation (e.g., time in nature, time with family), and those specific to fishing (e.g., catching fish, catching large fish).
4. Width of arrows reflects the relative strength of effect. Arrows pointing from one element to another imply direction of causality.

Fig. 1 Conceptual framework of the social-psychological dynamics of participation in fishing

Past studies examining recreational fisheries from the perspective of social psychology, often referred to as "human dimensions" (Hunt et al. 2013), have drawn on diverse research paradigms and concepts and focused on disparate elements of the fishing experience (Aas and Ditton 1998; Fenichel et al. 2013). This diversity of concepts might be confusing to those not formally trained in applied psychological sciences. To provide some structure and thereby help the social scientific study of recreational fishers and fisheries, we present a novel integrated conceptual framework of the fishing experience from the perspective of social psychology (Fig. 1). The model includes those factors that affect fishers' choices about fishing through to the benefits they derive from participation. The model outlines and organizes the key concepts in past studies from a social psychological research perspective (e.g., motives, beliefs, attitudes, preferences, constraints, satisfaction), shows their relationships, and accounts for the underlying mechanisms that explain them. We note that, like utility maximization as a guiding behavioural driver in microeconomics (see Melstrom et al. 2026; Parkkila et al. 2010), most theories underlying the social psychological concepts and their relations (Fig. 1) are fundamentally rational actor theories (Elster 1988, 1989). That is, a key assumption is that choices made by humans are conditioned on the desire to achieve satisfaction of preference and are rooted and influenced by a set of interrelated cognitions (e.g., values, beliefs, norms, attitudes) that individual people hold (Hausman 2011). For example, the cognitive hierarchy model (Homer and Kahle 1988) assumes that general values inform more specific beliefs, which in turn inform even more specific attitudes that ultimately direct behavioural intent and behaviour. Therefore, cognitions as conceptualized in social psychology are underlying reasons for behaviour and they form behavioural decisions in a rational way (Elster 1988). For example, when recreational fishers make a specific choice, they would assume to draw on their underlying values and beliefs that in turn rationally drive concrete attitudes and behaviours. In this interpretation, behaviour is (typically boundedly) rational in that fishers evaluate choice alternatives using an internal set of cognitions (e.g., values, beliefs, norms, and attitudes), and choose from among those alternatives available to them; fishers' behaviours correspond to anticipated outcomes, both positive and negative. Leading social psychological theories, such as the reasoned action model (Fishbein and Ajzen 2010), also shown (Fig. 1), adopt a similar theoretical mechanism as the cognitive hierarchy to explain individual behaviour, but this model emphasizes more factors that affect personal behaviour, such as the degree of behavioural control. We synthesize the various concepts that psychologists have put forward to explain human behaviour and assume the most direct predictor of behavioural intent is preference (Fig. 1), in a manner analogous to its use in resource economics (Hunt et al. 2019).

We structure our review of key concepts in social psychology around the model (Fig. 1), including:

(a) The cognitive antecedents of fishers' preferences (e.g., values, beliefs, norms, and attitudes), followed by
(b) The relationship between preference, intention, and behaviour
(c) Fishers' negotiation of conditions encountered in the field that affect decisions, and
(d) The outcomes of the fishing experience on the individual participant (e.g., benefits and satisfaction).

To understand the logic of the framework (Fig. 1), we first provide a brief review of three distinct, but related, theoretical frameworks applied to understand the cognitions and behaviours of recreational fishers and that are therefore reflected in our conceptual model:

1. The cognitive hierarchy
2. The norm activation and value-belief-norm models
3. The reasoned action model

We begin by reviewing these theories to demonstrate the similarities among the theories and the need for integration, and to situate our arguments in the broader literature.

2 Social Psychological Behavioural Theories and Constructs

2.1 The Cognitive Hierarchy

Implicit in many social psychological theories of the antecedents of behaviour is a hierarchy of cognitions. Drawing on this assumption, many applied human dimensions scholars in the field of fish and wildlife have adopted this position as an explicit organizing framework for their inquiry (e.g., Vaske and Donnelly 1999; Whittaker et al. 2006), often under a banner of "the cognitive hierarchy". Other theoretical frameworks are common in research on human-environment interactions (e.g., Klöckner 2013), including the norm activation model (Schwartz 1977), value-belief-norm model (Stern et al. 1999; Stern 2000), and the reasoned action model (Fishbein and Ajzen 2010). All these basic models of human thought and action are consistent with a commitment to the existence of a hierarchy of cognitions that affect intent and behaviour. We use the term "the cognitive hierarchy" to describe studies using the existence of a hierarchy of cognitions to justify their assumptions about the causes of behaviour, but note that "a cognitive hierarchy" is implied by the other theoretical models described.

As alluded to previously, the cognitive hierarchy suggests the existence of an ordered set of cognitions internal to the individual (Homer and Kahle 1988). These cognitions range from the extremely specific and direct antecedents of behaviour like attitudes or intentions, to the very abstract and acontextual like values. We provide a graphical depiction of the assumptions of the cognitive hierarchy (Fig. 2). Each of these cognitions has a specific meaning and function. As these constructs transcend theoretical models, and confusion exists regarding their meaning as applied across

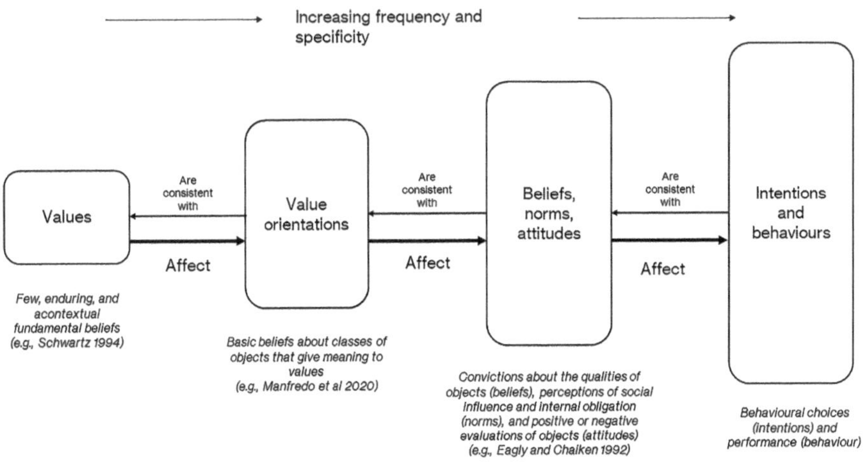

Fig. 2 Cognitive hierarchy theory. (Adapted from Homer and Kahle (1988), Fulton et al. (1996), and Milfont et al. (2010))

literatures, we pause to define them (values, value orientations, beliefs, norms, and attitudes) here, but will return to each in our subsequent description of the conceptual framework and their application to recreational fisher behaviour.

2.1.1 Values

At the foundation of the cognitive hierarchy are values (Fig. 2). Values are trans-situational beliefs about how to live one's life; they are not specific to any domain (e.g., fishing or the environment) and hold generally (Schwartz 1994). Values, in turn, shape individuals' evaluations of the more specific phenomena they encounter in their external environment. Values are few, strongly held, and formed early in life through socialization (Rokeach 1973) and are resistant to change (Manfredo et al. 2017).

Schwartz's refined values theory is perhaps the most comprehensive account of human values. This theory describes values as possessing six basic attributes. They (i) are beliefs tied to affect (e.g., emotion), (ii) refer to desirable goals, (iii) transcend specific actions and situations, (iv) serve as evaluative criteria, (v) are ordered by importance, and (vi) their relative importance drives actions (Schwartz 2012). Schwartz and colleagues hypothesize that there exists a conserved set of values (self-direction, stimulation, hedonism, achievement, power, security, conformity, tradition, benevolence, and universalism) from which broader scale patterns of ideology, preference, and culture derive. Variance exists in values between individuals within cultures, and in the dominant patterns found between cultures. Importantly, the relative priorities that individuals place on the values described by the theory influence their general approach to life and affect their behaviour, mediated by the other cognitions in the hierarchy.

2.1.2 Value Orientations

Unlike values that transcend objects and contexts, value orientations are beliefs about broad classes of objects and give meaning to values in a more specific context or domain, such as fisheries or wildlife (Fulton et al. 1996). Some also refer to value orientations as basic beliefs or worldviews, but all serve a similar psychological function. We separate value orientations from beliefs per se and suggest their primacy following application of the cognitive hierarchy, which has centred on value orientations as explanatory of higher-order cognitions like beliefs, norms, and attitudes (Whittaker et al. 2006). Examples of value orientations encountered in the fisheries and wildlife and conservation literatures include wildlife value orientations (Fulton et al. 1996; Manfredo et al. 2020), the new ecological paradigm as a measure of environmental concern (Dunlap et al. 2000), and ecological dominance orientation (Uenal et al. 2022). Value orientations operate like a lens that shapes how individuals see and respond to what they encounter in life as it pertains to certain issues, like one's relationship to the natural environment (Stern et al. 1995).

2.1.3 Beliefs

Beliefs take on many forms, but at their core are convictions that individuals hold to be true about the qualities of objects (Eagly and Chaiken 1993). For example, fishers can hold a functional belief that introducing fish through stocking can benefit and elevate fish stocks (Klefoth et al. 2023). Belief concepts applicable to behavioural prediction as operationalized in social psychological theories include beliefs about one's capacities to undertake behaviours (e.g., self-efficacy and perceived behavioural control), and beliefs about the outcomes of one's behaviour (e.g., behavioural beliefs). However, other beliefs play a prominent role in the broader social psychological literature (e.g., risk perceptions, Siegrist and Árvai 2020). Importantly, the term belief is used by some to describe many differentiated cognitions, including values and value orientations.

2.1.4 Norms

Like beliefs, norms are more specific situational constructs. Social psychologists differentiate between several types of norms, including social norms (descriptive and injunctive) (Cialdini et al. 1990) and personal norms (Schwartz 1977).

Social norms are shared standards of acceptable behaviour that imply an individual's obligation (Vaske et al. 2001; Vaske and Whittaker 2004; Manning 2007). Norms reflect what is considered "normal" in each context and are in part constructed by one's perception of the fit between the self and the context, including the expectations of others. Descriptive norms express an individuals' beliefs about what is "normal to do," or what is commonly done by one's peers, whereas injunctive norms represent what people "should do" (Cialdini et al. 1990; Schroeder et al. 2014). To further emphasize the influence of the individuals' perspective in evaluating standards of behaviour, some social psychological scholars have dubbed social norms "subjective norms" (e.g., Ajzen 1992).

Where social norms refer to shared expectations of behaviour in the social context, personal norms reflect the obligations that one ascribes to oneself to behave in a way consistent with their values (Schwartz 1977; Stern et al. 1999; Stern 2000). Thus, norms contain both elements of the self (personal norms) and the context (social norms) and imply constraint on preference and behaviour.

2.1.5 Attitudes

Though attitudes are commonly invoked in everyday language, the term has a more specific meaning in the academic literature. Social psychologists define an attitude as a tendency to evaluate an object with favour or disfavour (Eagly and Chaiken 1993). Attitudes are conceptualized as having a tripartite structure consisting of cognitive, affective (emotional), and conative (behavioural) dimensions. Typical attitude objects are people, situations, species, or management actions.

2.1.6 Intentions

Finally, intentions are the most proximal antecedent of behaviour in most social psychological theories. Intentions reflect a commitment to undertake a certain behaviour, at a certain time, and in a certain place. Some scholars (e.g., Fishbein and Ajzen 2010) consider intention analogous to a behavioural choice and suggest that an individual forms an intention to undertake a behaviour only after considering its alternatives.

Humans possess an innate need for consistency, and this need is thought to structure relationships among levels of cognitions (Homer and Kahle 1988). Confirmation through behaviour of deeply held cognitions like values results in positive affect (emotion), whereas disconfirmation results in negative affect. Thus, the anticipation of positive affect and avoidance of negative affect directs behaviour, and behaviour is rational. The strength of the hierarchy in directing behaviour depends on the strength of the individuals' commitment to the cognitions via identity and the context in which the behaviour occurs (Heberlein 2012). Formal behavioural theories in social psychology focus on behaviours with distinctive characteristics, and operationalize the constructs reviewed above in separate ways given their foci, but an implicit hypothesis of a hierarchy of cognitions from values to intention exists in most. One example of the application of the cognitive hierarchy in recreational fishing is the study of Riepe and Arlinghaus (2014) who reported that wildlife value orientation shaped specific beliefs about how human should treat animals and beliefs on whether fish and other animals have human-like traits, which in turn affect attitudes to recreational fisheries and the intention to act (e.g., signing a petition to ban fishing). Similarly, Bruskotter and Fulton (2008) explored fishers stewardship norms drawing on the cognitive hierarchy. These authors found that fishers' values orientations affected the appropriateness of technologies for fishing.

Next, we consider two additional theories, the norm activation model and associated value-belief-norm model, and the reasoned action model, and describe the cognitions hypothesized to affect behaviour therein, and their applications to recreational fisheries.

2.2 *Norm Activation Model and Value-Belief Norm Model*

Schwartz (1977) developed the norm activation model to describe the psychological antecedents of prosociality, or the donation of one's time, money, or physical being in the form of blood or organs. The norm activation model suggests that prosocial behaviours stem from an internalized obligation to act according to one's values. That obligation is known as a personal norm as described above. When a personal norm is activated in a context, it directs behaviour. Personal norms are activated when the individual is aware that their behaviours may impose negative consequences on objects they value, or their behaviours could alleviate negative consequences imposed by others, and that the individual ascribes some level of responsibility to themselves to act. Those later cognitions (ascription of responsibility and awareness of consequence) are beliefs.

Stern, Dietz, and colleagues (Stern et al. 1999; Stern 2000) extended this model as the value-belief-norm model to understand pro-environmental behaviours. Their extension includes one's environmental worldview (value orientation) and personal values as additional antecedents to an ascription of responsibility and awareness of consequence. Worldviews and values direct selective attention to stimuli pertinent to the self in context. The norm activation model and value-belief-norm model are consistent with a hierarchy of cognitions and suggest that one's internalized moral obligations in the form of personal norms affect behavioural choice.

There are several applications of the norm activation and value-belief-norm models in the recreational fisheries literature. Landon et al. (2018) drew on the norm activation model to explain the stewardship behaviours of fishers in Texas, USA. These authors extended the model to include one's identity as a fisher as an antecedent to personal norms and awareness of consequences. They found support for the norm activation model as an explanation for stewardship behaviours of fishers that were conducted outside the view of others, whereas identity directly predicted more public-facing stewardship activities. Riepe et al. (2021) found support for the value-belief-norm as a model of what compels European residents to engage in fish biodiversity conservation behaviours. Stensland et al. (2013) modelled fishers' intentions to release fish as a function of their personal norms and awareness of consequences derived from the norm activation model, among other constructs. These studies, and others, provide support for an effect of personal norms on the behavioural choice of fishers, and that personal norms are activated via a chain of hierarchical cognitions including beliefs, value orientations, and values.

2.3 *Reasoned Action Model*

The reasoned action model is perhaps the most applied theory of behaviour from the social psychological tradition in the environmental sciences and beyond (Cooke et al. 2009; St John et al. 2010), including especially the human dimensions literature in fish and wildlife (Miller 2017). Several iterations have existed since its first introduction in the late 1970s (e.g., Fishbein and Ajzen 1975—Theory of Reasoned Action; Ajzen 1992—Theory of Planned Behaviour). We focus on the most contemporary iteration in our discussion, known as the integrated behavioural model or "IM," but use the term reasoned action model to reflect the broader paradigm (e.g., Fishbein and Ajzen 2010). We provide a graphical depiction of the relationships among the constructs that comprise the IM (Fig. 3). The IM suggests that humans' volitional (meaning freely chosen, or not coerced) behaviours are predictable given an underlying set of cognitions including their intent or choice to undertake the behaviour (intention), their attitudes toward the behaviour (attitudes), their perceptions of the behaviours of peers and their expectations (perceived norms), and

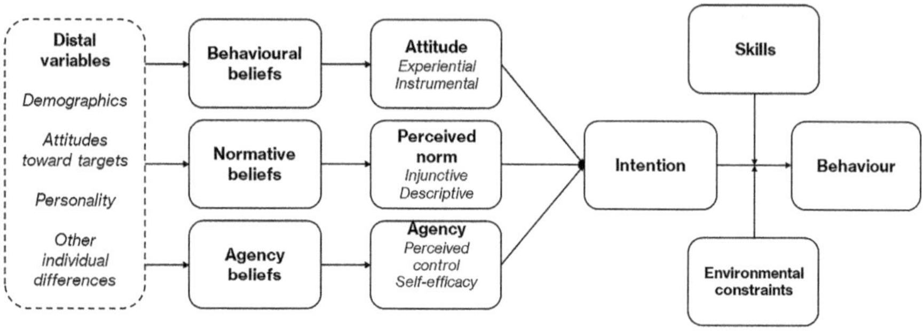

Fig. 3 Integrated model of behavioural prediction. (Adapted from Fishbein and Ajzen (2010))

perception of their ability to undertake the behaviour (self-efficacy). In turn, attitudes are a function of beliefs about the outcomes of the behaviour. Perceived norms, reflecting both a descriptive and injunctive component, are a function of beliefs about others' behaviour, compliance (or non-compliance), and one's motivation to comply. Self-efficacy, including both perceptions of one's control over their behaviour, referred to as perceived behavioural control (PBC), and capacities to undertake the behaviour, stems from associated beliefs. These cognitions are further conditioned on more abstract patterns of individual difference (e.g., personality, values, identities, and habits), sociodemographic variables, and exposure to persuasive communications not directly measured. Proponents of the model further hypothesize that one's skills and the environmental conditions of the context can moderate the relationship between intention and behaviour.

The reasoned action model is centred on attitude theory, specifically that humans' evaluations of, or attitudes about behaviours, depend on higher-order accessible beliefs about those behaviours (Ajzen 2012). The expectancy-value model provides a summation of an attitude based on the strength (unlikely-likely, probable-improbable) and evaluation (good-bad, wise-foolish) of beliefs about the attitude object or behaviour. This is similarly the case for one's normative and self-efficacy beliefs. Thus, the IM is consistent with our proposition that preference is a function of a hierarchy of cognitions and reflected in the behavioural choices of fishers (e.g., intentions), and that the satisfaction of preference is the primary motivator of behaviour conditioned on the context, including one's social relationships, and capacities.

The reasoned action model, predominantly in the form of the Theory of Planned Behaviour (TPB), has seen several applications in recreational fisheries. Crandall et al. (2018) employed the TPB to model how fisher's attitudes and subjective norms correlated with the use of venting tools and descending gear to mitigate barotrauma (i.e., major stressors on fish related to expansion of internal gases due to declining pressure during rapid ascent from depth) on released reef fish in Florida. Whereas attitudes and PBC influenced fishers' intentions to use barotrauma mitigation, subjective norms exerted the strongest influence, highlighting a common finding that norms are particularly behaviourally relevant (Arlinghaus et al. 2022). This, however, depends on the decision context. In a different study, attitudes and PBC—but not social norms—predicted intentions related to fish stocking among Swiss fishers (von Lindern and Mosler 2014). Actual fish stocking behaviour was predicted by both intentions and PBC, but also by the mental models about the functionality of stocking (von Lindern and Mosler 2014). Riepe et al. (2017) and Fujitani et al. (2020) also found that perceived functionality affected fish stocking decisions by managers, but psychological factors played a much smaller role in the context of decisions about stocking than about habitat management (Fujitani et al. 2020), indicating the context defines the salience of attitudes, norms, and efficacy in determining intention and behaviour.

The social psychological theories describe how an array of underlying cognitions affects fishers' evaluations of fishing opportunity. Those evaluations are manifest in an intention or choice to fish in a certain place, in a certain way, and at a certain time as a function of one's attitude toward that specific behaviour, perceived norms to engaging in that behaviour, and self-efficacy to do so. Attitude, norms, and self-efficacy are conditioned on underlying beliefs, which are in turn a function of more abstract cognitions, including value orientations and values. Each of these theories and their associated constructs featured in research explain the attitudes and behaviours of recreational fishers, as do other concepts like identity, involvement and specialization, place attachment, coping, constraints and negotiation, motivation, and satisfaction. This diversity of topics has led to a fractured literature using different names for the same concepts and adopting different philosophical positions on the causes and consequences of behaviour. We introduce our conceptual framework to attempt to bring a unifying thread to this literature. As we describe our conceptual framework, we continue to introduce the remainder of these concepts and their applications in recreational fisheries research.

3 Conceptual Framework for Understanding Recreational Fisheries from a Psychological Perspective

3.1 Cognitive Antecedents of Preference for Fishing Opportunity

Our framework is centred around a fisher's choice to participate in fishing in a certain place, in a certain way, and at a certain time. In making that choice, fishers must consider the many alternatives they have before them. Which species to target? What tackle to use? What lake or river to fish on and where? Fish at night or during the day? Bring the kids? Fish alone or with friends? Whether a fisher realizes it, they are considering the many alternatives before them when they make a choice to fish. We refer to this process as evaluation of fishing opportunity and suggest that it occurs via the formation of concrete preferences. Preferences refer to the order of importance individuals hold for a set of alternatives before them (Hausman 2005,

2011). For example, a harvest regulation type A might be preferred over a harvest regulation type B or night fishing over day fishing. As mentioned previously, participation in fishing involves numerous choices, and implied in those choices are trade-offs among alternatives with distinctive characteristics. One's inclination toward one alternative fishing opportunity over another, given its characteristics, reflects the immediate influences of preferences and the trade-offs fishers are willing to make before making a decision (Hunt et al. 2019). Elements of the choice context, including time, financial resources, and rules and regulations necessary to carry out a behaviour, limit the alternatives available to individuals despite their preference (Gintis 2007). Accounting for the constraints of the context, fishers exhibit patterns of choice corresponding to attributes of the experience and its setting (Fenichel et al. 2013; Hunt et al. 2019). Thus, on average, fishers make choices about fishing opportunities that satisfy their preferences (Beardmore et al. 2013), provided the options are available (and people are aware of them) and constraints can be negotiated (Stensland et al. 2016; TenHarmsel et al. 2021).

Fishers may exhibit preferences for various elements of the experience, but three categories capture the majority pertinent to understanding their behaviour: activity, species, and site preferences. Activity preferences refer to those preferences for the way one fishes (Sutton and Ditton 2005; Sutton and Oh 2015). At the most fundamental level, it is about choosing fishing versus doing something else with free time (Sutton and Oh 2015). Once a decision on fishing has been made, the process follows to think about the choice of a site, bait (e.g., live bait versus artificial lures), tackle (e.g., spinning versus fly), and within site location (e.g., from shore or a boat), among many others. Species preferences, as the name suggests, reflect preference for the target of the fishing activity. Site preferences include those preferences for the context of the fishing activity, including the number of other fishers (crowding), the expected catch rate at the location, the size of the fish at the location, fishing regulations at the location, and numbers of non-fishing users at the site, among many others. Scholars (e.g., Hunt 2005; Hunt et al. 2019) have examined the role of these attributes, and others, in fishers' choice of fishing location. Preferences for these attributes in many cases cannot be separated from one another, as, for instance, one's choice of tackle may stem from their preferred species, and site preference is conditioned on the presence of that species and quality of its stock. The combination of activity, species, and site preferences captures the trade-offs fishers make in evaluating fishing opportunity.

Yet, the question remains: why do fishers prefer one fishing opportunity over another? The question of where preferences come from remains a steadfast pursuit and source of debate in the literature on economics and social psychology (Sen 1977; Hausman 2011; Engelen 2017). Most scholars accept that preferences relate to individuals' cognitions, including awareness and habit, identities, values, beliefs (including self-efficacy), norms, and attitudes (Gintis 2007). Economists often refer to this diverse set of cognitions collectively as beliefs (Hausman 2011), whereas social psychologists differentiate carefully among them as we have previously described (Homer and Kahle 1988; Stern et al. 1999; Fishbein and Ajzen 2010; Vaske and Manfredo 2012). Where resource economics offers one leading theoretical framework of choice (utility theory), social psychology offers rich descriptions of the cognitive processes that underpin choice, and the consequences of choice once the behaviour is undertaken. Unlike resource economics, where a single theoretical account dominates most inquiry, social psychological scholarship is fragmented into several traditions despite some unifying assumptions (e.g., the reasoned action model, norm-activation model, the cognitive hierarchy, Klöckner 2013). The different cognitions in these frameworks are similar and related, and our framework attempts to synthesize them. We assume that fishers' preferences are a function of their multiple underlying cognitions, and that cognitions and preferences work together to determine behaviour within the confines of the social, ecological, and institutional context. In the sections to follow, we describe the set of cognitions that underpin fishers' preferences, beginning with the most encompassing, enduring, and abstract (identities and values), and ending with the most specific, dynamic, and fleeting (attitudes). In each section, we introduce the theoretical foundation of the construct and describe its application in recreational fisheries.

3.2 *Identity and the Social Context*

Fishers, regardless of where they may live, exist in highly differentiated social contexts. Examples of differentiation include gender, occupation, and race, among others. Social psychologists explain the causes and consequences of differentiation through the lens of social identity (Tajfel and Turner 1985; Lalonde 2002). Social identities are those internalized self-meanings that arise through social interaction and link to those roles that individuals occupy in society. Fishers may possess numerous identities (see next subsection on involvement and specialization), each with differing salience, prominence, and commitment in myriad contexts (Stryker and Burke 2000). Identities shape and are shaped by the sets of values, beliefs, norms, and attitudes that describe the dominant pattern exhibited by the archetype representative of that social group (Hogg et al. 2017). Thus, identities possess motivational properties and preferences (Bryan 1977) as individuals seek to think and behave in a manner consistent with expectations of other members of the group, or with meanings ascribed to the self (Stryker and Burke 2000). Identities also exist in a dialectic relationship with the meanings with which they associate and serve to reify between-group differences in exhibited patterns of preference and behaviour. "Recreational fisher," as a social role, is an example of an identity that has implications for the behaviours of individuals that have internalized it (Landon et al. 2018). Within the recreational fishing social world (Ditton et al. 1992), other identities exist, such as specialized and less specialized fishers (Bryan 1977, see next section). Fishers may possess more specific identities beyond general involvement and tied to styles of participation like "fly fisher," "ice fisher," or "saltwater fisher," among many others (Ditton et al. 1992), and each with possible unique constructions across social fields (e.g., the totality of their person-person relationships). Each identity has a bearing on the patterns of values, beliefs, norms, and attitudes internalized by individuals, their preferences, intentions, and behaviours (Stets and Biga 2003). That is to say that variance in recreational fishers' cognitions and behaviours stems from the consequences of differentiation and can in part account for differences among groups. These processes are depicted at the bottom of the left-hand side of figure (Fig. 1) as the "fisher identity." As noted in the figure, other salient identities can have a bearing on the evaluation of fishing opportunity. In a hypothetical example, one's identification as an environmentalist might influence one's choice of fishing location, given beliefs about the carbon costs of travel to alternative locations.

Research exploring concepts of identity in recreational fisheries falls under two predominant and related traditions: involvement (Havitz and Dimanche 1990; Kyle et al. 2007a) and specialization (Bryan 1977; Ditton et al. 1992; Scott and Shafer 2001; Hunt et al. 2023). It is suggested that these concepts and their associated measures reflect recreational fishers' "personalities" and deeply held cognitions about their relationship to fishing (Ditton et al. 1992; Scott and Shafer 2001; Koemle et al. 2024). Involvement in recreational fisheries is defined as an individual's "...unobservable state of motivation, arousal or interest..." in the activity (Havitz and Dimanche 1997, p. 246). Specialization as a concept was developed in trout recreational fishing in the USA and was described as a behaviour that moves from the "general to the particular," reflected in preferences, catch attitudes, and behaviours (Bryan 1977). Specialization has three dimensions—behavioural commitment, psychological commitment, and cognition or skill (Scott and Shafer 2001); it captures the diversity of behaviours exhibited toward fishing and their cognitive correlates. Like involvement, specialization is further defined as reflective of one's affective attachment to the activity (Ditton et al. 1992). Importantly, involvement and specialization do not proscribe fishers' means of engagement in fishing but relate to their intensity of engagement, or the importance people ascribe to it. People with varying involvement or specialization systematically vary in norms and attitudes to fishing, management preferences, and behaviours (e.g., Bryan 1977; Chipman and Helfrich 1988; Fisher 1997; Schroeder et al. 2008a, b; Johnston et al. 2010; Hunt et al. 2023; Koemle et al. 2024).

Hunt et al. (2023) argue that the concept of involvement is deeply grounded in attitudinal concepts and hence social psychology, whereas specialization has an origin in qualitative, inductive reasoning in sociology (see Bryan 1977). In measures of specialization, attitudes and behaviours are often mixed, which is conceptually problematic. Although specialization as a way to measure fisher identity and diversity has been around for more than 40 years, there is a large variation in how the concept, or sub-dimensions, have been measured in outdoor recreation studies, including fishing (Scott and Shafer 2001). This is of ongoing concern (Hunt et al. 2023). Though standard scales exist to measure involvement (and sub-dimensions such as attraction and identity, Kyle et al. 2007a), no such standard exists for specialization. The only recommendation is that specialization is a multidimensional construct with three sub-dimensions (behavioural commitment, psychological commitment, and skill). These sub-dimensions exert different relationships on attitudes and management preferences (Slaton et al. 2023). If researchers only measure one or two of the sub-dimensions, the term specialization should not be used, and measures should be kept separately. Involvement may also be used as a sub-concept of specialization, especially psychological commitment. Some researchers have dubbed the relationship between the self and the activity "ego-involvement" (Selin and Howard 1988); we will use that term in the following text to reflect both traditions but want to highlight that specialization is the prevalent term in recreational fishing literature. In short, both are conceptually a measure of "fisher identity," and empirical studies support that they relate to other concepts, such as beliefs, attitudes, preferences, and behaviour.

Measurement and operationalization of involvement and specialization differ, but some elements are consistent across the literature. Researchers have consistently looked to the centrality of fishing in one's life as an indicator of one's degree of ego-involvement, although centrality is only one component of involvement, which also has measures of identity or attraction (Kyle et al. 2007a). Centrality is, however, an often-used measure of the psychological relevance of fishing in the lifestyle of a fisher. Beardmore et al. (2013) found that centrality influenced fishers' general fishing preferences in Germany. Sutton (2006) provides evidence for centrality as an antecedent to participation in public fisheries consultation meetings by fishers. Van den Heuvel et al. (2022) suggest that centrality, mediated by place attachment, relates to fishers' fishing destination loyalty. Sutton and Ditton (2001) evaluated the effects of centrality on fishers' catch-and-release behaviours in the Atlantic bluefin tuna (*Thunnus thynnus*) fishery. These authors found the likelihood of releasing fish on a 1-day trip increased with increasing centrality. Koemle et al. (2024) found that ego-involvement, operationalized as centrality, influenced the decision rules used by fishers when choosing among alternative regulatory options for northern pike (*Esox lucius*) in Germany. These authors further suggest that one's degree of ego-involvement in the activity influences the availability of some choice alternatives, such that regret minimization rather than utility maximization captures their cognitive process in deciding among them. Schroeder et al. (2018) and Arlinghaus and Mehner (2005) found that measures of ego-involvement predicted fishers' attitudes toward habitat restoration and stocking. Slaton et al. (2023) compared different measures of specialization and found that centrality explained management attitudes related to habitats and harvest regulations, and behavioural commitment was a better predictor of attitudes to commercial fisheries.

Collectively, these studies suggest that the more ego-involved a fisher is in fishing, the more sensitive they are to fishing-related stimuli, the more crystallized their attitudes about fishing, and the more likely they are to engage in behaviours surrounding fishing. This includes acceptance of regulations or aversion to certain fishery management tools such as protected areas (Oh and Ditton 2006; Slaton et al. 2023). Involvement and related concepts such as commitment (Buchanan 1985) or specialization have implications for long-term participation in recreation (Havitz and Howard 1995) and the ability to negotiate constraints (Jun and Kyle 2011; Mueller et al. 2019). We will return to these points in later sections. Ego-involvement in fishing is an important aspect of the cognitions that influence preference, but in and of itself, it does not fully account for the diversity of preferences and behaviours exhibited by fishers. In addition to identity, concrete preferences are reflective of myriad other cognitions that can be correlated to ego-involvement or may also vary from culture to culture. One must therefore be critical of broad categorization and not conflate identity with specific attitudes and behaviours. For example, highly specialized fishers are often assumed to be less consumptive and engage more in catch-and-release (Bryan 1977), but the very same fishers may be highly consumptive in other cultures (Dorow et al. 2010). Hunt et al. (2023) propose that measures of specialization or involvement must be separated from the traits (e.g., consumptiveness) that fishers may express in a particular context. This position is further supported by social identity theory (Tajfel and Turner 1985) as discussed above.

3.3 *Cognitive Antecedents of Preference*

We suggest that the cognitive antecedents of preference exist in a hierarchy as we have described above, and that the relationships among the cognitions in that hierarchy are moderated by identity (Kyle and Landon 2023). Though it is beyond the scope of this review to describe every cognition pertinent to fishers' preferences, we endeavour to review the constructs that have featured prominently in the literature. The cognitive hierarchy has become an often-used template for understanding the relationships between cognitions, specifically in fisheries and wildlife (Fulton et al. 1996). However, cognitions at levels of the hierarchy do not fully mediate one another, and some conceptual overlap may exist between definitions of forms of cognitions, especially as they are used in applied fields like recreational fisheries. Here we discuss several forms of cognitions, including values, beliefs (including value orientations, control beliefs, and beliefs about the outcomes of fishing), norms, and attitudes (including place attachment) as they relate to fishers' preferences and behaviour. The cognitions that underpin preference are key attributes of the literature on the human dimensions of recreational fisheries from the social psychological perspective, specifically, beliefs and attitudes stand out as quantified in many studies.

3.3.1 Values

Few studies in recreational fisheries have examined basic values like those reflective of Schwartz's theory, as these values are causally distal from most outcomes of interest to management. However, values have been used to explain the chain of cognitive reasoning as applied to selected constructs (e.g., anti-fishing attitudes, Riepe and Arlinghaus 2014). Despite this paucity of research, it is the variation in basic personal values that manifests in higher-order evaluations of fisheries-relevant attitude objects, such as how fishers or the public position themselves in relation to certain management tools (Riepe et al. 2021), among other topics. For these reasons, we retain values as the foundation of the cognitive antecedents of preference (Fig. 1) but suggest that additional research on personal values could enhance understanding of fisher behaviour. This may be especially true for the diversity of fishers and their identities.

3.3.2 Beliefs

Given the ubiquity of the belief concept in both psychology and economics, numerous conceptualizations exist. Social psychologists, as noted previously, define beliefs as convictions that individuals hold to be true about the qualities of certain objects or domains (Eagly and Chaiken 1993). Importantly, beliefs are also implied in values as mentioned above; here we mean more specific beliefs related to a specific domain (e.g., wildlife or fishing environment). Here we examine three types of beliefs germane to understanding recreational fishers' preferences and behaviours: basic beliefs (i.e., value orientations or worldviews), control beliefs, and beliefs about the outcomes of behaviour.

Value orientations relevant to fisheries management include wildlife value orientations (Fulton et al. 1996) and the new ecological paradigm (Dunlap et al. 2000), but others exist. Wildlife value orientations are basic beliefs about the relationship between humans and wildlife (including fish). Manfredo and colleagues have argued that wildlife value orientations in western societies can be characterized along two broad dimensions, domination and mutualism, with a trend toward increasing mutualism in the twenty-first century in some western countries (Manfredo et al. 2017, 2020). Those with beliefs in line with the domination dimension believe wildlife exists for human use and that consumptive uses of wildlife are appropriate. Those with beliefs reflective of mutualism believe wildlife is a part of a broader community to which humans belong and deserve rights similar to humans (Teel and Manfredo 2010). These value orientations are relevant to recreational fisheries and wildlife management in that they are highly predictive of support or opposition to fishing and hunting, as well as management activities such as lethal control of species and stewardship of resources (Riepe and Arlinghaus 2014). An increasing proportion of Western societies holding a mutualism orientation may also result in declining participation in activities like hunting and fishing, and as such, it is a potentially important element of fishers' cognitions. Wildlife value orientations also structure what actions of fish and wildlife management agencies are deemed acceptable and unacceptable (Whittaker et al. 2006; Jacobs et al. 2014).

Research in recreational fisheries has drawn on the concept of value orientations to explain broad patterns of attitudes and behaviours among fishers (Bruskotter and Fulton 2008; Arlinghaus et al. 2012; Riepe and Arlinghaus 2014; Schroeder et al. 2018). For example, researchers grouped recreational fishers into three value orientations termed utilitarian, dominance, and protection, which they determined from analysing the pattern of belief statements of recreational fishers (Bruskotter and Fulton 2008). The utilitarian value orientation was comprised of beliefs about the appropriate uses of fisheries resources and their utility to human beings (e.g., fish are valuable only if people get to use them in some way). Individuals with utilitarian orientations believed that fisheries should be used in a way that maximizes their utility to people. In this study, Minnesota, USA fishers displayed low levels of utilitarianism. The dominance value orientation focused on the rights and importance of human beings versus those of non-human organisms (e.g., fish have as much right to exist as people [reverse coded]; humans have a right to change the natural world to suit their needs). Individuals who hold dominance orientations believe in human mastery over the natural world, and they hold that humans have the "right" to modify fisheries and other resources to accommodate human wants and needs. Dominance orientations were less prominent among Minnesota, USA fishers than utilitarian orientations. The protection value orientation dealt with prioritizing environmental protections over peoples' wants (e.g., protecting the environment is more important than providing fishing opportunities.) Minnesota, USA fishers displayed a strong protection orientation. Using the measurement approach from Bruskotter and Fulton (2008) to define relevant value orientations, Schroeder et al. (2018) examined trade-offs for funding fish stocking versus habitat management. They found that protection value orientations predicted decreased allocation of funds for fish stocking, and that stronger utilitarian values predicted increased allocation of funds for stocking. The intensity of value orientations among the constituency of a fisheries management institution can explain priorities for alternative approaches to management, like stocking versus ecological restoration. Understanding values and value orientations has utility in predicting beliefs, norms, attitudes, and behaviours.

Wildlife value orientation is just one type of value orientation. In the environmental psychological sciences, other constructs have been developed, such as the New Ecological Paradigm (Dunlap et al. 2000), to measure humans' beliefs about their relationship to the environment broadly and applied in the recreational fisheries context (Mackay et al. 2020). The compendium associated with chapter "Measure ment Instruments" by Kyle et al. (2026) summarizes key constructs, which have been applied in recreational fishing studies in the past (e.g., Riepe and Arlinghaus 2014; Riepe et al. 2021).

3.3.3 Control Beliefs

Fishers possess beliefs about their ability to undertake fisheries-related behaviours in certain places and at certain times. The reasoned action model, as reviewed previously, suggests that fishers' intentions and behaviours stem in part from their beliefs about their capacities to undertake those behaviours. Other theories, like Bandura's (1977) self-efficacy theory of motivation, suggest self-efficacy is the primary motivator of behaviour and arises from experience. Individuals form expectations about the outcomes of their behaviour by modelling observations of others and from their own past behaviours (Bandura 1997). Control beliefs play a role in shaping individuals' preferences and evaluations of recreation opportunity and the ability to participate (Smith et al. 2022). Thus, we suggest that control beliefs (self-efficacy and perceived behavioural control) are an important part of the cognitions that affect fishers' preferences. Fishers may, for instance, evaluate fishing opportunity based on their perceptions of efficacy. Hypothetically, one's beliefs about their abilities to paddle a river safely and effectively, or traverse open water in a boat, or climb to a river fishing site, could all influence preference for fishing sites, tackles, and species.

3.3.4 Beliefs About the Outcomes of Fishing

Fishers hold beliefs about specific objects in time and space, with implications for their evaluations of fishing opportunity. Fishers may also have beliefs about the benefits that fishing will bring to them and act upon that expectation. Outcome beliefs are also important in terms of beliefs that fishers' hold regarding the expected outcomes of fishing experience or of certain actions (e.g., belief that released fish survive; Stensland et al. 2013). At the most fundamental level, leisure researchers have dubbed outcome beliefs as recreation experience preferences or motives (Driver and Tocher 1970; Driver and Brown 1975; Driver and Cooksey 1980; Manfredo et al. 1996), and the concept has seen widespread use in research on recreational fisheries to help explain why fishers participate in fishing and what psychological benefits they seek (Fedler and Ditton 1994). Scholars and practitioners have conflated experience preferences with motivation per se and the recreational fisheries sector is not immune. Rooted in the benefits to leisure paradigm (Hendee 1974; Driver 1985), measures of recreation experience preference capture those elements of recreation experience that fishers find important, and thus seek to obtain from the experience. Under this conceptualization, the expectation of obtaining preferred benefits (e.g., achievement, self-determination, escape) motivates behaviour. Scholars have subsequently dubbed these beliefs motives or motivations. We contend that although recreation experience preferences reflect convictions that fishers hold to be true about the qualities of the experiences they endeavour to engage in (e.g., beliefs), and are therefore reasons for engagement (Hausman 2011), motivation stems from the pursuit of those goals via the satisfaction of preference. Therefore, recreation experience preferences are one set of the many cognitions that may shape fishers' preferences with implications for the evaluation of fishing opportunity. Importantly, however, beliefs about the expected outcomes of experiences serve a role in individuals' evaluations of the quality of those experiences post hoc but may not relate strongly to their behavioural choices (Schramm et al. 2003). We will elaborate on this point further in the section to follow and return to it in our discussion of satisfaction under hedonic well-being and the outcomes of behaviour. Despite our qualification regarding the conceptualization of recreation experience preferences in the existing literature, and to remain consistent with that literature, we will use the term motivations to reference the construct in the remainder of our discussion in this section.

As early as the 1960s, researchers explored why people participate in fishing and found multiple motivations for their trips (Bultena and Taves 1961). In the 1970s, Driver and colleagues began building a conceptual framework for studying motivations and benefits associated with outdoor recreation (Driver and Tocher 1970; Driver and Brown 1975; Driver 1976; Driver and Brown 1978), following a multiple benefits approach to recreation (Hendee 1974; Driver 1985). This research was instrumental in measuring (Manfredo et al. 1996) and identifying the motivations and outcomes that encourage fishing participation very generally (Knopf et al. 1973; Bryan 1974; Driver and Knopf 1976; Fedler and Ditton 1994; Ditton 2004; Beardmore et al. 2011).

Fedler and Ditton (1994) summarized motivations for fishing from US studies available at the time to include: (i) general psychological and physiological, (ii) natural environment, (iii) social, (iv) fisheries resource, and (v) skill and equipment. It is worth noting that the evolving application of the recreation experience preference measures to recreational fisheries (e.g., Fedler and Ditton 1994; Beardmore et al. 2011) deviates from the original intent and item lists of its authors, specifically by the addition of catch-related motives. However, in the original conceptualization by Driver and colleagues, which was general to outdoor recreation, catching fish was not present in any of the items used to measure motives. This is because Driver and colleagues' original work emphasized fundamental benefits of leisure that transcend outdoor activities. The addition of fisheries-specific measures ("to catch fish") occurred later, specifically in recreational fishing studies (Fedler and Ditton 1994) and without the theoretical foundation of the original work. Catching fish is perhaps not an instrumental motive per se, but it serves to contribute to other expected benefits that are more generally salient (e.g., achievement). Being aware of this conceptual departure, item lists to measure fishing motives now commonly include so-called activity specific (related to catch or consumption of fish) and activity general measures that are unrelated to catch (Fisher 1997; Arlinghaus 2006; Beardmore et al. 2011; see compendium associated with chapter "Measure ment Instruments" for more detail on measurement). However, the relevance of catch can perhaps be better assessed not by a motivational, but by an attitudinal concept, which is catch orientation (Anderson et al. 2007, see below for details).

Motivations, or expected benefits, for recreational fishing differ by fishers engaged in different settings such as stream, bank, lake, freshwater, saltwater, wild vs. put and take (Fedler and Ditton 1994; Manning 2022), demographic background such as gender, race (Schroeder et al. 2006; Toth Jr and Brown 1997; Witter et al. 1982), and other contextual and situational elements of the activity including duration, number, and characteristics of other trip participants, and regulations governing the harvest at the site for that species and season. Motivations also vary by fisher "personality" or identity as elaborated above, such as level of specialization and involvement (Ditton et al. 1992; Beardmore et al. 2011). Similarly, the orientation to catch varies strongly with different settings, fisher types, and identities (Dorow et al. 2010).

Findings that show catching fish is not an important motive for fishing should not be misinterpreted to mean that catch is unimportant to fishers, quite the opposite is true (Arlinghaus 2006; Beardmore et al. 2011). Motivations should not be confused with satisfactions because they hold more predictive power to explain general behaviours (e.g., which type of fishing to engage in or whether to fish) than specific behaviours or attitudes (Arlinghaus 2006). Many early studies of recreational fishers' motivations emphasized activity-general recreation motivations (e.g., enjoying nature, spending time with family and friends) over activity-specific motivations related to catching fish or the challenge of catching fish (Fedler and Ditton 1994). When motives are assessed at a general level, it is a recurring finding that activity-general motives associated with recreation are generally more important to fishers than specific motives related to catch or consumption (Ditton 2004). This must not be misinterpreted as if catch is not important because activity-general components of fishing are more easily satisfied than catch expectations (Birdsong et al. 2021). The relevance of catch to fishers is not well revealed by classical motivation research as commonly operationalized in the literature (e.g., the item list by Fedler and Ditton (1994)). Instead, assessing attitudes to catch, known as consumptive or catch orientation (Anderson et al. 2007), has been found to better represent the importance of catch to fishers and better explain behaviour (Schramm et al. 2003). This is an attitudinal concept, not an outcome belief. We will return to this important concept under the section on attitudes. However, beliefs about the outcomes of behaviour as operationalized by the recreation experience preference scales may shape fishers' general preferences for fishing opportunity.

3.3.5 Norms

The norm activation and value-belief-norm models and the reasoned action model suggest there are normative influences on preference and behaviour. Studies reviewed previously suggest that these constructs (i.e., personal norms, injunctive and descriptive social norms) play a role in fishers' behavioural choices. We refer to these in figure (Fig. 1) collectively as norms. Norms link preference to the social context (Kimbrough and Vostroknutov 2016) and to individuals' deeply held convictions like values (Schwartz 1977) and limit the availability of some behavioural alternatives (Hausman 2011). Norms, therefore, comprise an element of fishers' cognitions pertinent to their preferences for fishing opportunity. Norms also make choices boundedly rational, as the mechanisms that link norms and behaviour include the minimization of negative affect (Schwartz 1977). Thus, without norms, fishers may form different concrete preferences and make different choices (Mackay et al. 2020). As reviewed previously, research from Riepe et al. (2017) and Landon et al. (2018) highlights the importance of norms in the cognitions and behaviours of managers and fishers, respectively. Stensland et al. (2013) provide further evidence for the role of norms in shaping fisher behaviour. These authors describe the perceived social pressures that fishers feel to participate in catch-and-release behaviour. Thus, norms are an important antecedent to fishers' preferences.

3.3.6 Attitudes

Recreational fishers' attitudes are frequently studied, both as independent (predictor of some other construct) or as dependent variables (aspects to be explained), and examples are numerous (e.g., Chipman and Helfrich 1988; Arlinghaus and Mehner 2005; Slaton et al. 2023). For instance, a manager might wish to stock fish in a waterbody they manage and, before doing so, would like to know the level of support or opposition to stocking those fish among a population of fishers that use the resource (Arlinghaus et al. 2014). Fishers' support or opposition for stocking is an example of an attitude; a positive or negative evaluation (support, opposition) of an object (stocking). Those same fishers might possess numerous beliefs (e.g., stocking usually works, or stocking is a harm to local genetic strains) about stocking that inform and co-affect their attitude. A belief that stocked fish are inferior to naturally produced fish might lead a fisher to oppose stocking, whereas a belief that the stock in that waterbody is low and that stocking helps might lead a fisher to support stocking (von Lindern and Mosler 2014). Other recreational fisheries studies have gauged attitudes and beliefs about many different regulations, including creel and size limits, protected seasons, catch-and-release regulations, and catch-reporting programs (e.g., Hicks et al. 1983; Midway et al. 2020; van den Heuvel et al. 2020; Slaton et al. 2023). Attitudes are critical outcome variables in policy research on recreational fisheries. Many regulatory schemes require consulting populations of stakeholders to assess their preferences or attitudes before implementing change in accordance with best governance practice (Decker et al. 2016). Critically, fishers' beliefs about fisheries resources, and their attitudes toward elements of those resources are situated in a hierarchy of cognitions, and may influence behaviours such as voting behaviours, illegal releases of fishes, or in voicing of complaints to new policies (Hunt et al. 2010; von Lindern and Mosler 2014; Arlinghaus et al. 2022). Although attitudes can be considered an intermediate construct that explains behaviours of fishers, in practical applications attitude is perhaps the most often studied concept when used as the dependent variable (i.e., as an end in itself) by helping to understand the variation of attitudes to salient topics (such as management measures) in populations of fishers. We note that attitudes are, however, also a predictor of behaviour, but that behaviour is affected by many more aspects, including values, beliefs, and norms. Nevertheless, attitudes by themselves are highly relevant concepts to understand from a management perspective, precisely because they tie to concepts of acceptability and because they are antecedents of behaviour (Heberlein 2012).

3.3.7 Catch Orientation as an Attitude

Fishers have a tendency to evaluate certain classes of attitude objects positively or negatively as a function of the underlying constructs previously reviewed. Fishers' evaluation of the catch and retention aspects of fishing, known as consumptive orientation or catch orientation, is an important and commonly studied attitude in recreational fisheries research (Anderson et al. 2007; Aas and Vitterso 2000). Whereas some have conceptualized the importance of catch as a catch motivation in the recreation experience preference tradition as discussed previously, we agree with Anderson et al. (2007) that catch orientation is an attitudinal construct, and a quite specific one. By contrast, motives are general tendencies to act in a given outdoor situation or when viewed as expected benefits, general expected psychological benefits (i.e., beliefs).

Graefe (1980) was the first to introduce a scale designed to measure fisher catch orientation along four subdimensions of catch something, catching numbers, catching trophy fish, and keeping (or releasing) fish (Anderson et al. 2007; Hutt and

Jackson 2008; Schroeder and Fulton 2013). A growing body of evidence supports the validity and reliability of these dimensions of catch orientation, but item wording and item lists have changed over time (Aas and Vitterso 2000; Sutton and Ditton 2001; Anderson et al. 2007; Kyle et al. 2007b, 2026). Studies have demonstrated the importance of catch-related attitudes to fishers' satisfaction (Graefe and Fedler 1986; Fedler and Ditton 1994; Finn and Loomis 2001; Beardmore et al. 2011) catch and release versus consumption behaviour (Sutton and Ditton 2001; Sutton 2003), and in fishers' preferences for management (Arlinghaus and Mehner 2005; Carlin et al. 2012).

Catch orientation is not only a relevant concept as a predictor of other variables (e.g., support for management, Slaton et al. 2023), but has utility as a tool of market segmentation. Kyle et al. (2007b) used catch orientation to group fishers at a South Carolina recreational sport fishery into market segments. They found that fishers who were least concerned with catching fish expressed the strongest social motivations and place attachment to the lake system being studied, whereas fishers who were most concerned with catching large fish were least motivated by non-angling factors (i.e., escape, social) and had comparatively low attachment to the lake system. The fishers who were least concerned with catching and consuming fish scored low on measures of motivation but expressed the strongest satisfaction and likelihood of returning to the fishery (Kyle et al. 2007b). Using samples of walleye (*Sander vitreus),* northern pike, largemouth bass (*Micropterus salmoides*), and smallmouth bass (*Micropterus dolomieu*) fishers, Schroeder and Fulton (2013) examined how catch orientation related to management preferences for these species. Results suggested that the orientation to keep fish related to reduced support for harvest slot limits or restrictive creel limits, whereas an orientation to catch big fish predicted increased support for slot limits or more restrictive creel limits for certain species. They emphasized that examining how catch orientation relates to management preference in specific fisher groups could enable managers to provide desirable fishing opportunities that satisfy and retain fishers. By knowing the correlates of managerial concerns (e.g., attitudes to regulations) with catch orientation, applicants can do rapid assessments of the variation and types of catch orientations in a local fishery using just an abbreviated item list and use this information to "foresee" how the population of fishers thinks and feels about tangible issues such as dropping catch rates or proposed restrictions on harvest, among others. Although the same value might lie in an original motivation assessment using items summarized by Manfredo et al. (1996) and more specifically Fedler and Ditton (1994), we believe the catch orientation construct may hold greater promise for recreational fishing studies than the original experience preference scales (Manfredo et al. 1996). This is particularly so if the investigator wants to predict responses to specific actions, which specific constructs (such as catch orientation) are better at explaining than general constructs (such as motives). The original experience preference scale, instead, is perhaps best suited to compare different recreationists and their expected benefits rather than guiding management responses in fishing specifically. Clearly, more research on fishing motives using alternative conceptualizations is necessary.

3.3.8 Attachment to Setting (Place) as an Attitudinal Construct

Through sustained interaction, humans tend to develop connections to the physical environments around them. Scholars refer to this connection as place attachment. Where involvement and specialization reflect ego-involvement in fishing, place attachment reflects ego-involvement in its setting, which is often conceptualized as an evaluation of place (i.e., an attitude to place). Several conceptualizations of place attachment exist in the literature (Hidalgo and Hernandez 2001). We limit our review to the dominant tradition used in quantitative studies of outdoor recreation, including fishing (Kyle et al. 2005). In this tradition, place attachment is defined as the cognitive and affective bonds between people and the physical environment, especially as applied to settings of outdoor recreation like fishing. Scholars conceptualize place attachment as an attitude possessing cognitive, emotional, and behavioural dimensions (Jorgensen and Stedman 2000). Kyle et al. (2005) conceptualize aspects of one's attachment to place using four dimensions, including place identity, affective attachment, place dependence, and social bonding. Place identity refers to one's incorporation of the physical setting into their sense of self (Prohansky 1978). Affective attachment includes those emotions elicited by the setting. Place dependence refers to one's reliance on the setting to provide for recreational opportunities, and social bonding refers to a setting's ability to afford interactions with people important in one's life.

Few studies have investigated the relationship between place attachment and fishers' preferences and behaviours, but place attachment has the potential to influence their evaluations of fishing opportunity as the activity occurs in settings with which fishers have varying degrees of connection. van den Heuvel and Rönnbäck (2023) found that fishing tourists exhibit higher attachment to fishing tourism sites and have higher loyalty to those destinations. Koemle et al. (2024) did not find evidence for a relationship between place attachment and fishers' preferences for northern pike management in Germany in a choice model. Similarly, Hunt (2008) found a tenuous link between place attachment and fishers' site choice behaviour. Slaton et al. (2023) determined that some measures of place attachment affected attitudes to spatial and other types of management in a sample of German pike fishers. More research is needed to fully explicate the relevance of place to fishers' evaluation of fishing opportunity as the construct has been shown to have effects on the cognitions and behaviours of recreationists in other contexts. For instance, previous research based on Sherif and Hovland's (1961) social judgment theory concluded the more strongly an individual is attached to a setting, the more sensitive they are to conditions they encounter in that setting, including the density of other recreationists (Kyle and Landon 2021) and other environmental conditions (Kyle et al. 2004b). This sensitivity also relates to preferences for management (Kyle et al. 2004a, b), and propensity to undertake behaviours to protect that place (Devine-Wright 2009; Halpenny 2010). We will return to these points in the discussion of fishers' negotiation of conditions encountered in the field.

In summary, research on the social psychology of recreational fisheries suggests that fishers' evaluations of opportunity are a function of their preferences, especially species, site, and activity preferences. These more specific preferences emerge from myriad underlying cognitive antecedents that exist in a hierarchy from specific to general levels that contain few abstract, yet enduring values and many more concrete and dynamic beliefs (including motives), norms, and attitudes (including catch orientation and place attachment). Those cognitions shape and are shaped by social identities embedded in a broader context of one's relationships to other humans. Recreational fishers' degree of ego-involvement in fishing, a psychological measure of fisher personality and engagement with fishing, and the setting of the activity have implications for their fisheries-related cognitions and evaluations of fishing opportunity. These processes are depicted on the left-hand side of figure (Fig. 1).

3.4 The Relationship Between Preference, Intention, and Behaviour, Including Constraints

Preferences, as we defined previously, are total subjective evaluations of the available alternatives (Hausman 2011). Fishers' preferences for fishing opportunity are a function of the cognitions reviewed above, among others that may be relevant to the individual. From among those behavioural alternatives available to fishers, they will choose ones near the top of their preference hierarchy. Their behavioural choices are boundedly rational as they choose alternatives that satisfy their preferences. We contend that the process of behavioural choice is analogous to the development of an intention to fish in a certain place, at a certain time, and in a certain manner. This process is depicted at the top of figure (Fig. 1) where preference → intention.

Following the development of an intention, fishers must follow through on their behaviour, and several factors can impede their intention. Thus, the intention → behaviour relationship is not a given. First, intentions vary in strength, in part conditioned on fishers' ego-involvement. The stronger one's commitment to the identity, the greater their motivation to perform behaviours that verify that identity (Burke 1991). This has implications for the negotiation of constraints to participation, and we will return to this in a section to follow. As we have alluded, a fishers' context has bearing on both their cognitions and behaviours. The hierarchical model of leisure constraints (Crawford et al. 1991) is one conceptualization of the role of the context in the processes that underpin fishers' preferences, and on the relationship between intention and behaviour. Though research to date has not conceptualized leisure constraints in a model of this nature, we believe the implications are clear and relate to the proposition within the IM that environmental conditions moderate the intention-behaviour relationship. Given the role of self-efficacy in determining intention and behaviour, there are also clear links between elements of constraints on cognitions.

Research has classified leisure constraints as: intrapersonal, interpersonal, and structural (Crawford and Godbey 1987; Crawford et al. 1991). Intrapersonal constraints involve the individual's internal psychological processes that affect preferences toward activities (e.g., perception of skill or appropriateness of the activity). Interpersonal constraints relate to other people (e.g., not having a social group with whom to participate, or their approval). Structural constraints involve elements of the context of the activity itself. We differentiate these dimensions of constraints as having different effects on the psycho-behavioural process of fishing participation. Given that constraints are beliefs about one's capacities, the social environment, and the context, they have the potential to effect one's evaluations of fishing opportunity via preference. Constraints may manifest in fishers' attitudes toward, perceived norms about, and perceptions of efficacy related to undertaking a particular fishing behaviour, at a particular place, and at a particular time, and ultimately one's intent to do so. Constraints may also influence the process of engagement once a choice has been made to do so. This is especially the case for structural constraints relating to the context of behaviour.

Several studies have examined constraints on recreational fishing participation (Ritter et al. 1992; Aas 1995; Fedler and Ditton 2001; Schroeder et al. 2008a, b; Freudenberg and Arlinghaus 2009). Consistent with other research, Aas (1995) found that interested fishers reported higher levels of constraints to fishing compared to less-interested fishers and non-fishers, and that different constraints inhibit participants versus nonparticipants (Sutton 2007). Sutton (2007) found that a large majority of Queensland, Australia fishers reported constraints to fishing participation. Schroeder et al. (2008a, b) found that people of colour faced unique constraints to their fishing participation, including selective law enforcement, discriminatory behaviour, safety concerns related to lack of swimming ability, and language barriers. Across studies, structural factors appear to be the constraints fishers perceived most strongly to inhibit fishing participation (Ritter et al. 1992; Aas 1995; Fedler and Ditton 2001; Sutton 2007; Schroeder et al. 2008a, b; Freudenberg and Arlinghaus 2009).

3.5 Coping and Conditions Encountered in the Field

Once fishers have made choices about where, when, how, and with whom to fish, and attempt to do so, they are faced with the realities of the conditions they encounter in the field. These conditions may or may not correspond to their expectations for the experience and elicit cognitive and behavioural reactions with potential implications for well-being. These processes are depicted on the right-hand side of figure (Fig. 1) and include fishers' appraisal of the social, environmental, and managerial conditions, and the cognitive and behavioural coping strategies employed to maintain the well-being fishers derive from the experience.

Lazarus and Folkman's (1984) stress coping model has been applied widely in the leisure sciences to explain the coping mechanisms of recreationists (Schneider and

Hammitt 1995), including fishers (Schroeder and Fulton 2010). The model suggests that the coping process begins with an appraisal of conditions in the field. One's propensity to interpret conditions as a threat to preferred outcomes is a function of their underlying values, beliefs, norms, and attitudes, and ego-involvement in the setting and activity (Kyle et al. 2004a, b). This process is depicted along the bottom of figure (Fig. 1). If fishers appraise setting conditions negatively, they can employ strategies to maintain participation and enjoyment.

Miller and McCool (2003) describe seven forms of coping employed by recreationists under two broad categories: cognitive and behavioural. Behavioural coping, as the name suggests, refers to changes in behaviour made by recreationists in the face of negatively appraised encountered conditions. Examples of behavioural coping include changing the location or time that one fishes (temporal substitution), changing the waterbody one fishes (resource substitution), choosing to fish in a different way (activity substitution), choosing not to fish at all (absolute displacement), or seeking out authorities to make changes to the negatively appraised condition (direct action). There are two main forms of cognitive coping. The first is known as product shift, and references recreationists' change in their expectations from the activity. For example, a fisher encountering more other fishers in the field than they expected, may reevaluate their expectations from the experience. An unsuccessful fisher may also downrate expected catch rates in the future (Birdsong et al. 2021, 2022). This cognitive coping process allows the participant to maintain enjoyment and satisfies needs for competence following one's choice to partake in the behaviour. The second type of cognitive coping is known as rationalization and references the participant's reappraisal of the encountered condition. A fisher may, for instance, rationalize encountering a higher-than-expected number of other fishers by viewing it as a challenge, or by blaming an authority for failing to manage fisher numbers, among others. Both forms of cognitive coping allow the fisher to navigate stress, maintain competence, and continue participation.

3.6 *Outcomes of the Fishing Experience*

Ultimately, fishing will result in a flow of benefits to participants. Social psychologists, among others, refer to the benefits obtained from life experience as well-being and may use proxies for their measurement, such as satisfaction (Manning 2022). Following (Ryff 1989) and Ryan and Deci (2001), we consider two predominant forms of well-being experienced by fishers: hedonia (hedonic well-being) and eudaimonia (eudaimonic well-being). We also note that fishing generates flows of material benefits, including fish retained for consumption and physiological changes to the body that occur through nature exposure (Kaplan 1995; Ulrich 1983), but because our focus is on the (rational) social psychology of fishing experience, we do not discuss them further. These benefits are depicted in the centre of figure (Fig. 1), between behaviour and associated coping of conditions encountered in the field and the cognitive antecedents of preference. The benefits obtained from participation are partly dependent on and may affect the cognitive antecedents of preference.

3.6.1 Hedonic Well-Being

Hedonic well-being, in relation to recreational fisheries, includes those positive emotional outcomes of participation. Thus, hedonic well-being is synonymous with economic notions of utility (Kahneman et al. 1999). Among the concepts most strongly associated with hedonic well-being is satisfaction. Satisfaction is a measure of the psychological quality of fishing experiences, and cultivating satisfactory experiences is a major stated goal for managers (Birdsong et al. 2021). For example, the Kansas Department of Wildlife and Parks reports fisher satisfaction garnered through creel surveys to state legislators as a performance measure of its fisheries division (S. Steffen, personal communication).

It is hypothesized that hedonic well-being, in the form of satisfaction, arises through the confirmation of expectations for recreational experiences, which are a function of the expected psychological benefits in relation to the actual fishing outcomes (Arlinghaus 2006). Factors that may affect fisher satisfaction include all key aspects that are expected benefits of fishing, including physical (e.g., location, site characteristics and amenities), social (e.g., social group, crowding), managerial (e.g., management objectives, regulations) and ecological attributes (e.g., species richness and diversity, forage and prey dynamics, turbidity). Yet, abundant research underscores the clear importance of catch, access, crowding, and cost to fisher satisfaction (Arlinghaus 2006; Birdsong et al. 2021, 2022; Gundelund et al. 2022). Clearly, how experiences are judged depends on the fisher type (Beardmore et al. 2015). For example, centrality-to-lifestyle, which reflects one's meaningful bond to fishing, influences satisfaction (Beardmore et al. 2015; van den Heuvel et al. 2022). Consistent with theoretical expectations about the relationship between motivations and satisfaction, catch provides greater benefit to fishers with stronger activity-specific motivations compared to those with more activity-general motivations (Gundelund et al. 2022). Satisfaction ultimately derives from expectations in light of experiences, which evolve from motivations, fisher type (e.g., level of centrality), context, and management actions (Gale 1987; Spencer and Spangler 1992; Arlinghaus 2004; van Poorten et al. 2011; Birdsong et al. 2021, 2022; Gundelund et al. 2022). Satisfactory experiences occur when expectations are fulfilled or exceeded (Tang 2014). Fishers' satisfaction with the number of fish caught predicts management preferences, in particular stocking (Arlinghaus and Mehner 2005), so efforts to encourage reasonable fisher expectations and maximize opportunities to catch fish may increase satisfaction (Schroeder et al. 2018).

Whereas research demonstrates a relationship between catch and key non-catch issues (e.g., access and cost) and satisfaction, the relationship is not explicitly linear and varies according to elements of the catch. Birdsong et al. (2022) found that German recreational fishers exhibited diminishing marginal utility for increasing numbers of fish caught, but that the fish size—utility relationship was exponential and positive. Stated another way, fishers derived less utility for each additional fish they caught, but the larger the fish they caught, the more they were satisfied. Managing for fisher satisfaction is thus a difficult prospect, but these findings have some important implications. It is possible, for instance, to identify catch rates that meet a satisfaction threshold beyond which there is diminishing marginal utility for fishers. Managing for larger fish also has the potential to generate positive welfare outcomes, given the positive and exponential relationship described by Birdsong et al. (2022) and Beardmore et al. (2015). However, ecologically driven trade-offs in management strategies may exist for catch rates and age and size structure of the fishery (Ahrens et al. 2020), and especially given the ecological context. These findings also require confirmation in different social-ecological contexts, but the approach offers promise in how fisheries managers can operationalize satisfaction as an outcome of fisheries management and relate it to strategies.

These propositions are depicted under the box labelled hedonic well-being in the centre of figure (Fig. 1). Fishers possess beliefs about the expected outcomes of the fishing experience as a function of their values, beliefs, norms, and involvement in fishing. These cognitions shape fishers' evaluations of fishing opportunity via preferences and past expectations. Once a fisher has followed through on their intention to participate in a fishing opportunity, and they have negotiated the conditions in the field, they re-appraise the actual experience given their expectations. The confirmation of expectations yields positive affect, often measured as satisfaction (but other measures could apply), the disconfirmation of expectations thwarts the experience of positive affect, often measured as dissatisfaction.

3.6.2 Eudaimonic Well-Being

Well-being in humans is multifaceted, and not all facets of well-being are adequately captured by the positive affective state characteristic of hedonia (Ryff 1989; Ryan and Deci 2001). Eudaimonic well-being refers to those elements of experience that support individuals' growth, development, and realization of their true self (Ryan and Deci 2001; Huta and Ryan 2010). Eudaimonic well-being has implications for understanding motivation and engagement, particularly in recreation (which does not involve meeting essential physiological needs), and the reasons for participation in fisheries beyond expected utility as operationalized in contemporary treatments of satisfaction (Smith et al. 2024). One of the predominant frameworks for the operationalization of aspects of Eudaimonic well-being is that of self-determination theory (Ryan and Deci 2001).

Self-determination theory suggests that humans have an innate tendency toward growth and development (Ryan and Deci 2000). Scholarship in self-determination theory further suggests that it is in the satisfaction of basic psychological needs for autonomy, competence, and relatedness that humans experience higher forms of motivation, including intrinsic motivation, that support growth, development, and optimal psychological function (Ryan 1995). Recreation is a particularly relevant playing field to sensing such "non-essential," yet highly relevant outcomes. In fact, contemporary definitions of leisure include the expression of autonomy as a constituent part (Walker et al. 2019). Contexts that support the satisfaction of "higher order" needs are more likely to foster higher forms of motivation. Research in leisure contexts further suggests that experiencing the satisfaction of psychological needs in a setting or while participating in a leisure activity can foster attachment to that setting (Landon et al. 2020), or identification as an activity participant (Smith et al. 2023). These observations have implications for the cognitive antecedents of preference as we have depicted (Fig. 1). Whereas fishing may generate emotional happiness and other rewards (e.g., fish for dinner) for participants, it may also support their development as agentic human beings and give meaning to their life including in how they define and see themselves (Smith et al. 2023). In fact, we position that these self-reflective outcomes may be of higher importance and better explain why fishing is so attractive and results in people investing tremendously more time and money than any catch would generate in the commercial market if sold. The reason must lie in the self-determination properties of fishing, a concept economists measure under net willingness-to-pay, i.e., benefits beyond those that cost money to fishers (e.g., licenses or costs).

Eudaimonic well-being is an emerging area of scholarship in recreational fisheries, but it is not without precedent. For instance, many measures of experience preferences used in recreational fisheries studies reflect aspects of basic psychological needs, including spending time with family and friends (relatedness) and experiencing challenge and achievement (competence). Importantly, these elements of the experience are often rated highly by participants, but do not correlate strongly with satisfaction as mentioned previously (Arlinghaus 2006). More work is needed to understand how fishing participation may support eudaimonic well-being, the relationship between hedonic and eudaimonic well-being, and the dynamic relationship between experience and fishers' psychological involvement with the activity. Given the importance of involvement to the evaluation of fishing opportunity, appraisal of environmental conditions, and expenditure of negotiation effort, further research on the factors that support the development of involvement is warranted.

4 A Summary View of Our Integrated Framework

In this chapter, we reviewed the social psychological approach to understanding fisher behaviour. We presented a new integrated conceptual framework of the fishing experience. In this model, we suggested that fishers' evaluations of fishing opportunity reflect their preferences, and that those preferences are a function of their underlying cognitions, including identification and involvement as a fisher, and their values, beliefs, norms, and attitudes toward the activity and the setting where it occurs. Fishers develop intentions to undertake specific behaviours following their attitudes toward the outcomes of those behaviours, perceptions of norms surrounding those behaviours, and self-efficacy. Intention is the most proximate antecedent of behaviour. Fishers face constraints to participation even after they have decided to participate, and those constraints can be negotiated using coping strategies, and as a function of their ego-involvement in fishing. Participation in fishing can yield hedonic, eudaimonic, and material well-being. Hedonic well-being in the form of satisfaction depends on the relationship between fishers' beliefs about the expected benefits of fishing and their experience, whereas eudaimonic well-being stems from the satisfaction of basic psychological needs. Fishers' experiences can dynamically affect their cognitions. This model presents numerous assumptions about fishers and the factors that affect fisher behaviour, not all of which are shared across the social sciences, or even within social psychology. However, the model we propose tries to unite a separate literature stream by proposing how key concepts of social psychology are related and how they could be studied together. We hope our framework helps to clarify some of the existing confusion and proposes a united way forward to the next decade of social-psychological research to understand recreational fishing thought and action regarding fisheries resources.

Acknowledgments We thank the anonymous reviewers and Len Hunt (editor) for their feedback on drafts of this manuscript.

References

Aas Ø (1995) Constraints on sportfishing and effect of management actions to increase participation rates in fishing. N Am J Fish Manag 15(3):631–638

Aas Ø, Ditton RB (1998) Human dimensions perspective on recreational fisheries management: implications for Europe. In: Hickley P, Tompkins H (eds) Recreational fisheries: social, economic, and management aspects. Fishing New Books, pp 153–164

Aas Ø, Vitterso J (2000) Re-examining the consumptiveness concept: some suggestions from a confirmatory factor analysis. Hum Dimens Wildl 5:1–18

Ahrens RN, Allen MS, Walters C, Arlinghaus R (2020) Saving large fish through harvest slots outperforms the classical minimum-length limit when the aim is to achieve mulitple harvest and catch-related fisheries objectives. Fish Fish 21:483–510

Ajzen I (1992) The theory of planned behavior. Organ Behav Hum Decis Process 50:179–211

Ajzen I (2012) Martin Fishbein's legacy: the reasoned action approach. Ann Am Acad Polit Soc Sci 640:11–27

Anderson DK, Ditton RB, Hunt KM (2007) Measuring angler attitudes toward catch-related aspects of fishing. Hum Dimens Wildl 12:181–191

Arlinghaus R (2004) A human dimensions approach towards sustainable recreational fisheries management. Turnshare, London

Arlinghaus R (2006) On the apparently striking disconnect between motivation and satisfaction in recreational fishing: the case of catch orientation of German anglers. N Am J Fish Manag 26: 592–605

Arlinghaus R, Mehner T (2005) Determinants of management preferences of recreational anglers in Germany: habitat management versus fish stocking. Limnologica 35:2–17

Arlinghaus R, Schwab A, Riepe C, Teel T (2012) A primer on anti-angling philosophy and its relevance for recreational fisheries in urbanized societies. Fisheries 37(4):153–164

Arlinghaus R, Beardmore B, Riepe C, Meyerhoff J, Pagel T (2014) Species-specific preferences of German recreational anglers for freshwater fishing experiences, with emphasis on the intrinsic utilities of fish stocking and wild fishes. J Fish Biol 85(6):1843–1867

Arlinghaus R, Riepe C, Theis S, Pagel T, Fujitani M (2022) Dysfunctional information feedbacks cause the emergence of management panaceas in social-ecological systems: the case of fish stocking in inland recreational fisheries. J Outdoor Recreat Tour 38:100475

Bandura A (1977) Toward a unifying theory of behavioral change. Psychol Rev 84(2):191–215

Bandura A (1997) Self-efficacy: the exercise of control. W.H Freeman/Times Books/Henry Holt & Co, New York

Beardmore B, Haider W, Hunt LM, Arlinghaus R (2011) The importance of trip context for determining primary angler motivations: are more specialized anglers more catch-oriented than previously believed? N Am J Fish Manag 31(5):861–879

Beardmore B, Haider W, Hunt LM, Arlinghaus R (2013) Evaluating the ability of specialization indicators to explain fishing preferences. Leis Sci 35(3):273–292

Beardmore B, Hunt LM, Haider W, Dorow M, Arlinghaus R (2015) Effectively managing anglers satisfaction in recreational fisheries requires understanding the fish species and the anglers. Can J Fish Aquat Sci 72(4):500–513

Birdsong M, Hunt LM, Arlinghaus R (2021) Recreational angler satisfaction: what drives it? Fish Fish 22(4):682–706

Birdsong M, Hunt LM, Beardmore B, Dorow M, Pagel T, Arlinghaus R (2022) Does the relevance of catch for angler satisfaction vary with social-ecological context? A study involving angler cultures from West and East Germany. Fish Res 254:106414

Bruskotter JT, Fulton DC (2008) Minnesota anglers' fisheries-related value orientations and their stewardship of fish resources. Hum Dimens Wildl 13:207–221

Bryan RC (1974) The dimensions of a saltwater sport fishing trip; or what do people look for in a fishing trip besides fish? PAC/T-74-1. Fish Mar Serv, So Operations Branch, Pacific Reg, Vancouver

Bryan H (1977) Leisure value systems and recreational specialization: the case of trout fishermen. J Leis Res 9:174–187

Buchanan T (1985) Commitment and leisure behavior: a theoretical perspective. Leis Sci 7(4): 401–420

Bultena GL, Taves MJ (1961) Changing wilderness images and forestry policy. J For 59:167–171

Burke P (1991) Identity processes and social stress. Am Sociol Rev 56(6):836–849

Carlin C, Schroeder SA, Fulton DC (2012) Site choice among Minnesota walleye anglers: the influences of resource conditions, regulations and catch orientation on lake preference. N Am J Fish Manag 32(2):299–312

Chipman BD, Helfrich LA (1988) Recreation specialization of Virginia river anglers. N Am J Fish Manag 8(4):390–398

Cialdini RB, Reno RR, Kallgren CA (1990) A focus theory of normative conduct: recycling the concept of norms to reduce littering in public places. J Pers Soc Psychol 58:1015–1026

Cooke IR, Queenborough SA, Mattison EH, Bailey AP, Sandars DL, Graves AR, Morris J, Atkinson PW, Trawick P, Freckleton RP, Watkinson AR, Sutherland WJ (2009) Integrating socio-economics and ecology: a taxonomy of quantitative methods and a review of their use in agro-ecology. J Appl Ecol 46(2):269–277

Crandall CA, Garlock TM, Lorenzen K (2018) Understanding resource-conserving behaviors among fishers: barotrauma mitigation and the power of subjective norms in Florida's reef fisheries. N Am J Fish Manag 38:271–280

Crawford DW, Godbey G (1987) Reconceptualizing barriers to family leisure. Leis Sci 9:119–127

Crawford DW, Jackson EL, Godbey G (1991) A hierarchical model of leisure constraints. Leis Sci 13:309–320

Decker D, Smith C, Forstchen A, Hare D, Pomeranz E, Doyle-Capitman C, Schuler K, Organ J (2016) Governance principles for the 21st century. Conserv Lett 9(4):290–295

Devine-Wright P (2009) Rethinking NIMBYism: the role of place attachment and place identity in explaining place-protective action. J Community Appl Soc Psychol 19(6):426–441

Ditton RB (1996) Human dimensions in fisheries. In: Ewert A (ed) Natural resource management: the human dimension. Westview Press, Boulder, pp 73–90

Ditton RB (2004) Human dimensions of fisheries. In: Manfredo MJ, Vaske JJ, Bruyere BL, Field DR, Brown PJ (eds) Society and natural resources: a summary of knowledge. Modern Litho, Missouri, pp 199–208

Ditton RB, Loomis DK, Choi S (1992) Recreation specialization: reconceptualization from a social worlds perspective. J Leis Res 24:35–51

Dorfman PW (1979) Measurement and meaning of recreation satisfaction: a case study in camping. Environ Behav 11(4):483–510

Dorow M, Beardmore B, Haider W, Arlinghaus R (2010) Winners and losers of conservation policies for Europen eel, *Angulla anguilla* an economic welfare analysis for differently specialized eel anglers. Fish Manag Ecol 17:106–125

Driver BL (1976) Quantification of outdoor recreationists' preferences. In: Research, camping and environmental education. Penn State series II. Penn State University, Pennsylvania

Driver BL (1985) Specifying what is produced by management of wildlife by public agencies. Leis Sci 7(3):281–295

Driver BL, Brown PJ (1975) A sociopsychological definition of recreation demand, with implications for recreation resource planning. In: Assessing demand for outdoor recreation. National Academy of Sciences, Washington, DC

Driver BL, Brown PJ (1978) The opportunity spectrum concept and behavioral information in outdoor recreation resource supply inventories: a rationale. In: Integrated inventories of renewable natural resources: proceedings of the workshop, USDA Forest Service Gen. Tech. Rep. RM-55. Rocky Mountain Forest and Range Experiment Station, Fort Collins

Driver BL, Cooksey RW (1980) Preferred psychological outcomes of recreational fishing. In: Barnhart RA, Rodlofs TD (eds) Catch and release fishing as a management tool: a national sport fishing symposium. Humboldt State University, Arcata, pp 27–40

Driver BL, Knopf RC (1976) Temporary escape: one product of sport fisheries management. Fisheries 1(2):21–29

Driver BL, Tocher SR (1970) Toward a behavioral interpretation of recreation engagements with implications for planning. In: Driver BL (ed) Elements of outdoor recreation planning. University Microfilms Michigan, Michigan

Dunlap RE, Van Liere KD, Mertig AG, Jones RE (2000) New trends in measuring environmental attitudes: measuring endorsement of the new ecological paradigm: a revised NEP scale. J Soc Issues 56(3):425–442

Eagly AH, Chaiken S (1993) The psychology of attitudes. Wadsworth, Belmont

Elster J (1988) The nature and scope of rational-choice explanation. In: Ulmann-Margalit E (ed) Science in reflection, the Israel colloquium: studies in history, philosophy, and sociology of science, vol 3. Kluwer Academic Publishers

Elster J (1989) Social norms and economic theory. J Econ Perspect 3(4):99–117

Engelen B (2017) A new definition of and role for preferences in positive economics. J Econ Methodol 24(3):254–273

Fedler AJ, Ditton RB (1986) A framework for understanding the consumptive orientation of recreational fishermen. Environ Manag 10:221–227

Fedler AJ, Ditton RB (1994) Understanding angler motivations in fisheries management. Fisheries 19(4):6–13

Fedler AJ, Ditton RB (2001) Dropping out and dropping in: a study of factors for changing recreational fishing participation. N Am J Fish Manag 21:283–292

Fenichel EP, Abbott JA, Huang B (2013) Modeling angler behavior as part of the management system: synthesizing a multi-disciplinary literature. Fish Fish 14:137–157

Finn KL, Loomis DK (2001) The importance of catch motives to recreational anglers: the effects of catch satiation and deprivation. Hum Dimens Wildl 6:173–187

Fishbein M, Ajzen I (1975) Beliefs, attitude, intention, and behavior: an introduction to theory and research. Addison-Wesley, Reading

Fishbein M, Ajzen I (2010) Predicting and changing behavior: the reasoned action approach. Psychology Press, New York

Fisher MR (1997) Segmentation of the angler population by catch preference, participation, and experience: a management-orientated application of recreation specialization. N Am J Fish Manag 17(1):1–10

Freudenberg P, Arlinghaus R (2009) Benefits and constraints of outdoor recreation for people with physical disabilities: inferences from recreational fisheries. Leis Sci 32(1):55–71

Fujitani ML, Riepe C, Pagel T, Buoro M, Santoul F, Lassus R, Cucherousset J, Arlinghaus R (2020) Ecological and social constraints are key for voluntary investments into renewable natural resources. Glob Environ Chang 63:102125

Fulton DC, Manfredo MA, Lipscomb J (1996) Wildlife value orientations: a conceptual and measurement approach. Hum Dimens Wildl 1(2):24–47

Gale RP (1987) Resource miracles and rising expectations: a challenge to fishery managers. Fisheries 12:8–13

Gintis H (2007) A framework for the unification of the behavioral sciences. Behav Brain Sci 30:1–61

Graefe AR (1980) The relationship between level of participation and selected aspects of specialization in recreational fisherman. Dissertation, Texas A&M University, College Station, TX

Graefe AR, Fedler AJ (1986) Situational and subjective determinants of satisfaction in marine recreational angling. Leis Sci 8:275–295

Gundelund C, Arlinghaus R, Birdsong M, Flávio H, Skov C (2022) Investigating angler satisfaction: the relevance of catch, motives and contextual conditions. Fish Res 250:106294

Halpenny EA (2010) Pro-environmental behaviours and park visitors: the effect of place attachment. J Environ Psychol 30:409–421

Hausman DM (2005) Sympathy, commitment, and preference. Econ Philos 21(1):33–50

Hausman DM (2011) Preferences, value, choice and welfare. Cambridge University Press, Cambridge

Havitz ME, Dimanche F (1990) Propositions for guiding the empirical testing of the involvement construct in recreational and tourist contexts. Leis Sci 12(2):179–196

Havitz ME, Dimanche F (1997) Leisure involvement revisited: conceptual conundrums and measurement advances. J Leis Res 29(3):245–278

Havitz ME, Howard DR (1995) How enduring is involvement? A seasonal examination of three recreational activities. J Consum Psychol 4(3):255–276

Heberlein TA (2012) Navigating environmental attitudes. Oxford University Press, New York

Hendee JC (1974) A multiple-satisfaction approach to game management. Wildl Soc B 2(3): 104–113

Hicks CE, Belusz LC, Witter DJ, Haverland PS (1983) Application of angler attitudes and motives to management strategies at Missouri's trout parks. Fisheries 8(5):2–7

Hidalgo HC, Hernandez B (2001) Place attachment: conceptual and empirical questions. J Environ Psychol 21:273–281

Hogg MA, Abrams D, Brewer MB (2017) Social identity: the role of self in groups processes and intergroup relations. Group Process Intergroup Relat 20(5):570–581

Homer PM, Kahle LR (1988) A structural equation test of the value-attitude-behavior hierarchy. J Pers Soc Psychol 5:638–646

Hunt LM (2005) Recreational fishing site choice models: insights and future opportunities. Hum Dimens Wildl 10:153–172

Hunt LM (2008) Examining state dependence and place attachment within a recreational fishing site choice model. J Leis Res 40(1):110–127

Hunt KM, Ditton RB (1997) The social context of site selection for freshwater fishing. N Am J Fish Manag 17(2):331–338

Hunt LM, Gonder D, Haider W (2010) Hearing voices from the silent majority: a comparison of preferred fish stocking outcomes for Lake Huron by anglers from representative and convenience samples. Hum Dimens Wildl 15:27–44

Hunt LM, Sutton SG, Arlinghaus R (2013) Illustrating the critical role of human dimensions research for understanding and managing recreational fisheries within a social-ecological systems framework. Fish Manag Ecol 20:111–124

Hunt LM, Camp E, van Poorten B, Arlinghaus R (2019) Catch and non-catch related determinants of where anglers fish: a review of three decades of site choice research in recreational fisheries. Rev Fish Sci Aquac 27(3):261–286

Hunt LM, Arlinghaus R, Scott D, Kyle G (2023) Diversity of anglers: drivers and implications for fisheries management. In: Neal JW, Lang TJ, Krogman RM, Kurzawski K, Hunt KM, Taylor J (eds) Angler recruitment, retention, and reactivation: the future of fisheries and aquatic conservation. American Fisheries Society, Bethesda

Huta V, Ryan RM (2010) Pursuing pleasure or virtue: the differential and overlapping well-being benefits of hedonic and eudaimonic motives. J Happiness Stud 11(6):735–762

Hutt CP, Jackson JR (2008) Implications of angler motivations and preferences for urban fisheries management. In: Eades RT, Neal JW, Lang TJ, Hunt KM, Pajak P (eds) Urban and community fisheries programs: development, management, and evaluation, American Fisheries Society symposium, vol 67. American Fisheries Society, Bethesda, pp 63–76

Jacobs MH, Vaske JJ, Sijtsma MTJ (2014) Predictive potential of wildlife value orientations for acceptability of management interventions. J Nat Conserv 22(4):377–383

Johnston FD, Arlinghaus R, Dieckmann U (2010) Diversity and complexity of angler behavior drive socially optimal input and output regulations in a bioeconomic recreational-fisheries model. Can J Fish Aquat Sci 67(9):1507–1531

Jorgensen BS, Stedman RC (2000) Sense of place as an attitude: lakeshore owners' attitudes toward their properties. J Environ Psychol 21(3):233–248

Jun J, Kyle GT (2011) Gender identity, leisure identity, and leisure participation. J Leis Res 44(3): 353–378

Kahneman D, Diener E, Schwartz N (1999) Well-being: the foundations of hedonic psychology. Russell Sage, New York

Kaplan S (1995) The restorative benefits of nature: toward and integrative framework. J Environ Psychol 15(3):169–182

Kimbrough EO, Vostroknutov A (2016) Norms make preferences social. J Eur Econ Assoc 14(3): 608–638

Klefoth T, Wegener N, Meyerhoff J, Arlinghaus R (2023) Do anglers and managers think similarly about stocking, habitat management and harvest regulations? Implications for the management of community governed-recreational fisheries. Fish Res 260:106598

Klöckner CA (2013) A comprehensive model of the psychology of environmental behaviour – a meta-analysis. Glob Environ Chang 23(5):1028–1038

Knopf RC, Driver BL, Bassett JR (1973) Motivations for fishing. Trans N Am Wildl Nat Resour Conf 38:191–204

Koemle D, Gassler B, Kyle G, Meyerhoff J, Arlinghaus R (2024) How involvement drives rules behind stated preferences for recreational-fisheries management. J Environ Manag 349:119604

Kyle G, Landon A (2021) Shifting setting densities and normative evaluations of recreation experiences over time. Landsc Urban Plan 208:104043

Kyle G, Landon A (2023) Modernity and the western value-action paradox: contributions from conservation psychology. Front Conserv Sci 4:1092259

Kyle G, Graefe A, Manning R, Bacon J (2004a) Effect of activity involvement and place attachment on recreationists' perceptions of setting density. J Leis Res 36(2):209–231

Kyle G, Graefe A, Manning R, Bacon J (2004b) Effects of place attachment on users' perceptions of social and environmental conditions in a natural setting. J Environ Psychol 24:213–225

Kyle G, Graefe A, Manning R (2005) Testing the dimensionality of place attachement in recreational settings. Environ Behav 37(2):153–177

Kyle G, Absher J, Norman W, Hammitt W, Jodice L (2007a) A modified involvement scale. Leis Stud 26(4):399–427

Kyle G, Norman W, Jodice L, Graefe A, Marsinko A (2007b) Segmenting anglers using their consumptive orientation profiles. Hum Dimens Wildl 12:115–132

Kyle G, Landon AC, Pilgreen D, Golebie E (2026) Measurement instruments. In: Pope KL, Arlinghaus R, Hunt LM, Lynch AJ, van Poorten BT (eds) Understanding recreational fishers: disciplinary and interdisciplinary approaches for fisheries management. Springer Nature, Cham, pp 355–385

Lalonde RN (2002) Testing the social identity-intergroup differentiation hypothesis. Br J Soc Psychol 41:611–630
Landon AC, Kyle GT, van Riper CJ, Schuett MA, Park J (2018) Exploring the psychological dimensions of stewardship in recreational fisheries. N Am J Fish Manag 38:579–591
Landon AC, Woosnam KM, Kyle GT, Keith SJ (2020) Psychological needs satisfaction and attachment to natural settings. Environ Behav 53(6):661–683
Lazarus RS, Folkman S (1984) Stress, appraisal and coping. Springer Publishing Company, New York
Mackay M, van Putten EI, Yamazaki S, Jennings S, Sibly H (2020) Me and my behavior: an experiment on individual compliance behavior in recreational fishing. Front Mar Sci 7:579213
Manfredo MJ, Vaske JJ, Decker DJ (1995) Human dimensions of wildlife management: basic concepts. In: Knight RL, Gutzwiller KJ (eds) Wildlife and recreationists. Island Press, Washington, DC
Manfredo MJ, Driver BL, Tarrant MA (1996) Measuring leisure motivation: a meta-analysis of the recreation experience preference scales. J Leis Res 28(3):188–213
Manfredo MJ, Bruskotter JT, Teel TL, Fulton D, Schwartz SH, Arlinghaus R, Oishi S, Uskul AK, Redford K, Kitayama S, Sullivan L (2017) Why social values cannot be changed for the sake of conservation. Conserv Biol 31(4):772–780
Manfredo MJ, Teel TL, Don Carlos AW, Sullivan L, Bright AD, Dietsch AM, Bruskotter J, Fulton D (2020) The changing sociocultural context of wildlife conservation. Conserv Biol 34(6): 1549–1559
Manning RE (2007) Parks and carrying capacity: commons without tragedy. Island Press, Washington, DC
Manning RE (2022) Studies in outdoor recreation. Oregon State University Press, Corvallis
Melstrom RT, Jungers B, Bi X, Abbott JK (2026) Resource economics for understanding recreational fishers and fisheries. In: Pope KL, Arlinghaus R, Hunt LM, Lynch AJ, van Poorten BT (eds) Understanding recreational fishers: disciplinary and interdisciplinary approaches for fisheries management. Springer Nature, Cham, pp 123–156
Midway SR, Adriance J, Banks P, Haukebo S, Caffey R (2020) Electronic self-reporting: angler attitudes and behaviors in the recreational Red Snapper fishery. N Am J Fish Manag 40(5): 1119–1132
Milfont TL, Duckitt J, Wagner C (2010) A cross-cultural test of the value-attitude-behavior hierarchy. J Appl Soc Psychol 40:2791–2813
Miller ZD (2017) The enduring use of the theory of planned behavior. Hum Dimens Wildl 22(6): 583–590
Miller TA, McCool SF (2003) Coping with stress in outdoor recreation settings: an application of transactional stress theory. Leis Sci 25:257–275
Mitchell JP (2009) Social psychology: a natural kind. Trends Cogn Sci 13(6):256–251
Mueller JT, Landon AC, Graefe AR (2019) Modeling the role of social identity in constraint negotiation for ultra-endurance gravel cycling. J Leis Res 50(2):81–106
Murphy R Jr, Scyphers S, Gray SH, Grabowski J (2019) Angler attitudes explain disparate behavioral reactions to fishery regulations. Fisheries 44(1):475–487
Oh C-O, Ditton RB (2006) Using recreation specialization to understand multi-attribute management preferences. Leis Sci 28(4):369–384
Oh C-O, Ditton RB, Gentner B, Riechers R (2005) A stated preference approach to understanding angler preferences for management options. Hum Dimens Wildl 10(3):173–186
Parkkila K, Arlinghaus R, Artell J, Gentner B, Haider W, Aas Ø, Barton D, Roth E, Sipponen M (2010) Methodologies for assessing socio-economic benefits of European inland recreational fisheries. Food and Agricultura Organization of the United Nations, European Inland Fisheries Advisory Commission, Occasional Paper 46, Ankara
Prohansky HM (1978) The city and self-identity. Environ Behav 10(2):147–169
Riepe C, Arlinghaus R (2014) Explaining anti-angling sentiments in the general population of Germany: an application of the cognitive hierarchy model. Hum Dimens Wildl 19:371–390

Riepe C, Fujitani M, Cucherousset J, Pagel T, Buoro M, Santoul F, Lassus R, Arlinghaus R (2017) What determined the behavioral intention of local-level fisheries managers to alter fish stocking practices in freshwater recreational fisheries in two European countries? Fish Res 194:173–187

Riepe C, Liebe U, Fujitani M, Kochalski S, Aas Ø, Arlinghaus R (2021) Values, beliefs, norms, and conservation-oriented behaviors toward native fish biodiversity in rivers: evidence from four European countries. Soc Nat Resour 34(6):703–724

Ritter C, Ditton RB, Riechers RK (1992) Constraints to sport fishing: implications for fisheries management. Fisheries 17(4):16–19

Rokeach M (1973) The nature of human values. Free Press, New York

Ryan RM (1995) Psychological needs and the facilitation of integrative processes. J Pers 63:397–427

Ryan RM, Deci EL (2000) Self-determination theory and the facilitation of intrinsic motivation, social development, and well-being. Am Psychol 55(1):68–78

Ryan RM, Deci EL (2001) On happiness and human potentials: a review of research on hedonic and eudaimonic well-being. Annu Rev Psychol 52:141–166

Ryff CD (1989) Happiness is everything, or is it? Explorations on the meaning of psychological well-being. J Pers Soc Psychol 57(6):1069–1081

Salz RJ, Loomis DK (2005) Recreation specialization and anglers' attitude towards restricted fishing access. Hum Dimens Wildl 10(3):187–199

Schneider IE, Hammitt WE (1995) Visitor response to on-site recreation conflict. J Appl Recreat Res 20(4):249–268

Schramm HL, Gerard PD, Gill DA (2003) The importance of environmental quality and catch potential to fishing site selection by freshwater anglers in Mississippi. N Am J Fish Manag 23(2):512–522

Schroeder SA, Fulton DC (2010) The land of 10,000 lakes and 2.3 million anglers: problems and coping response among Minnesota anglers. J Leis Res 42(2):291–315

Schroeder SA, Fulton DC (2013) Comparing catch orientations among walleye, northern pike and bass anglers. Hum Dimens Wildl 18:1–18

Schroeder SA, Fulton DC, Currie L, Goeman T (2006) He said, she said: gender and angling specialization, motivations, ethics, and behaviors. Hum Dimens Wildl 11:301–315

Schroeder SA, Fulton DC, Nemeth ML, Sigurdson RE, Walsh RJ (2008a) Fishing in the neighborhood: understanding motivations and constraints to angling among Twin Cities metro residents. In: Eades RT, Neal JW, Lang TJ, Hunt KM, Pajak P (eds) Urban and community fisheries programs: development, management, and evaluation, American Fisheries Society symposium, vol 67. American Fisheries Society, Bethesda, pp 77–96

Schroeder SA, Nemeth ML, Sigurdson RE, Walsh RJ (2008b) Untangling the line: barriers to fishing participation in communities of color. In: Eades RT, Neal JW, Lang TJ, Hunt KM, Pajak P (eds) Urban and community fisheries programs: development, management, and evaluation, American Fisheries Society symposium, vol 67. American Fisheries Society, Bethesda, pp 97–112

Schroeder SA, Cornicelli L, Fulton DC, Grund MD (2014) What predicts support for antler point restrictions? Hum Dimens Wildl 19:301–318

Schroeder SA, Fulton DC, Altena E, Baird H, Dieterman D, Jennings M (2018) The influence of angler values, involvement, catch orientation, satisfaction, agency trust, and demographics on support for habitat protection and restoration versus stocking in publicly managed waters. Environ Manag 62:665–677

Schwartz SH (1977) Normative influences on altruism. Adv Exp Soc Psychol 10:221–279

Schwartz SH (1994) Are there universal aspects in the structure and contents of human values? J Soc Issues 50(4):19–45

Schwartz SH (2012) An overview of the Schwartz theory of basic values. Online Read Psychol Cult 2(1):11. https://doi.org/10.9707/2307-0919.1116

Scott D, Shafer CS (2001) Recreational specialization: a critical look at the construct. J Leis Res 33: 319–343

Selin SW, Howard DR (1988) Ego involvement and leisure behavior: a conceptual specification. J Leis Res 20(3):237–244
Sen AK (1977) Rational fools: a critique of the behavioral foundations of economic theory. Philos Public Aff 6(4):317–344
Sherif M, Hovland CI (1961) Social judgement: assimilation and contrast effects in communication and attitude change. Yale University Press
Siegrist M, Árvai J (2020) Risk perception: reflections on 40 years of research. Risk Anal 40(S1): 2191–2206
Slaton C, Koemle D, Birdsong M, Arlinghaus R (2023) Explaining attitudes to management actions and beliefs about other user groups and conservation with angler characteristics: a case study in a coastal pike (*Esox lucius*) fishery in the southern Baltic Seas, Germany. Fish Res 263:106669
Smith AP, Metcalf EC, Nesbitt HK, Leonard HJ, Cummins TM, Metcalf AL, Graefe AR (2022) Confidence, community, and conservation: exploring the relationship between self-efficacy and experience in female hunters. J Outdoor Recreat Tour 40:100568
Smith K, Landon AC, Fulton DC (2023) A self-determination approach to understanding leisure identity salience among lapsed hunters. Leis Sci:1–26. https://doi.org/10.1080/01490400.2023.2222738
Smith K, Landon AC, Fulton DC et al (2024) Self-determination theory as an alternate conceptual foundation for motivation in natural resource recreation. Hum Dimens Wildl:1–9. https://doi.org/10.1080/10871209.2023.2285017
Spencer PD, Spangler GR (1992) Effect that providing fishing information has on angler expectations and satisfaction. N Am J Fish Manag 12:379–385
St John FAV, Edwards-Jones G, Jones JPG (2010) Conservation and human behavior: lessons from social psychology. Wildl Res 37(8):658–667
Stensland S, Aas Ø, Mehmetoglu M (2013) The influence of norms and consequences on voluntary catch and release behavior. Hum Dimens Wildl 18(5):373–385
Stensland S, Aas Ø, Mehmetoglu M (2016) Understanding constraints and facilitators to salmon angling participation: insights from structural equation modeling. Hum Dimens Wildl 22(1): 1–17
Stern PC (2000) New environmental theories: toward a coherent theory of environmentally significant behavior. J Soc Issues 56(3):407–424
Stern PC, Kalof L, Dietz T, Guagnano GA (1995) Values, beliefs, and proenvironmental action: attitude formation through emergent attitude objects. J Appl Soc Psychol 25(18):1611–1636
Stern PC, Dietz T, Abel T, Guagnano GA, Kalof L (1999) A value-belief-norm theory of support for social movements: the case of environmentalism. Hum Ecol Rev 6(2):81–97
Stets JE, Biga CF (2003) Bringing identity theory into environmental sociology. Soc Theory 21(4): 398–423
Stryker S, Burke PJ (2000) The past, present, and future of an identity theory. Soc Psychol Q 63(4): 284–297
Sutton SG (2003) Personal and situational determinants of catch-and-release choice of freshwater anglers. Hum Dimens Wildl 8(2):109–126
Sutton SG (2006) Understanding recreational fishers' participation in public consultation programs. Hum Dimens Wildl 11(5):329–341
Sutton SG (2007) Constraints on recreational fishing participation in Queensland, Australia. Fisheries 32:73–83
Sutton SG, Ditton RB (2001) Understanding catch-and-release behavior among U.S. Atlantic bluefin tuna anglers. Hum Dimens Wildl 6:49–66
Sutton SG, Ditton RB (2005) The substitutability of one type of fishing for another. N Am J Fish Manag 25(2):536–546
Sutton SG, Oh C-O (2015) How do recreationists make activity substitution decisions? A case of recreational fishing. Leis Sci 37(4):332–353
Tajfel H, Turner JC (1985) The social identity theory of intergroup behavior. In: Worchel S, Austin WG (eds) Psychology of intergroup relations. Nelson-Hall, Chicago, pp 7–24

Tang Y (2014) Travel motivation, destination image and visitor satisfaction of international tourists after the 2008 Wenchun earthquake: a structural equation modeling approach. Asia Pac J Tour Res 19(11):1260–1277

Teel T, Manfredo MJ (2010) Understanding the diversity of public interests in wildlife conservation. Conserv Biol 24(1):128–139

TenHarmsel HJ, Boley BB, Irwin BJ, Jennings CA (2021) Perceived constraints and negotiations to trout fishing in Georgia based on angler specialization. N Am J Fish Manag 41:115–129

Toth JF Jr, Brown RB (1997) Racial and gender meanings of why people participate in recreational fishing. Leis Sci 19:129–146

Uenal F, Sidanius J, Maertens R, Hudson SK, Davis G, Ghani A (2022) The roots of ecological dominance: assessing individual preference for an anthropocentric and hierarchically organized world. J Environ Psychol 81:101783

Ulrich RS (1983) Aesthetic and affective response to natural environment. In: Altman I, Wolhill JF (eds) Human behavior and the natural environment. Plenum Press, New York, pp 85–125

van den Heuvel L, Rönnbäck P (2023) What you see isn't always what you get: on how anglers' fish stock perceptions are influenced by motivations, satisfaction and engagement. Fish Res 258: 106519

van den Heuvel L, Blicharska M, Blyth S, Rönnbäck P (2020) Catch reporting in recreational fishing: Swedish anglers' attitudes and preferences, and the effect of social factors. Fish Res 223:105444

van den Heuvel L, Blicharska M, Stensland S, Rönnbäck P (2022) Been there, done that? Effects of centrality-to-lifestyle and experience use history on angling tourists' loyalty to a Swedish salmon fishery. J Outdoor Recreat Tour 39:100549

van Poorten BT, Arlinghaus R, Daedlow K, Haertel-Borer SS (2011) Social-ecological interactions, management panaceas, and the future of wild fish populations. Proc Natl Acad Sci USA 108: 12554–12559

Vaske JJ, Donnelly MP (1999) A value-attitude-behavior model predicting wildland preservation voting intentions. Soc Nat Resour 12:523–537

Vaske JJ, Manfredo MJ (2012) Social psychological considerations in wildlife management. In: Decker DJ, Riley SJ, Siemer WF (eds) Human dimensions of wildlife management. Johns Hopkins University Press, Baltimore, pp 43–57

Vaske JJ, Whittaker D (2004) Normative approaches to natural resources. In: Manfredo MJ, Vaske JJ, Bruyere BL, Field DR, Brown PJ (eds) Society and natural resources: a summary of knowledge. Modern Litho, Jefferson, pp 283–294

Vaske JJ, Donnelly MP, Williams DR, Jonker S (2001) Demographic influences on environmental value orientations and normative beliefs about national forest management. Soc Nat Resour 14: 761–776

von Lindern E, Mosler H-J (2014) Insights into fisheries management practices: using the theory of planned behavior to explain fish stocking among a sample of Swiss anglers. PLoS One 9(12): e115360

Walker GJ, Kleiber DA, Mannell RC (2019) A social psychology of leisure. Sagamore-Venture, Champaign

Whittaker D, Vaske JJ, Manfredo MJ (2006) Specificity and the cognitive hierarchy: value orientations and the acceptability of urban wildlife management actions. Soc Nat Resour 19: 515–530

Witter DJ, Haverland PS, Belusz LC, Hicks CE (1982) Missouri trout park anglers: their motives and opinions of management. University of Minnesota Agricultural Experiment Station, St. Paul

Political Ecology for Understanding Recreational Fishers and Fisheries

Noëlle Boucquey, Jessie Fly, Wiebren J. Boonstra, Lauren Drakopulos, Kirsten M. Leong, Meghna N. Marjadi, Sarah Wise, and Luke Fairbanks

Abstract Political ecology, and its core concern with how power shapes social–ecological relationships, has much to offer recreational fisheries analyses. Political ecologists bring critical questions about how different fishers may have uneven access to resources, how particular policy narratives affect fishers, and how fishing communities are entangled with broader social, economic, and ecological processes. We explain the origins, key theoretical tenets, and methodological approaches of the field, describe the ways political ecology theory has been applied to understand recreational fisheries, and explore ways it could be applied in future research and management. With its focus on making visible the often hidden, power-laden relationships that shape the character of recreational fishing in specific places, political ecology investigations reveal the ways recreational fishing grapples with its own role in shifting ecologies and is also interwoven with resource-use conflicts and political movements. We bring our diverse perspectives as academics and

N. Boucquey (✉)
Department of Environmental Studies, Eckerd College, St. Petersburg, FL, USA
e-mail: boucqun@eckerd.edu

J. Fly
Department of Anthropology, Eckerd College, St. Petersburg, FL, USA

W. J. Boonstra
Department of Earth Sciences, Uppsala University, Uppsala, Sweden

L. Drakopulos
Department of Global Development, Cornell University, Ithaca, NY, USA

K. M. Leong
Pacific Islands Fisheries Science Center, NOAA Fisheries, Honolulu, HI, USA

M. N. Marjadi
Department of Fisheries Oceanography, School for Marine Sciences and Technology, University of Massachusetts Dartmouth, New Bedford, MA, USA

S. Wise
Alaska Fisheries Science Center, NOAA Fisheries, Seattle, WA, USA

L. Fairbanks
Institute at Brown for Environment and Society, Brown University, Providence, RI, USA

K. L. Pope et al. (eds.), *Understanding Recreational Fishers*, Fish & Fisheries Series 45, https://doi.org/10.1007/978-3-031-99739-6_7

fisheries managers to illustrate key moments when the central themes of political ecology have helped us to better understand recreational fisheries dynamics. Finally, we offer a set of best practices for integrating a political ecology perspective into recreational fishing studies.

Keywords Access · Discourse · Political ecology · Power · Scale · Social-ecological systems

1 Introduction

As this book highlights, recreational fishing is studied from a wide variety of perspectives (see Hunt et al. 2026). In this chapter, we introduce the exciting field of political ecology and discuss how we can study recreational fishing using the theoretical concepts and practical tools of the field. We begin by explaining the origins and key theoretical tenets of political ecology. We then describe the ways political ecology theory has been applied to understand recreational fisheries thus far, using specific examples from our own and others' research. Then we delve into how political ecology research is conducted, using specific cases to highlight the variety of social scientific methods used in political ecology. Further, we introduce readers to the ways this young field is evolving and responding to critiques. Finally, we outline some best practices for applying political ecology in future research and management of recreational fisheries.

2 What Is Political Ecology?

In this section, we briefly review the history of the field, then discuss in more detail its key theoretical tenets. We use specific examples from fisheries and other issues to show how political ecologists approach researching social–ecological relationships.

Political ecology is a relatively new field, coalescing in the late 1970s and early 1980s as scholars in a variety of areas (e.g., cultural ecology, hazards research, environmental science, anthropology, institutional economics, and human geography) began to question apolitical explanations for the relationship between socio-economic and environmental changes. Many of these scholars critiqued their own fields for offering overly simplistic explanations of environmental degradation (Painter and Durham 1995; Robbins 2019). Others pushed back against the neo-Malthusian work then popular in the environmental movement (e.g., Ehrlich 1968) as insufficiently addressing issues of power (Watts 1983). Indeed, the field's "originality and ambition lay in its efforts to link social and physical sciences through an explicitly theoretical approach" to thinking through the role of power and politics in shaping social-ecological relationships (Paulson et al. 2003, p. 206). In the most frequently cited definition, political ecology is described as a field that "combines the concerns of ecology and a broadly defined political economy"

(Blaikie and Brookfield 1987, p. 17). Whereas topics and approaches have proliferated since the field's origins, most political ecology work continues to share several key commitments, including: a theoretical commitment to critical social theory that questions the role of power in social relations including the production of knowledge; a methodological commitment to "in-depth, direct observation;" and a normative political commitment to research in the service of social–ecological justice (Perreault et al. 2015, p. 7).

Though the field of political ecology is dynamic and varied, most practitioners share a set of core concerns. Political ecologists are interested in how social–ecological relationships can be better understood by analysing: (i) the power dynamics at work (who holds political, financial, or social power? What are the tactics and decisions of those with more and less power? Who controls the creation and dissemination of environmental knowledge, and who decides what knowledge is considered legitimate?) (Neumann 2014; Ahlborg and Nightingale 2018; Robbins 2019); (ii) how socioeconomic and environmental processes are interconnected and both shape and are shaped by particular institutions (what historic political–economic decisions shaped resource use in particular ways? How do cultural norms affect the ways people think about and act toward nonhuman actors? How do current environmental conditions affect people's thinking, and what actions flow from those understandings?) (Blaikie and Brookfield 1987; Blaikie 1999; Escobar 1998; Neumann 1998; Schroeder et al. 2006); and (iii) the role of discourse in shaping political–economic and environmental policies and practices (how do collections of texts and images help shape narratives, scientific and political choices, and ultimately material actions that affect human-environment relationships?) (McCarthy 2002; Campbell 2007; Campbell and Meletis 2011; Hawkins and Silver 2017; Leslie-Bole and Perramond 2017). Whereas the diversity of political ecology means that others may focus on complementary key themes (e.g., equity, space, scale, historical processes, commons), we highlight these three here as being central to most work in the field.

2.1 Power and Politics

Political ecology is based on the proposition that social–ecological relationships are political. This means that the ways in which people use, enjoy, and experience benefits as well as harm from their environments always involve social interdependencies. These interdependencies can be more or less equal depending on the degrees of power that people have (Elias 2012 [1970]). Political ecology is thus driven by the wish to understand how power relations shape people's relationships with the nonhuman world. One way that political ecologists do this in the fisheries world is to reveal the power dynamics behind dominant narratives and practices in science and policymaking, and to illustrate how these may silence alternative narratives (May 2016; Loring 2017; Boucquey 2020). At the same time, political ecology strives to move beyond critique to provide productive ways forward (Robbins 2019).

Power is defined and understood in myriad ways (Boonstra 2016; Ahlborg and Nightingale 2018; Svarstad et al. 2018). A central distinction, however, is the difference between "power to," "power over," and "power-with" (Pansardi and Bindi 2021). Power-to denotes all the different capacities, such as physical, economic, cultural, social, and technological, that people have to lead a life, and to achieve outcomes that contribute to their own and others' wellbeing. A classic source of this understanding of power is found in the work of Baruch Spinoza (Haugaard 2014) and furthered by Morriss (2002 [1988]). Power-over refers to situations where the use of power-to impacts other people's lives. Power-over is often associated with domination and violence (Arendt 1970) and has roots in the work of Thomas Hobbes (Haugaard 2014). The final aspect of power is power-with, which refers to the capacities that people have as a group (i.e., when they act in concert). This aspect of power is often associated with empowerment (Arendt 1970; Allen 1999). In recent years, some scholars have argued that these three aspects, dimensions, or types of power are not mutually exclusive but rather presuppose each other and that the distinction between them is analytical. Nevertheless, different scholarly traditions tend to emphasise one over the other (Haugaard 2014) (see Table 1). Scholars continue to debate under which conditions power should be understood as domination or as empowerment, and when people can be held responsible for outcomes of power.

All these traditions inform the attention that political ecologists pay to power. What is distinct for this discipline is the context: power that shapes and stems from

Table 1 Types and definitions of power

Type of power	Definition	Tradition of study (founding scholar)	Recreational fishing example
Power-to	The capacities (creative, social, technical, economic, etc.) people have to shape their own lives (and potentially affect others)	Consensual (Talcott Parsons) Constitutive (Michel Foucault)	The power to catch a fish depends on multiple capacities (level of skill, ability to afford fishing gear, ability to travel to fishing location).
Power-over	The ability people have to force others to change their behaviours	Conflictual (Karl Marx) Constitutive	The state exerts power over fishers by setting and enforcing fisheries regulations via fines.
Power-with	The abilities that emerge when people work collectively	Consensual Constitutive	Indigenous fishing groups have collectively lobbied for their right to fish within historic waters.

social interdependencies in the use and experience of natural environments. Power here refers to human as well as human-nature interdependencies. Thus, power is used to understand struggles between more and less powerful groups over access and control of ecological resources. These struggles occur through direct confrontation but also via people trying to influence the socioeconomic and political contexts in which others live. These contexts include the norms, cultures, institutions, and structures that regulate how people relate to nature, others, and themselves (e.g., the "more-than-human" world (Braun 2005; Whatmore 2006; Collard and Dempsey 2013; Todd 2014)). Powerful actors derive their power from how these contexts currently operate and are thus typically trying to maintain the status quo. Less powerful groups often try to unsettle and change these structures of power. Political ecology attends to both the actions of contemporary more- and less-powerful groups as well as the historical legacies that have (re)produced unequal interdependencies between groups (Nixon 2011).

For example, several fisheries studies have highlighted how certain groups can disproportionately influence the development of institutions related to the distribution of fish catch. In some cases, these groups are large international or national fishing operations (e.g., Greenberg 2006; Nolan 2019), but they can also be government officials (Tan-Mullins 2007) or non-governmental organisations (Petersson et al. 2019). These studies show how powerful actors utilise their economic, social, and cultural capital to gain access to and control over fish stocks ("power to"). The studies also indicate how, in these efforts, powerful actors exert "power over" others who, in turn, depend on fish stocks for their wellbeing. These others include small-scale fishers and coastal communities, but also the human population at large, non-human others, and even future human generations.

Though specific theories of power may be used alone or in combination (Svarstad et al. 2018), political ecologists recognize that power is multifaceted and is created and enacted in a variety of ways. In line with their normative commitments to increasing justice in the world, political ecologists often try to illuminate channels of power that are hidden or taken for granted. Moreover, political ecologists work to show how agency and power are exercised by the seemingly less powerful (e.g., marginalised peoples, non-humans subject to exploitation). In general, political ecologists seek to produce careful analyses of the interplay between individual agency and social discourses and structures in any particular resource conflict. In some of our own work, for example, NB has examined how different narratives by commercial and recreational fishing interests have interacted with coastal land use change, scientific information, and prescribed policy performances to exacerbate power struggles among fishers both on and off the water (Boucquey 2016, 2017). Others have shown how ecological data themselves and the ways they are collected, organised, and deployed can play a powerful role in shaping how fisheries and ocean environments are understood (St. Martin and Hall-Arber 2008b; Knol 2011; Ascui et al. 2018).

2.2 Material and Discursive Social–Ecological Relationships

Another key element of political ecology work is interrogating the characteristics of specific interconnections between ecological and socioeconomic processes. Political ecologists are interested in both the ecological and socioeconomic effects of natural resources decision making (Neumann 2014; Doolittle 2015; Robbins 2019). For example, in an analysis of Florida's gill-net ban, Loring (2017) traced the ban's interlinked social and ecological consequences and their relationship to the political power of the sportfishing lobby. He found that sportfishing groups, interested in a greater allocation of fish, succeeded in convincing Florida voters to pass the net ban using scientifically unsupported statements about the environmental effects of commercial fishing. Although some finfish stocks improved after the ban, significant commercial effort shifted to shellfishing, leading to declining stone crab (*Menippe mercenaria*) populations. Simultaneously, the ban caused multiple deleterious economic and social effects on commercial fishers, including decreasing food security and increasing rates of depression (Loring 2017, pp. 96–97). The case illustrates that "ecological concerns, even spurious ones, can be used as a veil for advancing political and economic agendas that may not be socially just or ecologically justified" (Loring 2017, p. 99). In its attention to these interconnected issues and interest in the relative equity of the policy outcomes, Loring's study demonstrates core elements of political ecology research.

In the interest of better discerning how social norms evolve and political decisions are made and ultimately influencing what natural resources are used and who is empowered to access them, political ecologists often pay close attention to discourse. As used by political ecologists, the term differs from its colloquial meaning as conversation or debate. Rather, *discourses* are sets of ideas expressed by groups of people via a variety of means (e.g., traditional spoken or written narratives, and also visual images, art, and social media). However, as Neumann (2004, p. 816) argues, "discourses are not simply a set of ideas. By defining categories and creating and limiting possibilities, a discourse has historical, material consequences." Discourses play an important role in environmental decision-making. For example, conventional ideas about the causes of environmental decline are circulated as discourses, shaping the sphere of possible solutions and foreclosing the possibility of other causes and contexts.

In a study of sea turtle (*Chelonioidea*) conservation practices, for example, Campbell (2007) explored how global conservationists' beliefs about the ecology of and rights of access to turtles by locals, tourists, scientists, or managers are articulated at different sociopolitical scales. By examining these discourses, she showed that even in areas where sea turtles were not threatened and successful limited egg harvests were ongoing, conservationists used narratives about the global ecology of turtles to argue for implementing ecotourism instead. As she explains, "the global distribution and migrations of turtles. . .are invoked by sea turtle experts to override local rights of withdrawal, management, and exclusion, and to assign these rights at other sociopolitical scales" (Campbell 2007, p. 327). In this case, a particular scientific and conservation discourse was used to transfer power away from local resource users.

In its focus on power relationships and their effects on human–environment interactions, political ecology is interested in human and more-than-human behaviours. Unlike several of the other disciplines profiled in this volume (e.g., Landon et al. 2026; Mackay et al. 2026; Melstrom et al. 2026), political ecology does not model human behaviours in the traditional sense. Rather, political ecology conceptualises more-than-human agency as conditioned by particular constellations of historical and contemporary cultural, political, environmental, and economic processes. These features shift in space and time, thus political ecologists generally reject standardised models of behaviour and often very explicitly reject rational-actor models that fail to account for the importance of context (Peet and Watts 2004). Political ecology's strong emphasis on behaviours in context allows researchers to "embrace complexity and uncertainty in their analyses" (Doolittle 2015, p. 527). When focusing on recreational fishers, then, the only assumption political ecologists might make is that any individual fisher's behaviour is likely influenced by their intersectional identity, the history and geography of the space in which they are fishing, the political economy of the fisheries they are targeting, and the governance practices of the fishery.

A recent review demonstrates that there is a small, though steadily growing, literature of political ecology research within the general scientific literature on ocean, marine, and coastal environments (Bennett 2019). We suspect the same to be true of political ecologists' attention to recreational fisheries specifically. Why is it important for scientists, policymakers, and managers to pay attention to this field? In the remainder of this chapter, we aim to answer this question by highlighting the key questions that political ecologists ask, the theories and methods they use, and the knowledge they obtain about recreational fisheries.

3 How Is Political Ecology Used to Understand Recreational Fisheries?

Here, we begin by discussing how political ecology's attention to the importance of context and discourse means that studies of recreational fishing must often start by probing the term itself. Using US legislative examples, we show how a political ecology perspective can highlight the challenges and consequences of defining recreational fishing in particular ways. We then turn our attention to some recent examples of political ecology studies that have focused on fishing and discuss key insights gleaned by applying political ecology theories.

Political ecology has been described as both a "hatchet" and a "seed," a way of thinking about social–ecological systems that uses critique to question and denaturalize what has often been accepted as common knowledge, and then focuses on repair and regrowth (Robbins 2019). As introduced above, political ecologists perform this critique and repair by focusing on power dynamics, socioeconomic and environmental context, and the role of discourse. Regardless of the specific data-gathering methods used (discussed in Sect. 4), case study research is common in political ecology because it allows for an assessment of context: "the close examination of complex processes specific to space and time to gain empirical insights into the social relations and power hierarchies that have shaped environmental change" (Helmcke 2022, p. 267). Although literature on the political ecology of recreational fisheries is not abundant, case study research on the interactions and relations between humans and freshwater and marine ecosystems is fertile ground for the critique of "common knowledge," whether that of the public, conservation biologists, or the fishers themselves, and the fruitful rebuilding of relationships.

3.1 Analysing Discourses

Political ecologists often begin by critically examining the very words used to describe a group of people and their activities vis-à-vis their social–ecological context. For instance, US law (the Magnuson-Stevens Fishery Conservation and Management Act) categorises and defines commercial and recreational fishing separately (16 U.S.C. § 1801–1891(d)). A political ecologist might ask how ideas about fishing as either a work or leisure activity, an essential or non-essential activity, and the discourses through which these ideas are circulated (e.g., policy and fishing regulations) shape norms and understandings about people's motivations for fishing. For example, in their research on commercial fishing, Poe et al. (2015) determined that classifying fishing as either work (e.g., commercial fishing performed for money) or leisure implies that fishers' motivations and behaviours are driven solely by the capitalist economy. In practice, the distinction between the two may be blurred, and a capitalist point of view overlooks and undervalues motivations derived from other important forms of social value and exchange (e.g., gifting or trading fish) (Fig. 1) (Leong et al. 2020).

Similarly, "recreational" is a term often used to classify fishing performed "with leisure as the primary motivation" (FAO 2012, see also Lynch et al. 2026). The Magnuson-Stevens Fishery Conservation and Management Act defines recreational fishing as "for sport or pleasure" (16 USC § 1802). Behind these definitions lie powerful discourses about recreational fishing in which norms and understandings about fishers' motivations are articulated, categorised, and defined. Moreover, they can have real material impacts on people's livelihoods or ways of life (Jenkins 2015). Recreational fishing as a management category may include a broad suite of motivations and practices that go beyond sport or pleasure (e.g., food, cultural, social, spiritual, and dietary (Leong et al. 2020; Nieman et al. 2021)). It is also often monetized, for example, with the charter fishing industry. Concurrently, it has been well documented that commercial fishers frequently find great pleasure in their livelihood (Pollnac et al. 2015; Holland et al. 2020). By defining recreational fishing

Fig. 1 Fishing often serves an important role in maintaining social ties and cultural traditions and providing for community events (in this photograph, bottomfish captain Ed Ebisui III and crew Bryce Whittaker share an onaga [*Etelis coruscans*] they caught and prepared ikizukuri style for a New Year's party in Hawaii). (Photograph courtesy of Robyn Ebisui)

to exclude other motivations for fishing, these discourses also shape, and are shaped by, decisions about how fishing is regulated. As a result, this can ignore or obscure other kinds of non-commercial fishing practices. In some contexts, these practices are included under the term subsistence, for example, as defined in the Alaska National Interest Land Conservation Act (Public Law 96-487, December 2, 1980), which applies to Indigenous and non-Indigenous peoples. When and how these definitions are deployed is strategic and shapes how specific actions are perceived and regulated. For example, as described in Iwane et al. (2021), the Presidential Proclamation designating the core Papahānaumokuākea Marine National Monument stated that non-commercial fishing may be allowed. The National Oceanic and Atmospheric Administration promulgated regulations that allow for sustenance fishing in federal waters, where fish may be harvested under a permit for another primary purpose (e.g., Native Hawaiian cultural practices, scientific research), but it must be eaten within the Monument boundary. This differentiates the activity from subsistence fishing, which is allowed in state waters and defined as "…harvesting for direct personal or family consumption and not for commercial purposes" (Iwane et al. 2021, p. 3).

Political ecology studies of fishers' motivations and practices (Ebbin 2017), and the impacts of policies on subsistence fishers (Nieman et al. 2021), challenge dominant discourses on recreational fishing. A growing body of political ecology scholarship is further questioning conventional beliefs about where recreational and subsistence fishing are practised and the challenges of classifying fishing types and behaviours, particularly in urban environments (Emery and Pierce 2005; Quimby et al. 2020; Boucquey and Fly 2021; Joosse et al. 2021; Marjadi et al. 2021; Boonstra et al. 2022). This can matter in real-world ways. For instance, if fishing is classified as a solely "recreational" activity, it can be easier to close historic fishing locations when there are conflicts with neighbours or other activities (Boucquey and Fly 2021).

3.2 Analysing Context

Political ecologists understand that human behaviours are conditioned by particular constellations of historical and contemporary cultural, political, environmental, and economic processes. Thus, understanding specific contexts and their intersection with individuals and social groups with varying identities is vital to any political ecology study (e.g., Rocheleau et al. 1996; Carney 2014; Cole 2017). In recreational fisheries, attention to context matters for understanding the behaviours and choices of varying social groups. Such attention also matters for deconstructing conventional wisdom, to plant the seeds of better management and communication. Important factors in a recreational fishing context might include the ecological and regulatory histories of a fishery, tensions between different user groups, expansions or enclosures of access to fishery resources, or shifting ecologies under climate change. Political ecologists would think about how these contexts might differentially affect people of varying economic status, gender, and racial identities, fishing experience, or tenure in a community.

One piece of so-called conventional wisdom that has received much attention, especially from fisheries scientists and policymakers, is Garrett Hardin's (1968) "tragedy of the commons." This idea posits that any resources not governed by formal access and use rules, or privatised, will be overexploited by users. This assertion, which was long held to influence important policymaking around fisheries and forestry, has been critiqued by institutional economists (Ostrom 1990, 2009) and political ecologists as ignoring the role of community self-determination and collective action for common benefits (Berkes et al. 1989; Ostrom et al. 2002; McCay and Jentoft 2007). Numerous ethnographic examples exist of successful commons management among resource users, many from fishing communities (McEvoy 1987; Acheson 1988; Feeny et al. 1990; Acheson and Brewer 2003). This is also a prime example of the importance of context. Mansfield (2004, 2010), for instance, considers explanations of overfishing as a tragedy of the commons facile and overstated, and calls for greater attention to sociopolitical context. Longo et al. (2015) noted how this and other critiques of the tragedy of the commons have helped to better understand how social conditions, including norms, organisations and property regimes influence how fishers fish, and whether this leads to overfishing (e.g., as seafood has become a global commodity, international industrial fishing fleets encroach nearer to historic coastal fishing communities, affecting fish populations and forcing local fishers to fish nearshore waters more heavily to maintain a livelihood). This literature, in which political ecology is prominent (see, for recent examples, Cinner et al. (2022) and Loring (2022)), has enriched scientific understanding of commons and user cultures, moving beyond the essentialist and individualist approach of Hardin's initial idea.

In an example from our own work, NB and JF (Boucquey and Fly 2021; Fly and Boucquey 2023) explored how piers and other shore fishing spaces are sometimes painted as victims of the tragedy of the commons, especially by those wanting to use the coastal space for another activity (e.g., biking, jogging) or to exclude strangers from their neighbourhoods. Letters to the editor of the local newspaper decried shore fishers' lack of regard for cleanliness, road safety, and environmental health. Yet, using data from a quantitative survey and several years of ethnographic interviews and observations, we discovered that many shore fishers express pro-social and pro-environmental attitudes, and exhibit "commoning" (Gibson-Graham et al. 2016) behaviours (e.g., behaviours that show care and community-making). Contextual factors contribute to fishers' abilities and desires to effectively manage their fishing spaces and target species and their definition of "effective management" for these bridges, seawalls, and piers (Fig. 2). Some of these contextual factors include the existence of amenities (e.g., trashcans, restrooms) and "social infrastructure" (i.e., space conducive to community-building (Klinenberg 2018)), the legality of fishing in the space, fishers' complex internal and social calculations of the costs of violating regulations, and the constantly shifting fishing rules and regulations and the means by which they are communicated.

Fig. 2 Contextual factors (such as the characteristics of shore fishing infrastructures [here offering lowered pier railings and a fish cleaning station]) are important in influencing whether fishers exhibit commoning behaviours. (Image courtesy of the Tampa Bay Urban Fishing Project)

3.3 Analysing Power Dynamics

Finally, political ecologists focus on the power dynamics at play and their effect on policymaking, coastal land use, and the health and well-being of fishers and fisheries. For instance, in some places, recreational fishing has crowded out and replaced subsistence or small-scale fishing, especially where shore space has been privatised or otherwise commodified (Kalina et al. 2019). Mas (2015) describes the process of marine and coastal gentrification of the Balearic Islands of Spain. Tourism has begun to dominate the local economy; thus, coastal space has been converted to accommodate recreational fishing. However, recreational catch is not included in official tallies of total catch, and recreational fishing is often unregulated (Mas 2015, p. 225). Traditional small-scale fishing is dwindling in this example because of the enclosure of public marine access space and competition for the same target species.

Power dynamics are not always in favour of recreational fishers, however, particularly given the diversity of practices and motivations subsumed by that categorization. As described above, we detail struggles for and restrictions on access to shore space among different groups of recreational users (e.g., fishers, bikers, joggers), with notable cases of recreational fishers "losing" (Boucquey and Fly 2021). Additionally, political ecologists often examine what and whose practices are legitimised through environmental decision-making and codified through legal and other institutional processes. In another example from our work, LD and MM have shown that fish testing for safe consumption in the United States of America does not always reflect the fishing preferences and practices of recreational and subsistence fishers, leaving out the culturally important species often preferred by fishers with minoritized or otherwise marginalised ethnic and cultural identities (Drakopulos et al. 2020; Marjadi et al. 2021).

Thus, the work of political ecologists on issues related to recreational fisheries and beyond is meant to question the received wisdom and dominant narratives that direct public behaviour, social–ecological data collection, and policymaking. This questioning provides historical and cultural contextual information that allows for a more holistic understanding of the subject of study. It may also uncover power dynamics that may have silenced alternative narratives or oversimplified a complex social–ecological system. However, this work should not end with a critique. Instead, political ecology aims to provide a productive way forward with new definitions, classifications, and understandings. Political ecologists use the methodologies described below to act as both critiques and seeds for change (Robbins 2019).

4 How Is Political Ecology Research Conducted?

Political ecologists have employed a variety of methods to examine interconnected social–ecological systems and how socioeconomic and political power influence how such systems operate. Here, we first provide a short overview of the rationales in political ecology for using different approaches, then we use some specific examples from our own and others' work to show how political ecology research can be conducted.

4.1 Research Practices

The array of research methods used in political ecology stems from the "productive tension" between more or less realist and critical epistemologies (e.g., theories of legitimate knowledge production, see also Hunt et al. (2026)) within political ecology (Doolittle 2015, p. 517). At the same time, political ecologists often take a strong normative stance, critiquing dominant narratives and favouring increased ecological, economic, political, and social equity (Forsyth 2003; Bryant 2015; Perreault et al. 2015; Robbins 2019). Central to political ecology is the deep and nuanced exploration of the ways cultural practices are shaped, contested, and negotiated (Watts 2000). Thus, many political ecology studies will integrate qualitative methods, which are well suited to capture the kind of nuanced, context-dependent data on the "less visible social, political, or economic structures and processes" that may be at work (Doolittle 2015, p. 516). Quantitative methods can allow for a broad assessment of differences in experience between groups with different social and economic positions, whereas qualitative methods contribute to understanding the underlying forces leading to these differences and the ways that individuals experience them. In addition, political ecologists often incorporate participatory, feminist, or arts-based methods into their projects (Harris 2015; Osborne 2017; Batterbury 2018; Springett 2022). Depending on the situation and guided by the relevant research questions, political ecologists may draw on a variety of methods to conduct their investigations. Here, we discuss examples to help illustrate this methodological diversity.

As noted above, an important contribution of political ecology work is foregrounding the role of power in resource management frameworks. In the realm of fisheries, attention is often focused on how power is exercised via management structures and processes of rulemaking (Bennett 2018). Investigating how power works is challenging, and political ecologists often draw on multiple methods to better understand how power functions in particular cases such as in fisheries management. In his work with commercial fisheries in New England, United States of America, St. Martin (2001) creatively combined a discourse analysis of regional fisheries management documents with in-person interviews with fishers and participatory mapping of their fishing spaces (e.g., home ports, typical fishing grounds, prominent environmental features). The discourse analysis helped reveal a powerful bioeconomic narrative driving fishery management decisions, one that imagined the fishery as open access and the fishermen as individual profit maximisers, a "tragedy of the commons" scenario as described above. Such assumptions were "institutionalised in data collection, fishery analysis, and policy recommendations" and consequently left "no place for community in New England fisheries"

(St. Martin 2001, pp. 124, 125). However, in their interviews and mappings, fishers themselves painted a very different picture of their reality, one that emphasised shared physical and social spaces. In making these spaces visible through his research, St. Martin demonstrated how a process of literal and metaphorical "re-mapping" could provide alternatives to standard management practices, in this case proposed participatory management based on community-linked fishing areas.

In another interesting example of mixed-methods work in marine political ecology, Chiaravalloti (2017) investigated a complex conflict between artisanal fishers, policymakers, and tourism-focused recreational fishing businesses in a Brazilian freshwater fishery. In this conflict, local artisanal fishers were subject to increasingly stringent regulations as tourism declined, and the artisanal fishers were blamed for overfishing. Chiaravalloti (2017) employed a multi-pronged approach for collecting both qualitative and quantitative data: conducting participant observation and semi-structured interviews with the artisanal fishers to better understand their perspectives, outfitting a subset of fishers with GPS devices to track their fishing routes and target species, and collecting local records on hydrologic events, tourism visitation, and permitted recreational fishing sizes and quotas to examine the linkages between fishing activities, local policymaking, and tourism. With this diversity of data, Chiaravalloti painted a clearer picture of the political–ecological dynamics at work in this fishery. The qualitative interview and GPS data both revealed that fishers engaged in several customary management practices in contrast to policy narratives painting fishers as blindly overfishing. These findings led Chiaravalloti to conclude that the fishery "exemplifies a type of dynamic socio-ecological system which calls for flexible tenure and access and adaptive management practices" and to urge the local policymakers to change their existing top-down restrictions that deny local knowledge and practice (Chiaravalloti 2017, pp. 118–119).

Although political ecology research can often benefit from mixed methodologies, the most important criterion to consider for any research project is whether the methods provide suitable answers to the questions asked. In our own work, we use a variety of methods from the political ecology "toolkit" to address different types of questions. For example, NB once examined why fisheries policymaking was an arena of intense conflict between recreational and commercial interests, even where fishery resources were *increasing* in abundance (Boucquey 2016). To study this question, she focused on the policymaking process itself, and people's performances therein. She collected several years of public meeting records, conducted detailed observations of meetings in person, and interviewed policymakers and both types of fishers. These data formed the basis for a close narrative and policy performance analysis, where she found that a combination of the meeting structure and the language used associated with fisheries science established "some players as dominant and put others in a defensive mode prior to any actual discussion" (Boucquey 2016, p. 15). These findings allowed her to recommend that more deliberative policymaking processes be pursued to rearrange "the relationships between policy actors and the spaces they perform [in], in order to promote genuine collaboration and improve acceptance of compromises" (Boucquey 2016, p. 17). Using qualitative methods, this study embodied several key political ecology tenets in its focus on how power and knowledge are employed in decision making around fishery resources, and in its normative concern with more equitable material outcomes.

In their work in Bristol Bay, Alaska, Donkersloot et al. (2020) explored the role of diverse economies within rural and small-scale fisheries. This research examined how institutional structures support and constrain participation in fisheries and determine legitimate rights holders through resource allocation. Through extensive ethnographic work (Donkersloot is a small-scale commercial fisher herself), the authors problematized dominant narratives of fishers as market-driven, self-interested actors, and reinforced the importance of community and social networks.

In our own work in Bristol Bay on Alaska Native women fishers, SW and colleagues explored the lived experience of Indigenous women engaged in commercial and subsistence fishing (Lavoie et al. 2019). Using oral history methods, we conducted a series of open-ended interviews with women who have lived and fished for generations. Secondary data were used to develop quantitative metrics of salmon harvests and permit holdings over time. Although women were primary permit holders in the area, there was a steep decline in permits held by Bristol Bay residents. Women fishers acted as knowledge holders and community keystones. As such, women were critical in fisheries management decisions, community resilience, and social–ecological sustainability.

Oral history methods allow for complex narratives and limit additional layers of interpretation and reinterpretation. In the Bristol Bay study, because women told their own stories in their own words, they became the authors of their experience, which illuminated their engagement with power and institutional frameworks. In this way, oral history is a way for people to disrupt dominant power structures. Feminist research has long embraced oral history for these reasons. Anderson et al. (2019, p. 92) argue that oral history allows for the exposure of "the dialectical nature of female consciousness," reflecting the authored experience as positioned within social structures. Key to our work in Bristol Bay was our commitment to partnership, both with the women themselves, as well as with the Bristol Bay Native Association. In line with the principles of political ecology, the goal was to upend the normalised researcher-subject relationship and make room for authorship and ownership of the story. The Bristol Bay Native Association was involved in recording the oral histories and maintained ownership of the research to support equity in the research and any resulting products.

Interdisciplinary and transdisciplinary methods provide another avenue to address complex research questions in fisheries that span disparate fields. Transdisciplinary research is research that (1) considers the complex system of factors that explain an issue's current state, (2) addresses both science and society's diverse perceptions of an issue, and (3) sets aside the idealised context of science to produce practically relevant knowledge (Pohl 2005). Transdisciplinary approaches may be used to investigate questions about power dynamics in the implementation of environmental policies among responsible stakeholders. For example, decisions about access to safe fishing areas and safety recommendations for consumption of fish, including those caught by recreational and subsistence fishers, are developed through collaboration between state departments, which may have different priorities and resources.

MM and LD used a transdisciplinary approach to elucidate relationships between urban development, socio-ecological vulnerability, and subsistence fisheries in the Gulf Coast, United States of America (Drakopulos et al. 2020; Marjadi et al. 2021). Qualitative analysis of relevant planning documents allowed us to evaluate how local governments prioritised access to fishing sites for urban marine fishers. Quantitative analysis of recreational fishing activity from a national fisheries survey, along with Census data, allowed an assessment of the socioeconomic makeup of the subsistence community (i.e., a subset of the larger recreational fishing group focused more on consumption), along with their preferred catch. Finally, quantitative methods were used to assess toxicology data from local agencies to compare testing practices to fish consumption practices. Overall, policy practices and planning failed to reflect fisher needs, especially for those in marginalised or minoritized communities (Marjadi et al. 2021). The interplay between fields and methods was necessary to this finding, as was the consideration of planning documents and socioeconomic and toxicology data to answer questions about the power dynamics around decisions about access to safe urban fishing sites and reliable fish testing (Fig. 3). Further, respectful collaboration across disciplines (e.g., social science, ecology, and public health) was necessary to conduct a range of qualitative (i.e., content analysis) and quantitative (i.e., spatial analysis, regression analysis) analyses, which allowed for a more nuanced understanding of the issues (Drakopulos et al. 2020).

Fig. 3 A reduction in legal points of water access for urban fishers means increasing precariousness for those who often continue to fish in popular locations like local bridges (here, a "no fishing" sign tries to prevent fishing from a neighbourhood bridge in St. Petersburg, Florida). (Photograph courtesy of the Tampa Bay Urban Fishing Project)

4.2 From Research to Action

These examples demonstrate a growing body of literature on the political ecology of recreational fishing. However, this type of research has yet to be regularly included as a standard practice in natural resource governance. Whereas dialogue-based approaches to stakeholder engagement have emerged over the past few decades that allow for greater understanding of competing discourses and opportunities for increased power sharing (Leong et al. 2012), systemic barriers have prevented them from becoming institutionalised in practice (Leong et al. 2011). Recognition of this shortcoming is reflected in the recent Executive Order on Advancing Racial Equity and Support for Underserved Communities Through the Federal Government (Biden 2021). This situation is due in part to managers and scientists often not recognizing their own biases towards ecological science as the dominant form of knowledge based on a more positivist worldview, which leaves little space for other worldviews in the decision-making process (IPBES 2022). Natural resource management has historically prioritised ecological data collection, whereas the conservation social sciences, including political ecology, have been largely misunderstood and underutilised (Bennett et al. 2017) (see our discussion of best practices in Sect. 6 for ways to begin addressing this challenge).

Aligning with the field's normative commitments, most political ecology work suggests changes that might be made to increase the social–ecological wellbeing and equity in a particular system. For example, Alaska contains language in its state constitution that requires "equity in how natural resources are allocated," which contributed to a more deliberative policy process than in Florida (Loring 2017, p. 102). Still, after examining the effects of fishing gear bans in Florida and Alaska as discussed earlier in this chapter, Loring (2017) found that questions of social justice were rarely part of the policy conversations, despite the clear socioeconomic effects of gear bans. Overall, he suggests that because all fisheries decisions have a political dimension, it is essential to reorganise fisheries governance in ways that take better account of people's "needs and values" in resource allocation decision making (Loring 2017, p. 102). In a different context, Karrow and Thompson (2016) noted that management decisions around the recreational Bahamas bonefish (*Albula vulpes*) industry often benefited travel companies, large non-governmental organisations, and certain tourism sectors while marginalising local residents. They suggest that "consultation, education, and co-management decision making are critical" to reduce conflict and improve communication in communities that cater to bonefishing tourism (Karrow and Thompson 2016, p. 49). Even if these types of recommendations are a small step toward change, it is part of the political ecology ethos to strive to give back to a community that is participating in research.

Some political ecologists also advocate for collaborative community research or community-based participatory research (Radonic and Kelly-Richards 2015). In their call for attention to power dynamics, political ecologists must not forget to address the researcher–participant power differentials inherent in most projects and instead enter into projects as "co-learners" (Wiber et al. 2009, p. 173). Participatory research is “a process of sequential reflection and action, carried out with and by local people rather than on them” (Cornwall and Jewkes 1995, p. 1667). Participatory research not only accesses and records “situated knowledge” through its methodology, but it also includes the possessors of that situated knowledge in research design, data collection, data analysis, and application of results (Radonic and Kelly-Richards 2015, p. 397). Participatory research in political ecology is particularly focused on planting productive seeds and has been useful in developing fisheries resource conservation strategies (Wiber et al. 2004), securing land tenure in contentious areas (Herlihy 2005), bolstering community food security (Carney et al. 2012), and mapping community use of fishing territories (St. Martin and Hall-Arber 2008a).

A specific example of this type of approach is embodied in Mai Ka Pō Mai: A Native Hawaiian Guidance Document for Papahānaumokuākea Marine National Monument (Office of Hawaiian Affairs et al. 2021). The monument is co-managed by the National Oceanic and Atmospheric Administration, US Fish and Wildlife Service, the State of Hawai‘i Department of Land and Natural Resources, and the State of Hawai‘i Office of Hawaiian Affairs. Mai Ka Pō Mai is the result of over 10 years of collaboration between the co-managers and the Native Hawaiian community and is based on conceptual representations of Hawaiian cosmology. Non-commercial fishing may be allowed under the Presidential Proclamations that established the core monument and expansion area (3 CFR 8031 (2006), 3 CFR 9478 (2016)). Further, ongoing regulatory discussions, while always challenging, are considering guiding principles outlined in Mai Ka Pō Mai.

5 How Is Political Ecology Evolving?

As a field with diverse origins, methods, and analytical approaches, political ecology is constantly evolving. Although the field originally focused on rural, third-world contexts (e.g., Blaikie and Brookfield 1987; Bassett 1988; Bryant 1992), political ecology perspectives have since been applied to issues around the globe at a variety of spatial and political scales (e.g., Adger et al. 2001; McCarthy 2005; Schroeder et al. 2006; Bennett 2010; Gabriel 2014). With its openness to diverse methodologies, political ecology also lends itself well to integration with other research perspectives, assuming such perspectives do not undermine political ecology’s normative outlook in favour of increased equity. However, political ecology has been critiqued for failing to live up to its interdisciplinary roots and for wielding critique more often than identifying actionable ways forward. In this section, we discuss these critiques and highlight how integration and collaboration with other disciplines could enhance political ecology research, especially as applied to recreational fishing. We also identify some exciting new areas of research that political ecology could productively engage with in relation to recreational fishing.

5.1 Critiques and Responses

Several internal and external critiques have been levelled at political ecology as it has evolved. For a field with the term "ecology" in it, some have suggested that political ecology should engage more deeply with the biophysical processes of the systems it studies (Vayda and Walters 1999). Moreover, these scholars cautioned political ecologists against assuming politics (or any other a priori process) as the main driver of ecological outcomes (Vayda and Walters 1999, 2011). Political ecologists themselves have debated the role of ecology in the field, with some of these debates relating to differing underlying epistemologies (Paulson et al. 2003). Many political ecologists take a pragmatic route in deciding when to engage in such work, agreeing that "levels of engagement with ecology…should stem from the questions being asked" (Turner 2016, p. 2). The arena of recreational fishing offers many intriguing opportunities for political ecologists to grapple more directly with the ecological elements of their work or to collaborate with ecologists. For example, to better understand the potentially uneven distributions of risks and benefits experienced by urban fishers, political ecology researchers studying patterns of resource access could collaborate with fishery biologists to test for contaminant levels in harvested resources.

Another set of critiques has focused on the other half of the term political ecology, pointing out that political ecologists could be more effective at addressing the real-world problems they study by increasing engagement with policymakers and politics more broadly. This can be challenging, as practitioners often feel a tension between a desire to further the theoretical goals of the field and engaging in activism. Nevertheless, political ecologists have pointed out that a desire "to make politics not only a research subject but a *practice* has long been an explicit and central goal of political ecology" (Peet and Watts 2004; Walker 2007, p. 363; Perreault et al. 2015; Robbins 2019). Recent work has suggested that political ecology can be a model for practising engaged scholarship in the Anthropocene era (Osborne 2017). In working toward this goal, clear narratives are critical for effecting change (Roe 1994). Here again, the realm of recreational fishing research offers opportunities for remedying this disconnect between academic and governance work.

The good news is that, both within fisheries research and the field of political ecology more broadly, progress is being made in relation to these critiques. In our own experience, addressing these issues has meant speaking with people outside of the usual academic social science forums and seeking cross-collaborations when possible. For example, NB and JF have discussed the primarily social results from their urban fishing work in Tampa Bay with regional organisations focused on ocean acidification and red tide issues. LF has served on several coastal and marine government and non-governmental organizations' advisory boards, where he has worked to integrate a political ecology perspective. MM and LD worked directly with state and regional agencies to collect relevant data for their analysis of social–ecological feedback loops facing subsistence fishers, and SW continues to partner with tribal associations for research projects in western Alaska. In this way, we and many others are working to "plant the seed" of more useful collaborations. Further, there is growing recognition that multiple knowledge systems (particularly Indigenous Knowledge) are important to develop improved understanding of the effects of and responses to climate change, as well as community resilience (Obermeister 2017). Indigenous Knowledge is embedded in people and shared through oral exchange and engagement with the natural environment; therefore, strong, enduring, and equitable partnerships are crucial.

Methodological flexibility will remain important to political ecology in its evolving applications to recreational fisheries research and management. As discussed above, the arena of recreational fishing provides an ideal place to address the most prominent critiques in political ecology relating to integrating more ecological work and better interfacing with resource managers and policymakers. That said, there remains a variety of potentially productive collaborations with other disciplines that could enhance political ecology research on recreational fishing. In terms of those featured in this volume, for instance, collaborations with environmental history, historical ecology, and social psychology could be particularly useful (see Landon et al. 2026; Thurstan et al. 2026). Political ecology work often benefits from a historical component, and the more detailed social–ecological histories that can be pieced together about a particular fishery, the better we can understand current challenges. In collaboration with social psychologists or behavioural economists, it would be interesting to learn more about the internal and shared motivations different groups of fishers have, and how such motivations might interact with the political–economic aspects of a fishery.

5.2 New Directions

Political ecology is also evolving in other ways that are applicable to recreational fishing research. Political ecologists have long been interested in how social relations are shaped by nature's material and biophysical properties (Bakker and Bridge 2006). In recent years, this interest in materiality has evolved in exciting ways with particular attention to more-than-human agency (i.e., the ability of non-human entities such as animals, plants, fungi, rocks, and others to have material effects in the world) (Bakker and Bridge 2021). More-than-human approaches employ a relational perspective that understands humans as "one actor in relation to a multiplicity of beings and entities without whom the human could not even live, let alone accomplish anything" (Durand and Sundberg 2022, p. 191). In other words, instead of a fixed world with humans as the sole drivers of action, human and more-than-human entities make a dynamic world through their interactions.

More-than-human political ecology is as diverse as the theoretical traditions from which it draws, including feminist theory, posthumanism, science and technology studies, post-colonial and decolonial thought, queer and trans theories, and Indigenous scholarship. For example, using posthumanist theory, (Durand and Sundberg 2022) examine how the unique qualities, or "plantliness," of a species endemic to

Chiapas, Mexico, have shaped environmental politics, knowledge production, and identity in the region. Hazard's (2022) work on river justice draws on queer and trans theory to advocate for a relational approach to river science and management that prioritises Indigenous worldmaking while disrupting colonial environmental practices. In fisheries, more specifically, political ecologists have examined how science and management practices interface with the slippery materialities (i.e., physical properties) of fish. This work is similarly wide-ranging in terms of environments (e.g., freshwater and marine) and contexts (e.g., Indigenous fisheries, aquaculture, commercial fisheries) (for a substantive review of more-than-human engagements with fisheries, see Drakopulos 2020). For example, Todd's (2014) work calls attention to how Indigenous ways of knowing and being with salmon could inform processes of Indigenous-state reconciliation. Havlick and Biermann (2021) demonstrate and critique how the unique qualities of fish genetics are leveraged through science to reduce freshwater cutthroat trout (*Oncorhynchus clarkii*) to their "genetic bodies" to be more easily managed and regulated through discourses of purity and nativeness, restructuring long-standing conservation practices in the process.

Beyond these more-than-human engagements, political ecology is also expanding in new directions that can be productively applied to recreational fishing questions. For instance, some political ecologists are taking an "experimental turn" to move beyond the field's reliance on critique (Hinchliffe et al. 2005; Landström et al. 2011; Lane et al. 2011; Braun 2015). This "experimental turn" does not refer to laboratory or field experiments following the scientific method, but rather to exploring and making visible ways of knowing, being, and governing that differ from the status quo (Braun 2015). For researchers and resource managers in the realm of recreational fishing, engaging with this "experimental turn" might mean thinking outside of typical management and research practices. For example, bringing together community members in ways that go beyond typical management meeting formats by perhaps holding a series of more casual meetings on board fishing boats, or by organising fishing policy "design charrettes" borrowed from participatory action techniques in public planning (Howard and Somerville 2014; Roggema et al. 2014). Using filmmaking as a participatory research method and link between community members and resource managers has further experimental potential (Cumming and Norwood 2012; Cumming et al. 2021; Acott et al. 2022). The hope is that more creative and experimental thinking can generate new practices "through which difference can be identified and new knowledges and political possibilities generated" (Braun 2015, p. 112).

Finally, another important development in political ecology is its "increasingly international and polylinguistic" character (Perreault et al. 2015, p. 624). For many years, political ecology was predominantly conducted by researchers from so-called developed countries studying issues in developing countries (Bryant 1992; Zimmerer and Bassett 2003). In the last two decades, this has increasingly changed as political ecologists both expanded their analyses to include "first world" problems (McCarthy 2002, 2005; Schroeder et al. 2006; Galt 2013) and simultaneously increased collaborations with scholars around the world (Bryant 2015). In the realm of fishing, exciting work is being done by global scholars investigating

Fig. 4 Urban fishing is common globally (here fishers try their luck at Nacka Strand, Stockholm, Sweden). (Photograph courtesy of Tessa Joosse)

transitions from commercial fishing to tourism (Chen and Chang 2017; Fabinyi 2020), the politics of rapid social–ecological regime changes (Nayak et al. 2016), implications of an emerging "blue economy" (Andriamahefazafy and Kull 2019; Childs and Hicks 2019; Cohen et al. 2019), questions surrounding recreational fishing in marine reserves (Engie and Quiroga 2014; Kayal et al. 2020), issues of urban fishing (Fig. 4) (Kalina et al. 2019; Boucquey and Fly 2021; Joosse et al. 2021; Marjadi et al. 2021; Boonstra et al. 2022; Carlson et al. 2022), and recreational fisheries governance challenges (Mmopelwa et al. 2013; Gómez et al. 2021; Gregory and Grant-Smith 2022). Political ecologists have only just begun to turn attention toward recreational fishing questions, and opportunities abound for further work.

6 Best Practices

Here, we highlight best practices for fisheries managers and researchers wanting to apply a political ecology lens to their work. Fundamentally, approaching a recreational fishing question from a political ecology perspective means thinking critically about how power may operate in any particular context. Besides a sensitivity to the workings of power, close attention to how fishers interact with the other social–ecological elements of a recreational fishery (e.g., the fish themselves, fishing infrastructures, accessibility, the macro- and micro-politics of the fishery regulations, cultural attitudes) is essential to analyse a fishery's political ecology. Finally, as we have emphasised throughout this chapter, findings from political ecology work will ideally result in some practical suggestions for changes to fisheries management practices that may increase social–ecological wellbeing and equity for all humans and non-humans involved in a fishery. With these key tenets in mind, in this section, we offer best practices for integrating political ecology in recreational fishing research and fisheries governance and management practices.

To organise this discussion and illustrate how interconnections between research and governance practices can be furthered, we use the four categories that (Bennett et al. 2017) identified as barriers to better incorporating social sciences in conservation: ideological, institutional, knowledge, and capacity.

6.1 Addressing Ideological Challenges

Ideological barriers in research and governance stem from the differing worldviews or paradigms that people bring to their work (Bennett et al. 2016). As discussed in Hunt et al. (2026), these paradigms can contain dramatically different sets of ontologies (theories of the nature of reality), epistemologies (theories about what constitutes valid knowledge), and methodologies (beliefs about the best ways to obtain valid knowledge). These differing paradigms matter in research projects because they influence how different disciplines structure data collection and analysis, and can affect how well a project team functions, depending on the extent to which differences in worldviews are acknowledged and discussed at the outset (Campbell 2005; Eigenbrode et al. 2007; Miller et al. 2008; Moon et al. 2021). Similar challenges can occur in governance with even greater implications because managers' acknowledged or taken-for-granted paradigms can affect what information is deemed legitimate or relevant for policymaking decisions.

Political ecologists, many of whom hold worldviews on the spectrum between *critical realism* and *constructivism*, may clash with researchers or policymakers holding more *positivist* worldviews (typically biophysical scientists or economists) (see Table 1 in Hunt et al. 2026). This can lead to fundamental disagreements about the ontological character (the material and social realities) of recreational fisheries, the issues of greatest relevance to address, and the best approaches to researching and rulemaking. To address such challenges, political ecologists and their research and governance partners must all approach their work with humility, empathy, and open-mindedness. Beyond that, some concrete steps that can be taken to increase the success of interdisciplinary collaborations include involving all perspectives in the scoping and problem definition phases of a project in addition to the later phases, addressing differences in worldviews explicitly rather than implicitly, balancing membership on a project team to avoid a single worldview dominating the work, maintaining frequent communications throughout a project, and allowing for divergent interpretations of project results (Campbell 2005; Miller et al. 2008; Turner et al. 2015; Bennett et al. 2016; Moon et al. 2021).

Their central interest in power relationships means that political ecologists should go beyond collaborations with others working with paradigms based on social or natural science. Recreational fishing communities have long histories based on many types of relationships, which can be as diverse as viewing nature as resources or kin (IPBES 2022). Regular inclusion of methodologies that respect and honour stakeholders' multiple relationships, worldviews, and knowledge systems in relation to marine ecosystems is needed (e.g., participatory methods from the inception of a project) (Reed 2008). Rather than approaching collaborations in terms of stakeholder knowledge that can be added to established ecological models, researchers can approach their work from the participants' perspective—What would make the process rewarding, worthwhile, credible, fair, inclusive, and create shared ownership? (Ballou et al. 2016; Wondolleck and Yaffee 2017).

6.2 Addressing Institutional Challenges

Institutions, with their norms, cultures, and particular histories, along with more formal rules, laws, or processes, can serve as either roadblocks or gateways to more progressive research and policymaking (Bennett et al. 2016). Particularly in the so-called "first world," commercial fisheries management is often deeply institutionalised, with reliance on specific bioeconomic data inputs and a conceptualization of fishers as individualized rational actors. This has historically left little room for creativity or community in fisheries policymaking (St. Martin 2001, 2005; Mansfield 2004, 2007). Though some large, institutionalised data-gathering programs exist with respect to recreational fisheries (e.g., the Marine Recreational Information Program surveys in the United States of America (NOAA Fisheries 2022)), these often suffer from similar concerns and tend to artificially smooth over important variabilities across regions and communities. With their more diffuse and difficult-to-track character, recreational fisheries present an opportunity to destabilise the status quo and insert some of political ecology's key concerns. Recreational fisheries may be more amenable to flexibility in management because of their diversity (Cooke et al. 2013). Further, the deeply interlinked social and ecological issues and challenges in recreational fisheries (Arlinghaus et al. 2016; Arlinghaus and Cooke 2009) are just the sort of questions that political ecology can help to address. We suggest that political ecologists should advocate for further inclusion in established and emerging recreational fisheries governance institutions.

6.3 Addressing Knowledge Challenges

The creation and deployment of knowledge is another key issue that can be a hidden barrier to integrating political ecology into recreational fishing research and management. Namely, that political ecology may be unfamiliar to many of those already working in this area. As Bennett et al. (2016, p. 61) writes about the social sciences more broadly, "the value of the range of social science methods (e.g., qualitative, quantitative, spatial, planning, evaluative, historical, meta-analytical, arts-based, and participatory methods) and related analytical techniques may not be immediately apparent to natural scientists, practitioners, or policymakers." If we are interested in extending political ecology work, we need to (i) do quality political ecology research, and (ii) communicate its value effectively. If research is co-produced (conducted in equitable partnerships with communities), these goals may even be accomplished simultaneously.

Here, we briefly review some best practices for creating knowledge with political ecology. Though there is no single prescription for conducting political ecology research, quality work in political ecology will generally include the following characteristics:

- A study design that makes methodological choices that flow from the key questions at hand (e.g., employing surveys when broadly descriptive data are needed, conducting in-depth interviews when "how" and "why" questions need to be answered, gathering ecological data for better understanding social–ecological connections) (Doolittle 2015; Turner 2016).
- A study design that explicitly considers and seeks to further understanding of power relationships: (i) between the researchers themselves and the populations to be studied, (ii) amongst the various human groups in the study, (iii) between these groups and other institutions, policies, or practices, and (iv) how such relationships interact with and affect the nonhuman actors relevant to the study. Importantly, particular power relationships should not be assumed at the outset, but their discovery should emerge from the evidence collected (e.g., it may be discovered that marginalised fishers or fish themselves have more agency than previously thought).
- A study design that examines both the discursive and material aspects of the social–ecological system being studied, and how they are interrelated. In doing so, the study should seek to develop a fuller understanding of the social–ecological relationships in the system.
- A study design that allows for time to be spent investigating the larger historical, socioeconomic, and political context of the particular case, and that follows "webs of relations across scales" (Helmcke 2022, p. 268). This type of design explicitly remains open to new, unexpected findings.
- A study design that seeks to link empirical fieldwork with an effort to engage with and further political ecology theory.
- A study design that engages with research participants as more than research subjects, for instance, as reviewers, co-designers, or other forms of partnership.
- Study outputs that address how wellbeing, equity, and justice might be improved in relation to the more-than-human actors involved in the study. For example, with respect to justice, important questions exist about the social interdependencies through which people exert power over human and non-human others.

It is also critical for political ecologists to communicate the substance and value of their research beyond academia. If political ecologists are to make a difference in recreational fisheries policymaking, they must learn to speak the language of their colleagues in fisheries science and management. As Bennett et al. (2016, p. 64) note,

> To connect and gain traction, social scientists may need to reflect on their outreach strategies...and grapple with how their work links to conservation biology and ecological outcomes throughout the research process. This does not mean the theory and language of social science should be abandoned, rather, it means social scientists need to learn to communicate for different audiences and purposes.

Although communicating and collaborating across the academia-policy boundary can be messy and challenging, it has long been an effort that political ecologists value (Walker 2006; Blaikie 2008; Robbins 2019). Despite (and because of) the field's many critiques of existing policymaking structures and processes, it is still important for political ecologists to be "useful" whenever possible (Blaikie 2012).

6.4 Addressing Capacity Challenges

Finally, the overall capacity in terms of human capital and skills and financial and temporal resources (Bennett et al. 2016) of recreational fisheries science and management to support political ecology work is currently limited. In an ideal world, more funding would be allocated to fisheries management agencies to support staff positions in the social sciences beyond economics, such as political ecology, anthropology, human geography, sociology, and others, and more time allocated to conduct the in-depth research that quality social science demands. In a resource-limited world, there are still ways to increase the capacity of managers to incorporate key concepts and principles from political ecology into existing management processes.

For instance, many external observers and internal agency professionals have noted the often acrimonious relationship that can develop between communities of natural resource users and the people in charge of regulating those resources, particularly during the policymaking process (Reed 2008; Ballou et al. 2016; Walker and Daniels 2019; Leong et al. 2020). With its focus on the role of power in shaping relationships, a political ecology perspective applied to the planning and implementation of public engagement processes could improve these processes' equity, and perhaps their success in terms of overall stakeholder and manager satisfaction. This could mean communicating key findings to resource managers to help them proactively recognize and address the specific power dynamics that might undermine fair and equitable engagement processes. For example, a common issue in fisheries management is that only those with the available time and funds are typically able to engage in most public input processes (Ballou et al. 2016). Attending to the basic needs of participants, such as providing childcare and reimbursing participants for travel costs, missed work time, and meals, could be an important first step to increase diversity in public participation.

Beyond these practical items, managers could be encouraged to shift the structure of both public engagement processes and fisheries governance more broadly to build greater capacity for thinking about recreational fisheries from a political ecology perspective. In terms of public engagement processes, managers could do more to think about creative ways to meet communities where they are, designing context-specific collaborative learning opportunities instead of applying the same types of engagements (e.g., public meetings) for different issues and communities (Leong et al. 2007, 2009; Reed 2008; Walker and Daniels 2019). In terms of broader governance structures, bureaucratic barriers could be removed (such as those that prevent cost reimbursements to participants), and more frequent and sustained collaborations between the natural and social scientists charged with fisheries management could be built into typical processes such as fisheries management plan updates or regional rebuilding plans.

7 Chapter Summary

We hope you, the reader, have learned from this chapter more about political ecology and how to apply its concepts to the study and management of recreational fisheries. Key takeaways include:

1. Whereas the field of political ecology is heterogeneous, a central thread running through its diverse topics is a concern with power relationships and how they affect both human and nonhuman actors. Related areas of focus include the complex interconnections between socioeconomic and ecological processes, and the types of visual and verbal discourses that help shape such processes via specific policies and practices.
2. With specific examples, we have shown how political ecologists working in fisheries question the received wisdom of dominant narratives, provide historical and cultural contextual information that allows for a more holistic understanding, and uncover power dynamics in complex socio-ecological systems. Further, this work moves beyond critique to provide ways forward with new understandings of what might be possible to shift the system of study toward more socially and ecologically just conditions.
3. Political ecologists use a variety of empirical research methods that flow both from the theoretical and normative commitments of the field and the specific questions of a particular study. These may include qualitative, quantitative, or mixed methods, and often include participatory components that engage community partners in the research.
4. As a young field, political ecology is evolving in interesting ways. Practitioners continue to work toward weaving together social–ecological insights, attending to more-than-human agency, incorporating more diverse and indigenous ways of knowing, engaging in creative methodological experimentation, and collaborating internationally. Thus far, political ecologists have studied recreational fishing in a variety of contexts, including in relation to livelihood transitions, the growth of fishing tourism, marine spatial planning, the "blue economy," urban provisioning, and protected areas work.
5. We have made suggestions for best practices in conducting political ecology work. These include addressing epistemological, ontological, and methodological differences with collaborators. Quality political ecology studies carefully examine power relationships, the discursive and material ways social–ecological systems are linked, and the historic, economic, and political contexts affecting

such systems. Finally, political ecology endeavours to make positive improvements in the distribution of justice and wellbeing among the more-than-human elements of social–ecological systems.

We hope that our suggestions in this chapter will inspire more political ecology work in the realm of recreational fisheries research and management. We look forward to engaging further in these efforts as this work grows.

Acknowledgments We are grateful for the helpful comments of two peer reviewers, a National Oceanic and Atmospheric Administration reviewer, and the editors of this volume. We also thank the many research participants we have worked with, whose kindness and knowledge have made our work possible.

References

Acheson J (1988) The lobster gangs of Maine. University Press of New England, Hanover

Acheson JM, Brewer J (2003) Changes in the territorial system of the Maine lobster industry. In: Dolšak N, Ostrom E (eds) The commons in the New Millenium: challenges and adaptation. Massachusetts Institute of Technology, Cambridge, MA, pp 37–59

Acott TG, Willis C, Ranger S, Cumming G, Richardson P, O'Neill R, Ford A (2022) Coastal transformations and connections: revealing values through the community voice method. People Nat 5(2):403–414

Adger WN, Benjaminsen TA, Brown K, Svarstad H (2001) Advancing a political ecology of global environmental discourses. Dev Change 32(4):681–715

Ahlborg H, Nightingale AJ (2018) Theorizing power in political ecology: the where of power in resource governance projects. J Polit Ecol 25:381–401

Allen A (1999) The power of feminist theory: domination, resistance, solidarity. Westview Press, Boulder

Anderson K, Armitage S, Jack D, Wittner J (2019) Beginning where we are: feminist methodology in oral history. In: Nielsen JM (ed) Feminist research methods: exemplary readings in the social sciences. Routledge, New York, pp 91–109

Andriamahefazafy M, Kull CA (2019) Materializing the blue economy: tuna fisheries and the theory of access in the Western Indian Ocean. J Polit Ecol 26(1):403–424

Arendt H (1970) On violence. Harcourt Brace, New York

Arlinghaus R, Cooke SJ (2009) Recreational fisheries: socioeconomic importance, conservation issues and management challenges. In: Dickson B, Hutton J, Adams WA (eds) Recreational hunting, conservation and rural livelihoods: science and practice. Wiley, Oxford, pp 39–58

Arlinghaus R, Cooke SJ, Sutton SG, Danylchuk AJ, Potts W, Freire KM, Alós J, Da Silva ET, Cowx IG, Van Anrooy R (2016) Recommendations for the future of recreational fisheries to prepare the social-ecological system to cope with change. Fish Manag Ecol 23(3–4):177–186

Ascui F, Haward M, Lovell H (2018) Salmon, sensors, and translation: the agency of big data in environmental governance. Environ Plan D 36(5):905–925

Bakker K, Bridge G (2006) Material worlds? Resource geographies and the "matter of nature". Prog Hum Geogr 30(1):5–27

Bakker K, Bridge G (2021) Material worlds redux: mobilizing materiality within critical resource geography. In: Himley M, Havice E, Valdivia G (eds) The Routledge handbook of critical resource geography. Routledge, New York, pp 43–56

Ballou LM, Albritton N, Horowitz LS (2016) Enforcement encourages participation in resource management: explaining a fisheries management paradox in Hawaii. Asia Pac Viewp 57(3): 379–392
Bassett TJ (1988) The political ecology of peasant-herder conflicts in the northern Ivory Coast. Ann Assoc Am Geogr 78(3):453–472
Batterbury S (2018) Affirmative and engaged political ecology: practical applications and participatory development actions. Nord Geogr Publ 47(5):111–131
Bennett J (2010) Vibrant matter: a political ecology of things. Duke University Press, Durham
Bennett NJ, Roth R, Klain SC, Chan KM, Clark DA, Cullman G, Epstein G, Nelson MP, Stedman R, Teel TL, Thomas RE, Wyborn C, Curran D, Greenberg A, Sandlos J, Verissimo D (2017) Mainstreaming the social sciences in conservation. Conservation Biology, pp 31(1): 56–66
Bennett NJ (2018) In political seas: engaging with political ecology in the ocean and coastal environment. Coast Manag 47(1):67–87
Bennett NJ (2019) In political seas: engaging with political ecology in the ocean and coastal environment. Coastal Management 47(1):67–87
Bennett NJ, Roth R, Klain SC, Chan KM, Clark DA, Cullman G, Epstein G, Nelson MP, Stedman R, Teel TL (2016) Mainstreaming the social sciences in conservation. Conserv Biol 31(1):56–66
Berkes F, Feeny D, McCay BJ, Acheson JM (1989) The benefits of the commons. Nature 340:91–93
Biden J (2021) Executive Order 13985 of January 20, 2021: advancing racial equity and support for underserved communities through the Federal Government. Fed Regist 86(14):7009–7012
Blaikie P (1999) A review of political ecology: issues, epistemology and analytical narratives. Z Wirtschaftsgeogr 43(3–4):131–147
Blaikie P (2008) Epilogue: towards a future for political ecology that works. Geoforum 39(2): 765–772
Blaikie P (2012) Should some political ecology be useful? The inaugural lecture for the Cultural and Political Ecology Specialty Group, annual meeting of the Association of American Geographers, April 2010. Geoforum 43(2):231–239
Blaikie P, Brookfield H (1987) Land degradation and society. Methuen and Co., York
Boonstra W (2016) Conceptualizing power to study social-ecological interactions. Ecol Soc 21(1): 21
Boonstra WJ, Boucquey N, Carlson AK, Drakopulos L, Fly J, Joosse S, Panchang S, Marjadi MN, Rieser A, Wernersson HC (2022) Urban fishing reveals underrepresented diversity. Nat Food 3: 295
Boucquey N (2016) Actors and audiences: negotiating fisheries management. J Environ Policy Plan 18(4):426–446
Boucquey N (2017) "That's my livelihood, it's your fun": the conflicting moral economies of commercial and recreational fishing. J Rural Stud 54:138–150
Boucquey N (2020) The "nature" of fisheries governance: narratives of environment, politics, and power and their implications for changing seascapes. J Polit Ecol 27(1):169–189
Boucquey N, Fly J (2021) Contested commoning: urban fishing spaces and community wellbeing. Int J Commons 15(1):305–319
Braun B (2005) Environmental issues: writing a more-than-human urban geography. Prog Hum Geogr 29(5):635–650
Braun B (2015) From critique to experiment? Rethinking political ecology for the Anthropocene. In: Perreault TA, Bridge G, McCarthy J (eds) The Routledge handbook of political ecology. Routledge, London, pp 102–114
Bryant R (1992) Political ecology: an emerging research agenda in third world studies. Polit Geogr 11(1):12–36
Bryant R (ed) (2015) The international handbook of political ecology. Edward Elgar Publishing, Cheltenham

Campbell LM (2005) Overcoming obstacles to interdisciplinary research. Conserv Biol 19(2): 574–577

Campbell LM (2007) Local conservation practice and global discourse: a political ecology of Sea Turtle conservation. Ann Assoc Am Geogr 97(2):313–334

Campbell LM, Meletis ZA (2011) Agreement on water and a watered-down agreement: the political ecology of contested coastal development in Down East, North Carolina. J Rural Stud 27(3): 308–321

Carlson AK, Boonstra WJ, Joosse S, Rubenstein DI, Levin SA (2022) More than ponds amid skyscrapers: urban fisheries as multiscalar human-natural systems. Aquat Ecosyst Health Manag 25(1):49–58

Carney MA (2014) The biopolitics of "food insecurity": towards a critical political ecology of the body in studies of women's transnational migration. J Polit Ecol 21(1):1–18

Carney PA, Hamada JL, Rdesinski R, Sprager L, Nichols KR, Liu BY, Pelayo J, Sanchez MA, Shannon J (2012) Impact of a community gardening project on vegetable intake, food security and family relationships: a community-based participatory research study. J Community Health 37(4):874–881

Chen CL, Chang YC (2017) A transition beyond traditional fisheries: Taiwan's experience with developing fishing tourism. Mar Policy 79:84–91

Chiaravalloti RM (2017) Overfishing or over reacting? Management of fisheries in the Pantanal Wetland, Brazil. Conserv Soc 15(1):111–122

Childs J, Hicks CC (2019) Securing the blue: political ecologies of the blue economy in Africa. J Polit Ecol 26(1):323–340

Cinner JE, Zamborain-Mason J, Maire E, Hoey AS, Graham NA, Mouillot D, Villéger S, Ferse S, Lockie S (2022) Linking key human-environment theories to inform the sustainability of coral reefs. Curr Biol 32(12):2610–2620

Cohen PJ, Allison EH, Andrew NL, Cinner J, Evans LS, Fabinyi M, Garces LR, Hall SJ, Hicks CC, Hughes TP, Jentoft S, Mills DJ, Masu R, Mbaru EK, Ratner BD (2019) Securing a just space for small-scale fisheries in the Blue Economy. Front Mar Sci 6:171

Cole S (2017) Water worries: an intersectional feminist political ecology of tourism and water in Labuan Bajo, Indonesia. Ann Tour Res 67:14–24

Collard RC, Dempsey J (2013) Life for sale? The politics of lively commodities. Environ Plan A 45(11):2682–2699

Cooke S, Suski CD, Arlinghaus R, Danylchuk AJ (2013) Voluntary institutions and behaviours as alternatives to formal regulations in recreational fisheries management. Fish Fish 14(4):439–457

Cornwall A, Jewkes R (1995) What is participatory research? Soc Sci Med 41(12):1667–1676

Cumming G, Norwood C (2012) The community voice method: using participatory research and filmmaking to foster dialog about changing landscapes. Landsc Urban Plan 105(4):434–444

Cumming G, Campbell L, Norwood C, Ranger S, Richardson P, Sanghera A (2021) Putting stakeholder engagement in its place: how situating public participation in community improves natural resource management outcomes. GeoJournal 87:209–221

Donkersloot R, Coleman J, Carothers C, Ringer D, Cullenberg P (2020) Kin, community, and diverse rural economies: rethinking resource governance for Alaska rural fisheries. Mar Policy 117:103966

Doolittle A (2015) The best of many worlds: methodological pluralism in political ecology. In: Bryant R (ed) The international handbook of political ecology. Edward Elgar Publishing, Cheltenham, pp 515–529

Drakopulos L (2020) New materialist approaches to fisheries: the birth of "bycatch". Environ Soc 11(1):100–114

Drakopulos L, Robertson D, Panchang S, Marjadi M, Koehn Z, Guo L (2020) Transdisciplinary synthesis research in unruly environments: reflecting on a case study of vulnerability and urban dishing in the American Gulf Coast. Projections 15:1–29. https://doi.org/10.1162/00c13b77.0c274823

Durand L, Sundberg J (2022) Monster plants: the vegetal political ecology of Lacandonia schismatica. J Polit Ecol 29(1):189–207
Ebbin SA (2017) Fishing for food: piloting an exploration of the invisible subsistence harvest of coastal resources in Connecticut. Agriculture and Food Security 6(1):12
Ehrlich P (1968) The Population Bomb. Ballantine Books, New York
Eigenbrode SD, O'Rourke M, Wulfhorst JD, Althoff DM, Goldberg CS, Merrill K, Morse W, Nielsen-Pincus M, Stephens J, Winowiecki L (2007) Employing philosophical dialogue in collaborative science. Bioscience 57(1):55–64
Elias N (2012) What is sociology?, vol 5 (trans: Morrissey G, Mennell S, Jephcott E). University College Dublin Press, Dublin
Emery MR, Pierce AR (2005) Interrupting the telos: locating subsistence in contemporary US forests. Environment and Planning A 37(6):981–993
Engie K, Quiroga D (2014) The emergence of recreational fishing in the Galápagos Marine Reserve: adaptation and complexities. In: Denkinger J, Vinueza L (eds) The Galapagos Marine Reserve: a dynamic social-ecological system. Springer, New York, pp 203–226
Escobar A (1998) Whose knowledge, whose nature? Biodiversity, conservation, and the political ecology of social movements. J Polit Ecol 5:53–82
Establishment of the Northwestern Hawaiian Islands Marine National Monument, 3 CFR 8031 (2006)
Fabinyi M (2020) The role of land tenure in livelihood transitions from fishing to tourism. Marit Stud 19(1):29–39
FAO (2012) Recreational fisheries. FAO technical guidelines for responsible fisheries, no. 13. FAO, Rome
Feeny D, Berkes F, McCay BJ, Acheson JM (1990) The tragedy of the commons: twenty-two years later. Hum Ecol 18(1):1–19
Fly J, Boucquey N (2023) Flows of care in "third places": the role of shore fishing spaces in collective wellbeing. Wellbeing Space Soc 4:100128
Forsyth T (2003) Critical political ecology: the politics of environmental science. Routledge, New York
Gabriel N (2014) Urban political ecology: environmental imaginary, governance, and the non-human. Geogr Compass 8(1):38–48
Galt RE (2013) Placing food systems in first world political ecology: a review and research agenda. Geogr Compass 7(9):637–658
Gibson-Graham JK, Cameron J, Healy S (2016) Commoning as a postcapitalist politics. In: Amin A, Howell P (eds) Releasing the commons: rethinking the futures of the commons. Routledge, New York, pp 192–212
Gómez S, Carreño A, Lloret J (2021) Cultural heritage and environmental ethical values in governance models: conflicts between recreational fisheries and other maritime activities in Mediterranean marine protected areas. Mar Policy 129:104529
Greenberg JB (2006) The political ecology of fisheries in the Upper Gulf of California. In: Biersack A, Greenberg JB (eds) Reimagining political ecology. Duke University Press, Durham, pp 121–148
Gregory K, Grant-Smith D (2022) Engaging recreational fishers in fisheries' stewardship: a case of responsibilisation in natural resource management. J Environ Plan Manag 66(7):1563–1578
Hardin G (1968) The tragedy of the commons. Science 162:1243–1248
Harris LM (2015) A quarter century of knowledge and change: pushing feminism, politics, and ecology in new directions with feminist political ecology. In: Buechler S, Hanson AM (eds) A political ecology of women, water, and global environmental change. Routledge, New York
Haugaard M (2014) Power. In: Gibbons MT, Coole D, Ellis E, Ferguson K (eds) The encyclopedia of political thought. Wiley, Oxford, pp 2965–2978
Havlick DG, Biermann C (2021) Wild, native, or pure: trout as genetic bodies. Sci Technol Hum Values 46(6):1201–1229

Hawkins R, Silver JJ (2017) From selfie to #sealfie: nature 2.0 and the digital cultural politics of an internationally contested resource. Geoforum 79:114–123

Hazard CW (2022) Underflows: queer trans ecologies and river justice. University of Washington Press, Seattle

Helmcke C (2022) Ten recommendations for political ecology case research. J Polit Ecol 29(1): 266–277

Herlihy PH (2005) Participatory research mapping of Indigenous lands in Darién, Panama. Hum Organ 62(4):315–331

Hinchliffe S, Kearnes MB, Degen M, Whatmore S (2005) Urban wild things: a cosmopolitical experiment. Environ Plan D 23(5):643–658

Holland DS, Abbott JK, Norman KE (2020) Fishing to live or living to fish: job satisfaction and identity of west coast fishermen. Ambio 49(2):628–639

Howard Z, Somerville MM (2014) A comparative study of two design charrettes: implications for codesign and participatory action research. CoDesign 10(1):46–62

Hunt LM, Boucquey N, Beardmore B, Christensen J, Fulton DC, Mackay M, Melstrom RT, Post JR, Schroeder SA, Thurstan RH, van Putten EI, Arlinghaus R (2026) Introducing disciplinary perspectives and interdisciplinary possibilities for understanding recreational fishers and fisheries. In: Pope KL, Arlinghaus R, Hunt LM, Lynch AJ, van Poorten BT (eds) Understanding recreational fishers: disciplinary and interdisciplinary approaches for fisheries management. Springer Nature, Cham, pp 57–94

IPBES (2022) Summary for policymakers of the methodological assessment of the diverse values and valuation of nature of the Intergovernmental Science-Policy Platform on Biodiversity and Ecosystem Services. https://doi.org/10.5281/zenodo.7075892. Accessed 9 July 2022

Iwane M, Leong KM, Kleiber D (2021) Non-commercial fishing in policy, practice, and culture: insights from the Papahanaumokuakea Marine National Monument. U.S. Dept. of Commerce, NOAA technical memorandum NOAA-TM-NMFS-PIFSC-114. https://doi.org/10.25923/r365-2f47

Jenkins D (2015) Impacts of neoliberal policies on non-market fishing economies on the Yukon River, Alaska. Mar Policy 61:356–365

Joosse S, Hensle L, Boonstra WJ, Ponzelar C, Olsson J (2021) Fishing in the city for food—a paradigmatic case of sustainability in urban blue space. NPJ Urban Sustain 1(1):41

Kalina MR, Mbereko A, Maharaj B, Botes A (2019) Subsistence marine fishing in a neoliberal city: a political ecology analysis of securitization and exclusion in Durban, South Africa. J Polit Ecol 26(1):363–380

Karrow T, Thompson T (2016) Political ecology of the flats fishing industry in the Bahamas. In: Nepal S, Saarinen J (eds) Political ecology and tourism. Routledge, London, pp 63–77

Kayal M, Cigala M, Cambra E, Soulat N, Mercader M, Lebras A, Ivanoff P, Sébési L, Lassus-Debat A, Hartmann V (2020) Marine reserve benefits and recreational fishing yields: the winners and the losers. PLoS One 15(12):e0237685

Klinenberg E (2018) Palaces for the people: how social infrastructure can help fight inequality, polarization, and the decline of civic life. Broadway Books, New York

Knol M (2011) Mapping ocean governance: from ecological values to policy instrumentation. J Environ Plan Manag 54(7):979–995

Landon AC, Schroeder SA, Stefen SF, Arlinghaus R (2026) Social psychology for understanding recreational fishers and fisheries. In: Pope KL, Arlinghaus R, Hunt LM, Lynch AJ, van Poorten BT (eds) Understanding recreational fishers: disciplinary and interdisciplinary approaches for fisheries management. Springer Nature, Cham, pp 157–196

Landström C, Whatmore SJ, Lane SN, Odoni NA, Ward N, Bradley S (2011) Coproducing flood risk knowledge: redistributing expertise in critical "participatory modelling". Environ Plan A 43(7):1617–1633

Lane SN, Odoni N, Landström C, Whatmore SJ, Ward N, Bradley S (2011) Doing flood risk science differently: an experiment in radical scientific method. Trans Inst Br Geogr 36(1):15–36

Lavoie A, Lee J, Sparks K, Hoseth G, Wise S (2019) Engaging with women's knowledge in Bristol Bay Fisheries through oral history and participatory ethnography. Fisheries 44(7):331–337

Leong KM, McComas KA, Decker DJ (2007) Commentary: matching the forum to the fuss: using coorientation contexts to address the paradox of public participation in natural resource management. Environ Pract 9(3):195–205

Leong KM, Forester JF, Decker DJ (2009) Moving public participation beyond compliance: uncommon approaches to finding common ground. George Wright Forum 26(3):23–39

Leong KM, Emmerson DP, Byron R (2011) The new governance era: implications for collaborative conservation and adaptive management in Department of the Interior Agencies. Hum Dimens Wildl 16(4):236–243

Leong KM, Decker DJ, Lauber B (2012) Stakeholders as beneficiaries of wildlife management. In: Decker DJ, Riley SJ, Siemer WF (eds) Human dimensions of wildlife management. Johns Hopkins University Press, Baltimore, pp 26–40

Leong KM, Torres A, Wise S, Hospital J (2020) Beyond recreation: when fishing motivations are more than sport or pleasure. Administrative report H-20-05. Pacific Islands Fisheries Science Center. https://doi.org/10.25923/K5HK-X319

Leslie-Bole H, Perramond EP (2017) Oyster feuds: conflicting discourses and outcomes in Point Reyes, California. J Polit Ecol 24(1):144–166

Longo SB, Clausen R, Clark B (2015) The tragedy of the commodity: oceans, fisheries, and aquaculture. Rutgers University Press, New Brunswick

Loring PA (2017) The political ecology of gear bans in two fisheries: Florida's net ban and Alaska's Salmon wars. Fish Fish 18(1):94–104

Loring PA (2022) Coral reefs: moving beyond Malthus. Curr Biol 32(12):R569–R571

Lynch AJ, Hunt LM, Beardmore B, van Poorten BT, Pope KL, Arlinghaus R (2026) Complexity and integration of recreational fisheries. In: Pope KL, Arlinghaus R, Hunt LM, Lynch AJ, van Poorten BT (eds) Understanding recreational fishers: disciplinary and interdisciplinary approaches for fisheries management. Springer Nature, Cham, pp 3–19

Mackay M, van Putten EI, Richter A, Abatayo AL, Chuah S-H, Dankel DJ, Schadeberg A, Snyder H (2026) Behavioural economics for understanding recreational fishers and fisheries. In: Pope KL, Arlinghaus R, Hunt LM, Lynch AJ, van Poorten BT (eds) Understanding recreational fishers: disciplinary and interdisciplinary approaches for fisheries management. Springer Nature, Cham, pp 261–284

Mansfield B (2004) Neoliberalism in the oceans: "rationalization," property rights, and the commons question. Geoforum 35(3):313–326

Mansfield B (2007) Property, markets, and dispossession: the Western Alaska Community Development Quota as neoliberalism, social justice, both, and neither. Antipode 39(3):479–499

Mansfield B (2010) Modern industrial fisheries and the crisis of overfishing. In: Peet R, Robbins P, Watts MJ (eds) Global political ecology. Routledge, New York, pp 98–113

Marjadi MN, Drakopulos L, Guo LW, Koehn JZ, Panchang SV, Robertson D (2021) Negative socio-environmental feedback loop may foster inequality for urban marine subsistence fishers. Environ Sci Pol 121:68–77

Mas IM (2015) The fishing footprint of a tourism-based economy: displacing seafood consumption from local to distant waters in the Balearic Islands. J Polit Ecol 22(1):211–238

May CK (2016) Visibility and invisibility: structural, differential, and embedded power in collaborative governance of fisheries. Soc Nat Resour 29(7):759–774

McCarthy J (2002) First World political ecology: lessons from the wise use movement. Environ Plan A 34:1281–1302

McCarthy J (2005) First World political ecology: directions and challenges. Environ Plan A 37(6): 953–958

McCay B, Jentoft S (2007) Market or community failure? Critical perspectives on common property research. Hum Organ 57(1):21–29

McEvoy AF (1987) Toward an interactive theory of nature and culture: ecology, production, and cognition in the California fishing industry. Environ Rev 11(4):289–305

Melstrom RT, Jungers B, Bi X, Abbott JK (2026) Resource economics for understanding recreational fishers and fisheries. In: Pope KL, Arlinghaus R, Hunt LM, Lynch AJ, van Poorten BT (eds) Understanding recreational fishers: disciplinary and interdisciplinary approaches for fisheries management. Springer Nature, Cham, pp 123–156

Miller TR, Baird TD, Littlefield CM, Kofinas G, Chapin FS III, Redman CL (2008) Epistemological pluralism: reorganizing interdisciplinary research. Ecol Soc 13(2):46

Mmopelwa G, Ngwenya BN, Kgathi DL (2013) Contending claims over access to fisheries: a case study of the Okavango Delta Panhandle, Botswana. J Sustain Dev 6(3):58–69

Moon K, Cvitanovic C, Blackman DA, Scales IR, Browne NK (2021) Five questions to understand epistemology and its influence on integrative marine research. Front Mar Sci 8:574158

Morriss P (2002 [1988]) Power: a philosophical analysis. Manchester University Press, Manchester

Nayak PK, Armitage D, Andrachuk M (2016) Power and politics of social-ecological regime shifts in the Chilika lagoon, India and Tam Giang lagoon, Vietnam. Reg Environ Chang 16(2): 325–339

Neumann RP (1998) Imposing wilderness: struggles over livelihood and nature preservation in Africa. University of California Press, Berkeley

Neumann RP (2004) Moral and discursive geographies in the war for biodiversity in Africa. Polit Geogr 23(7):813–837

Neumann RP (2014) Making political ecology. Routledge, New York

Nieman CM, Rudman AN, Chory ML, Murray GD, Fairbanks LW, Campbell LM (2021) Fishing for food: values and benefits associated with coastal infrastructure. PLoS One 16(4):e0249725

Nixon R (2011) Slow violence and the environmentalism of the poor. Harvard University Press, Cambridge

NOAA Fisheries (2022) Recreational fishing data. https://www.fisheries.noaa.gov/topic/recreational-fishing-data. Accessed 7 Oct 2022

Nolan C (2019) Power and access issues in Ghana's coastal fisheries: a political ecology of a closing commodity frontier. Mar Policy 108:103621

Obermeister N (2017) From dichotomy to duality: addressing interdisciplinary epistemological barriers to inclusive knowledge governance in global environmental assessments. Environ Sci Pol 68:80–86

Office of Hawaiian Affairs, NOAA, U.S. Fish and Wildlife Service, and State of Hawai'i (2021) Mai Ka Pō Mai: a native Hawaiian guidance document for Papahānaumokuākea Marine National Monument. https://dlnr.hawaii.gov/ahamoku/files/2022/04/OHA-NOAA-USFWS-and-State-of-Hawaii.-2021.-Mai-Ka-Po-Mai-A-Native-Hawaiian-Guidance-document-for-Papahanaumokuakea-Marine-National-Monument.-Honolulu-HI-Office-of-Hawaiian-Affairs.pdf. Accessed 30 June 2023

Osborne T (2017) Public political ecology: a community of praxis for Earth stewardship. J Polit Ecol 24(1):843–860

Ostrom E (1990) Governing the commons: the evolution of institutions for collective action. Cambridge University Press, Cambridge

Ostrom E (2009) A general framework for analyzing sustainability of social-ecological systems. Science 325:419–422

Ostrom E, Dietz T, Dolšak N, Stern PC, Stonich S, Weber EU (eds) (2002) The drama of the commons. National Academy Press, Washington, DC

Painter M, Durham WH (1995) The social causes of environmental destruction in Latin America. University of Michigan Press, Ann Arbor

Pansardi P, Bindi M (2021) The new concepts of power? Power-over, power-to and power-with. J Political Power 14:51–71

Papahānaumokuākea Marine National Monument Expansion, 3 CFR 9478 (2016)

Paulson S, Gezon LL, Watts M (2003) Locating the political in political ecology: an introduction. Hum Organ 62(3):205–217

Peet R, Watts M (eds) (2004) Liberation ecologies: environment, development, social movements. Routledge, London

Perreault TA, Bridge G, McCarthy J (eds) (2015) The Routledge handbook of political ecology. Routledge, London

Petersson MT, Dellmuth LM, Merrie A, Österblom H (2019) Patterns and trends in non-state actor participation in regional fisheries management organizations. Mar Policy 104:146–156

Poe MR, Levin PS, Tolimieri N, Norman K (2015) Subsistence fishing in a 21st century capitalist society: from commodity to gift. Ecol Econ 116:241–250

Pohl C (2005) Transdisciplinary collaboration in environmental research. Futures 37(10): 1159–1178

Pollnac RB, Seara T, Colburn LL (2015) Aspects of fishery management, job satisfaction, and well-being among commercial fishermen in the Northeast Region of the United States. Soc Nat Resour 28(1):75–92

Quimby B, Crook SE, Miller KM, Ruiz J, Lopez-Carr D (2020) Identifying, defining and exploring angling as urban subsistence: pier fishing in Santa Barbara, California. Mar Policy 121:104197

Radonic L, Kelly-Richards S (2015) Pipes and praxis: a methodological contribution to the urban political ecology of water. J Polit Ecol 22(1):389–409

Reed MS (2008) Stakeholder participation for environmental management: a literature review. Biol Conserv 141(10):2417–2431

Robbins P (2019) Political ecology: a critical introduction. Wiley, Oxford

Rocheleau DE, Thomas-Slayter BP, Wangari E (1996) Gender and environment: a feminist political ecology perspective. In: Rocheleau DE, Thomas-Slayter BP, Wangari E (eds) Feminist political ecology: global issues and local experience. Routledge, London, pp 3–24

Roe E (1994) Narrative policy analysis: theory and practice. Duke University Press, Durham

Roggema R, Martin J, Vos L (2014) Governance of climate adaptation in Australia: design charrettes as a creative tool for participatory action research. In: Van Buuren A, Eshuis J, Van Vliet M (eds) Action research for climate change adaptation. Routledge, London, pp 110–127

Schroeder RA, St. Martin K, Albert KE (2006) Political ecology in North America: discovering the Third World within? Geoforum 37(2):163–168

Springett S (2022) Art-making for political ecology: practice, poetics and activism through enchantment. Continuum 36(3):478–494

St. Martin K (2001) Making space for community resource management in fisheries. Ann Assoc Am Geogr 91(1):122–142

St. Martin K (2005) Mapping economic diversity in the First World: the case of fisheries. Environ Plan A 37:959–979

St. Martin K, Hall-Arber M (2008a) Creating a place for "community" in New England Fisheries. Hum Ecol Rev 15(2):161–170

St. Martin K, Hall-Arber M (2008b) The missing layer: geo-technologies, communities, and implications for marine spatial planning. Mar Policy 32(5):779–786

Svarstad H, Benjaminsen TA, Overa R (2018) Power theories in political ecology. J Polit Ecol 25(1):350–363

Tan-Mullins M (2007) The state and its agencies in coastal resources management: the political ecology of fisheries management in Pattani, southern Thailand. Singap J Trop Geogr 28(3): 348–361

Thurstan RH, Chong-Montenegro C, Christensen J, Clark A, Gaynor A (2026) Environmental history and historical ecology for understanding recreational fishers and fisheries. In: Pope KL, Arlinghaus R, Hunt LM, Lynch AJ, van Poorten BT (eds) Understanding recreational fishers: disciplinary and interdisciplinary approaches for fisheries management. Springer Nature, Cham, pp 233–259

Todd Z (2014) Fish pluralities: human-animal relations and sites of engagement in Paulatuuq, Arctic Canada. Étud Inuit 38(1–2):217–238

Turner MD (2016) Political ecology II: engagements with ecology. Prog Hum Geogr 40(3): 413–421

Turner VK, Benessaiah K, Warren S, Iwaniec D (2015) Essential tensions in interdisciplinary scholarship: navigating challenges in affect, epistemologies, and structure in environment-society research centers. High Educ 70(4):649–665
Vayda AP, Walters BB (1999) Against political ecology. Hum Ecol 27(1):167–179
Vayda AP, Walters BB (2011) Causal explanation for social scientists: a reader. AltaMira Press, Plymouth
Walker P (2006) Political ecology: where is the policy? Prog Hum Geogr 30(3):382
Walker P (2007) Political ecology: where is the politics? Prog Hum Geogr 31(3):363
Walker GB, Daniels SE (2019) Collaboration in environmental conflict management and decision-making: comparing best practices with insights from collaborative learning work. Front Commun 4:2
Watts MJ (1983) Silent violence: food, famine, and peasantry in Northern Nigeria. University of California Press, Berkeley
Watts M (2000) Political ecology. In: A companion to economic geography, Sheppard E and Barnes T, Eds Blackwell, Malden, p 257–274
Whatmore S (2006) Materialist returns: practising cultural geography in and for a more-than-human world. Cult Geogr 13(4):600–609
Wiber M, Berkes F, Charles A, Kearney J (2004) Participatory research supporting community-based fishery management. Mar Policy 28(6):459–468
Wiber M, Charles A, Kearney J, Berkes F (2009) Enhancing community empowerment through participatory fisheries research. Mar Policy 33(1):172–179
Wondolleck JM, Yaffee SL (2017) Marine ecosystem-based management in practice: different pathways, common lessons. Island Press, Washington, DC
Zimmerer KS, Bassett TJ (2003) Political ecology: an integrative approach to environment-development studies. The Guilford Press, New York

Environmental History and Historical Ecology for Understanding Recreational Fishers and Fisheries

Ruth H. Thurstan, Carolina Chong-Montenegro, Joseph Christensen, Anna Clark, and Andrea Gaynor

Abstract The concept of fishing primarily for leisure, personal fulfilment or sport has existed for centuries. In contrast, the application of history and the use of historical sources and approaches to understanding recreational fisheries and fisher behaviour remains an emerging area of research. This chapter explores the variety of ways in which historical sources and perspectives have been used to expand our understanding of recreational fishing, past and present. We describe how historical approaches can shed light on the origins of recreational fishing, changing patterns in recreational fishing behaviour, the drivers and outcomes of behavioural change, the extent to which recreational fishing is bound up in people's values, identities and cultures, and how fishers have treated and viewed the natural environment, and their place within it, through time. Historical approaches help us to understand how and when the fishing activities, behaviours, values, cultures and identities we observe today were formed, and how they change across time. The chapter also surveys the range of sources that have been used to study the past ecological impacts of recreational fishing, and some of the challenges involved in this kind of research. We conclude with an overview of future opportunities for research in this area.

Keywords Archival research · Artefacts · Environmental history · Historical ecology · Narrative · Oral history

R. H. Thurstan (✉)
Centre for Ecology and Conservation, University of Exeter, Cornwall, UK
e-mail: r.thurstan@exeter.ac.uk

C. Chong-Montenegro
School of Biological Sciences, The University of Queensland, St Lucia, QLD, Australia

J. Christensen · A. Gaynor
School of Humanities, The University of Western Australia, Crawley, WA, Australia

A. Clark
Australian Centre for Public History, University of Technology Sydney, Sydney, NSW, Australia

K. L. Pope et al. (eds.), *Understanding Recreational Fishers*, Fish & Fisheries Series 45, https://doi.org/10.1007/978-3-031-99739-6_8

1 Introduction

Recreational fishing, that is, fishing for leisure, relaxation or competition, may have existed for as long as people have fished (Pitcher and Hollingworth 2002). Historical sources—albeit often highly fragmented—provide indications of the behaviours and cultures of people fishing for leisure or competition in the past, including who took part, the norms and practices associated with recreational fishing, the species targeted, locations fished and the gears used. It is sometimes difficult to define—and therefore confidently identify—past forms of recreational fishing. The earliest known images of hook, line and pole fishing, discovered in the tombs of wealthy Egyptians, date from 4000 BC (Pitcher and Hollingworth 2002), and treatises and texts that mention fishing for leisure or sport are known across Europe from 1200 AD, with descriptions of the practice originating independently in multiple countries (Hoffman 1985).

As historians, it is difficult to know if subsistence fishers in the distant past continued to fish beyond their nutritional needs for the "rush" of the hunt, or if they chose to fish at times primarily as a form of escapism, or meditation, for that matter. The Food and Agricultural Organisation of the United Nations' (FAO) definition of recreational fisheries, "fishing of aquatic animals [. . .] that do not [. . .] meet basic nutritional needs and are not generally sold or otherwise traded [. . .]" indicates a separation between recreational and other forms of fishing (FAO 2013). However, diverse forms of fishing are experienced as enjoyable, and may or may not be perceived as leisure or recreational activities. These activities include fishing by people who do not perceive themselves as commercial fishers but would barter or sell some or all their catch, subsistence fishing and Indigenous fishing practices, among others (Nyboer et al. 2022). Blurry definitions of what constitutes "recreation"—especially in times and places with less formal notions of "work"—in addition to a typical lack of formal written evidence about individuals' leisure activities and perspectives on fishing, means that identifying historical recreational fishing practices and behaviours is not always simple or clear cut (e.g., Paulin 2007).

In this chapter, we explore how history and historical approaches have been applied to understanding recreational fisheries. We describe the discipline of history, how historical sources have been used to understand and measure past recreational fishing activity and fisher behaviour, sources and key methods of historical analysis, and developments that have the potential to fill gaps in knowledge. While attempting to be as inclusive as possible, we recognise that our own disciplinary and cultural biases necessarily limit the approaches and perspectives put forward in this chapter. The history of recreational fishing remains scattered and incomplete, but as historical research continues we are beginning to better understand how and where recreational fisheries were conducted in the past, by whom and for what purposes, the species targeted and the environmental impacts of their removal, and the past technological, societal and cultural developments that influence recreational fishing practices and attitudes today.

2 What Is History as a Discipline?

The practice of recording and interpreting past events has been practised in one form or another for millennia (e.g., Lateiner 1989; Locker 2014; Nunn and Reid 2016). However, it was not until the nineteenth century that history was formalised as an academic discipline concerned with interpreting and understanding past events (Harrison et al. 2004). Over time, the discipline of history would come to be understood as the critical analysis of oral, written, and material objects recovered from past human societies (Claus and Marriott 2017). Initially, the discipline of history was modelled upon the natural sciences, that is, emphasis was placed upon the objective gathering and analysis of text-based historical evidence, and the employment of empirical observation to uncover the "truth" of the past (Tyrrell 2005; Woolf 2011; Donnelly and Norton 2020). This approach prevailed until the 1960s, when the rise of social history produced dissenting voices that argued the difficulties (even impossibility) of separating historians' interpretation of sources from their own cultural and social experiences and biases (Donnelly and Norton 2020). Social histories also registered the importance of evidence that existed outside of the written archive—including oral, embodied, material and environmental records—as vital historical sources which allowed for more inclusive historical analysis and narrative (e.g., Allen 1986; Thompson 1963). These differing perspectives contributed to an increasing recognition in the discipline (if not universal acceptance) that there is no single truth when it comes to reconstructing the past, but that understanding of the past can be improved and explanations of change developed and refined through interpretation and critique of historical sources (Claus and Marriott 2017; Donnelly and Norton 2020). Simply put, the creation of history is shaped by multiple factors, including the kinds of evidence preserved, the methods employed and the historian's experience and assumptions. Plural interpretations of the past exist, but careful and critical study of historical sources can help us draw conclusions and create new accounts of the past and how we got to where we are today.

Sub-disciplines of history are numerous and include social history, maritime history, economic history and political history, among others. In tandem with the emergence of the environmental movement of the 1960s and 1970s, environmental history coalesced as a new sub-discipline (White 1985; Hunt et al. 2026). Environmental history focuses on how humans interacted with the natural world, the outcomes of these interactions for people and nature, how nature has been thought about and valued, how environmental change influences human actions and behaviours and the historical roots of contemporary environmental problems (Worster 1990; MacEachern and Turkel 2009). Environmental historians have charted the history of human impact on and interaction with the non-human world across numerous times and places, including the rise of wilderness preservation, the influence of climate on human affairs, the growth of cities and their ecological footprints, and human-environmental interactions within marine and freshwater systems (Melosi 1993; White 1996; Bogue 2001; Maohong 2004; Williamson 2013; MacNeil 2014).

As the sub-discipline of environmental history was being developed, other disciplines started to adopt historical sources and methods to understand the scale, direction and drivers of ecosystem change more universally through decadal to centennial timescales (Jackson et al. 2001; Hunt et al. 2026). With origins in the disciplines of history, ecology, geography and anthropology (Szabó 2015), these attempts started to coalesce and be termed "historical ecology," which today is defined as a research framework or community of practice that integrates historical approaches to understand past human influences on ecosystems, and their social and ecological outcomes through time (Armstrong et al. 2011; Kittinger et al. 2015; Crumley 2021).

Historical ecology and environmental history have since developed complementary and collaborative approaches to studying human–environment interactions and their social and ecological outcomes in marine and aquatic environments. Both make use of often diverse and fragmentary historical sources, which present methodological challenges to their analysis and interpretation. It is primarily the questions, sources and methods of environmental history and historical ecology that lie at the core of this chapter, although we note that the history of recreational fisheries and fisher behaviours is frequently embedded in regional and social histories (e.g., Kirby 2008; Coopey 2010; Henshaw 2011; McPhee 2017), and museum exhibits and collections. Some museums are specifically dedicated to recreational fishing (e.g., the International Game Fishing Association Fishing Hall of Fame and Museum in Florida, USA, the Recreational Fishing Museum of Finland and the River Tweed Salmon Fishing Museum in Kelso, Scotland), whereas others host exhibitions or displays on recreational fishing as part of their wider collections.

3 How Has History Been Used to Understand Recreational Fishers?

Despite increased interest in historical environmental agendas across diverse disciplines from the 1960s, research into aquatic ecosystem change and human–environment interactions has not been expressed uniformly across ecosystems and sectors. The greater accessibility and awareness of commercial fishery records, which are frequently collated and archived by government bodies, has led to this sector dominating fishery-related enquiry in environmental history and historical ecology (e.g., Engelhard et al. 2016). Yet, as highlighted above, historical scholarship on recreational fisheries exists across numerous sub-disciplines of history (Pears 1982; Kirby 2008; Henshaw 2011; Snyder et al. 2016; McPhee 2017), and as the economic, social and environmental significance of this sector becomes increasingly appreciated by environmental historians and historical ecologists, scholarship continues to expand (e.g., Chiarappa and Szylvian 2003; Brennan 2004; Franklin 2011; Frawley 2015; Gaynor et al. 2016; Clark 2017).

Archaeological deposits have yielded spearpoints from 50,000 to 100,000 years ago, which possibly included fishing activities, and ancient murals record the existence of hook, pole and line fishing from as early as 4000 BC in Egypt (Pitcher and Hollingworth 2002). Whether these activities took place primarily for leisure is unknown, but the existence of recreational fishing activities and behaviours can be inferred from early publications from the late mediaeval and early modern eras in Europe (synthesised in part by Hoffman (1985), see also Pitcher and Hollingworth (2002) for examples of ancient sources). These often took the form of instructions on angling materials and techniques, as well as the habits and food preferences of target species. An example of this is *The Treatyse of Fysshynge wyth an Angle from the book of Saint Albans*, originally published in 1496 by Wynkyn de Worde, although the true authorship is commonly attributed to a female prioress, Juliana Berners (Fig. 1). This and similar texts (e.g., Izaac Walton's *The Compleat Angler*, published in 1653) now form valuable sources for historical inquiry into recreational fishing activities, values and fisher behaviours during these periods. Variations of recreational fishing almost certainly occurred historically in cultures all around the world, and the pursuit of fishing for leisure (likely also for subsistence, we cannot always be sure) appears in historical records as people migrated from Europe and colonised other countries (e.g., Senior 1890; Manning 1908; Dane 1921).

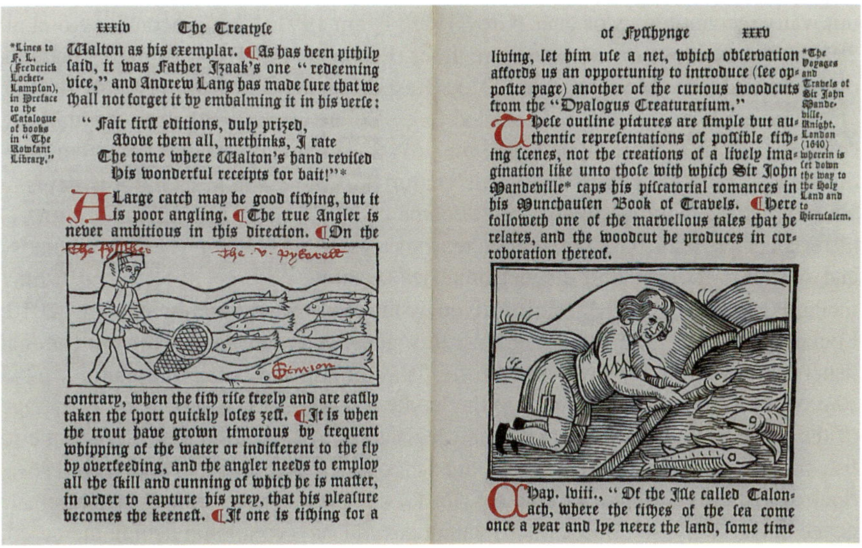

xxxiv The Treatyse

Walton as his exemplar. ¶ As has been pithily said, it was Father Izaak's one "redeeming vice," and Andrew Lang has made sure that we shall not forget it by embalming it in his verse:

"Fair first editions, duly prized,
Above them all, methinks, I rate
The tome where Walton's hand revised
His wonderful receipts for bait!"*

*Lines to F. L. (Frederick Locker-Lampson), in Preface to the Catalogue of books in "The Rowfant Library."

A Large catch may be good fishing, but it is poor angling. ¶ The true Angler is never ambitious in this direction. ¶ On the contrary, when the fish rise freely and are easily taken the sport quickly loses zest. ¶ It is when the trout have grown timorous by frequent whipping of the water or indifferent to the fly by overfeeding, and the angler needs to employ all the skill and cunning of which he is master, in order to capture his prey, that his pleasure becomes the keenest. ¶ If one is fishing for a

of Fysshynge xxxv

living, let him use a net, which observation affords us an opportunity to introduce (see opposite page) another of the curious woodcuts from the "Dyalogus Creaturarium."

These outline pictures are simple but authentic representations of possible fishing scenes, not the creations of a lively imagination like unto those with which Sir John Mandeville* caps his piscatorial romances in his Munchausen Book of Travels. ¶ Here followeth one of the marvellous tales that he relates, and the woodcut he produces in corroboration thereof.

*The Voyages and Travels of Sir John Mandeville, Knight. London (1640) wherein is set down the way to the Holy Land and to Hierusalem.

Chap. lviii., "Of the Isle called Calonach, where the fishes of the sea come once a year and lye neere the land, some time

Fig. 1 An extract from *The Treatyse of Fysshynge wyth an Angle from the book of Saint Albans* (originally published in 1496 by Wynkyn de Worde, with authorship commonly attributed to Juliana Berners, this version was edited and published in 1903 by William Loring Andrews). (Source: Juliana Berners, CC0, via Wikimedia Commons)

From the eighteenth century onwards, written and material sources on recreational fishing become more diverse and numerous, enabling historians and historical ecologists to better understand how the diversity of recreational fishing we see today developed, the changing values held by past recreational fishers and their fishing behaviours, and how fishing activities have influenced the ecosystems and fish populations targeted (e.g., Locker 2014). These questions have been investigated using more conventional documentary evidence such as fishing diaries, government reports, newspapers, logbooks, catch records and magazine articles, as well as material artefacts such as fishing gears, boats and artefacts lost around structures such as jetties, imagery such as photographs, mosaics, paintings, participatory mapping and oral history (Rodrigues 2002; Casasola 2007; Paulin 2007; Monteagudo 2007; McClenachan 2009; Beaudreau and Whitney 2016; Buckley et al. 2017; Chong-Montenegro et al. 2022a; Obregón et al. 2022). In the following paragraphs we provide more details on how some of the above sources have been used to understand past recreational fishing and fisher behaviour.

3.1 Behavioural Responses to Change

Historical research has shown how particular communities of recreational fishers responded to distinct types of change in the past, such as the introduction of innovative technologies or gear (Grey 1928; Dunn 1991; Chong-Montenegro et al. 2022b), regulations or management regimes (Parenteau 2004) and changing patterns of fish abundance (McClenachan 2013). Responses to societal movements such as environmentalism are also documented (e.g., Young et al. 2014). In industrialising countries, scholars have investigated the growth of recreational fishing sub-cultures during the nineteenth and twentieth centuries. For example, within Britain—a country with a comparatively well-documented history of recreational fishing—historians have argued that one of the reasons recreational fishing grew in popularity and started to diverge into distinct branches targeting different species and fishing methods, was due to societal shifts driven by the industrial revolution (Coopey 2010; Locker 2014). In freshwater angling, this generated two classes of recreational fishers: the wealthy landowners who fly-fished for species such as salmon (*Salmo salar*) and trout (*Salmo trutta*), and the landless working class who took advantage of the growth in public waterways and legislation of recreational or "free" time to fish for a variety of species using baited hooks (Locker 2014; Thompson 1967). Societal patterns differed by country (for an example of recreational fishing emergence in Germany, see Haase 2000), but in general, as recreational fishing increased, so did specialisation and competitive elements (McClenachan 2013; Locker 2014), and the development of fishing clubs and associations further facilitated access to recreational fishing (Chong-Montenegro et al. 2022b; Haase 2000). Historical research also allows us to better understand how fishers' responses to ecological or broader social changes led to shifting norms in recreational fishing communities, such as community-driven codes of practice that encouraged recreational fishers to conserve fish, engage in catch and release, or otherwise distinguish them from their commercial or subsistence-driven counterparts (Sjovold 1999; Frawley 2015).

Recreational fishers' responses to change can be highly variable (Beaudreau and Whitney 2016). This is partly due to the differing motivations driving participation in the activity but may also be influenced by the multiple histories of recreational fisheries development and the influence of culture upon fishing behaviour in any given context. Technological innovations have undoubtedly influenced where people choose to fish and the targeting abilities of recreational fishers (Thurstan et al. 2018; Cooke et al. 2021). However, changes in locations fished and preferred target species often reflects a combination of individual preferences and motivation, as well as regulatory, technological and environmental factors (Rothlisberger et al. 2010; Beaudreau and Whitney 2016). Historically popular fishing grounds often continue to be fished even when catch rates decline and further grounds are discovered (Beaudreau and Whitney 2016; Thurstan et al. 2018; Chong-Montenegro et al. 2022a), for example, demonstrating that maximising catch rates is not necessarily the primary goal of many recreational fishers. In contrast, and indicative of the diversity of motivations held by recreational fishers, studies have also shown historical expansion in the areas fished in response to improved technology and fishers' increased knowledge (Thurstan et al. 2018; Chong-Montenegro et al. 2022b).

Technological innovations, changing abundances of fish and regulatory and cultural shifts also influence the target species and the quantities of species (or proxies of quantity i.e. catch per unit of fishing effort) reported through time. Historical perspectives show how new target species grew in popularity as technology developed (e.g., the emergence of game fishing and the targeting of large pelagic fish from the 1920s (Grey 1928)). Regardless of new technologies and regulations, studies have uncovered the long-term popularity of certain species and explanations as to why: these species may be abundant and easily caught, be readily identifiable, or provide a specific leisure or culinary experience, for example, the targeting of freshwater carp (typically *Cyprinus carpio*) in coarse fishing (Locker 2018). History can also discern the roots of strong cultural attachments to such varieties of fish and the locations in which they are typically captured and help elucidate why such vehement opposition to—or acceptance of—regulatory changes might occur (e.g., Arlinghaus 2007; Christensen and Jackson 2014; Harris 2018).

The implementation of restrictive regulations on the catching or landing of a species, or the emergence of new cultural norms (e.g., voluntary catch and release fishing), can heavily influence fisher behaviour (Arlinghaus et al. 2007). Historical research has explored the mechanisms by which cultural norms in fishing communities interact and influence fishing behaviours through time, often heavily influenced by class distinctions and societal or gendered expectations (Sjovold 1999; Parenteau 2004). Although the regulation of recreational fishing became increasingly common in the twentieth century, formal restrictions do not always account for changing patterns of behaviour. For example, analysis of magazine articles over multi-decadal periods showed that recreational fishing communities can self-regulate in the face of species decline by encouraging changes in fishing practices years before formal legislation is enacted, as occurred in Australia when certain species underwent rapid declines from spearfishing (Young et al. 2014, 2015). Historical studies have also explored when and why new social norms are fostered in the absence of legislation and how these have contributed to changing fishing practices, such as shifts in the numbers or sizes of fish acceptable to catch and display, moves away from take-all to catch and release practices, or in the case of boat-based fishing, switching fishing effort from more heavily-exploited demersal stocks to more abundant pelagic species (Aas et al. 2002; Frawley 2015; Thurstan et al. 2016). Studies of illegal behaviours such as the use of nets or traps, breaching closed seasons, fishing in closed waters or exceeding limits on the legal size or total number of fish caught, are not as frequently published upon by historians. This largely reflects the difficulties of locating evidence of such behaviour in historical archives unless individuals were detected and prosecuted at the time.

3.2 Fishing Cultures and Practices

Present-day fishing cultures and practices are heavily grounded in the past. Historical research can help us understand when and where particular cultures of fishing developed, the identities and values associated with these, and how recreational fishing heritage influences fishing practices today. The cultures, identities and values attached to recreational fisheries can be diverse and conflicting, and this is also reflected in discourses within the historical record. For example, recreational fishing is (and has been) lauded as both highly competitive and a relaxing leisure pursuit (Grey 1928; Clark 2017). Recreational fishing has (and continues to be) represented as highly masculine and gendered, often rendering women largely absent or with highly specific roles, yet is also strongly family orientated (Sjovold 1999; McKenzie 2000; Gaynor et al. 2016). Recreational fishing can be an individual or a group activity, and is portrayed as variously dangerous, cavalier and nurturing (Sjovold 1999; McKenzie 2000; Clark 2017). These diverse cultures are in part due to the highly diversified nature of recreational fishing, which is conducted in both highly managed and entirely unregulated, "wild" environments, thus attracting people with a diverse range of interests and motivations.

As briefly highlighted in the previous section, recreational fishing cultures and practices are also driven by broader cultural phenomena and change within society (such as immigration), meaning cultural shifts in recreational fishing cannot be explored in isolation from wider trends. For example, using multiple historical sources and foci, Frawley (2015) demonstrated that the shift from take-all to catch and release within Australia was likely due to the convergence of multiple influences that resulted in broad-scale behavioural changes and a shift in dominant fishing practice. These influences included the popularisation of catch and release through television media and the concurrent development of discourses within recreational fishing communities about the negative impacts of recreational fishing on inshore environments and fish populations (Frawley 2015). This in itself was likely influenced by the burgeoning conservation movement and government campaigns aimed at preventing the landing of undersized fish (Clark 2017). The concept of catch and release is not limited to wild stocks of fish, as shown by the prevalence of this practice in inland coarse angling for non-native species such as carp, with large and evasive individuals particularly prized and repeatedly caught and released (Locker 2014). Contemporaneous research also demonstrates the diversity of opinion and contrasting discourses around catch and release within fishing communities, and their influences (Aas et al. 2002; Arlinghaus et al. 2007).

3.3 Conflict and Collaboration

Historical perspectives help us to understand the basis and emergence of conflicts and collaboration among recreational fishing groups and with other sectors, as well as the strategies employed to protect or advance recreational sectors' interests. This can encompass the formation of recreational fishing associations and campaign groups, the forging of alliances with commercial fishing sectors and conservation organisations when objectives align, public and political campaigns to protect recreational fishers' interests, and shifts or transformations in governance (Parenteau 2004; Szylvian 2004; Christensen 2009; Daedlow et al. 2013). In countries where recreational fishing is a popular activity with ties to tourism industries and regional economies, historical research has illuminated the factors shaping a present-day sector that can wield significant, cross-party political influence (McClenachan 2013).

Conflicts, either among recreational fishing sectors or with wider sectoral interests such as commercial fishing and conservationists, have been observed in both official (i.e. government) and popular (i.e. print media) historical documents (Sjovold 1999; McClenachan 2013). Sometimes conflicts arise between commercial and recreational fishers over "fishing rights" to a particular space, such as a bay or estuary, or over perceptions of unjustifiable levels of environmental impact by a particular sector or gear type (e.g., Loring 2017). Examples of archival documents detailing these conflicts include signed petitions and counter-petitions in local government archives aimed at preventing net fishers from operating in areas popular with recreational anglers (Szylvian 2004; Christensen 2009), and concerns voiced by recreational fishers about the actions of their peers (e.g., concerns voiced about the taking of undersized fish by other recreational fishers). Historical research has likewise shown how recreational fishing movements sometimes aligned closely, and facilitated, the emergence of broader conservation movements. For example, starting in the nineteenth century and continuing to the present, American and British angling associations have successfully lobbied to enhance the protections of target species and their habitats (Reiger 1975; Bate 2001). Another source of historical evidence is the emergence of published community-agreed rules and norms within the sector which reflected and amplified these concerns (Johnstone and Field 1998). The growth in political representation of recreational fishing sectors and the subsequent influences upon policies also leave a documentary trail, particularly when these place recreational fishing groups in opposition to government policy or other sectors or lobbying for action such as legislative change or habitat restoration (Szylvian 2004; McClenachan 2013; Christensen and Jackson 2014; Clark 2019). Where such events are within living memory, oral histories can help to illuminate the political manoeuvrings and drivers of these actions (Loring 2017).

4 What Methods Are Typically Used to Study Recreational Fishers Using a History Lens?

The process of interpreting the past through the study of contemporaneous documentary, material and oral evidence allows us to identify and describe past fisher behaviours, and their drivers and outcomes, subject to the availability of historical sources. Methods to interrogate and analyse historical materials include the study of sources within archives (Box 1), palaeography (the study and interpretation of historical writing), chronology (where historians establish an order and dates to past events), codicology (the study of manuscripts—distinct from palaeography in that it focuses upon the broader cultural context of a manuscript, i.e. how and when it was made, who published or owned it), epigraphy (the study of ancient inscriptions), material culture (the study of objects used by past societies) and oral history (Box 2), among others (Sreedharan 2007; Roberts and Robinson 2020). Methods such as lexicometry and content analysis are used to determine the frequency and presence of specific themes within sources (Kochalski et al. 2026), with quantitative approaches applied to analyse numerical data extracted from source materials (Sreedharan 2007). In both environmental history and historical ecology, researchers identify historical sources, synthesise the information within them (using qualitative and quantitative approaches) and use this information to generate historical explanations of change. The sources used, types of data collected, their spatial and temporal scales, the focus upon the individual or population, and the ways in which researchers measure outcomes are highly heterogeneous. This heterogeneity reflects the questions being asked, the diversity of historical sources available to study, and the embodied and experiential nature of recreational fishing itself.

Box 1 Archival Research

Archival research involves identifying, extracting, and interpreting primary sources such as written documents and material objects to generate new accounts and draw conclusions about the past. What sources are available for study will depend upon the archive being used and its purpose (is it a local library, government, special-interest community or academic archive, is it a museum or private collection?), what sources were preserved, the historical

(continued)

Box 1 (continued)

context provided alongside these sources (e.g., from where and when they were sourced, their uses or functions, and how they came to be stored in the archive), and how accessible and findable those sources are within the archive. Archival collections—both written documents and artefacts—are increasingly being digitised, which enhances their accessibility and searchability. However, not all archives or sources are digitised, and much archival research involves identifying and locating less readily findable sources. Relevant archival sources may not be individually catalogued but instead exist within an unprocessed collection of documents. Hence, archival searches must involve careful selection of keywords to systematically search archival catalogues, as well as knowledge of key figures or collections that may hold information of interest; knowledgeable archivists or curators are invaluable as they often know the details of collections beyond what has been formally catalogued. Not all archive materials are readily accessible. For example, archival collections may be private or require special requests to access if material is deemed fragile or classified. Materials may be hand-written, in a different language or contain highly technical terms or language that is highly specific to the period and events around the time of publication. What's more, they may not be text-based at all—instead, material cultural archives might consist of archaeological remains located in landscapes (e.g., Rodrigues 2002). Uncovering key historical sources thus requires perseverance, skill, and a bit of luck.

Once relevant sources have been located within archives, the interpretation of their content involves additional challenges, as covered in the *source criticism* and *analytical approaches* sections of this chapter. Guidelines or protocols for using and interpreting culturally hurtful or insensitive materials, such as those originally collected without the consent of their owners or which contain content that is offensive or will cause distress to individuals or communities, increasingly exist, and should be identified and followed at all stages of research (e.g., AIATSIS 2020).

Box 2 Oral History

Oral history is a method of preserving oral testimony that is produced with the intention of creating a record of a person's or group's experience of the past (Oral History Association 2023). Researchers then use this record, alongside other oral histories, or alternative sources, to conduct historical enquiry. As oral history deals with living subjects, ethical guidelines must be followed at all stages of the research and researchers must be empathetic and alert to the present and past contexts in which interviewees tell their stories, as well as

(continued)

Box 2 (continued)
who ultimately "owns" the oral history and its outputs (i.e. recordings, transcripts, or research resulting from access to and study of these sources). Oral history is not the only form of oral testimony used by historical researchers, but unlike other interview forms, oral histories specifically aim to deliver highly personalised, in-depth reflections on a particular topic in the past (Ritchie 2014). Oral histories also tend to identify—with consent—the interviewee by name and other identifying characteristics, with the aim to provide context for the interview narrative and its subsequent interpretation (Oral History Association 2023). Unlike other forms of interview, the recording, and any related documents (i.e. transcripts, photographs, or objects, created from or related to the oral history interview) are typically preserved—again with the interviewee's consent—in an archive for future research. Restrictions on access and ownership are set in agreement with the interviewee, and in line with copyright and national law. The researcher must be aware of how existing laws apply to oral history outputs, and that such laws and ethical norms, as well as the recognised practice and methods of oral history itself, may differ across jurisdictions and cultures. The researcher should also realise that the practices of oral history continue to be debated and to evolve, as, for example, Indigenous-led perspectives, traditions, and embodied knowledge are increasingly centred within oral history research (e.g., Mahuika 2019).

Once archived and consents around access and use for research are clarified, oral history content is identified and sourced using similar methods to archival research. Analysis of oral histories tell us about how interview subjects understand the past and place it into a narrative. Oral histories highlight voices and experiences of past events that may not otherwise exist in archives, or otherwise lend additional depth or perspectives to existing historical sources. Who the researcher chose to interview, who was available or agreed to be interviewed, the interviewer's and interviewee's personal agendas and cultural backgrounds, and the questions asked by the interviewer, are all necessary to consider when interpreting these versions of the past. Collaborative and contesting accounts across multiple oral histories or compared to other forms of primary sources e.g., the written record, help the historical researcher to critically examine and draw conclusions about the past.

4.1 Identifying Sources

Historical sources pertaining to recreational fisheries are highly diverse and differ greatly in the information provided, which in turn informs the analyses conducted. Within historical study, the distinction between primary and secondary sources is critical. Primary sources are those written at the time of or shortly after an event, either by direct witnesses or those who present a view and values that were expressed or otherwise existed at the time. Secondary sources are accounts written after the event and which are based on the collation of primary sources and often include significant interpretation by the author.

A broad range of sources have been applied to fill gaps in our historical knowledge of recreational activities. Examples of documentary records used to understand past recreational fisheries include newspaper articles, fishing magazines, angling guidebooks, published books, personal logbooks and fishing club records (Buckley et al. 2017; Santos et al. 2017; Clark 2017; Thurstan et al. 2017; Chong-Montenegro et al. 2022a; Boucek et al. 2023). Newspapers and magazines are important primary sources of cultural history as they provide written information on past social, sports and entertainment subjects that were deemed of significance to their readership, including recreational fishing (Young et al. 2015). Preceding the publication of special interest magazines were angling guidebooks (e.g., Anonymous 1714; Alexander 1860; Aflalo 1896) and newspaper columns, which published peoples' accounts of and opinions on recreational fishing (Thurstan et al. 2016). Material collections, logbooks and fishing club records may not record peoples' perceptions or values directly, but they can provide information about where people were fishing, the gears used and species targeted. Such sources may be found within public or private collections, including government, academic, national, or local archives. Archival research (Box 1) leverages such sources to help historians examine past values in terms of preferred target species, motivations and changes in what was or was not socially acceptable and infer possible environmental changes over time (McClenachan 2009, 2013). New archives can also be created through oral histories and other forms of interviews with recreational fishers, by focusing on their past experiences and observations or perceptions of change (Box 2) (e.g., Darcey 1989/90).

4.2 Source Criticism and Source Comparison

Important practices in historical enquiry include source criticism and source comparison. Source criticism is the practice of critically examining the content and context of the historical sources being used—whether documentary, material or oral—and determining their value and limitations as historical evidence. As well as examining the content for meaning, this also includes considering who created the source and their purpose in creating this source, the social or political context at the time of source creation, the intended use of the source and who the intended audience was. To move towards a fuller understanding of past change, multiple sources must be consulted. As evidence is collated, source comparison is required, where multiple sources are cross analysed for consistency, that is, what information is corroborated or contradicted across sources? When sources seem to contradict each other, the reasons for this should be closely examined as far as possible. Do these sources differ because the authors genuinely had different experiences or perceptions of an event (and if so, why might this be)? Or do they have other reasons for providing conflicting accounts? Are we missing specific information within the historical record or nuances that would have been clear in the culture and context of the time for which we are drawing these conclusions, for example, how an object was used or what it might have meant to somebody? For both contemporaneous sources and recollections collated long after an event—such as those recorded in oral histories—is there the possibility that an individual's description of events has had information omitted, embellished, or biased in any way? For example, newspaper articles are often published to support or contest a specific outcome or agenda. Finally, if there are clear contradictions or inconsistencies across sources, what does it mean for the trustworthiness and value of these sources in drawing conclusions about the past? Within recreational fisheries, differing accounts may be due to real differences in experience, biases in our present-day interpretation of the uses or intent of historical objects or documents, or a reflection of the highly diverse motivations, worldviews and values of recreational fishers.

Source criticism and comparison do not just stop at the sources in hand, they also consider why we may not have certain types of evidence available to us and what this could mean for the conclusions being drawn. Historical sources are typically incomplete, and the reasons for this must be considered and the potential influence on conclusions evaluated. For example, why were documents saved and archived in the first place? This could be because they were viewed as significant and worth saving at the time of their publication, records were kept due to statutory requirements, or perhaps documents were saved by chance and eventually archived as part of a wider collection, for example, from a person who bequeathed papers or belongings to a library or museum. Documentary records are fragile and entire collections have been lost to fire, floods, or poor storage conditions. In the case of oral histories, some communities are chosen for interviews but not others. In the case of archaeological and cultural collections, certain sections of society (typically the wealthiest) may be better represented (e.g., ancient murals of fishing observed in the tombs of wealthy Egyptians (Pitcher and Hollingworth 2002)). What could the loss or exclusion of certain sources or societies mean for the accuracy or completeness of the historical explanations being developed? Often, the dearth of evidence in the historical record provides us with important clues as to the past values of specific sections of society, i.e. what information was deemed important enough to keep, record or preserve, whose experiences are privileged in the historical record, and hence how (in)-complete or biased our interpretations of the past might be.

4.3 Analytical Approaches and Interpretation

As described earlier, the discipline of history is about understanding—as far as we can ascertain with the sources available to us and within the bounds of our (socially and culturally conditioned) comprehension—what happened in the past, how these events unfolded and why, and what the outcomes were for humans and the world they live in. Historical researchers must develop and refine explanations based on their comparison, evaluation and critical analysis of multiple historical sources, and their understanding of the historical context in which these sources were produced.

Researchers have used multiple analytical approaches to investigate the history of recreational fishing. Depending on the focus of the historical enquiry and the sources being used, the questions asked, and analytical approaches will differ. For example, historical ecologists may ask how recreational fishing activities have impacted marine ecosystems or fish populations, whereas environmental historians may ask what the outcomes of these interactions were for people and nature, or how fishers' values have changed over time and why. These foci are not mutually exclusive: historical ecologists include humans as a component of ecosystems and may focus upon both ecological and social outcomes (Kittinger et al. 2015), whereas some environmental historians have chosen to centre nature—rather than humans and their interactions with environments—in their research (O'Gorman and Gaynor 2020).

Statistical methods are commonly used for quantitative and semi-quantitative data, such as historical catch or landings records. In using these sources researchers may be seeking to identify or present additional proxies of environmental or population change, with the aim to extend the timescale beyond that of scientific monitoring data. Sometimes these historical records are already known to scientists but may not be used due to a lack of clarity regarding the utility, historical context, or trustworthiness of the data. In other cases, they may be hidden in archives until uncovered by historical researchers. For example, quantitative data extraction from historical sources has been used to build recreational catch reconstructions (e.g., Moutopoulos et al. 2013), whereas descriptions of fishing locations from historical records have been used to quantitatively derive spatial changes in fishing activities several decades prior to scientifically collated data (Chong-Montenegro et al. 2022b). Together, these sources can provide information on the changing experience of recreational fishers and their behaviours through time, as well as indirect insights into ecological and environmental changes, including relative species abundance and size, and habitat degradation or modification (Coverdale et al. 2013; Sharpe et al. 2017; Lybolt and Pandolfi 2019; Chong-Montenegro et al. 2022b; Stamp et al. 2022).

Similar to catch or landings records, newspapers and magazine articles can provide paired quantitative and qualitative data associated with fishing activities over longer or deeper timescales to existing scientific data, including detailed descriptions of the intended target species, the number of fish caught, number of fishers per outing, total weight of catch, species composition, the locations and hours fished and the types of gears used. Such data have been used to reconstruct catch and catch rate trends and evaluate the spatial and temporal development of recreational fishing activities over extended periods of time, if the historical records cover appropriate spatial scales (Gartside et al. 1999; Young et al. 2014; Thurstan et al. 2016; Chong-Montenegro et al. 2022b; Boucek et al. 2023). Typically, the competitive nature of recreational fishing played a significant role in ensuring past catches and catch rates entered the archival record, as fishing clubs published results of competitions in regional papers and charter businesses published the outcomes of past trips as a form of advertisement (Richardson et al. 2006; Thurstan et al. 2017). Conversely, fishing experiences of individuals or families, or of communities marginalised from mainstream cultural reporting (e.g., Indigenous cultures), are typically underrepresented in these types of records.

Oral histories and similar approaches such as semi-structured interviews have been used to generate new accounts of the past, including asking interviewees to provide quantifiable data such as recalled changes in fish catch through time, in addition to being used to characterise fishers' attitudes, or as information on fishers' responses to cultural or environmental changes over time (Beaudreau and Whitney 2016; Young et al. 2016; Gaynor and McCann 2017). For example, Beaudreau and Whitney (2016) documented the spatial expansion of recreational fishing efforts in Puget Sound, USA, and assessed variations in species-specific fishing intensity over five decades. Kroloff et al. (2019) used fishers' knowledge to document the historical drivers of bonefish (*Albula vulpes*) declines in Florida and their implications for the fishery over 40 years. Retrospective accounts thus provide valuable information on fishers' perceptions of environmental and behavioural changes, and may fill gaps in historical knowledge where alternative, documentary historical records are lacking (Kerr 1985, Tull 1999, but see Fowles and Gaynor (2011) for some of the challenges of applying retrospective accounts in this way).

Quantitative data can also be extracted from historical paintings and photographs, such as the quantity of fish of varied species, their sizes relative to a person or item in the image and—if comparable timeseries of images exist—be compared through time. Similar to the above sources, with appropriate knowledge of the historical context within which these pictures were taken and preserved, these may extend understanding of ecological changes over longer time series, or they can allow questions to be asked about changes in target species or gears, or the social and cultural norms and values surrounding fishing (McClenachan 2009; Shiffman et al. 2014; Jiménez-Alvarado et al. 2019). The more sporadic nature of archaeological finds of fishing hooks and gears, and images from ancient times, may not allow us to draw conclusions such as the above, but they can still indicate fishing for leisure, either by the size or quantity of hooks used (i.e. smaller or fewer hooks, or those of poor workmanship, could be associated with leisure or subsistence fishing (Casasola 2007)), the species targeted, and who in ancient societies accessed fish resources in leisure or subsistence contexts (Pitcher and Hollingworth 2002; Monteagudo 2007).

Historical qualitative sources present valuable information on people's perceptions and opinions as they related to fishing at the time, or—in the case of oral histories—can provide reflections upon the past through a present-day lens. For example, evidence of fishers' attitudes towards fishing and prescribed norms can be found within opinion or editorial pieces in newspapers and magazines, and fishing guides and government inquiries record concerns expressed by recreational fishers, such as perceptions of declining stocks or worrying behaviour by their peers or other sectors. Calls for greater regulation and fisheries management can be seen in the evidence provided to government fisheries inquiries in the mid- to late-nineteenth century. The 1880 Commission of Inquiry into the New South Wales fisheries, for example, lamented how fish "are followed up every creek and cranny by their relentless human enemies, and so perpetually harassed and hunted" (Commission 1880). In an 1882 study, fisheries scientist Julian Tenison-Woods described how declines in the recreational and commercial fisheries in Sydney harbour were considered to be due to commercial netting (Tenison-Woods 1882). Narrative-driven analyses that tie together multiple independent archival sources are particularly important for informing cause and effect and understanding the underlying context of present-day situations. For example, the concern expressed by recreational fishers about declining stocks and recorded by Tenison-Woods (1882) contributed to growing calls for increased fisheries regulation. These concerns appear multiple times in the historical record (e.g., Champley 1910) and understanding the context of these concerns in tandem with observations of the past that can be compared to today can help place current ecosystem health, as well as current sectoral conflicts, into historical context.

5 How Might This Discipline Evolve in the Future?

There is much still to learn about past recreational fisher behaviour and the links to the present. Growing interest in recreational fisheries histories focusses attention on new sources and approaches that can contribute to our understanding. The emergence of "digital humanities," which promote the intersection of digital technologies and historical approaches, presents exciting opportunities to expand our understanding of past recreational fishing activities and the people that partook in these. This includes the curation of traditional archives into digital forms, novel visualisations of archives and geographic information systems, and new methods of interrogating historical sources (Travis et al. 2022), as well as citizen science and history projects (Williams et al. 2015). Together, these have the potential to not only promote greater accessibility of historical archives, including to people who may not previously have been able to access these sources, they also open up history to new forms of interpretation and communication.

Research into past recreational fishing has made use of a range of documentary evidence and retrospective accounts, but there remain opportunities to further explore past recreational fisheries that not only further our understanding of historical fisheries and fishing, but also what materials and forms of knowledge should contribute to this. For example, fishing experiences, the transmission of that experience to later generations, and social and cultural norm shifts through time, ties into place-based and other knowledge forms such as traditional and Indigenous knowledge (Martinussen 2007; Lam 2016). Fishing practices are themselves a form of "embodied archive," that is, understanding is passed through community and intergenerational transmission. Study of such "embodied archives," their transmission and interpretation could therefore be a way to engage more deeply with and understand past cultures and fishing experiences, with large samples across generations enabling the generalisation of these insights at a broader scale. As part of this, the creation of new archives by conducting oral history, recording historical practices, or making private materials publicly accessible with the full consent of their owners, is vital to our understanding of historically overlooked communities and their activities, as well as informing contemporary sustainable fisheries management and practice. In considering how we conduct history, we also need to consider who holds applicable knowledge: accessing fishing histories does not need to be limited to fishing communities; for instance, those tasked with understanding or regulating fisheries are also knowledge holders.

There remain many knowledge gaps in our understanding of past recreational fisheries and fisher behaviour. Deeper understanding requires an increased research focus on regions beyond North America, Europe and Australasia and particularly non-Western cultures (e.g., Kadagi et al. 2021). For example, over the past decades recreational game fishing has become an increasingly popular activity in several countries of the Tropical Eastern Pacific and Western Pacific (e.g., Costa Rica, Panama). Luxury fishing resorts have been operating in these regions since the 1960s (Harvey 2010) and contribute to local economies by employing small-scale fishers as fishing guides (CCM pers. obs.). However, there is limited research on the social, ecological and cultural changes associated with the establishment and operation of big game fishing resorts in local and traditional fishing communities. In addition, there remains a lack of scholarly understanding of the perspectives, behaviours and motivations of customary or traditional fishing cultures in non-Western and Indigenous communities and the overlap with or distinctiveness from leisure forms of fishing (e.g., Paul-Burke et al. 2020; Waiti and Wheaton 2022). We note that angling or recreational histories may be readily available in the published record in countries (e.g., China) with a long history of administrative record keeping. There is growing recognition of oral history and interviews as important tools and methods for documenting the past experiences and practices of marginalised individuals and cultures whose histories, identities and perspectives were excluded or destroyed by colonialism, in addition to elucidating the connections between practices, knowledges, values and place in relation to recreational and wider forms of fishing (e.g., Smyth et al. 2018). Equally, Indigenous-collated/owned archival records and material collections are increasingly available for study, including online repositories developed by and with Indigenous communities that aim to convey past and present Indigenous cultures, practices, connections and perspectives, including for activities such as fishing. For example, the Australian Institute of Aboriginal and Torres Strait Islander Studies website brings together oral, documentary and material collections and Indigenous knowledges from community members and elders on fishing cultures, values and fishing rights, among other topics (AIATSIS 2022).

Beyond making history more inclusive in terms of acknowledging the differing experiences of multiple cultures, and how such histories can and should be told, there is the opportunity to generate knowledge to inform decision making. Recreational fishing culture and identity are grounded in the (often very personal) past, which presents both difficulties and opportunities when attempting to manage recreational fisher behaviour through regulation or other means, or when attempting to predict behavioural responses to change. This provides an opportunity for historical perspectives to identify the heritage and cultures that surround and influence recreational fishers today and articulate how these factors might be better acknowledged and accounted for in decision making. Historical insight and knowledge can also provide a powerful trigger for conservation action by drawing attention to the scale of ecological change in marine and freshwater environments, thus addressing the problem of "shifting baseline syndrome," or the loss of environmental awareness and understanding on an inter-generational basis (Pauly 1995; Kittinger et al. 2015).

Finally, taking its cue from developments in geography, anthropology and political ecology, "more-than-human" histories endeavour to displace the human from the centre of our narratives and foreground the entanglement of human and non-human that has inextricably shaped our shared pasts (Demuth 2019; O'Gorman and Gaynor 2020). Though it may not be immediately clear as to how such histories may be applied to understanding recreational fishing, this approach would foreground the agency and (inferred) interests of diverse types of fish—or even individual fish—as well as the diversity of human actors in multispecies worlds. For example, Mordue and Wilson (2022) use ethnography of coarse angling practice to focus "on the simultaneity of coarse fish as victims of human play and as biosocial actors" and explore what they term a "care/cruelty paradox" in catch and release fishing by examining both how fishers affect fish they catch and vice versa. Such approaches would enable historians—and their audiences—to reflect more deeply upon the relationships and knowledges that our histories make visible, as well as those that are obscured.

6 Best Practices When Applying the Disciplinary Perspective of History

When working with fishers and other stakeholders, it is necessary to communicate that historical sources and methods reflect the (oft-times problematic) social and cultural norms of the time. Historians follow national, institutional and discipline-established protocols when it comes to accessing, using and interpreting culturally insensitive materials (e.g., human ethics procedures and Indigenous research guidelines (AIATSIS 2020)). Protocols for best practice continue to evolve and are now relatively robust in terms of how we use and communicate the findings from historical materials today, but it is important to note that best practices may not have been followed during the original acquisition of source materials, and that in turn what is considered "best practice" will inevitably change. For example, some historical sources in the public domain are displayed without the explicit agreement—and sometimes with explicit disagreement—of the original owners of that source and their ancestors (The Guardian 2019). This applies to artefacts, pictures, interviews, Indigenous community oral histories and documents. The use of these historical sources for research may therefore be contested or cause great hurt and should take place with the full knowledge and consent of all parties.

Published standards, guidelines and codes of conduct now exist for methods such as oral history and other forms of interview to ensure ethical procedures are followed (e.g., AIATSIS 2020; Canadian Institutes 2022; Oral History Association 2023). Researchers are increasingly required to comply with relevant standards and codes in oral history research, and these must be identified and consulted prior to the start of research. These differ by country and the community or communities being interviewed, which makes understanding of local knowledge and cultural norms, and alignment with laws around copyright and consent, important and necessary. Guidelines on the oral history process typically do not just include the questions that are asked, they also include how people are identified for an interview and how they are approached, the level of information required for them to give true consent, how the imparted information is used within research and beyond, what happens to it afterwards, and who owns the outputs or associated intellectual property (e.g., Ritchie 2014; AIATSIS 2020; Canadian Institutes 2022). Consulting with and including local collaborators and knowledge holders from the very beginnings of the research process, and appropriately acknowledging their contributions and ownership, is a vital part of the process and is also included in published standards and codes of conduct (AIATSIS 2020, Canadian Institutes 2022).

Studies of past fishing activity often raise basic methodological problems surrounding the use of archaic weights and measures, the appearance of colloquial names for fish that are either no longer in use, or which vary even within the same region or country, and the fact that historical archives are often highly fragmentary, with large temporal and spatial gaps in the information they contain. As a result, there is often a need to standardise units of measurement, to standardise the names of key species or to group taxa, and for extrapolating data into long-run series, and thus, for clear hypotheses and processes that can guide analysis. These challenges remind us that all historical sources, whether they relate to fishers or to fish, inevitably reflect human as much as (if not more than) natural variables. Therefore, it is necessary to understand as much as possible about the relevant cultural, political and social contexts within which the source was produced, and to embed any interpretation within this (van Sittert 2001).

7 Conclusions

Historical sources and approaches continue to shed new light on recreational fishers' behaviours through time, the social and ecological outcomes of these, and the origins of the identities and cultures bound up in recreational fishing today. As interdisciplinary researchers increasingly engage with the discipline of history, it is vital that the formation of this discipline and its highly varied epistemologies and practices are not overlooked. Historians and history-facing researchers are increasingly embracing modern technologies, methods and perspectives. These are not without significant challenges to historical analysis and interpretation, but if we are to understand the full breadth of past recreational fisheries and how the history of recreational fishing links to today, while ensuring the creation of this history is inclusive of all, then we must embrace these challenges and continue to break new disciplinary and interdisciplinary ground.

Acknowledgments We are grateful to two anonymous reviewers, Dr Brett van Poorten, and Dr Robert Arlinghaus for suggestions and critical comments that significantly improved this chapter.

References

Aas Ø, Thailing CE, Ditton RB (2002) Controversy over catch-and-release recreational fishing in Europe. In: Pitcher TJ, Hollingworth C (eds) Recreational fisheries: ecological, economic and social evaluation. Blackwell Science, Oxford, pp 95–106

Aflalo FG (1896) A sketch of the natural history of Australia with some notes on sport. MacMillan and Co., London

AIATSIS (2020) AIATSIS code of ethics for Aboriginal and Torres Strait Islander research. Australian Institute for Aboriginal and Torres Strait Islander Studies, Canberra

AIATSIS (2022) Fishing. https://aiatsis.gov.au/explore/fishing. Accessed 20 May 2023

Alexander SJE (1860) Salmon-fishing in Canada. Longman, Green, Longman and Roberts

Allen J (1986) Evidence and silence: feminism and the limits of history. In: Pateman C, Grosz E (eds) Feminist challenges: social and political theory. Northeastern University Press, Boston, pp 173–189

Anonymous (1714) The whole art of fishing. Being a collection and improvement of all that has been written upon this subject: with many new experiments, Shewing the different ways of angling, and the best methods of taking fresh-water fish. To which is added, the laws of angling. E. Curll, London

Arlinghaus R (2007) Voluntary catch-and-release can generate conflict within the recreational angling community: a qualitative case study of specialised carp, *Cyprinus carpio*, angling in Germany. Fish Manag Ecol 14:161–171

Arlinghaus R, Cooke SJ, Lyman J, Policansky D, Schwab A, Suski C, Sutton SG, Thorstad EB (2007) Understanding the complexity of catch-and-release in recreational fishing: an integrative synthesis of global knowledge from historical, ethical, social, and biological perspectives. Rev Fish Sci 15:75–167

Armstrong CG, Shoemaker AC, McKechnie I, Ekblom A, Szabó P, Lane PJ, McAlvay AC, Boles OJ, Walshaw S, Petek N, Gibbons KS, Morales EQ, Anderson EN, Ibragimow A, Podruczny G, Vamosi JC, Marks-Block T, LeCompte JK, Awâsis S, Nabess C, Sinclair P, Crumley CL (2011) Anthropological contributions to historical ecology: 50 questions, infinite prospects. PLoS One 12(2):e0171883

Bate R (2001) Saving our streams. The role of the anglers' conservation association in protecting English and Welsh rivers. Institute of Economic Affairs

Beaudreau AH, Whitney EJ (2016) Historical patterns and drivers of spatial changes in recreational fishing activity in Puget Sound, Washington. PLoS One 11:e0152190

Bogue MB (2001) Fishing the Great Lakes: an environmental history, 1783–1933. University of Wisconsin Press

Boucek RE, Rehage JS, Castillo NA, Dwoskin E, Lombardo SM, Santos R, Navarre C, Larkin M, Adams AJ (2023) Using recreational tournament records to construct a 53-year time series of the Florida Keys recreational bonefish fishery. Environ Biol Fish 106:279–291

Brennan C (2004) Imperial game: a history of hunting, society, exotic species and the environment in New Zealand and Victoria 1840–1901. PhD thesis, University of Melbourne

Buckley SM, Thurstan RH, Tobin A, Pandolfi JM (2017) Historical spatial reconstruction of a spawning-aggregation fishery. Conserv Biol 31:1322–1332

Canadian Institutes (2022) Canadian Institutes of Health Research, Natural Sciences and Engineering Research Council of Canada, and Social Sciences and Humanities Research Council of Canada. Tri-council policy statement: ethical conduct for research involving humans, December 2022. https://ethics.gc.ca/eng/documents/tcps2-2022-en.pdf. Accessed 30 Oct 2023

Casasola DB (2007) Fishing tackle in Hispania: reflections, proposals and first results. In: Bekker-Nielsen T, Casasola DB (eds) Ancient nets and fishing gear. Proceedings of the international workshop on "Nets and fishing gear in Classical Antiquity: a first approach", Cádiz, November 15–17, 2007. Aarhus University Press

Champley JT (1910) Australian angler's guide and sea fisher's manual. G Robertson, Melbourne

Chiarappa M, Szylvian KM (2003) Fish for all: an oral history of multiple claims and divided sentiment on Lake Michigan. Michigan State University Press, East Lansing

Chong-Montenegro C, Thurstan RH, Campbell AB, Cunningham ET, Pandolfi JM (2022a) Historical reconstruction and social context of recreational fisheries: the Australian east coast Barramundi. Fish Manag Ecol 29:44–56

Chong-Montenegro C, Thurstan RH, Pandolfi JM (2022b) Quantifying the historical development of recreational fisheries in Southeast Queensland, Australia. Mar Ecol Prog Ser 696:135–149

Christensen J (2009) Recreational fishing and fisheries management: a HMAP Asia project paper. Working paper no. 157, Murdoch Business School and National Library of Australia, ISSN: 1037-4612

Christensen J, Jackson G (2014) Shark Bay snapper: science, policy, and the decline and recovery of a marine recreational fishery. In: Christensen J, Tull M (eds) Historical perspectives of fisheries exploitation in the Indo-Pacific. Springer Netherlands, Dordrecht

Clark A (2017) The catch: the story of fishing in Australia. National Library of Australia, Canberra

Clark A (2019) Rebuilding reefs, restoring memory: at work in the waters of history. Griffith Rev 63:107–113

Claus P, Marriott J (2017) History: an introduction to theory, method and practice. Routledge

Commission (1880) Report of the Royal Commission appointed on the 8th January, 1880, to inquire into and report upon the fisheries of this colony: together with the minutes of evidence and appendices. Government Printer, Sydney

Cooke SJ, Venturelli P, Twardek WM, Lennox RJ, Brownscombe JW, Skov C, Hyder K, Suski CD, Diggles BK, Arlinghaus R, Danylchuk AJ (2021) Technological innovations in the recreational fishing sector: implications for fisheries management and policy. Rev Fish Biol Fish 31:253–288

Coopey R (2010) A river does indeed run through it: angling and society in Britain since 1800. In: Lykke Syse KV, Oestigaard T (eds) Perceptions of water in Britain from early times to the present: an introduction. University of Bergen

Coverdale TC, Herrmann NC, Altieri AH, Bertness MD (2013) Latent impacts: the role of historical human activity in coastal habitat loss. Front Ecol Environ 11:69–74

Crumley CL (2021) Historical ecology: a robust bridge between archaeology and ecology. Sustainability 13:8210

Daedlow K, Beckmann V, Schlüter M, Arlinghaus R (2013) Explaining institutional persistence, adaptation, and transformation in East German recreational-fisheries governance after the German reunification in 1990. Ecol Econ 96:36–50

Dane R (1921) Sport in Asia and Africa. Andrew Melrose Ltd

Darcey J (1989/90) Oral history of the Australian fishing industry. https://www.frdc.com.au/sites/default/files/products/1987-021.90-DLD.pdf. Accessed 23 June 2023

Demuth B (2019) Floating coast: an environmental history of the Bering Strait. WW Norton and Company

Donnelly M, Norton C (2020) Doing history. Routledge, Oxon

Dunn B (1991) Angling in Australia: its history and writings. David Ell Press, Balmain

Engelhard GH, Thurstan RH, MacKenzie BR, Alleway HK, Bannister RC, Cardinale M, Clarke MW, Currie JC, Fortibuoni T, Holm P, Holt SJ, Mazzoldi C, Pinnegar JK, Raicevich S, Volckaert FAM, Klein ES, Lescrauwaet A-K (2016) ICES meets marine historical ecology: placing the history of fish and fisheries in current policy context. ICES J Mar Sci 73:1386–1403

FAO (2013) Recreational fisheries. FAO technical guidelines for responsible fisheries, no. 13. Food and Agriculture Organisation of the United Nations

Fowles B, Gaynor A (2011) The challenge of creating a scientifically-robust historical description of changing finfish populations in the Ningaloo Marine Park. In: Gaynor A, Davis J (eds) Environmental exchanges: studies in Western Australian history, vol 27. Centre for Western Australian History, University of Western Australia, Crawley, pp 99–123

Franklin AS (2011) Performing acclimatisation: the agency of trout fishing in postcolonial Australia. Ethnos 76:19–40

Frawley J (2015) Kissing fish: Rex Hunt, popular culture, sustainability and fishing practices. J Aust Stud 39:307–325

Gartside DF, Harrison B, Ryan BL (1999) An evaluation of the use of fishing club records in the management of marine recreational fisheries. Fish Res 41:47–61

Gaynor A, Mccann J (2017) "I've had dolphins...looking for abalone for me": oral history and the subjectivities of marine engagement. Oral Hist Rev 44:260–277

Gaynor A, Frawley J, Máñez KS (2016) "Slim female records the same old story": newspapers, gender, and recreational fishing in Australia, 1957-2000. Geoforum 77:114–123

Grey Z (1928) Big game fishing in New Zealand seas. Sci Am 139:116–118

Haase H (2000) Faszination fisch. Geschichtliches zum fisch und seinem fang. Findling, Buch und Zeitschriftenverlag, Neuenhagen

Harris D (2018) A slow catastrophe? Fishing for sport and commerce in colonial Victoria. In: Cushing N, Frawley J (eds) Animals count: how population size matters in animal-human relations. Routledge, London, pp 155–167

Harrison R, Jones A, Lambert P (2004) The institutionalisation and organisation of history. In: Lambert P, Schofield P (eds) Making history: an introduction to the history and practices of a discipline. Routledge, Oxford, pp 9–25

Harvey G (2010) Panama paradise: a tribute to Tropic Star Lodge. Derrydale Press

Henshaw RE (2011) Environmental history of the Hudson River. State University of New York Press, Albany

Hoffman RC (1985) Fishing for sport in medieval Europe: new evidence. Speculum 60:877–902

Hunt LM, Boucquey N, Beardmore B, Christensen J, Fulton DC, Mackay M, Melstrom RT, Post JR, Schroeder SA, Thurstan RH, van Putten EI, Arlinghaus R (2026) Introducing disciplinary perspectives and interdisciplinary possibilities for understanding recreational fishers and fisheries. In: Pope KL, Arlinghaus R, Hunt LM, Lynch AJ, van Poorten BT (eds) Understanding recreational fishers: disciplinary and interdisciplinary approaches for fisheries management. Springer Nature, Cham, p 57–94

Jackson JB, Kirby MX, Berger WH, Bjorndal KA, Botsford LW, Bourque BJ, Bradbury RH, Cooke R, Erlandson J, Estes JA, Hughes TP, Kidwell S, Lange CB, Lenihan HS, Pandolfi JM, Peterson CH, Steneck RS, Tegner MJ, Warner RR (2001) Historical overfishing and the recent collapse of coastal ecosystems. Science 293:629–638

Jiménez-Alvarado D, Sarmiento-Lezcano A, Guerra-Marrero A, Tuya F, Santana Del Pino Á, Sealey MJ, Castro JJ (2019) Historical photographs of captures of recreational fishers indicate overexploitation of nearshore resources at an oceanic island. J Fish Biol 94:857–864

Johnstone EE, Field D (1998) Amateur Fishermen's Association of Queensland: official history 1904–1997. Amateur Fishermen's Association of Queensland, Bribie Island

Kadagi NI, Wambiji N, Fennessy ST, Allen MS, Ahrens RN (2021) Challenges and opportunities for sustainable development and management of marine recreational and sport fisheries in the Western Indian Ocean. Mar Policy 124:104351

Kerr GJ (1985) Craft and craftsmen of Australian fishing, 1870–1970: an illustrated oral history. Mains'l Books

Kirby JT (2008) Mockingbird song: ecological landscapes of the South. The University of North Carolina Press

Kittinger JN, Mcclenachan L, Gedan KB, Blight LK (2015) Marine historical ecology in conservation: applying the past to manage for the future. University of California Press

Kochalski S, Barz F, Pita P, Harrison HL (2026) Quantitative and qualitative content analysis of text and images. In: Pope KL, Arlinghaus R, Hunt LM, Lynch AJ, and van Poorten BT (eds) Understanding recreational fishers: disciplinary and interdisciplinary approaches for fisheries management. Springer Nature, Cham, Switzerland, p 507–532

Kroloff EK, Heinen JT, Braddock KN, Rehage JS, Santos RO (2019) Understanding the decline of catch-and-release fishery with angler knowledge: a key informant approach applied to South Florida bonefish. Environ Biol Fish 102:319–328

Lam ME (2016) Aboriginal freshwater fisheries as resilient social-ecological systems. In: Craig JF (ed) Freshwater fisheries ecology. Wiley, pp 422–437

Lateiner D (1989) The historical method of Herodotus. University of Toronto Press

Locker A (2014) The social history of coarse angling in England AD 1750-1950. Anthropozoologica 49:99–107

Locker A (2018) Freshwater fish in England: a social and cultural history of coarse fish from prehistory to the present day. Oxbow Books

Loring PA (2017) The political ecology of gear bans in two fisheries: Florida's net ban and Alaska's salmon wars. Fish Fish 18:94–104

Lybolt MJ, Pandolfi JM (2019) Holocene history of Moreton Bay reef habitats. In: Tibbetts IR, Rothlisberg PC, Neil DT, Homburg TA, Brewer DT, Arthington AH (eds) Moreton Bay Quandamooka and catchment: past, present, and future. The Moreton Bay Foundation, Brisbane, pp 153–162

MacEachern A, Turkel WJ (eds) (2009) Method and meaning in Canadian environmental history. Nelson Education, Toronto

MacNeil JR (2014) The great acceleration: an environmental history of the Anthropocene since 1945. Harvard University Press

Mahuika N (2019) Rethinking oral history and oral tradition: an Indigenous perspective, Oxford oral history series. Oxford University Press, New York

Manning DW (1908) Trout fishing in the Cape Colony. Argus Printing and Publishing

Maohong B (2004) Environmental history in China. Environ Hist 10:475–499

Martinussen AO (2007) Fishing gear, open boats and preserving skills. In: Bekker-Nielsen T, Casasola DB (eds) Ancient nets and fishing gear. Proceedings of the international workshop on "Nets and fishing gear in Classical Antiquity: a first approach", Cádiz, November 15–17, 2007. Aarhus University Press, pp 307–322

McClenachan L (2009) Documenting loss of large trophy fish from the Florida Keys with historical photographs. Conserv Biol 23:636–643

McClenachan L (2013) Recreation and the "Right to Fish" movement: anglers and ecological degradation in the Florida Keys. Environ Hist 18:76–87

McKenzie C (2000) The British big-game hunting tradition, masculinity and fraternalism with particular reference to the "The Shikar Club". Sports Hist 20:70–96

McPhee D (2017) Environmental history and ecology of Moreton Bay. CSIRO Publishing, Clayton

Melosi MV (1993) The place of the city in environmental history. Environ Hist Rev 17:1–23

Monteagudo GL (2007) Nets and fishing gear in Roman mosaics from Spain. In: Bekker-Nielsen T, Casasola DB (eds) Ancient nets and fishing gear. Proceedings of the international workshop on "Nets and fishing gear in Classical Antiquity: a first approach", Cádiz, November 15–17, 2007. Aarhus University Press, pp 157–182

Mordue T, Wilson S (2022) More-than-human encounters with fish in the city: from careful angling practice to deadly indifference. Leis Stud 42:709–722

Moutopoulos DK, Katselis G, Kios K, Tsotskou A, Tsikliras AC, Stergiou KI (2013) Estimation and reconstruction of shore-based recreational angling fisheries catches in the Greek Seas (1950-2010). J Biol Res 20:376

Nunn PD, Reid NJ (2016) Aboriginal memories of inundation of the Australian coast dating from more than 7000 years ago. Aust Geogr 47:11–47

Nyboer EA, Embke HS, Robertson AM, Arlinghaus R, Bower S, Baigun C, Beard D, Cooke SJ, Cowx IG, Koehn JD, Lyach R, Milardi M, Potts W, Lynch AJ (2022) Overturning stereotypes: the fuzzy boundary between recreational and subsistence inland fisheries. Fish Fish 23:1282–1298

O'Gorman E, Gaynor A (2020) More-than-human histories. Environ Hist 25:711–735

Obregón C, Christensen J, Zeller D, Hughes M, Tweedley JR, Gaynor A, Loneragan NR (2022) Local fisher knowledge reveals changes in size of blue swimmer crabs in small-scale fisheries. Mar Policy 143:105144

Oral History Association (2023). https://oralhistory.org/about/principles-and-practices-revised-2009/. Accessed 20 May 2023

Parenteau B (2004) A "very determined opposition to the law": conservation, angling leases, and social conflict in the Canadian Atlantic salmon fishery, 1867-1914. Environ Hist 9:436–463

Paul-Burke K, O'Brien T, Burke J, Bluett C (2020) Mapping Māori knowledge from the past to inform marine management futures. N Z Sci Rev 76:32–41

Paulin CD (2007) Perspectives of Māori fishing history and techniques. Ngā āhua me ngā pūrākau me ngā hangarau ika o te Māori. Tuhinga 18:11–47

Pauly D (1995) Anecdotes and the shifting baseline syndrome of fisheries. Trends Ecol Evol 10:430

Pears N (1982) Familiar aliens: the acclimatisation societies' role in New Zealand's biogeography. Scott Geogr Mag 98:23–34

Pitcher TJ, Hollingworth CE (2002) Fishing for fun: where's the catch? In: Pitcher TJ, Hollingworth CE (eds) Recreational fisheries: ecological, economic and social evaluation, Fish and aquatic resources series, vol 8. Blackwell Sciences, pp 1–16

Reiger JF (1975) American sportsmen and the origins of conservation. Winchester Press

Richardson EA, Kaiser MJ, Edwards-Jones G, Ramsay K (2006) Trends in sea anglers' catches of trophy fish in relation to stock size. Fish Res 82:253–262

Ritchie DA (2014) Doing oral history: a practical guide. Oxford University Press

Roberts J, Robinson P (2020) Paleography and codicology. In: Eliot S, Rose J (eds) A companion to the history of the book: methods, materials and readers. John Wiley & Sons Ltd, pp 51–64

Rodrigues JA (2002) Jetty-site artifacts from Holdfast Bay, South Australia: tracing the State's historical development from its early years, based on a functional analysis of submerged material culture. Int J Naut Archaeol 31:108–128

Rothlisberger JD, Lodge DM, Cooke RM, Finnoff DC (2010) Future declines of the binational Laurentian Great Lakes fisheries: the importance of environmental and cultural change. Front Ecol Environ 8:239–244

Santos RO, Rehage JS, Adams AJ, Black BD, Osborne J, Kroloff EK (2017) Quantitative assessment of a data-limited recreational bonefish fishery using a time-series of fishing guides reports. PLoS One 12:e0184776

Senior W (1890) Near and far: an angler's sketches of home sport and colonial life. Sampson Low, Marston, Searle & Rivington

Sharpe DMT, De León LF, González R, Torchin ME (2017) Tropical fish community does not recover 45 years after predator introduction. Ecology 98:412–424

Shiffman DS, Gallagher AJ, Wester J, Macdonald CC, Thaler AD, Cooke SJ, Hammerschlag N (2014) Trophy fishing for species threatened with extinction: a way forward building on a history of conservation. Mar Policy 50:318–322

Sjovold C-P (1999) "An angling people": nature, sport and conservation in nineteenth-century America. PhD thesis, University of California

Smyth L, Egan H, Kennett R (2018) Livelihood values of Indigenous customary fishing. Australian Institute of Aboriginal and Torres Strait Islander Studies. Final report to the Fisheries Research and Development Corporation, Canberra

Snyder S, Borgelt B, Tobey E (eds) (2016) Backcasts: a global history of fly fishing and conservation. University of Chicago Press

Sreedharan E (2007) A manual of historial research methodology. The Centre for South Indian Studies, Trivandrum

Stamp T, West E, Robbins T, Plenty S, Sheehan E (2022) Large-scale historic habitat loss in estuaries and its implications for commercial and recreational fin fisheries. ICES J Mar Sci 79: 1981–1991

Szabó P (2015) Historical ecology: past, present and future. Biol Rev 90:997–1014

Szylvian KM (2004) Transforming Lake Michigan into the "World's greatest fishing hole": the environmental politics of Michigan's Great Lakes sport fishing, 1965-1985. Environ Hist 9: 102–127

Tenison-Woods JE (1882) Fish and fisheries of New South Wales. Thomas Richards Government Printer, Sydney

The Guardian (2019) "They're not property": the people who want their ancestors back from British museums. 23 April 2019. https://www.theguardian.com/culture/2019/apr/23/theyre-not-property-the-people-who-want-their-ancestors-back-from-british-museums. Accessed 20 May 2023

Thompson EP (1963) The making of the English working class. Pantheon Books, New York

Thompson EP (1967) Time, work-discipline, and industrial capitalism. Past Present 38:56–97

Thurstan RH, Campbell AB, Pandolfi JM (2016) Nineteenth century narratives reveal historic catch rates for Australian snapper (*Pagrus auratus*). Fish Fish 17:210–225

Thurstan RH, Game E, Pandolfi JM (2017) Popular media records reveal multi-decadal trends in recreational fishing catch rates. PLoS One 12:e0182345

Thurstan RH, Buckley SM, Pandolfi JM (2018) Trends and transitions observed in an iconic recreational fishery across 140 years. Glob Environ Chang 52:22–36

Travis C, Dixon DP, Bergmann L, Legg R, Crampsie A (eds) (2022) Routledge handbook of the digital environmental humanities. Routledge, London

Tull M (1999) Oral history of the Australian fishing industry. Fisheries Research and Development Corporation

Tyrrell I (2005) Historians in public: the practice of American history, 1890–1970. University of Chicago Press, Chicago

Van Sittert L (2001) The South African fisheries: a preliminary survey of historical sources. In: Holm P, Smith TD, Starkey DJ (eds) The exploited seas: new directions for marine environmental history. Liverpool University Press, pp 167–180

Waiti J, Wheaton B (2022) Indigenous Māori knowledges of the ocean and leisure practices. In: Peters K, Anderson J, Davies A, Steinberg P (eds) The Routledge handbook of ocean space. Routledge, pp 85–100

White R (1985) Environmental history: the development of a new historical field. Pac Hist Rev 54: 297–335

White R (1996) The organic machine: the remaking of the Columbia river, Hill and Wang critical issues. Hill and Wang, New York

Williams SM, Holmes BJ, Pepperell JG (2015) The novel application of non-lethal citizen science tissue sampling in recreational fisheries. PLoS One 10:e0135743

Williamson T (2013) An environmental history of wildlife in England 1650–1950. Bloomsbury, London

Woolf D (2011) A global history of history. Cambridge University Press, Cambridge

Worster D (1990) Transformations of the earth: toward an agroecological perspective in history. J Am Hist 76:1087–1106

Young MA, Foale S, Bellwood DR (2014) Impacts of recreational fishing in Australia: historical declines, self-regulation and evidence of an early warning system. Environ Conserv 41:350–356

Young MA, Foale S, Bellwood DR (2015) Dynamic catch trends in the history of recreational spearfishing in Australia. Conserv Biol 29:784–794

Young MA, Foale S, Bellwood DR (2016) Why do fishers fish? A cross-cultural examination of the motivations for fishing. Mar Policy 66:114–123

Behavioural Economics for Understanding Recreational Fishers and Fisheries

Mary Mackay, E. Ingrid van Putten, Andries Richter, Anna Lou Abatayo, Swee-Hoon Chuah, Dorothy J. Dankel, Amanda Schadeberg, and Hunter Snyder

Abstract Fisher behaviour is a key consideration and source of uncertainty for recreational fisheries management. All human behaviour is complex and thus difficult to understand, predict, and manage. In recreational fisheries, classical economic theories of profit maximization or rational choice tend to poorly explain fisher behaviour. Consequently, policy interventions that fail to understand and anticipate behavioural responses of fishers tend to fail their intended objectives. The topic of

---

M. Mackay (✉) · E. I. van Putten
Commonwealth Scientific and Industrial Research Organisation Environment, Hobart, TAS, Australia

Centre for Marine Socioecology, University of Tasmania, Hobart, TAS, Australia
e-mail: Mary.Mackay@csiro.au

A. Richter
Environmental Economics and Natural Resources group, Wageningen University, Wageningen, The Netherlands

Centre for Ecological and Evolutionary Synthesis, Department of Biosciences, University of Oslo, Oslo, Norway

A. L. Abatayo
Environmental Economics and Natural Resources group, Wageningen University, Wageningen, The Netherlands

S.-H. Chuah
Centre for Marine Socioecology, University of Tasmania, Hobart, TAS, Australia

Tasmanian Behavioural Lab and Centre for Marine Socioecology, University of Tasmania, Hobart, TAS, Australia

D. J. Dankel
Climate and Sustainability Research Group, Department of Climate and Environment, SINTEF Ocean, Bergen, Norway

A. Schadeberg
Environmental Economics and Natural Resources Group & Environmental Policy Group, Wageningen University, Wageningen, The Netherlands

H. Snyder
Thayer School of Engineering, Dartmouth College, Hanover, NH, USA

K. L. Pope et al. (eds.), *Understanding Recreational Fishers*, Fish & Fisheries Series 45, https://doi.org/10.1007/978-3-031-99739-6_9

behavioural economics offers an opening by enriching our understanding of fisher behaviour by studying how psychological mechanisms such as behavioural preferences, beliefs, and cognitive biases affect behaviour. The behavioural economic literature can be leveraged to better understand the impacts of fishery management systems. Identifying behavioural regularities and mechanisms can help to inform and improve the current monitoring, control, and surveillance model of management. Potential applications for novel recreational fisheries management build upon a rich theoretical literature and practice. In this chapter, we introduce behavioural economics and its key theories, describe the types of methods used to study recreational fisher behaviour, explain past applications in recreational fisheries research and management, provide some best practices when applying behavioural economics in research and in the field, and consider future prospects.

Keywords Behavioural insights · Behavioural mechanisms · Cognitive bias · Experimental economics · Social psychology · Sociology

1 What Is Behavioural Economics?

In the absence of any sort of intervention by managers, recreational fisheries can be characterized as open-access resources, allowing unrestricted use by anyone. This unrestricted access can lead to overexploitation and resource depletion, a situation commonly referred to as the "Tragedy of the Commons" (Hardin 1968). Hardin proposed privatization (i.e., making resources private properties) and passing laws restricting public access to resources as solutions to the Tragedy of the Commons. Real-world examples of such regulations include, for example, marine recreational fishing in Portugal that used to be open access but, since 2006, has had restrictions to control recreational harvest (Veiga et al. 2013).

However, Elinor Ostrom, who received the Nobel Memorial Prize in Economics in 2009, challenged this approach. She demonstrated that, under specific conditions, communities could effectively manage and sustainably govern their resources without resorting to privatization or government control (Ostrom 2015). Ostrom emphasized that successfully managing the commons requires bottom-up organization and adherence to social norms. Key to its success are behavioural factors such as internal trust and reciprocity (Ostrom 2009), effective communication (Ostrom 2012), active involvement in creating governing rules that leads to greater acceptability of the rules (Abatayo and Lynham 2016, 2020; Janssen et al. 2008), and enforceability of regulations (Bergstrom 2010), among others. There is ample evidence that those factors are also highly relevant for recreational fishing. In the case of the saltwater recreational fishers in Portugal, for instance, Veiga et al. (2013) find that many recreational fishers were against the regulations currently in place. The lack of acceptance of the regulations among recreational fishers, along with their detailed knowledge of the regulations, has resulted in their lack of compliance and hence, the failure in managing the fish stocks in Portugal.

Indeed, although many management tools (quota allocation, bag and size limits, season closures, gear restrictions, etc.) have been effective in managing recreational fisheries, the absence of a direct profit motivation among recreational fishing foreshadows the difficulty in neatly fitting their decision-making into traditional economic analysis. The behavioural responses of fishers to diverse types of available management approaches and restrictions should be considered to ensure the effectiveness of recreational fisheries policies. Behavioural economics can help understand how recreational fishers are likely to respond and behave (Wieczorek et al. 2021) by combining economic principles with knowledge from behavioural sciences, especially psychology, which broadly examines human behaviour and mental processes, and social psychology, which focuses on how behaviour is affected by the social and physical environment.

1.1 Behavioural Economics Versus Traditional Economics

The neoclassical economist Adam Smith famously argued in *The Wealth of Nations*, that economic behaviour is motivated by self-interest and that by intending one's own security and gain, one is "led by an invisible hand" (Smith 1776, p. 540) to advance the interest of society. In this world, individuals are viewed as *Homo economicus*: rational and self-interested. This notion of rationality and self-interest is often based on two key assumptions: (i) that individuals have stable preferences and can rank different options consistently (i.e., transitive) and (ii) that individuals have access to all relevant information to compare and evaluate different outcomes based on a measurable utility scale (i.e., complete) (Mas-Colell et al. 1995). According to this view, rational individuals are expected to have consistent preferences and make choices that will lead to the best possible outcome for themselves (i.e., in terms of maximizing their utility or satisfaction), given their constraints and available information (Jentoft et al. 1998). The baker bakes and the brewer brews not out of benevolence but out of self-interest (i.e., because they profit from doing so). As such, in this "rational" world, the only conceivable way to change decisions is to change the costs and benefits of available options.

Reality, however, is far from this simplified version of human interaction and economic behaviour. In practice, individuals tend to make decisions that are vastly different from what is predicted by traditional economic theory, as various other factors influence their decision-making. This, however, is not to say that individuals make irrational or random choices, but rather that the traditional definition of selfish rationality is too narrow to characterize actual individual decision-making. For instance, an individual may decide to follow certain rules not because they directly benefit from doing so but because they observe that the majority of society is following the rules. Also, losses may be weighted more heavily than gains, making individuals pursue strategies that are not necessarily yielding the highest expected payoff (known as loss aversion). Further, the concept of framing, filtering a communication to inspire a certain interpretation, can yield desired behaviour:

individuals are likely to accept a policy that is positively framed ("Help to keep the environment clean") compared to the same policy that is negatively framed ("Do not leave your trash lying around"). These quite different examples show that behavioural economics can enter the decision-making process in several ways.

To better understand the drivers of individual behaviour and decision-making, it is important to go beyond assumptions of selfish rationality and to consider psychological elements in individual behaviour and decision-making. People's behaviour is influenced by their environment and their circumstances and they do not always do things that are in their own self-interest. Rather, a wider set of motivations and psychological factors plays a role, such as sympathy for others, altruism, fairness, trust, loss aversion, and social status.

Many ideas core to behavioural economics have roots dating back centuries. For instance, though Adam Smith is famously the father of neoclassical economics in the eighteenth century, he did not have a simple-minded view of human nature but rather had a deep interest in the psychological underpinnings of behaviour (Angner and Loewenstein 2012). He recognized that individuals sympathize with other individuals and act altruistically towards them for no other reason than that it gives them personal pleasure. Before publishing *The Wealth of Nations*, Smith proposed a theory of individual behaviour that is a constant struggle between "passions" (i.e., drives, emotions, motivational feelings) and the "impartial spectator" (i.e., self-denial, self-government, conscience) (Smith 1759). According to Smith, the struggle between these two determines individual perceptions and preferences.

Nevertheless, behavioural economics is often viewed as a relatively fresh alternative to "traditional" or "neoclassical" economics and is seen by economists and non-economists alike as a young subfield of economics. Its contribution to economics is its understanding of economic decision-making through realistic psychological foundations (Camerer and Loewenstein 2004). It extends, rather than abandons, traditional economic theory (Ho et al. 2006) to better understand what was previously seen as irrational behaviour. Despite this, some economists prefer to remain agnostic about behavioural underpinnings of individual decision-making, mostly because they are difficult to observe and quantify (Angner and Loewenstein 2012).

It is also worth emphasizing that although many insights from behavioural economics are novel to economics, they are well-established and often borrowed from fields as social psychology or sociology. Indeed, many behavioural mechanisms seem very intuitive and based on common sense to people on the street. Therefore, one of the major contributions of behavioural economics is introducing insights generated by other social sciences into the traditional field of economics. One can wonder how insightful the label "behavioural economics" is, but it is noticeably clear that most of the work originates in psychology, sociology, or behavioural sciences more generally, and would not exist without cross-pollination from these.

The emergence of behavioural economics as a field is due mostly to the recognition that individuals are *Homo sapiens,* rather than merely *Homo economicus.* There are many ways in which human decision-making diverges from the assumption of traditional rationality (Mullainathan and Thaler 2001), and the academic pursuit of behavioural economics started with the realization—that traditional economic models are not able to predict actual consumer choices. This makes it clear that social and behavioural sciences should incorporate aspects that are not immediately directly observable, such as an individual's cognitive and affective states (Angner and Loewenstein 2012).

1.2 The Importance of Behavioural Mechanisms

The "Rational actor" model remains the main workhorse in economics, even if empirical support can be feeble—often to the surprise of non-economists. However, the persistent popularity is not because of plausibility (economists know the shortcomings very well), but convenience. This is because "optimal" behaviour can be modelled mathematically with some basic calculus. In contrast, behavioural economics is often perceived as a potpourri of different anomalies, biases, and peculiarities of human behaviour, making it more difficult to use in formal modelling or indeed making its use in welfare analysis controversial. Indeed, compiling a list of only the most important behavioural mechanisms easily results in a catalogue of different mechanisms (McFadden 1999; Shogren and Taylor 2008; Wieczorek et al. 2021), but no unified way exists to classify behavioural economic mechanisms (Carlsson and Johansson-Stenman 2012).

A broad classification of behavioural economics and how it contrasts with traditional economics is shown (Fig. 1). Traditional economics assumes that individuals make decisions that simply maximize private welfare, whereas behavioural economics allows for a richer set of explanations. First, individuals hold behavioural preferences (e.g., avoiding losses or caring more about how outcomes relate to the performance of peers, rather than absolute outcomes). Second, individuals hold behavioural beliefs, for example, regarding what is appropriate or acceptable behaviour. Third, contextual factors—cognitive biases—tend to affect our decisions, even if they do not affect outcomes. According to traditional economics, an inconsequential framing of a certain message or presenting information in a certain order should not matter for our decisions—but in reality, they do.

The first important acknowledgement of *behavioural preferences* was made by the *Prospect theory* of Kahneman and Tversky (1979). Prospect theory shows that people are disproportionally motivated by their fear of losses (loss aversion) and place more weight on losses than equivalent gains (Tversky and Kahneman 1981). More broadly, individuals do not judge potential outcomes based on their "expected value" but put weight on the probabilities of different outcomes. Moreover, people are not good at systematically evaluating probabilities correctly and acquiring insight in an asymmetric fashion. Low probability events may be over-weighted (putting more stress on extreme outcomes) or under-weighted (neglecting extreme events) (Diekert 2017). In addition, individuals tend to hold social preferences, which implies that our happiness is affected by what happens to others, as well as how our outcomes relate to those of others. Another example of a behavioural

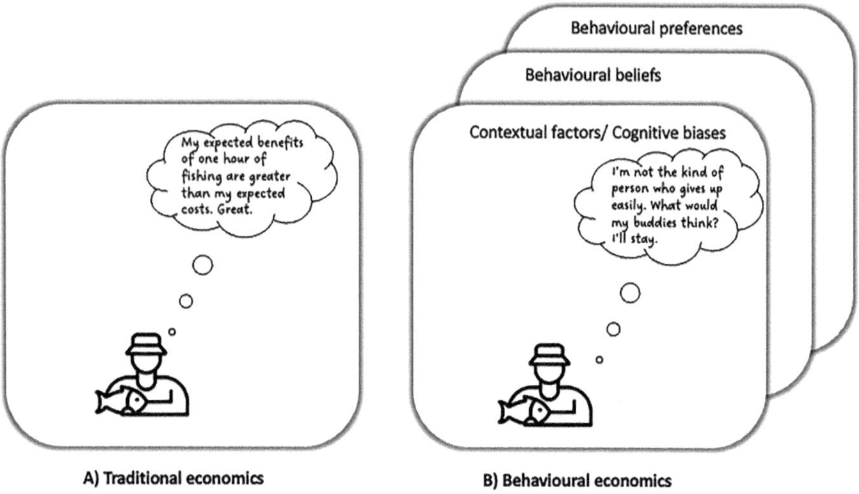

Fig. 1 Behavioural Economics in Contrast with Traditional Economics (**a**: traditional economics assumes that individuals pursue decisions that maximize own payoffs; **b**: behavioural economics considers a richer set of mechanisms to explain decisions) Highlighting that Process of Decision-making is Multi-dimensional in Which Behavioural Preferences, Beliefs, Contextual Factors, and Cognitive Biases Play Key Roles (*behavioural preferences* imply that we do not necessarily follow decisions that are optimal and give highest expected payoff, for example because we avoid choices that could lead to loss; *behavioural beliefs* reflect that individuals have certain expectations about what others are doing, or what is appropriate to do, both also affecting decisions; *cognitive biases* reflect all processes that systematically affect our decisions [e.g., the framing or cultured understanding of a situation] even though they do not matter for outcomes)

preference is how individuals may evaluate outcomes differently depending on the timing of the outcome. Often, individuals tend to have present biases and may strongly prefer immediate outcomes over outcomes that are in the future, leading to seemingly inconsistent choices over time.

Behavioural economics also considers how our cultural and social norms influence our decisions through *behavioural beliefs* (Wilson 2020). If behavioural preferences are our moral value compass, beliefs are the needle that is changing with our environment. Together, they tend to guide our decisions. It is natural for individuals to compare themselves to others, and this is an example of how people's decision making is influenced by social norms. Social norms are people's perceptions of what is common behaviour in specific situations (Ajzen 1991). Social norms reflect the reality that decisions are not taken in a vacuum, but in a social context (Bicchieri 2017; Nyborg 2018). This implies that actors are not only self-interested but also consider how decisions affect others. Also, we infer appropriate actions and the success of actions by observing what people around us are doing. Social norms were central in the work of Elinor Ostrom, who showed that evolutionary theory, including the study of cultural evolution, can explain the propensity to cooperate. This observation was corroborated by the development and growth of social norms in natural resource management (Ostrom 2000). People can cooperate to avoid the tragedy of the commons (Wilson 2020).

Cognitive bias means individuals diverge from rational decision making by making systematic thinking errors influenced by *contextual factors*, such as emotion and opinions. A lengthy list of cognitive biases exists that affect behaviour. Thaler (2016) aptly refers to these as "supposedly irrelevant factors" as they do not matter for the outcomes. The framing effect is an example of a cognitive bias that influences decision making. Other examples are the "Ostrich effect" (i.e., selectively avoiding negative information) or confirmation bias (e.g., selecting information that confirms earlier beliefs). Decisions to take or avoid certain risks depend on how messages are framed. The framing effect means that when an individual chooses from a set of options, their choice can be influenced more by the way the information is presented than the substance of the information (Plous 1993).

If one compares the role of preferences, beliefs, and contextual factors, it is worth highlighting that they often interact. The tendency to follow social norms, for example, comes from a desire to conform (a preference), in combination with an observation of what others do (a belief). Also, framing and loss aversion go together because individuals are more likely to desire risks when the information is framed negatively but seek to avoid risks when the information is framed positively (Tversky and Kahneman 1981).

Finally, the decision-making process itself is not necessarily optimizing, but may be subject to some other decision-making rule or heuristic. Heuristics are mental shortcuts or rules of thumb that simplify decisions under risk and uncertainty. For example, people will use heuristics (these psychological rules of thumb) to make decisions that rely on information that is easy to remember and access (availability heuristics). Another heuristic that is often encountered is termed "anchoring," which is the tendency to fixate on a reference point that may have no direct relevance to a decision. Hence, a basic premise of behavioural economics is that people have bounds to their rationality; they tend not to always do things in their own best interest. People's rationality is bounded because they have limited cognitive abilities, information, and time. Bounded rationality describes "rational choice that takes into account the cognitive limitations of the decision-maker—limitations of both knowledge and computational capacity" (Simon 1990, p. 1). Bounded rationality aims to understand the actual cognitive processes that explain the discrepancy between human reasoning and economic decision making (Gigerenzer and Selten 2001).

1.3 Behavioural Economics for Policy

Policy makers use behavioural economics to improve the design and delivery of public policy through improved prediction of fisher responses to management (Wieczorek et al. 2021). Using behavioural economics techniques can improve the effectiveness, efficiency, and acceptability of policies. Nevertheless, the role of behavioural economics in welfare economics, drawing normative conclusions, and developing policy prescriptions, is something that is currently being actively debated (Angner and Loewenstein 2012).

In their book *Nudge* (2008), Richard Thaler and Cass Sunstein argue that being aware of how people behave and the psychology underlying their behaviour will serve to improve policy design. “Nudges” are an example of policies informed by behavioural economics. Nudges are interventions where the decision context (or choice architecture) is designed to steer people’s behaviour in a certain direction (and build on knowledge of cognitive processes). In contrast, command-and-control policy instruments bring about behaviour change via mandates and prohibitions. Economic instruments, such as taxes or subsidies, give financial incentives by manipulating costs and benefits. Nudging is becoming increasingly prominent in different arms of government and policy making (Gane 2021).

Effective nudging requires policymakers to be aware of how people behave as opposed to how they are assumed to behave as a rational response to traditional policies. Examples of nudges are pop-up alerts that need to be actively clicked away when a certain number of games has been reached to help addicted online gamblers break out of the “zone.” Also, enrolling employees automatically into a retirement savings plan unless they opt-out has been shown to be effective.

Initially, Thaler and Sunstein (2003) described nudges as libertarian paternalism (Thaler 2015) that operates without coercion but instead steers individuals to make better choices (Gane 2021). Unlike mandates and prohibitions, nudges do not force or remove choice. Unlike taxes and subsidies, nudges do not change prices by making products or services more expensive or cheaper. Instead, nudges appeal to people’s psychology (e.g., their cognitive processes, heuristics, biases) by making it easier and more attractive for people to adopt the recommended behaviour (Thaler and Sunstein 2008). Because they retain choice, a strength of nudges is that they are a lighter touch compared to more heavy-handed traditional policies. Nudges steer towards private welfare-promotion or steer behaviour to promote public goods (Barton and Grüne-Yanoff 2015), which is critically important to their normative justifications.

Economics traditionally addresses normative questions with respect to policy interventions and whether they are good or bad, using, for instance, cost-benefit analysis (Bernheim 2016). Welfare economics studies the allocation of resources to promote welfare and is a normative branch of economic theory. Behavioural economics is playing an increasingly important role in policy development and analysis, but policies underpinned by behavioural economics can be a unique challenge for welfare analysis (Just 2017) and welfare economics. Some argue that insights from behavioural economics develop paternalistic policies and therefore, create a paradox for welfare economics and undermine its foundations in relation to free choice (Just 2017).

The role of behavioural economics in normative policies has not been sufficiently addressed in the literature (Baldwin 2014; Just 2017). Moreover, there are many concerns about using these (paternalistic) behavioural economic tools to manipulate behaviour. Evidence of the effectiveness of nudges and refining their normative justification is needed (Benartzi et al. 2017; Congiu and Moscati 2022). The cost-effectiveness of nudges compared to measures of traditional policy tools, such as tax incentives and other financial inducements, is favourable (Benartzi et al. 2017), although sometimes disputed (Baldwin 2014).

It is important to note that interventions based on behavioural economics are not without limitations. First, they are not suited for all circumstances. Nudges, for example, will likely not work on people who have already consciously and deliberately made the decision to flaunt regulations. Second, some have argued that interventions in the form of nudges are manipulative as they exploit people's psychology. However, a large study conducted by Sunstein et al. (2019) showed that people were generally supportive of nudges if they were transparent, and their outcomes were consistent with people's values. In the case of monitoring and controlling the behaviour of recreational fishers, no widely shared consensus exists in society about whether these are the right things to do (unlike, for example, recycling or volunteering). Due to their contested nature, nudging may not be widely supported (Cotterill et al. 2012). Third, nudges do not address the structural impediments and the cultural and social institutions that give rise to policy challenges. Responsibility for these lies with the government and should not be shifted to individuals. Finally, the effectiveness of behavioural economics interventions over the longer term and at scale still needs to be evaluated. Some interventions may be effective for decisions that only need to be made once, but for sustained compliance, repeated interventions may be required.

Despite these limitations, behavioural economics offers a new important interdisciplinary research lens to recreational fisheries. A strong promise exists for new breakthroughs for sustainability, trust, and policy collaboration when best practices of behavioural economics and best practices of interdisciplinary research come together for a deeper understanding of the human aspects of recreational fishing.

2 Behavioural Economics and Understanding Recreational Fishing

The average recreational fisher does not fit the neoclassical characterization of the *Homo economicus.* Recreational fishing is a leisure activity and sport in contrast to other types of fishing for pure subsistence or for profit. Hence, the personal choices and behavioural patterns associated with recreational fisheries can be vastly different between artisanal or small-scale fisheries and large-scale commercial fisheries. Attempts to regulate these distinct types of fisheries using one-size-fits-all policies could result in negative consequences such as overfishing and irreversible ecological damage.

Behavioural economics, with its focus on the (social) psychology underlying people's behaviour, has an important role to play in supporting more effective management of recreational fisheries and supporting economic and ecological objectives (Greiner et al. 2013). Recreational fishing behaviour is influenced by the complex interactions of preferences, beliefs, and contextual factors. Whereas preferences are often portrayed in a somewhat stylized way, they can be multidimensional and context-dependent and encompass motivation, identity, and a wider value set (Birdsong et al. 2021). In recreational fishing, those deeper motivational factors are expected to be particularly strong as fishing does not happen for monetary reasons. Recreational fishing is an inherently social activity (despite many cliches portraying recreational fishers otherwise), fishers have a strong intrinsic motivation to go fishing (and decisions may be driven by passion as well as reason) and fishing takes place in a highly stochastic environment.

The above-mentioned characteristics suggest that behavioural economic approaches can provide valuable insights to understand and then to address policy challenges in recreational fishing. For example, a passionate recreational fisher caught in the moment may be tempted to flaunt regulations and keep a certain type of rare fish. In this "hot" mental state, their emotions take over. In such cases, it may be more effective to apply behavioural insights to de-bias or temper such emotions rather than to quote a "cold" regulation to them. Even though behavioural insights can be leveraged to develop policies and regulatory measures that are more effective to support compliance and sustainable recreational fishing, only few research papers have applied behavioural economics to recreational fishing and many of the insights are not directly attributable to behavioural economics but instead to related fields such as psychology or other social sciences. The following examples will illustrate how social and motivational factors may play a role in recreational fishing, and these motivational factors link to behavioural economics, even if they have not been considered in that light.

An individual may fish for social reasons, such as spending time with friends and family. In a survey by Young et al. (2016), they found that more than half of their sample of recreational fishers in Australia and the Solomon Islands emphasized the importance of fishing to build relationships with friends and other recreational fishers. They believe that recreational fishing facilitates social cohesion and improves relationships. Recreational fishers can derive a sense of accomplishment that comes from catching a fish, a feeling of relaxation and groundedness from being one with nature, and feelings of pleasure as they reach a state of "flow" (i.e., a psychological state characterized by complete absorption in an activity). Many recreational fishers have financial support and resources but choose to go fishing because they "love it—it's a hobby" (Young et al. 2016). Non-catch aspects play a role for many recreational fishers in how satisfied they are with the fishing experience. Whether these non-catch aspects are motivators or contribute to satisfaction with the fishing trip is disconnected (Arlinghaus 2006). Satisfaction may indeed be catch-related (Connelly and Brown 2000; Birdsong et al. 2021), and it is important for managers to know what drives satisfaction with a fishing experience when developing management actions.

Some motivations are more salient in decision making than others. Individuals may participate in recreational fishing to improve individual skill, engage in competitions, show off, or to prove their ability to catch a certain type or size of fish.

Recreational fishing behaviour can be influenced by cultural norms and values, such as the importance placed on traditional fishing practices or the perceived value of catching certain types of fish. There is currently, for example, a renaissance of traditional fishing practices in the Hawaiian Islands, which was eroded during Western contact (Friedlander et al. 2013). Traditionally, fishing activities were based on identifying strictly enforced times and places where fishing could occur to encourage the regrowth of resources. Unlike the past, where violators were punished by blinding or death, present-day violations are more likely to lead to unfavourable social perceptions from others.

It is important to note that these motivations are neither all-encompassing nor mutually exclusive. People may engage in recreational fishing for a combination of reasons. Understanding these preferences and motivations can help managers and conservationists to develop more effective policies and regulations, as well as education and awareness campaigns that consider the motivations and decision-making processes of recreational fishers (see Landon et al. 2026). If, for example, recreational fishers fish for social reasons, they may be more likely to participate in group fishing trips or join fishing clubs, which can lead to increased fishing pressure on certain areas or types of fish. Similarly, if people are fishing for social status, they may be more likely to participate in tournaments or to keep the fish they can catch, which can also lead to increased fishing pressure.

The motivational differences between fishers and between fisheries provide an avenue for behavioural economic studies to investigate how behaviour might be influenced. For example, messaging and appropriate framing of messages may have an impact on participation and compliance (Martin et al. 2017). Simple short SMS messages can be an effective means to increase compliance (Snyder et al. 2021). Framing studies that influence social norms in recreational fishing are not common, but recreational fishing education campaigns that seek to influence social norms to improve compliance offer opportunities to inform new frames for decision-making (Bova et al. 2017).

In recreational fishing, compliance with rules and regulations is of crucial importance. In contrast to what one may think, strong enforcement and penalties may be much less important for compliance behaviour than social norms (Bova et al. 2017, 2022). This may not be unique to recreational fishing, as this has also been reported for commercial fisheries (Bisack and Clay 2021; Diekert et al. 2021). Interventions to reduce illegal fishing can (amongst other things) influence “shame” if apprehended (Wilcox and Bergseth 2021). Influencing the social norm can be a behavioural economic approach to understanding compliance (Mackay et al. 2018). The potential role of social norms and peer pressure on enforcement and self-governance is an area where insights from behavioural economics can help improve management. Social norms have an enormous influence on fishers’ compliance behaviour. In the recreational blue cod (*Parapercis colias*) fishery in the Marlborough Sounds in New Zealand, behavioural interventions that target social norms (instead of increasing enforcement) were recommended (Thomas et al. 2016).

The attributes that characterize a typical recreational fishing trip (other than cost) are stochastic, like the amount of catch and the weather. Behavioural economics stresses how challenging it is for individuals to deal with probabilities. Furthermore, even if the relevant probability distributions can be estimated, they may not be available to recreational fishers and the spatial resolution that is relevant to them. Likewise, when considering the substantial number of site options recreational fishers typically choose from, one must wonder how fishers resolve the complexity associated with those many choices. In other words, bounded rationality and limited information are at the centre of a fisher's trip choices. Bounded rationality is an important driver of fisher behaviour and vessel search patterns (Carrella et al. 2020) .

Bounded rationality may also lead to resistance to change and repeated behaviour. It can contribute to technological lock-in because fishers only look for information on technology that aligns with what they already use (related to anchoring and status quo biases). Moreover, fishers may have present bias and prefer immediate outcomes over delayed future outcomes, leading to management preferences at odds with, for example, environmental management objectives.

Some empirically based studies that provide insight into aspects of fisher behaviour include, for example, Klefoth et al. (2023), who highlight the importance of understanding preferences, especially because they found that managers and fishers differ with respect to their pro-ecological beliefs and their behavioural intentions. Making a management decision based on misperceptions or misalignment of beliefs can create conflicts. Though that is true, even more concerning is the possibility that it may lead to over-fishing, stock depletion, and ultimately to management failure. Understanding the subtleties in beliefs and behavioural intentions has implications for many recreational fishing aspects, such as support for climate adaptation and making climate-resilient fisheries (Gregory and Grant-Smith 2023; Varade et al. 2023).

3 Behavioural Economics and Experimental Economics?

The way theories are tested using empirical data is essentially the same in traditional and behavioural economics. However, the way data are collected in behavioural economic studies differs. The norm in behavioural economics research is the application of experimental approaches to test theories, but many distinct types of experiments exist. The experiments can be laboratory-based (e.g., with the experiment subject participating in a "game," discussed further below), which is a controlled setting using participants or volunteers. Experiments can also be field-based, which is a less controlled setting usually using specific actors, like fishers (or lab in the field). One key component of behavioural economic experiments is that participants are paid for their participation, which differs from other experimental disciplines such as psychology. There is a trade-off regarding external and internal validity between lab and field experiments. In a laboratory setting, it is easier to control the impact of other external variables. But because a lab environment is different from the outside world, the external validity (the ability to generalize to the general population) can diminish (see van Putten et al. 2026).

Experimental and behavioural economics are terms often used together, but they are not the same. Experimental economics is a methodological approach. It uses controlled, scientific experiments to test people's choices and decisions under certain conditions. It is one way to approach a behavioural economic question—like the contribution of social norms to compliance behaviour (Mackay et al. 2020; Thomas et al. 2016). Experimental economics does not exclusively get implemented in behavioural economics but has been used to answer traditional economic questions such as the law of supply and demand. In its simplest form, this laboratory experiment was used as a tool in empirical economic analysis as it addresses how a single buyer and seller reach, or fail to reach, mutually agreeable terms of trade. The study of alternative market mechanisms was at the heart of the Nobel prize being awarded to Vernon Smith (1962), the founder of experimental economics.

Economic experiments have been carried out to establish the influence of different behavioural mechanisms on decision making, but they have not been so commonly explored in a recreational fishing context (Bergseth et al. 2017; Mackay et al. 2020). Moreover, aspects such as cognitive biases that play a role in recreational fisher decisions are not frequently formally tested in a laboratory setting with either recreational fishers (the stakeholders) or students.

Nevertheless, many laboratory-based experimental approaches have been developed. Behavioural economic experiments are often referred to as games or economic game experiments. Games are used by behavioural and traditional economists and social scientists to develop and test theories. There are numerous games, but perhaps the best known is the *public goods game* (Rojas et al. 2021). Public good games are relevant to recreational fishing because they can explore cooperation (Aswani et al. 2013) (information sharing), trust, self-interest, free-riding, and injustices (Maldonado and Moreno-Sanchez Rdel 2016).

These games can help explore and gain insight into cultural and social norms. It is well known that there are cross-cultural differences that will impact the outcomes of a public good game. For example, in public goods' games with fishers and reindeer herders in Kamchatka, Russia, Gerkey (2013) highlighted their unconditional generosity posited to emerge from economic interdependence. Their average contributions were substantially higher than reported elsewhere. However, the framing of the game shifted the participants' strategy toward conditional cooperation.

The *prisoner's dilemma* is game -theoretic approach that investigates the problem of cooperating in a situation where individuals have an incentive to renege (Haynie et al. 2009). The game represents a situation where different parties have a choice between cooperating or not. In this game, an optimal solution (the highest reward) is not possible if the parties (individuals) behave rationally and make decisions in their own self-interest. Only when both parties choose to co-operate can the highest reward be achieved. This approach has been applied to open access fisheries issues, such as in the Arctic, to assist decision-makers in resolving conflicts (Cole et al. 2014). Cole et al. (2014) find that access to different information often explains undesirable outcomes and "a reduced probability of reaching agreement, delays in making decisions or decisions that turn out to be poor from a social perspective. These outcomes underscore the importance of strong governing institutions in providing and distributing credible information, for example, about the size of fish stocks, the consequences of oil spill or, more generally, the value of ecosystem services" (p. 11). Also Stoop et al. (2012) noted in a laboratory setting that group earnings are greater when group members catch fewer fish, which is consistent with classical economic theory. However, they found in a field experiment that when cooperation requires reducing fishing effort, individuals are not cooperative.

Trust plays a crucial role in the many behavioural economic experiments (Lopez et al. 2012). The *coordination game* can explore this to some degree. In the coordination game, players will earn a higher payoff if they all do the same thing. In a fisheries context, a dynamic coordination game was implemented to analyse the dynamics of a common-property fishery and the non-cooperative behaviour of fishers (Vosooghi 2013, 2019). If fishers face uncertainty about the other players' behaviour because they cannot obtain an accurate assessment of the current stock, their coordination is influenced by "fear of miscoordination." Incentivized coordination games can also be used to investigate the influence of social norms, showing that players are willing to sacrifice money to behave in a socially appropriate way (Krupka and Weber 2013).

Another game that investigates sharing behaviour and self-interest as a driver of behaviour is the *dictator game* (Sorokowski et al. 2017). It questions the standard economic assumption that individuals will act solely out of self-interest, and many experiments have shown that people are more willing to share goods than what a self-centred exchange would predict. This is done by presenting real-life data of otherwise rational individuals acting in a manner that runs counter to the predictions of economic assumptions. The *ultimatum game* is somewhat related in that, in this take-it-or-leave-it bargaining game, it investigates the players' willingness or unwillingness to accept injustices (Harsanyi 1977). Very simply, in the game, one player has a sum of money, and they must share it with another player. It becomes apparent in this game that people tend to refuse small offers that are perceived as unfair (i.e., less than half or what they consider is a fair amount). Rationally, one would accept any amount of money, even if it is an unfair small amount, but the refusal of an unfair offer can be, for example, because the receiver wants to punish the giver or as a honour-related tactic to avoid being attributed an inferior status (Yamagishi et al. 2012).

The above-mentioned games have many possible permutations and variants where various aspects (such as culture or social norms or gender or race, or trust or so many other aspects) are tested or where the games are altered to have, for example, incomplete information, tipping points, or diverse levels of competitiveness. But again—many opportunities exist for researchers to apply experimental games in the context of recreational fishing.

4 What Are the Best Practices When Applying This Disciplinary Perspective?

4.1 Theoretical Best Practice

Behavioural economics was born out of the need to extend traditional economic theory of selfishly rational individuals to one that is closer to reality. Although doing so is a step in the right direction from the much-oversimplified economic theories, it does make theoretical work more complicated and possibly less tractable. It is important not to lose track of what is important to be as close to reality as possible. The advantage of the traditional way of doing economics is that it provides "analytical tractability," so that one can say that an element x affects an element y, ceteris paribus (i.e., all things held constant). Given this, two potential ways of moving forward exist. First, economists can incorporate relevant behavioural mechanisms into their traditional mathematical (i.e., equation-based) models. There have been several successful cases. For instance, the role of identity and self-image (Bulte and Horan 2010), social pressure (Richter et al. 2013; Tavoni et al. 2012), or social norms more generally (Nyborg 2018) have all been formally incorporated in economic models. However, despite these successes, there are behavioural mechanisms that are simply incompatible with traditional economic models. Hence, and as a second potential way of moving forward, economists can consider more theoretically descriptive works or other forms of modelling (e.g., agent-based models) that are process-based, rather than equation-based.

In any case, one needs to carefully evaluate which mechanisms are useful additions to the behavioural economics toolbox by considering the robustness and order of importance of different mechanisms (Shogren and Taylor 2008). Robustness implies that not every random deviation from rational behaviour is important to consider. Given limited cognitive abilities, attention span, and information, actors may make mistakes that can be better characterized as noise rather than as a systematic deviation. Some deviations may also only occur in extremely specific contexts, limiting its relevance. Moreover, some deviations from rationality may be systematic, but have little additional power in explaining how humans respond to different incentives, policies, and social factors compared to the rational actor model. In the context of recreational fishing, it seems particularly important to understand how fishers respond to regulation (e.g., catch limits, licences), to the social environment (e.g., social norms to conduct certain practices), and to environmental change (e.g., a stock collapse).

4.2 Methodological Best Practice

Behavioural economics is a fast-moving field with other exciting emerging fields developing alongside behavioural economics. For example, neuroeconomics look at the role of emotions (affect) in economic behaviour. Below, we discuss several areas for methodological best practices when applying behavioural economics.

Ethics Approval, Consent, and Fairness: Research should be carried out ethically. Usually, this implies getting ethical approval before conducting the research. Participants must provide their consent to participate in research and have the ability to withdraw from the experiments at any time. Participants are often compensated for participating in an experiment, so that such an experiment is not only to the benefit of the research team. This may also involve sharing results and following up with participants.

Deception. Economic experiments, as opposed to psychology experiments, must always be truthful to their participants. The basis for this assertion is the idea that psychological reaction to suspected manipulation may jeopardize experimental control and validity and contaminate the subject pool (Bonetti 1998; Charness et al. 2022). Hence, any form of deception is considered a red flag in economics research and is usually considered unpublishable. Examples of deception include giving false information about the behaviour of other people in the experiment or the nature of the experiment. However, the omission of information is tolerated. For instance, informing participants that they will participate in a "decision-making" experiment while leaving out the more specific research question (e.g., violation of rules) is acceptable (Charness et al. 2022).

Consequentiality of Decisions. In economics research, the quality of information derived from hypothetical questions is always seen as inferior to decisions that are either incentivized or a real-world decision-making context. Indeed, asking someone whether they *would* respect a certain regulation or act cooperatively may lead to an affirmative answer. But in reality, people do not necessarily "put their money where their mouth is."

External Validity. Many economic experiments raise the question of external validity, the extent to which the experimental results apply to a wider group of participants and decision contexts (i.e., will the results of experiments conducted with university students hold with similar experiments conducted with Australian recreational fishers). Establishing external validity is exceedingly difficult, and it is often outside the scope of the research. Research done by Henrich et al. (2010), for instance, shows how varied social norms can be across different participant pools. Nevertheless, discussing the external validity of a research project is seen as a key ingredient in any behavioural economics research. One should be careful in simply assuming that results from a Western, Educated, Industrialized, Rich, and Democratic (WEIRD) sample (Henrich et al. 2010) can simply be extrapolated across socio-cultural contexts.

Socio-Cultural Sensitivity. Experiments should reflect the socio-cultural context of participants. Steps should be taken to ensure proper understanding of the experiment instructions, such as offering instructions in the language most accessible to experiment participants (Abatayo and Lynham 2020; Gneezy and Imas 2017).

Cultural and religious restrictions should also be considered. For instance, the standard incentivized elicitation of risk preferences may be considered gambling, and hence, frowned upon, in some cultures. Experiment participants should not be forced into situations that belie their cultural and religious beliefs and practices, which may result in any form of post-experimental social ostracism. On a similar note, unintended detrimental social dynamics after the experiment may result from singling out a person for payment. As such, although done in practice in several laboratory experiments, one should be careful in implementing such in the field.

Open Science. In the wake of the replication and reproduction crisis in the social sciences (Camerer et al. 2016; Chen et al. 2022), researchers have advocated for a range of new approaches to improve scientific practices in behavioural economics. The general criterion is best described by Palfrey and Porter (1991): "the relevant criterion is that enough detail be provided to enable another researcher to replicate the results in a manner that the original author or authors would accept as being valid." But what is considered "enough detail"? Chen et al. (2022) identify three areas where details should be provided when conducting experiments for any potential replicators: (i) recruitment, (ii) pre-experimental procedures, and (iii) experimental procedures. In the recruitment stage, the experimenters should specify who the subjects are, what their cultural norms are, and what method was used to recruit them up for the experiment. In the pre-experimental procedures stage, the experimenters should report the circumstances that the recruited participants experienced (i.e., the availability of a waiting room, communication among participants and between experimenter and participants, etc.). Finally, in the experimental procedures stage, the experimenters should again report the circumstances that the recruited participants experienced (i.e., delivery of instructions, questions asked and answered, measures of participant comprehension). A prevalent best practice is to upload the experiment instructions to an open-access archive.

Research Is Not a Fishing Trip. Often, economic experiments deliver extensive data. It is considered inappropriate to search in this ocean of data for surprising and exciting relationships and develop a story that explains those, often counterintuitive, relationships. When one is statistically testing the significance of 100 different relationships, it is expected that some of these will be statistically significant by chance. To circumvent this, the standard practice is to submit a pre-analysis plan, in which the researchers specify the theories, hypotheses, and statistical tests that they will use prior to carrying out the data collection (Burlig 2018; Olken 2015). Note, however, that pre-analysis plans come at a cost. In practice, most research is more complicated than a single, simple hypothesis with a single, simple outcome variable of interest. A hypothesis could be conditional on the realizations of other previous hypothesis tests and specifying the entire chain of logic for every possible realization of the data can quickly become overwhelming even for the most ardent pre-specifier (Burlig 2018). Hence, the researcher should aim for a balance between the two extremes.

5 How Might Behavioural Economics Evolve in the Future?

One of the main challenges of using behavioural economics approaches is the diversity of different mechanisms, biases, and anomalies that do not lend itself to guide research in a very meaningful way (McFadden 1999; Shogren and Taylor 2008; Wieczorek et al. 2021). Here, we offer some suggestions for how future research can make behavioural economics meaningful for recreational fishing and vice versa. An obvious fruitful avenue for further research is the identification of key mechanisms for (recreational) fisheries. Such inventory may imply a prioritization for practitioners and researchers alike to make sure that the key social processors are not missed, helping to avoid getting bogged down in a list of mechanisms that are neither relevant for understanding behaviours of recreational fishers, nor relevant for designing management. It seems plausible that—given the "leisure" context of recreational fishing—the more important mechanisms are the ones relating to social norms and also intrinsic motivation, though there may be other relevant mechanisms as well (Magee et al. 2018). It is also important that behavioural economists engage with the extensive literature around recreation studies, including what has been done specifically to understand recreational fisheries.

Second, recreational fishing is often open access but can also be a tightly regulated activity with entrance limits, licences, fees, catch limits, and temporal and spatial closures used by management authorities. A question of key importance is how those regulations may interact with human behaviour. In the worst case, recreational fishers may respond in unanticipated, and perhaps opposite ways to the regulation at hand, making the outcome worse with the regulation than without. For example, in the absence of regulation, recreational fishers may still be guided by social norms or intrinsic motives of stewardship that regulate how to fish. Fishers will likely respond to norms or intrinsic motivations to fish even in the presence of regulations, and regulators may need to consider how to best harness and leverage these non-economic incentives. A new regulation may crowd out or replace such norms, and if the new regulation is poorly enforced, the new behaviour can undermine sustainability goals compared to when the regulation was absent (Eikeset et al. 2011; Ostrom 2015; Richter et al. 2013). In the best case, regulations may crowd-in (i.e., reinforce nature-positive practices) and help uphold such norms, though the pertinent question is what determines which policies crowd in, and which crowd out sustainable practices.

Third, an important avenue for further research will be around the persistence of behavioural change. Currently, we know extraordinarily little about the durability and scalability of interventions to promote sustainable practices. A small intervention (e.g., a nudge) may generate an immediate and sizable effect on behaviour, though such an effect may not last over time and may also disappear when repeated. In contrast, a social norm intervention may take some time to spread, but once social norms have reached a certain level in a society, they are likely to reinforce themselves and spread more quickly.

More attention also needs to be paid to "boost" (Hertwig and Grüne-Yanoff 2017) and "nudge-plus" (Banerjee and John 2021) measures, which are the next generation of nudges. With these, behaviour change comes from interventions that stimulate reflection and awareness of the part of the person nudged, rather than just forcing a fast reaction from them. As such, the interventions are more likely to result in sustained effects if the outcome aligns with the person's values. They are also viewed as more respectful of the person's agency and autonomy, and therefore more palatable.

A highly promising strand of literature will be to use randomized control trials to measure potential impacts of management interventions. Now, the impacts of specific management interventions are often not properly measured, leading to missed learning opportunities. Together, the future of behavioural economics and recreational fisheries is one where key social and behavioural processes are widely known and considered in fisheries management and policy design. Moreover, the social and behavioural processes are included in the models (currently mostly single-species fisheries assessment models) used for setting future sustainable catch levels (Nieman and Solomon 2022).

Finally, behavioural economics already combines knowledge from different social disciplines, and we have illustrated how it can contribute to management and policy processes. It can grow its contribution to the interdisciplinary research community (Arlinghaus et al. 2013), where recreational fishing is increasingly represented as a social-ecological system (Solomon et al. 2020). Recreational fishing will be more evidence-based and behaviourally inspired to avoid unanticipated and unintended effects prior to policy implementation and will be set up such that each intervention can be carefully evaluated ex post.

Acknowledgments We acknowledge and thank the contribution of our reviewers, who provided constructive criticism that has greatly improved our writing. Also, many thanks to past, present, and future participants of behavioural economics experiments and studies around the world.

References

Abatayo AL, Lynham J (2016) Endogenous vs. exogenous regulations in the commons. J Environ Econ Manag 76:51–66

Abatayo AL, Lynham J (2020) Risk preferences after a typhoon: an artefactual field experiment with fishers in The Philippines. J Econ Psychol 79:102195

Ajzen I (1991) The theory of planned behavior. Organ Behav Hum Decis Process 50(2):179–211

Angner E, Loewenstein G (2012) Behavioral economics. In: Mäki U (ed) Handbook of the philosophy of science: philosophy of economic. Elsevier, Amsterdam, pp 641–690

Arlinghaus R (2006) On the apparently striking disconnect between motivation and satisfaction in recreational fishing: the case of catch orientation of German anglers. N Am J Fish Manag 26(3): 592–605

Arlinghaus R, Cooke SJ, Potts W (2013) Towards resilient recreational fisheries on a global scale through improved understanding of fish and fisher behaviour. Fish Manag Ecol 20(2–3):91–98

Aswani S, Gurney GG, Mulville S, Matera J, Gurven M (2013) Insights from experimental economics on local cooperation in a small-scale fishery management system. Glob Environ Change 23(6):1402–1409

Baldwin R (2014) From regulation to behaviour change: giving nudge the third degree. Mod Law Rev 77(6):831–857

Banerjee S, John P (2021) Nudge plus: incorporating reflection into behavioral public policy. Behav Public Policy 8:69–84

Barton A, Grüne-Yanoff T (2015) From libertarian paternalism to nudging—and beyond. Rev Philos Psychol 6(3):341–359

Benartzi S, Beshears J, Milkman KL, Sunstein CR, Thaler RH, Shankar M, Tucker-Ray W, Congdon WJ, Galing S (2017) Should governments invest more in nudging? Psychol Sci 28(8):1041–1055

Bergseth BJ, Williamson DH, Russ GR, Sutton SG, Cinner JE (2017) A social–ecological approach to assessing and managing poaching by recreational fishers. Front Ecol Environ 15(2):67–73

Bergstrom TC (2010) The uncommon insight of Elinor Ostrom. Scand J Econ 112(2):245–261

Bernheim BD (2016) The good, the bad, and the ugly: a unified approach to behavioral welfare economics. J Benefit-Cost Anal 7(1):12–68

Bicchieri C (2017) Norms in the wild: how to diagnose, measure, and change social norms. Oxford University Press

Birdsong M, Hunt LM, Arlinghaus R (2021) Recreational angler satisfaction: what drives it? Fish Fish 22(4):682–706

Bisack KD, Clay PM (2021) Behavioral responses to competing incentives and disincentives: compliance with marine mammal protection. Mar Policy 132(2021):104674

Bonetti S (1998) Experimental economics and deception. J Econ Psychol 19(3):377–395

Bova CS, Halse SJ, Aswani S, Potts WM (2017) Assessing a social norms approach for improving recreational fisheries compliance. Fish Manag Ecol 24(2):117–125

Bova CS, Stephens J, Aswani S, Potts WM (2022) Is the instrumental approach a 'silver bullet' for addressing non-compliance in recreational fisheries: a South African case study. Fish Res 255: 106439

Bulte E, Horan RD (2010) Identities in the commons: the dynamics of norms and social capital. B E J Econ Anal Policy 10(1):1–33

Burlig F (2018) Improving transparency in observational social science research: a pre-analysis plan approach. Econ Lett 168:56–60

Camerer CF, Loewenstein G (2004) Chapter one. Behavioral economics: past, present, future. In: Colin FC, George L, Matthew R (eds) Advances in behavioral economics, vol 1. Princeton University Press, pp 3–51

Camerer CF, Dreber A, Forsell E, Ho T-H, Huber J, Johannesson M, Kirchler M, Almenberg J, Altmejd A, Chan T, Heikensten E, Holzmeister F, Imai T, Isaksson S, Nave G, Pfeiffer T, Razen M, Wu H (2016) Evaluating replicability of laboratory experiments in economics. Science 351(6280):1433–1436

Carlsson F, Johansson-Stenman O (2012) Behavioral economics and environmental policy. Annu Rev Resour Econ 4(1):75–99

Carrella E, Saul S, Marshall K, Burgess MG, Cabral RB, Bailey RM, Dorsett C, Drexler M, Madsen JK, Merkl A (2020) Simple adaptive rules describe fishing behaviour better than perfect rationality in the US West Coast Groundfish Fishery. Ecol Econ 169:106449

Charness G, Samek A, van de Ven J (2022) What is considered deception in experimental economics? Exp Econ 25(2):385–412

Chen Z, Liu P, Schultz ET, Kasper JM, Swallow SK (2022) The influence of projected outcomes on preferences over alternative regulations: evidence from a recreational fishery. Land Econ 98(4): 599–617

Cole S, Izmalkov S, Sjöberg E (2014) Games in the Arctic: applying game theory insights to Arctic challenges. Polar Res 33(1):23357

Congiu L, Moscati I (2022) A review of nudges: definitions, justifications, effectiveness. J Econ Surv 36(1):188–213

Connelly NA, Brown TL (2000) Options for maintaining high fishing satisfaction in situations of declining catch rates. Hum Dimens Wil 5(1):18–31
Cotterill S, Moseley A, Richardson L (2012) Can nudging create the big society? Experiments in civic behaviour and implications for the voluntary and public sectors. Volunt Sect Rev 3(2): 265–274
Diekert FK (2017) Threatening thresholds? The effect of disastrous regime shifts on the non-cooperative use of environmental goods and services. J Public Econ 147:30–49
Diekert F, Nøstbakken L, Richter A (2021) Control activities and compliance behavior—survey evidence from Norway. Mar Policy 125:104381
Eikeset A, Richter MA, Diekert FK, Dankel DJ, Stenseth NC (2011) Unintended consequences sneak in the back door: making wise use of regulations in fisheries management. In: Belgrano A, Fowler CW (eds) Ecosystem based management for marine fisheries: an evolving perspective. Cambridge University Press, pp 183–218
Friedlander AM, Shackeroff JM, Kittinger JN (2013) Customary marine resource knowledge and use in contemporary Hawai'i. Pac Sci 67(3):441–460. 420
Gane N (2021) Nudge economics as libertarian paternalism. Theory Cult Soc 38(6):119–142
Gerkey D (2013) Cooperation in context: public goods games and post-soviet collectives in Kamchatka, Russia. Curr Anthropol 54(2):144–176
Gigerenzer G, Selten R (2001) Rethinking rationality. In: Gigerenzer G, Selten R (eds) Bounded rationality: the adaptive toolbox. The MIT Press
Gneezy U, Imas A (2017) Lab in the field: measuring preferences in the wild. In: Duflo E, Banerjee AV (eds) Handbook of field experiments, vol 1. Elsevier, North-Holland, pp 439–464
Gregory K, Grant-Smith D (2023) Engaging recreational fishers in fisheries' stewardship: a case of responsibilisation in natural resource management. J Environ Plan Manag 66(7):1563–1578
Greiner R, Franklin DC, Gregg D (2013) Towards an improved understanding of angler tourism in northern Australia. Fish Manag Ecol 20(2–3):161–173
Hardin G (1968) The tragedy of the commons. Science 162(3859):1243–1248
Harsanyi JC (1977) Rule utilitarianism and decision theory. Erkenntnis 11(1):25–53
Haynie AC, Hicks RL, Schnier KE (2009) Common property, information, and cooperation: commercial fishing in the Bering Sea. Ecol Econ 69(2):406–413
Henrich J, Heine SJ, Norenzayan A (2010) The weirdest people in the world? Behav Brain Sci 33(2–3):61–83
Hertwig R, Grüne-Yanoff T (2017) Nudging and boosting: steering or empowering good decisions. Perspect Psychol Sci 12(6):973–986
Ho TH, Lim N, Camerer CF (2006) Modeling the psychology of consumer and firm behavior with behavioral economics. J Mark Res 43(3):307–331
Janssen MA, Goldstone RL, Menczer F, Ostrom E (2008) Effect of rule choice in dynamic interactive spatial commons. Int J Commons 2(2):288–312
Jentoft S, McCay BJ, Wilson DC (1998) Social theory and fisheries co-management. Mar Policy 22(4):423–436
Just DR (2017) The behavioral welfare paradox: practical, ethical and welfare implications of nudging. Agric Econ Res Rev 46(1):1–20
Kahneman D, Tversky A (1979) Prospect theory: an analysis of decision under risk. Econometrica 47(2):263–291
Klefoth T, Wegener N, Meyerhoff J, Arlinghaus R (2023) Do anglers and managers think similarly about stocking, habitat management and harvest regulations? Implications for the management of community-governed recreational fisheries. Fish Res 260:106589
Krupka EL, Weber RA (2013) Identifying social norms using coordination games: why does dictator game sharing vary? J Eur Econ Assoc 11(3):495–524
Landon AC, Schroeder SA, Stefen SF, Arlinghaus R (2026) Social psychology for understanding recreational fishers and fisheries. In: Pope KL, Arlinghaus R, Hunt LM, Lynch AJ, van Poorten BT (eds) Understanding recreational fishers: disciplinary and interdisciplinary approaches for fisheries management. Springer Nature, Cham, p 157–196

Lopez MC, Murphy JJ, Spraggon JM, Stranlund JK (2012) Comparing the effectiveness of regulation and pro-social emotions to enhance cooperation: experimental evidence from fishing communities in Colombia. Econ Inq 50(1):131–142

Mackay M, Jennings S, van Putten EI, Sibly H, Yamazaki S (2018) When push comes to shove in recreational fishing compliance, think 'nudge'. Mar Policy 95:256–266

Mackay M, van Putten EI, Yamazaki S, Jennings S, Sibly H (2020) Me and my behavior: an experiment on individual characteristics and compliance behavior in recreational fishing [original research]. Front Mar Sci 7:740

Magee C, Voyer M, McIlgorm A, Li O (2018) Chasing the thrill or just passing the time? Trialing a new mixed methods approach to understanding heterogeneity amongst recreational fishers based on motivations. Fish Res 199:107–118

Maldonado JH, Moreno-Sanchez Rdel P (2016) Exacerbating the tragedy of the commons: private inefficient outcomes and peer effect in experimental games with fishing communities. PLoS One 11(2):e0148403

Martin VY, Weiler B, Reis A, Dimmock K, Scherrer P (2017) 'Doing the right thing': how social science can help foster pro-environmental behaviour change in marine protected areas. Mar Policy 81:236–246

Mas-Colell A, Whinston MD, Green JR (1995) Microeconomic therory. Oxford University Press, New York

McFadden D (1999) Rationality for economists? J Risk Uncertain 19(1):73–105

Mullainathan S, Thaler R (2001) Behavioral economics. In: Smelser PBN (ed) Int J Behav Sci, vol 20. Oxford University Press, pp 1094–1100

Nieman CL, Solomon CT (2022) Slow social change: implications for open access recreational fisheries. Fish Fish 23(1):195–201

Nyborg K (2018) Social norms and the environment. Annu Rev Resour Econ 10(1):405–423

Olken BA (2015) Promises and perils of pre-analysis plans. J Econ Perspect 29(3):61–80

Ostrom E (2000) Collective action and the evolution of social norms. J Econ Perspect 14(3): 137–158

Ostrom E (2009) A general framework for analyzing sustainability of social-ecological systems. Science 325(5939):419–422

Ostrom E (2012) Experiments combining communication with punishment options demonstrate how individuals can overcome social dilemmas. Behav Brain Sci 35(1):33–34

Ostrom E (2015) Eight design principles for successful commons. In: Bollier D, Helfrich S (eds) Patterns of commoning. Commons Strategies Group, pp 47–48

Palfrey T, Porter R (1991) Guidelines for submission of manuscripts on experimental economics. Econometrica 59:1197–1198

Plous S (1993) The psychology of judgment and decision making. Mcgraw-Hill Book Company

Richter A, van Soest D, Grasman J (2013) Contagious cooperation, temptation, and ecosystem collapse. J Environ Econ Manag 66(1):141–158

Rojas CA, Cinner J, Lau J, Ruano-Chamorro C, Contreras-Drey FJ, Gelcich S (2021) An experimental look at trust, bargaining, and public goods in fishing communities. Sci Rep 11(1):20798

Shogren JF, Taylor LO (2008) On behavioral-environmental economics. Rev Environ Econ Policy 2(1):26–44

Simon HA (1990) Bounded rationality. In: Eatwell J, Milgate M, Newman P (eds) Utility and probability. Palgrave Macmillan, pp 15–18

Smith A (1759) The Theory of Moral Sentiments, History of Economic Thought Books, McMaster University Archive for the History of Economic Thought, number smith1759

Smith A (1776) An inquiry into the nature and xauses of the wealth of nations. McMaster University. Archive for the history of economic thought

Smith VL (1962) An experimental study of competitive market behavior. J Polit Econ 70(2): 111–137

Snyder HT, Cox ME, Bork Hansen S, Connors C, Eckstein S (2021) Deterrents and nudges improve compliance in Greenland's Atlantic salmon (Salmo salar) fishery. ICES Mar Sci Symp 78(8): 2809–2817

Solomon CT, Dassow CJ, Iwicki CM, Jensen OP, Jones SE, Sass GG, Trudeau A, van Poorten BT, Whittaker D (2020) Frontiers in modelling social–ecological dynamics of recreational fisheries: a review and synthesis. Fish Fish 21(5):973–991

Sorokowski P, Oleszkiewicz A, Niemczyk A, Marczak M, Huanca T, Velasco EC, Sorokowska A (2017) Money, food, and daily life objects are similarly shared in the dictator game. a study among Poles and Tsimane' [original research]. Front Psychol 8:554

Stoop J, Noussair CN, Soest D (2012) From the lab to the field: cooperation among fishermen. J Polit Econ 120(6):1027–1056

Sunstein CR, Reisch L, Kaiser M (2019) Trusting nudges? Lessons from an international survey. J Eur Public Policy 26(10):1417–1443

Tavoni A, Schlüter M, Levin S (2012) The survival of the conformist: social pressure and renewable resource management. J Theor Biol 299:152–161

Thaler RH (2015) Misbehaving: the making of behavioral economics. W W Norton & Co, New York

Thaler RH (2016) Behavioral economics: past, present, and future. Am Econ Rev 106(7): 1577–1600

Thaler RH, Sunstein CR (2003) Libertarian paternalism. Am Econ Rev 93(2):175–179

Thaler RH, Sunstein CR (2008) Nudge: improving decisions about health, wealth, and happiness, vol 19. Yale University Press, pp 356–360

Thomas AS, Milfont TL, Gavin MC (2016) A new approach to identifying the drivers of regulation compliance using multivariate behavioural models. PLoS One 11(10):e0163868

Tversky A, Kahneman D (1981) The framing of decisions and the psychology of choice. Science 211:453–458

van Putten EI, Trudeau A, Mackay M, Obregón C, Post JR, Richter A, Schuch E, Solomon CT (2026) Economic experiments in recreational fishing studies. In: Pope KL, Arlinghaus R, Hunt LM, Lynch AJ, van Poorten BT (eds) Understanding recreational fishers: disciplinary and interdisciplinary approaches for fisheries management. Springer Nature, Cham, p 411-442

Varade ML, Choi F, Helmuth B, Scyphers S (2023) Catching versus counting: comparing the pro-environmental attitudes, behaviors, and climate concerns of recreational fishers and citizen scientists. Sustainability 15(1):307

Veiga P, Pita C, Leite L, Ribeir J, Ditton RB, Goncalves JMS, Erzini K (2013) From a traditionally open access fishery to modern restrictions: Portuguese anglers' perceptions about newly implemented recreational fishing regulations. Mar Policy 40:53–63

Vosooghi S (2013) Uncertainty and coordination in common-property fisheries. University of Edinburgh

Vosooghi S (2019) Panic-based overfishing in transboundary fisheries. J Assoc Environ Resour Econ 73(4):1287–1313

Wieczorek AM, Schadeberg A, Krogh Hallin J, van Putten I, Kraak SBM, Richter A, Clay PM, Goti Aralucea L, Pedreschi D, Hamon KG, Dankel DJ, Mackay M (2021) Behavioural economics in fisheries: a systematic review protocol. PLoS One 16(8):e0255333

Wilcox C, Bergseth BJ (2021) Effectiveness of interventions to shift drivers of roving banditry and reduce illegal fishing by Vietnamese blue boats. Conserv Lett 14(5):e12823

Wilson AC (2020) Behavioral economics in context: applications for development, inequality & discrimination, finance, and environment. Global Development Policy Center, Boston University, Boston

Yamagishi T, Horita Y, Mifune N, Hashimoto H, Li Y, Shinada M, Miura A, Inukai K, Takagishi H, Simunovic D (2012) Rejection of unfair offers in the ultimatum game is no evidence of strong reciprocity. PNAS 109(50):20364–20368

Young MAL, Foale S, Bellwood DR (2016) Why do fishers fish? A cross-cultural examination of the motivations for fishing. Mar Policy 66:114–123

Part III
Engaging Fishers to Support Effective Social Science

Guidelines for Writing and Constructing Recreational Fisher Surveys

Jerry J. Vaske, Mark D. Needham, Craig A. Miller, and Carly C. Sponarski

Abstract Surveys are a useful tool for understanding natural resource stakeholders such as fishers. Surveys can be used to measure support for an agency's decisions and policy, motivations to fish and satisfaction in fishing, as well as barriers to participation. Survey design is a skill that requires training and knowledge to craft a questionnaire that is useful, as well as sequential steps taken in the design process. Like any research endeavour, a research problem, goal, objective, or question must be articulated before designing a questionnaire. Decisions around which concepts and questions should be included in the survey to answer the research question must be made, along with deciding the methods used to deploy the survey. Taking time to draft, revise, and pre-test a survey is key to making sure your research goals are addressed. Planning how your survey data will be analysed to answer the research question is critical to make sure you will collect the appropriate data, given the needs of the project. In this chapter, we provide 37 guidelines to help you design a survey and the questions within that survey to meet your needs.

Keywords Guidelines for constructing questionnaires · Guidelines for writing questions · Survey research

---

J. J. Vaske (✉)
Human Dimensions of Natural Resources Department, Colorado State University, Fort Collins, CO, USA
e-mail: jerryv@colostate.edu

M. D. Needham
Department of Forest Ecosystems and Society, Oregon State University, Corvallis, OR, USA

C. A. Miller
Illinois Natural History Survey, University of Illinois, Champaign, IL, USA

C. C. Sponarski
Northern Forestry Centre, Canadian Forest Service, Natural Resources Canada, Edmonton, AB, Canada

K. L. Pope et al. (eds.), *Understanding Recreational Fishers*, Fish & Fisheries Series 45, https://doi.org/10.1007/978-3-031-99739-6_10

1 Introduction to Surveys

Surveys are commonly used for examining topics such as what motivates people to participate in fishing or measuring support for managing agency decisions and policies. Survey research involves administering questionnaires to a sample of respondents selected from a particular population. Although some individuals mistakenly believe that designing and implementing a questionnaire is a simple task of quickly writing questions and asking people to respond, several broad steps should be followed to ensure that it is conducted scientifically:

1. Identify the problem, goals, objectives, and research questions or hypotheses that the project is trying to solve, and the necessary information needed to make informed management decisions. Note, there is a fundamental difference between a research question presented in a theoretical or academic proposal and a question asked of respondents in a questionnaire; these often differ.
2. Decide what should be included in the questionnaire, such as what specific concepts are of interest to the study (e.g., motivations, satisfaction, attitudes, behavioural intentions).
3. Identify any existing banks of proven (valid) questions and appropriate scales that others have used for measuring these concepts. These are usually found in the methods sections of social science journal articles. For example, Milfont and Duckitt (2010) provided an inventory of questions and scales measuring attitudes in natural resource contexts, Dunlap et al. (2000) and Manfredo et al. (2020) discussed questions commonly used for measuring environmental beliefs and value orientations, and Manfredo et al. (1996) listed questions often used for measuring motivations for participating in a recreation activity such as fishing. The Kyle et al. (2026) chapter in this book also discusses this in more detail.
4. Select an appropriate survey method (e.g., mail, telephone, onsite, internet). See the other Vaske et al. (2026) chapter in this book, Dillman et al. (2014), and Vaske (2019).
5. Write draft questions as part of constructing a first draft of the questionnaire. This is discussed in more detail in this chapter in the sections below titled "Guidelines for Writing Questions" (Guidelines 1 through 28) and "Guidelines for Constructing Questionnaires" (Guidelines 29 through 37).
6. Ask experts to review the questionnaire and revise it based on their input.
7. Pretest the questionnaire on a small sample (e.g., $n \sim 100$) of people from the population of interest and revise it accordingly based on pretest results. This pretest can be conducted using a variety of methods (e.g., survey, interview, focus group).
8. Establish procedures for coding responses and analysing data.
9. Administer the final questionnaire. This is discussed in more detail in the other Vaske et al. (2026) chapter in this book.
10. Conduct a nonresponse bias check if the response rate is low, and then weight the data if necessary. Note that the definition of what constitutes "low" varies. See Dillman et al. (2014), Vaske (2019), and Vaske et al. (2026) for further discussion.
11. Enter and analyse responses, and then report the results of the analyses.

Writing good questions and constructing a professional questionnaire that is easy to follow and complete are not trivial tasks. Taking time to translate ideas that motivated the survey into good questions is key (Salant and Dillman 1994). This chapter offers 28 guidelines (Guidelines 1 through 28) for how to write specific questions to be included in a questionnaire. This chapter also offers nine guidelines (Guidelines 29 through 37) for how to construct an attractive and professional questionnaire (these nine guidelines are not focused on specific questions). These guidelines were based on various sources (e.g., Pollock et al. 1994; Salant and Dillman 1994; Tourangeau et al. 2000; Harkness et al. 2010; Dillman et al. 2014; Johnson et al. 2019; Vaske 2019). Readers are also encouraged to review these sources for more information on the 11 broad steps outlined above and all the guidelines listed below (in no particular order) that are critical to conducting a scientific survey.

2 Guidelines for Writing Questions

2.1 *Guideline 1: Identify What Information Respondents Should Provide*

The first step to writing questions is to clarify the information needed to meet project goals and the concepts (e.g., motivations, satisfaction, attitudes, behavioural intentions) that need to be measured. It can be useful to map the proposed survey questions against the overall goals to show which question or questions address which goal or goals. Participation in fishing, for example, is often of managerial interest. If the goal is to find out *whether* people participated in fishing, the following question may work:

Did you go fishing at Smith Lake in the last 12 months? (Please check one).

☐ No
☐ Yes

If the objective is to estimate *how many days* people went fishing, the following question may be more suitable:

About how many days did you go fishing at Smith Lake in the last 12 months? (Please write response. If you did not fish at Smith Lake in the last 12 months, please write 0)

I went fishing at Smith Lake about ______ days in the last 12 months.

The word "about" suggests to the respondent that an accurate and precise number may not be possible, and providing an estimate is acceptable. A researcher could sum the number of 0 days and the number of people who gave values greater than 0 to answer the No versus Yes categories in the first question.

2.2 Guideline 2: Choose Between Close-Ended (i.e., Fixed-Scale) Versus Open-Ended Questions, and Use Open-Ended Sparingly

Open-ended questions do not provide response alternatives from which to select a particular answer. Rather, respondents write their answers however they choose. Open-ended questions are easy to construct because all that is required is blank space or a box where people respond in their own words. There is also a richness in open-ended responses and comments, and qualitative text-based analytical methods (e.g., content analysis) are available to handle qualitative data (Kochalski et al. 2026)

Open-ended questions can provide highly useful information if done properly, but they require personnel time to sort responses and code answers into themes, and this can be complicated by poor spelling, messy handwriting, and use of abbreviations that require time and effort to discern (Dillman et al. 2014; Vaske 2019). Open-ended questions can also produce a wide variety of responses depending on sample size. For example, an open-ended question such as "Where do you live?" may yield responses such as Colorado, Larimer County, Fort Collins, or Red Feather Lakes. These locations could be coded into one category labelled Colorado because Larimer County, Fort Collins, and Red Feather Lakes are all in Colorado.

Open-ended questions also increase respondent burden, which can increase survey completion time and decrease response rates. Also, when people are asked to write sentences about unfamiliar topics, there may be a greater likelihood that questionnaires will be returned incomplete (Dillman et al. 2014; Vaske 2019). Open-ended questions also limit the ability to estimate the reliability and validity of concepts in a questionnaire, and do not easily allow for comparisons among groups. Some respondents also write long and detailed answers to a question, whereas others write a brief reply. Open-ended questions also create problems when completing questionnaires on mobile devices (e.g., smartphone, tablet) where typing can be challenging, so people often only provide short responses (Vasja et al. 2022). Research has shown high levels of nonresponse to open-ended questions when using mobile devices (Peytchev and Hill 2010).

Open-ended questions can sometimes be useful for informing close-ended questions and their responses. For example, answers from an elicitation survey with open-ended questions can then be used for developing fixed-scale close-ended questions to include in a questionnaire that is administered later to a larger sample. This is useful when researchers have little knowledge about a topic and are unable to specify response choices. Open-ended questions are also commonly used at the end of questionnaires to allow respondents to provide additional comments, such as "Do you have any other comments about these issues? If yes, please write your comments below." These types of open-ended questions can give respondents the feeling of really being asked to provide more context and depth to their thoughts and opinions on a topic. Open-ended questions can also be used following a fixed scale question to ask respondents to explain, elaborate, or clarify responses, although some respondents may avoid these types of questions because they can increase respondent burden (Geer 1991; Dillman et al. 2014). Furthermore, open-ended questions can be used in conjunction with fixed response categories where an exhaustive list of all possible responses is not possible. In this situation, a list of the most likely responses is provided, with the final response being "Other (please specify ______________)" presented at the end of the categories. Finally, open-ended questions are useful when asking for estimates of a personal characteristic or behaviour where precise information is needed and can be recalled without a list or scale of answer choices (Geer 1991). Examples include:

What is your age? _____ years old
About how many years have you fished in your life? _____ years
In what state do you live? ______________________ state

Taken together, whether researchers choose to use open-ended or close-ended questions largely depends on the goals of their project. If the main goal is to do formal statistical analyses, then close-ended is usually best. However, this requires the researcher to know exactly what questions to ask, response options to include, and how to format these response options (e.g., scales). There might be ample space in a questionnaire to include both open-ended and close-ended questions.

Close-ended (i.e., fixed-scale) questions are preferred, especially when using formal statistical or quantitative analyses. These types of questions include: (i) close-ended with ordered response choices, (ii) close-ended with unordered response choices, and (iii) partially close-ended response choices (Salant and Dillman 1994). *Close-ended questions with ordered response choices* present responses in a particular order or gradation. These questions are easy to code and less demanding for respondents. Examples include:

Overall, how dissatisfied or satisfied were you with your fishing trip to Smith Lake today? (Please check one).

☐ Very dissatisfied
☐ Dissatisfied
☐ Neither dissatisfied nor satisfied
☐ Satisfied
☐ Very satisfied

To what extent do you oppose or support each of the following possible management actions at Smith Lake? (Please circle one number for each)

	Strongly oppose	Oppose	Neither	Support	Strongly support
Improve road access	1	2	3	4	5
Provide more parking	1	2	3	4	5
Provide more trash cans	1	2	3	4	5
Provide more toilets	1	2	3	4	5
Prohibit motorboats	1	2	3	4	5
Require dogs be kept on leash	1	2	3	4	5

Questions that are *close-ended with unordered response choices* are similar to fixed-order items with one exception — responses are not in an order or a gradation; people choose from discrete unordered responses. With these response options, respondents usually read the entire list, compare response options, and then select an answer. These tasks become more challenging as more choices are added. An example is:

What agency is responsible for managing marine reserves in Oregon? (Please check one).

- ☐ Oregon Marine Board
- ☐ Oregon Department of Fish and Wildlife
- ☐ National Oceanic and Atmospheric Administration
- ☐ U.S. Fish and Wildlife Service
- ☐ U.S. Coast Guard
- ☐ Unsure

Questions with *partially close-ended response choices* offer a compromise between open-ended and close-ended questions. These questions give respondents flexibility to choose from a set of researcher-determined responses or create their own answers by including, for example, an "other" category and asking respondents to write their answer. If the researcher has provided a relatively exhaustive list of options, few people will identify other responses, but at least they retain the ability of not being forced into a predefined category (Salant and Dillman 1994). For example:

What one main activity did you participate in today at this wildlife refuge? (Please check one)

- ☐ Hunting
- ☐ Fishing
- ☐ Wildlife viewing
- ☐ Wildlife photography
- ☐ Other (Please write response) ____________

Deciding on open-ended, close-ended, or partially close-ended questions is never easy, as each has strengths and weaknesses, and should be evaluated within the context of project objectives and how other studies have measured identical concepts. Note that many statistical techniques (e.g., reliability, regression, analysis of variance) require variables to be measured using close-ended questions with ordered response choices on continuous scales (Vaske 2019). Five, seven, or nine-point scales (e.g., strongly disagree [1], disagree [2], neither [3], agree [4], strongly agree [5]) allow for these techniques and are considered continuous in the human dimensions literature. It is possible to collapse these scales by recoding them as categorical (e.g., disagree, neither, agree) or dichotomous (e.g., agree or not) later. Continuous scales provide researchers with more options when analysing data and reporting results.

2.3 *Guideline 3: Use Simple Words and Phrases, and Different Languages If Necessary*

Language and literacy issues limit some people's ability to participate in both national and international surveys (Dillman et al. 2014). If portions of the target population do not speak or struggle with the native language, questionnaires should

Table 1 Examples of Simple Words and Phrases Instead of Specialized Words (modified from Dillman et al. 2014)

Worse	Better
Words	
Exhausted	Tired
Highest priority	Most important
Employment	Work
Leisure	Free time
Rectify	Fix
Phrases	
Occupants of this residence	People who live here
Your responses in reply to this survey	Your answers
Post-school extracurricular activities	What you do after school
Work related employment concerns	Job concerns

be translated into other languages and back-translated by a speaker of the native language to ensure accuracy. Some strategies recommend a team approach to survey translations (Harkness et al. 2010). With this methodology, different individuals collaborate to maximize the available expertise (Johnson et al. 2019). The TRAPD model (Translation, Review, Adjudication, Pretesting, Documentation) is an illustration of the team approach (Harkness et al. 2010). Other team approaches are described at the Cross-Cultural Survey Guidelines website (http://ccsg.isr.umich.edu).

It is also important to use short, simple, and easy-to-understand words. Pretesting the instrument is important and will almost certainly identify problems with words and phrases (Table 1).

2.4 Guideline 4: Provide Important Information in the Stem of a Question, But Use as Few Words as Possible

The stem (i.e., words forming the query) is the most important part of the question (Dillman et al. 2014). The question stem must provide explicit information about what is being asked, how people are to respond (e.g., "Please check one"), and definitions, examples, or additional instructions. The type of response in the stem should not be vague (e.g., "In an average month, how often do you go fishing?"), but clear (e.g., "In an average month, how many *days* do you go fishing?"). This is especially important for telephone surveys where respondents do not see the response options that provide visual cues for how to answer.

Extra verbiage is not necessary. Long questions confuse respondents and may cause them to not respond or even stop the questionnaire altogether (Dillman et al. 2014). Brevity and clarity are important when writing questions. For example, not all response options need to be included because extra words make it challenging to understand what is being asked. An example of a poor question stem would be:

To what extent do you strongly oppose, oppose, neither oppose nor support, support, or strongly support the U.S. Fish and Wildlife Service charging people fees to enter and visit wildlife refuges? (Please check one).

☐ Strongly oppose
☐ Oppose
☐ Neither oppose nor support
☐ Support
☐ Strongly support

A more succinct version of this question for self-administered modes would be:

Do you oppose or support the U.S. Fish and Wildlife Service charging entrance fees at wildlife refuges? (Please check one).

☐ Strongly oppose
☐ Oppose
☐ Neither oppose nor support
☐ Support
☐ Strongly support

2.5 *Guideline 5: Avoid Vague Quantifiers So More Precise Estimates Can Be Obtained*

When selecting responses for close-ended questions, be as precise as possible without being so specific that it is nearly impossible to answer (Schaeffer 1991). Although broad categories make it easier for respondents and researchers, categories that are too broad provide little useful information. Consider the following examples:

How often did you go fishing during the past year? (Please check one).

☐ Never
☐ Rarely
☐ Occasionally
☐ Regularly

How often did you go fishing during the past year? (Please check one).

☐ Never
☐ 1 to 12 times
☐ 13 to 24 times
☐ 25 to 52 times
☐ More than 52 times

The first question is problematic because “rarely,” “occasionally,” and “regularly” mean different things to different people. For example, for one person, “rarely” might mean 1–3 times, whereas for another it might mean 3–5 times. The second question is problematic because responses are so precise that it may be impossible to correctly answer the question. This level of precision may provide more detail than the researcher needs.

A problem with both questions is what is meant by "the past year?" If questions were answered in June 2022, it is unclear if the last year refers to January to June 2022 or the last 12 months (June 2021 to June 2022). A revision could be:

How often did you go fishing during the past 12 months? (Please check one).

☐ Not at all
☐ A few times
☐ About once a month
☐ About two or three times a month
☐ About once a week
☐ More than once a week

2.6 Guideline 6: Use an Equal Number of Positive and Negative Responses (i.e., Bipolar) for Scale Questions, But Recognize that an Unequal Number of Responses (i.e., Unipolar) Are Sometimes Appropriate

Bias can occur when ordered response categories are weighted in one direction, causing an imbalance. A problematic question would be:

Overall, how dissatisfied or satisfied were you with your fishing guide today? (Please check one)

☐ Dissatisfied
☐ Neither dissatisfied nor satisfied
☐ Somewhat satisfied
☐ Mostly satisfied
☐ Completely satisfied

This question has only one negative ("dissatisfied") compared to three positive responses ("somewhat satisfied," "mostly satisfied," "completely satisfied"). The following revised question provides a more balanced or *bipolar* response scale:

Overall, how dissatisfied or satisfied were you with your fishing guide today? (Please check one)

☐ Very dissatisfied
☐ Dissatisfied
☐ Neither dissatisfied nor satisfied
☐ Satisfied
☐ Very satisfied

There is an exception to this guideline. If most respondents are likely to check one extreme, or a middle point (e.g., neither, neutral) does not make sense for the question, unequal numbers of positive and negative responses may be appropriate. These *unipolar* scales measure gradation along one dimension. The 9-point perceived crowding scale (see Guideline 20 below and Shelby et al. 1989; Shelby and Vaske 2007; Vaske and Shelby 2008), for example, ranges from "not at all crowded" to "extremely crowded." Also, questions measuring motivations for engaging in a recreation activity are often measured on a unipolar scale of "not important" to "extremely important."

2.7 Guideline 7: Consider How "Neither," "Neutral," "No Opinion," and "Not Sure" or "Do Not Know" Responses Are Used

"Neither," "neutral," "no opinion," and "not sure" or "do not know" do not always mean the same thing (Blasius and Thiessen 2001; Dillman et al. 2014). A question may ask how many respondents oppose or support a fishing regulation on a specific lake. One respondent may never have given careful thought to the regulation, but had at least heard about the topic. For this individual, a correct answer is likely "no opinion." A second respondent who researched the topic and weighed pros and cons of the regulation without coming to any resolution may say "neutral" or "neither oppose nor support." A third respondent might have no idea what the regulation means and had never heard of it, so "not sure" or "do not know" would be appropriate.

Some researchers provide many or all these response options (e.g., "neither," "neutral," "no opinion," "not sure," "do not know"), suggesting this is the only way for people who cannot provide a positive or negative response to accurately answer; without these options, individuals would give an untrue answer or skip the question (Dillman et al. 2014). Others, however, suggest that providing these options allows respondents to *satisfice* by giving them an easy out of answering the question without doing any of the cognitive thinking necessary to report a true response (Krosnick 2002; Roberts et al. 2019). Although studies have supported both sides of this issue (see Dillman et al. 2014 for a review), most suggest that "neither" or "neutral" are substantive responses that should form the midpoint of bipolar scales, whereas "no opinion" and "not sure" or "do not know" are non-substantive responses that should be placed after the scale and visually separate (e.g., extra space between response categories) from the substantive options, such as the following (Dillman et al. 2014):

To what extent do you oppose or support fishing restrictions on the Colorado River? (Please check one).

☐ Strongly oppose
☐ Oppose
☐ Neither oppose nor support
☐ Support
☐ Strongly support

☐ No opinion
☐ Don't know

Some research (e.g., Sturgis et al. 2014) has cast doubt on this advice by showing that many individuals who select a scale midpoint of "neither" or "neutral" actually have no opinion or do not know much about the issue (called "face-saving don't knows"). Reallocating these individuals from the substantive midpoint response to a non-substantive "no opinion," "not sure," or "do not know" response alters distributions, inferences, and conclusions. In addition, researchers often agonize over whether to include a scale midpoint ("neither," "neutral"), even though research has shown it "has little effect on the resulting data quality and conclusions drawn from the data" (Dillman et al. 2014, p. 154). Researchers may consider using scales that do not include the midpoint ("neither," "neutral"), but still offer one or two visually separated options ("no opinion," "do not know" or "not sure") when it is likely that several respondents will not have an opinion or adequate knowledge:

To what extent do you oppose or support fishing restrictions on the Colorado River? (Please check one).

☐ Strongly oppose
☐ Oppose
☐ Support
☐ Strongly support

☐ No opinion
☐ Don't know

For questions where it is highly likely that all individuals in the sample should know about or have an opinion regarding the issue, it may be useful to format the question without a midpoint and any non-substantive options:

To what extent do you oppose or support fishing restrictions on the Colorado River? (Please check one).

☐ Strongly oppose
☐ Oppose
☐ Support
☐ Strongly support

Research has yet to determine a single correct approach for addressing these issues. It is important that researchers keep these considerations in mind and proceed thoughtfully when designing questionnaire items and scales (Sturgis et al. 2014). In addition, it may be wise to keep the same number of response options consistent throughout sections of a questionnaire to avoid respondent confusion.

2.8 *Guideline 8: Avoid Double Negatives or Asking People to Say "Yes" to Mean "No"*

Double negatives confuse respondents. Although this seems obvious, such questions are frequently asked. For example, the question "Do you oppose or support not requiring catch-and-release fishing at Smith Lake?" is a double negative, which will confuse many. If respondents quickly read or skim this question, they are likely to miss the word "not." And the mental connection of supporting a "not" is challenging (Dillman et al. 2014). A revision would be:

Do you oppose or support requiring all fishers to catch and release their fish at Smith Lake? (Please check one).

☐ I oppose requiring catch-and-release at Smith Lake
☐ I support requiring catch-and-release at Smith Lake

2.9 Guideline 9: Do Not Write Double-Barrelled Questions; Ask One Question at a Time

A double-barrelled question contains two different questions or components, but asks for only one answer. Respondents may feel differently about each component. For example, the question "Do you agree or disagree that fish and other marine species (e.g., crabs, sea stars, oysters) will benefit from the new fishing restrictions?" is double-barrelled because a respondent may agree that fish will benefit, but disagree that other marine species will benefit, or vice versa. It is unclear what part of the question a response of "agree" or "disagree" refers. In this instance, the question should be revised and separated into two questions, with one about fish and the second about other marine species. Double-barrelled questions are a common mistake in survey research (Dillman et al. 2014; Vaske 2019). Researchers should be cautious when using coordinating conjunctions such as "and," "nor," "but," and "so" in questions because they may result in double-barrelled questions.

2.10 Guideline 10: Use Appropriate Timeframes for Questions and Responses

Surveys often ask fishers how frequently they fish for a particular species. Dillman et al. (2014) identified several issues that should be considered when using time referents in questionnaires. First, when something is so common (e.g., watching television), it is hard for a person to quantify exactly how often they engage in the behaviour. It may be more appropriate to ask for a general estimate (e.g., about once a day) rather than specific counts.

Second, memory diminishes over time, so it is difficult to categorize information by precise periods (e.g., month, year). This can cause recall bias in a study, leading to overestimates or underestimates in responses. For example, it would be difficult for avid fishers to remember how often they went fishing during the last five years. Depending on project objectives, it is often useful to shorten the timeframe.

Third, some studies focus on short periods, such as engaging in a behaviour in the last few days or one week. Participation in the behaviour may differ substantially by day of week, so different responses may be obtained depending on when respondents complete the questionnaire. Finally, avoid vague quantifiers (e.g., "during the past year," "this year") if more precise estimates can be obtained (e.g., "in the last 12 months"). Please refer to *Guideline 5* for further information.

2.11 Guideline 11: Reduce the Effect of Sensitive or Objectionable Questions

Respondents are reluctant to answer sensitive or potentially objectionable questions. This is especially problematic if the nonresponse reduces sample sizes and inflates social desirability bias (i.e., respondents give answers that are consistent with perceived societal norms or shared viewpoints, Tourangeau and Yan 2007). Asking respondents about their exact annual total income is one potentially objectionable question, and few people can easily recall their income to the precise monetary amount. Using broad categories in a close-ended question format can reduce nonresponse:

Which of the following broad categories best describes your approximate annual household income before taxes in 2022? (Please check one).

☐ Less than $10,000
☐ $10,000 – $29,999
☐ $30,000 – $49,999
☐ $50,000 – $69,999
☐ $70,000 – $89,999
☐ $90,000 – $109,999
☐ $110,000 – $129,999
☐ $130,000 – $149,999
☐ $150,000 – $169,999
☐ $170,000 or more

Researchers should consider questions and their response options in relation to study goals and the target population. For example, if the study is concerned with the impact of recreation fee increases on low-income individuals, finer distinctions at the low end of the response scale for this income question may be necessary. More broadly, however, it is important for researchers to make sure that they absolutely need to ask these potentially objectionable questions (e.g., income) to meet the overall goals of the study. If a question does not map to a specific study goal, then it would be wise to leave it out of the questionnaire to reduce respondent burden and minimize the potential for objectionable questions.

Asking questions about behaviours in which people should engage (but many do not) can also be objectionable. An objectionable question would be:

Did you vote in the 2020 national election? (Please check one).

☐ No
☐ Yes

People who did not vote may feel offended by this question because voting is a socially desirable behaviour. These people may say "yes" even though they did not vote, or skip this question without answering. One way to soften the impact is:

For various reasons, some people did not have a chance to vote in the last national election. Did you happen to vote in the 2020 national election? (Please check one).
☐ No
☐ Yes
☐ Prefer not to say

When asking sensitive or objectionable questions, it is useful to provide short and simple explanations for why responses are important, how the information will be protected (e.g., anonymity, confidentiality), and softer wording in requests for personal information (Dillman et al. 2014).

2.12 *Guideline 12: Define Terms and Use Terminology that Makes Sense to Respondents*

Some terminology may be unclear and not make sense, causing respondents to misunderstand questions, and thereby produce bias and error in responses (Fowler 1992). Questions examining fishing participation, for example, may ask "how many fishing trips" people have taken, but what constitutes a "trip" is unclear and can be confusing:

How many fishing trips have you taken in the last three months?
(Please write response).
I have taken _________ fishing trips.

For some people, a trip may involve walking a few blocks from their house to a nearby lake or stream. For others, a trip might constitute a multi-day excursion to hike-in lakes in another region. Potentially unclear terms should be clarified and defined, such as:

How many fishing trips have you taken in the last three months?
A fishing trip is defined as a trip of more than 50 miles from home
(Please write response).
I have taken ________ fishing trips.

2.13 *Guideline 13: Define Abbreviations and Acronyms*

Undefined abbreviations and jargon should never be used (Salant and Dillman 1994). There is an exception to this rule. Some abbreviations and jargon may be appropriate if the questionnaire is being completed by a specific group of people who use certain phrases or abbreviations frequently and will not be confused. For example, members of a fishery management council are likely to know that ABC means Acceptable Biological Catch or EFH means Essential Fish Habitat. These abbreviations would not be acceptable to use in a survey of the public. In case of doubt, always define abbreviations and avoid jargon.

2.14 Guideline 14: Avoid Bias from Unequal Comparisons

When asking close-ended questions with unordered response categories, it is possible to create bias when responses are skewed. Consider the following question:

Who do you feel is most responsible for the litter at Smith Lake? (Please check one).

☐ Careless fishers
☐ Hikers
☐ Campers

The word "careless" adds a value connotation for fishers, which is not present for hikers or campers. This question should be revised:

Who do you feel is most responsible for the litter at Smith Lake? (Please check one)

☐ Fishers
☐ Hikers
☐ Campers

2.15 Guideline 15: Avoid Using Slanted or Leading and Loaded Introductions and Questions

Questions can be biased when they are written in a way that makes it appear as though everyone shares the same opinion or participates in a behaviour, and as a result, the respondent should too. Slanted or leading introductions create social desirability bias. An example of a slanted introduction is:

More Americans participate in recreational fishing now than 10 years ago.

Do you fish? (Please check one).

☐ No
☐ Yes

This question should be revised to be simple, objective, and neutral:

Do you participate in recreational fishing? (Please check one).

☐ No
☐ Yes

2.16 Guideline 16: State Both Sides of Scales in Question Stems or Introductions

As stated in Guideline 4, researchers should reduce the length of questions, but one problem with reducing length is to mention only one side of a response scale:

To what extent do you support using nets to take fish? (Please check one).

- ☐ Strongly oppose
- ☐ Oppose
- ☐ Neither oppose nor support
- ☐ Support
- ☐ Strongly support

Not mentioning disagreement or opposition in the question stem can bias responses (Salant and Dillman 1994). Mentioning "oppose or support" (or vice versa) conveys a greater range of responses and that opposition or disagreement is acceptable. This question stem might be revised as: "To what extent do you oppose or support using nets to take fish?"

2.17 *Guideline 17: Minimize "Check All That Apply" Questions to Avoid "Satisficing" and Both "Primacy" and "Recency" Effects*

For some topics, "check all that apply" questions are essential. For example, people can participate in many activities in a day, so the following is often necessary:

Please check all activities in which you are participating at Smith Lake today. (Please check all that apply).

☐ Sunbathing	☐ Boating (with a motor)
☐ Swimming	☐ Boating (without a motor, such as kayak, canoe)
☐ Fishing	☐ Camping
☐ Waterskiing	☐ Hiking around the lake

There are, however, problems with "check all that apply" questions (Rasinski et al. 1994). Respondents may *satisfice* by checking answers and proceeding down the list until they believe that they have provided a "satisfactory" answer (Dillman et al. 2014). Although almost all respondents will read the first few options, some will not read the entire list. Options listed first are more likely to be checked. This is analogous to a *primacy effect* or selecting from the first few answers, which is common in mail, internet, and onsite surveys. These types of questions can also cause a *recency effect* by selecting from the last few answers mentioned, which is common in telephone and interviewer-completed onsite surveys (Dillman et al. 2014). In internet surveys, it is possible to address this issue by randomizing the order of response options. To minimize primacy and recency effects from "check all that apply" questions, researchers could also use scale responses instead:

People visit Smith Lake for many activities. Please indicate how important each of the following are in influencing you to visit Smith Lake today.

(Please circle one number for each statement).

I visited Smith Lake today to:	Not important	Slightly important	Moderately important	Extremely important
Sunbathe	1	2	3	4
Swim	1	2	3	4
Fish	1	2	3	4
Waterski	1	2	3	4
Go boating (with a motor)	1	2	3	4
Go boating (without a motor, such as kayak or canoe)	1	2	3	4
Camp	1	2	3	4
Hike around the lake	1	2	3	4

2.18 Guideline 18: Use Response Categories That Are Mutually Exclusive and Exhaustive

All answers to questions should be *mutually exclusive*, meaning they must not overlap. If response categories overlap, there is potential for a respondent to select two or more answers for a "check one" question. This confuses respondents and complicates data entry and analysis. For example:

How did you first learn about the fishing boat accident on March 17, 2022 that killed two people? (Please check one).

☐ From a friend or family member
☐ From my spouse
☐ From the radio
☐ While at home
☐ While at work
☐ While attending a meeting

Problems with this question are: (i) choices combine sources and locations, so someone could have heard about the accident from their spouse while at home, or from a friend who they work with while attending a meeting; (ii) sources are not mutually exclusive because a spouse is also a friend and family member; and (iii) locations are not mutually exclusive because a person could have heard about the accident while at work and attending a meeting. One option for revising this question would be to separate it into two questions with revised choices:

From whom or what did you first learn about the fishing boat accident on March 17, 2022 that killed two people? (Please check one).

☐ From my spouse
☐ From another friend or relative
☐ From the radio

Where were you when you first learned about the fishing boat accident on March 17, 2022 that killed two people? (Please check one).

☐ At home
☐ At work
☐ At a meeting at a place other than my work

A further improvement could include more sources (e.g., television, internet) or locations (e.g., in car) to make options more exhaustive. Items such as "at home" and "at work" may still not be mutually exclusive because some people work from home. Researchers could reconsider these types of lists and use scales or partially close-ended response choices (i.e., an "other" open-ended category) instead.

A more subtle, but common example of responses not being mutually exclusive is for questions with numerical response categories. Consider the following:

Which of the following broad categories best describes your current annual household income before taxes? (Please check one).

☐ Less than $10,000
☐ $10,000 – $30,000
☐ $30,000 – $50,000
☐ $50,000 – $70,000
☐ $70,000 – $90,000
☐ $90,000 – $110,000
☐ $110,000 – $130,000
☐ $130,000 – $150,000
☐ $150,000 – $170,000
☐ $170,000

The first problem is almost all these response categories overlap (e.g., $30,000 is listed in two response categories). The second problem is that the response categories are not *exhaustive* (i.e., all possible answers must be provided) because there is no option for people whose income exceeds $170,000 (Dillman et al. 2014). One approach for solving these problems would be:

Which of the following broad categories best describes your current annual household income before taxes? (Please check one).

☐ Less than $10,000
☐ $10,000 – $29,999
☐ $30,000 – $49,999
☐ $50,000 – $69,999
☐ $70,000 – $89,999
☐ $90,000 – $109,999
☐ $110,000 – $129,999
☐ $130,000 – $149,999
☐ $150,000 – $169,999
☐ $170,000 or more

2.19 *Guideline 19: Make Sure Questions and Statements Are Technically Accurate*

Statements, questions, or words that are inaccurate can reduce credibility of the survey, the researcher, and the organization funding the study. Errors in accuracy range from spelling or grammar mistakes to inaccurate information. An example of an *inaccurate* question is:

The ivory-billed woodpecker is considered by the National Conservation Union as "threatened," meaning that it is vulnerable to extinction in the distent future. Should the U.S. federal government provide more funding for research on this species? (Please check one).

☐ No
☐ Yes
☐ Unsure

There are three major problems with this question. First, the ivory-billed woodpecker is actually classified as "critically endangered" and possibly extinct. Second, the organization that classifies species (e.g., threatened, endangered) is actually named the International Union for Conservation of Nature (IUCN), not the National Conservation Union. Third, "distent" is misspelled; it should be spelled "distant."

2.20 *Guideline 20: Use Questions and Responses that Allow Comparisons with Existing Information or Previously Collected Data*

One benefit of survey research is the ability to compare results to other studies. To compare data, questions must be asked in the same way as earlier studies. Measurement consistency can allow for statistical approaches such as meta-analysis to examine generalizability, and longitudinal analysis to assess change over time. Comparisons can also ensure that samples are representative of the larger population from which they were drawn and inform any necessary weighting procedures.

An example of measurement consistency is the 9-point scale often used for measuring perceived crowding in recreation settings (Vaske and Shelby 2008):

To what extent did you feel crowded by other fishers while you were fishing at Smith Lake today? (Please circle one number).

1	2	3	4	5	6	7	8	9
Not At All Crowded		Slightly Crowded			Moderately Crowded			Extremely Crowded

This scale has been used in hundreds of studies (see Vaske and Shelby 2008 for a review), which allows for comparisons across time, settings, activity groups, geographical regions, and other contexts. As stated earlier, it is important for researchers to consult the literature when designing questionnaires because there are banks of proven (valid) questions and appropriate scales that others have used for measuring concepts. For example, Milfont and Duckitt (2010) provided an inventory of questions and scales measuring attitudes, Dunlap et al. (2000) and Manfredo et al. (2020) discussed questions used for measuring beliefs and value orientations, and Manfredo et al. (1996) listed questions commonly used for measuring motivations. The Kyle et al. (2026) chapter in this book provides more examples. Although it is important to consult the literature prior to survey research, it is equally important to scrutinize how others have measured issues and concepts, as their questions and response options may not follow the guidelines for rigor outlined here.

2.21 Guideline 21: Provide Information Instead of Assuming Respondent Knowledge

Researchers should never assume that all respondents know enough to answer every question (Salant and Dillman 1994). When asking about a specific topic of which some respondents may not be aware, it is useful to provide background information. An example of a question that assumes too much knowledge is:

To what extent do you oppose or support the Oregon Department of Fish and Wildlife's new regulation on fishing for both salmon and steelhead in the Santiam River? (Please check one).

☐ Strongly oppose
☐ Oppose
☐ Neither oppose nor support
☐ Support
☐ Strongly support

This question assumes that all respondents are aware of the new regulation, which is unlikely. The question could be improved by: (i) providing more information about the regulation, (ii) asking if respondents were aware of this regulation, and then (iii) asking if they support or oppose the regulation:

The Oregon Department of Fish and Wildlife has introduced a new regulation requiring all fishers to release wild (nonhatchery) salmon and steelhead caught in the Santiam River. Prior to receiving this survey, were you aware of this regulation? (Please check one).

☐ No, I was not aware of this regulation
☐ Yes, I was aware of this regulation

Now, do you oppose or support this new regulation? (Please check one).

☐ Strongly oppose
☐ Oppose
☐ Neither oppose nor support
☐ Support
☐ Strongly support

2.22 Guideline 22: Avoid Questions That Are Too Difficult or Require Calculations

Asking respondents to report percentages or make other calculations should be avoided because errors are common, especially in onsite or telephone surveys where people are pressured to answer quickly (Dillman et al. 2014, Salant and Dillman 1994). An example of an unnecessary calculation is:

What percent (%) of days that you fished in the last three months were spent at Smith Lake? (Please write response).

____________ % of days fishing in last three months spent at Smith Lake.

This calculation is challenging, especially for avid fishers who may have fished many days at multiple locations. Asking for approximates and separating into two questions (approximate number of days fishing in the last three months, followed by approximate number of days fishing in the last three months at Smith Lake) reduces respondent burden and leaves the calculations to the researcher, not the respondent.

Some people will attempt to report a precise number, but many respondents will offer numbers ending in 0 or 5. This digit preference can be an issue, but researchers have offered suggestions for correcting the problem (Vaske et al. 1996; Beaman et al. 1998, 2015).

2.23 *Guideline 23: Use Multiple Questions to Measure Complex Concepts*

For questions about topics such as demographics (e.g., age) and behaviour (e.g., fishing participation), respondents usually attempt to give accurate estimates or correct answers. If asked the identical question in a questionnaire administered a day or two later, people are likely to provide similar responses because they possess many of these attributes. Questions about cognitions such as beliefs, norms, attitudes, or intentions, however, are more prone to measurement error and imprecision because people do not possess these characteristics in the same way as they possess things such as age and behaviour (for details, see Kyle et al. 2026). Opinions can change from day to day and may not be well thought-out before questionnaire completion (Salant and Dillman 1994).

In addition, no single question is a perfect measure of a complex concept such as an attitude or a belief. To improve the reliability and validity of measures of concepts derived from a questionnaire, researchers typically use scales and ask multiple questions about a concept or issue. Statistical approaches such as reliability analysis (e.g., Cronbach's alpha, McDonald's omega) can be employed to examine patterns in how people answer similar questions, which can then be used to justify combining answers into a single numerical index estimating the latent (not directly observed) concept (see Kyle et al. 2026, chapter "Measurement Instruments", for details).

The following example lists multiple questions measuring the concept of specialization in fishing. Specialization was originally defined as "a continuum of behaviour from the general to the particular, reflected by equipment and skills used in the sport and activity setting preferences" (Bryan 1977, p. 175). At one end of this continuum are novices or infrequent participants who do not consider the activity to be a central life interest or show strong preferences for equipment and technique. The other end includes more avid participants who are committed to the activity and may use sophisticated methods (Needham et al. 2009). The first five items listed in the example measure the affective or psychological sub-dimension of specialization called "centrality to lifestyle," the next five measure the cognitive sub-dimension of "skill level," and the last five measure the behavioural sub-dimension of "equipment" (Needham et al. 2009). These 15 items and other related items are commonly used in questionnaires measuring this multi-dimensional concept of specialization (see Manning 2022; Needham et al. 2009; Scott and Shafer 2001 for reviews). If researchers choose banks of items and questions from established scales such as this, it is important to either stick to them or have a good argument for leaving specific items out and then run reliability (e.g., alpha, omega) and validity (e.g., confirmatory factor analysis) statistical analyses. The Kyle et al. (2026) chapter in this book discusses this issue in more detail.

In addition, as discussed later in this chapter and shown in this example, numbers and "circle a number" commands have been recommended for these close-ended questions with ordered choices that are part of a list of items on a scale in grid format where items are arranged in rows and responses in columns (e.g., Mitra and Lankford 1999; Vaske 2019). Some others, however, have shown examples using check boxes or repeating the response labels instead of using numbers, but also claimed that there is no standardized format that is widely accepted among researchers (e.g., Dillman et al. 2014).

To what extent do you agree or disagree with each of the following statements about fishing? There are no right or wrong answers; just state your opinion. (Please circle a number for each).

	Strongly disagree	Somewhat disagree	Neither	Somewhat agree	Strongly agree
If I stopped fishing, an important part of my life would be missing.	1	2	3	4	5
I would rather go fishing than do most anything else.	1	2	3	4	5
Participation in fishing is a large part of my life.	1	2	3	4	5
Most other recreation activities do not interest me as much as fishing.	1	2	3	4	5
Fishing is becoming a more central part of my life each year.	1	2	3	4	5
Given the fishing skills I have developed, it is important that I continue to participate in fishing.	1	2	3	4	5
I feel that I am more skilled in fishing than other anglers in general.	1	2	3	4	5
Testing my fishing skills is very important to me.	1	2	3	4	5
In general, I am becoming more skilled in fishing each year.	1	2	3	4	5
I would rate my skill level in fishing as advanced or expert.	1	2	3	4	5
I have accumulated a lot of fishing equipment.	1	2	3	4	5
I have invested a lot of money in fishing equipment.	1	2	3	4	5
I feel that I have more fishing equipment than other anglers in general.	1	2	3	4	5
I often spend time learning about newest fishing equipment available.	1	2	3	4	5
In general, I am obtaining more fishing equipment each year.	1	2	3	4	5

Note that it is generally recommended that individuals should circle their responses instead of using crosses in paper-and-pencil questionnaires. Crossing out a response can have an ambiguous meaning.

2.24 Guideline 24: Make Sure Questions Apply to Respondents

It is common for people to receive questionnaires about topics with which they are unfamiliar. Imagine receiving a questionnaire at a local park and being asked "How many hours did you fish on the lake in this park today?" Likely, some visitors will not have fished, so the question is not applicable.

One methodological issue that is challenging is distinguishing between people who choose not to respond because a question does not apply to them versus those who are simply unmotivated and decide to skip the question. In addition, questions can be problematic for surveys (e.g., internet) that can be formatted to require responses to every question and could cause respondents to provide false information or quit answering altogether (Dillman et al. 2014). A useful approach for addressing these issues is to use skip patterns that act as filter questions (these filters are easy and useful in internet surveys), such as:

Did you fish on the lake during your visit to this park today? (Please check one).

☐ No

☐ Yes → If yes, about how many hours did you spend fishing here today?
(Please write number) _________ hours

2.25 Guideline 25: Provide Extra Motivation for Challenging or Important Questions

Some questions necessitate substantial time or effort to answer. Many open-ended questions, for example, require respondents to think about and formulate answers in their own words. A question for fishers, such as "Please tell us what you might do if a fish disease was found in the location where you fish most often" may require substantial thought and time to answer. Given the effort required for responding to challenging questions, some people may be inclined to skip these questions, so they may need motivation to respond (Dillman et al. 2014).

One approach for maintaining motivation throughout the questionnaire is to keep the number of these types of questions to a minimum. A second approach is to provide motivational text clarifying the importance of these questions. Question stems such as "This question is extremely important for understanding this issue, so please take your time answering" can improve compliance and the substance of responses (Dillman et al. 2014).

2.26 *Guideline 26: Limit Use of Ranking Questions and Ask Respondents to Rank Only a Few Items at a Time Instead of a Long List*

One method that some researchers use for asking questions involves having respondents rank several items, such as the following question:

Please rank each of the following 10 reasons for visiting Smith Lake today, where 1 is your most important reason for visiting and 10 is your least important. No ties, so please do not write the same number twice.

_____ Be in the outdoors close to nature
_____ Have fun or excitement
_____ Be with friends or family
_____ Get exercise
_____ Rest or relax
_____ Escape crowds of people
_____ Develop skills or get better at an activity
_____ Go fishing
_____ Feel safe away from risks
_____ Experience nice weather or temperatures

Although data from rankings can provide useful results, these types of questions can be challenging for respondents to understand and complete (Dillman et al. 2014). When these items are verbally read in telephone surveys or interviewer-completed onsite surveys, it is even more difficult for respondents to keep track of the items and their rankings, often leading to primacy and recency effects (see Guideline 17 above). Ideally, these questions should be placed on scales, such as the following, which represent a small subset of commonly used measures (e.g., the Recreation Experience Preference [REP] scales) for measuring motivations to participate in an activity such as fishing (REP items are often measured on 4-point scales of 1 "not at all important" to 4 "very important;" Manfredo et al. 1996):

How important was each of the following reasons for visiting Smith Lake today? (Please circle one number for each)

I visited today to:	Not at all important	Slightly important	Moderately important	Very important
Be in the outdoors close to nature.	1	2	3	4
Have fun or excitement.	1	2	3	4
Be with friends or family.	1	2	3	4
Get exercise.	1	2	3	4
Rest or relax.	1	2	3	4
Escape crowds of people.	1	2	3	4
Develop skills or get better at an activity.	1	2	3	4
Go fishing.	1	2	3	4
Feel safe away from risks.	1	2	3	4
Experience nice weather or temperatures.	1	2	3	4

If ranking questions must be used, the goal should be to reduce respondent burden. One option would be to simplify the task by asking for a ranking of only the top three. A second option would be to present paired comparisons where respondents compare only two options at a time until they have compared all options against each other (Dillman et al. 2014). A third option is to rank only a few items at a time in separate groups. The example below is an excerpt from one common approach for measuring values, which involves ranking four different items at a time in several separate groups (Inglehart 1990, 1997; Manfredo et al. 2003; Manfredo et al. 2009).

Below are separate groups of goals people might prioritize differently.

For each group, rank the four goals in order of importance to you. That is:

1 = the goal that is most important to you
2 = the 2nd most important goal
3 = the 3rd most important goal
4 = the least important goal

Group 1 Rank these four goals from 1 = most important to 4 = least important. *No ties* (Do not give any items the same rank).

	Rank
Maintain a high level of economic growth.	______
See that people have more to say about how things are in their communities.	______
Make sure this country has strong defense forces.	______
Try to make our cities and countryside more beautiful.	______

Group 2 Repeat for this next set of four goals (1 = most important, 4 = least important). *No ties* (Do not give any items the same rank).

	Rank
Maintain order in the nation.	______
Give people more to say in important governmental decisions.	______
Fight rising prices.	______
Protect freedom of speech.	______

2.27 *Guideline 27: Limit Scale Questions to Four or Five Response Options*

Continuous scales (e.g., 1 "strongly disagree" to 5 "strongly agree") are widely used and accepted (Vaske 2019). The challenge, however, is choosing a scale length that provides enough range and variance for respondents to be able to place themselves along the scale, but not so many options that respondents get confused or distinctions between options are ambiguous (Dillman et al. 2014). People can only remember a limited number of response options at a time, so fewer options reduce complexity and burden. Dillman et al. (2014) suggested that for bipolar scales, the optimal number of response options ranges from five to seven when there is a scale midpoint (e.g., "neither," "neutral"). Other research has suggested that researchers should offer only five options for these questions rather than seven, nine, or more because additional options increase variability in personal interpretation of scales, leading to more method effects and lower validity and data quality (Revilla et al. 2014). For unipolar scales, the optimal scale length is four or five options (Dillman et al. 2014). Examples of bipolar and unipolar scales were shown earlier.

For continuous scales, it is often advisable to associate higher numbers with more positive responses (e.g., 1 "very dissatisfied," 10 "very satisfied") because people tend to expect this based on their upbringing and questions commonly asked in mass media and public opinion polls (e.g., "On a 10-point scale where 10 is the highest or most positive response, how do you feel about…"). Some research has shown no differences in responses when scales were presented with the most positive response option presented first or the most negative presented first (Christian et al. 2009). What is important is ensuring that scales are presented in a consistent direction across all questions in a questionnaire to avoid respondent confusion (Dillman et al. 2014).

For telephone surveys or interviewer-completed onsite surveys, researchers should break scale questions into two separate questions. For example, instead of verbally asking "Overall, how dissatisfied or satisfied are you with your visit to Smith Lake: very dissatisfied, slightly dissatisfied, neither dissatisfied nor satisfied, slightly satisfied, or very satisfied?" cognitive burden can be reduced by separating this into two questions. First, "Overall, are you dissatisfied or satisfied with your visit to Smith Lake?" Then, "Would you say you are slightly or very satisfied" (or dissatisfied, depending on their response to the first question)? Although this approach increases reliability and validity and reduces completion time, it also increases the number of skip patterns and questionnaire length and clutter (Malhotra et al. 2009). It can be avoided in respondent self-completed surveys (e.g., mail and internet surveys) where respondents can see response options.

2.28 Guideline 28: Use Construct-specific Labels Instead of Always Agree-Disagree Scales When Possible

Disagree to agree scales are common in survey research (Höhne and Lenzner 2018) and there is nothing inherently wrong with them, but they are sometimes used in place of more appropriate construct-specific scales such as:

To what extent do you disagree or agree that you are satisfied with your fishing trip to Smith Lake today? (Please check one).

☐ Strongly disagree
☐ Disagree
☐ Neither disagree nor agree
☐ Agree
☐ Strongly agree

The challenge with this question is that respondents first must decide how satisfied they are and then convert this judgment into a different construct and evaluation scale (i.e., disagree to agree). A more appropriate construct-specific format ensures consistency between the question stem and response options, such as:

To what extent are you dissatisfied or satisfied with your visit to Smith Lake today? (Please check one).

☐ Very dissatisfied
☐ Dissatisfied
☐ Neither dissatisfied nor satisfied
☐ Satisfied
☐ Very satisfied

Many researchers use agree-disagree scales out of habit because they are widely used, or they borrow them from other studies using these scales (Dillman et al. 2014). Research has shown that more construct-specific scales (e.g., oppose—support, do not trust—highly trust) reduce bias and respondent burden, and increase reliability and validity compared to agree-disagree scales (Saris et al. 2010; Dillman et al. 2014). As a result, researchers are encouraged to use scales that directly measure the construct of interest rather than generic scales such as agree-disagree scales.

Taken together, these 28 guidelines do not represent a fully exhaustive list of all suggestions for writing questions. There are a few other suggestions (e.g., use complete sentences and avoid cryptic phrases when asking questions, make every question count to avoid wasting space and increasing respondent burden) provided by various sources (e.g., Salant and Dillman 1994; Dillman et al. 2014; Vaske 2019).

3 Guidelines for Constructing Questionnaires

Once questions have been written and problems have been corrected, the next step is to construct an attractive and professional questionnaire that is easy to follow and complete in a timely manner. Well-designed questionnaires take time to construct, but the payoff is that people are more likely to respond and the resulting responses will be valid.

Questionnaires should minimize respondent burden, which means: (i) keeping completion time to a minimum, (ii) decreasing the time that respondents need to think about questions, and (iii) respecting respondents by ensuring they will not be embarrassed by not understanding what is expected (Salant and Dillman 1994). This section offers guidelines for constructing questionnaires that minimize burden, in no particular order.

3.1 Guideline 29: Start with an Interesting, Easy, and Relevant Question

The first few questions should receive more scrutiny than the others. The first question is crucial because it influences whether people will continue to participate. This question should not take long to answer, require an open-ended response, involve long scale responses, be something that does not apply to all respondents, or ask people about themselves (e.g., personal characteristics, demographics). This question should be simple to answer, short, related to the topic of study, relevant to all respondents, and motivate respondents to complete the rest of the questionnaire. This question should catch the interest of respondents and give them a context for more challenging questions later. This is especially important for telephone or internet surveys because respondents cannot look through the rest of the questions as they can in a paper questionnaire (Dillman et al. 2014, Vaske et al. 2023). An example starting question would be:

Fishing has not been allowed at Smith Lake since the early 1990s.

Do you think fishing should be allowed at Smith Lake again? (Please check one).

☐ No
☐ Yes
☐ Unsure

3.2 Guideline 30: Never Start with Demographic Questions; Place These and Other Sensitive Questions at the End

Some respondents consider demographic questions (e.g., age, marital status, income) to be no one's business. Respondents who find such questions objectionable may be reluctant to answer them and any remaining questions, which reduces sample sizes and response rates. These questions should never be at the beginning of a questionnaire and should be grouped together near the end. If these questions go unanswered, respondents have likely already answered the earlier questions, so useful information can be salvaged. In addition, asking too many demographic questions may jeopardize the anonymity of respondents and their answers (e.g., knowing a licensed recreational angler's age, gender, residence, marital status, income, and other personal characteristics could possibly identify an individual).

3.3 Guideline 31: Group Similar Questions into Sections, But Be Mindful of Potential Order Effects

Questionnaires should be arranged logically. Questions at the *beginning* should be simple and motivate respondents. Questions at the *end* should be those most likely to be objectionable (e.g., demographics). Questions in the *middle* should be the most difficult, but important to answer.

Questions examining the same specific subject or concept should be grouped together on the same page or in the same section. In a mail survey of fishers, for example, questions about their motivations could be on one page, those examining trust in agencies could be on another page, and questions measuring demographics should be on a separate page near the end. This improves readability and flow, and reduces respondent burden because they are not constantly switching between topics.

The order of questions can influence how people answer. Although researchers may want questions to build on each other, it is not desirable to have questions that may bias responses to other questions positioned near each other. Questions that might influence responses to other questions later in the questionnaire should be separated as much as possible to avoid unintended order effects (Dillman et al. 2014). For example, an open-ended question should be placed before a close-ended question about the same topic if the wording of this close-ended question can influence responses to the open-ended question. Another approach for addressing order effects is to construct half of the questionnaires with one order and half with another order. Different questionnaire versions can also be used to limit content for everyone, thus reducing respondent burden (e.g., some questions are only asked to one half and others are asked to the other half of the sample). Researchers can also randomize the sequence of items in a list. These techniques are easy in telephone, internet, and interviewer-completed onsite surveys (e.g., interviewers can start with different items when reading out a list of statements; Dillman et al. 2014).

Despite these order effects of specific questions, respondents will develop ideas about how the researcher wants them to answer certain questions. Survey questions can guide the thought processes inside a respondent's mind and stimulate their thinking about the subjects being addressed in the questionnaire. This adds another important aspect to the issue of how questions are sequenced and framed in a questionnaire, including any introductory and explanatory text. This is impossible for researchers to avoid, but they should be aware of this issue.

3.4 Guideline 32: Use Transitions to Guide Respondents

Organize questionnaires so that respondents can proceed smoothly from question to question and section to section. Flow is improved by using transitions (e.g., "Now, we would like to ask you…;" "Finally, we would like to ask you…") that provide a conversational tone and act like a roadmap signalling changes in topics or types of questions. Transitions are especially beneficial in telephone and onsite surveys, but can also be useful in other surveys (e.g., mail, internet).

3.5 Guideline 33: Short Questionnaires Are Not Always Better; Use "White Space" to Avoid Visual Clutter

Questionnaire layout can influence response rates and data quality, especially for mail, onsite, and internet surveys. Instruments read to respondents by interviewers (e.g., telephone) are less susceptible to layout. Some researchers believe that short questionnaires (e.g., 1–2 pages) are always best. This is not true. Well-designed and interesting questionnaires, even if they are lengthy, will be positively received, although response rates to surveys have been dropping as discussed by Vaske et al. (2026) in the chapter "Fundamentals of Sampling and Implementing Human Dimensions of Fishing Surveys" of this book (Dillman et al. 2014; Stedman et al. 2019; Vaske et al. 2023). Importantly, condensing questions on pages can look cluttered and make it challenging to read and respond, so it is important to work on the visual appeal of surveys in a professional manner.

There are several things to consider when structuring the layout of questionnaires. First, questions should not appear on one page with their response categories on the next page. If items in a series must be split across pages, headings (e.g., questions, response scales) should be repeated on each page. Second, questionnaires should never contain pages with large blank spaces while other pages are crammed with questions. Third, use adequate "white space" between questions to allow respondents to easily see each question and progress relatively quickly. Fourth, do not add extra or irrelevant questions just to fill space. Too many questions decrease response rates and increase respondent burden and item nonresponse (Dillman et al. 2014; Vaske 2019).

Reducing the number of questions on a page minimizes visual clutter, ensures adequate spacing between questions, creates an attractive-looking questionnaire that is easy to read, and encourages people to respond. There are, however, a few things to consider before adding more "white space." First, longer questionnaires in terms of pages rather than questions can be perceived as "too long" by respondents, leading to lower response rates (Dillman et al. 2014; Vaske 2019). Second, outlets such as text messaging and Twitter (now known as "X") have become commonplace and require condensing ideas into few words in a minimal amount of space. People have become accustomed to reading less and seeing shorter requests. Attention spans are also more limited as people use computers, smartphones, and tablets to get their information. Researchers, therefore, need to exercise caution in the length of questionnaires (Dillman et al. 2014). Third, adding pages to mail and onsite questionnaires increases costs to purchase more paper and print more pages. Fourth, more pages add thickness and weight to mail surveys, increasing postage costs.

3.6 *Guideline 34: Minimize Skip Patterns*

Skip patterns ask respondents who provided a particular answer to one question to skip over and not answer one or more questions that do not apply to them. Skip patterns tell respondents where to go next, such as:

Are you fishing at Smith Lake today? (Please check one).

☐ No → If no, please skip to question 3

☐ Yes

Skip patterns are sometimes unavoidable and serve as filters for questions that do not apply to everyone (Guideline 24 earlier). Skip patterns, however, should be minimized because they cause clutter and respondent confusion. These issues can be problematic for paper surveys (e.g., mail, onsite), but are not a concern for telephone and internet surveys where skip patterns can be automated. Skip patterns are also time-consuming and cumbersome when entering data because the analyst must distinguish answers purposefully not provided as directed by skip patterns ("does not apply") from those voluntarily not answered ("missing response").

3.7 *Guideline 35: Sequentially Number Questions Throughout the Questionnaire or Within Sections*

Number questions so respondents know which question to read and answer next. Numbering is also required for skip patterns. Questions should be numbered sequentially throughout the entire instrument, but questions should be numbered within sections or particular pages for long questionnaires because respondents may be reluctant to complete a questionnaire where the last question is a high number (e.g., Question 172). There is no rule for numbering, but researchers should be aware of hidden messages that may be communicated by providing numbers or a progress bar (which, according to Dillman et al. 2014, should be avoided) about questionnaire length.

3.8 Guideline 36: Format Responses Consistently

Response formats should be consistent throughout the questionnaire. This is especially important for respondent-completed surveys (e.g., mail, internet) and less important for interviewer-completed surveys (e.g., telephone; Vaske et al. 2023). Salant and Dillman (1994) initially suggested using numbers for all close-ended responses rather than boxes or lines, such as:

Did you go fishing in 2022 in Colorado? (Please circle one number)

1. No
2. Yes

Their rationale was that it "precodes" answers so they can be quickly entered as numbers into software. More recently, however, Dillman et al. (2014) suggested the following approach for formatting these questions with close-ended responses by using a combination of boxes and numbers:

Did you go fishing in 2022 in Colorado? (Please check one).

☐ $_1$ No
☐ $_2$ Yes

In the authors' studies using mail and onsite questionnaires, respondents have commented that it is easiest to answer questions using: (i) *lines* (i.e., fill-in-the-blank) and "write response" commands for open-ended items (see example in Guideline 12 earlier), (ii) *boxes* and "check one" or "check all that apply" commands for close-ended questions with responses that are not part of a list of items on a scale (see example in Guideline 7 earlier), and (iii) *numbers* and "circle a number" commands for close-ended questions with ordered choices that are part of a list of items on a scale in grid format where items are arranged in rows and responses in columns (see example in Guideline 23 earlier). Others have also recommended using numbers (e.g., Mitra and Lankford 1999; Vaske 2019). Dillman et al. (2014), however, showed examples using check boxes or repeating the response labels instead of using numbers in these scales or lists, but also claimed that there is no standardized format that is widely accepted among researchers. Their approach, however, adds visual clutter and makes data entry slower and more prone to errors. Most internet and onsite (intercept) surveys completed using electronic devices use radio buttons instead of numbers for these scales. There is no "perfect" way to format responses. It is crucial, however, that researchers maintain consistency throughout the entire questionnaire to minimize respondent confusion and maximize ease of questionnaire completion and data entry.

3.9 Guideline 37: Questionnaires Should Be Visually Appealing and Allow Easy Progress Through Questions

Questionnaires should be visually appealing and allow easy progression between questions. This is important for respondent-completed surveys (e.g., mail, internet), but less so for interviewer-completed surveys (e.g., telephone). Dillman et al. (2014) offered the following suggestions:

- Organize and show questions consistently within and across pages or screens.
- Avoid placing questions side-by-side. A single vertical navigational path should be established to encourage responses in the intended order.
- Consider darker and larger print for the question stem and lighter and smaller print for answer choices, and spaces to create contrast and separation between the question stem and responses.
- Use spacing to create subgroups within a question. Items close together will be perceived as belonging together, and those farther apart will be perceived as separate (e.g., indent answer spaces or responses underneath the question stem, use "white space" effectively to separate content). Researchers, however, should use caution with this recommendation if a concept is measured with multiple items, especially if the goal is to estimate the reliability of these items. In this case, intermixing items belonging to different concepts can sometimes be beneficial.
- Visually standardize answer spaces and response options (e.g., same font, size, spacing).
- Use visual design properties to emphasize important elements (e.g., bold, italics, underline, capitalization).
- Select font, size, and line length to ensure readability (e.g., Dillman et al. 2014 suggested using 10–12-point font for most populations or larger for older populations; use sans serif fonts such as Calibri or Arial instead of serif fonts such as Times New Roman). In addition to considering respondent age, researchers should also strive to make their surveys accessible to people with various disabilities.
- Integrate special instructions into questions where they are more likely to be read, rather than as free-standing entities such as shaded boxes.
- Use colour and contrast to help respondents recognize important elements and navigate questions (e.g., lightly shaded or coloured [e.g., blue, green] background with white answer spaces, black font colour).
- If a questionnaire is formatted as a booklet (e.g., mail survey), then include a title page with an attractive graphic and logos or names of agencies or organizations funding the study. These could also be used in online or internet surveys, but are not necessary for telephone surveys or short in-person or on-site surveys. Title pages and cover letters should also include statements about respondent confidentiality and anonymity, what will happen to the data (e.g., where results will be

published), and contact information of the researchers. More information on these topics is found in Vaske et al. (2026).

Regardless of what layout and design elements are used, ensure consistency to minimize confusion and error.

4 A Note on Levels of Measurement and Types of Analyses

Stevens (1951) described a hierarchy of four levels of measurement based on how questions in a questionnaire are designed (nominal, ordinal, interval, ratio). Nominal data are labelled and classified into mutually exclusive categories, but are not ordered in any meaningful way (e.g., fishers who use bait, fishers who use lures, fishers who use artificial flies). Ordinal data are the same as nominal data, but the categories are ordered in some kind of ranked or hierarchical manner (e.g., fishers who *rarely* go fishing, fishers who *often* go fishing, fishers who *always* go fishing). Interval data are the same as ordinal data, but are measured along a numerical scale that has equal distances between values, and there is no true zero point (e.g., credit scores and both Fahrenheit and Celsius temperatures are often cited examples). Ratio data are the same as interval data, but there is a true zero point (e.g., 0 days fishing, 1 day fishing, 2 days fishing).

In addition to introducing these four levels of measurement, Stevens (1951) also recommended that these levels were necessary for informing the selection of appropriate statistics (e.g., parametric statistics, such as means and *t*-tests, can only be used with interval or ratio data). Since the 1950s, these recommendations and levels of measurement have been discussed in most statistics and methods books, and have led to considerable debate and confusion (e.g., Baker et al. 1966; Binder 1984; Velleman and Wilkinson 1993). At the heart of this controversy is the nature of the relationship between ordinal and interval data, and the appropriate statistics to use with these types of data (Borgatta and Bohrnstedt 1980).

In Stevens' (1951) view, many of the measures discussed in this chapter, such as satisfaction with a fishing trip (see guideline 28 above), perceived crowding while fishing (see guideline 20 above), specialization in fishing (see guideline 23 above), or motivations for fishing (see guideline 26 above), would be considered ordinal. As such, Stevens (1951) contended that transformations (e.g., computing means and standard deviations) and parametric statistics (e.g., *t*-tests) that are appropriate for interval and ratio data are not appropriate for ordinal data. For example, computing a grade point average (a mean usually on a scale from 0 "F" to 4 "A") from several discrete letter grade categories (e.g., A, B, C, D, F [ordinal data according to Stevens 1951]) would not be appropriate using Stevens' (1951) logic. Not all statisticians, however, have agreed about the utility of using Stevens' traditional levels (e.g., Borgatta and Bohrnstedt 1980; Vaske 2008, 2019).

Borgatta and Bohrnstedt (1980), for example, argued that most of the scales that have been used and published widely in the social sciences are not truly ordinal scales, especially when they are measured on numerical scales (e.g., 1 "strongly disagree" to 5 "strongly agree;" 1 "not at all crowded" to 9 "extremely crowded;" 1 "very dissatisfied" to 7 "very satisfied"). This is particularly the case when multiple questionnaire items on these scales are designed to measure a latent (unobserved) concept (e.g., attitude, crowding, satisfaction), tested for reliability using statistics such as Cronbach's alpha or McDonald's omega, and then combined into composite indices measuring these concepts (see guideline 23 above and Kyle et al. 2026). In fact, most concepts of interest to human dimensions researchers (e.g., attitudes, beliefs) are conceptualized to be on scales at the latent (unobserved) level, even though they are usually measured (observed) as discrete variables (e.g., questions in a questionnaire with response options such as a 7-point disagree-agree scale).

Vaske (2008, 2019) introduced alternative levels of measurement (dichotomous, categorical, continuous) that are, from our perspective, more appropriate and useful than Stevens' (1951) terminology for the selection and interpretation of statistics. A dichotomous variable has only two response options (e.g., lake fishers vs. river fishers, support vs. oppose a policy). A categorical variable has three or more *unordered* and mutually exclusive response options (e.g., lake fishers vs. river fishers vs. ocean fishers; fishers who live in Illinois vs. Colorado vs. Oregon vs. Florida). A continuous variable has an *ordered* set of response options (typically 5 or more levels) with a distribution that is approximately normally distributed in the population sampled. Many human dimensions variables, such as age (in years), days of fishing participation, and amount of money spent on fishing, are continuous variables. Variables with ordered and numerical response options (e.g., 1 "strongly disagree" to 5 "strongly agree;" 1 "highly unacceptable" to 7 "highly acceptable;" 1 "not at all crowded" to 9 "extremely crowded") are also considered to be continuous variables. As such, parametric measures for these variables, such as means and standard deviations, can be computed and are appropriate. Different types of statistical analyses can be used (Table 2) depending on how the survey data are coded (i.e., the levels of measurement of the data). See Vaske (2019) for additional details on this topic.

5 Conclusion

Writing questions and constructing questionnaires are complex tasks that researchers and funding agencies should not treat as afterthoughts. Good questionnaires cannot be completed in final form in a single sitting. Even experienced social science researchers often need multiple drafts and revisions, including pre-tests, of a questionnaire before the guidelines discussed here have been addressed, and it is ready to administer. This chapter discussed 28 guidelines for writing questions. Once

Table 2 Guidelines for selecting an appropriate analysis strategy (Vaske 2019; "Cases" refers to the sample of respondents)

Type of variable		Number of variables			
Independent variable	Dependent variable	Independent variable	Dependent variable	Analysis strategy	Test statistic
Dichotomous Categorical	Dichotomous Categorical	1	1	Chi-square	χ^2
Dichotomous	Continuous	1	1	*t*-test (groups)	*t*
	Continuous	Cases	2	*t*-test (paired)	*t*
Dichotomous Categorical	Continuous	1	1	1-way ANOVA	*F*
Dichotomous Categorical	Continuous	2 +	1	*n*-way ANOVA	*F*
Dichotomous Categorical	Continuous	2 +	2 +	MANOVA	*F*
	Continuous	Cases	2 +	Repeated measures ANOVA	*F*
Continuous	Continuous	1	1	Correlation	*r*
Continuous	Continuous	1 +	1	Regression	*F*
Dichotomous Continuous	Dichotomous	1 +	1	Logistic regression	Wald χ^2
Dichotomous Continuous	Dichotomous Categorical	1 +	1	Discriminant	*F*
Dichotomous Categorical	Dichotomous Categorical	1 +	1 +	Log linear	χ^2

questions are written, the instrument must be constructed. This chapter then offered nine guidelines for constructing questionnaires. Taken together, these are the most common guidelines provided in the literature, but they are not fully exhaustive. Researchers should consult additional sources to ensure they design and administer the best possible questionnaire to meet project objectives (e.g., Salant and Dillman 1994; Dillman et al. 2014; Vaske 2019).

References

Baker BO, Hardyck CD, Petrinovich LF (1966) Weak measurements vs. strong statistics: an empirical critique of S. S. Stevens's proscriptions on statistics. Educ Psychol Meas 26:291–309

Beaman J, Vaske JJ, Grenier M (1998) A prototype model for estimating and correcting bias in number preference. Tour Anal 2(2):77–89

Beaman J, Vaske JJ, Schmidt JI, Huan TC (2015) Measuring and correcting response heaping arising from the use of prototypes. Hum Dimens Wildl 20(2):167–173

Binder A (1984) Restrictions on statistics imposed by method of measurement: some reality, some myth. J Crim Just 12:467–481

Blasius J, Thiessen V (2001) The use of neutral responses in survey questions: an application of multiple correspondence analysis. J Off Stat 17(3):351–367
Borgatta EF, Bohrnstedt GW (1980) Level of measurement: once over again. Sociol Method Res 9(2):147–160
Bryan H (1977) Leisure value systems and recreational specialization: the case of trout fishermen. J Leisure Res 9:174–187
Christian LM, Parsons NL, Dillman DA (2009) Designing scalar questions for web surveys. Sociol Method Res 37:393–425
Dillman DA, Smyth JD, Christian LM (2014) Internet phone mail and mixed-mode surveys: the tailored design method. Wiley, Hoboken
Dunlap RE, Van Liere KD, Mertig AG, Jones RE (2000) New trends in measuring environmental attitudes: measuring endorsement of the new ecological paradigm: a revised NEP scale. J Soc Issues 56:425–442
Fowler FJ Jr (1992) How unclear terms affect survey data. Public Opin Quart 56:218–231
Geer JG (1991) Do open-ended questions measure "salient" issues? Public Opin Quart 55:360–370
Harkness JA, Braun M, Edwards B, Johnson TP, Lyberg L, Mohler PP, Pennell B-E, Smith TW (2010) Survey methods in multinational multiregional and multicultural contexts. Wiley, Hoboken
Höhne JK, Lenzner T (2018) New insights on the cognitive processing of agree/disagree and item-specific questions. J Surv Stat Methodol 6:401–417
Inglehart R (1990) Culture shift in advanced industrial societies. Princeton University Press, Princeton
Inglehart R (1997) Modernization and postmodernization. Princeton University Press, Princeton
Johnson TP, Pennell B-E, Stoop AL, Dorer B (2019) Advances in comparative survey methods. Wiley, Hoboken
Kochalski S, Barz F, Pita P, Harrison HL (2026) Quantitative and qualitative content analysis of text and images. In: Pope KL, Arlinghaus R, Hunt LM, Lynch AJ, van Poorten BT (eds) Understanding recreational fishers: disciplinary and interdisciplinary approaches for fisheries management. Springer Nature, Cham, 507–532
Krosnick JA (2002) The causes of no-opinion responses to attitude measures in surveys: they are rarely what they appear to be. In: Groves RM, Dillman DA, Eltinge JL, Little RJA (eds) Survey nonresponse. Wiley, New York, pp 87–100
Kyle G, Landon AC, Pilgreen D, Golebie E (2026) Measurement instruments. In: Pope KL, Arlinghaus R, Hunt LM, Lynch AJ, van Poorten BT (eds) Understanding recreational fishers: disciplinary and interdisciplinary approaches for fisheries management. Springer Nature, Cham, p 355–385
Malhotra N, Krosnick JA, Thomas RK (2009) Optimal design of branching questions to measure bipolar constructs. Public Opin Quart 73:304–324
Manfredo MJ, Driver BL, Tarrant MA (1996) Measuring leisure motivation: a meta-analysis of the recreation experience preference scales. J Leisure Res 28(3):188–213
Manfredo MJ, Teel TL, Bright AD (2003) Why are public values toward wildlife changing? Hum Dimens Wildl 8(4):287–306
Manfredo MJ, Teel TL, Henry KL (2009) Linking society and environment: a multilevel model of shifting wildlife value orientations in the western United States. Soc Sci Quart 90:407–427
Manfredo MJ, Urquiza-Haas EG, Don Carlos AW, Bruskotter JT, Dietsch AM (2020) How anthropomorphism is changing the social context of modern wildlife conservation. Biol Conserv 241:108297
Manning RE (2022) Studies in outdoor recreation: search and research for satisfaction, 4th edn. Oregon State University Press, Corvallis
Milfont TL, Duckitt J (2010) The environmental attitudes inventory: a valid and reliable measure to assess the structure of environmental attitudes. J Environ Psychol 30:80–94
Mitra A, Lankford S (1999) Research methods in park, recreation, and leisure services. Sagamore, Illinois

Needham MD, Sprouse LJ, Grimm KE (2009) Testing a self-classification measure of recreation specialization among anglers. Hum Dimens Wildl 14:448–455

Peytchev A, Hill S (2010) Experiments in mobile web survey design: similarities to other modes and unique considerations. Soc Sci Comput Rev 28:319–335

Pollock KH, Jones CM, Brown TL (1994) Angler survey methods and their application in fisheries management. American Fisheries Society, Special Publication 25, Bethesda

Rasinski KA, Mingay D, Bradburn NM (1994) Do respondents really "mark all that apply" on self-administered questions? Public Opin Quart 58:400–408

Revilla MA, Saris WE, Krosnick JA (2014) Choosing the number of categories in agree-disagree scales. Sociol Method Res 43:73–97

Roberts C, Gilbert E, Allum N, Eisner L (2019) Research synthesis: satisficing in surveys: a systematic review of the literature. Public Opin Quart 83(3):598–626

Salant P, Dillman DA (1994) How to conduct your own survey. Wiley, New York

Saris WE, Revilla M, Krosnick JA, Schaeffer EM (2010) Comparing questions with agree/disagree response options to questions with item-specific response options. Surv Res Methods 4:61–79

Schaeffer NC (1991) Hardly ever or constantly: group comparisons using vague quantifiers. Public Opin Quart 55:395–423

Scott D, Shafer CS (2001) Recreational specialization: a critical look at the construct. J Leisure Res 33:319–343

Shelby LB, Vaske JJ (2007) Perceived crowding among hunters and anglers: a meta-analysis. Hum Dimens Wildl 12(4):241–261

Shelby B, Vaske JJ, Heberlein TA (1989) Comparative analysis of crowding in multiple locations: results from fifteen years of research. Leisure Sci 11:269–291

Stedman RC, Connelly NA, Heberlein TA, Decker DJ, Allred SB (2019) The end of the (research) world as we know it? Understanding and coping with declining response rates to mail surveys Society and Natural Resources 32:1139–1154

Stevens SS (1951) Mathematics, measurement and psychophysics. In: Stevens SS (ed) Handbook of experimental psychology. Wiley, New York, pp 1–49

Sturgis P, Roberts C, Smith P (2014) Middle alternatives revisited: how the neither/nor response acts as a way of saying "I don't know." Soc Res Methods 43:15–38

Tourangeau R, Yan T (2007) Sensitive questions in surveys. Psychol Bull 133(5):859–883

Tourangeau R, Rips LJ, Rasinski K (2000) The psychology of survey response. Cambridge University Press, Cambridge, UK

Vasja V, Couper MP, Cehovin G (2022) Alternative layouts for grid questions in PC and mobile web surveys: an experimental evaluation using response quality indicators and survey estimates. Soc Sci Comput Rev 41(6):1–23

Vaske JJ (2008) Survey research and analysis: applications in parks, recreation and human dimensions. Venture Publishing Inc, State College

Vaske JJ (2019) Survey research and analysis, 2nd edn. Sagamore-Venture Publishing LLC, Urbana

Vaske JJ, Shelby LB (2008) Crowding as a descriptive indicator and an evaluative standard: results from 30 years of research. Leisure Sci 30:111–126

Vaske JJ, Beaman J, Manfredo MJ, Covey CD, Knox R (1996) Response strategy recall frame and digit preference in self-reports of angling participation. Hum Dimens Wildl 1(4):54–68

Vaske JJ, Don Carlos AW, Manfredo MJ, Teel TW (2023) Evaluating alternative survey methodologies in human dimensions of wildlife research. Hum Dimens Wildl 28:320–334. https://doi.org/101080/10871209202222057622

Vaske JJ, Needham MD, Miller CM, Sponarski C (2026) Guidelines for writing and constructing recreational fisher surveys. In: Pope KL, Arlinghaus R, Hunt LM, Lynch AJ, van Poorten BT (eds) Understanding recreational fishers: disciplinary and interdisciplinary approaches for fisheries management. Springer Nature, Cham, p 287–325

Velleman PF, Wilkinson L (1993) Nominal, ordinal, interval, and ratio typologies are misleading. Am Stat 47(1):65–72

Fundamentals of Sampling and Implementing Human Dimensions of Fishing Surveys

Jerry J. Vaske, Mark D. Needham, and Craig A. Miller

Abstract Once a questionnaire has been constructed, the next steps include developing sampling procedures, administering the questionnaire, calculating response rates, and conducting a nonresponse bias check or comparing data to census information about the population (and weighting data if necessary). We compared and contrasted various sampling strategies (e.g., random, systematic, stratified, cluster), and then discussed steps for implementing different survey methods (e.g., on-site, telephone, mail, email, Internet, Internet panels). Each method has advantages and disadvantages. For example, when researchers are interested in fishers' perceptions of specific experiences and locations, on-site surveys may be the only alternative. Telephone surveys offer rapid data collection, but are costly, often have low response rates, and cannot accommodate complex questions or long questionnaires. Mail surveys allow for longer and more complex questions, but are also costly and take time to complete. Email and Internet surveys are attractive because of the low cost and rapid data collection, but tend to have low response rates and difficulty in obtaining email addresses for general population surveys. Internet panels involve self-selected individuals who are usually paid to complete questionnaires, but these are usually based on nonprobability samples. We also discuss strategies for increasing response rates, performing nonresponse bias checks, and weighting data.

Keywords Nonresponse bias check · Response rates · Sampling strategies · Survey methods · Weighting data

J. J. Vaske (✉)
Human Dimensions of Natural Resources Department, Colorado State University, Fort Collins, CO, USA
e-mail: jerryv@colostate.edu

M. D. Needham
Department of Forest Ecosystems and Society, Oregon State University, Corvallis, OR, USA
e-mail: mark.needham@oregonstate.edu

C. A. Miller
Illinois Natural History Survey, University of Illinois, Champaign, IL, USA
e-mail: craigm@illinois.edu

K. L. Pope et al. (eds.), *Understanding Recreational Fishers*, Fish & Fisheries Series 45, https://doi.org/10.1007/978-3-031-99739-6_11

1 Implementing Human Dimensions Surveys

This chapter discusses approaches for implementing human dimensions (i.e., social science) of fishing surveys, including various approaches for how to sample fishers (e.g., random, systematic, stratified) using diverse types of survey methods (e.g., on-site, mail, Internet). Other chapters in this book (e.g., Kyle et al. 2026, Vaske et al. 2026) discuss procedures for writing and designing questionnaires, as well as how to develop measurement instruments. Once a questionnaire has been constructed and pretested, steps that follow include developing sampling procedures, administering the questionnaire, entering the data into software, calculating response rates, conducting a nonresponse bias check, and weighting the data if necessary. This chapter discusses many of these steps.

2 Sampling

The first step to collecting survey data is defining the population or populations of interest (Babbie 2003; Tourangeau 2004). Populations may be broad (e.g., public in a country) or specific (e.g., fishers on a specific section of a river). Once the population has been defined, the researcher should, if possible, obtain or else generate (e.g., via random telephone surveys or other address collection) a list of the entire population from which to select a sample. Fisheries researchers, for example, might use customer lists maintained by their state or local fish and wildlife agencies. In the United States of America, however, state license databases are legally protected and may not be available simply upon request. It is important to obtain agreement with agencies to access license databases prior to any other planning process (e.g., developing questionnaires) and this permission takes time to develop and negotiate.

In other countries that lack fishing licenses or license addresses are not collated (e.g., Germany), complete license frames may not exist and must then be created via other means (e.g., random telephone surveys or address collection via creel surveys). If complete address lists are not available, researchers might still engage in close-to-representative sampling using quota sampling or weigh survey respondents (discussed later in this chapter), but this also demands knowledge about characteristics in the population or populations of interest. If the research is designed to generate insights about population-level metrics, it is important to sample representatively from the population of interest and avoid inferences from biased samples (discussed later in this chapter). Snowball samples (i.e., where research participants are asked to assist researchers in identifying other potential participants) or responses from open-access surveys on social media or the Internet, for example, may provide biases in the answer patterns that weighing responses might be impossible to rectify. It can be tempting to offer a survey link on the Internet and advertise it on social media, but the responses may be highly biased and not represent the population or populations of interest, unless the population of interest is avid social media users who respond to survey links. Taken together, if the researcher is not sampling rigorously from the population or populations of interest, the resulting data should never be reported as representing the population or populations, and only certain types of analyses (e.g., testing hypotheses about correlations among variables) are warranted.

Depending on the project and what data an agency maintains, license sale records may be incomplete, too broad, or unavailable. For example, if the study is specific to attitudes of fishers on Lake Michigan, sampling a statewide database of all Illinois fishers may not yield the desired sample, as residents in southern Illinois are less likely to have fished Lake Michigan compared to those living in the northern areas of this state. When population lists are unavailable, on-site (intercept) survey methods, door-to-door household surveys, telephone surveys, or address collection via creel surveys may be the only options, but some of these approaches (e.g., on-site surveys) may draw from the population of trips and not the population of recreational fishers, so the samples (e.g., on-site sample) may contain an avidity bias if not adjusted during analysis.

Once the population has been defined, researchers must choose a method for selecting a representative sample. *Sampling* is the process of selecting population units (e.g., people) and collecting data that describe the larger population based on a selection of that population (Vaske 2019). *Representativeness of the sample selection* occurs when all constituents in the population have a known chance of being selected (Salant and Dillman 1994). The sample is representative of the population if characteristics of the sample closely approximate and resemble the same characteristics in the population for issues of interest (Cochran 1977). This is also known as *scope of inference*. Unless the sample is representative of the population, it can be impossible to make accurate predictions about the population from which the sample was selected (Mitra and Lankford 1999). Researchers should avoid the temptation to report statements about a population when the sample is clearly not representative (e.g., "percentage of fishers in area X who hold an attitude of the strength Y"). With non-representative samples, the researcher is bound to rely on only descriptions of the sample or testing hypotheses about relationships between concepts rather than any accurate descriptions of the entire population of fishers.

Sampling designs that strive for representativeness are called probability samples (Yeager et al. 2011). *Probability sampling* usually involves random selection of units, requires relatively few observations, and allows results to be inferred to the population from which it was selected, if all members of the population have a known nonzero chance of being selected (Morgan and Harmon 1999; Dillman et al. 2014). Probability sampling has two advantages (Vaske 2019). First, probability theory allows researchers to estimate the precision (i.e., random *sampling error*) of estimates derived from a sample, using design-based estimators where the randomness is introduced by the sampling design (Jessen 1978). Some sampling error always occurs because only a sample of the population is asked to participate rather than everyone in the population. Every sample that could be drawn from the population can produce slightly different estimates (Peytcheva and Groves 2009; Dillman et al. 2014). Sampling error can be avoided by conducting a census (i.e.,

Table 1 Completed sample sizes needed for various population sizes and characteristics at three levels of precision

	Sample size for the 95% confidence level					
Population	± 3% sampling error		± 5% sampling error		± 10% sampling error	
Size	50/50 split	80/20 split	50/50 split	80/20 split	50/50 split	80/20 split
100	92	87	80	71	49	38
200	169	155	132	111	65	47
400	291	253	196	153	78	53
600	384	320	234	175	83	56
800	458	369	260	188	86	57
1000	517	406	278	198	88	58
2000	696	509	322	219	92	60
4000	843	584	351	232	94	61
6000	906	613	361	236	95	61
8000	942	629	367	239	95	61
10,000	965	640	370	240	95	61
20,000	1013	661	377	243	96	61
40,000	1040	672	381	244	96	61
100,000	1056	679	383	245	96	61
1000,000	1066	683	384	246	96	61
1000,000,000	1067	683	384	246	96	61

Sources: Salant and Dillman 1994, Dillman et al. 2014, Vaske 2019

surveying the entire population), but this is usually not practical or feasible due to the costs and time required (Dillman 2007).

Sampling estimates based on random sampling are typically close to the true value, on average, but this closeness depends on sample size, lack of non-response bias, and how the sample is selected (discussed later in this chapter). It is often desirable to obtain a sample large enough to be 95% confident that the data are within ±5% of the population of interest (i.e., margin of error) if one wants to describe the population of interest. In other words, chances are 19 out of 20 that the population value is within 5% of the estimate in either direction. Different sample sizes are needed for various population sizes and levels of sampling error (Table 1). A 50/50 split is the most conservative and means that the population is divided with 50% expected to answer one way (e.g., support, agree) and 50% the other way (e.g., oppose, disagree). An 80/20 split means that most people respond one way or have a certain characteristic, whereas a few do not. If researchers have little knowledge about the population, the 50/50 split is recommended (Vaske 2019). For simple random samples, a sample size of approximately 400 (rounded from 384) is considered suitable for generalizing to a population at a 95% confidence level with a ± 5% margin of error (Salant and Dillman 1994; Dillman et al. 2014).

Second, probability samples are representative because selection biases are minimized as much as possible (Groves and Peytcheva 2008). *Bias in population estimates*, for example, occurs when individuals in the sample are not representative of the population. Bias would likely occur, for example, if the researcher avoided including certain people in the sample (e.g., only selected male fishers in a study of all fishers; did not include those fishing without a license or those exempt from needing a license), or if a substantial number of people contacted had systematically different characteristics to the rest of the sample and they refused to participate by not answering the questionnaire (nonresponse). These biases are examples of coverage errors, which occur with a discrepancy between the target population or populations and a subset of individuals who are included in the sampling frame. Coverage errors occur, for example, when the list of individuals from which the sample is selected is not inclusive of all elements of the population or populations of interest and does not accurately represent the population characteristics to be estimated (Dillman et al. 2014). If there are errors in coverage, then all elements of the target population do not have an equal or known chance of being included in the sample.

If the researcher makes a subjective judgment to include or exclude individuals based on certain characteristics, the integrity of the selection is compromised, and inferring results to the population is difficult at best and usually impossible. This is called *nonprobability sampling*. Convenience samples, snowball samples, and focus groups are examples of nonprobability samples where some members of the population have a high chance of being included (e.g., social media users if links to questionnaires are provided on the Internet) and others have little or no chance of being selected (e.g., non-users of social media). Nonprobability samples are useful for exploratory studies that may generate innovative ideas to be rigorously tested later via probability sampling or for hypothesis testing, but are questionable if a researcher wants to describe the population or populations of interest without any significant effort to eliminate bias (Vaske 2019).

After collecting data from nonprobability samples, researchers may examine demographic or other (e.g., avidity) characteristics of their sample and conclude that respondents are like those of the population. Although this is an attempt at checking for representativeness, similarities do not prove the sample is representative of the population, and researchers should generally not use nonprobability samples to make inferences about the larger population, at least without carefully weighting the data, which is discussed later in this chapter (Morgan and Harmon 1999; Salant and Dillman 1994). However, some recent research has started using big data from nonprobability samples (e.g., from patterns in social media data, or electronic citizen science platforms such as smartphone applications) and even some automated analyses of these large datasets (e.g., artificial intelligence, machine learning) to measure issues such as population-level participation rates in a fishery, with some results showing promise for supplementing data from traditional probability samples (e.g., Gundelund et al. 2021; Venturelli et al. 2017). This chapter focuses primarily on probability samples, but recognizes that non-probability samples are becoming common, especially in the areas of big data and digital data, and that such data may yield useful conclusions if any biases are known and corrected. The Venturelli et al. (2026) chapter in this book discusses this issue in more detail.

There are several probability sampling methods (for reviews, see Cochran 1977, Jessen 1978, Fowler Jr. 1993, Scheaffer et al. 1996, Lohr 1999, Dillman et al. 2014).

Single-stage sampling (i.e., a single set of primary sampling units) includes simple random, systematic, and stratified sampling. *Multistage sampling* uses both primary (e.g., households) and secondary (e.g., individuals within households) sampling units and a combination of sampling techniques (e.g., a cluster sample combined with a stratified sample). Deciding on an approach depends on the study purpose, population surveyed, and available resources (e.g., personnel, time, cost).

2.1 Simple Random

The most basic probability sampling method is *simple random sampling*, which involves selecting people so that each member of the population has an equal chance of being selected (Scheaffer et al. 1996). Members are selected randomly, and their selection is independent of one another (Dillman et al. 2014). For example, if a target population consists of 100,000 fishers, a random sample of 1000 fishers means that every member of this population has a 1 in 100 chance of selection. Random samples are reasonably unbiased, but require all members of the population to be included in the list; if some members are missing, they do not have an equal chance of being selected (Morgan and Harmon 1999).

To select a simple random sample, researchers might use a random numbers table, which is a long set of integers generated so the table contains all 10 integers (i.e., 0, 1, 2, . . . 9) in relatively equal proportions with no trends in the pattern in which digits were generated. If a number is selected from any point on the table, it has a relatively equal chance of being any digit from 0 through 9 (Scheaffer et al. 1996). Sampling and statistics books often provide tables of random numbers and step-by-step directions for selecting random samples from these tables (e.g., Salant and Dillman 1994; Mitra and Lankford 1999; Dillman et al. 2014). Software packages can also generate random samples (e.g., Microsoft Excel, SPSS).

For surveys where only indirect sample frames are available for selecting individuals (e.g., household surveys, intercept surveys of boat fishers), commonly used approaches for generating approximate random samples (secondary sampling units) within clusters of individuals (primary sampling units) include the Kish (1949) and birthday methods. The *Kish method* "involves listing all members of a household, by gender and age (or sometimes name and age), and then selecting one member from the list using previously developed random selection tables" (Dillman et al. 2014, p. 84). The *birthday method* involves selecting the adult who either had the most recently passed birthday or has the most recent upcoming birthday (Rizzo et al. 2004). The birthday method is among the most easily executed methods, especially for on-site surveys where groups of individuals are encountered (Dillman et al. 2014).

2.2 Systematic

Random sampling can be cumbersome when applied to on-site surveys (Salant and Dillman 1994). An alternative is *systematic sampling*, which involves randomly selecting the first person in a population, choosing an appropriate fixed interval, and then systematically selecting subsequent individuals (Fowler 1993; Morgan and Harmon 1999). For an on-site study such as a survey of fishers along the entire bank of a river, the first person can be selected at random (using a randomly generated number) to guard against any bias, followed by every n^{th} (e.g., 15th) person along the riverbank. Compared to random samples, systematic approaches to on-site surveys are easier to perform in the field and less subject to selection errors by researchers (Thompson 1992; Babbie 2003). Systematic sampling can provide more information per unit cost than random sampling because the sample tends to be uniformly spread over the population (Scheaffer et al. 1996).

Two terms are common in systematic sampling (Vaske 2019). First, a *sampling interval* is the distance between people selected in the sample (e.g., five fishers along a stream). The calculation is:

$$\text{sampling interval} = \frac{\text{population size}}{\text{sample size}} \tag{1}$$

Second, *sampling ratio* is the proportion of units in the population that are selected (e.g., 1/5 fishers along a stream). The calculation is:

$$\text{sampling ratio} = \frac{\text{sample size}}{\text{population size}} \tag{2}$$

To select a systematic sample, the researcher must choose the sampling ratio needed for obtaining a specific sample size. The ratio could, for example, be 1 in 5, 1 in 10, 1 in 100, or in general 1 in k. For a systematic sample of n people from a population of size N, k must be less than or equal to N/n (Kish 1965). For example, if the population was $N = 30{,}000$ and the sample size for a $\pm$ 5% margin of error at the 95% confidence level was $n = 400$, k would be 75. The researcher would select every 75th person after randomly selecting the first individual. If $k < 75$, the sample size would be greater than 400 people.

Determining a value for k when the population size is unknown is challenging (Cochran 1977). In recreational fishing, for example, agencies may not have accurate data on use levels at a location. In this case, researchers must determine a desired sample size and margin of error, and then estimate the value of k needed for reaching this sample size (see Vaske 2019 for details). Researchers should select a conservative value for k because if it is too large, the required sample size may not be obtained through a systematic sample (Scheaffer et al. 1996).

2.3 Stratified Random

Stratified random sampling involves: (i) dividing the population into nonoverlapping groups (i.e., strata) that are of interest because of project objectives or hypotheses, and then (ii) selecting a random sample from each stratum or group (Scheaffer et al. 1996; Morgan and Harmon 1999; Dillman et al. 2014). The resulting stratification helps researchers to determine differences among subgroups and can increase precision for a given sample (i.e., cost-efficient). Example strata in recreational fishing include fishing locations (e.g., river, lake), equipment types (e.g., fly fishing, artificial lures, bait casting), and target species (e.g., bass, trout).

Stratified samples are useful for two reasons (Vaske 2019). First, many populations contain subgroups that are of interest. For example, if a manager is interested in the extent to which fishers would support and be willing to pay a new user fee to enter a water-based resource (e.g., a specific lake), it would be useful to examine people in distinct income brackets (e.g., high, low). Opinions of fishers in the high-income bracket may differ from those in the lower bracket.

Like many recreation activities, fishers may form groups with greater homogeneity based on day of the week, time of the day, or type of day (e.g., holidays, stocking schedule) combined with other characteristics (e.g., skill or experience, target species). Such consideration requires temporally stratified sampling. For example, avid fishers may avoid fishing a particular lake during a holiday weekend due to crowding. Conversely, an occasional fisher may fish during the same weekend because fishing may be combined with other activities (e.g., boating, picnicking). Weekdays, evenings, and weekends will appeal to different fishers based on their schedules. Retired workers may fish during weekdays to avoid crowding, whereas evenings may be favoured by fishers who work during the day. Bass fishers may be best sampled during the day, but people who fish for catfish may be best sampled during late evenings. Such temporal distinctions must be taken into consideration when developing sampling schedules based on factors of interest, such as demographic characteristics and target species.

Second, stratification with more homogeneous populations within strata can greatly reduce sampling error. If 99% of the population agrees with an issue, for example, it is unlikely that a probability sample will misrepresent the extent of agreement. If the population is split (e.g., 50% agree, 50% disagree), however, sampling error will be larger. Rather than selecting a sample from the total population at large, the researcher could select appropriate samples from more homogeneous subgroups of the population (e.g., annual license, lifetime license, senior license buyers).

To obtain accurate information about populations of interest, researchers need to ensure an adequate sample size for each subgroup (Jessen 1978; Thompson 1992). Stratified samples assume that sample sizes of subgroups are relatively equal and allow for tests of statistical differences among groups (Mitra and Lankford 1999). For example, if the stratification variable has two levels (e.g., trout fishers, bass fishers) and both levels involve large population sizes, a final sample size of approximately 800 (i.e., 400 trout fishers, 400 bass fishers) would be required for findings to be discussed in terms of the 95% confidence level with a ± 5% margin of error (Dillman et al. 2014).

Sometimes it is useful to estimate population proportions of each group in advance and use these estimates to determine the proportion of the sample needed for each stratum. This involves *proportionate sampling* (within a stratified random sample) of subgroups or strata at rates equal to their size in the population. This eliminates the possibility of drawing a sample where groups are not accurately represented. By comparison, *disproportionate sampling* (within a stratified random sample) involves deliberately sampling subgroups at rates unequal to their size in the population (i.e., over- or under-sampling groups). Disproportionate sampling is useful when researchers need to draw conclusions about uncommon or low-incidence populations (e.g., lifetime license buyers). This helps to ensure that enough members of these groups will be surveyed (Dillman et al. 2014).

2.4 Cluster

Cluster sampling involves conducting a random sample of groups or clusters (e.g., locations along a river) and then sampling people within the selected clusters (Kelly and Cumberland 1990; Scheaffer et al. 1996). This approach differs from stratified sampling, where a random sample of people is drawn within each group; cluster sampling involves a random sample of groups. For cluster sampling, in contrast to stratification, it is most effective (if feasible) to have clusters that are at least somewhat heterogeneous to each other (i.e., with variation between clusters similar to the general population).

Cluster sampling can be effective when: (i) a good sample list of population units is unavailable or expensive to obtain, but a listing of potential clusters is obtainable; and (ii) the cost of obtaining completed responses increases as distance separating population and sampling units increases (Scheaffer et al. 1996; Rocco 2003). A goal of cluster sampling is identifying groups of similar types of people. It is assumed that in some geographical areas, a greater probability exists for sampling desired individuals. If these areas can be identified in advance, the cost and time for data collection can be less than other approaches (e.g., random sample). For example, if a study examines fishing participation among Hispanic residents in an urban area where only 5% of the population is Hispanic, and the researcher wants a sample size of 2000 Hispanics, 40,000 questionnaires would need to be completed for a sample of 2000 Hispanics (i.e., 40,000 * 0.05 = 2000). Given that most urban areas contain neighbourhoods that are largely populated by specific ethnic or racial populations (Mitra and Lankford 1999), it may be more efficient for researchers to use cluster sampling by identifying areas that are predominantly Hispanic, conducting a random sample of these areas, and then sampling within the selected areas.

The first step in cluster sampling is identifying appropriate clusters. People within a cluster are often physically close together and may share characteristics. Once clusters have been specified, the second step is developing a sampling frame listing the clusters and then conducting a random or stratified random sample of clusters. The last step involves selecting a sample from within each cluster. To illustrate, assume that to conduct a cluster sample, the researcher identified six predominantly Hispanic communities. For simplicity, these communities will be named A, B, C, D, E, and F. Assume that communities A, C, and F were selected through a random sample of the six communities. The researcher then randomly selected streets within each of these three clusters (i.e., A, C, F) and then randomly selected a certain number of households along these streets.

Cluster samples are often used for on-site surveys conducted in person because grouping sampling units into geographical areas reduces costs and decreases distances that interviewers need to travel (Dillman et al. 2014; Vaske 2019). For mail, Internet, and telephone surveys, geographical issues such as distances between respondents have minimal effect on costs and responses, so clustering is usually not needed for these modes (Dillman et al. 2014).

2.5 *More Complex Multistage Sampling Approaches*

Sampling can be complex, especially when several types of sampling approaches are combined into a more complex multistage design. A stratified-cluster sampling approach (i.e., cluster combined with stratification), for example, can be useful for on-site studies when a list of potential respondents in the population is unavailable and researchers do not know in advance who is in the population. Readers should consult resources focusing on sampling for more information about these and other more complex sampling approaches (e.g., Kish 1965; Cochran 1977; Kelly and Cumberland 1990; Thompson 1992; Scheaffer et al. 1996; Rocco 2003; Dillman et al. 2014; Vaske 2019).

3 Implementation

Surveys should be designed to encourage high response rates and secure adequate sample sizes. Strategies such as multiple contacts, design and content of a questionnaire, appearance of accompanying materials (e.g., envelopes), sponsorship, and personalization should all be considered. Dillman et al. (2014) provided several suggestions for addressing these issues.

Research has also shown consistently that sending a *token cash incentive* with survey requests is one of the most effective ways for increasing trust and response rates (for a review, see Dillman et al. 2014). A few dollars (e.g., $2–$5) are usually best because higher amounts may create more suspicion than trust. Including cash with the invitation is usually more effective than promising larger payments after respondents complete the questionnaire, such as being included in a drawing for a prize (Dillman et al. 2014). Prepaid cash incentives are most useful and easiest to administer for mail surveys, but are difficult to implement for other modes (e.g., Internet, telephone; Schouten et al. 2013). Research has shown, however, that mailing a letter with a cash incentive and a request to complete a questionnaire on the Internet improves response rates more than a cash incentive with a request to complete and mail back a questionnaire (Messer and Dillman 2011). Approaches such as PayPal payments, gift certificates, and gift cards are also possible, but increase response rates only minimally and not as effectively as cash payments (Vaske 2019). Other incentives, such as prize drawings, also have little, if any, substantive impact on response rates (Dillman et al. 2014). Although cash sent with the request may increase response rates, it also increases the overall project budget, and some university institutional review boards (i.e., human subjects) prohibit cash incentives.

3.1 Mail Surveys

With mail surveys, respondents receive a questionnaire by mail, answer questions, and return the questionnaire by mail. Given that no direct interaction occurs between the researcher and respondent, people may see little incentive to cooperate. To obtain a reasonable response rate, it is imperative to produce a well-written and attractive questionnaire, use personalized correspondence, and conduct repeat mailings as reminders (Kwak and Radler 2002; Dillman et al. 2014).

Before administering a mail survey, the researcher needs to have the sampling list in electronic format so that names and addresses can be quickly transferred to software to print mailing labels. Names and addresses will be printed several times because multiple mailings are necessary for achieving a high response rate (Vaske 2019). The researcher needs to sort through the list between mailings and delete information about people who have responded and do not need to be contacted again. Electronic information streamlines the process and reduces cost (e.g., by allowing barcode scanning to remove respondents from the sample). Researchers should prepare survey materials in advance. For example, print materials for multiple mailings (e.g., letters, questionnaires, postcards, envelopes) and address envelopes for the first waves of mailings before mailing the initial packets. All these tasks can also be contracted to a private company or other organization to perform (e.g., a printing and mailing service).

Multiple contacts are more effective than any other technique for increasing mail survey responses (Dillman et al. 2014; Vaske 2019). The content of contacts (e.g., prenotifications, full questionnaire packets, postcard reminders) is varied, and each differs from one another. Dillman et al. (2014) suggested a strategy where contacts are tailored to the study and "multiple carefully designed contacts that are strategically timed are more important than using an exact system of contacts" (p. 373). For example, prenotification letters are now less effective and may not be needed.

Replacing postcards with formal letters may induce more trust and higher responses (Dillman et al. 2014).

If a *prenotification letter* is used, it should be less than one page and sent to the sample a few days or a week before the questionnaire is mailed, notifying people: (i) about the study purpose, (ii) that the questionnaire will arrive shortly, and (iii) that their participation is appreciated. Letters should contain letterhead or logos of organizations conducting or sponsoring the study, signatures and contact information of lead researchers, personalized addresses, and a date. Prenotification letters should be brief, personally addressed, positively worded, emphasize the study's importance, and build anticipation; precise details about conditions for participation are not needed (e.g., anonymity, confidentiality, voluntary). Providing a unique identification code (PIN) and a URL (i.e., website) for respondents to complete the questionnaire on the Internet instead of by mail may increase participation (this is a type of mixed-mode survey, discussed later in this chapter). Offering a mail questionnaire as a follow-up to an Internet questionnaire request can increase responses (Dillman et al. 2014). These same guidelines apply to harvest records distributed pre-season to sampled individuals.

Dillman et al. (2014) suggested that if it is possible and financially feasible, letters should be mailed using first-class and manually stamped postage with the recipient being referred to by their name, not "Resident of...," if the sample list contains this information. First-class postage allows items to be processed faster by the mail service, returned or forwarded for invalid addresses, and perceived as more important by respondents (Dillman et al. 2014).

Some researchers opt for postcards instead of letters as prenotifications to reduce postage costs and personnel time. Dillman et al. (2014) emphasized that prenotifications should be sent as a letter instead of a postcard because a letter is more formal, contains more detail and trust-inducing elements (e.g., personalized address, ink signature, letterhead), and takes longer for people to open and attend to compared with postcards that can be glanced at and discarded.

Prenotifications can be beneficial because they are less expensive to print and mail (i.e., postage) than full questionnaire packets. Postage costs associated with undeliverable questionnaires (e.g., invalid addresses, moved) are also lower for this mailing compared with sending full packets. Prenotifications do increase costs, and for large samples, this cost may be a consideration. In 2023, in the United States of America, a prenotification letter for a sample of 5000 individuals could increase costs approximately USD $3000 or more when costs for letters, envelopes, and postage are factored into the budget.

The *first questionnaire packet* typically contains the: (i) envelope where documents are contained, (ii) personalized and signed cover letter on letterhead, (iii) questionnaire, (iv) addressed and personally stamped postage-paid return envelope, and (v) sometimes a token incentive (Dillman et al. 2014). The mailing envelope will vary in size depending on the size of paper used for the questionnaire, but it should contain logos and addresses of organizations or agencies conducting the study, be labelled with the respondent's name and address, and mailed using manually stamped first-class postage if it is possible and financially feasible. Envelopes should never include statements such as "your response required," "open now," or "important documents" and should not be brightly coloured (e.g., yellow, blue) because these convey a marketing approach and recipients may treat such envelopes accordingly (Dillman et al. 2014). Before mailing a survey, contact the postal service, agency, or university mailing services for information about postage rates and procedures that avoid complications with format for electronic address files, automated scanning, and sorting equipment (e.g., proper label placement).

The cover letter is important for persuading people to answer the questionnaire and take the project seriously. Cover letters should contain letterheads or logos of agencies conducting and sponsoring the survey (the same as prenotification letters, if used), signatures and contacts of lead researchers, personalized addresses, and a date. Salutations such as "Dear Bill" or "Dear Mrs. Jones" help to personalize letters, whereas statements such as "Dear Resident" or "Dear Colorado Fishing Participant" convey that a form letter was sent, and people are likely to discard these impersonalized packets (Dillman et al. 2014). If it is difficult to determine gender from names, researchers can avoid offending recipients by not using salutations such as Mr. or Mrs.

Letters should not exceed one page and should start by stating the purpose of the study, information requested, that they are among a limited group selected, and the importance of the request (e.g., Porter and Whitcomb 2003; Dillman et al. 2014). Statements about the survey being voluntary and anonymous or confidential should then be included because they are often required by human subject and regulatory compliance protocols (e.g., university institutional review boards). Surveys are typically *voluntary* because people must be allowed to skip questions or not participate. Researchers should convey the study's importance and value of their contribution, but must not be coercive or offensive. *Confidentiality* means that responses will remain private by removing identification numbers, destroying mailing lists after the study ends, and ensuring that people remain *anonymous* so that personal and contact information cannot be associated with specific responses (Salant and Dillman 1994). The concluding section of the letter should provide directions for returning the questionnaire, contact information for inquiries, and a statement of appreciation. Letters should be personable, professional, and free of jargon.

The main component of the packet is the questionnaire (Vaske et al. 2026). It is important to provide an ID code on the questionnaire and mailing labels that correspond to the mailing list. This code tracks returned questionnaires and who should receive follow-up mailings. This procedure reduces mailing and administrative costs, and avoids assembling and mailing duplicate packets to respondents, but can create concerns with respondents about the perceived lack of anonymity. The importance of these ID codes for tracking purposes and how respondent anonymity will be protected should be clearly articulated in the cover letter accompanying the questionnaire. These ID codes have had minimal negative effects on response rates (Salant and Dillman 1994), but should be placed in a corner of the questionnaire so that if respondents tear them out, they do not tear through questions and answers. Trying to hide ID numbers (e.g., incorporate into an approval processing number such as "Approved Form 426") is unethical and discouraged because it conveys that the researcher is trying to hide something or trick people (Dillman et al. 2014). However, it can be useful to add ID numbers to both the questionnaire and return envelope because respondents seldom notice these numbers in multiple locations (Vaske 2019). Barcodes can be used instead of numbers, with macrocodes compatible with mailing list platforms (e.g., Excel) that automatically delete individuals from the list once their returned questionnaire is scanned.

Ideally, a stamped return envelope must be provided so respondents can return completed questionnaires. This envelope should provide first-class postage and be printed with the researcher's or agency address. Expecting respondents to pay postage decreases responses (Salant and Dillman 1994). Adding stamps to return envelopes, however, increases postage costs and personnel time when assembling packets. For large projects or studies constrained by budget, business reply envelopes may be more practical because postage is only paid for returned envelopes, but at a higher charged rate. Response rates from studies using business replies may be up to 10% lower than studies using manually stamped first-class postage (Armstrong and Lusk 1987; Dillman et al. 2014).

When selecting a mailing date, avoid holiday periods because extra postal service volume during these times increases the likelihood of delivery mistakes (Weeks et al. 1980; Dillman et al. 2014). Mailings should be tailored to the population. If a study is interested in fisher participation during the most recent fishing season, questionnaires should be mailed shortly after the season so respondents can easily recall specific trips. January to March appears to be one of the most conducive periods for mail questionnaires in the northern hemisphere, but little evidence suggests that time of week or month of year has a dramatic effect on response rates (Dillman et al. 2014).

The first follow-up is often a *thank-you or reminder* expressing gratitude to those who responded and encouraging nonrespondents to complete and return the questionnaire. Most people who answer questionnaires respond within a few days to a week, and as each day passes, the questionnaire becomes a lower priority. A carefully timed contact can remind people to complete the questionnaire before it is forgotten, lost, or discarded (Dillman et al. 2014). Reminders can be a letter or postcard, but postcards are less expensive and a different stimulus than opening another envelope (Heberlein and Baumgartner 1981; Dillman et al. 2014).

Reminders state when the questionnaire was sent, state why it is important, thank those who responded, and ask those who have not responded to do so. For postcards, the text is on the back. Respondent names and addresses appear on a mailing label on the front. First reminders should be sent using first-class postage and arrive a few days to 1 week after the first packet. This timing allows the first mailing to have its major effect (Dillman et al. 2014). Reminder letters or postcards should be printed and labelled before the first mailing. Cost savings from doing so are minimal compared to the risk of sending a late reminder (Vaske 2019).

Two to four weeks after this first reminder, a second complete mailing should be sent to nonrespondents. This *replacement questionnaire packet* should contain a: (i) reminder letter, (ii) questionnaire, and (iii) addressed and stamped postage-paid return envelope. This packet should be sent using first-class postage. The timing for sending this packet can be longer than earlier mailings to allow for more questionnaires to be received, especially those that have been delayed in the mail. This timing also reduces costs associated with another complete mailing.

This replacement packet is identical to the first except that a different letter encourages nonrespondents to complete and return the instrument. This letter should convey a tone of insistence that previous mailings lacked, but remain professional (Dillman et al. 2014). The most important text in this letter is, "A few weeks ago, we sent you a questionnaire asking about (include brief text here). We have not received your questionnaire. Perhaps you misplaced it or forgot to mail it back. We are providing you another questionnaire..." (Dillman et al. 2014). This phrasing attempts to convince people that their responses are important. A replacement questionnaire and stamped return envelope must be included because many people will have misplaced or discarded the first packet.

If the number of responses is still low after these mailings, a *final reminder* or *contact* may be necessary using priority mail or courier (e.g., FedEx, UPS) shipping instead of regular mail, or telephone (if the sample list contains telephone numbers) to contact nonrespondents (Keeter et al. 2017). This contact emphasizes the importance and legitimacy of the request using different packaging and appearance. When the budget allows, another questionnaire should be included (Dillman et al. 2014). Final contacts using priority mail or courier can be cost-prohibitive, but do increase responses.

Other details for implementing mail surveys include the type and size of stationery used. Larger envelopes for outgoing mailings, for example, require no folding of materials and draw attention when they appear in a person's mailbox. These considerations increase postage costs. Readers should consult other texts and articles that discuss these issues in more detail and provide example figures illustrating cover letters, envelopes, and questionnaires (e.g., Salant and Dillman 1994; Mitra and Lankford 1999; Dillman et al. 2014; Vaske 2019).

3.2 Telephone Surveys

Telephone surveys rapidly generate data, allow researchers to control the sequence in which questions are asked, and ensure all questions are answered (an ideal goal that does not always happen). These surveys, however, often have low response rates (e.g., people not answering or hanging up, do-not-call lists, call display and screening technology) and present challenges when trying to ask complex or long questionnaires. Telephone surveys also have higher administration costs than other types of surveys, except for on-site surveys. Telephone surveys are less sensitive to design layout because respondents rarely see the instrument.

The first step in conducting telephone surveys is selecting the sample. In countries where fishing licenses are common, fishers usually record their telephone numbers on license applications. Other sources of telephone numbers are problematic for assorted reasons (for reviews, see Dillman et al. 2014 and Hammarstedt et al. 2017).

Telephone directories, for example, are now largely non-existent. Cellular telephone numbers present coverage issues and people often keep their cellular numbers after they move to a new location. An even larger issue is many people talk using platforms such Zoom or use video messaging on social media platforms (e.g., Facebook and Instagram Messenger), and accessing contact information for these platforms can be challenging, presenting a barrier to surveying some individuals.

Random digit dialling (RDD) may seem acceptable for overcoming problems, but it must be used with caution because any randomly generated number could be a: (i) working household number, (ii) working non-residential number, (iii) nonworking number, or (iv) household with landline and cellular numbers (which could lead to coverage overlap issues). With RDD, the Internet or other directories are used for determining area codes and prefixes (Yeager et al. 2011). The next four digits are selected randomly using a random numbers table or software.

One alternative is the Waksberg (1978) method, which involves a two-stage cluster design to reduce unproductive calls. In the first stage, a listing of all area codes and prefix numbers in the area of interest is obtained. For each area code and prefix combination, all possible choices for the next two numbers are added (e.g., 970–491-00 through 970–491-99); a list of all possible eight-digit numbers of the 10 digits in telephone numbers (in the United States of America) is constructed. These eight-digit combinations represent primary sampling units. Software or a random numbers table is used for assigning two additional digits to a randomly selected primary sampling unit. After the number is dialled, the primary sampling unit is retained if it is a residential or personal cellular number. If the number is not a working residential or personal cellular number, the primary sampling unit is eliminated from sampling. In the second stage, additional last two digits are selected randomly within each valid primary sampling unit until a set number of telephones is reached, and this process is repeated until the desired sample size is reached. A disadvantage of this method is the amount of personnel time needed.

Once a telephone sample has been selected, arrangements for facilities, equipment, and personnel must be made. These surveys are most successful when administered from a centralized location because it allows interviewers and supervisors to work together, and allows supervisors to support and monitor interviewers (Salant and Dillman 1994). Computer-assisted telephone interviewing (also known as CATI) software can reduce errors and speed up interviewing and data entry. Interviewers must also be hired and trained before placing calls. The success of telephone surveys depends on interviewers being able to read fluently, communicate in a clear and pleasant manner without hesitating or stumbling with pronunciation, and respond well to inquiries (e.g., Oksenberg et al. 1986; Groves and McGonagle 2001). Interviewers must convey professionalism and legitimacy. Some interviewer training issues include: (i) keeping a record or log of calls (e.g., busy, disconnected, call back); (ii) dealing with abusive or aggressive respondents (e.g., be nice, keep cool, be professional); (iii) tone and pace (e.g., clear, neutral, enthusiastic, loud enough, slow pace, sensitive to respondents); (iv) reading every question in order and reading the entire question before accepting responses, and (v) reading response scales (e.g., 5 or 7 points).

Once interviewers are trained and a sample has been selected, numbers can be called. Given that it may take several calls to reach people, it is important to maintain a call record so different interviewers know what occurred during previous calls and the best time to reach each individual. Making too many attempts in a brief time, however, can frustrate people rather than increase cooperation (Dillman et al. 2014). The length of the call period (e.g., 1 week, several months) impacts decisions about the number of attempts and how long to wait between attempts (Kalsbeek et al. 1994; Dillman et al. 2014). Outcomes of the previous attempts also influence the number of attempts because more attempts are usually made to numbers that were busy, no one answered, or a voicemail was encountered. The disposition of each call is important for calculating response rates.

Calling on weekday or weekend evenings during 5:00–9:00 p.m. and occasional afternoons on weekends are most productive (Salant and Dillman 1994; Dillman et al. 2014). Daytime hours on weekdays are less productive because many individuals will be working. Days and times of calls should vary because people have different schedules. One consideration, however, is that cellular numbers are mobile, and people can be in a different location and time zone than their number would suggest.

3.3 On-Site Surveys

On-site (or intercept) surveys are common in fisheries management. Systematic, cluster, and more complex multistage sampling designs can be effective when conducting on-site surveys because population lists of fishers may be unavailable. Pollock et al. 1994 noted that population lists of fishers are hardly ever available in advance, but for well-defined access-points, sampling frames can be based on site-days. Clusters of fishers within samples of site-days can then be subsampled using various methods such as on-site surveys. Although on-site questionnaires are typically shorter in length (e.g., one sheet of paper printed on both sides), the number of interviews completed per unit of time varies by the use level at the research site. In addition, the sampling frame is usually from the population of trips, not necessarily the population of unique fishers.

Given that interviewers typically administer questionnaires independently on-site, it is important that they be trained properly. Interviewers must also be prepared to address inquiries about the project or questionnaire. If the questionnaire is provided on a tablet or smartphone instead of paper, the interviewer must be able to assist respondents with these devices and troubleshoot hardware or software issues.

Before collecting data, public officials and resource managers should be informed of when and where questionnaires will be administered and the names and contact information of interviewers and supervisors. Interviewers should carry a formal letter from managers of agencies to demonstrate the study's legitimacy to respondents (Salant and Dillman 1994). Interviewers should also wear a name tag identifying their name and organization, and showing a photograph of the individual. Interviewers should carry clipboards and pens if the questionnaire is printed on paper, or electronic devices (e.g., tablets, smartphones) if it will be completed on these devices. Paper questionnaires, clipboards, and pens should still be available in case these electronic devices get damaged or people would prefer paper versions.

Fisheries management often includes creel surveys. Creel surveys offer valuable opportunities for on-site surveys. In addition to the biological data collected, human dimensions data can also be collected along with weather and other physical information. For many studies, such as fisher satisfaction, coupling the biological and social data can prove valuable. Moreover, creel surveys with on-site surveys can be linked to mail or other surveys as a follow-up for more detailed questions and items addressed to the entire season. Care must be made in planning this mixed-mode approach (more on this topic later in this chapter) so that creel survey data are linked to social data. This connection is easily accomplished by linking creel, on-site, and follow-up surveys through numerical codes.

3.4 Email and Internet Surveys

Electronic surveys using email contacts to obtain Internet responses are popular (Dillman et al. 2014; Vaske et al. 2023a). Their low cost and fast speed of data collection are attractive, but these surveys present challenges. Low response rates and issues associated with coverage are common, especially for public surveys (Stedman et al. 2019; Vaske et al. 2023a, b). Response rates for Internet surveys of the public are often in the single digits because the technology that makes it possible to achieve unprecedented speedy access also provides the means for easily avoiding or ignoring requests (Dillman et al. 2014). In addition, it is unlikely that everyone in the population of interest has a computer or smartphone with Internet access. Email addresses are also difficult to obtain or unavailable for most general population surveys, and there is often no straightforward process for drawing samples of emails where all members of the larger population have a known, nonzero chance of inclusion (Dillman et al. 2014). Many people, especially younger individuals, are also replacing email with other communication forms (e.g., texting, social media), which makes it difficult to reach them and request participation. Given these issues, researchers have been using Internet surveys in mixed-mode designs, such as contacting people by another mode (e.g., mail, telephone, text messages when this contact information is available) and asking them to complete the questionnaire on the Internet or in another format (Wallen et al. 2016). Although these approaches may increase access to potential respondents, they may result in nonprobability samples where the resulting data may be biased and not representative of the population or populations of interest, especially if the data are not weighted (discussed later in this chapter).

To minimize email requests landing in an individual's spam or junk folder, multiple personalized contacts with varied and different messages are essential for Internet surveys (Porter and Whitcomb 2003). Personalizing contacts (e.g., "Dear [First Name]") also increases responses and makes invitations less likely to be considered as spam or junk. Sending requests individually instead of in mass bulk emails (e.g., electronic mailing list posts, one email to multiple recipients at once) also helps. There are a number of other techniques that can be used for minimizing the chance that email notifications are screened as spam or junk (e.g., building your list of email addresses rather than using a third-party commercial list that is known to spam or junk filters; using online software to verify recipient email registration and include double opt-in tools; use methods to formally authenticate your email as a trusted source; regularly maintain and clean up your email list; use real-time email address validation and implement a sunset policy to remove confirmed unengaged subscribers; comply with Internet privacy laws; monitor email engagement metrics; utilize online spam checker tools; carefully proofread emails to avoid typographical and grammatical errors). However, institutional firewalls and software packages are always adapting, which increases the chance that email notifications are considered as spam or junk.

According to Dillman et al. (2014), a prenotification email is rarely used. Instead, the first email starts with a survey invitation, which is then followed by reminders. If each reminder generates substantially more completed questionnaires, additional reminders may be worthwhile. Conversely, if reminders yield few new responses, additional reminders may be unwise and irritate people. It is important that message content varies across emails because repeatedly emailing the same invitation is not effective and can get filtered as spam or junk. If emails are long, they often go unread. Emails should be short and professional (e.g., two to four short paragraphs to the point) expressing what the study is about, why they were selected, what is being asked of respondents, how to access the questionnaire (i.e., URL link), who to contact with questions, and appreciation for participating. The link to the questionnaire should be placed early in the email so recipients understand what to do next; distinguishing the link by using a box or other device makes it stand out from the email text. The questionnaire website can be designed and operated by researchers and their institutions or commercial packages that provide relatively inexpensive, prepackaged, and customizable instrument delivery services that do not require substantial technological or programming experience (e.g., Qualtrics, SurveyMonkey, Alchemer; Brandon et al. 2014). Grammar in emails should be checked carefully. Placing text in capital letters should be avoided because it gives the impression of yelling, links to "unsubscribe" should not be included because they convey the message is spam or junk, and acronyms, abbreviations, symbols, and special characters should be avoided. Emails should fit on most screens, so readers are not required to scroll far down to read the message. This is especially important for mobile devices such as smartphones and tablets.

Time intervals between contacts (e.g., emails) should be shorter than for other modes (e.g., mail), although timing will depend on the goals of the study and the population being surveyed (Kwak and Radler 2002). Contact intervals can be pre-set in many packages (e.g., Qualtrics). Pacing of contacts is quicker (e.g., three emails in two weeks, four to six emails in a month), but annoying rapid-fire reminders should be avoided (Dillman et al. 2014). Emails should be sent on weekdays in the morning before working hours because those receiving emails later in the day and on weekends are less likely to respond (Trouteaud 2004).

Each time a message is sent, a replacement website link for the questionnaire should be provided in the email. This approach is somewhat different than mail surveys, where reminder letters or postcards are sent without a replacement questionnaire. Each sample member should be assigned a unique ID access code (PIN) that is listed in every email. This number allows the researcher to delete respondents from the active email list and allows only reminder emails to nonrespondents. The code protects the integrity of the sample and ensures that only individuals in the sample can access and complete the questionnaire (Dillman et al. 2014). Surveys should be programmed so respondents who stop answering before completing the entire questionnaire and then return to it later are returned to where they stopped. Once a respondent submits the completed questionnaire, their code is deactivated, preventing multiple responses from the same individual or code (Dillman et al. 2014).

Researchers should be cautious when using advanced tools for designing and implementing these surveys because some techniques make it difficult to respond. For example, hardware and software can cause performance or incompatibility issues, or computer literacy may differ among respondents. Researchers should opt for simpler online questionnaires that can be accessed by respondents, especially given that many people now use devices with smaller screens (e.g., tablets, smartphones). Several books and articles provide example email letters and website layouts for these types of surveys (e.g., Dillman et al. 2014; Vaske 2019).

3.5 Internet Panels

One type of survey gaining popularity is *Internet panels* (Brandon et al. 2014, Vaske et al. 2023a). These usually involve self-selected individuals who willingly join panels, provide their contact information, and complete questionnaires on assorted topics several times per year (Dillman et al. 2014). Some members are recruited through prize drawings (e.g., opportunities to win gift cards) and are paid to complete questionnaires. Examples of these participant recruitment services include *Qualtrics Panel*, *SurveyMonkey Audience*, *Prolific,* and *Amazon Mechanical Turk*. Some panels contain thousands or millions of members, which is useful given that contact information (e.g., emails) for Internet surveys can be difficult to obtain or unavailable, especially for the public.

There are some challenges with online panels (Vaske 2019; Wardropper et al. 2021). First, most panels rely on nonprobability samples, which makes it difficult for the data to be representative of the population. Some studies, however, show minimal differences between responses from panel members and those from more traditional probability samples (e.g., Berinsky et al. 2012). A few panels have been created to serve as probability samples (e.g., *LISS Panel, Ipsos KnowledgePanel*). Some of the more common panels (e.g., *Prolific, Qualtrics Panel*) do permit sampling that matches population characteristics (e.g., some demographics, geographical areas), but sometimes there are not enough panel members who meet certain characteristics. Second, pricing is often based on questionnaire length and complexity, which can prohibit using some panels. Panels created via probability samples tend to be more expensive (e.g., USD $50 or more for each completed survey) than nonprobability panels (e.g., $5–10 each). Third, depending on the study location (e.g., rural vs. urban), there may not be enough panel members in the area to fill the desired quotas. This issue is especially problematic in rural areas where there are often not enough members of most panels. Fourth, like some other methods (e.g., telephone survey), panel members often have a low tolerance for difficult or lengthy questionnaires. Fifth, some people skip questions or always select the low or high ends of scales (i.e., straight lining) to speed up questionnaire completion (Kaminska et al. 2010).

Procedures are available for addressing these issues. For example, a soft launch should always be conducted and response times for the questionnaire and each question should be scrutinized. If, for example, an individual answers 30 questions in less than a minute or two, they likely did not carefully read each question. Some companies include a survey duration screener to exclude anybody who answers too quickly. In addition, attention filters and quality check questions should be embedded in the questionnaire. An example is "This is an attention filter. Please select "strongly disagree" with this statement." Anybody who answers differently than how they are directed should be screened out and not included in the data. Two or three of these filters should be included in a questionnaire.

The use of Internet panels is increasing given the potential cost savings and ability to generate numerous responses in a brief time (Dillman et al. 2014). Researchers must be careful in how they design and implement panel studies, exercise caution when reporting results from these panels, and not assume data are representative of a larger population (Brandon et al. 2014).

3.6 Mixed-Mode Surveys

Guidelines presented here for implementing surveys have focused on a single type of contact for each mode, such as letters for mail surveys, emails for Internet surveys, or calls for telephone surveys. These approaches can reduce responses because people increasingly delete emails, screen calls, or discard mail they do not recognize (Dillman et al. 2014). Mixed-mode surveys use various modes, making them less likely to be ignored, such as using random digit dialing on the telephone (to ask people their mail and email addresses) to create the sample frame and then using mail, email, or Internet for administering the questionnaire (e.g., Riepe et al. 2019). Complicating the benefits of mixed-mode surveys is the issue that many people now use a single device (e.g., smartphone, tablet) for answering the telephone, reading and responding to emails, and browsing the Internet (Dillman et al. 2014). Implementing these surveys can also be challenging because they require knowing how to implement various survey modes (e.g., mail, telephone, Internet). The details discussed above for mail, telephone, on-site, and Internet surveys apply to implementing mixed-mode surveys.

An important requirement for mixed-mode surveys is access to contact information for more than one survey mode and preferably as many modes as possible. If a general population sample is obtained from a private firm such as Dynata or Marketing Systems Group, multiple types of contact information may be available (e.g., postal addresses, telephone numbers, email addresses). This information can be used for contacting sample members via a mode (e.g., mail) that is different from the mode they can use for responding (e.g., Internet). Sending a letter in the mail, for example, can build trust and increase response rates for subsequent telephone and Internet surveys. Likewise, a telephone call reminder for a questionnaire that was mailed can encourage responses and convey legitimacy and importance.

When implementing mixed-mode surveys, avoid offering a choice of response modes simultaneously (i.e., concurrent design). Offering multiple response modes (e.g., mail, Internet) in the same contact at the same time and allowing people to choose a response mode sometimes decreases response rates compared with only one mode (Medway and Fulton 2012). Offering multiple modes can confuse people and consequently decrease responses (Dillman et al. 2014).

One of the most successful mixed-mode surveys involves a mail response option after an Internet option (i.e., web-first design; Dillman et al. 2014). A prenotification letter contains an Internet response option, followed a week or two later by a mail survey packet containing a paper questionnaire. In some cases, however, this web-first mixed-mode design can generate fewer responses compared to single-mode mail surveys (i.e., mail only; Dillman et al. 2014), but the mixed-mode approach provides more rapid responses because Internet surveys generate faster responses than mail surveys. This mixed mode also reduces printing and mailing costs because fewer questionnaire packets are needed. Survey costs in mixed-mode designs can be reduced by using more expensive modes later in the implementation process.

4 Response Rates

One reason for following these implementation strategies is to achieve the highest possible response rate. A *response rate* is the proportion of completed questionnaires to the total number of eligible respondents, and indicates success in gaining cooperation from people (Vaske 2019). A poor response rate increases the likelihood of introducing *nonresponse error*, which occurs when "a significant number of people in the survey sample do not respond to the questionnaire and are different from those who do in a way that is important to the study" (Salant and Dillman 1994, p. 20).

When a response rate for a sample from a population is low, the researcher may not know whether those who completed a questionnaire are like those who did not. Low response rates, therefore, weaken the representativeness of the sample to the population and can compromise inference (Medway and Fulton 2012; Keeter et al. 2017). On the other hand, a high response rate is only a good survey indicator if the sample was representative. A major mistaken belief by inexperienced survey researchers or those lacking formal training is that a high sample size is a good indicator of survey quality when it is not a good indicator of whether the sample is biased or not representative.

Disagreement abounds on what constitutes acceptable response rates. Two decades ago, for example, Babbie (2003) considered 50% adequate, 60% good, and 70% very good. Unfortunately, response rates have been strongly declining for all survey modes (e.g., Connelly et al. 2003; Stedman et al. 2019; Vaske et al. 2023a, b). Response rates for surveys of the public are generally low and now sometimes in the single digits, especially those conducted solely via telephone or the Internet (Dillman et al. 2014, Stedman et al. 2019, Vaske et al. 2023a). Issues such as saliency of the topic to respondents, number and personalization of contacts, time of survey administration, question complexity, modes used for administering the survey (e.g., mail, Internet, on-site, telephone), and questionnaire design all influence response rates (Dillman et al. 2014; Stedman et al. 2019). Readers should consult other sources for equations to calculate response rates for various data collection modes (e.g., Vaske 2019).

5 Nonresponse Bias Checks, Census Information, and Weighting

Individuals who completed a questionnaire may be different from those who did not, so respondents may not be representative of the population or sample of which they are a member. One way to address this issue is to conduct a nonresponse bias check, which involves contacting a sample of nonrespondents and asking them a short number of questions from the questionnaire. The number of questions will depend on the length of the original questionnaire. If a mail questionnaire were 12 pages, for example, it would be advisable to only ask a subset of important questions from each major section to minimize respondent burden. If a questionnaire was only one or two pages, it might be feasible to ask all questions in the nonresponse bias check.

It can be useful to use a different type of survey mode for nonresponse bias checks (Dillman et al. 2014; Yao and Xu 2021). One way to do this is to provide a different survey mode and appearance. After conducting multiple mailings for a mail survey, for example, it can be effective to conduct a nonresponse bias check by using telephone contacts of a sample of nonrespondents and asking them questions from the mail questionnaire (e.g., Manfredo et al. 2003; Needham et al. 2007). Another form of contact may not always be possible because full contact information may not be available for all members of the sample (Mitra and Lankford 1999). A telephone nonresponse bias check for a mail survey, for example, assumes that telephone numbers are available for all sample members. Alternating modes of surveys also requires having training in each survey mode.

According to Babbie (2003) and Dillman et al. (2014), a lack of response bias is more important than a high response rate. Nonresponse bias may exist even with a high response rate, such as 70%, whereas a low response rate such as 30% may be acceptable with no differences between respondents and nonrespondents (Krosnick 1999; Crompton and Tian-Cole 2001). Given that some studies of relatively homogeneous groups have found few differences among multiple contacts (e.g., first contact, postcard, final contact), some researchers have argued that going to the expense of obtaining high response rates and extensive contacts may not be necessary to avoid nonresponse error (e.g., Hammitt and McDonald 1982; Dolson and Machlis 1991). Other studies of heterogeneous samples (e.g., public), however, have found substantial differences between respondents and nonrespondents (e.g., Woodside and Ronkainen 1984; Manfredo et al. 2003). A nonresponse bias check is important and may be more important than high response rates for allowing confidence in the representativeness of results (Crompton and Tian-Cole 2001).

Researchers should also compare characteristics of respondents to recent census information if it is available, especially: (i) for general population surveys, (ii) when there is also nonresponse bias within a sample of original nonrespondents (e.g., people who did not respond to the nonresponse bias questionnaire), or (iii) if a nonresponse bias check is not conducted (Groves 2006; Groves and Peytcheva 2008). If a nonresponse bias check or census information differs from respondents, the data may need to be statistically weighted for sample results to be representative of the larger population from which the sample was drawn. Assume, for example, that the response rate for a random mail survey of residents in an area was only 10% and the data from this sample showed that 30% of respondents had not earned a college degree and 70% had earned a college degree. However, both census information for the population of this area and a follow-up telephone nonresponse bias check with a random sample of those who did not respond to the initial mail survey showed that 60% had not earned a college degree and 40% had earned a college degree. Although the original sample was selected at random, this suggests that responses to the mail survey may not be truly representative of the population. Given the low response rate and this substantial disparity in results between the mail survey and both census information and the telephone nonresponse bias check, it may be prudent to weight the sample data from the mail survey, as follows:

$$\text{Weight} = \frac{\text{Population\%}}{\text{Sample\%}} \tag{3}$$

Therefore, Weight (no college degree) = 0.60/0.30 = 2.0
Weight (college degree) = 0.40/0.70 = 0.57.

Multiple characteristics known from the sample population can be used to calculate more complex weights (Vaske 2019). For example, it is often most useful for weighting to be based on variables and questions that are of most interest and importance to the study (e.g., attitudes, motivations, behaviour) rather than just demographic characteristics. These weights are then inputted at the individual level (e.g., weighing an individual up or down) into statistical software for analysing the data and are activated when running analyses. Length constraints of this chapter prohibit a more detailed discussion of weighting, but several books and articles provide more examples and information on the calculations and steps required for weighting data (e.g., Vaske 2019; Vaske et al. 2011).

Acknowledgments All the authors are based in the United States of America. Some examples used in this chapter should be viewed from this country's context. For instance, when discussing mail surveys, reference is occasionally made to the United States Postal Service. Postal services in other countries may not function in an analogous manner.

References

Armstrong JS, Lusk EJ (1987) Return postage in mail surveys: a meta-analysis. Public Opin Q 51: 233–248

Babbie E (2003) The practice of social research with infotrac, 10th edn. Wadsworth Publishing Co, Belmont

Berinsky AJ, Huber GA, Lenz GS (2012) Evaluating online labor markets for experimental research: Amazoncom's mechanical Turk. Polit Anal 20:351–368

Brandon DM, Long JH, Loraas TM, Mueller-Phillips J, Vansant B (2014) Online instrument delivery and participant recruitment services: emerging opportunities for behavioral accounting research. Behav Res Account 26:1–23

Cochran WG (1977) Sampling techniques, 3rd edn. Wiley, New York NY

Connelly NA, Brown TL, Decker DJ (2003) Factors affecting response rates to natural resource-focused surveys: empirical evidence of declining rates over time. Soc Nat Resour 16:541–547

Crompton JL, Tian-Cole S (2001) An analysis of 13 tourism surveys: are three waves of data collection necessary? J Travel Res 39:356–368

Dillman DA (2007) Mail and internet surveys: the tailored design method, 2nd edn. Wiley, Hoboken

Dillman DA, Smyth JD, Christian LM (2014) Internet phone mail and mixed-mode surveys: the tailored design method. Wiley, Hoboken

Dolson DE, Machlis GE (1991) Response rates and mail recreation survey results: how much is enough? J Leis Res 23(3):272–277

Fowler FJ Jr (1993) Survey research methods, 2nd edn. Sage, Newbury Park

Groves RM (2006) Nonresponse rates and nonresponse bias in household surveys. Public Opin Q 70(5):646–675

Groves RM, McGonagle KA (2001) A theory-guided interviewer training protocol regarding survey participation. J Off Stat 17(2):249–265

Groves RM, Peytcheva E (2008) The impact of nonresponse rates on nonresponse bias: a meta-analysis. Public Opin Q 72(2):167–189

Gundelund C, Venturelli P, Hartill B, Hyder K, Olesen H, Skov C (2021) Evaluation of a citizen science platform for collecting fisheries data from coastal sea trout anglers. Can J Fish Aquat Sci 78:1576–1585

Hammarstedt JE, Redmond JM, Gupta A, Dunne KF, Vemula SP, Domb BG (2017) Survey mode influence on patient-reported outcome scores in orthopedic surgery: telephone results may be positively biased. Knee Surg Sports Traumatol Arthrosc 25(1):50–54

Hammitt WE, McDonald CD (1982) Response bias and the need for extensive mail questionnaire follow-ups among selected recreation samples. J Leis Res 14(3):207–216
Heberlein TA, Baumgartner R (1981) Is a questionnaire necessary in a second mailing? Public Opin Q 45(1):102–108
Jessen RJ (1978) Statistical survey techniques. Wiley, Hoboken
Kalsbeek WD, Botman SL, Massey JT, Liu PW (1994) Cost-efficiency and the number of allowable call attempts in the National Health Interview Survey. J Off Stat 10(2):133–152
Kaminska O, McCutcheon AL, Billiet J (2010) Satisficing among reluctant respondents in a cross-national context. Public Opin Q 74:956–984
Keeter S, Hatley N, Kennedy C, Lau A (2017) What low response rates mean for telephone surveys. Pew Research Center, Washington DC
Kelly EJ, Cumberland WG (1990) Prediction theory approach to multistage sampling when cluster sizes are unknown. J Off Stat 6(4):437–449
Kish L (1949) A procedure for objective respondent selection within the household. J Am Stat Assoc 44:380–387
Kish L (1965) Survey sampling. Wiley, New York
Krosnick JA (1999) Survey research. Annu Rev Psychol 50:537–567
Kwak N, Radler B (2002) A comparison between mail and web surveys: response pattern respondent profile and data quality. J Off Stat 18(2):257–273
Kyle G, Landon A, Pilgreen D, Golebie E (2026) Measurement instruments. In: Pope KL, Arlinghaus, R, Hunt LM, Lynch AJ, and van Poorten BT (eds) Understanding recreational fishers: disciplinary and interdisciplinary approaches for fisheries management. Springer Nature, Cham, p 355–385
Lohr SL (1999) Sampling: design and analysis. Duxbury Press, Pacific Grove
Manfredo MJ, Teel TL, Bright AD (2003) Why are public values toward wildlife changing? Hum Dimens Wildl 8:287–306
Medway RL, Fulton J (2012) When more gets you less: a meta-analysis of the effect of concurrent web options on mail survey response rates. Public Opin Q 76(4):733–746
Messer BL, Dillman DA (2011) Surveying the general public over the internet using address-based sampling and mail contact procedures. Public Opin Q 75:429–457
Mitra A, Lankford S (1999) Research methods in park recreation and leisure services. Sagamore, Champaign
Morgan GA, Harmon RJ (1999) Sampling and external validity. J Am Acad Child Psy 38(8): 1051–1053
Needham MD, Vaske JJ, Donnelly MP, Manfredo MJ (2007) Hunting specialization and its relationship to participation in response to chronic wasting disease. J Leisure Res 39(3):413–437
Oksenberg L, Coleman L, Cannell CF (1986) Interviewers' voices and refusal rates in telephone surveys. Public Opin Q 50(1):97–111
Peytcheva E, Groves RM (2009) Using variation in response rates of demographic subgroups as evidence of nonresponse bias in survey estimates. J Off Stat 25(2):193–201
Pollock KH, Jones CM, Brown TL (1994) Angler survey methods and their applications in fisheries management. American Fisheries Society. Bethesda MD. ISBN 0-913235-88-1
Porter SR, Whitcomb ME (2003) The impact of contact type on web survey response rates. Public Opin Q 67:579–588
Riepe C, Meyerhoff J, Fujitani M, Aas Ø, Radinger J, Kochalski S, Arlinghaus R (2019) Managing river fish biodiversity generates substantial economic benefits in four European countries. Environ Manag 63(6):759–776
Rizzo L, Brick JM, Park I (2004) A minimally intrusive method for sampling persons in random digit dial surveys. Public Opin Q 68:267–274
Rocco E (2003) Constrained inverse adaptive cluster sampling. J Off Stat 19(1):45–57
Salant P, Dillman DA (1994) How to conduct your own survey. Wiley, New York
Scheaffer RL, Mendenhall W, Ott L (1996) Elementary survey sampling. Duxbury Press, Belmont
Schouten B, van den Brakel J, Buelens B, van der Laan J, Klausch T (2013) Disentangling mode-specific selection and measurement bias in social surveys. Soc Sci Res 42(6):1555–1570

Stedman RC, Connelly NA, Heberlein TA, Decker DJ, Allred SB (2019) The end of the (research) world as we know it? Understanding and coping with declining response rates to mail surveys. Soc Nat Resour 32:1139–1154
Thompson SK (1992) Sampling. Wiley, New York
Tourangeau R (2004) Survey research and societal change. Annu Rev Psychol 55:775–801
Trouteaud AR (2004) How you ask counts: a test of internet-related components of response rates to a web-based survey. Soc Sci Comput Rev 22:385–392
Vaske JJ (2019) Survey research and analysis, 2nd edn. Sagamore-Venture Publishing LLC, Urbana
Vaske JJ, Jacobs MH, Sijtsma MTJ, Beaman J (2011) Can weighting compensate for sampling issues in online surveys? Hum Dimens Wildl 16(3):200–215
Vaske JJ, Don Carlos AW, Manfredo MJ, Teel TW (2023a) Evaluating alternative survey methodologies in human dimensions of wildlife research. Hum Dimens Wildl 28(4):320–334
Vaske JJ, Miller CA, Williams BD (2023b) Declining hunter and trapper mail survey response rates between 1976 and 2020 in Illinois. Unpublished manuscript. Colorado State University, Fort Collins
Vaske JJ, Needham MD, Miller CM, Sponarski C (2026) Guidelines for writing and constructing recreational fisher surveys. In: Pope KL, Arlinghaus R, Hunt LM, Lynch AJ, van Poorten BT (eds) Understanding recreational fishers: disciplinary and interdisciplinary approaches for fisheries management. Springer Nature, Cham, p 287–325
Venturelli P, Hyder K, Skov C (2017) Angler apps as a source of recreational fisheries data: opportunities, challenges and proposed standards. Fish Fish 18:578–595
Venturelli P, Skov C, Audzijonyte A, Belmaker J, Ana Cardoso AS, Gordon L, Gundelund C, Hyder K, Jarić I, Lennox RJ, Potts WM, Vacura K, Vaz AS, Sbragaglia V (2026) Digital data mining. In: Pope KL, Arlinghaus R, Hunt LM, Lynch AJ, and van Poorten BT (eds) Understanding recreational fishers: disciplinary and interdisciplinary approaches for fisheries management. Springer Nature, Cham, p 443–479
Waksberg J (1978) Sampling methods for random digit dialing. J Am Stat Assoc 73:40–46
Wallen KE, Landon AC, Kyle GT, Schuett MA, Leitz J, Kurzawski K (2016) Mode effect and response rate issues in mixed-mode survey research: implications for recreational fisheries management. N Am J Fish Manag 36(4):852–863
Wardropper CB, Dayer AA, Goebel MS, Martin VY (2021) Conducting conservation social science surveys online. Conserv Biol 35(5):1650–1658
Weeks MF, Jones BL, Folsom RE, Benrud CH (1980) Optimal times to contact sample households. Public Opin Q 44(1):101–114
Woodside AG, Ronkainen IE (1984) How serious is non-response bias in advertising conversion research? J Leis Res 22(4):34–37
Yao M, Xu Y (2021) Method bias mechanisms and procedural remedies. Sociol Methods Res 53(1): 235–278. https://doi.org/10.1177/00491241211043141
Yeager DS, Krosnick JA, Chang L, Javitz HS, Levendusky MS, Simpser A, Wang R (2011) Comparing the accuracy of RDD telephone surveys and internet surveys conducted with probability and non-probability samples. Public Opin Q 75(4):709–747

Measurement Instruments

Gerard Kyle, Adam C. Landon, Daniel Pilgreen, and Elizabeth Golebie

Abstract The development of measures designed to reveal heterogeneity among anglers and predict their behaviour has been central to scholarship within the human dimensions of recreational anglers for over 40 years. Measures of these constructs — such as values, attitudes, preferences, and norms–are developed using instrumentation that permits their inference based on subjects' responses to statements that are manifestations of their definition. The development of tools for establishing the validity and reliability of instrumentation has been a century-long pursuit for researchers in the social sciences. Two fundamental questions lie at the heart of the challenge: (1) how can we be sure we a measuring what we claim to measure (validity)? and (2) how can we establish consistency in measurement (reliability)? With the increasing incidence of their application by both scholars with and without training in psychometric measurement, it is worthwhile to review the methods quantitative social scientists use to evaluate measurement instruments, and to help ensure that these normative standards are upheld. We provide a step-by-step guide for the protocols and metrics used to evaluate the psychometric adequacy of latent

G. Kyle (✉)
Department of Rangeland, Wildlife and Fisheries Management, Texas A&M University, College Station, TX, USA
e-mail: gkyle@tamu.edu

A. C. Landon
Minnesota Department of Natural Resources, St. Paul, MN, USA

Department of Fisheries, Wildlife and Conservation Biology, University of Minnesota, St. Paul, MN, USA
e-mail: adam.landon@state.mn.us

D. Pilgreen
Department of Recreartion, Park and Tourism Sciences, Texas A&M University, College Station, TX, USA
e-mail: dpilgreen@tamu.edu

E. Golebie
Department of Natural Resources and Environmental Sciences, University of Illinois at Urbana-Champaign, Urbana, IL, USA
e-mail: golebie2@illinois.edu

K. L. Pope et al. (eds.), *Understanding Recreational Fishers*, Fish & Fisheries Series 45, https://doi.org/10.1007/978-3-031-99739-6_12

variable instrumentation. Ultimately, the ability to correctly identify significant relationships among these constructs and their influence on human action is directly tied to our ability to adequately measure the constructs. Studies employing flawed instrumentation will lead to erroneous conclusions and likely undermine the authors' goal—an understanding of recreational anglers' heterogeneity and behaviour that can inform the sustainable management of the resource.

Keywords instrument development · Latent variables · Measurement · Psychometric assessment

1 Introduction

The human dimensions of fisheries often rely on measurement instruments (e.g., questionnaires) to collect data directly from fishers regarding their characteristics, demographics, fishing experience, knowledge, beliefs, attitudes, and behaviour. This information supports fisheries managers in understanding and forming positive relationships with stakeholders and informing management decisions with an awareness of stakeholder preferences and needs. Interest in collecting survey data from fishers is growing among managers and implementing surveys is seemingly easier than ever, given the widespread accessibility of the internet and advances in technology. However, collecting survey data that is reliable and valid is not as simple as drafting questions as they come to mind and uploading them to a survey platform. In this chapter, we provide an overview of the rich literature on scale development for a class of variables that is growing increasingly important for understanding recreational anglers. For our work, "instrument" refers to the measurement tool (i.e., questionnaire) and "scale" refers to a subset of questions or measures imbedded within the questionnaire.

Many psychometrically tested scales have been developed through past research (e.g., see https://doi.org/10.18738/T8/LFWX64). Thus, regardless of whether the investigator is adapting an existing scale or considering the development of new measures, questionnaire design should begin with a review of past work to identify scales that may be implemented or adapted in a new study. Drawing on past work is essential for several reasons. First, using established scales increases the likelihood that the scales will provide a reliable and valid measure. Reliability and validity (elaborated later in this chapter) require data to assess. Thus, it can be risky to implement new measures because if they are shown to be psychometrically questionable, the data may be unusable. Established scales have been tested and refined throughout many studies and are more likely to produce data that accurately and robustly measure the concept of interest. Second, using scales from the literature allows the researcher to draw direct comparisons with past research and contribute to that body of literature. For instance, a researcher who wants to study fisher satisfaction would typically start by reviewing past work. Utilizing a scale from that body of work will allow the researcher to comment on, for instance, whether satisfaction was higher or lower than in other studies, whether it predicts behaviour positively or negatively, and how satisfaction relates to other variables. Replication is also the means through which science progresses. Continued interrogation of the same phenomena in new contexts probes the boundary conditions where theory holds, where it does not, and feeds the flame of innovation. New explanations of observed patterns of cognition and behaviour arise when old ones fail. This systematic progress is only possible when the measurement is valid, replicable, and consistent. Otherwise, every study is merely an idiosyncratic description of a population. If the researcher simply constructed their own questions to measure satisfaction, they would not be certain that they were actually measuring the concept and would, thus, not be able to draw conclusions beyond their own study.

Although researchers typically draw from past work when developing scales, the items often need to be adapted for the study site or context. Thus, understanding how to design psychometrically sound scales is an important skill for many researchers working within the human dimensions of recreational fisheries. A psychometrically sound scale measures what it intends to measure, consistently over time, in a specific setting and population. This chapter sets out a stepwise approach for the development and psychometric testing of scales designed to measure abstract social psychological constructs relevant for research in the human dimensions of recreational fisheries. Much of the focus of this chapter is structured around the design and adaptation of existing scales to fit the context and needs of an investigation. However, many of the guidelines we present are relevant to the development of new scales designed to measure phenomena not previously explored. For comprehensive reviews on the design of survey instruments more broadly, we direct readers to the work of Dillman et al. (2014) and Vaske (2019). Numerous concepts are of interest to human dimensions researchers and studies using instrumentation to measure them are becoming common in the literature. Their popularity is attributable to the insight they provide on fisher behaviour, heterogeneity, and management preferences.

Regarding the compendium of measures that we have assembled (https://doi.org/10.18738/T8/LFWX64), these are examples of measures of latent variables that have been shown to influence behaviour in an array of contexts. Each construct and associated measure's utility for understanding behaviour will vary depending on the question being asked and the cultural context. Decisions on which measure is most appropriate rest on the researcher's understanding of the literature and their capacity to adequately analyse the data.

2 Measures of Latent Variables in Recreational Fisheries Research

In social psychology, a construct is considered latent when it is hidden in the mind of the respondent. For example, an individual's attitude toward some object (e.g., a fisheries-management tool or the fisheries manager) is not something that is immediately apparent to an observer. It can be inferred through a variety of ways, but it is not something that has a defined shape, weight, or area. Latent constructs commonly used in recreational fisheries research include concepts such as values, attitudes, satisfaction, motivation, norms, specialization, psychological involvement, and so on. Alternately, observed variables can be directly reported or measured. Examples include socio-demographic characteristics such as age, education, race, household income, and behavioural indicators such as frequency of participation, the number of fish caught, and species caught. The accurate and reliable measurement of observed variables is relatively straightforward, in contrast to the complex process of measuring latent constructs. The researcher cannot peer directly into the mind of the respondent to determine what they think about the issue being addressed in the instrument. Consequently, single indicators are often imperfect measures of complex latent phenomena. Researchers typically do not measure latent constructs directly but are able to infer their existence using a series of questions designed to measure the construct. Using multiple questions to measure a construct enables the researcher to assess the adequacy of the measures by employing several statistical tools (elaborated later in the chapter).

The series of questions to measure latent variables are known as observed measures or "items." Researchers present respondents with a question or statement and request an accompanying response. The response format or scale (e.g., strongly disagree to strongly agree) quantifies the respondent's interpretation of the item. An example is presented (Table 1). The six items displayed measure two dimensions of individuals' involvement with fishing (see Kyle et al. 2007); (i) respondents' *attraction* to fishing and (ii) the *centrality* of fishing within the context of the respondent's life. When constructing a questionnaire, the researcher would not display construct

Table 1 Multi-item scales

Below is a series of statements that capture how you might feel about fishing. Please indicate the extent to which you agree with each state by checking the appropriate circle.	Strongly disagree	Disagree	Neither agree nor disagree	agree	Strongly agree
Attraction					
Angling is one of the most enjoyable things I do	O	O	O	O	O
Angling is very important to me	O	O	O	O	O
Angling is one of the most satisfying things I do	O	O	O	O	O
Centrality					
I find a lot of my life is organized around angling	O	O	O	O	O
Angling occupies a central role in my life	O	O	O	O	O
To change my preference from angling to another recreation activity would require major rethinking	O	O	O	O	O

labels (i.e., *attraction* and *centrality*) and would randomize items to avoid the possibility of respondents mindlessly checking circles without fully processing the information presented in each statement. Most electronic data collection platforms have numerous options to randomize the order of questions presented to respondents.

Much has been written about item construction and how people respond to survey items (for reviews, see Krosnick 2018 and Vaske 2019). Well-constructed items enable respondents to (i) understand the intent of the item, (ii) recall knowledge from memory, (iii) formulate a summary judgment, (iv) translate their judgment to the provided response options, and (v) minimize cognitive burden. When these conditions are met, it is more likely that responses provided reflect the latent construct of interest.

3 Reflective Versus Formative Measurement Models

Because many aspects of the social-psychological dimensions of fishing are latent, particularly those related to individuals' thoughts and feelings about fishing and the contexts in which it is experienced, efforts to measure these phenomena have relied on multi-indicator scales that have been shown to be valid and reliable. When developing or evaluating scales to measure latent variables, the researcher must consider the relationship each indicator shares with the latent variable. This part of the model, when nested with a broader set of path or structural equation models, is referred to as the measurement model. Two broad possibilities for conceptualizing latent variable measurement models are depicted (Fig. 1). Following a tradition in

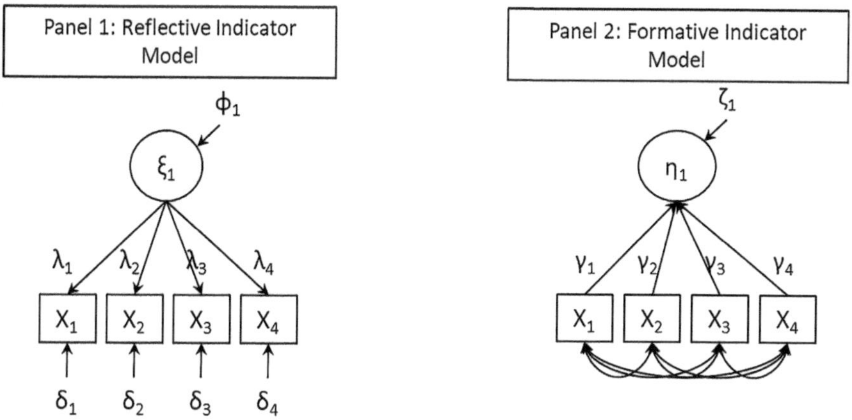

$$X_1 = \lambda_1\xi + \delta_1$$
$$X_2 = \lambda_2\xi + \delta_2$$
$$X_3 = \lambda_3\xi + \delta_3$$
$$X_4 = \lambda_4\xi + \delta_4$$

$$\xi = \gamma_1 X_1 + \gamma_2 X_2 + \gamma_3 X_3 + \gamma_4 X_4 + \zeta$$

ig. 1 Reflective and formative measurement models

quantitative social psychology and in accordance with classical test theory (Lord et al. 1968), human dimensions researchers have almost exclusively conceptualized their latent variables to be reflective (see Panel 1 in Fig. 1). In so doing, several key assumptions are made. First, as noted, the direction of causality trails from the latent construct to the measured indicator. As such, variation in the latent variable precedes variation in the indicators (see equations in Panel 1, Fig. 1). Consequently, the indicators ought to share a common theme and are interchangeable. This interchangeability, theoretically, enables researchers to measure a specific construct by sampling a few relevant indicators from the domain underlying the construct (Churchill 1979; Nunnally and Bernstein 1994). Inclusion or exclusion of one or more indicators from the domain does not materially alter the content or validity of the construct (Bollen 1989). Because the indicators share a common theme, they are also assumed to be strongly and positively correlated (Brown 2015) and have the same antecedents and consequences (Cenfetelli and Bassellier 2009).

Alternately, for formative measurement models where the indicators define the construct, directionality flows from the measured items to the latent construct (Bollen and Lennox 1991, see Panel 2, Fig. 1). Unlike the reflective model, this model does not assume that the measures are caused by a single underlying construct. Rather, it assumes that the measures have an impact on the construct. As such, no assumptions are made about inter-item collinearity. In fact, Jarvis et al. (2003, p. 202) have suggested that it would be "entirely consistent for formative indicators to be completely uncorrelated." Because the items are not necessarily correlated, measures of internal consistency reliability are not appropriate for formative models (Bollen and Lennox 1991). Also, the consequences of dropping one of the indicators are potentially serious, with the potential to omit an important element of the composite latent construct and change the meaning of the variable. Thus, item selection should be guided by decisions that attempt to capture the domain of interest (Coltman et al. 2008). Finally, although formative indicators are not necessarily correlated, they do have the same antecedents and consequences (Cenfetelli and Bassellier 2009).

Though formative measurement models have yet to be widely used in human dimensions research, Kyle and Jun (2015) suggested that the leisure constraints measurement model likely follows formative structure and provided an empirical examination. Constraints to fishing has long been a topic in the recreational fisheries literature (Aas 1995; Fedler and Ditton 2001; Schroeder et al. 2008; Sutton 2007; Sutton et al. 2009; TenHarmsel et al. 2021). Kyle and Jun suggest other constructs relevant to the human dimensions of recreation fishing might also be formative, e.g., satisfaction, fishing quality, agency trust (Jarvis et al. 2003). Rather than factor analysing multi-item scales that follow this form, researchers have tended to construct indices that provide respondents a "score" related to the phenomenon of interest. In these instances, principal components analysis (PCA) is often used as the data reduction tool. It is important to note that the latent constructs we have been discussing are reflective and emerge from the common factors model (Thurstone 1947). For reflective latent constructs, factor analysis (discussed later in this chapter) is most often used to (i) identify *common* variance accounted for by the laten construct, and (ii) *unique* variance comprised of both reliable variance specific to the indicator or item and random error variance. Alternately, PCA aims to account for the variance in the observed measurements rather than explain the correlations among them (Brown 2015).

The remainder of this chapter will focus on the development and analyses of latent constructs comprised of reflective indicators. As noted, most human dimensions research has treated the measurement of latent variables as reflective. Item development has followed the definitions of the latent constructs. As such, variance shared among items can be attributed to the latent construct. In the example we presented above for fisher involvement, the items measuring attraction and centrality were developed with the definitions of attraction and centrality at the fore. Although the use of formative measures has been a controversial issue within the structural equation modelling and measurement community (see Bollen and Diamantopoulos 2017 for review and defence), we do see some utility despite the paucity of research appearing in the human dimensions literature. An example was presented by Kyle and Jun (2015) within the context of leisure constraints; a construct human with which dimensions researchers have had a long-standing interest.

4 Scale Development—Key Concepts

4.1 Operationalization

Scale development requires selecting a response scale, defined as the continuum of categories from which responses to each indicator are selected. Three types of response scales are commonly used in human dimensions research: (i) Likert-type scaling, (ii) semantic differential, and (iii) visual analogue scales. Response scales affect the way respondents understand the questionnaire and have implications for data analysis and, consequently, researchers must carefully consider the response scale selected. Though they produce ordinal level data, treating the indicators as continuous in subsequent analyses (e.g., factor analyses) is common and supported by the literature under certain conditions (Carifio and Perla 2008; Robitzsch 2020).

First, a Likert-type response scale requires respondents to evaluate the indicator on a bipolar spectrum (e.g., strongly disagree to strongly agree). An example of a type of Likert-type response scale is presented (Table 1) where respondents express their level of agreement with each statement by checking a circle (or box, or number, depending on the platform, i.e., web-based vs. pencil and paper) best matching their sentiment.

Semantic differential scales, on the other hand, involve the construction of a series of opposing adjective pairs. One of the most popular scales for measuring emotions toward wildlife has been Mehrabian and Russell's (1974) pleasure, arousal, dominance (PAD) model. Respondents select a point between each adjective pair to indicate the strength with which they associate the adjective and attitudinal object (Coolican 2014). By including >5 bipolar items, these scales can produce interval data (McCroskey 2006) because the same binary responses are used, removing ambiguity or item-specific differences in the interpretation of relative scale positions. Nonetheless, the potential for position response bias may be greater than for the Likert-type scale because of the lack of descriptors at points within the response scale, and the ambiguity of the central position remains challenging to interpret (Coolican 2014).

Visual analogue scales (e.g., slider scales on electronic platforms) can be a convenient display format that can overcome the limitations of predetermined categories by providing a line to represent bipolar continuums (e.g., high to low, good to bad). These are particularly useful when dealing with abstract concepts such as emotions or moods (Coolican 2014; Jacobs et al. 2014); the increased scale sensitivity achieved by being able to measure precise distance indicated on the continuum (as opposed to a priori demarcated points) may be important for constructs that are not dichotomous, e.g., stakeholder perceptions of wildlife rights (Whitehouse-Tedd et al. 2020). Moreover, respondents define their unique position on the continuum and therefore generate interval data. Betella and Verschure (2016) provide an illustrative example of using emoticons with a digital slider such that the expression of the emoticon alters as the respondent shifts the slider across from left to right (see Fig. 2).

Decisions based on the type of response scale to utilize are typically based on prior research and the nature of the construct being measured. Even in the context of new scale development, prior research can inform given that construct classes (e.g., attitudes, beliefs, evaluations, management preferences, emotions, etc.) often employ similar response scales. For Likert-type questions, the type of response anchor is guided by the construct and question stem. For scales measuring latent constructs, question stems often present statements that require respondents to express their level of agreement (e.g., strongly agree through strongly disagree).

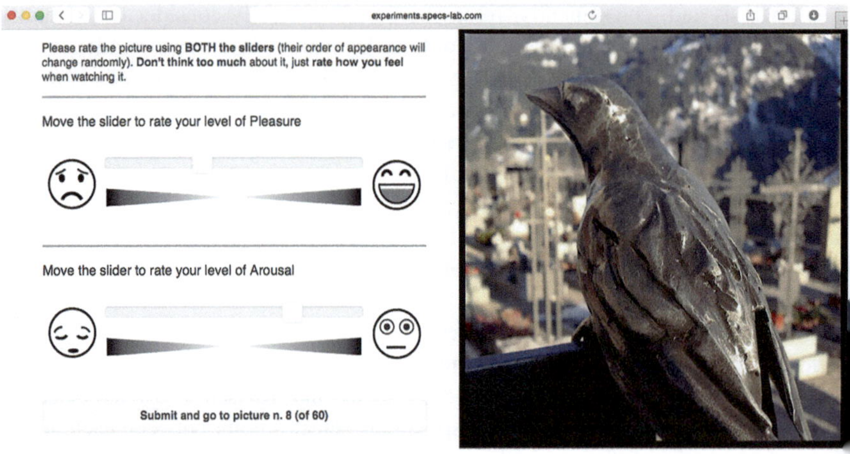

Fig. 2 Betella and Verschure's (2016) affective slider

For management-focused constructs, there are many more options that can have polarity or express levels of the particular phenomena of interest. For example:

(a) Familiarity with an issue of interest—Not at all familiar through very familiar.
(b) Behavioural frequency—Not at all through all the time, never through always.
(c) Behavioural intent—Very unlikely though very likely.
(d) Perceived effectiveness—Not at all effective through very effective.
(e) Perceived importance—Not at all important through extremely important.
(f) Policy and management support—Strongly oppose through strongly support.

For semantic differential scales, the choice of paired adjectives is also driven by the construct of interest. Of principal concern is the need to cover the conceptual breadth of the construct.

Other considerations for response format include:

(a) The number of response options—For Likert-type scales, five and seven categories are most used (Malhotra and Peterson 2006). Increasing categories can reduce skewness and kurtosis, but also increase the cognitive load on respondents as they attempt to discriminate between response options (DeCastellarnau 2018).
(b) Should responses be numerically labelled, e.g., 1 = strongly agree through 5 = strongly disagree? Should response categories each receive a descriptor? For a five-point Likert-type scale, it is common for each response category labelled, e.g., strongly agree, agree, neither agree nor disagree, disagree, and agree. Should numeric values be used in conjunction with descriptors? The literature is not definitive beyond suggesting not to make the instrument visually challenging by including too much information.
(c) The inclusion of a mid-point, e.g., "neither agree nor disagree", "no opinion" or "I don't know"? Researchers need to be careful forcing respondents to take a side, e.g., agree or disagree. Respondents not having an opinion or neither agreeing nor disagreeing with the statement presented are potentially genuine responses. They are distinct from "I don't know". Their indecisiveness may be contextual. In some contexts, the statement might read true to them and in other contexts not. It is not that they are uninformed. "I don't know" options are appropriate in a context where the researcher might be requesting the respondent to express an opinion about a policy recommendation. If they are uninformed about the policy and its implications, it is a genuine response category. Note, however, that it is not treated as a mid-point along a continuum. In subsequent analyses, it could be treated as a missing value for the respondent. For additional guidance on when not to use a mid-point, see Chyung et al. (2017).

4.2 Item Wording

Though researchers designing survey instruments with the intent to utilize latent variables ought to begin with a thorough review of the extant literature, there are occasions when societal change exposes voids within the existing literature. These voids sometimes expose the inadequacy of existing measures to capture the phenomena of interest. Shifting values, climates, and pandemics are altering the way humans relate to nature, giving rise to the possibility of new social-psychological phenomena. These contexts may necessitate the development of new scales, items, or the modification of existing scales. For the construction of new scales and the development of item wording, some basic tenets need to be observed (DeVillis 2017; Dillman et al. 2014; Vaske 2019; Schaeffer and Dykema 2011):

(a) Beyond an extensive review of the extant literature, past qualitative work can be used to identify salient constructs around which items can be developed (e.g., grounded theory; Glaser and Strauss 1967).
(b) Items should be written in clear, unambiguous language.
(c) Items are most commonly positively worded (positive valence) for optimal comprehension by the respondent.
(d) If there is concern that respondents may not be adequately considering each item response, consideration for the use of items with opposite polarity to the scale may be warranted. There is evidence, however, that items with opposing valences often perform poorly in subsequent analyses (i.e., reliability and validity assessments; DeVillis 2017).
(e) Item wording should be neutral and not lead the respondent to answer in a particular manner.
(f) Items should only reference a single construct or issue to ensure clarity (Marsden and Wright 2010).

4.3 *Multiple Indicators*

As noted, because latent constructs are complex and challenging to measure, researchers utilize multiple items to capture the breadth of their meaning. But how many items provide an adequate assessment? When determining the number of items to be included within an item pool, there is a tension between striving for parsimony and developing enough items to cover the conceptual breadth of the construct. Items will or should have a degree of shared variance, given that they will have been constructed around a narrow definition of the concept of interest. However, at what point does redundancy emerge from the inclusion of too many items? If adapting an existing scale where the dimensionality of the construct is known, a priori, for statistical reasons, we outline below (i.e., model identification; Brown 2015) that three to five items per construct or dimension is sufficient. If the researcher is developing their own scale, it would be prudent to develop additional items (five to seven items per construct) in anticipation that not all items will perform adequately and some items will eventually be omitted throughout the scale purification process. Researchers should be mindful that when collecting data and assessing multiple latent constructs that will be tested within full path or structural equation models, questionnaire length and respondent burden can become a concern.

Lengthy surveys contribute to both respondent attrition and measurement error (Dillman et al. 2014).

4.4 *Pretesting and Pilot Testing*

Once a scale has been developed to measure a construct, it is important to move through several stages of testing prior to collecting data. First, researchers solicit a review of the survey from experts, which may include other professionals or members of the sample population. This review may identify points of confusion or clarification within the instrument that should be addressed. Second, researchers may conduct interviews with several members of the sample population. During the interviews, the researcher may ask participants to "think out loud" while reading through the measurement instrument and selecting answers. This process enables the researcher to understand how each question is being perceived and where revisions might be needed. Finally, researchers conduct a pilot test, in which they disseminate to a small percentage of the sample population. The pilot test generates preliminary data, which can be used to conduct preliminary tests of the reliability and validity of the scale and allow the researcher an opportunity to make any changes to the scale. After pilot and pre-testing have provided evidence of the scales' psychometric adequacy, the researcher proceeds with formal data collection (see Vaske et al. 2026 for guidance on selecting a sample population and survey method).

5 Assessing the Psychometric Properties of Latent Variable Measures

Once formal data collection is complete, it is necessary to evaluate the psychometric properties of each scale. That is, the researcher must test whether all items do indeed measure the construct of interest and whether items need to be removed, in a process known as factor analysis. Employing a reflective measurement model, factor analysis assumes that the observed measures (i.e., indicators) are manifestations of a theoretically defined latent variable. Factor analysis identifies the number and nature of latent variables (i.e., factors or constructs) that account for the variation among a set of observed measures (Thurstone 1947). The goal of factor analysis is to develop a parsimonious understanding of the covariation among a set of indicators because the number of factors is often substantially less than the number of observed variables (Brown 2015).

There are two broad classes of factor analyses: exploratory (EFA) and confirmatory (CFA). Exploratory factor analysis is used when the links between the observed and latent variables are unknown, a priori. As an inductive approach, EFA reveals associations between the observed measures and underlying factors based on empirical evidence. The goal is to identify the minimum number of factors that account for covariation among the observed measures. Alternately, CFA is used when the researcher has an a priori understanding of the underlying factor structure, which stems from theory, past empirical evidence, or both. Confirmatory factor analysis examines the associations between the observed measures and their underlying latent factors (i.e., CFA tests a hypothesized factor structure). When conducting CFA, the researcher specifies both the number of latent factors and their associations with the observed measures. For both EFA and CFA, relationships between observed measures and latent variables are represented by factor loadings. In the discussion that follows, we provide a step-by-step guide to the decision framework that drives each of the analyses. For guidance on the software interface and program syntax, there are several online and published resources available. For EFA, SPSS has accessible online guides (see www.spss-tutorials.com/). For CFA, Brown (2015) provides program syntax for LISREL and Mplus in addition to the analysis conceptual underpinnings.

Although EFA and CFA are both based on the common factors model, CFA is driven by theory, empirical evidence, or both. Unlike EFA, CFA tests a priori hypotheses about the number of factors, the pattern of factor loadings, and errors associated with the latent constructs. The acceptability of the specified model is evaluated relative to goodness of fit and the strength of the resulting parameter estimates. Goodness of fit refers to the extent to which the hypothesized model reproduces the observed data (i.e., the variance or covariance matrix). Model fit is a necessary but insufficient condition for establishing the psychometric fitness of latent measures. The output associated with CFA provides information for the preliminary evaluation of the adequacy of the measures and guidance on how to improve the model.

5.1 Exploratory Factor Analyses

The general steps for conducting EFA are outlined below (Osborne 2015). Each step requires substantive interpretation:

1. Determining data suitability.
2. Deciding upon an extraction method.
3. Deciding on the method of rotation.
4. Deciding on how many factors to retain.
5. Scale purification.
6. Interpretation of results (return to #3 if the solution is not ideal).
7. Replication.

5.1.1 Determining Data Suitability

Before conducting EFA, it is important to determine whether this type of analysis is appropriate for the data collected, given sample size, power, sampling adequacy, and sphericity. First, the minimum sample size is determined based on the number of items. Although there is a lack of consensus as to what constitutes a minimum sample size, a common way to determine the minimum required sample size is using the subject-to-item ratio (Costello and Osborne 2005; Field 2000; Yong and Pearce 2013), although there is a lack of consensus as to what value should be considered the minimum ratio. One common suggested minimum subject-to-item ratio is five to one, although higher ratios are likely to result in superior metrics, with a twenty-to-one ratio being suggested by some analyses (Costello and Osborne 2005). Second, the calculation of statistical power (Cohen 1990) can assist with decision-making. Statistical power is one minus the probability of type II error, i.e., the probability of not identifying an existing effect—false negative (Brown 2015). Based on the work of Cohen (1988, 1992), the cutoff most frequently used to determine acceptable power is 0.80. This value implies the researcher has an 80% chance of rejecting a false null hypothesis. Both Brown (2015) and Kyriazos (2018) provide thorough reviews and guidance on how to implement. Third, the Kaiser-Meyer-Olkin (KMO) measure represents sampling adequacy for each variable and the complete scale or model. Values range between zero and one, with values $\geq$0.80 considered adequate (Dziuban and Shirkey 1974). Finally, Bartlett's test of sphericity compares the observed correlations among items to an identity matrix (a matrix where items share no common variance). Values range between −1.0 and 1.0. The higher the proportion (farther from zero), the more suited the data is to factor analysis. If the correlation matrix is significantly (χ^2, $p \leq 0.05$) different from the identity matrix, the dataset is appropriate for factor analysis.

5.1.2 Deciding Upon an Extraction Method

Most software packages offer several extraction methods. For instance, SPSS has six options for extraction methods: unweighted least squares, generalized least squares, maximum likelihood, principal axis factoring, alpha factoring, and image factoring. If data are relatively normally distributed, maximum likelihood is the best choice because "it allows for the computation of a wide range of indexes of the goodness of fit of the model [and] permits statistical significance testing of factor loadings and correlations among factors and the computation of confidence intervals." (Fabrigar et al. 1999, p. 277). If the assumption of multivariate normality is "severely violated," principal axis factoring is recommended (Fabrigar et al. 1999). Although alternative extraction techniques may be useful in certain circumstances, maximum likelihood (when data are normally distributed) or principal axis factoring (when data are non-normal) will generally provide the best results (Costello and Osborne 2005).

5.1.3 Deciding on the Method of Rotation

Though it is possible to conduct EFA without rotation, interpretation of the derived factors is more straightforward with rotation (Osborne 2015). There are two broad classes of rotation methods: orthogonal and oblique. Orthogonal rotations (varimax, quartimax, and equamax) produce factors that are uncorrelated, whereas oblique rotational methods (direct oblimin, quartimax, promax) produce factors that are correlated. The choice of method is guided by past work and the researcher's conceptual understanding of the concept of interest. Costello and Osborne (2005) caution that though conventional wisdom might suggest researchers ought to use orthogonal methods because they produce more easily interpretable results, ignoring the possibility of factor correlations can result in the loss of important information. Because multidimensional latent variables often address an overarching theme (second-order factor), there is likely shared variance owing to some level of shared meaning. In these situations, the choice of an oblique rotation should theoretically render a more accurate and perhaps more reproducible solution. If the factors are truly uncorrelated, orthogonal, and oblique rotations produce nearly identical results. Oblique rotation output is slightly more complex than orthogonal rotation output. In SPSS output, when using oblique rotation, the pattern matrix is examined for factor or item loadings, and the factor correlation matrix reveals any correlation between the factors. The substantive interpretations are essentially the same. There is no widely preferred method of oblique rotation; all tend to produce similar results (Fabrigar et al. 1999), and it is fine to use the default delta (0) or kappa (4) values in the software packages. Manipulating delta or kappa changes the amount the rotation procedure "allows" the factors to correlate, and this appears to introduce unnecessary complexity for the interpretation of results (Costello and Osborne 2005).

5.1.4 Deciding How Many Factors to Retain

Once the initial analysis is generated, the researcher needs to decide on how many factors to retain, guided by one or more statistical procedures. The most common selection procedures include (i) the Kaiser-Guttman rule, (ii) the scree test, and (iii) parallel analysis. Each of these procedures relies on eigenvalues, which provide insight into the variance accounted for by the factor solution within the variance-covariance matrix of analysed indicators (Tabachnick and Fidell 2013). The simplest procedure is applying the Kaiser-Guttman rule, which states that when an eigenvalue is less than 1.0, the variance explained by the factor is less than the variance of a single indicator (Brown 2015). There is broad consensus in the literature, however, that this test is among the least accurate methods for selecting the number of factors to retain (Velicer and Jackson 1990).

The scree test involves examining the graph of the eigenvalues (available via every software package) and looking for the natural bend or break point in the data

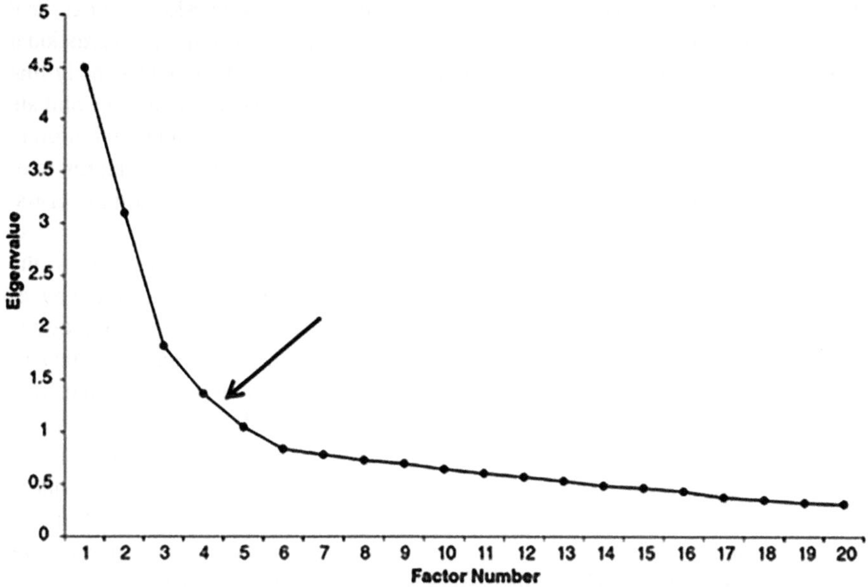

Fig. 3 Scree plot

where the curve flattens out (see Fig. 3). The number of data points above the "break" (i.e., not including the point at which the break occurs) is usually the number of factors to retain. When data points are clustered together near the bend, additional tests may be needed: the researcher may run multiple factor analyses and set the number of factors to retain manually—once at the projected number based on the a priori factor structure, again at the number of factors suggested by the scree test if it is different from the predicted number, and again at numbers above and below those numbers. For example, if the predicted number of factors is six and the scree test suggests five, then run the data four times, setting the number of factors extracted at four, five, six, and seven. After rotation, compare the item loading tables taking into consideration the one with the "cleanest" factor structure; item loadings above 0.40 (Hair et al. 2018; Tabachnick and Fidell 2013), no or few item cross-loadings (i.e., items loading strongly on more than one factor), no factors with fewer than three items, and has the best fit to the data.

In instances of weak loadings or cross-loading, items may need to be removed from the analysis (one at a time) and the model re-run. If all loading tables are uninterpretable, then there is likely a problem with the data that cannot be resolved by manipulating the number of factors retained. Sometimes dropping problematic items (ones that are low-loading, cross-loading, or freestanding) and rerunning the analysis can solve the problem, but the researcher must consider if doing so compromises the integrity of the data. If the factor structure still fails to clarify after multiple test runs, there is a problem with item construction, scale design, or the hypothesis itself, and the researcher may need to throw out the data as unusable and start from scratch. One other possibility is that the sample size was too small, and more data needs to be collected before running the analyses. A final issue to consider is the indeterminacy of the number of factors, i.e., there's no salient bend in the elbow. As depicted (Fig. 3), there is some uncertainty between factors five and six when using data containing 20 items. This is a case where the researcher will need to make an assessment based on their knowledge of the subject area and the range of empirical indicators discussed below (e.g., the strength of factor loadings, cross-loading, number of items per factor, parallel analyses).

A third tool to aid decision-making is parallel analysis, which uses a randomly generated sample of the same observations from the original data and creates eigenvalues based on this random data set. The researcher then compares the eigenvalues of the data set with the eigenvalues generated from parallel analysis. If the eigenvalue generated from the true data set is larger than that generated from the random data set, then the factor is retained. Once the random eigenvalues exceed those generated from the true data set, then the factors are no longer retained. The rationale of parallel analysis is that retained factors should capture more variance than is expected by chance. Given the tendency of both the Kaiser-Guttman rule and the scree plot to over-dimensionalize, it is recommended that parallel analysis be used for all EFA studies. This issue is somewhat evidenced in the scree plot (Fig. 3). If relying on the scree plot alone, the researcher would be challenged to choose between five to six factors. Findings from the parallel analysis suggest the presence of four factors in the data. Although the researcher's professional judgment is still required, the analysis rules out the possibility of a sixth factor. A practical drawback of the procedure is that it is not available in popular statistical packages like SAS and SPSS. The syntax for running the analysis in SPSS has been published by Hayton et al. (2004); however, the user must save and summarize the eigenvalues generated from the random data. Parallel analysis is an option in Mplus and STATA software and various shareware programs on the internet, e.g., the user-written "paran" and "EFAtools" packages for program R (Fig. 4).

5.1.5 Scale Purification

Every item in a scale will load onto every factor when analysed using EFA. However, the strengths of these loadings will differ. In EFA, items can load onto factors strongly or weakly. Poorly performing items (Hair et al. 2018; Tabachnick and Fidell 2013) behave in one of two ways: (i) fail to load meaningfully on any factor, i.e., all loadings ≤ 0.4, or (ii) load strongly on more than one factor, i.e., cross-loading. Another metric used to assess item performance is its communality. This reflects the amount of variance being accounted for in the item by factors emerging in the EFA. Brown (2015, p. 77–78) provides an overview of how the metric is computed. MacCallum et al. (2001) have suggested item communalities should be greater than 0.6 or an average of 0.7 across all items for EFAs with small sample sizes. Ideally, a clean solution would have an item loading strongly on one factor (≥ 0.7) and weakly (<0.3) on all other factors. If an item or items are performing

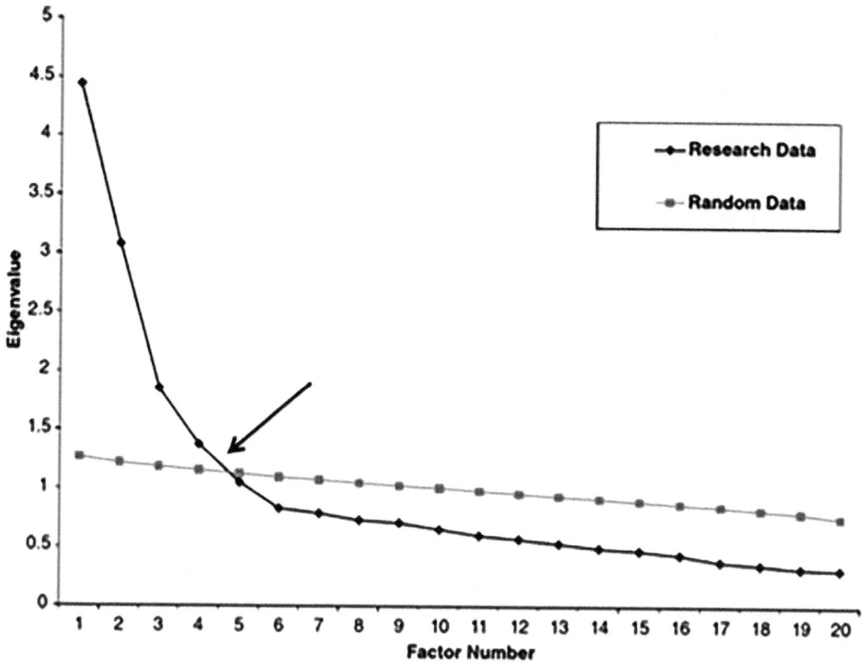

Fig. 4 Parallel analysis

poorly, the researcher must make a decision related to item removal. This decision is tempered by the number of items loading one factor. Again, it is desirable to have three or more for concerns over reliability and for model identification if the researcher plans to use the solution in a structural equation model. There are also other considerations we discuss below that are related to the validity and reliability of the scale and model identification when switching CFA in subsequent analyses (Brown 2015). As we note, these metrics also assist in decision-making related to scale purification.

5.2 *Confirmatory Factor Analysis*

Researchers use CFA in contexts where they know, in advance, the underlying factor structure of their scale, given its use in previous research, or they are attempting to validate their scale and its factor structure that emerged from EFA. In EFA, the researcher can only specify the number of factors, not the specific relationship between items and their latent constructs. By contrast, the CFA researcher tests a more parsimonious solution by specifying the number of factors, the pattern of factor loadings (and cross-loading usually fixed to zero), and an appropriate error theory (unique or correlated). Every aspect of the CFA model is specified in advance. The acceptability of the hypothesized model is evaluated by its goodness of fit (discussed below) and by the interpretability and strength of the parameter estimates. Similar to EFA, the answer to the question, "what's an adequate sample size?" is, "it depends". Issues to consider include statistical power (Cohen 1988), bias in parameter estimates and standard errors, and solution propriety where sample size can impact model convergence, generating improper solutions or impossible parameter estimates (Gagné and Hancock 2006; Wolf et al. 2013). Improper solutions often produce fit indices well outside acceptable guidelines or parameter estimates that are implausible (e.g., standardized factor loadings >1.5). Though we are hesitant to report rules of thumb owing to these considerations and their reported fallibility, Muthén and Muthén's (2002) Monte Carlo simulation revealed that for normally distributed data without missing values, a sample of 150 produced reliable parameter estimates.

When testing CFA measurement models, the researcher is faced with several important decisions focused on:

1. Model estimation—Choosing the appropriate algorithm.
2. Model identification—Ensuring that there is adequate data available (variances and covariances) to solve the equations generating parameter estimates.
3. The handling of missing data—Evaluating the nature of the sample's missing data and deciding how to address.
4. Assessing model fit—Evaluating how well the hypothesis model structure aligns with the structure evidenced in the data.
5. Model revision and comparison—Should the model be revised to better align with the data?

5.2.1 Model Estimation

The default model estimation procedure in most software programs (e.g., LISREL, Mplus, EQS, AMOS) used to test CFA measurement models with normally distributed data is maximum likelihood (ML). For data with non-normal distributions (kurtosis or skewness), robust estimation procedures that include the asymptotic covariance matrix in addition to the correlation or covariance matrices are analysed. The use of robust ML estimation procedures provides standard error and model fit indices (χ^2) that accommodate non-normality. Similarly, for latent variables with categorical indicators (i.e., dichotomous, polytomous, ordinal), different estimation procedures are required along with alternate matrices to be analysed. The most used estimator for categorical data is weighted least squares (WLS). The WLS estimator does, however, require a substantially larger sample size. Brown (2015) provides a thorough discussion of estimating CFA measurement models with non-normal and categorical data. Byrne (1998) also provides helpful insight on constructing matrices (polyserial, polychoric) that accompany different combinations of categorial and continuous indicator latent variable models on the LISREL/PRELIS software platform.

5.2.2 CFA Model Identification

To estimate the parameters in CFA (e.g., factor loadings, indicator variances, factor variance, factor covariance), the measurement model must be "identified". Model identification pertains to the difference between the "knowns" in the sample input matrix (variance and covariances) and the "unknowns" being estimated (e.g., factor loadings, indicator variances, etc.) with the identification status for the model solution requiring the knowns to be equal to or greater than the unknowns. Though the issue of identification is complex and difficult to discuss in non-technical terms, there are some general guidelines that researchers follow to ensure model identification. First, the latent variable must have its scale identified. This can be done in one of two ways: (i) most common is to set the scale (factor loading) of a marker indicator at 1.0, or (ii) set the variance of the latent factor to 1.0. Second, each latent factor should have three indicator variables (items). This is also advisable for addressing issues of validity and reliability (see below). Algebraically, it ensures we have more knowns than unknowns. Using the formula, p(p + 1)/2, where p is the number of data points in the input matrix, we can count the number of knowns in each model. For example, for the model being estimated in panel 1 (Fig. 1), we would be attempting to estimate eight parameters. These include four error variances for each indicator (δ_1- δ_4), three-factor loadings (λ_2- λ_4); one-factor loading is not estimated, given it was selected to be the scaling indicator, and one latent factor variance term (φ_1). Using the formula above $4(4 + 1)/2 = 10$ data points. Given that the number of knowns (reflected in the data points) exceeds the parameters to be estimated, the model is said to be "over-identified." A three-item factor would yield six data points and would require estimating six parameters. In this instance, it would be "just-identified." A two-item factor would be "under-identified" because it only contains three data points while estimating four parameters. The exception to the two-item per latent factor requirement arises when the two-item factor is correlated with another latent factor. Algebraically, the model will solve. Though algebraic identification is a requirement for testing the CFA measurement model, it is not necessarily sufficient. For example, in a three-item model, if one or more factor loadings are very low (<0.3), empirically, it could be performing like a two-item (or even one item) factor. In this case, the model will likely be empirically under-identified and not fit the sample data. Most introductory texts examining CFA and the structural equation model offer detailed guidance on addressing issues of identification (See Byrne 1998; Brown 2015).

5.2.3 Handling Missing Data

Missing data is an issue for both EFA and CFA. It can result in the loss of critical data and, for the reasons discussed above related to the number of items required per indicator, can be critical when analysing small samples. Most software packages that estimate CFA measurement models, however, allow for superior ways of managing missing data over methods such as pair- or listwise deletion. Additionally, mean or regression-based imputation is also problematic because they underestimate variance and overestimate correlations (Brown 2015). There are three broad classes of missing data: (i) missing completely at random (MCAR) where the missing data are randomly distributed across a variable and unrelated to other variables, (ii) missing at random (MAR) where the missing data are not randomly distributed but are accounted for by other variables, and (iii) missing not at random (MNAR) where the missing data systematically differ from the observed variables. The missing data approaches we discuss here are suitable for missingness that follows MCAR or MAR distributions. Under MCAR and MAR, the imputation process can increase the efficiency and power of the estimation, but when these assumptions do not hold, estimates may be biased given patterns of response and non-response. Missingness with MNAR data is said to be nonignorable, given that the missingness is related to values that would have been observed. They contain cases with missing data on a given variable that would have higher or lower values on that variable than cases with data present after all other variables in the data set have been controlled for (Brown 2015). Though this is not a common occurrence in survey research related to the human dimensions of recreational fisheries, there are contexts where it is conceivable. For example, in the context of the United States, white men dominate recreational fishing participation (RBFF 2022). Many states will conduct surveys of licensed anglers to identify constraints to access and engagement with the objective of trying to identify barriers that other segments of the population (e.g., Hispanic, Black, women) might face. An inherent bias in surveying licensed anglers, however, is that many have already negotiated the constraints that latent segments of participants may have yet to overcome. Consequently, the missingness introduced by this approach is not at all random. Their absence from the sample is tied to the variable of concern, i.e., constraints to participation. We direct readers to the work of Graham and Schafer for further guidance on how to address this form of missingness (Graham 2012; Schafer and Graham 2002).

Regarding MCAR and MAR missingness, two commonly used approaches are employed: full information maximum likelihood (FIML; also referred to as direct ML) and multiple imputation. FIML is readily available in latent variables software packages. When using raw data in LISREL, FIML is the default estimation option when missing data are encountered in the raw data input file. Broadly, the approach uses each case's available data to compute ML estimates. The ML estimate of a parameter is the value of the parameter that is most likely to have resulted in the observed data. When data are missing, we can factor the likelihood function. The likelihood is computed separately for those cases with complete data on some variables and those with complete data on all variables. These two likelihoods are then maximized together to find the estimates. Like multiple imputation, this method provides unbiased parameter estimates and standard errors (Brown 2015). Other statistics programs like MPLUS, ANOS, Stata, and the "lavaan" package for program R have capabilities for estimating CFA models using the FIML method.

An alternative to FIML is multiple imputation (MI; Rubin 1987). Multiple imputation possesses statistical properties that closely approximate FIML (Schafer

1997). Unlike FIML, MI is a data-based process that occurs as a separate step before CFA. Where FIML provides direct estimates in the CFA model, MI in LISREL is a two-step procedure. First, using the expectation-maximization (EM) procedure, MI produces ML estimates of a mean vector and a covariance matrix for analyses to follow. This is advantageous because it can be used to estimate both EFA and CFA models in addition to being utilized in other analyses of interest. It can provide a complete dataset. As a two-step iterative procedure, the EM algorithm first replaces missing values by constructing a series of regression equations based on the means and covariance matrix obtained from the observed data (the E or expectation step). Through the dataset, each missing variable is predicted by the remaining observed variables in each case. When all the missing values are filled, a new mean vector and covariance matrix are calculated by using both the observed and imputed values (Allison 2002).

5.2.4 Assessing the CFA Model

Before examining parameter estimates generated from the CFA, an assessment of model fit is required. Ultimately, the goal of model estimation is to obtain estimates for each parameter in the measurement model generated by the CFA (i.e., factor loadings, factor variances and covariances, indicator variance, and possible covariance). These parameters provide insight into the indicators' relationship with the parent factor (i.e., latent variable), item and factor associations, and construct validity and reliability (discussed below). Before doing so, however, an examination of the model's fit indices is required to determine the extent to which the predicted variance-covariance matrix reflected in the hypothesized measurement model (symbolized as Σ) resembles the sample variance-covariance matrix (symbolized as S). Fit indices provide metrics that illustrate how closely or how distant $\Sigma = S$. Over the past 50 years, since the first generation of LISREL and the development of a broad range of software programs, several model fit indices have been developed. Accompanying this development has been a lively discussion surrounding cut-off values for each of the indices (Hair et al. 2017; Hu and Bentler 1999). Brown (2015) provides a review of three classes of fit indices based on their popularity in the applied literature and performance in Monte Carlo simulations: absolute fit, fit adjusting for model parsimony, and comparative or incremental fit. When reporting fit indices, convention typically requires the reporting of one or more indices from each class.

For absolute fit indices, the viability of the hypothesis that $\Sigma = S$ without consideration of other aspects of the model. In other words, what counts is how well the hypothesized covariances from the specified model's fixed and free parameters match the observed covariances estimated from the free parameters in the model (Hoyle 1995). A foundational absolute fit index is the chi-square (χ^2) statistic. It is foundational in the sense that it was used to assess model fit in early iterations of structural equation modelling using ML estimation but also because it is used in the calculation of most of the other fit indices that have emerged in the years since. Though it is still common to report the χ^2 value of the measurement model, the statistic is sensitive to sample size and often yields a statistically significant value (meaning $\Sigma \neq S$) when samples begin to climb beyond two to three hundred cases. With this understanding, researchers more often rely on other classes of fit indices. Another index is the standardized root mean residual (SRMR), which is calculated from the average discrepancy observed between the residuals of the sample covariance matrix and the hypothesized model for the covariance. A SRMR value of zero indicates a perfect fit, but values less than 0.05 are considered acceptable (Diamantopoulos and Siguaw 2000).

Another class of fit indices is parsimony correction indices, which differ from absolute fit indices in that they include a penalty function for poor model parsimony, i.e., the number of freely estimated parameters (factor loadings, error variances, etc.) as evidenced in the model's degrees of freedom. A widely recommended index in this class is the root mean square error of approximation (RMSEA). It provides an assessment of "how far" the hypothesized model is from the "perfect model". Like SRMR, a RMSEA value of zero indicates a perfect fit, but values less than 0.05 are considered acceptable.

The third class of fit indices is comparative fit (or incremental fit indices; Hu and Bentler 1998). These indices evaluate model fit by comparing the hypothesized model to a null model (also called the independence model) where all the covariances of the input model are fixed to zero. Two popular indices from this class are the comparative fit index (CFI; Bentler 1990) and the Tucker-Lewis index (TLI; Tucker and Lewis 1973; also called the non-normed fit index, NNFI). There are similarities in the calculation of these indices, so they behave similarly regardless of sample size and model complexity. As such, it is common to only see one of the indices reported when assessing model fit. Values range between zero and one, with values greater than 0.95 being considered acceptable.

5.2.5 Model Revision and Comparison

Having run the CFA, researchers often need to revise their hypothesized model in light of poor model fit, weak parameter estimates (low factor loadings), or both. Though decisions on model specification ought to be firmly anchored in theory or past empirical evidence, contemporary software programs provide an array of metrics that can help to inform decision-making related to specification. Model respecification, post hoc, does reflect a departure from past empirical evidence or theory and will require substantive justification. Nearly all CFA measurement models impose some kind of restriction on the number of parameters to be estimated. Usually, some parameters are set to zero (and thus not estimated at all), but sometimes restrictions come in the form of equality constraints (e.g., constrain factor loadings to be equal) or other kinds of structured relations among parameters. The model fit indices reflect the extent to which these imposed restrictions impede the ability of the model to reproduce the means, variances, and covariances that were observed in the sample. Broadly, the fit indices indicate the extent to which the estimated model can adequately reproduce the observed sample statistics.

Alternately, a poorly fitting model indicates that some aspect of the hypothesized model is inconsistent with the characteristics of the observed sample. Brown (2015) noted that the main sources of model misspecification (and poor model fit) are the number of factors (too many or too few), the indicators (choice of indicators, pattern of factor loadings), and error theory (uncorrelated vs. correlated error terms).

For poor model fit, information about how to improve model fit can be found in the modification indices (MI; also called the Lagrange Multiplier or Score Test, depending on the software being used). An MI is an estimate of the amount by which the χ^2 would be reduced if a single parameter restriction were to be removed from the model. There are as many Mis as imposed restrictions in the model. Most commonly, an MI reflects the improvement in model fit that would result if a previously omitted parameter were to be added and freely estimated. MI values greater than 3.84 (with 1 degree of freedom) reflect a potential statistically significant improvement in model fit. If a parameter is added based on a large MI, this represents a data-driven modification of the original hypothesized model. It is not uncommon in practice for researchers to consult Mis to suggest model modifications that lead to a "better" fitting model, but over-dependence on them also carries risks. First, they are completely determined by the data and, thus, devoid of theory. The largest Mis might be associated with parameters that are unsupported by theory and instead represent idiosyncratic characteristics of the data. Second, simulation research has suggested that using Mis to guide model specification rarely leads to the true population model. MacCallum et al. (1992. p. 502) conducted a comprehensive study of Mis and concluded "In summary, our results bring us to a position of considerable scepticism concerning the validity of the model modification process as it is often used in practice." The point in all of this, the use of Mis can be beneficial if utilized thoughtfully and judiciously.

6 Assessing Validity and Reliability

The final step in the scale development process is to empirically assess its psychometric adequacy, which usually focuses on two components, i.e., the scale's validity and reliability. Where validity refers to the extent to which a scale measures what it claims to measure, reliability represents the extent to which the scale scores are repeatable. Two recent papers in the conservation literature by Kyle et al. (2020) and Whitehouse-Tedd et al. (2020) provide a thorough review of the topic and the criteria for assessment within the context of human dimensions research. Although they provide an overview of the multiple forms and procedures that can be used for assessment validity and reliability, we restrict our discussion to procedures most reported in human dimensions work that assess latent variables.

Two forms of construct validity need to be established before the scale can be deemed psychometrically adequate and utilized in further analyses (e.g., path modelling): convergent and discriminant validity. Evidence of convergent validity is displayed when scale items designed to measure the same theoretical construct are related to one another. Because they have been constructed around the theoretical definition of the construct of interest, they ought to display some association with one another. Conversely, discriminant validity is evidenced when different measures of theoretically distinct constructs are not highly correlated. Landon et al. (2018) reported three statistics for assessing the convergent validity of their latent constructs using data collected from recreation anglers in Texas. Their indices focused on the strength of the factor loadings emerging from their CFA, the statistical significance of these loadings, and the average variance explained (AVE; Fornell and Larcker 1981). For the strength of the factor loadings, Fornell and Larcker suggested that values of 0.707 or greater are desired, given that values less than this indicate that the latent factor is capturing at least 50% of the variation in the indicator. In applied research, this threshold is demanding. Although Hair et al. (2018) have offered less restrictive thresholds ($\geq$0.40), we recommend that the preponderance of items load on their theoretical construct at or above 0.707. Constructs containing a preponderance of items falling below 0.707 will likely run into other issues related to the model solution (e.g., empirical under-identification), discriminant validity, and reliability. For the statistical significance of factor loadings, output from the CFA analyses will guide assessment. Last, for the estimate of AVE, Fornell and Lacker suggest a value greater than 0.5. The AVE provides an assessment of the average variance captured by the latent construct in the indicators concerning the amount of variance due to measurement error. Values greater than 0.5 indicate the latent construct is capturing more than 50% of the variation among its indicator variables.

Similarly, for discriminant validity, multiple indices are available for researchers to utilize for assessment. With the goal of empirically distinguishing constructs, three commonly used procedures include:

1. Assessing the confidence intervals around latent variable estimates. If the confidence interval (CI with 2 SEs) around the correlation estimate between the two factors includes 1.0, it could suggest that the measures reflect the same variable (Anderson and Gerbing 1988).
2. Calculated AVEs that are greater than the squared correlations between each of the latent constricts. This test compares the relative amount of variance explained by the latent variable and the latent variable's relationship with other variables in the measurement model. When a latent variable's association with other variables in the measurement model is stronger than its relationships with its manifest indicators, it points to a lack of discrimination between the two variables (i.e., they are measuring the same underlying construct; Fornell and Larcker 1981).
3. Constraining factor correlations (one pair at a time) to be 1.0 and examining the constraint's impact on the model fit indices. If the χ^2 value with 1 degree of freedom increases by more than 3.84 (a statistically significant increase), the researcher can assume the constructs are not perfectly correlated and has further evidence of discriminant validity. However, very highly correlated latent factors may still present issues of cross-loading of indicators.

Finally, for reliability, commonly utilized metrics rely on the assessment of internal consistency (i.e., Cronbach's alpha, α (Cronbach 1951)) and the composite reliability (i.e., Jöreskog's rho; ρ (Raykov 1997)) of the latent variables. Research has shown that coefficient alpha is a good estimate of reliability under conditions of tau equivalence (i.e., factor loadings are equal) and error terms are not correlated (Novick and Lewis 1967; Raykov 1997; McDonald 1999). In applied research, however, this assumption often does not hold, leading to an alpha that provides a lower bound estimate of reliability (Sijtsma 2009). In the context of congeneric measures (i.e., inequality among the factor loadings for a unidimensional latent factor), an alternate estimator such as composite reliability or McDonald's omega (ω; McDonald 1999) is recommended. The lower bound cutoff for alpha is 0.70 (Nunnally 1951) and 0.60 for composite reliability (Bagozzi and Yi 1998).

7 Adapting Existing Scales to Align with Your Study Context

For an example of the iterative process underlying scale development, adaption, and refinement we use the construct, place attachment, to illustrate how the measurement of a latent construct has evolved over 30-plus years of use in an array of outdoor recreation contexts. Williams et al.'s (1992) much cited place attachment scale was among the first published scales to measure recreationists' attachments to the landscape. An evolution from their earlier work (Williams and Roggenbuck 1989), the scale has since been validated in several studies (e.g., Moore and Graefe 1994, Williams and Vaske 2003) and extended to include other facets that reflect recreationists' attachment to an array of landscapes. For example, Williams et al.'s original conceptualization included two dimensions of attachment, *place identity* and *place dependence*. Where place dependence focused on recreationists' functional attachment to the setting (i.e., it's value for facilitating desired leisure experiences), place identity referenced recreationists' emotional and symbolic attachments to the setting (i.e., the landscape has become part of the recreationists' identity). Drawing on theory related to the structure of attitudes, Jorgensen and Stedman (2001) later distinguished between the identity and emotion-based indicators of Williams et al.'s measure of place identity to derive two distinct sub-dimensions titled *place identity* (cognition) and *attachment* (emotion), respectively. Kyle et al. (2004) further refined the scale, extending Jorgensen and Stedman's work with the addition of a fourth dimension titled *social bonding*. They cited phenomena where recreationists' attachment to the setting is often anchored in the experiences (and memories) shared with family and friends. This process of refinement reflects both a reconceptualization of Williams et al.'s original work both in terms of conceptualization and associated measurement. Importantly, these authors' decisions underpinning refinement have been anchored in theory and past empirical evidence.

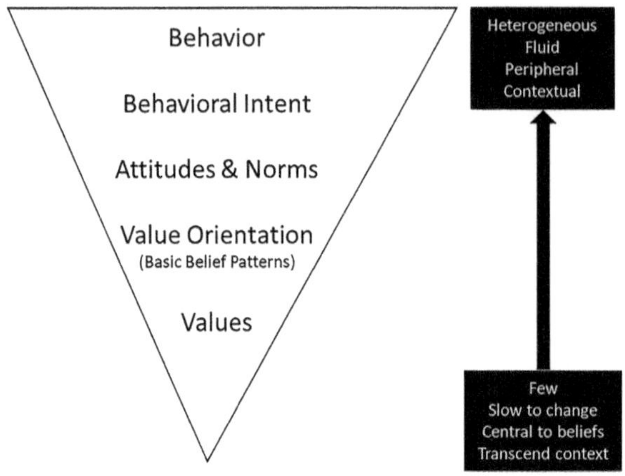

Fig. 5 Cognitive hierarchy model of human behaviour (adapted from Fulton et al. 1996)

8 Understanding the Nomological Network

Having adequately measured the latent construct or constructs, decisions then need to be made concerning its relationship or their relationships with other variables of interest. The conceptual and methodological orientation in which this chapter has been crafted is anchored in the ontology and epistemology of quantitative social psychology. Conceptually, the latent constructs we reference in the compendium can be situated within the nomological network defined (Cronbach and Meehl 1955) by the cognitive hierarchy (Fulton et al. 1996, Homer and Kahle 1988; see Fig. 5).

Broadly, the cognitive hierarchy portrays a model of human behaviour driven by constructs that vary in their specificity and proximity to behaviour. Though human values undoubtedly shape action, their influence is mediated by beliefs, attitudes, norms, and intent. Other constructs referenced in our compendium can also be situated along the hierarchy. There is an established literature that provides some insight on their roles as mediator or moderator (e.g., Iwasaki and Havitz 2004, Stern et al. 1999, Kyle et al. 2023). It is important that the researcher understands where within the hierarchy variables are situated to enable valid decisions related to their status as exogenous or endogenous indicators. These decisions are based on theory (e.g., theory of planned behaviour, norm activation model, values-beliefs-norm theory) and past empirical evidence.

9 Concluding Comments

Researchers interested in understanding fishers' cognition and affect for the purpose of behavioural prediction should consider these recommendations a starting point for a foray into psychometrics. Considering the process for scale development we have outlined, and the criteria for the psychometric adequacy of measures of latent variables, it could be assumed that the best approach to measuring latent variables is to utilize existing scales that have established validity and reliability (Shelby 2011). However, validation of any scale is likely to include some degree of sample specificity (Whitehouse-Tedd et al. 2020). Accordingly, it must have its continued validity for a different sample confirmed. This is especially important when respondents speak a language different from that in which the scale was originally constructed and validated (thereby requiring validation of the translation process) or when species have multiple common names or even no name in local dialects or languages (e.g., Stevens et al. 2014).

There are numerous constraints to implementing the recommendations described here. Managers commonly collect data on fishers' attitudes and behaviours in the field, and under constraints for time, labour, and cost. Many scales for the measurement of psychological constructs of interest to human dimensions inquiry are long, multidimensional, and difficult to implement in the field. It is important to understand the trade-offs that come with choices made about operationalizing latent constructs like those described here in ways that deviate from past work. Managers may choose, for instance, to reduce a multi-item scale to a single item to measure a construct of interest. Doing this (1) ignores the uncertainty inherent in the measurement of latent constructs, and (2) increases the odds of that item failing to capture the latent construct of interest, or only capturing a part of its character and variance. In either case, the veracity of claims about, for instance, the effect of an incomplete measure of a latent psychological construct on fishers' behaviour is questionable. These are certainly limitations worth considering when designing research on fishers' cognition and affect.

Commenting on the increasing volume of social science-informed research appearing in outlets that have traditionally published work from the life sciences, St. John et al. (2014, p. 2) noted the increasing incidence (and failure) of "ecologically trained scientists adding social science research to their mainly ecological studies, with greater or lesser success." They contend that low-quality science may be published under these circumstances, given editors' and reviewers' training in the life sciences. They warn of the emergence of a potentially sinister problem: "because researchers read publications in high-impact journals which validate lower quality social science… journals risk institutionalizing poor social science" (p. 2). Teel et al. (2018) echo this sentiment, noting the need for social science to be evaluated by the disciplinary norms from which the work draws. Although the full impact of mismeasurement remains unknown, its pervasiveness and persistence may erode trust and perpetuate misconceptions.

References

Aas Ø (1995) Constraints on sportfishing and effect of management actions to increase participation rates in fishing. N Am J Fish Manag 15(3):631–638

Allison PD (2002) Missing data. Sage Publications, Thousand Oaks

Anderson JC, Gerbing DW (1988) Structural equation modeling in practice: a review and recommended two-step approach. Psychol Bull 103:411–423

Bagozzi RP, Yi Y (1998) On the evaluation of structural equation models. J Acad Market Sci 16(1): 74–94
Bentler PM (1990) Comparative fit indexes in structural models. Psychol Bull 107(2):238–246
Betella A, Verschure P (2016) The affective slide: a digital self-assessment scale for the measurement of human emotions. PLoS One 11(2):e0148037
Bollen KA (1989) Structural equations with latent variables. Wiley, Malden
Bollen KA, Diamantopoulos A (2017) In defense of causal-formative indicators: a minority report. Psychol Methods 22:581–596
Bollen K, Lennox R (1991) Conventional wisdom on measurement: a structural equation perspective. Psychol Bull 110(2):305–314
Brown TA (2015) Confirmatory factor analysis for applied research, 2nd edn. Guildford Press, New York
Byrne BM (1998) Structural equation modeling with LISREL, PRELIS, and SIMPLIS: basic concepts, applications, and programming. Lawrence Erlbaum Associates, Mahwah
Carifio J, Perla R (2008) Resolving the 50-year debate around using and misusing Likert scales. Med Educ 42(12):1150–1152
Cenfetelli RT, Bassellier G (2009) Interpretation of formative measurement in information systems research. Manag Inform Syst Q 33:689–707
Churchill GA (1979) A paradigm for developing better measures of marketing constructs. J Mark Res 16(2):64–73
Chyung SY, Roberts K, Swnason I, Hankinson A (2017) Evidence-based survey design: the use of a midpoint on the Likert scale. Perform Improv Q 56(10):15–23
Cohen J (1988) Statistical power analysis for the behavioral sciences, 2nd edn. Lawrence Erlbaum Associates
Cohen J (1990) Things I have learned (so far). Am Psychol 45(12):1304–1312
Cohen J (1992) A power primer. Psychol Bull 112:155–159
Coltman T, Devinney TM, Midgley DF, Venaik S (2008) Formative versus reflective measurement models: two applications of formative measurement. J Bus Res 61(12):1250–1262
Coolican H (2014) Psychological tests and measurement scales. In: Coolican H (ed) Methods and statistics in psychology, 6th edn. Psychological Press, New York, pp 193–229
Costello AB, Osborne JW (2005) Best practices in exploratory factor analysis: four recommendations for getting the most from your analysis. Pract Assess Res Eval 10(7):1–9
Cronbach LJ (1951) Coefficient alphas and the internal structure of tests. Psychometrika 16(3): 297–334
Cronbach LJ, Meehl PE (1955) Construct validity in psychological tests. Psychol Bull 52:281–302
DeCastellarnau A (2018) A classification of response scale characteristics that affect data quality: a literature review. Qual Quant 52:1523–1559
DeVillis RF (2017) Scale development, 4th edn. Sage, Thousand Oaks
Diamantopoulos A, Siguaw JA (2000) Introducing LISREL. Sage, London
Dillman DA, Smyth JD, Christian LM (2014) Internet, phone, mail, and mixed-mode surveys: the tailored design method, 4th edn. Wiley
Dziuban CD, Shirkey EC (1974) When is a correlation matrix appropriate for factor analysis? Some decision rules. Psychol Bull 81(6):358–361
Fabrigar LR, Wegener DT, MacCallum RC, Strahan EJ (1999) Evaluating the use of exploratory factor analysis in psychological research. Psychol Methods 4(3):272–299
Fedler AJ, Ditton RB (2001) Dropping out and dropping in: a study of factors for changing recreational fishing participation. N Am J Fish Manag 21(2):283–292
Field A (2000) Discovering statistics using SPSS for windows. Sage Publications, Los Angeles
Fornell C, Larcker DF (1981) Evaluating structural equation models with unobservable variables and measurement error. J Mark Res 1(1):39–50
Fulton DC, Manfredo MJ, Lipscomb J (1996) Wildlife value orientation: a conceptual and measurement approach. Hum Dimens Wildl 1(2):24–47

Gagné P, Hancock GR (2006) Measurement model quality, sample size, and solution propriety in confirmatory factor models. Multivar Behav Res 41:65–83

Glaser B, Strauss A (1967) The discovery of grounded theory: strategies for qualitative research. Sociology Press, New York

Graham JW (2012) Missing data: analysis and design. Springer, New York

Hair JF, Babin BJ, Krey N (2017) Covariance-based structural equation modeling in the journal of advertising: review and recommendations. J Advert 46:163–177

Hair JF, Black WC, Babin BJ, Anderson RE, Black WC (2018) Multivariate data analysis, 8th edn. Cengage, Boston

Hayton JC, Allen DG, Scarpello V (2004) Factor retention decisions in exploratory factor analysis: a tutorial on parallel analysis. Organ Res Methods 7(2):191–205

Homer PM, Kahle LR (1988) A structural equation test of the value-attitude-behavior hierarchy. J Pers Soc Psychol 54:638–646

Hoyle RH (1995) Structural equation modeling: concepts, issues, and applications. Sage, Thousand Oaks

Hu LT, Bentler PM (1998) Fit indices in covariance structure modeling: sensitivity to underparameterized model misspecification. Psychol Methods 3(4):424

Hu LT, Bentler PM (1999) Cutoff criteria for fit indexes in covariance structure analysis: conventional criteria versus new alternatives. Struct Equ Modeling 6:1–55

Iwasaki Y, Havitz ME (2004) Examining relationships between leisure involvement, psychological commitment and loyalty to a recreation agency. J Leis Res 36(1):45–72

Jacobs MH, Vaske JJ, Dubois S, Fehres P (2014) More than fear: role of emotions in acceptability of lethal control of wolves. Eur J Wildlife Res 60:589–598

Jarvis CB, MacKenzie SB, Podsakoff PM (2003) A critical review of construct indicators and measurement model misspecification in marketing and consumer research. J Consum Res 30(2): 199–218

Jorgensen BS, Stedman RC (2001) Sense of place as an attitude: lakeshore owners attitudes toward their properties. J Environ Psychol 21(3):233–248

Krosnick JA (2018) Questionnaire design. In: Vannette DL, Krosnick JA (eds) The palgrave handbook of survey research. Springer, New York, pp 439–455

Kyle GT, Jun J (2015) An alternate conceptualization of the leisure constraints measurement model: formative structure? J Leis Res 47:337–351

Kyle GT, Landon AC (2023) Modernity and the value-action paradox: contributions from conservation psychology, vol 4. Front Conserv Sci, p 1092259

Kyle GT, Mowen AJ, Tarrant M (2004) Linking place preferences with place meaning: an examination of the relationship between place motivation and place attachment. J Environ Psychol 24(4):439–454

Kyle GT, Absher J, Norman W, Hammitt W (2007) A modified involvement scale. Leisure Stud 26: 399–427

Kyle GT, Landon AC, Vaske JJ, Wallen K (2020) Tools for assessing the psychometric adequacy of latent variables in conservation research. Conserv Biol 34(6):1353–1363

Kyriazos TA (2018) Applied psychometrics: sample size and sample power considerations in factor analysis (EFA, CFA) and SEM in general. Psychology 9:2207–2230

Landon AC, Kyle GT, van Riper CJ, Schuett MA, Park J (2018) Exploring the psychological dimensions of stewardship in recreational fisheries. N Am J Fish Manag 38:579–591

Lord FM, Novick MR, Birnbaum A (1968) Statistical theories of mental test scores. Addison-Wesley

MacCallum RC, Roznowski M, Necowitz LB (1992) Model modifications in covariance structure analysis: the problem of capitalization on chance. Psychol Bull 111:490–504

MacCallum RC, Widaman KF, Preacher KJ (2001) Sample size in factor analysis: the role of model error. Multivar Behav Res 36:611–637

Malhotra NK, Peterson M (2006) Basic marketing research: a decision-making approach, 2nd edn. Pearson, Upper Saddle River

Marsden PV, Wright JD (2010) Handbook of survey research, 2nd edn. Emerald Group Publishing, United Kingdom
McCroskey JC (2006) Reliability and validity of the generalized attitude measure and generalized belief measure. Commun Q 54(3):265–274
McDonald RP (1999) Test theory: a unified treatment. Lawrence Erlbaum Associates, Mahwah
Mehrabian A, Russell JA (1974) An approach to environmental psychology. MIT Press, Cambridge, MA
Moore RL, Graefe AR (1994) Attachments to recreation settings: the case of rail-trail users. Leis Sci 16(1):17–31
Muthén LK, Muthén BO (2002) How to use a Monte Carlo study to decide on sample size and determine power. Struct Equ Modeling 9(4):599–620
Novick MR, Lewis C (1967) Coefficient alpha and the reliability of composite measurements. Psychometrika 32:1–13
Nunnally JC (1951) Coefficient alpha and the internal structure of tests. Psychometrika 16(3): 297–334
Nunnally JC, Bernstein IH (1994) Psychometric theory, 3rd edn. McGraw-Hill, New York
Osborne JW (2015) What is rotating in exploratory factor analysis? Pract Assess Res Eval 20(2): 1–7
Raykov T (1997) Scale reliability, Cronbach's coefficient alpha, and violations of essential tau equivalence with fixed congeneric components. Multivar Behav Res 32:329–353
Recreation Boating and Fishing Foundation (RBFF) (2022) 2022 special report on fishing. Recreation Boating and Fishing Foundation, Alexandria
Robitzsch A (2020) Why ordinal variables can (almost) always be treated as continuous variables: clarifying assumptions of robust continuous and ordinal factor analysis estimation methods. Front Educ 5:589965
Rubin DB (1987) Multiple imputation for nonresponse in surveys. Wiley
Schafer JL (1997) Analysis of incomplete multivariate data. Chapman Hall, Boca Raton
Schaeffer NC, Dykema J (2011) Questions for surveys: current trends and future directions. Public Opin Q 75:909–961
Schafer JL, Graham JW (2002) Missing data: our view of the state of the art. Psychol Methods 7: 147–177
Schroeder SA, Fulton DC, Nemeth ML, Sigurdson RE, Walsh RJ (2008) Fishing in the neighborhood: understanding motivations and constraints for angling among Minneapolis-St. Paul, Minnesota metro residents. Am Fish Soc Symp 67:1–19
Shelby LB (2011) Beyond Cronbach's alpha: considering confirmatory factor analysis and segmentation. Hum Dimens Wildl 16(2):142–148
Sijtsma K (2009) On the use, the misuse, and the very limited usefulness of Cronbach's alpha. Psychometrika 74:107–120
St. John FAV, Keane AM, Jones JPG, Milner-Gulland EJ (2014) Robust study design is as important on the social as it is on the ecological side of applied ecological research. J Appl Ecol 51:1479–1485
Stern PC, Dietz T, Abel T, Guagnano GA, Kalof L (1999) A value-belief-norm theory of support for social movements: the case of environmentalism. Hum Ecol Rev 6(2):81–97
Stevens SS, Amulike B, Ndaga S, Organ JF, Serfass TL (2014) The confusion of common names: a methodological challenge. Hum Dimens Wildl 19:191–199
Sutton SG (2007) Constraints on recreational fishing participation in Queensland, Australia. Fisheries 32(2):73–83
Sutton SG, Dew K, Higgs J (2009) Why do people drop out of recreational fishing? A study of lapsed fishers from Queensland, Australia. Fisheries 34(9):443–452
Tabachnick BG, Fidell LS (2013) Using multivariate statistics, 6th edn. Pearson, Boston
Teel TL, Anderson CB, Burgman MA, Cinner J, Clark D, Estévez RA, Jones JPG, McClanahan TR, Reed MS, Sandbrook C, St. John FAV (2018) Publishing social science research in conservation biology to move beyond biology. Conserv Biol 32:6–8

TenHarmsel HJ, Boley BB, Irwin BJ, Jennings CA (2021) Perceived constraints and negotiations to trout fishing in Georgia based on angler specialization level. N Am J Fish Manag 41(1):115–129
Thurstone LL (1947) Multiple factor analysis. University of Chicago Press, Chicago
Tucker LR, Lewis C (1973) A reliability coefficient for maximum likelihood factor analysis. Psychometrika 38:1–10
Vaske JJ (2019) Survey research and analysis: application in parks, recreation, and human dimensions, 2nd edn. Venture, State College
Vaske JJ, Needham MD, Miller CA (2026) Fundamentals of sampling and implementing human dimensions of fishing surveys. In: Pope KL, Arlinghaus R, Hunt LM, Lynch AJ, van Poorten BT (eds) Understanding recreational fishers: disciplinary and interdisciplinary approaches for fisheries management. Springer Nature, Cham, p 327–353
Velicer WF, Jackson DN (1990) Component analysis and common factor analysis: some issues in selecting and appropriate procedure. Multivar Behav Res 25(1):1–18
Whitehouse-Tedd K, Abell J, Dunn AK (2020) Evaluation of the use of psychometric scales in human-wildlife interaction research to determine attitudes and tolerance toward wildlife. Conserv Biol 35(2):533–547
Williams DR, Roggenbuck JW (1989) Measuring place attachment: some preliminary results. Paper presented at the NRPA Symposium on Leisure Research, San Antonio
Williams DR, Patterson ME, Roggenbuck JW, Watson AE (1992) Beyond the commodity metaphor: examining emotional and symbolic attachment to place. Leisure Sci 14(1):29–46
Williams DR, Vaske JJ (2003) The measurement of place attachment: validity and generalizability of a psychometric approach. Forest Science 49(6): 830–840
Wolf EJ, Harrington KM, Clark SL, Miller MW (2013) Sample size requirements for structural equation models: an evaluation of power, bias, and solution propriety. Educ Psychol Meas 76: 913–934
Yong AG, Pearce PA (2013) A beginner's guide to factor analysis: focusing on exploratory factor analysis. Tutor Quant Methods Psychol 9(2):79–94

Optimizing Creel Surveys

Derrick de Kerckhove, Caroline M. Tucker, Lee Gutowsky, George Morgan, Matthew Guzzo, and Vianey Leos Barajas

Abstract The standardization of monitoring protocols across jurisdictions is a common theme in fisheries management because it provides a straightforward method for unbiased comparisons among different systems (i.e., across watersheds, lakes, states, or countries) and through time. However, in fisheries, one size rarely fits all, and as such, standard protocols must sometimes be optimized to fit the conditions of the study or management priorities. Creel surveys (i.e., the collection of socio-economic, fish, and fisheries information related to fishing trips and activities) are no exception. Further, over a long enough time, a creel survey may need modifications as the fish stock or societal conditions surrounding the fishery change. This chapter offers advice on how to measure the effectiveness of traditional creel survey protocols (e.g., stratified single-system angler interview and count surveys) and incorporate newer methods to gain additional insight into creel survey design and angler behaviour. First, we offer some practical advice on linking the mandate of fishery management with the outputs of creel surveys. Second, we introduce some approaches to improving the efficiency of the creel survey if the analyst deems the general protocol unbiased. Third, we present some approaches to identify, mitigate,

D. de Kerckhove (✉) · C. M. Tucker · G. Morgan
Aquatic Research and Monitoring Section, Ontario Ministry of Natural Resources, Peterborough, ON, Canada
e-mail: dak.dekerckhove@ontario.ca; george.morgan@ontario.ca

L. Gutowsky
Freshwater Institute, Fisheries and Oceans Canada, Winnipeg, MB, Canada
e-mail: lee.gutowsky@dfo-mpo.gc.ca

M. Guzzo
Department of Biology, University of Toronto Mississauga, Toronto, Toronto, ON, Canada

V. L. Barajas
Department of Statistical Sciences, University of Toronto, Toronto, ON, Canada
e-mail: vianey.leosbarajas@utoronto.ca

K. L. Pope et al. (eds.), *Understanding Recreational Fishers*, Fish & Fisheries Series 45, https://doi.org/10.1007/978-3-031-99739-6_13

and quantitatively correct biased data. Finally, we discuss how a Bayesian-based joint estimation modelling framework can combine multiple survey protocols (e.g., roving and waterbody access point, vehicle counts, and aerial angler counts) into one creel survey design.

Keywords Avidity · Biases · Efficiency · Management needs · Standardization

1 Introduction

The act of documenting the effort, catch, and socioeconomics associated with all types of fisheries using both on- and off-site methodologies such as counts, interviews, catch-sampling, or mail-in post cards can collectively be called "creel surveys." The data that creel surveys provide fisheries managers has been instrumental for setting fishing regulations, managing quotas, or understanding the effects of external pressures and new regulations on the behaviour of fishers. There is a long history of creel surveys being the only source of data for fisheries managers, such that quotas were often based on the previous year's yields or landings alone. Today, creel surveys are often integrated in a broader set of fisheries tools, including independent surveys and remote sensing. In addition, creel surveys remain a useful tool for evaluating fisher behaviour and specialization (Beardmore et al. 2013; Hunt et al. 2019), and the social-ecological dynamics of fisheries (Nieman et al. 2021). In this chapter, we focus on how to optimize creel surveys to improve the precision of fishery indicators, handle changing social and ecological conditions, or become generally more efficient in data collection. In general, "creel optimization" refers to changes to the protocol or analysis a manager might make to an ongoing creel survey program to leverage a desired outcome that is not currently being well addressed. More specifically, we define the analyst's role in creel optimization as loosely encompassing practices intended to improve the efficiency, effectiveness, and utility of creel surveys. As we demonstrate, creel optimization is desirable for a variety of reasons and achievable using strategies tailored to specific project objectives. Here we discuss commonplace rationales for optimization and several approaches that fisheries managers can take.

The needs of fisheries management are in a constant state of change. Standard methodologies for creel surveys have been established for decades (Malvestuto 1983; Pollock et al. 1994), and though these resources can provide sources of advice for designing or optimizing creels (e.g., see the many case studies described in American Fisheries Society Symposium - Guthrie et al. 1991), the type of information these survey designs yield could be less relevant now than they once were (Nieman et al. 2021). The information provided by creel surveys should be revisited over time to address these shifting priorities and societal norms. Evidence of such concern was clear on Lake Opeongo (Ontario, Canada), site of the oldest continuous recreational fishery creel survey (ca. 1930), where the adoption of the family car and paving of highways was predicted to substantially change the lake trout (*Salvelinus namaycush*) fishery (Shuter et al. 1987). However, the next few decades brought unanticipated disruptions, including World War II and the total loss of guided American tourists, which was cause for a modified creel survey design for Lake Opeongo, and even today the increased use of advanced fish finders is changing our perception of angler activity metrics. Since the 1991 symposium, the fishing landscape has continued to change, with the advent of the internet, mobile communications, and advanced fishing technology, as well as public sensibilities around outdoor recreation. Further, social-economic shocks, including the 2008 global financial crisis to the start of the 2020 COVID-19 global pandemic, had strong and immediate impacts on fishers and fisheries. This backdrop of constant societal change makes the optimization of creel survey protocols inevitable, because time will catch up to even the best designs.

A creel surveyor can choose to design a new protocol from scratch, circumventing the need for optimizing existing protocols. A new protocol would allow a fresh look at the changing fishing landscape and let the designer apply new techniques. However, implementing a new protocol is generally impractical as most resource managers desire a level of continuity with past creel surveys to be able to detect trends through time, even if collected from a different water body. Therefore, aspects of past but imperfect surveys often need preserving in the creel survey design process. This chapter focuses on such a scenario, where an existing creel survey protocol was designed following standards of time-honoured advice and methods (Carlander et al. 1958; Malvestuto 1983; Pollock et al. 1994), but is now in need of modernization.

In general, scrutiny of an existing creel survey design arises from a perceived change to fisheries and their management. In our experience and others (Soupir et al. 2006; Vølstad et al. 2006; McCormick and Meyer 2017), a common driver of change is the budgets of resource managers, who must optimize creel surveys to serve the same function with less resources. However, there are many other possible drivers of change. Those related to the fishery could include an external change to the fishing society (e.g., economic recession, a pandemic, or war), a new management action to the stock (e.g., a regulation that changes the efficacy of fishing effort) or the emergence of a new fishery within the system (e.g., stocking or invasive species). Those related to the protocol itself could include a change to the logistics of implementing the creel survey (e.g., increased bad weather days) or new Health and Safety regulations that affect the timing of the crew's shifts. Finally, those related to the creel survey analysis could include a recognition that internal management needs were not being met (e.g., trend detection is overwhelmed by uncertainty) or that the existing analytical framework had become antiquated (e.g., old computer software may no longer be compatible with new operating systems). In all these scenarios, and the many others that exist, designers will be faced with the decision of what aspects of creel optimization will most likely address the driver of change while reflecting various constraints.

In this chapter, we present four general approaches to optimizing an existing creel survey based on some driver of change. We start with evaluating the survey protocol based on the direct linkages between outputs and management actions. Second, we present basic approaches to improve the efficiency of a survey protocol that is assumed to be unbiased. Third, we present approaches that could be used to address potential bias or uncertainty in survey outputs. And last, we suggest a modern, flexible analytical approach for the joint estimation of creel survey parameters from a range of potentially biased or unbiased surveys. We will draw from published and unpublished sources to present optimization strategies and provide context from our experience with the management of Ontario, Canada's fisheries.

2 Optimizations in Response to Changing Management Needs

Given sufficient time, creel survey programs may span the tenure of multiple management regimes, political climates, and sometimes even stocking priorities. In Ontario, Canada, many creel surveys have been more-or-less continuous since the 1960s when management expectations were more focused on effort and catch-rates. Prior to the development of formal Biological Reference Points (Smith et al. 1993), fisheries were largely monitored for drastic changes in fisher-focused metrics such as effort and catch rate. As the theory of maximum sustainable yield became prevalent in management systems, associated harvest estimates became the primary indicator of interest (e.g., Shuter et al. 1998). Today in Ontario, most heavily exploited fish stocks are managed based on fishery-independent data (e.g., scientific surveys; Bonar and Huber 2002). As such, the role of creel surveys has shifted from stock assessment to a tactic for tracking the socioeconomic and demographic aspects of fisheries, e.g., fisher satisfaction and economic expenditures. Sweeping changes to management objectives, such as those witnessed in Ontario, have been the central impetus to creel design optimization. For example, if a fishery experiences a reduction in exploitation, the precision in estimates of fishing activity will decline without increased intensive and targeted sampling. Managers may prefer to shift the priority to allocating limited resources to conduct comprehensive interviews that ascertain fisher satisfaction, rather than chase a difficult fishery metric.

Optimizations in response to changing management needs first require assessing and reassessing the mandate of the creel survey and how it aligns with current fisheries management objectives: e.g., Do managers need new or different information? What are the budgetary limitations? Do known design and statistical limitations of existing creel surveys need addressing? For older creel programs, it is useful to consider management actions at the time the creel survey was designed, and to determine how the decision-making process was informed in the past and will be informed going forward. This requires both a quantitative and qualitative lens. Some indicators are associated with clear quantitative requirements of creel surveys, such as sufficient sample sizes to ensure there is adequate statistical power to detect an effect (Parkinson et al. 1988; Peterman 1990). Other indicators–for example, fisher satisfaction—may be qualitative measures meant to assure managers or stakeholders that important characteristics of the fishery are recognized by decision-makers.

Sometimes the same indicator may be used both quantitatively and qualitatively. For example, a declining catch rate might have once been an important quantitative indicator but has now become a verification tool for fisher complaints. The direction of change for an indicator may also be relevant. For example, assessing a recovering fishery is difficult if low rates of fisher activity and harvest result in an abundance of zeros in an indicator (McCormick and Meyer 2017). Rather than greatly increasing survey effort or turning to zero-inflated or hurdle models to match the statistical distribution (Zuur et al. 2009; Hofstetter et al. 2016), decision makers may temporarily accept low statistical precision and accuracy until the statistical power of the survey improves as the fishery is managed to achieve targets.

A second step is to revisit and re-evaluate the practicality of the protocol, including elements of the creel survey design. One of the most obvious is to determine the ways in which surveys and the associated datasets are stratified. Stratification is a means to group similar categories or strata (e.g., time of day, depth) and reduce the overall variance of a survey estimate. In practice, stratification will minimize variation within a stratum by maximizing it among strata (e.g., the stratification of a busy tournament weekend from the rest of the season). However, in our experience survey stratification may be conducted out of convenience, based on the available technology or historical fishing patterns. For example, spatial strata for large lakes may be defined by the distance a field crew could travel within one temporal stratum (e.g., AM-PM shifts). Should the mode of surveyor transportation change (e.g., from walking to ATV or snow machine), such spatial strata can change or become irrelevant. A similar example includes within-season strata, which sometimes were based solely on reporting timelines for within-season closures or the behaviour of fishers relative to nearby popular fisheries (e.g., fishers moving from closed to open fisheries), both of which are prone to change.

The statistical uncertainty between levels of stratification in creel surveys is commonly measured by the coefficient of variation (CV). The CV is a unitless measure of relative standard deviation meant for assessing the variability or dispersion of data, particularly among datasets. This simple metric is useful for comparing the relative variation of creel surveys with differing strata and by extension, informing the optimal survey design given management objectives. For example, reductions in the CVs of fishing effort have been achieved through seasonal stratifications associated with opening fishing weeks (Sztramko 1991) or periods of good fishing (Weithman and Haverland 1991). Malvestuto and Knight (1991) used this approach to show that day-type (i.e., workday vs non-workdays) reduced the CV of fishing effort, but AM-PM strata had no benefit. The authors further demonstrated that though frequent sampling reduced the CV of monthly estimates, the strategy failed to affect the precision of the annual estimates (Knight and Malvestuto 1991) that are likely more relevant for managing exploitation on the stock. A similar analysis can be conducted for catch- or harvest-based outputs, as Crone and Malvestuto (1991) demonstrated with Harvest-Per-Unit-Effort. Exploring a range of options in stratification or primary sampling units may reveal higher levels of precision with a simple change in survey design.

It is worth evaluating whether alternative survey methodologies available through technological advancements could more efficiently yield the same observations as traditional methods. Boat counts are a good example, and many creel surveys depend on mobile or access point creel (i.e., the count and interview of fishers as they come off the water or ice at the completion of their trip, herein referred to as access creels) crews to document the number of the boats on the water at a given time. The same count may be collected through traffic counters (Hartill et al. 2020), shoreline cameras (Stahr and Knudsen 2018), or plane- or drone-based aerial counts (Askey et al. 2018). This methodological change does little to affect the quantitative results of the creel survey, but potentially makes data collection faster, more extensive, more frequent, or cheaper. It is worth noting that though technological advancements may save money, they will require verification from experimental studies to test for bias, e.g., traffic counters (see Simpson 2018) and aerial counts (see Soupir et al. 2006; Vølstad et al. 2006).

Although creel surveys have the capacity to address a much wider range of questions than those immediately required by resource managers (Brownscombe et al. 2019), they are susceptible to "scope creep" as additional questions or protocols are added. Unfortunately, these additions can often neglectfully become enshrined in long-term protocols and should be identified and considered for removal. For example, in the Ontario Broad-scale Monitoring Program, 700 lakes were surveyed through netting and aerial fisher count across a 5-year period (Lester et al. 2021). Over the first 10-years, complex flight lines were necessary to ensure that the lakes being netted were also flown in the same year. These paired flights were a temporary and undocumented choice as part of a study to link fisheries-independent (i.e., index netting) and fisheries-dependent (i.e., fisher activity) indices, yet the complex flight paths continued long after the study was completed. Decoupling flight lines from the netting proved a simple design change that saved significant annual operating costs.

3 Optimizations to Increase Efficiency

When a creel program is optimized with the primary goal of improving efficiency—i.e., the monetary or labour cost associated with a survey—it becomes necessary to first confirm that the underlying assumptions are robust, and as such, managers are not compelled to expend resources to examine possible sources of bias. Managers may feel confident in this decision if indicator precision appears relatively stable across time, results match with stakeholder expectations, or results are verified through independent and associated surveys (e.g., similar exploitation rates are derived from index netting surveys). The need for optimization can arise from an external factor that disrupts the ease of implementing the survey. Perhaps expenses have grown, logistical challenges for the crews have emerged, development has brought new access points for the fishery, or a new fishery has emerged through stocking, species invasion, or changes in fisher preferences. The goal of optimizing creel survey efficiency is to reduce survey effort while maintaining the existing robust assumptions and statistical precision.

A simple preliminary analysis is to check whether the sample size (i.e., the numbers of counts, interviews, or fish samples per strata) result in the desired precision of the survey. Most creel surveys did not have a robust statistical power analysis at the time of their design, because the within- and among-strata variances were unknown ahead of time unless there was a pilot study or robust prior knowledge. However, historical long-term datasets can provide good estimates of variance. For example, Lester et al. (1991) calculated statistical power using the variance in fishing activity, catch-per-unit-effort, and mean party size from long-term datasets for 47 lakes in Ontario, Canada. They developed sample size guidelines for a prescribed institutional threshold of 20% relative standard error (or equivalently, a coefficient of variation), and showed that 20–30 survey days were required to detect catch rates of at least 1 fish per 10 h of fishing in active fisheries.

McCormick et al. (2015) present a robust methodology with accompanying programming code to generate sample size estimates to achieve a desired level of precision for effort and catch in a multi-stage design with days as the primary unit, and AM-PM and workday-non-workday stratifications. The authors present an analytical (i.e., solving the variance equations for sample size; Schaeffer et al. 2006) and an empirical (i.e., bootstrapping to achieve sampling distributions of effort, catch-rate, and catch) methodology to derive a relative confidence interval. 40% of the mean was selected as a suitable threshold for precise creel survey outputs (note that this is roughly equivalent to the 20% relative standard error referenced above). The sample size (n) for fishing effort (E) is analytically calculated as:

$$n = \frac{N^2 \left(Z_{\alpha/2}\right)^2}{CI^2_{\widehat{E}}} \sum_{k=1}^{L} \left(\frac{N_k}{N}\right)^2 \frac{s^2_{\widehat{E_k}}}{w_k.}$$

where L is the number of strata, k is a particular stratum, N is the number of days in all strata (unless with a k subscript in which case it is the number of days within a stratum), w is the proportion of the sample that will be allocated to the stratum, $Z_{\alpha/2}$ is the critical value for the confidence interval (i.e., 1.96 for a 95% CI interval), $CI^2_{\widehat{E}}$ is the confidence interval width, and $s^2_{\widehat{E_k}}$ is the sample variance, which is calculated as:

$$s^2_{\widehat{E_k}} = \frac{\sum_{d=1}^{n_k} \left(\widehat{E}_d - \underline{E}_k\right)}{n_k - 1}$$

where d is the day, E_d is the effort over that day, and E_k is average effort over the stratum. Note that this sample variance is an approximation of within-sample variance, which could be directly estimated if creel survey crews could perform multiple counts within one shift in one stratum. If a creel optimization is a planned event, resource managers may want to consider adding activities such as extra crews for multiple counts per shift during the year preceding the analytical work. This formulation was used to estimate a minimum average sampling size for nine fisheries in Idaho and Oregon (McCormick et al. 2015).

Statistical power analyses (see Fabrizio et al. 1991 for a change detection example for Lake Michigan, USA) and sample size estimators illustrate the law of diminishing returns such that improvements in precision at low relative standard errors require a tremendous increase in survey effort, especially for low activity or low catch fisheries. Conversely, creel surveys will fluctuate between under- and over-sampling fishers if the crew's effort fails to scale with the characteristics of the fishery. It can therefore be instructive to revisit annual data from a long-running creel survey program, calculate the historical ideal sample sizes (Lester et al. 1991; McCormick and Meyer 2017), and derive a relationship of creel survey intensity with the fluctuating fishery. As a demonstration, we present such an approach for a winter walleye (*Sander vitreus*) fishery on Lake Nipissing (Ontario, Canada), for which 25 years of annual creel survey data revealed catch rates ranging by an order of magnitude. Here, we empirically derived a relationship between the relative standard error of the number of walleye caught and actual sector-days that were surveyed (Fig. 1, panel A), as well as the effect of the catch rate on the estimated sector-days required (Fig. 1, panel B). This exercise showed that fisheries assessments routinely over-surveyed the lake (~120 sector days annually) and could relax creel survey sampling effort by threefold (~40 sector days) when the stock was considered healthy. Bayley et al. (1991) used this same type of approach on 33 lakes in Illinois, USA, to show that 30 and 60% of the available days in a fishing season needed to be surveyed to achieve a reasonable precision for estimates of fisher effort and harvest.

Although instructive to examine long-term datasets, creel survey designers must consider whether surveyor mobility has increased through time. Indeed, the mode and manner of on-site crew movement have considerable implications for the efficiency and precision of a survey. This fact becomes most apparent when creel designers are tasked with surveying vast, complex systems (e.g., the Laurentian Great Lakes in North America or long marine coastlines), or a logistically challenging distribution of access locations (see Vølstad et al. 2006) or linked lake systems (see Soupir et al. 2006; Chizinski et al. 2014). Advice from the aforementioned research articles and the non-uniform bus-route creel survey protocols described in Pollock et al. (1994, 1997) can be extremely helpful in optimizing existing protocols as access points or fisheries change. For example, Roop et al. (2018) discuss the need to "coarsen" (i.e., accept a lower level of resolution on lake-specific catch rates, sensu Chizinski et al. 2014) their creel estimates as the logistics of visiting multiple lakes within a Georgia (USA) lake-chain became untenable. In Ontario, fisheries assessment programs have begun including the genetic composition of certain stocks. Over time, fisheries managers learn how stocks are distributed across multiple lakes such that there may be a benefit in "coarsening" the estimate to a lake-chain value, and substantially reducing creel survey sampling effort. Another approach is to shift from a two-stage strategy (i.e., count first, interview next; Lester et al. 1991) to "creel-as-you-go" strategy (described in Pollock et al. 1994). For assessing angler effort, encounter rate with fishing parties must be applicable system-wide without bias. Robson (1991) demonstrated, using a simple geometric

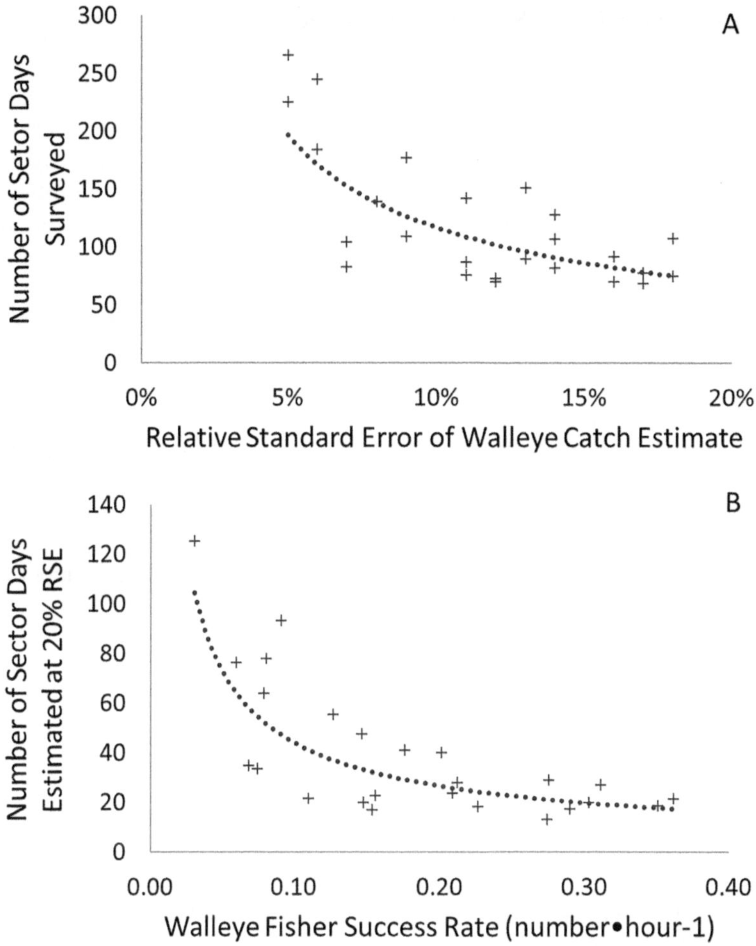

Fig. 1 Relationship between the relative standard error of the number of walleye caught and actual sector-days that were surveyed (panel A), and effect of the catch rate on the estimated sector-days required for the winter fishery on Lake Nipissing (years; panel B) (following Lester et al. 1991)

model, how the length and timing of the creel survey within the fishing day could be used to generate the probabilities of encountering any given angler. If designed properly, angler abundances could be calculated based on encounter rates without the need for area-based counts. Further, these models have inherent levers for optimization, as they can be used to demonstrate how the length of any single interview will affect the overall encounter rates by lowering the active search times (Robson 1991; Wade et al. 1991). Finally, protocols that force counts without interviews at particular sections of the route not only maximize the area covered in one shift, but also have the most accurate and precise results.

4 Optimizing Creel Surveys to Reduce Bias

Biases in creel surveys can result from systematic issues in survey design, data collection, and analytical approaches and more generally, from an imperfect understanding of the fishery. Identifying bias from observational data is challenging, but managers may become suspicious of bias if their conclusions greatly differ from their stakeholders' impressions of the fishery or if creel survey results are not robust to perturbations in protocol or external conditions. Cataloguing all sources of bias is challenging because creel methodologies are diverse and generally contain different errors for on-site vs off-site surveys. Further, though less discussed in the creel survey literature, common errors associated with the processing of scientific data also arise during data collection, quality control, coding, and preparation for analysis (Mazzi et al. 2019). Essig and Holliday (1991) provide a good example of specific sources of bias during various types of surveys (reproduced in Table 1), including sampling errors (e.g., imprecise protocols or instructions), non-coverage (e.g., missing a portion of the angling population), avidity bias (e.g., preferentially surveying avid anglers), and a set of response and non-response errors. In this section, six types of bias are presented, including coverage, response, nuanced avidity, temporal, and spatial biases.

4.1 *Coverage*

Given that financial resources limit the scope of creel surveys, they must be thoughtfully designed to best capture the full spatiotemporal extent and inherent variability of the fishery. Issues arising within the sampling scheme are likely to be closely tied to logistical limitations and can make optimizing coverage difficult. Before altering sampling effort to improve coverage-related biases, managers should carefully consider the major sources of spatial and temporal variation in the fishery to identify necessary changes in sampling effort. For example, if between-day variation in activity is especially high, reducing or shuffling sampling effort (e.g., by reducing the number of days sampled) will disproportionately affect creel estimates. If within-day variation is higher than among-day variation, it would be most valuable to maintain multiple daily time slots. If spatial variation is high, maintaining sampling across the spatial extent of the system should be a priority, potentially at the expense of some relatively superfluous strata (see below for more discussion of spatial structure).

4.2 *Response Bias*

Biases associated with angler interviews include prestige (providing answers to enhance their self-image), avidity (where keen anglers are more likely to respond to surveys), recall (the accuracy of a response is dependent on the time since the

Table 1 Specific forms of bias arising during fisheries creel surveys. (Reproduced from Essig and Holliday 1991; errors are rated as having low [L], medium [M], or high [H] probability of occurring in a survey)

	On-site survey methods					Off-site surveys methods			
Error type	Roving	Access point	Catch cards	Log-book	Aerial	Mail	Known phone	Random phone	Door-to-door
Sampling errors									
Improper sample selection	L	M	H	H	L	M	H	L	M
Noncoverage	L	M	H	H	H	H	H	L	H
Avidity bias	H	H	H	H	H	H	M	L	L
Responses errors									
Recall bias	L	L	M	M	L	H	M	M	M
Prestige bias	M	M	H	H	NA	H	M	M	M
Digit bias	L	L	H	H	L	H	H	H	H
Intentional misreporting	L	L	M	M	NA	M	M	M	M
Question misinterpretation	L	L	H	H	NA	H	M	M	M
Species misidentification	L	L	H	H	NA	H	H	H	H
Incorrect lengths-weight	L	L	H	H	NA	H	H	H	H
Nonresponse errors									
Refusals	L	L	H	H	NA	H	M	M	L
Language barrier	L	L	L	L	NA	L	L	L	L
Physical-mental impediment	L	L	L	L	NA	L	L	L	L

event occurred), and misreporting biases (errors in self-reporting including exaggeration, and misidentification; Sullivan 2003) (Table 1). These biases are typically believed to worsen when data are self-reported (e.g., catch cards, logbooks) and collected off-site (mail, phone) (Essig and Holliday 1991; Pollock et al. 1994; Connelly et al. 2000). Novel interview technologies such as smartphone apps (see Venturelli et al. 2017, 2026; Gutowsky et al. 2013) and e-licencing systems (Riggers and Jones 2022) may render surveys easy to distribute, but remain prone to self-reporting biases, e.g., avidity and age demographics will affect response rates. Self-reporting methods have been found to under- or overestimate catch and harvest depending on the context of the fishery and the creel methodology. For example, Weithman and Haverland (1991) found that catch rates for a variety of species were underestimated by ~20% when using a telephone survey, whereas McCormick et al. (2015) found a springtime underestimate of 16% and an otherwise 24% overestimate of steelhead trout (*Oncorhynchus mykiss*) harvest when using a mail or internet survey. Even in-person interview methods are susceptible to responder biases, e.g., access creels can underestimate harvests by 1.1× to 1.5× (Claytor and O'Neil 1991). Approaches to reduce responder biases include encouragement of rapid responses (e.g., provide anglers with no time to consider their exaggeration), offer incentives, carefully construct questions to avoid easy rounding errors, and avoid relying on scaling one factor by another (e.g., using time fished scaled with catch-rate to estimate harvest, rather than simply ask for harvest) (see Steffe et al. 2008; Lawson 2015). Consider how Kokel et al. (1991) found that avid anglers can still bias off-site mail or phone interviews, at least doubling estimates of "how long do you fish". Ultimately, self-reported data is most reliable when paired with other forms of data collection for validation and the development of bias correction factors, including follow-up phone calls (Hogg et al. 2010). Bias-corrected estimators of effort per fisher can further be used, or survey methods can be used which target fishers (e.g., phone survey) and not trips (e.g., roving or access creels) (Thomson 1991). For instance, the Marine Recreational Information Program combines their Access Point Angler Intercept Survey (APAIS) results with two off-site survey methods to correct coverage biases. The Fishing Effort Survey (FES) involves surveys mailed to all addresses within the study states; APAIS captures both out-of-state and state fishers (U.S. focus), therefore it can be used to correct for the absence of out-of-state fishers interviewed in the FES.

4.3 Nuanced Avidity

Avidity bias (Thomson 1991), where dedicated fishers are most likely to participate in a survey, is a well-established issue for off-site creels. However, there are nuanced aspects of avidity bias for on-site creels as well. Roving or access creels are intended to sample fishers at random, `but as a result, particularly avid fishers are more likely to be sampled (or repeatedly sampled). This mismatch between the intended design of creel surveys to sample population-level fishing effort and desire for individual-level data skews fisher demographics, under-represents more casual users of the fishery (Hunt et al. 2010), and can over-inflate estimates of per-capita effort (Thomson 1991). On the other hand, describing effort requires a representative sample of fishing trips, including the length and frequency of trips by avid fishers. Thus, the purpose of the survey will determine whether accounting for avidity is necessary. The variation between infrequent and frequent users of a fishery can be large. For example, Cowie and Ridgway (2023) conducted an access creel survey to gain insights into fisher motivations, methods, and demographics at Lake Opeongo, which supports the most popular lake trout fishery in Algonquin Provincial Park, Ontario, Canada. The authors applied a Cormack-Jolly-Seber mark-recapture model on boat registration numbers and found that most fishers are not avid or frequent users of the fishery. In fact, 70% of fishers were recorded only a single time in 14 years, whereas the most active participant was detected 128 times (<1% were observed every year). These data were relatively stable through time, although travel distance and, to a lesser extent, gas prices affected the probability of participation. Further, they demonstrated that observing "fishers" and not "trips" as the target measurement can be a useful exercise to control for such bias (Cowie and Ridgway 2023). A simple approach is to ask fishers how many times they have been interviewed during the current season. If fishers have been interviewed at least once, surveyors may ask the same demographic questions on repeated sampling. There exists a rich literature on assessing the fisher's avidity in terms of the centrality of the activity to their lifestyle. Beardmore et al. (2013) offer a list of useful metrics to assess this type of avidity. A practical, on-site approach used by Ontario's creel survey technicians is to ask whether fishers know the current regulations and classify them as such for qualitative and quantitative post-interview evaluations.

4.4 Temporal Biases

Biases associated with time, such as trip length, and survey design aspects such as timing, length, and trip timing of surveys, must be considered when designing a creel survey program. For example, roving creel surveys typically intercept fishers during their trip and therefore require an extrapolated estimate of fisheries metrics based on incomplete trip data. Though beyond the focus of this chapter, it is worth noting that common solutions to such biases are well described. For example, there are persistent biases when comparing mean-of-ratios and ratio-of-means estimators (e.g., Alexiades et al. 2015). Alexiades et al. (2015) suggest these arise from measurement error and can be addressed by using model II regression rather than linear regression because both the response and predictor contain measurement error. Other approaches rely on self-reports after a trip has concluded, even though post-trip self-reporting is also sensitive to bias (as described above). For example, ice fishers were given a self-reporting mail-in card during a winter creel survey program on Lake Nipissing, Ontario, Canada. On-ice creel survey ends at 5:30 pm local time, which precedes civil twilight and safe travel conditions for surveyors. Given that the roving creel was collecting incomplete trip data, fishers were asked when they expected to finish fishing, and asked to later record their actual fishing end time before mailing the survey card. Findings indicated that fishers generally predicted the end of their trip within a 30 min window of their actual completion ($r = 0.8$), suggesting that self-reported estimates of fishing end times could be a useful alternative to estimate effort when completed trips cannot be observed (D.T. de Kerckhove, Ontario Ministry of Natural Resources and Forestry, unpublished data).

Variation in fishery activity, effort, and catch rate within a single day can be an important but under-considered source of bias. For example, differences in the ecology of the target species may produce different daily patterns of fishing behaviour that are not well-described by uniform sampling effort (Parker et al. 2006) within a broad time strata. Parker et al. (2006) estimated diel activity patterns of fishers on Ontario lakes, demonstrating significantly different daily time-based distributions of fishers occurrence on the waterbody depending on the season (open water vs. winter), target species (walleye, lake trout, or bass), the region of Ontario (NW, NE, or S), and the distance of the lake to a road. A similar bias was uncovered on Lake Nipissing, in which a large proportion of interviewed fishers (41%) reported a fishing end time after the winter creel survey crews were off the ice. The main targeted species here is walleye, which is particularly active in low light conditions (Lester et al. 2004), and yellow perch. Using mail-in cards handed to the fishers at the time-of-interview, walleye catch rates were low when estimated via roving data (0.258 fish/hour, 0.122–0.368, 95% CI) and pre-5:30 pm reports from mail-in cards (0.290 fish/hour, 0.194–0.373, 95% CI), but were almost double for mail-in card reports calculated after 5:30 pm (0.473 fish/hour, 0.342–0.589 95% CI). To address such a temporal bias in activity, simultaneous access creel survey times on Lake Nipissing have been extended later into the evening to now provide more accurate estimates of fishing effort, catch, and harvest (0.554 fish/hour, 0.514–0.592, 95% CI).

4.5 *Spatial Biases*

Simulating potential survey paths over known spatial distributions of fishing parties can be used to improve estimates of fishing effort in creel surveys, but less often for catch rates (Wade et al. 1991). However, in "creel-as-you-go" protocols, it is important to note that if fisher success is spatially correlated to a particular set of habitat features, understanding the spatial autocorrelation of fisher success may allow for unbiased survey paths and count strategies to be developed. For example, on the Canadian side of Lake of the Woods (Ontario), which supports a large recreational walleye fishery, a "creel-as-you-go" strategy was implemented to solve the logistical challenges of completing counts in large waterbodies. Moreover, knowing the distribution of fishers can help direct creel surveys and provide valuable spatial data for specifying statistical prior probability distributions in future Bayesian estimation models. For example, data from the Lake of the Woods creel survey program was recently used to evaluate factors explaining fisher success (catch >0) during spring, summer, and fall fishing seasons. To model the counts for each season, we used a Poisson distribution where the expected number of catches on the log-scale is a function of five explanatory variables: time spent fishing, fishing start time, party size, and each two-way interaction. We further included a spatially correlated random effect component to account for spatial dependence in the model (Gutowsky et al. *in prep*). This approach is described well using an Integrated Nested Laplace Approximation and is often referred to as the INLA model (Gomez-Rubio 2020), or the R-INLA model using a common statistical programming platform (Bakka et al. 2018). In this approach, a normally distributed spatially dependent random intercept is estimated as a Gaussian Markov random field for which the parameters are estimated by a Matérn correlation function using the continuous domain stochastic partial differential equation approach (Lindgren et al. 2011). The INLA approach has shown great utility in separating the effects of spatial autocorrelation and a spatial random field from the direct effects of important covariates. As such, in the Lake of Woods analysis, we can test whether space should be considered at all as an important factor in designing our future surveys, and whether our covariates are likely to exhibit any spatial autocorrelation.

A spatial component improved the model's parsimony based on an information theoretic approach (i.e., each spatial model had a DIC at least 2 less than the next non-spatial model, Spiegelhalter et al. 2002) and illustrated spatial autocorrelation in catch would decrease to <10% at 62 km, 43 km and 55 km in the spring, summer, and fall, respectively. Overlaying the GMRF with existing sector-based stratifications showed no substantial pattern in fisher catch, which reinforced the hypothesis that sectors were designed for logistical rather than socio-ecological reasons. In planning a "creel-as-you-go" methodology with insight into spatial behaviour on Lake of the Woods, crews can focus effort on broader areas, perhaps based on boat accessibility, separated by 40–60 km in distance to avoid coverage bias.

5 Integrating Multiple Data Sources to Address Individual Biases

If target data are too sparse (e.g., infrequent visits to remote lakes or cost limitations), unbalanced, or missing too many values (e.g., missing days due to poor weather conditions), addressing the problem through expanded or altered sampling activities may be impractical. An increasingly common approach is to collect auxiliary data that provides additional information related to, but distinct from, more typical targets such as access activity counts. Approaches include aerial counts of fishers, ice huts, or fishing boats, traffic records, counts of boat trailers or vehicles parked in lots near access points, and remote cameras. Importantly, auxiliary data may be less expensive, logistically easier to implement, or faster to collect. Optimally, the principal survey data and auxiliary data should be gathered in tandem to allow for validation and identification of biases. A typical analytical practice is to derive a conversion ratio or mean expansion factor from auxiliary data that can be summed among strata and the conversion factor applied to the total (Steffe et al. 2008).

As an alternative to performing independent analyses on creel survey data, methods exist to jointly estimate target measurement by integrating multiple data types and models (Hoyle and Maunder 2004). Rather than fitting separate models to multiple sources (or types) of data, model parameters are estimated from a shared, multivariate distribution. Joint estimation reduces measurement bias by increasing replication (e.g., by integrating data from multiple years or different collection methods). Further, it allows several related data sets (i.e., for access creel and aerial survey data) to be fit with sub-models within the same larger model. Joint estimation improves model specification and precision and reduces bias (Maunder et al. 2013 and references within). This analytical practice is increasingly prevalent in fisheries stock assessments where, for example, catch-at-age models often integrate multiple types of data (catch, effort, age, or size from multiple sources) and models (Aeberhard et al. 2018, Maunder et al. 2013 and references within). Despite these benefits, joint estimation methods are relatively uncommon in creel survey analysis (but see Trudeau et al. 2021, Tucker et al. 2024). Here we present joint estimation methods for imputing missing values and integrating multiple datasets under one model for creel assessment.

5.1 Joint Estimation to Impute Missing Values

Sparse datasets are a common challenge in fisheries assessment, especially across varied landscapes with multiple access points. To address the issue of sparse data, van Poorten et al. (2015) applied joint estimation on time-lapse camera-derived fisher counts for recreational fisheries across 91 lakes in British Columbia, Canada. Due to the inherent cost to survey the region, only a subset was selected for creel surveys. Concurrent access creels facilitated the calculation of a correction factor to account for a limited field of view for some cameras. The authors then built a zero-inflated model of total effort that incorporated both creel survey and camera effort. Once fit to validation data, the model was used to impute fishing effort on times when creel surveys were not conducted, so the seasonal fishing effort is simply the sum of creel observations of effort and imputed effort. Using the same models, expected total fishing effort for lakes where no creel surveys were conducted (only the camera data were available), fishing effort was imputed for the entire season, effectively providing a solution for missing data. It is worth noting that credible intervals around the resulting predictions were generally small (but with some variation), even when predictions were made for lakes without creel surveys. This provides an important tool to monitor trends in fishing activity even when the relevant lakes are unable to be surveyed regularly by staff.

5.2 *Joint Estimation to Integrate Several Data Types*

Lake Simcoe, a large (744 km^2) inland lake in southern Ontario, supports a popular recreational ice fishery for yellow perch (*Perca flavescens*), lake trout, and lake whitefish (*Corgonus clupeaformis*). Yearly winter roving creel surveys have been conducted since the 1970s, however, with ice conditions becoming unsafe for survey crews to traverse over the last few years, there has been an increasing reliance on access creel surveys. As a matter of practice, the roving surveys were considered more robust than the access surveys because the latter imposes limitations for fisheries assessment on Lake Simcoe due to the highly developed shoreline and large number of public and private access points. To improve upon the poor confidence in the access surveys, auxiliary data, including aerial counts of ice huts and vehicle counts at access point parking lots, have been collected since 2020. Counts of vehicles are an alternate method to estimate the number of fishing parties; however, the analysis requires the assumption that parked vehicles are associated with a fishing party. Aerial surveys provide a broad-scale overview of active huts, but party-specific information on effort and catch is unavailable (Fig. 2).

To reduce bias and improve the precision of fishing activity estimates, we jointly modelled all three forms of data (access creels, vehicle counts, and aerial surveys; Tucker et al. 2024). Basically, access to creel data is used to inform rates at which fishers arrive at, and leave, the fishery each day (sensu van Poorten and Brydle 2018). This data provides an estimate of total proportional effort, which is then scaled to total effort based on instantaneous counts of vehicle counts or aerial surveys, thus providing total estimates of fishing activity. At this time, appropriate stratifications can be easily applied to the analysis if they are expected to significantly affect the means and variances among strata. Important strata retained as covariates in the Lake Simcoe model included day type (workday and non-workday) and ice thickness. Last, covariates are treated as potential scalars that affect the average activity at a site, for which vehicle counts provide a suitable example. This general model was fit to the data sets collected by stationary creel survey attendants at access points, and mobile crews counting parked cars by truck or ice huts by air, using a Bayesian framework. We found that predictions for mean site activity on a given day type had much higher precision when at least one auxiliary variable was included in the model (as compared to using access data alone). The scalar parameters for vehicle or ice hut counts provided useful insights into the fishery and larger social system. There is a large difference between the number of vehicles counted and the associated number of parties observed at an access creel, with approximately three vehicles being counted for each party counted at the access point. This highlighted the importance of many of these parking areas for non-fishing activities, including dog walkers, hockey rink users, and other activities. For aerial surveys, five ice huts were counted for every party. We expect that commercial businesses and private property provide lake access to many parties who are otherwise not counted by the access creel. This joint estimation process allows estimates to move from relative to absolute, by allowing each data type to correct biases associated with all other data types.

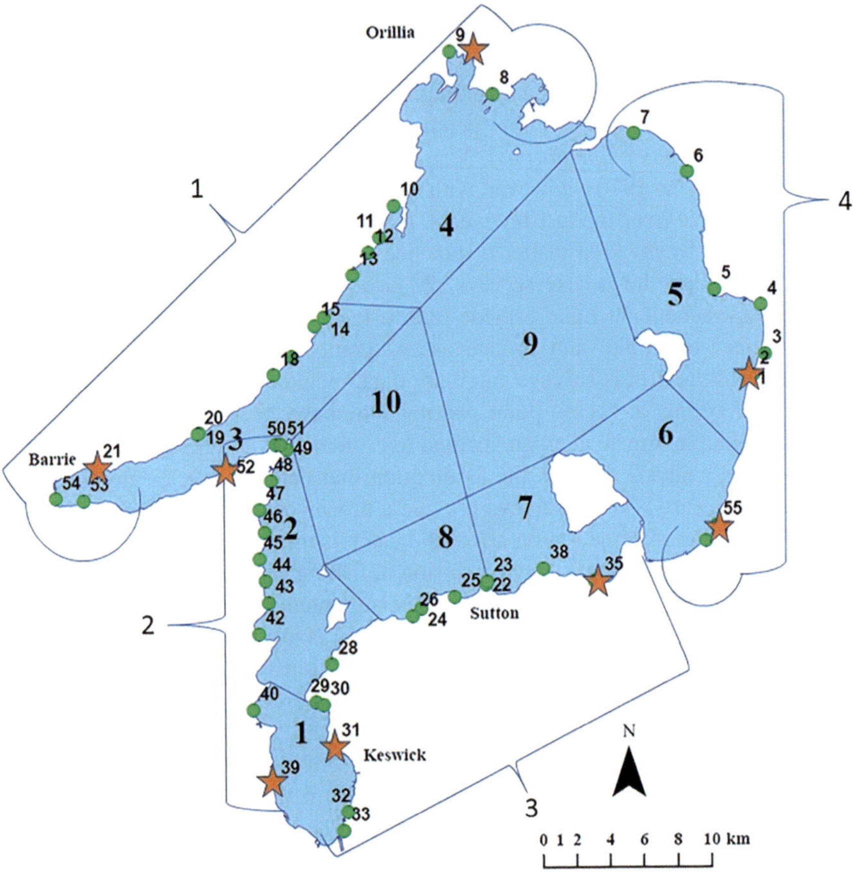

Fig. 2 Lake Simcoe creel design for roving creels and aerial surveys (numbering from 1 to 10 and blue lines delineate the ice sectors; access points for access creels shown numbered from 1 to 55; green circles are minor points and orange stars are major points; parentheses showing subsections of the shoreline are routes that can be covered by a creel team in a day)

6 Conclusion

Fisheries are complex and dynamic systems that demand considerable resources to sustainably manage. Although standardization remains crucial for robust, unbiased, and replicable monitoring protocols, the one-size-fits-all approach is increasingly recognized as an unsuitable approach. Improvements in technology and fisher mobility, changing demographics, and climate change (e.g., ice-on and ice-off dates in temperate climates) require constant adaptation to programs expected to simultaneously deliver on targets with adequate precision and reduce operational costs on limited budgets. Unless optimized in some manner, our experience indicates that expensive creel surveys are among the first components pruned from monitoring programs.

Creel survey optimization often begins with simple re-evaluations of the statistical power being achieved by the standard effort of the crew. We often find that legacy survey programs were designed at a time with greater funding capacity, or similarly, greater uncertainties surrounding the health of the fishery, and so were given a higher level of effort than is strictly necessary to achieve robust estimates of valued fishery metrics. For some systems in Ontario, we have been able to cut costs by as much as half when using this approach. Second, we have found that the re-evaluation of stratifications from a logical perspective, followed by statistical modelling (i.e., generalized linear models with strata being included as random variables) has allowed us to drop stratifications that were not helping our estimates enough to overcome the additional crew time to measure them. Thinking about the stratification logically first is necessary because, given high-effort creel surveys, one might find that a particular stratification has a statistically significant effect on the estimate, but that the magnitude of that effect is very small. If so, the insight to the fishery could be quite marginal, yet require a high cost to achieve. After these two quite straightforward types of optimizations, we have found that the next stage brings us to "blue-sky thinking" and a consideration of all the varied statistical and technologically advanced approaches only now at our disposal. For these approaches, we have always required some form of pilot study to either test a different counting method (i.e., drones vs ground crews) or new statistical frameworks (e.g., the joint estimation model). Fortunately, the gains that can be made through the first two optimization methods listed here have usually allowed the pilot study to be achieved without the need for additional creel survey crews. As such, we often plan our optimizations in this order, and don't discount the utility of small gains in efficiency or even small drops in precision.

The insights, recommendations, and approaches presented in this chapter hopefully provide a comprehensive guide for fisheries managers and researchers to navigate the challenging exercise of creel survey design. By harmonizing management objectives with survey outputs, enhancing protocol efficiency, addressing bias, and adopting innovative modelling techniques such as Bayesian joint estimation, creel designers can maximize the efficacy of creel surveys. An adaptive approach to creel design and optimization ensures that surveys remain a vital tool for informed decision-making as fisheries resources and stakeholders change over time. Finally, as analysts have found themselves firmly within the "big data era," we expect that machine learning applications on disparate datasets will be a new tool in optimizing creels soon. Our research team is actively participating in early pilot studies on applying random forest and neural networks to multiple survey datasets to look for conversion factors among established creel methodologies, and new digital insights such as website search metrics (e.g., the number of searches for a particular waterbody, or anonymized data from fisher mobile phone app). Whether these new approaches yield tangible results and can be replicated and then integrated into standardized monitoring protocols remains to be seen, but it seems inevitable given the current pace of progress.

Acknowledgments Thanks to the many members of the Ontario Ministry of Natural Resources creel survey crews and analysts who, over the decades, have designed and conducted comprehensive surveys under challenging conditions. In particular, the authors would like to acknowledge Kim Tremblay, Victoria Kopf, Victoria Danco, Michelle Gorrie, and Todd Langley for leading the operations of some of the creel surveys discussed in this chapter. More broadly, the authors give their thanks for support, advice, and analyses provided by Geoffrey Legault, Simone Collier, Nigel Lester, Henrique Giacomini, Mark Ridgway, and Blair Wasylenko.

References

Aeberhard WH, Mills Flemming J, Nielsen A (2018) Review of state-space models for fisheries science. Annu Rev Stat Appl 5:215–235

Alexiades AZ, Marcy-Quay B, Sullivan PJ, Kraft CE (2015) Measurement error in angler creel surveys. N Am J Fish Manag 35(2):253–261

Askey PJ, Ward H, Godin T, Boucher M, Northrup S (2018) Angler effort estimates from instantaneous aerial counts: use of high-frequency time-lapse camera data to inform model-based estimators. N Am J Fish Manag 38(1):194–209

Bakka H, Rue H, Fuglstad G-R, Riebler A, Bolin D, Illian J, Krainski E, Simpson D, Lindgren F (2018) Spatial modeling with R-INLA: A review. WIREs Computational Statistics 10: e1443. https://doi.org/10.1002/wics.1443

Bayley PB, Sobaski ST, Halter MJ, Austen DJ (1991) Comparisons of Illinois creel surveys and the precision of their estimates. In: Guthrie D, Hoenig JM, Holliday M, Jones CM, Mills MJ, Moberly SA, Pollock KH, Talhelm DR (eds) Creel and angler surveys in fisheries management. American Fisheries Society., Symposium 12, Bethesda, pp 206–211

Beardmore B, Haider W, Hunt LM, Arlinghaus R (2013) Evaluating the ability of specialization indicators to explain fishing preferences. Leis Sci 35(3):273–292

Bonar SA, Huber WA (2002) Standard sampling of inland fish: benefits, challenges, and a call for action. Fisheries (Bethesda) 27(3):10–16

Brownscombe JW, Hyder K, Potts W, Wilson KL, Pope KL, Danylchuk AJ, Cooke SJ, Clarke A, Arlinghaus R, Post JR (2019) The future of recreational fisheries: advances in science, monitoring, management, and practice. Fish Res 211:247–255

Carlander KD, Dicostanzo RCJ, Jessen J (1958) Sampling problems in creel census. Prog Fish C 20(2):73–81

Chizinski CJ, Martin DR, Pope KL (2014) Angler effort and catch within a spatially complex system of small lakes. Fish Res 154:172–178

Claytor RR, O'Neil SF (1991) Using small creek surveys and mark-recapture experiments to interpret angling statistics. In: Guthrie D, Hoenig JM, Holliday M, Jones CM, Mills MJ, Moberly SA, Pollock KH, Talhelm DR (eds) Creel and angler surveys in fisheries management. American Fisheries Society, Bethesda, pp 195–205

Connelly NA, Brown TL, Knuth BA (2000) Do anglers and fishery professionals think alike? Fisheries (Bethesda) 2:21–25

Cowie ED, Ridgway MS (2023) Angling party persistence and visitation in a recreational Lake trout fishery: relative influence of travel distance and fuel costs. N Am J Fish Manag 43(6): 1725–1734

Crone PR, Malvestuto SP (1991) Comparison of five estimators of fishing success from creel survey data on three Alabama reservoirs. In: Guthrie D, Hoenig JM, Holliday M, Jones CM, Mills MJ, Moberly SA, Pollock KH, Talhelm DR (eds) Creel and angler surveys in fisheries management. American Fisheries Society, Bethesda, pp 61–66

Essig RJ, Holliday MC (1991) Development of a recreational fishing survey: the marine recreational fishery statistics survey case study. In: Guthrie D, Hoenig JM, Holliday M, Jones CM, Mills MJ, Moberly SA, Pollock KH, Talhelm DR (eds) Creel and angler surveys in fisheries management. American Fisheries Society, Bethesda, pp 245–254

Fabrizio MC, Ryckman JR, Lockwood RN (1991) Evaluation of sampling methodologies of the Lake Michigan creel survey. In: Guthrie D, Hoenig JM, Holliday M, Jones CM, Mills MJ, Moberly SA, Pollock KH, Talhelm DR (eds) Creel and angler surveys in fisheries management. American Fisheries Society, Bethesda, pp 162–176

Guthrie D, Hoenig JM, Holliday M, Jones CM, Mills MJ, Moberly SA, Pollock KH, Talhelm DR (eds). American Fisheries Society Symposium, 1991. Creel and angler surveys in fisheries management. American Fisheries Society, Bethesda

Gutowsky LF, Gobin J, Burnett NJ, Chapman JM (2013) Smartphones and digital tablets: emerging tools for fisheries professionals. Fisheries (Bethesda) 38(10):455–461

Gomez-Rubio V (2020) Bayesian inference with INLA, 1st edition. Chapman and Hall, New York. 330 pg. https://doi.org/10.1201/9781315175584

Hartill B, Rush N, Payne G, Davey N, Bian R, Miller A, Armiger H, Spong K (2020) Camera and creel survey monitoring of trends in recreational effort and harvest from 2004–05 to 2018–19. Ministry for Primary Industries, Wellington

Hofstetter H, Dusseldorp E, Zeileis A, Schuller AA (2016) Modeling caries experience: advantages of the use of the hurdle model. Caries Res 50(6):517–526

Hogg SE, Lester NP, Ball H (2010) The effectiveness of the 2005 recreational fishing survey to deliver spatially explicit estimates of fishing effort and harvest: analysis for selected Ontario lakes. Government of Ontario, Peterborough

Hoyle SD, Maunder MN (2004) A Bayesian integrated population dynamics model to analyze data for protected species. Anim Biodiv Conserv 21(1):247–266

Hunt LM, Gonder D, Haider W (2010) Hearing voices from the silent majority: a comparison of preferred fish stocking outcomes for Lake Huron by anglers from representative and convenience samples. Human Dim Wild 15(1):27–44

Hunt LM, Camp E, van Poorten B, Arlinghaus R (2019) Catch and non-catch-related determinants of where anglers fish: a review of three decades of site choice research in recreational fisheries. Rev Fish Sci Aquac 27(3):261–286

Knight S, Malvestuto SP (1991) Comparison of three allocations of monthly sampling effort for the roving creel survey on West Point Lake. In: Guthrie D, Hoenig JM, Holliday M, Jones CM, Mills MJ, Moberly SA, Pollock KH, Talhelm DR (eds) Creel and angler surveys in fisheries management. American Fisheries Society, Bethesda, pp 97–101

Kokel WK, Stanovick JS, Nielsen LA, Orth DJ (1991) When to ask: angler responses at different times in the fishing trip and year. In: Guthrie D, Hoenig JM, Holliday M, Jones CM, Mills MJ, Moberly SA, Pollock KH, Talhelm DR (eds) Creel and angler surveys in fisheries management. American Fisheries Society, Bethesda, pp 102–107

Lawson A (2015) An investigative analysis of Queensland's statewide recreational fishing surveys Identifying bias in self-reported catch and effort data. Accessed 19 Jan 2023

Lester NP, Petzold MM, Dunlop WI (1991) Sample size determination in roving creel surveys. In: Guthrie D, Hoenig JM, Holliday M, Jones CM, Mills MJ, Moberly SA, Pollock KH, Talhelm DR (eds) Creel and angler surveys in fisheries management. American Fisheries Society, Bethesda, pp 25–46

Lester NP, Dextrase AJ, Kushneriuk RS, Rawson MR, Ryan PA (2004) Light and temperature: key factors affecting walleye abundance and production. T Am Fish Soc 133(3):588–605

Lester NP, Sandstrom S, de Kerckhove DT, Armstrong K, Ball H, Amos J, Dunkley T, Rawson M, Addison P, Dextrase A, Taillon D, Wasylenko B, Lennox P III, Giacomini HC, Chu C (2021) Standardized broad-scale management and monitoring of inland lake recreational fisheries: an overview of the Ontario experience. Fisheries (Bethesda) 46(3):107–118

Lindgren F, Rue H, Lindström J (2011) An explicit link between Gaussian fields and Gaussian Markov random fields: the stochastic partial differential equation approach. J R Stat Soc B 73(4):423–498

Malvestuto SP (1983) Sampling the recreational fishery. In: Nielson A, Johnson DL (eds) Fisheries techniques. American Fisheries Society, Bethesda, pp 397–419

Malvestuto S, Knight S (1991) Evaluation of components of variance for a stratified, two-stage roving creel survey with implications for sample size allocation. In: Guthrie D, Hoenig JM,

Holliday M, Jones CM, Mills MJ, Moberly SA, Pollock KH, Talhelm DR (eds) Creel and angler surveys in fisheries management. American Fisheries Society, Bethesda, pp 108–115

Maunder MN, Punt AE (2013) A review of integrated analysis in fisheries stock assessment. Fish Res 142:61–74

Mazzi GL, Mitchell J, Carausu F (2019) Data uncertainties: their sources and consequences. Eurostat Statistical Working Papers. https://cros-legacy.ec.europa.eu/system/files/ks-tc-19-005-en-n.pdf Accessed 19 Jan 2023

McCormick JL, Meyer KA (2017) Sample size estimation for on-site creel surveys. N Am J Fish Manag 37(5):970–980

McCormick JL, Whitney D, Schill DJ, Quist MC (2015) Evaluation of angler reporting accuracy in an off-site survey to estimate statewide steelhead harvest. Fish Manag Ecol 22(2):134–142

Nieman CL, Iwicki C, Lynch AJ, Sass GG, Solomon CT, Trudeau A, van Poorten B (2021) Creel surveys for social-ecological-systems focused fisheries management. Rev Fish Sci Aquac 29(4): 739–752

Parker S, Krausse M, Dunlop W (2006) Estimating angler effort from aerial surveys. Northwest Science and Information Technical Report, Peterborough

Parkinson EA, Berkowitz J, Bull CJ (1988) Sample size requirements for detecting changes in some fisheries statistics from small trout lakes. N Am J Fish Manag 8:181–190

Peterman RM (1990) Statistical power analysis can improve fisheries research and management. Can J Fish Aquat Sci 47(1):2–15

Pollock KH, Jones CM, Brown TL (1994) Angler survey methods and their application in fisheries management. In: special publication 25. American Fisheries Society, Bethesda, p 371

Pollock KH, Hoenig JM, Jones CM, Robson DS, Greene CJ (1997) Catch rate estimation for roving and access point surveys. N Am J Fish Manag 17(1):11–19

Riggers BL, Jones MK (2022) Paradigm shift: applying capture-recapture techniques to electronic licensing system data to estimate Chinook salmon harvest. N Am J Fish Manag 42:815–827

Robson DS (1991) The roving creel survey. In: Guthrie D, Hoenig JM, Holliday M, Jones CM, Mills MJ, Moberly SA, Pollock KH, Talhelm DR (eds) Creel and angler surveys in fisheries management. American Fisheries Society, Bethesda, pp 19–24

Roop HJ, Poudyal NC, Jennings CA (2018) Assessing angler effort, catch, and harvest on a spatially complex, multi-lake fishery in middle Georgia. N Am J Fish Manag 38(4):833–841

Schaeffer RL, Mendenhall III WM, Ott RL (2006) *Elementary survey sampling*, 6th edition. Duxbury Press, Belmont, California.

Shuter BJ, Matuszek JE, Regier HA (1987) Optimal use of creel survey data in assessing population behaviour: Lake Opeongo lake trout (*Salvelinus namaycush*) and smallmouth bass (*Micropterus dolomieui*), 1936–83. Can J Fish Aquat Sci 44(Suppl.2):229–238

Shuter BJ, Jones ML, Korver RM, Lester NP (1998) A general, life history based model for regional management of fish stocks: the inland lake trout (*Salvelinus namaycush*) fisheries of Ontario. Can J Fish Aquat Sci 55(9):2161–2177

Simpson G (2018) Use of a public fishing area determined by vehicle counters with verification by trail cameras. Nat Res 9(5):187–199

Smith SJ, Hunt JJ, Rivard D (1993) Risk evaluation and biological reference points for fisheries management, vol 120. National Research Council of Canada: Canadian Special Publication of Fisheries Aquatic Sciences, p 442

Soupir CA, Brown ML, Stone CC, Lott JP (2006) Comparison of creel survey methods on Missouri River reservoirs. N Am J Fish Manag 26(2):338–350

Spiegelhalter DJ, Best NG, Carlin BP, Van Der Linde A (2002) Bayesian measures of model complexity and fit. J R Stat Soc B 64(4):583–639

Stahr KJ, Knudsen R (2018) Evaluating the efficacy of using time-lapse cameras to assess angling use: an example from a high-use metropolitan reservoir in Arizona. N Am J Fish Manag 38(2):327–333

Steffe AS, Murphy JJ, Reid DD (2008) Supplemented access point sampling designs: a cost-effective way of improving the accuracy and precision of fishing effort and harvest estimates derived from recreational fishing surveys. N Am J Fish Manag 28(4):1001–1008

Sullivan MG (2003) Exaggeration of walleye catches by Alberta anglers. N Am J Fish Manag 23(2):573–580

Sztramko LK (1991) Improving precision of roving creel survey estimates: implications for fisheries with a closed season. In: Guthrie D, Hoenig JM, Holliday M, Jones CM, Mills MJ, Moberly SA, Pollock KH, Talhelm DR (eds) Creel and angler surveys in fisheries management. American Fisheries Society, Bethesda, pp 116–121

Thomson CJ (1991) Effects of the avidity bias on survey estimates of fishing effort and economic value. Am Fish Soc Symp 12:356366

Trudeau A, Dassow CJ, Iwicki CM, Jones SE, Sass GG, Solomon CT, van Poorten BT, Jensen OP (2021) Estimating fishing effort across the landscape: a spatially extensive approach using models to integrate multiple data sources. Fish Res 223:105768

Tucker CM, Collier S, Legault G, Morgan GE, de Kerckhove DK (2024) Estimating angler effort and catch from a winter recreational fishery using a novel Bayesian methodology to integrate multiple sources of creel survey data. Fish Res 272:106932

van Poorten BT, Brydle S (2018) Estimating fishing effort from remote traffic counters: opportunities and challenges. Fish Res 204:231–238

van Poorten BT, Carruthers TR, Ward HGM, Varkey DA (2015) Imputing recreational angling effort from time-lapse cameras using an hierarchical Bayesian model. Fish Res 172:265–273

Venturelli, P.A., Hyder, K., and Skov, C. (2017) Angler apps as a source of recreational fisheries data: opportunities, challenges and proposed standards. Fish and Fisheries 18:578–595.

Venturelli P, Skov C, Audzijonyte A, Belmaker J, Ana Cardoso AS, Gordon L, Gundelund C, Hyder K, Jarić I, Lennox RJ, Potts WM, Vacura K, Vaz AS, Sbragaglia V (2026) Digital data mining. In: Pope KL, Arlinghaus R, Hunt LM, Lynch AJ, and van Poorten BT (eds) Understanding recreational fishers: disciplinary and interdisciplinary approaches for fisheries management. Springer Nature, Cham, p 443–479

Vølstad JH, Pollock KH, Richkus WA (2006) Comparing and combining effort and catch estimates from aerial-access designs as applied to a large-scale angler survey in the Delaware River. N Am J Fish Manag 26(3):727–741

Wade D, Jones CM, Robson DS (1991) Computer simulation techniques to assess bias in the roving-creel-survey estimator. In: Guthrie D, Hoenig JM, Holliday M, Jones CM, Mills MJ, Moberly SA, Pollock KH, Talhelm DR (eds) Creel and angler surveys in fisheries management. American Fisheries Society, Bethesda, pp 40–46

Weithman AS, Haverland P (1991) Comparability of data collected by telephone and roving creel surveys. In: Guthrie D, Hoenig JM, Holliday M, Jones CM, Mills MJ, Moberly SA, Pollock KH, Talhelm DR (eds) Creel and angler surveys in fisheries management. American Fisheries Society, Bethesda, pp 67–73

Zuur AF, Ieno EN, Walker NJ, Saveliev AA, Smith GM (2009) Zero-truncated and zero-inflated models for count data. In: Mixed effects models and extensions in ecology in R. Springer, New York, NY, pp 261–293

Economic Experiments in Recreational Fishing Studies

E. Ingrid van Putten, Ashley Trudeau, Mary Mackay, Clara Obregón, John R. Post, Andries Richter, Esther Schuch, and Christopher T. Solomon

Abstract Experimentation in natural science is commonplace, and the reasons and methods for undertaking it are well understood. The use of experimental methods has risen remarkably over the last 25 years in the field of economics and has had a

E. I. van Putten (✉) · M. Mackay
Commonwealth Scientific and Industrial Research Organisation Environment, Hobart, TAS, Australia

Centre for Marine Socioecology, University of Tasmania, Hobart, TAS, Australia
e-mail: ingrid.vanputten@wur.nl; mary.mackay@csiro.au

A. Trudeau
Center for Limnology, University of Wisconsin-Madison, Madison, WI, USA
e-mail: aatrudeau@wisc.edu

C. Obregón
Australian National Centre for Ocean Resources and Security, University of Wollongong, Wollongong, Australia

Department of Biological Sciences, University of Calgary, Calgary, AB, Canada
e-mail: Clara.Obregon@murdoch.edu.au

J. R. Post
Environmental Economics and Natural Resources group, Wageningen University, Wageningen, The Netherlands
e-mail: jrpost@ucalgary.ca

A. Richter
Environmental Economics and Natural Resources group, Wageningen University, Wageningen, The Netherlands

Centre for Ecological and Evolutionary Synthesis, Department of Biosciences, University of Oslo, Oslo, Norway
e-mail: andries.richter@wur.nl

E. Schuch
Research Institute for Sustainability, Potsdam, Germany
e-mail: Esther.Schuch@iass-potsdam.de

C. T. Solomon
Cary Institute of Ecosystem Studies, Millbrook, NY, USA
e-mail: solomonc@caryinstitute.org

K. L. Pope et al. (eds.), *Understanding Recreational Fishers*, Fish & Fisheries Series 45, https://doi.org/10.1007/978-3-031-99739-6_14

tremendous impact on the way economic research is carried out. This raises the question; how can economic experiments advance our understanding of recreational fishing. Here, we argue that many opportunities and benefits exist for the application. There have already been several successful experiments (in the lab or the field) to understand the drivers of recreational fisher behaviour. Despite these efforts, there remains a raft of opportunities to use experiments to help understand cause and effect in terms of recreational fisher behaviour. In this chapter, we provide a taxonomy of economic experiments and highlight their application to recreational fisheries. We sketch how controlled and uncontrolled interventions in recreational fisheries are learning opportunities to better understand fisher behaviour and social-ecological dynamics. Further, we highlight practical aspects of experimentation that will help its implementation and suggest opportunities for future experiments in recreational fisheries.

Keywords Experimental economics · Lab-in-field · Laboratory experiments · Natural experiment · Social science

1 Experimental Approach to Investigation of Recreational Fisher Behaviour

Recreational fisheries can be understood as social–ecological systems with complex feedbacks within and between the ecological system (e.g., fish biomass) and social system (e.g., fisher or manager behaviour) (Arlinghaus et al. 2017). The complex interactions between the social and natural systems mean that emergent outcomes are not easily attributable to any one social–ecological system's component but arise from multiple and simultaneous feedbacks and system properties (Matsumura et al. 2019; Golden et al. 2022). Such emerging complexity makes it difficult to understand causes and effects in the real world (Ferraro et al. 2019). Experimental methods are fruitful approaches to establish causal pathways and corresponding impacts and help unpack some of the complexity inherent in coupled systems (Janssen et al. 2010; Poteete et al. 2010). Experiments are particularly useful to observe and understand behaviour of recreational fishers (Stoop et al. 2012; Ward et al. 2013). However, even though the use of experiments in ecological research of recreational fishing is relatively well developed, many opportunities remain to better understand the social aspects, causal mechanisms, and behavioural responses. For example, it is well known that there are differences in the motivation to participate in the commercial and recreational fishing sectors (Mackay et al. 2020). By using experiments, the difference between recreational and commercial fisher responses to traditional fisheries management interventions might be better understood, and unexpected management outcomes may be averted. This chapter discusses how insights from experiments can help inform a recreational fishing research agenda.

An experiment is a research method used to test hypotheses in a controlled setting. Defining features of an experiment are (i) a manipulation of key variables while controlling possible confounders and (ii) the presence of a control or reference group. Often, the researcher tries to manipulate a variable through an intervention (e.g., the introduction of a licence fee) and asks how that intervention affects an outcome of interest (e.g., days of fishing). The presence of a control group is needed to construct a credible counterfactual that tells the researcher how the outcome variable would have developed in the absence of the intervention. For example, if the introduction of a licence fee coincided with very bad weather conditions, it would not be possible to attribute a drop-in fishing activity to the intervention, as well as the change in weather. Therefore, it is crucial to have a design that creates a credible counterfactual, such as control sites without interventions sampled simultaneously and over time. Especially in complex field settings, counterfactuals are necessary to control for temporal effects unrelated to the intervention.

Depending on the hypothesis, there are different degrees to which a research design resembles reality and the amount of control one has over the setting. In other words, an intervention may be controlled and carefully designed or just "happen" outside the control of the research team studying it. For example, it may be the case that a management agency implements a licence fee without any involvement of a research team and without considering and measuring control sites. The challenge then lies for researchers to establish a credible counterfactual, e.g., a constructed "synthetic" control (Abadie 2021), that helps understand the impact of the intervention over time. If the intervention happens outside the control of the research team, it is known as a "natural experiment." These may be distinct events, brought about by policy change (e.g., access restrictions), natural disasters (e.g., floods), but also wider societal disruptions (e.g., COVID-19) that can inform how such events affected key variables of interest. Including natural experiments in our definition of "experiments" may be confusing to some—however, we believe they provide key opportunities for opportunistically inferring cause and effect in complex fishery systems. Confusion may arise because there is no agreed-upon definition of a natural experiment in social science (Leatherdale 2019), but it is generally accepted that it means that the conditions (i.e., treatment variables) surrounding an intervention are not, or cannot be controlled by the researcher (e.g., natural phenomena like COVID or a tsunami). Natural experiments also include, e.g., management interventions that are not controlled by the researcher, but where the impacts are observed and analysed (Fujitani et al. 2018). Because the treatment variable in natural experiments cannot be manipulated by the researcher, they can also be described as observational studies (Dunning 2008). However, a degree of experimental "control" remains possible through, e.g., site selection (Diamond 1983), and the term natural experiment applies when something fundamental changes that affects key variables of interest. Because this type of analysis of natural experiments is not uncommon in behavioural fisheries science (Pita et al. 2021; Ryan et al. 2021; Sawynok and Sawynok 2021), we include natural experiments in our definition of experiments.

In this chapter, we discuss the full range, from controlled experiments in a laboratory where people's choices (which can be students or the group or people of interest) are tested and observed under specific, often abstract, conditions (Smith 1982) to natural experiments (see Sect. 2) without any controlled manipulation of the research subjects, but which can capture more realistic decision making. This chapter provides an overview of the use of experiments in recreational fisheries and provides a taxonomy of experiments. Each of the experimental settings and the approaches has advantages and disadvantages. The chapter further provides best practices and mentions potential caveats for researchers. For example, when designing an experiment, it is crucial to ensure that the experiment itself is designed and communicated in a way that it is unambiguous to participants. The information provided to participants should be clear, and the context and problem should be understood by all in the same way. Also, experiments involving human subjects require special preparation. Of particular importance are ethical issues that can arise within the research.

1.1 What Is an Experiment?

Even though we have indicated above that we discuss the full spectrum of experiments from laboratory-based (controlled) to natural experiments, the boundary between what is an experiment and what is a "non-experiment" remains blurry. What distinguishes an experiment from other quantitative observational methods that are also testing hypotheses? Crucially, observational studies tend to use naturally occurring variation of key variables to infer relationships. For example, temperature or precipitation may have an impact on fisher participation at a given site. In contrast, experiments tend to measure the impacts of specific interventions on a certain group and compare the results with a control group. These specific interventions may also be naturally occurring events (a natural experiment), but often involve treatments that are applied to randomly selected treatment and control groups to measure the response. For example, in a field experiment where the effects of access fees on fishing effort are tested, different access fees may be randomly assigned to otherwise-similar waterbodies to test the hypothesis that higher access fees will reduce fishing effort. Fishing effort before and after the imposition of access fees can then be compared with those control sites using statistical tools to evaluate support for the hypothesis.

At the foundation of experiments lies the scientific method (Bacon 1818), which goes through several steps to progress from formulating one or more hypotheses to accepting or rejecting them. The hypotheses that are tested in the experiment must be solid (i.e., quantifiable and rejectable) because the experiment cannot make up for flawed assumptions or faulty premises (Webster and Sell 2014). If a hypothesis is tested using an experiment, this will lead to support for the hypothesis (i.e., rejection of the null hypothesis of no effect) or rejection of the hypothesis (i.e., support for the null hypothesis of no effect).

The key to experimental research is the principle of replication, which means that the same experiment should give the same results if conducted under identical conditions. In practice, this is often a question of internal and external validity. Internal validity implies that the results obtained in the experiment can be correctly

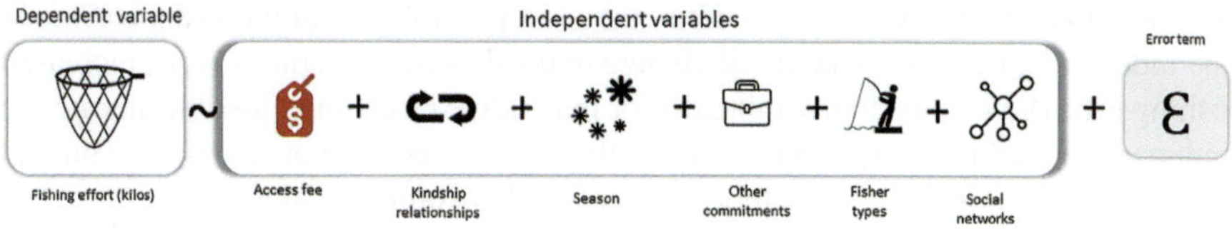

Fig. 1 Hypothetical example of a relationship that might be explored in an experiment (example shown is between fishing effort [the independent variable on the left-hand side of the ~ sign], and access fee and some social and demographic variables [the explanatory variables on the right-hand side of the ~ sign], plus an error term; in this example one of the independent variables [access fee] may be one that is manipulated; icons from https://www.flaticon.com)

attributed to the intervention, rather than a confounding factor. If a setting is very realistic, it may very well be that something happening outside the experiment (e.g., a public policy implemented at the same time) may affect behaviour in the experiment (e.g., the tendency to cooperate). External validity pertains to the question of how generalizable the results are. What do results obtained from an experiment conducted at a small lake in Germany say about recreational fishing in Germany more generally? And can we say anything about recreational fishing in Europe or globally?

Experiments are used to explore causal relationships and to see how manipulation in independent variables influences the dependent variable of interest. There can be more than one different independent variable (see hypothetical example in Fig. 1). Some of the independent variables (like access fee in Fig. 1) may be manipulated whilst the others are not. To test more than one independent variable, it is necessary to test the effect of each of the independent variables individually (e.g., the effect of a licence fee on fishing effort and the effect of season on fishing effort) before testing combinations of two independent variables (Kimmel et al. 2021). To establish causation, changes in the independent variable or variables of interest must be independent of confounding effects. Confounding effects are those other uncontrolled variables that influence the relationship between the dependent and independent variables. For example, comparisons of recreational fisher behaviour between stocked and unstocked lakes (Silwal et al. 2023) may be confounded by previously poor fishing conditions that led to stocking as a management intervention or by a particular fisher type heavily drawn to one lake type being more responsive than others to the intervention.

The gold standard for establishing this independence is through random assignment of subjects (e.g., recreational fishers or their communities) to treatment groups, ensuring all groups are comparable. In this experimental context, there is at a minimum one treatment and one control group, but ideally replication is necessary for each of the two. To avoid confounding effects, the subjects are (ideally) assigned randomly to the different treatment groups to avoid systematic differences between the groups. The treatment group receives a manipulated independent variable (e.g., a higher licence fee) than the control group. The more manipulations one wants to test, the more treatments are necessary. Only then is it possible to claim that the change in the independent variable causes the change in the dependent variable (e.g., increased fishing effort). Of course, multiplicative or interactive effects are possible among the independent variables (e.g., licence fees only affecting effort at otherwise low budget conditions), necessitating even more treatments that combine independent variables.

An example of a study from recreational fisheries that uses treatment groups is by (Fujitani et al. 2017) who randomly assigned recreational fishers organized in angling communities (clubs) to two groups: (i) the control group receives a passive lecture about the ecological effects of stocking, whereas (ii) the treatment group joined a participatory adaptive management program evaluating the effects of stocking in their fishery. A third placebo group received passive lectures, but not on the topic (stocking) of interest, being the control group. The use of this experimental setup allowed the authors to conclude that the treatment group (those who joined the participatory adaptive management program) retained more knowledge about ecological risks involved in stocking than the control group. Another example of the use of treatment groups can be found in the testing of enhanced enforcement actions for northern pike (*Esox lucius*) protected sizes in Alberta (Walker et al. 2007). In partnership with enforcement officers, the study compared deterrence of noncompliance at treatment lakes, which received enhanced enforcement actions, with control lakes subject to normal enforcement patrols. The crux of these experimental approaches is that assignment to treatment groups should be random for experiments to be carried out as randomized control trials (Banerjee 2020). Using the example from Fujitani et al. (2017), if participants can choose whether they want to be part of an adaptive management program, it would be natural to expect that the more motivated individuals will do so. In such cases, participants "self-selecting" into treatments would then drive most of the experimental results, rather than the intervention itself. Therefore, it is crucial that the selection of participants into treatments is carried out at random. It should be noted that such randomization, which is often very clear to researchers, is typically not intuitive for practitioners and at odds with how fisheries are usually managed. For example, organizers of a workshop are more inclined to invite committed users, rather than random ones. Also, management authorities hardly implement regulations at random but think carefully about which locations an intervention may be successful or most needed. The non-random placement of management regulations or area-based management tools (e.g., marine protected area) renders many of these actions quasi-experimental and non-random, with a high possibility for bias. Matching techniques or synthetic control methods may still be used to identify credible counterfactuals and allow causal inference of policy impacts (Ferraro et al. 2019; Medoff et al. 2022; Nowakowski et al. 2023).

In an experiment, researchers often manipulate one or more independent variables and measure the effect on one or more dependent variables. For example, the use of the catch-and-release method (the dependent variable) would be left free to vary, whereas the degree of information given to individual recreational fishers about the biological outcomes of catch-and-release is varied by the researcher (the independent variable). Though researchers may wish to study the effect of several variables at the same time, this reduces the ability to identify effects due to a loss of statistical power, as it is usually infeasible to increase treatment groups to a large degree. For experiments, the number of treatments increases with the number of variables one wants to vary. A full factorial design that can identify all interaction effects for three independent variables requires at least eight treatments ($2 \times 2 \times 2 = 8$), with multiple individual subjects (i.e., recreational fishers) in each treatment or control group. However, to identify the main effects would require varying each independent variable once from a common baseline, which would result in three additional treatments (see also discussion on Balanced Incomplete Block Design in choice experimentation, Young (2019)). The strength of experiments is to identify singular mechanisms, rather than simulating all the complexities of particular settings (Webster and Sell 2014). The topics analysed in experiments can therefore appear to be quite narrow by nature, but still insightful in terms of cause-and-effect inference.

The recreational fishery systems are often highly regulated and constantly exposed to various forms of interventions. Interventions provide learning opportunities, especially if experiments are carried out in a controlled way. Field-based experiments have taken advantage of new fish stocking initiatives to learn about the behavioural response of fisher effort to fish abundance (Lathrop et al. 2002; Mee et al. 2016) and whether fishers cooperate for the common good or not (Stoop et al. 2012). Also, experiments have helped understand the impacts of environmental factors (Golden et al. 2021) and fish population size (Askey et al. 2013; Dassow et al. 2020) on recreational fishers catch rates. Generally, experimentation provides excellent learning opportunities for adaptive management (Lester et al. 2003; Hansen et al. 2015).

Many examples exist of observational studies where fisher behaviour is observed across space and time, and site choice studies are an example of such natural experiments. Many fisheries are made up of multiple discrete patches across landscapes (i.e., lakes, streams, marine reefs, etc.), thus fishers must make site choice decisions involving travel time and travel costs (Post et al. 2008; Hunt et al. 2011; Camp et al. 2018; Wilson et al. 2020). Time series with a perturbation of site characteristics have identified site choice behaviour with aggregate measures of total fishing effort in experimentally opened fisheries (van Poorten and Post 2005) and regulation changes from harvest to catch-and-release fisheries (Johnston et al. 2011). In their review of where people fish, travel-related times and costs, as well as catch-related measures of fishing quality, were consistently demonstrated to be important in fisher site choice behaviour (Hunt et al. 2019). Other aspects of fisher behaviour have also been investigated through observational data. For example, harvest or release decisions, trip context (i.e., day or multi-day trips), and compliance with regulations was found to vary with travel time or costs and fisher typologies (Parkinson et al. 2004; Hunt et al. 2011; Ward et al. 2013; Kaemingk et al. 2019).

1.2 A Taxonomy of Experiments

When we consider the role of experiments in understanding the behaviour of recreational fishers, there are many different settings in which this can take place (i.e., experimentation in a laboratory setting, in the field, and the hybrid lab-in-the-field experiment, or a natural experiment) (Fig. 2 and Table 1). Experimental settings occur along opposite gradients of control and realism (Fig. 2). Therefore, research takes place along a gradient of controlled experimental methods situated in a laboratory to natural experiments (observational studies) undertaken in the "wild." Laboratory experiments allow the researcher complete control of the independent variable and are therefore internally valid (i.e., causation) can be supported. It can sometimes be challenging to generalize the results of lab experiments to the real world—even thoughtfully designed ones (e.g., Stoop et al. 2012). This is especially pertinent after finding that lab experiments involving human participants heavily rely on participants from "Western, Educated, Industrialized, Rich, and Democratic (WEIRD)" societies (Henrich et al. 2010). External validity, or generalizability to "real world" systems, is a strength of natural experiments. Real-world data sources

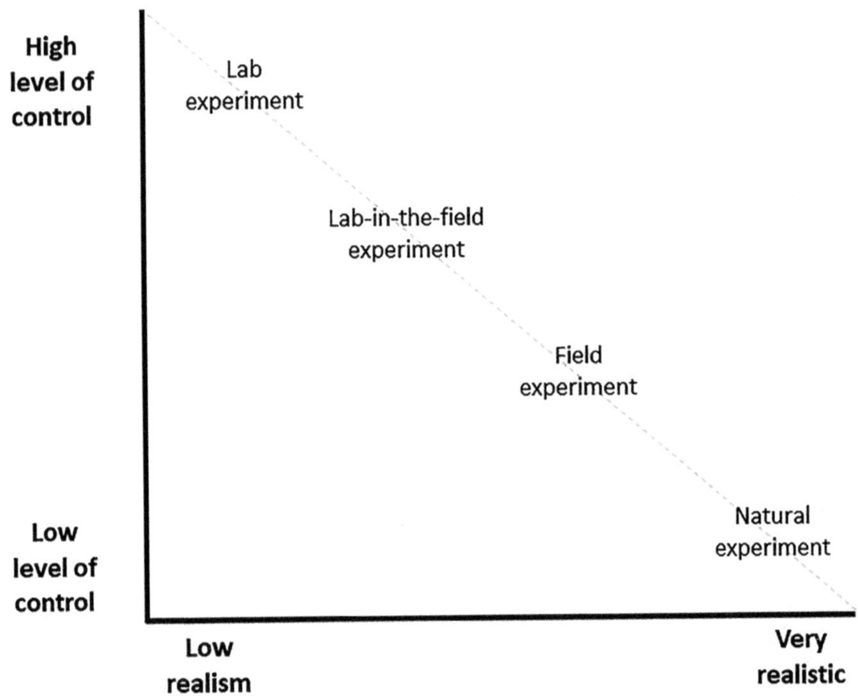

Fig. 2 Schematic illustrating the gradients of control and realism between different experimental settings

Table 1 Different types of experimental settings. (Based on Harrison and List 2004, Viceisza 2016, Gneezy and Imas 2017)

	Lab experiment[a]	Lab-in-the-field experiment[a]	Field experiment[a]	Natural experiment
Management (control)	Experiment conducted under highly controlled conditions	Control over conditions and ability to make direct comparisons between different populations and contexts	Experiment cannot control extraneous variables (different gradients of control)	No control over the independent variable as it occurs naturally in real life
Environment (realism)	Conducted in laboratories (i.e., environments that are abstract from the natural setting) Imposed	Conducted in the natural environment of the participants Imposed	Conducted in the everyday (i.e., real life) environment of the participants Imposed	Conducted in the everyday (i.e., real life) environment of the participants Natural
Participants (population)	Narrow set of participants, typically university students Artefactual = with a nonstandard subject pool Framed = field context in the information set	Targeting the theoretically relevant population but using a standardized, validated lab paradigm to perform tasks that are not part of their day-to-day environment	Typically use a non-student theoretically relevant population that is not aware that their decisions are being studied	Same as field experiments participants do not know that they are in an experiment
Comments	Results obtained for the respondent population may not be representative of behaviour in wider population	More confident that the results are applicable to the relevant context	Confidence that the results are applicable to the relevant context but extraneous variable influence	Natural experiments occur opportunistically and require random or "near-random" sorting into treatment groups.
Resources (design & implementation)	(Janssen et al. 2010) (Guala 2005) (van Soest 2013) (Poteete et al. 2010)	(Gangadharan et al. 2022) (Viceisza 2012, 2016)	(Glennerster and Takavarasha 2014) (Banerjee 2020)	(Dunning 2008, 2012) (Huntington-Klein 2021) (Cunningham 2021)

Note: Variation in control and realism is continuous. For instance, it is not that all field experiments are highly realistic and unable to control extraneous variables; rather, there is variation in realism and control

[a] Referred to elsewhere as a controlled experiment

are messy in that they are prone to multicollinearity, and causation, or internal validity, can only be established if a credible counterfactual can be constructed (e.g., Fujitani et al. 2012).

To test hypotheses in recreational fisheries and to make inferences that may be generally applicable to fisheries in many contexts, experiments and observational studies along the full spectrum of gradient of control and realism (Fig. 2) are necessary. Gold standard experiments with replication, controls, and randomization of treatments in a fisheries context have high internal validity, but these standards are frequently unachievable (Hilborn 2016). Experimental results, however, must demonstrate their external validity through effective choices of study participants and of explanatory variables that correspond with drivers of change in multiple fishery systems (Hilborn 2016). Controlled experiments benefit from being completed in conjunction with monitoring of "real world" systems (Lynch et al. 2021; Lennox et al. 2022). Other methods of inference, such as theoretical models or comparative studies, can also help to validate and provide context for experimental studies.

Natural experiments leverage uncontrolled changes that "naturally" assign random or pseudo-random treatment groups (Dunning 2008). Not all interventions can be studied through field experiments; e.g., local authorities may be reluctant to vary user fees or start charging users in the first place to enable different treatment groups. Also, some behaviours such as noncompliance are notoriously difficult to study in the field (Arlidge et al. 2023), because any attempts to measure non-compliance lead to behavioural change (Boonstra et al. 2017, Diekert et al. 2021, but see Walker et al. 2007). In such a case, a lab experiment may offer valuable insights. Laboratory experiments allow researchers full control of the study environment, the selection of treatment groups, and the application of treatments or independent variables. Many lab experiments in fisheries use experimental economics methods in which participants are presented with a task where they are incentivized to maximize a financial reward (often corresponding to a modest payout after the experiment) within specific rules and a highly controlled experimental setting (Guala 2005). The goal of this strict control is to avoid unknown confounding effects (omitted variables) introducing bias to the experimental results. In these experiments, the greatest unknown factor is the participants themselves. Participants who are distracted, confused, or even over-cooperative (i.e., by "helping" run the experiment) may bias the results. Therefore, in addition to careful recruitment of (to the extent possible) a random selection of participants, effectively introducing and explaining the rules of the experiment to participants is of critical importance. Mackay et al. (2020) present a laboratory experiment utilizing a common-pool resource game. Participants (who in this case were students—not fishers) were incentivized to harvest fish in the context of a social dilemma, where higher individual catches would increase their payment, but higher total group catches would decrease their payment. This experimental approach allowed the testing efficacy of "nudges" leveraging social norms versus the effects of low- and high-probability enforcement of regulations, which is difficult and expensive to test in the field (but see Walker et al. 2007). However, trade-offs of experimental approaches include a reliance on student participants in some studies (as in Mackay et al. 2020) and the unavoidable artificiality of a laboratory environment, potentially compromising the study's external validity.

One way to increase external validity is to combine a laboratory approach with a field approach (Stoop et al. 2012) or adopt a lab-in-the-field approach (Table 1). Though lab experiments sometimes rely on students as a study population, lab-in-the-field approaches take, as the name suggests, the lab to the field. This connects the controlled setting of the lab (or the less-controlled setting of a field site) with the study population of interest. One example is the multi-method lab-in-the-field approach to experiments conducted by Stoop et al. (2012) to observe whether classical economics could predict the cooperative behaviour of participants engaged in a social dilemma. For one lab experiment, participants were recreational fishers rather than students (unlike a typical laboratory experiment), and they were aware that they were taking part in an experiment (unlike a typical field experiment). The lab experiment consisted of a social dilemma game where, on pen and paper, participants could choose to cooperate by "harvesting" fewer fish to increase the group's payout (see Chapter on Behavioural Economics). In addition to economic experimental methods, lab-in-the-field approaches can also include (and study) recreational fishing behaviour in practice. For example, Stoop et al. (2012) additionally investigated cooperation behaviour when decisions were followed up by actual fishing behaviour at a pond with a known fish population, and they found no evidence of cooperative behaviour as was found in the lab experiment. By increasing the realism of the experiment, the researchers enhanced the study's external validity and detected decision patterns that are predicted by classical economics (Noussair et al. 2015) further expanded the experiment's realism by making the game dynamic and including feedback between overharvesting and reduced resource abundance. (Monk and Arlinghaus 2018) examined the relationship between fisher skill, their ability to locate fish, and their catch rates as recruited participants fished an experimental lake for perch (*Perca fluviatilis*), tracked by an acoustic telemetry system, and fisher behaviour using GPS on boats. Each of these experiments investigated recreational fishery dynamics that were relevant to a broad range of systems by implementing internally valid experimental methods within a naturalistic setting.

Field studies, in contrast to lab-based and lab-in-the-field experiments, can entirely avoid the trouble of recruiting a representative group of participants because they observe the response of fishers to experimental changes applied to their own fishery system. These fishers are unaware that they are part of an experiment and therefore present the greatest degree of realism in their responses. In field experiments, researchers apply interventions (Fig. 3) such as regulations, access fees, information, or stocking to selected waterbodies. Particularly critical in these studies is the appropriate selection of sites and the assignment of their treatments. Mee et al. (2016) estimated the responses of catch rate and fishing effort to a range of stocking densities in 34 lakes. As a result of this experimental design and the use of statistical models to account for temporal trends, regional effects, and differences in lake accessibility, this study could independently evaluate the effects of stocking density on fishing effort to identify optimal stocking densities. These sorts of experimental studies to evaluate the effects of management actions have been identified as a key research gap in recreational fisheries (Holder et al. 2020).

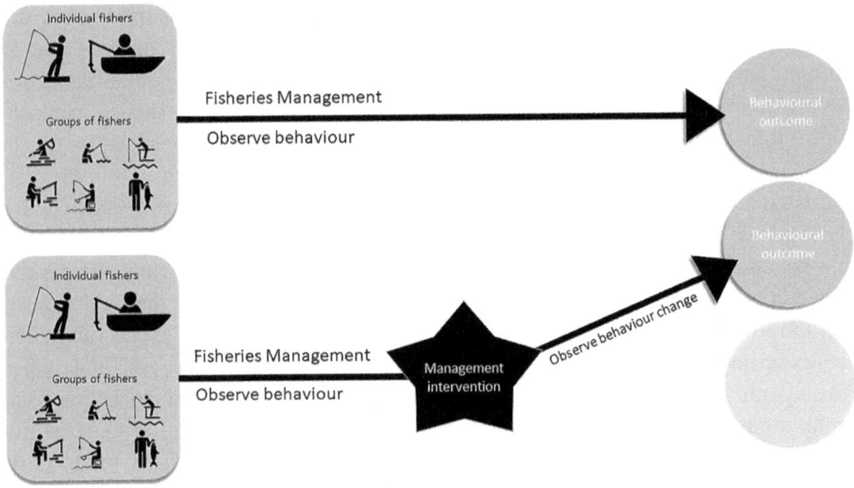

Fig. 3 Schematic diagram of fisher behaviour observed under "Normal" fisheries management (base case or counterfactual; panel a) and a management intervention is implemented, and fisher behaviour change is observed (panel b; icons from https://www.falticon.com)

Not all experiments are explicitly designed. In some cases, naturally occurring "interventions" can be opportunistically leveraged to investigate the effects of the naturally occurring change. Natural experiments can have different "treatments"—but unlike lab-in-the-field experiments, they do not have an active control and need a lot of care to statistically control for possibly confounding variables (Cunningham 2021; Huntington-Klein 2021). Correlation of independent variables may obscure mechanisms of causation. Ineffective statistical design can lead to confounding of the treatment with other site characteristics (Ferraro et al. 2019). Natural experiments are a common approach to learning about recreational fisher behaviour, and they can provide unique opportunities for deductive reasoning in natural, field settings. A controlled experiment faces very different challenges from a natural experiment. Design is key in controlled experiments; the experiment must be carefully set up to allow the desired causal claims, whereas the analysis is relatively simple. In natural experiments, there is no (apparent) design, but the researchers must find a convincing causal identification strategy. Definitions of natural experiments vary by author and field, spanning from studies leveraging naturally occurring change that is applied to participants or sites "as if" randomly (Dunning 2008), to observational studies utilizing careful selection of sites to maintain controls (Diamond 1983), to observing sites or participants before and after the effects of interventions that are not controlled by the researcher (Leatherdale 2019). For example, Klatt et al. (2014) investigated fishing participation before and after the introduction of viral hemorrhagic septicemia (VHS) in different management regions with different regulations. Similarly, studies of global events such as the COVID-19 pandemic have detected changes in harvest (Bunt and Jacobson 2022) and fishing effort (Ryan et al. 2021; Trudeau et al. 2022; Britton et al. 2023) before and during the height of the pandemic. Other studies have successfully investigated fisher behaviour by investigating how fishing effort responds to a variety of drivers, including regulations (Whitehead et al. 2011), facilities, travel time or cost (Post et al. 2008), stocking rate (Fayram et al. 2006), or regulatory changes (Beard et al. 2003).

2 Choosing the Appropriate Approach to Measure Preferences

Often, experimental research in recreational fishing tries to understand recreational fishers' behavioural choices and preferences. Such information may be obtained through stated or revealed preference methods (see Melstrom et al. 2026). Stated preference methods ask participants to make hypothetical choices between options that vary experimentally and systematically in their attributes. They are generally survey-based experiments.

Behaviour of recreational fishers can also be observed (visually, by site visits, or using digital apps or wildlife cameras). Observational methods "reveal" the preferences or behaviours of fishers without asking them directly. These revealed preference methods are not experimental in the true sense of the term and use the real-life decisions of fishers as dependent variables. For example, in a study by Bennear et al. (2005), the demand for state fishing licenses was used to infer the benefits of recreational fishing. Revealed preference approaches are more akin to natural experiments or observational studies and, by definition, use the most realistic and externally valid data of actual fishing behaviour. Revealed preference methods include surrogate market approaches such as travel cost and hedonic pricing. For example, Alvarez et al. (2014) use site choice by saltwater fishers in the Southeast US to estimate monetary compensation for recreational losses due to the Deepwater Horizon oil spill. Prayaga et al. (2010) used the travel cost methods to estimate the value of recreational fishing in a region in Queensland, Australia. Navarro et al. (2021) used phone diary surveys to develop Random Utility Models (RUMs) of site choice to predict the effect of management interventions (e.g., closed areas) on fisher behaviour.

There is a range of novel technologies that can measure recreational fisher preferences. These approaches include cameras, satellite imagery, radar, and hydrophones. Hartill et al. (2020) suggest that robustly designed digital camera monitoring systems are a cost-effective means of monitoring recreational fisher behaviour and that they can provide insight into fishery dynamics. The use of digital camera systems has been trialled in New Zealand (Hartill et al. 2016), Australia (Keller et al. 2016; Steffe et al. 2017), and Germany (Kaiser 2016). Satellite data has also been used to monitor recreational fisher behaviour in, e.g., Greece (Keramidas et al. 2018), where it helped estimate the number of recreational vessels in the Aegean and Ionian Seas. Satellite data has been used to evaluate the role of recreational fishing activities in transporting pest and disease vectors (Bacela-Spychalska et al. 2013).

Hydrophone, or acoustic data, has been used successfully in Australia to monitor fishing activity in MPAs (Kline et al. 2020). Other approaches to gain insight into the way fishers behave are using apps, providing an opportunity to exchange information that could change the way fishers fish and engage with management (Venturelli et al. 2017). Smartphone apps are now used to collect data on aspects of fisher behaviour in many different countries, but experts agree that they might not be a stand-alone method now (Skov et al. 2021). It is worth noting that even though the predictors in these revealed preference approaches on their own are unable to establish causation, they have external validity, in that the behavioural decisions of the fishers are real.

Some internal validity can be gained through combining different approaches. For example, combining autonomous cameras' information with creel surveys (both revealing fisher behaviour through observation) (Dutterer et al. 2020; Lai et al. 2021) to estimate, e.g., fishing effort. This dual approach was also used to test hypotheses of optimal management policy for rainbow trout (*Oncorhynchus mykiss*) fisheries in British Columbia (Askey et al. 2013; Mee et al. 2016; Carruthers et al. 2019).

Internal validity can be gained through additional surveys or questionnaires, where respondents are asked for their views, opinions, or preferences (Hoyos 2010; Hess and Daly 2014). For example, via a questionnaire, Pouso et al. (2018) found a relationship between ecological recovery and a change in recreational fishing behaviour. The potential impact and effect size of a particular type of regulation inconsistently impacting fishers can be picked up with the use of survey methods (Thomas et al. 2016). Surveys can be implemented on site (e.g., access point or boat ramp surveys) or off site (e.g., diary panels). Recreational diary panels are a common approach to gain insight into fisher behaviour, but a carefully designed recruitment process is key to ensuring the estimates of catch, harvest, release, and effort are not biased (Lewin et al. 2023) (see Chapter "Ecology for Understanding Recreational Fishers and Fisheries" in Sect. 3).

Many studies have investigated the implications of differences between fishers through parameters that are allowed to vary by individual (McFadden and Train 2000) or by including covariates that describe characteristics of the fishers themselves (Kamakura and Russell 1989). The goals of these studies vary widely. They include testing the effects of demographic variables or recreational fisher typology (such as specialized and committed fishers vs. casual fishers) on fishing preferences and site choice behaviour (Beardmore et al. 2013; Goldsmith et al. 2018), estimating the economic value of fishery resources (Lew and Larson 2012; Alvarez et al. 2014), and predicting spatial patterns in fishing effort (Post et al. 2008; Wilson et al. 2020). Some studies combine stated preference and revealed preference data, as in Hindsley et al. (2023), who combine data from a creel survey with a discrete choice experiment. They find that there may be hypothetical bias in the stated preference data and that the willingness to pay estimates for improved catch is higher than the revealed preference estimate. A review of revealed preference and stated preference studies in recreational fishing is found in Hunt et al. (2019).

Discrete choice experiments, a type of stated preference method, are quite popular in understanding recreational fishing behaviour (see Fig. 4 for an example). Choice experiments, e.g., may evaluate fisher response to regulations such as bag or size

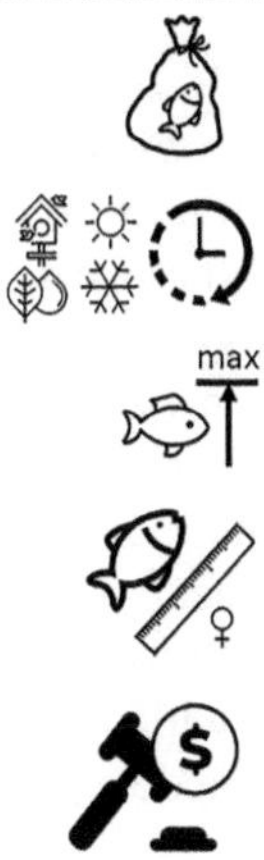

Management attribute	Status quo	Levels A	Levels B	Levels C
Daily bag limit	1/day	2/day	3/day	5/day
Season length	15 weeks	12 weeks	52 weeks	24 weeks
Maximum seasonal catch	15	25	10	30
Maximum size limit for females	105mm	115mm	110mm	95mm
Penalties for noncompliant act	$100	$150	$50	$250
Your choice	☐	☑	☐	☐

Fig. 4 Hypothetical example of question in a discrete choice experiment (loosely based on Mackay et al. 2020; icons from https://www.flaticon.com)

limits (Morey et al. 1991; Aas et al. 2000; Oh et al. 2005; Beardmore et al. 2011; Carr-Harris and Steinback 2020; Carter et al. 2020), catch outcomes (Koemle et al. 2022; Bronnmann et al. 2023), gear restrictions (Knoche and Lupi 2016), or the value of stocked versus wild fish (Arlinghaus et al. 2014). Alternatively, management attributes may describe particular management outcomes such as reductions in crowding (Lee et al. 2014), economic benefits to particular user groups (Wattage et al. 2005), or improvements in water quality (Kosenius 2010; Zhang and Sohngen 2017). Holzer and McConnell (2017) use a choice experiment for the Gulf of Mexico flounder (*Cyclopsetta chittendeni*) fishery to show that fishers exhibit disutility from randomness in the catch, and Koemle et al. (2022) found that fishers preferred ambiguity in expected catch over known probabilities of low catch rates. Risk aversion leads to different predictions on fisher participation and welfare from those assuming risk neutrality.

3 A Practical Guide to Doing an Experiment

In this section, we discuss the practical side of doing experiments to understand recreational fisher behaviour. There are many aspects that need to be considered, including hypothesis generation, selection of appropriate statistical models, recruitment of study participants, and ethical considerations of experiments involving human subjects.

3.1 Hypothesis Generation and Selection of Appropriate Statistical Models

The study of human behaviour within recreational fisheries draws from many disciplines. This means that researchers have many tools and theoretical frameworks available to them when formulating hypotheses and designing experiments. Although the interdisciplinary nature of recreational fisheries research generates a rich variety of perspectives on the relationships between humans and nature, the sheer variety of theoretical frameworks available in the literature means that researchers may have difficulty explicitly basing their recreational fishing experiments on theory. Before beginning the process of designing an experiment, a first key step is to identify the theoretical framework that is most appropriate for your study system and research question. The choice of theoretical framework will inform the scope of your hypothesis testing as well as your choice of statistical models. For example, random utility theory (McFadden 1974) is an effective framework for investigating and quantifying welfare changes for a population of fishers, but it would not be effective for understanding how, e.g., social norms influence fisher choice behaviour, which is more rooted in fields outside economics (Bicchieri 2005, 2017) (see Chapter "Social Psychology for Understanding Recreational Fishers and Fisheries" Social Psychology and Chapter "Behavioural Economics for Understand ing Recreational Fishers and Fisheries" Behavioural Economics). Explicitly rooting your hypothesis in theory should improve your experiments' potential applicability to other systems and avoid producing findings that cannot be replicated. If possible, collaborating with an expert in the discipline is an opportunity to identify unmet assumptions or "unknown unknowns" that may weaken or invalidate your results if unaddressed.

Before beginning any experiment, one should therefore identify an applicable theoretical framework, carefully review previous work to narrow the research gap, and develop a hypothesis describing the expected relationship between the predictor and response variables. The statistical model describes the relationship that the researcher is hypothesizing, i.e. that X change in a predictor, all else being equal, is expected to produce Y change in the response variable (e.g., Fig. 1). The fit of the final model and the effect size and uncertainty of the estimated coefficients are then used to evaluate the likelihood that the null hypothesis (that there is no effect or no relation among variables) should be rejected, given the data that were observed. The choice of appropriate statistical models will depend on whether the response variable is continuous or discrete, as well as its statistical distribution (e.g., Gaussian, binomial, Poisson, negative binomial). In addition, the model must account for any non-independence among observations. Non-independent observations may occur, e.g., when observations of individual fishers are collected across multiple waterbodies. Two observations of catch rates within a single waterbody will be more similar than two observations of catch rates from different waterbodies. Similarly, observations that are closer to each other in time or space may exhibit temporal or spatial autocorrelation. Both the experimental design and the statistical model must be formulated to either avoid or account for cases when observations are not independent. A rich set of statistical tools now exists to accommodate almost any data structure (Stevens 2012; Cunningham 2021; Huntington-Klein 2021).

Pre-testing the statistical model on simulated data is an effective strategy to (i) practice with the chosen statistical modelling tool (e.g., R packages, Python libraries, proprietary statistical software) before working with the "real" data and (ii) choose an appropriate sample size and experimental design. Simulated data may best be produced with a known relationship between the predictor and response variables. Different versions of the simulation can include smaller or larger effect sizes and narrower or wider variance of the error term to test the model's power to detect the known effect across a range of sample sizes and experimental designs (i.e., varying the number of treatment groups and replicates). These power analyses can be completed "manually" as described here or by using specific tools developed for that purpose. Pre-testing the experiment, e.g., by conducting an initial version of an economic experiment with a reduced sample size, provides another opportunity to practice fitting the statistical model and identify potential errors (e.g., non-convergence, suspicious residuals, or unexpected distribution of response variables) before beginning the energy- and resource-intensive process of conducting the experiment.

Finally, experiments that are not carefully linked to theories and hypotheses bear the risk that the research itself becomes a fishing trip, where researchers search for interesting and perhaps surprising results and try to come up with ex-post explanations for those. Therefore, it has become good practice to pre-register any experimental designs with hypotheses and expected results. This limits flexibility for researchers but takes away the pressure to find something interesting and makes it easier to publish null results.

3.2 Recruitment Processes

Recruiting representative recreational fishers or recreational managers for experimental research can be notoriously difficult, although in some jurisdictions, licence records provide a sample frame for probability-based sampling design. The fishers can be hard to reach simply because they are geographically dispersed (Moore et al. 2015). Often, fishers are hard to reach because there is no central way to contact them as a group. Increasingly, though, there are apps and social media sites that recreational fishers use, and these can provide a way for researchers to contact them and connect with them. Contacting fishers through fishing-specific media, however, may cause the recruitment of only highly engaged fishers (e.g., Venturelli et al. 2017). The risks of obtaining a biased sample of the overall population of fishers apply to almost any method of contacting, and researchers should try to understand and minimize or account for such biases and incentivize participation (e.g., using payments). Ideally, participants are a random sample of fishers, whereas in practice, the most motivated and committed fishers are those who are most likely to be responsive towards a request to participate in a study.

Participants may be recruited in person via or in different ways in a recreational fishing context. Access point surveys (where fishers are approached at boat launching ramps or popular fishing spots) are one way in which this has traditionally happened. A study by Griffiths and Fay (2015) addresses the challenge of reaching "hard-to-reach" longtail tuna (*Thunnus tonggol*) recreational fishers in Australian waters by trialling different methods. They compare the effectiveness (and other benefits) of time-location sampling (at e.g., tackle shops and trade shows) and online diaries to traditional access point surveys. Online diaries were inexpensive but showed signs of recruitment bias. Time-location sampling at key fisher infrastructure (such as tackle shops) was cost-effective and reached a broader set of fishers. Access point surveys were expensive and, perhaps surprisingly, were not able to reach the same breadth of fishers as time-location sampling. Regardless of how participation in the experiment is solicited, the researcher should be aware of any ethical issues around this.

In relation to recruiting individuals to the research, researchers must make sure that there is no undue pressure for people to participate. Even though this may seem unsurprising—undue pressure can be exerted in unexpected ways. To get participation in a survey, a researcher may send an email to the president of the local recreational fisher club and ask them to forward the email to their club members. Might it be that members feel pressured to do the survey because otherwise, there may be consequences from the president? Maybe—and maybe not—but this needs to be a thought that is considered when thinking about how to recruit participants. It also needs to be clear in the recruitment process that no false promises are made (e.g., if you do this survey, your fishing experience will certainly improve). In blinded experiments where individual people are allocated to specific treatments, e.g., in (Fujitani et al. 2017), it is important to be clear upfront that an experiment is being conducted and that all individuals are debriefed later about which treatment they belonged to, and which results were gathered. When interventions include sensitive issues, e.g., providing wrong messaging or lies, even greater standards apply for ethical approval. Selected interventions (e.g., providing information on catches) in real fisheries may substantially change fisher behaviour "in the real world," which can cause another set of ethical issues that must be clearly disclosed or avoided in the process.

3.3 Survey Biases

We recommend that researchers familiarize themselves with survey methods (Chapters "Fundamentals of Sampling and Implementing Human Dimensions of Fishing Surveys" and "Measurement Instruments") to ensure they are constructing clear, quality questions and prompts for participants. A variety of resources exist, including Survey Methodology by Groves et al. (2009). As mentioned earlier, it is important in all experiments that the respondents understand the question or task they are asked to carry out. If the questions are vague or poorly written, or the interviewer is impacting or influencing the respondents' answers, this can easily lead to response bias. There are many reasons the respondent may not answer the questions truthfully, for instance, if they are uncomfortable. Two related reasons they may not answer truthfully are confirmation bias and social desirability bias. The former is where the respondent's answer will conform to the direction of the question (Kunda and Sinclair 1999). For example, if fishers are asked if they are catch-and-release fishers, they may preferentially remember times that they released fish, which may bias their responses towards greater catch-and-release orientation. The latter is where the respondent second-guesses what is socially desirable and over-reports the behaviours that are desirable or answer to influence some fishery management outcome (Maccoby and Maccoby 1954; Fisher 1993). For example, if catch and release angling is perceived by the fisher as virtuous behaviour, they may over-report their catch and release activities. For activities that are socially unacceptable or illegal, such as noncompliance with fishing regulations, alternative approaches may be required, such as anonymous responses and indirect questions (Bova et al. 2022) or a randomized response technique (Lewis 2015; Thomas et al. 2015) where partial randomization of responses in a survey provides individual cover for illegal activity but allows population-level inference about rates of noncompliance (e.g., Blank and Gavin 2009).

Another important bias to be aware of is non-responder bias, where respondents refuse to answer the question or even return or participate in a survey (Tarrant et al. 1993). Non-response bias is essentially a form of selection bias. Selection bias may happen when the people targeted by the study are not included in the experiment. A good example of this is when, for a phone survey, only landlines are called, but many people no longer have landlines. Sampling bias is the same as selection bias and happens when the sampling frame does not represent the true population. This may be due to the sampling technique or the data collection method. It is good practice to report response rates and not to generalize to the population level if non-response or selection bias is present.

In lab and lab-in-the-field experiments, other biases apply that can create confounding effects. Participants may misunderstand the experimental procedures, or they may make incorrect inferences about their own preferences. Where participants interact with each other in an experiment, they may respond based on their subjective beliefs about the others and have false expectations (Cubitt et al. 2001).

When designing surveys, the information that is in the survey should be unambiguous and interpreted in the same way. This means that the researchers need to think carefully about how the participant will perceive and interpret the experiment and how to avoid divergence (Kawagoe and Takizawa 2019). For example, when asking fishers about their preferences for different management approaches (Mackay et al. 2020) using a stated choice experiment (which is a survey-based experiment), the description of the management attributes (that the respondent is weighing up and choosing between) needs to be clear and well understood.

All approaches are prone to participant and survey biases. Asking people for their preferences has an inherent risk that the respondent might not be answering truthfully or be self-aware of their preferences or behaviour. There may also be a lack of incentive to tell the truth. The disincentive may be created because the respondent thinks they should answer something to please the interviewer, or saying something that they think is right, gives them a warm glow, or it conforms with the social norm (Bova et al. 2022). Cognitive anomalies apply in the stated preference approach, e.g., the hypothetical setting may be confusing or difficult to relate to and thus causing problems. Similarly, there are issues with the revealed preference approach because it might be impossible to know which behavioural options were discarded in preference for the revealed behaviour. Neither the stated preference nor the revealed preference approach, when applied cross-sectionally, can account for preference changing over time. There is some consensus that revealed preference estimates are, in general, considered more reliable than stated preferences (List and Gallet 2001; Murphy et al. 2005; de Corte et al. 2021), this is thought to be due to the tendency for people to overstate their preferences, also known as "hypothetical bias" (Loomis 2011; Johansson-Stenman and Svedsäter 2012).

3.4 Counterfactual Thinking

In experimentation, a critical concept is not only to understand the causal relationships and the confounders but also to think in terms of the counterfactual or control (i.e., if X had not happened, then Y would not have occurred). In recreational fisheries, e.g., Dundas and von Haefen (2020) used counterfactuals in a study that identified and quantified the economic value of behavioural adaptation to climate change by fishers. A counterfactual simulation predicts welfare change under future climate scenarios. The behavioural response of fishers to high temperatures is to switch their recreational activities to night-time. This behavioural adaptation can save up to $27 million annually. In the counterfactual simulation, the fishers behavioural change to night-time fishing does not occur. Another example of the use of counterfactuals in recreational fisheries is presented by Fujitani et al. (2012) and their work on fishers' use of marine reserves in the Gulf of California. Authors compared visitation patterns from fishers to an area that was declared a marine reserve in 2022 and visitation patterns to that area if it had not been declared a marine reserve (i.e., counterfactual scenario). Findings showed that the marine reserve influenced fisher behaviour in the short term, as visits declined from the moment the marine reserve was announced. However, the strength of the reserve to hinder recreational fishing trips to that area declined over time as compliance was not adequately enforced, thus showing that the effect did not last without strong management implementation (Fujitani et al. 2018).

For experimental methods that use controlled interventions, the treatment measures the experimental effect, and the control serves as the counterfactual (Cunningham 2021). In choice experiments and welfare studies, the counterfactual is the status quo or base case. For natural experiments with observational data, we do not have a "parallel universe" that tells us what the world looked like, say, if COVID-19 never happened. Hence, it is the task of the researcher to construct a credible counterfactual. For example, in conservation research, the study of counterfactual scenarios can be used to celebrate success stories by evaluating the success of past conservation efforts as well as to predict the potential impact of a management approach (Grace et al. 2021). One key aspect in assessing counterfactual scenarios is to define the variable used to evaluate the impact of the treatment (which will result in a counterfactual outcome, different from the current outcome). For example, the past impact of a treatment (e.g., a conservation action) in a system where we know the outcome (e.g., current conservation status of a species), a counterfactual scenario might be where no conservation efforts were undertaken for that species (Grace et al. 2021).

One option is to compare *changes* in, e.g., a water body before and after treatment to changes in other systems (a diff-in-diff approach also referred to as a before-after-control treatment design). The difference-in-difference approach observes two groups over a certain period. If there were no treatments, the two groups would have similar trends (equal trend assumption), and there would be no time-varying differences. But then one of the groups (the treatment group) has a change in the independent variable over time (whereas the control group does not). For example, the introduction of benchmarking for fish stocks reduced behavioural biases in stock assessments in the treatment group when compared to the control group, where there was no benchmarking (Fujitani et al. 2017; Schuch and Richter 2022). The estimator of the difference in difference approach is literally the difference in outcome in the first period (before treatment) and second period (after treatment) in the treatment group minus the difference in average outcome in the control group before and after treatment (Fox and Swearingen 2021). There have been advances in statistical analysis towards causal analysis of multivariate data using structural causal and structural equation models (Pearl 2009; Pearl et al. 2016).

3.5 Ethical Considerations

Ethics are a key consideration when conducting experimental research involving human subjects. The extent of these considerations will depend on the types of data collected and the degree of local community engagement with the research. Although recreational fisheries experiments are unlikely to cause direct physical harm or mental distress to participants, harm to participants is possible through the breach of private data or the release of local, traditional, or Indigenous data to wider audiences without the consent of that community (Hudson 2009; Webster and Sell 2014). In addition, experiments where false information is provided, which is not allowed in experimental economics (see Chapter "Social Psychology for Understanding Recreational Fishers and Fisheries", Social Psychology), or the treatment is to expose illegal behaviour, can also risk harm. Some, but not all, of these potential harms are addressed through institution-specific human subjects research authorization protocols and reviews, which vary in degree from country to country. It is advisable to approach faculties that regularly engage in human research, e.g., psychology, to learn about ethical review needs in each culture or country.

Processes of authorization for human subject research depend on the location of the research and the policies of the research institution, and we will not present a full review of those policies here. It is therefore vital for researchers to verify local, national, and institutional policies before beginning any experiments involving interaction with and recruitment of human subjects, including administering surveys and conducting interviews. In a recreational fisheries context, researchers will likely need to address potential issues regarding participant privacy (i.e., avoiding breaches of private data), facilitating informed consent of participants, and ethical recruitment of study participants, including appropriate uses of incentives (Webster and Sell 2014). This process can be lengthy, even for experiments that pose little risk to participants. It is important to begin this process well in advance of the project's start date. Research carried out by non-governmental groups (NGOs) or independent consultants does not always have access to ethics committees to follow these same formal ethics approval pathways, but regardless, there is a need and opportunity for these organisations to consider the ethical aspects of their work.

Ethical considerations go beyond gaining institutional approval (Heimer and Petty 2010; Wilmer et al. 2021). Fisheries researchers increasingly adopt participatory research or knowledge co-production approaches to bridge the gap between scientific knowledge and policy implementation (Cooke et al. 2021). Regardless of institutional requirements, the integrity of the research design and the goal of maintaining equitable partnerships with fishery stakeholders and rights holders depend on obtaining informed consent and the avoidance of extractive research practices. These unethical research practices benefit academic researchers by extracting data, publications, and prestige without benefiting the community holding that knowledge (e.g., Trisos et al. 2021; Wilmer et al. 2021). Teams conducting research that relies on local, traditional, or Indigenous ecological knowledge should ensure that they are obtaining not just individual consent, but collective consent to broadly share this knowledge. Individual motivations for participating in research vary, so individuals should not be uncritically accepted as community representatives without broader community input (Hudson 2009). Consistent and clear communication between researchers and community members outside of typical academic modes (i.e., journal articles and formal presentations) is essential to ensure continued collective consent and to ensure that research goals are "credible, relevant, actionable, and impactful" (Nyboer et al. 2022). Different communities will have different needs, and a research protocol that is harmless in one community may cause significant harm elsewhere. It is essential to understand the history of the specific community with which you collaborate, and particularly their history of abuse by governmental entities and academics (e.g., Hatch et al. 1993; Smith 1999). However, every community is different, so teams should be prepared to listen to feedback and adjust their behaviour rather than relying exclusively on arbitrary or contextually ill-suited benchmarks or consider the job being done once ethical approval is obtained. Clearly, in an experiment, participating subjects need to be debriefed at the end and results fully disclosed and shared with participants.

4 Future Directions

As outlined in this chapter, experiments offer learning opportunities to study recreational fishing and they are currently underused in recreational fishing research, which is often only observational or cross-sectional (e.g., recreational fisher surveys). The social–ecological context of recreational fishing very much leans itself towards a "living laboratory" of constant learning and experimentation for at least two reasons. First, recreational fisheries are heavily regulated and tightly controlled in some parts of the world. Management interventions are implemented on a regular basis in recreational fisheries and each of those is an opportunity to learn. Second, recreational fishing is often an organized social activity with entrance licences, fishing clubs, and fixed fishing sites, which can serve as sampling units (e.g., Fujitani et al. 2017). Recreational fishers are also often a particularly motivated and engaged group who have a strong interest in their ecological system, which should make it easier to involve them in research activities than random citizens who lack such motivations. Together, this makes the economic experimental toolbox particularly useful for recreational fishing (e.g., Stoop et al. 2012).

One may argue that many of the ideas presented in this chapter are just old wine in new bottles. After all, the notion of "adaptive management" (Walters 1986) very much embraces the idea of experimentation, learning, and looking holistically at a social–ecological system. Yet, we think that recent developments in experimental economics can bring something to our profession and advance our understanding of recreational fishing. First, a rigorous application of "randomized control trials" where entire sites or lakes or recreational fishing groups or communities or clubs are "treated" and compared to control sites or groups holds the potential to answer big questions around recreational fishing. Studies using such an approach are clearly very costly to conduct, but the potential scientific impact is high as well; see, e.g., Radinger et al. (2023) who used a participatory habitat enhancement vs. stocking design applied to different angling clubs and implemented in partnership to learn ecologically how the entire system changed to interventions. Second, methods from economic experiments are increasingly useful to understand social dynamics and social norms (Andreoni et al. 2021), which are highly relevant for recreational fishing. Also, preferences of fishers are often unobserved, leaving practitioners guessing when it comes to the acceptability of, and response to policy interventions. Incentivized games (Diekert et al. 2023) or choice experiments (Rambonilaza and Kerouaz 2023) may offer excellent opportunities to learn about fisher preferences and behaviour. Third, recent developments around "natural experiments" and statistical models of causal inference (Pearl et al. 2016) offer opportunities to learn from the many interventions that are conducted by management authorities as "uncontrolled experiments." In such a case, establishing a credible counterfactual will always remain a challenge, because one needs to compare an observed outcome to one that would have happened in the absence of the intervention and therefore never happened. Yet, sufficiently long time series and the presence of potential control sites can help to construct a credible counterfactual.

Though there are reasons to be enthusiastic about the prospects of using experimental methods from economics for recreational fishing, we would like to end on a cautionary note. Conducting experiments in recreational fishing requires close coordination with stakeholders and time investment to build trust. Although this is ideal for genuine long-term collaborations, a quick and poorly designed experiment can be perceived as intrusive and shows a lack of respect for study participants and stakeholders. Therefore, there are strong ethical implications related to conducting such research. Also, methods from experimental economics are not necessarily easy to master, and those who are trained in those methods are not necessarily trained to work in interdisciplinary teams. So, there are certainly challenges when it comes to applying those methods in a field setting, and when it comes to bringing those field contexts to universities that teach those methods. Finally, the great strength of the approaches sketched in this chapter is the ability to generate quantitative insights, which is by no means the only and probably not even the best framework to study human experience and perceptions. Unlike natural scientists, social scientists hold the privilege of being able to talk to their study objects and ask questions. It is straightforward to recommend that qualitative and quantitative methods should be combined in a complementary way, but doing this in practice is much less straightforward. So here lies another challenge for future researchers to overcome.

Acknowledgments We would like to acknowledge the useful and insightful comments from the reviewers and the book editors. We thank them for their patience.

References

Aas Ø, Haider W, Hunt L (2000) Angler responses to potential harvest regulations in a Norwegian sport fishery: a conjoint-based choice modeling approach. N Am J Fish Manag 20(4):940–950

Abadie A (2021) Using synthetic controls: feasibility, data requirements, and methodological aspects. J Econ Lit 59(2):391–425

Alvarez S, Larkin SL, Whitehead JC, Haab T (2014) A revealed preference approach to valuing non-market recreational fishing losses from the Deepwater horizon oil spill. J Environ Manag 145:199–209

Andreoni J, Nikiforakis N, Siegenthaler S (2021) Predicting social tipping and norm change in controlled experiments. Proc Natl Acad Sci USA 118(16):e2014893118

Arlidge WNS, Arlinghaus R, Kurvers RHJM, Nassauer A, Oyanedel R, Krause J (2023) Situational social influence leading to non-compliance with conservation rules. Ecol Evol 38(12): 1154–1164

Arlinghaus R, Beardmore B, Riepe C, Meyerhoff J, Pagel T (2014) Species-specific preferences of German recreational anglers for freshwater fishing experiences, with emphasis on the intrinsic utilities of fish stocking and wild fishes. J Fish Biol 85(6):1843–1867

Arlinghaus R, Alós J, Beardmore B, Daedlow K, Dorow M, Fujitani M, Hühn D, Haider W, Hunt LM, Johnson BM, Johnston F, Klefoth T, Matsumura S, Monk C, Pagel T, Post JR, Rapp T, Riepe C, Ward H, Wolter C (2017) Understanding and managing freshwater recreational fisheries as complex adaptive social-ecological systems. Rev Fish Sci Aquac 25(1):1–41

Askey PJ, Parkinson EA, Post JR (2013) Linking fish and angler dynamics to assess stocking strategies for hatchery-dependent, open-access recreational fisheries. N Am J Fish Manag 33(3): 557–568

Bacela-Spychalska K, Grabowski M, Rewicz T, Konopacka A, Wattier R (2013) The 'killer shrimp' *Dikerogammarus villosus* (Crustacea, Amphipoda) invading Alpine lakes: overland transport by recreational boats and scuba-diving gear as potential entry vectors? Aquat Conserv 23(4):606–618

Bacon F (1818) Novum organum scientiarum, London

Banerjee AV (2020) Field experiments and the practice of economics. Am Econ Rev 110(7): 1937–1951

Beard TD, Cox SP, Carpenter SR (2003) Impacts of daily bag limit reductions on angler effort in Wisconsin Walleye Lakes. N Am J Fish Manag 23(4):1283–1293

Beardmore B, Dorow M, Haider W, Arlinghaus R (2011) The elasticity of fishing effort response and harvest outcomes to altered regulatory policies in eel (Anguilla Anguilla) recreational angling. Fish Res 110(1):136–148

Beardmore B, Haider W, Hunt LM, Arlinghaus R (2013) Evaluating the ability of specialization indicators to explain fishing preferences. Leis Sci 35(3):273–292

Bennear LS, Stavins RN, Wagner AF (2005) Using revealed preferences to infer environmental benefits: evidence from recreational fishing licenses. J Regul Econ 28(2):157–179

Bicchieri C (2005) The grammar of society: the nature and dynamics of social norms, Illustrated. Cambridge University Press, Cambridge

Bicchieri C (2017) Norms in the wild: how to diagnose, measure, and change social norms. Oxford University Press, New York, NY

Blank SG, Gavin MC (2009) The randomized response technique as a tool for estimating non-compliance rates in fisheries: a case study of illegal red abalone (*Haliotis rufescens*) fishing in northern California. Environ Conserv 36(2):112–119

Boonstra WJ, Birnbaum S, Björkvik E (2017) The quality of compliance: investigating fishers' responses towards regulation and authorities. Fish Fish 18(4):682–697

Bova CS, Stephens J, Aswani S, Potts WM (2022) Is the instrumental approach a 'silver bullet' for addressing non-compliance in recreational fisheries: a south African case study. Fish Res 255: 106439

Britton JR, Pinder AC, Alós J, Arlinghaus R, Danylchuk AJ, Edwards W, Freire KMF, Gundelund C, Hyder K, Jarić I, Lennox R, Lewin W-C, Lynch AJ, Midway SR, Potts WM, Ryan KL, Skov C, Strehlow HV, Tracey SR, Tsuboi J, Venturelli PA, Weir JL, Weltersbach MS, Cooke SJ (2023) Global responses to the COVID-19 pandemic by recreational anglers: considerations for developing more resilient and sustainable fisheries. Rev Fish Biol Fish 33(4): 1095–1111

Bronnmann J, Koemle D, Meyerhoff J, Weltersbach MS, Strehlow HV, Arlinghaus R (2023) Willingness to pay for harvest regulations and catch outcomes in recreational fisheries: a stated preference study of German cod anglers. Fish Res 259:106536

Bunt CM, Jacobson B (2022) The impact of the COVID-19 pandemic on a recreational rainbow trout (*Oncorhynchus mykiss*) fishery. Environ Biol Fish 105(4):499–507

Camp EV, Ahrens RNM, Crandall C, Lorenzen K (2018) Angler travel distances: implications for spatial approaches to marine recreational fisheries governance. Mar Policy 87:263–274

Carr-Harris A, Steinback S (2020) Expected economic and biological impacts of recreational Atlantic striped bass fishing policy. Front Mar Sci 6:814

Carruthers TR, Dabrowska K, Haider W, Parkinson EA, Varkey DA, Ward H, McAllister MK, Godin T, van Poorten BT, Askey PJ, Wilson KL, Hunt LM, Clarke A, Newton E, Walters C, Post JR (2019) Landscape-scale social and ecological outcomes of dynamic angler and fish behaviours: processes, data, and patterns. Can J Fish Aquat Sci 76(6):970–988

Carter DW, Lovell SJ, Liese C (2020) Does angler willingness-to-pay for changes in harvest regulations vary by state? Results from a choice experiment in the Gulf of Mexico. Mar Policy 121:104196

Cooke SJ, Nguyen VM, Chapman JM, Reid AJ, Landsman SJ, Young N, Hinch SG, Schott S, Mandrak NE, Semeniuk CAD (2021) Knowledge co-production: a pathway to effective fisheries management, conservation, and governance. Fisheries 46(2):89–97

Cubitt RP, Starmer C, Sugden R (2001) Discovered preferences and the experimental evidence of violations of expected utility theory. J Econ Methodol 8(3):385–414

Cunningham S (2021) Causal inference: the mixtape. Yale University Press, New Haven and London

Dassow CJ, Ross AJ, Jensen OP, Sass GG, van Poorten BT, Solomon CT, Jones SE (2020) Experimental demonstration of catch hyperstability from habitat aggregation, not effort sorting, in a recreational fishery. Can J Fish Aquat Sci 77(4):762–769

de Corte K, Cairns J, Grieve R (2021) Stated versus revealed preferences: an approach to reduce bias. Health Econ 30(5):1095–1123

Diamond JM (1983) Ecology: laboratory, field and natural experiments. Nature 304(5927):586–587

Diekert F, Nøstbakken L, Richter A (2021) Control activities and compliance behavior-survey evidence from Norway. Mar Policy 125:104381

Diekert F, Li Y, Nøstbakken L, Richter A (2023) Why do fishermen comply with regulations? The role of preferences. In: Bucciol A, Tavoni A, Veronesi M (eds) Behavioural economics and the environment: a research companion. Routledge, Oxon/New York, pp 83–106

Dundas SJ, von Haefen RH (2020) The effects of weather on recreational fishing demand and adaptation: implications for a changing climate. J Assoc Environ Resour Econ 7(2):209–242

Dunning T (2008) Improving causal inference: strengths and limitations of natural experiments. Polit Res Q 61(2):282–293

Dunning T (2012) Natural experiments in the social sciences: a design-based approach. Cambridge University Press, Cambridge, UK

Dutterer AC, Dotson JR, Thompson BC, Paxton CJ, Pouder WF (2020) Estimating recreational fishing effort using autonomous cameras at boat ramps versus creel surveys. N Am J Fish Manag 40(6):1367–1378

Fayram AH, Hansen MJ, Ehlinger TJ (2006) Influence of walleye stocking on angler effort in Wisconsin. Hum Dimens Wildl 11(2):129–141

Ferraro PJ, Sanchirico JN, Smith MD (2019) Causal inference in coupled human and natural systems. Proc Natl Acad Sci USA 116(12):5311–5318

Fisher RJ (1993) Social desirability bias and the validity of indirect questioning. J Consum Res 20(2):303

Fox HK, Swearingen TC (2021) Using a difference-in-differences and synthetic control approach to investigate the socioeconomic impacts of Oregon's marine reserves. Ocean Coast Manag 215: 105965

Fujitani ML, Fenichel EP, Torre J, Gerber LR (2012) Implementation of a marine reserve has a rapid but short-lived effect on recreational angler use. Ecol Appl 22(2):597–605

Fujitani M, McFall A, Randler C, Arlinghaus R (2017) Participatory adaptive management leads to environmental learning outcomes extending beyond the sphere of science. Sci Adv 3(6): e1602516

Fujitani ML, Fenichel EP, Torre J, Gerber LR (2018) Synthesizing ecological and human use information to understand and manage coastal change. Ocean Coast Manag 162:100–109

Gangadharan L, Jain T, Maitra P, Vecci J (2022) Lab-in-the-field experiments: perspectives from research on gender. Jpn Econ Rev 73(1):31–59

Glennerster R, Takavarasha K (2014) Running randomized evaluations: a practical guide. Princeton University Press, Princeton and Oxford

Gneezy U, Imas A (2017) Chapter 10-Lab in the field: measuring preferences in the wild. In: Banerjee AV, Duflo E (eds) Handbook of economic field experiments, vol 1, pp 439–464

Golden AS, Sivaram S, Batsaikhan G, Jensen OP (2021) Thunderstorms have species and gear-specific indirect effects on the catchability of Mongolian salmonids. Fish Manag Ecol 28(4): 373–382

Golden AS, van Poorten BT, Jensen OP (2022) Focusing on what matters most: evaluating multiple challenges to stability in recreational fisheries. Fish Fish 23(6):1418–1438

Goldsmith WM, Scheld AM, Graves JE (2018) Characterizing the preferences and values of U.S. recreational Atlantic bluefin tuna anglers. N Am J Fish Manag 38(3):680–697

Grace MK, Akçakaya HR, Bull JW, Carrero C, Davies K, Hedges S, Hoffmann M, Long B, Lughadha EMN, Martin GM, Pilkington F, Rivers MC, Young RP, Milner-Gulland EJ (2021) Building robust, practicable counterfactuals and scenarios to evaluate the impact of species conservation interventions using inferential approaches. Biol Conserv 261:109259

Griffiths SP, Fay G (2015) Integrating recreational fisheries data into stock assessment: implications for model performance and subsequent harvest strategies. Fish Manag Ecol 22(3):197–212

Groves RM, Fowler FJ, Couper MP, Lepkowski JM, Singer E, Tourangeau R (2009) Survey methodology. Wiley, Hoboken

Guala F (2005) The methodology of experimental economics. Cambridge University Press

Hansen GJA, Gaeta JW, Hansen JF, Carpenter SR (2015) Learning to manage and managing to learn: sustaining freshwater recreational fisheries in a changing environment. Fisheries (Bethesda) 40(2):56–64

Harrison GW, List JA (2004) Field experiments. J Econ Lit 42(4):1009–1055

Hartill BW, Payne GW, Rush N, Bian R (2016) Bridging the temporal gap: continuous and cost-effective monitoring of dynamic recreational fisheries by web cameras and creel surveys. Fish Res 183:488–497

Hartill BW, Taylor SM, Keller K, Weltersbach MS (2020) Digital camera monitoring of recreational fishing effort: applications and challenges. Fish Fish 21(1):204–215

Hatch J, Moss N, Saran A, Presley-Cantrell L, Mallory C (1993) Community research: partnership in black communities. Am J Prev Med 9(6):27–31

Heimer CA, Petty J (2010) Bureaucratic ethics: IRBs and the legal regulation of human subjects research. Annu Rev Law Soc Sci 6:581–599

Henrich J, Heine SJ, Norenzayan A (2010) The weirdest people in the world? Behav Brain Sci 33(2–3):61–83

Hess S, Daly A (2014) Handbook of choice modelling. Edward Elgar Publishing, Cheltenham

Hilborn R (2016) Correlation and causation in fisheries and watershed management. Fisheries (Bethesda) 41(1):18–25

Hindsley P, Landry CE, Schnier K, Whitehead JC, Zarei M (2023) Joint estimation of angler revealed preference site selection and stated preference choice experiment recreation data considering attribute non-attendance. J Environ Econ Policy 12(1):44–62

Holder PE, Jeanson AL, Lennox RJ, Brownscombe JW, Arlinghaus R, Danylchuk AJ, Bower SD, Hyder K, Hunt LM, Fenichel EP, Venturelli PA, Thorstad EB, Allen MS, Potts WM, Clark-Danylchuk S, Claussen JE, Lyle JM, Tsuboi J, Brummett R, Freire KMF, Tracey SR, Skov C, Cooke SJ (2020) Preparing for a changing future in recreational fisheries: 100 research questions for global consideration emerging from a horizon scan. Rev Fish Biol Fish 30(1):137–151

Holzer J, McConnell K (2017) Risk preferences and compliance in recreational fisheries. J Assoc Environ Resour Econ 4(S1):S1–S35

Hoyos D (2010) The state of the art of environmental valuation with discrete choice experiments. Ecol Econ 69(8):1595–1603

Hudson M (2009) Think globally, act locally: collective consent and the ethics of knowledge production. Int Soc Sci J 60(195):125–133

Hunt LM, Arlinghaus R, Lester N, Kushneriuk R (2011) The effects of regional angling effort, angler behavior, and harvesting efficiency on landscape patterns of overfishing. Ecol Appl 21(7):2555–2575

Hunt LM, Camp E, van Poorten BT, Arlinghaus R (2019) Catch and non-catch-related determinants of where anglers fish: a review of three decades of site choice research in recreational fisheries. Rev Fish Sci Aquac 27(3):261–286

Huntington-Klein N (2021) The effect: an introduction to research design and causality. CRC Press, Boca Raton

Janssen MA, Holahan R, Lee A, Ostrom E (2010) Lab experiments for the study of social-ecological systems. Science 328(5978):613–617

Johansson-Stenman O, Svedsäter H (2012) Self-image and valuation of moral goods: stated versus actual willingness to pay. J Econ Behav Organ 84(3):879–891

Johnston D, Harris D, Caputi N, Thomson A (2011) Decline of a blue swimmer crab (*Portunus pelagicus*) fishery in Western Australia—history, contributing factors and future management strategy. Fish Res 109(1):119–130

Kaemingk MA, Chizinski CJ, Allen CR, Pope KL (2019) Ecosystem size predicts social dynamics in recreational fisheries. Ecol Soc 24(2):17

Kaiser F (2016) Estimating German recreational salmon catches in the Baltic Sea. Thesis. University of Rostock, Rostock

Kamakura WA, Russell GJ (1989) A probabilistic choice model for market segmentation and elasticity structure. J Mark Res 26(4):379–390

Kawagoe T, Takizawa H (2019) Diversity of experimental methods in economics: an introduction. Springer, Singapore, pp 1–19

Keller K, Steffe AS, Lowry M, Murphy JJ, Suthers IM (2016) Monitoring boat-based recreational fishing effort at a nearshore artificial reef with a shore-based camera. Fish Res 181:84–92

Keramidas I, Dimarchopoulou D, Pardalou A, Tsikliras AC (2018) Estimating recreational fishing fleet using satellite data in the Aegean and Ionian seas (Mediterranean Sea). Fish Res 208:1–6

Kimmel K, Dee LE, Avolio ML, Ferraro PJ (2021) Causal assumptions and causal inference in ecological experiments. Ecol Evol 36(12):1141–1152

Klatt J, Lupi F, Melstrom RT (2014) A natural experiment identifying disease regulation effects on recreational fishing. In: 2014 Annual meeting. Agricultural and Applied Economics Association, Minneapolis

Kline LR, DeAngelis AI, McBride C, Rodgers GG, Rowell TJ, Smith J, Stanley JA, Read AD, Van Parijs SM (2020) Sleuthing with sound: understanding vessel activity in marine protected areas using passive acoustic monitoring. Mar Policy 120:104138

Knoche S, Lupi F (2016) Demand for fishery regulations: effects of angler heterogeneity and catch improvements on preferences for gear and harvest restrictions. Fish Res 181:163–171

Koemle D, Meyerhoff J, Arlinghaus R (2022) How catch uncertainty and harvest regulations drive anglers' choice for pike (*Esox lucius*) fishing in the Baltic Sea. Fish Res 256:106480

Kosenius AK (2010) Heterogeneous preferences for water quality attributes: the case of eutrophication in the Gulf of Finland, the Baltic Sea. Ecol Econ 69(3):528–538

Kunda Z, Sinclair L (1999) Motivated reasoning with stereotypes: activation, application, and inhibition. Psychol Inq 10(1):12–22

Lai EKM, Ryan KL, Mueller U, Hyndes GA (2021) Corroborating effort and catch from an integrated survey design for a boat-based recreational fishery in Western Australia. Fish Res 236:105865

Lathrop RC, Johnson BM, Johnson TB, Vogelsang MT, Carpenter SR, Hrabik TR, Kitchell JF, Magnuson JJ, Rudstam LG, Stewart RS (2002) Stocking piscivores to improve fishing and water clarity: a synthesis of the Lake Mendota biomanipulation project. Freshw Biol 47(12): 2410–2424

Leatherdale ST (2019) Natural experiment methodology for research: a review of how different methods can support real-world research. Int J Soc Res Methodol 22(1):19–35

Lee DE, Hosking SG, du Preez M (2014) A choice experiment application to estimate willingness to pay for controlling excessive recreational fishing demand at the Sundays River Estuary, South Africa. Water SA 40(1):39–48

Lennox RJ, Sbragaglia V, Vollset KW, Sortland LK, McClenachan L, Jarić I, Guckian ML, Ferter K, Danylchuk AJ, Cooke SJ, Arlinghaus R, Twardek WM (2022) Digital fisheries data in the internet age: emerging tools for research and monitoring using online data in recreational fisheries. Fish Fish 23(4):926–940

Lester NP, Marshall TR, Armstrong K, Dunlop WI, Ritchie B (2003) A broad-scale approach to management of Ontario's recreational fisheries. N Am J Fish Manag 23(4):1312–1328

Lew DK, Larson DM (2012) Economic values for saltwater sport fishing in Alaska: a stated preference analysis. N Am J Fish Manag 32(4):745–759

Lewin W-C, Weltersbach MS, Haase K, Riepe C, Strehlow HV (2023) Potential biases in angler diary data: the impact of the diarist recruitment process on participation rates, catch, harvest, and effort estimates. Fish Res 258:106551

Lewis SG (2015) Bags and tags: randomized response technique indicates reductions in illegal recreational fishing of red abalone (*Haliotis rufescens*) in Northern California. Biol Conserv 189:72–77

List JA, Gallet CA (2001) What experimental protocol influence disparities between actual and hypothetical stated values? Environ Resour Econ (Dordr) 20(3):241–254

Loomis J (2011) What's to know about hypothetical bias in stated preference valuation studies? J Econ Surv 25(2):363–370

Lynch AJ, Sievert NA, Embke HS, Robertson AM, Myers BJE, Allen MS, Feiner ZS, Hoogakker F, Knoche S, Krogman RM, Midway SR, Nieman CL, Paukert CP, Pope KL, Rogers MW, Wszola LS, Beard TD Jr (2021) The U.S. inland creel and angler survey catalog (CreelCat): development, applications, and opportunities. Fisheries (Bethesda) 46(11):574–583

Maccoby EE, Maccoby N (1954) The interview: a tool of social science. In: Lindzey G (ed) Handbook of social psychology, 1st edn. Addison-Wesley, Boston, pp 449–487

Mackay M, van Putten EI, Yamazaki S, Jennings S, Sibly H (2020) Me and my behavior: an experiment on individual characteristics and compliance behavior in recreational fishing. Front Mar Sci 7

Matsumura S, Beardmore B, Haider W, Dieckmann U, Arlinghaus R (2019) Ecological, angler, and spatial heterogeneity drive social and ecological outcomes in an integrated landscape model of freshwater recreational fisheries. Rev Fish Sci Aquac 27(2):170–197

McFadden D (1974) Conditional logit analysis of qualitative choice behavior. In: Zarembka P (ed) Frontiers in econometrics. Academic Press, pp 105–142

McFadden D, Train K (2000) Mixed MNL models for discrete response. J Appl Econ 15(5): 447–470

Medoff S, Lynham J, Raynor J (2022) Spillover benefits from the world's largest fully protected MPA. Science 378(6617):313–316

Mee JA, Post JR, Ward H, Wilson KL, Newton E, Cantin A (2016) Interaction of ecological and angler processes: experimental stocking in an open access, spatially structured fishery. Ecol Appl 26(6):1693–1707

Melstrom RT, Jungers B, Bi X, Abbott JK (2026) Resource economics for understanding recreational fishers and fisheries. In: Pope KL, Arlinghaus R, Hunt LM, Lynch AJ, van Poorten BT (eds) Understanding recreational fishers: disciplinary and interdisciplinary approaches for fisheries management. Springer Nature, Cham, pp 123–156

Monk CT, Arlinghaus R (2018) Eurasian perch, *Perca fluviatilis*, spatial behaviour determines vulnerability independent of angler skill in a whole-lake reality mining experiment. Can J Fish Aquat Sci 75(3):417–428

Moore A, Hall K, Giri K, Tracey S, Hansen S, Ward P, Stobutzki I, Andrews J, Nicol S, Brown P (2015) Developing robust and cost-effective methods for estimating the national recreational catch of southern bluefin tuna in Australia. 5 Fisheries Research and Development Corporation, Canberra ACT

Morey ER, Shaw WD, Rowe RD (1991) A discrete-choice model of recreational participation, site choice, and activity valuation when complete trip data are not available. J Environ Econ Manag 20(1):181–201

Murphy JJ, Allen PG, Stevens TH, Weatherhead D (2005) A meta-analysis of hypothetical bias in stated preference valuation. Environ Resour Econ (Dordr) 30(3):313–325

Navarro M, Hailu A, Langlois T, Ryan KL, Burton M, Kragt ME (2021) Combining spatial ecology and economics to incorporate recreational fishing into marine spatial planning. ICES J Mar Sci 79(1):147–157

Noussair CN, van Soest D, Stoop J (2015) Cooperation in a dynamic fishing game: a framed field experiment. Am Econ Rev 105(5):408–413

Nowakowski AJ, Canty SWJ, Bennett NJ, Cox CE, Valdivia A, Deichmann JL, Akre TS, Bonilla-Anariba SE, Costedoat S, McField M (2023) Co-benefits of marine protected areas for nature and people. Nat Sustain 6(10):1210

Nyboer EA, Reid AJ, Jeanson AL, Kelly R, Mackay M, House J, Arnold SM, Simonin PW, Sedanza MGC, Rice ED, Quiros TEAL, Pierucci A, Ortega-Cisneros K, Nakamura JN, Melli V, Mbabazi S, Martins MSL, Ledesma ABB, Obregón C, Labatt CK, Kadykalo AN, Heldsinger M, Green ME, Fuller JL, Franco-Meléndez M, Burnett MJ, Bolin JA, Andrade-Vera S, Cooke SJ (2022) Goals, challenges, and next steps in transdisciplinary fisheries research: perspectives and experiences from early-career researchers. Rev Fish Biol Fish 33(2):349–374

Oh C-O, Ditton RB, Gentner B, Riechers R (2005) A stated preference choice approach to understanding angler preferences for management options. Hum Dimens Wildl 10(3):173–186

Parkinson EA, Post JR, Cox SP (2004) Linking the dynamics of harvest effort to recruitment dynamics in a multistock, spatially structured fishery. Can J Fish Aquat Sci 61(9):1658–1670

Pearl J (2009) Causal inference in statistics: an overview. Stat Surv 3:96–146

Pearl J, Glymour M, Jewell NP (2016) Causal inference in statistics, a primer. Wiley

Pita P, Ainsworth GB, Alba B, Anderson AB, Antelo M, Alós J, Artetxe I, Baudrier J, Castro JJ, Chicharro B, Erzini K, Ferter K, Freitas M, García-de-la-Fuente L, García-Charton JA, Giménez-Casalduero M, Grau AM, Diogo H, Gordoa A, Henriques F, Hyder K, Jiménez-Alvarado D, Karachle PK, Lloret J, Laporta M, Lejk AM, Dedeu AL, Martín-Sosa P, Martínez L, Mira AM, Morales-Nin B, Mugerza E, Olesen HJ, Papadopoulos A, Pontes J, Pascual-Fernández JJ, Purroy A, Ramires M, Rangel M, Reis-Filho JA, Sánchez-Lizaso JL, Sandoval V, Sbragaglia V, Silva L, Skov C, Sola I, Strehlow HV, Torres MA, Ustups D, van der Hammen T, Veiga P, Venerus LA, Verleye T, Villasante S, Weltersbach MS, Zarauz L (2021) First assessment of the impacts of the COVID-19 pandemic on global marine recreational fisheries. Front Mar Sci 8:735741

Post JR, Persson L, Parkinson EA, van Kooten T (2008) Angler numerical response across landscapes and the collapse of freshwater fisheries. Ecol Appl 18(4):1038–1049

Poteete AR, Janssen MA, Ostrom E (2010) Working together: collective action, the commons, and multiple methods in practice, STU, Student edn. Princeton University Press, Princeton/Oxford

Pouso S, Uyarra MC, Borja Á (2018) Recreational fishers' perceptions and behaviour towards cultural ecosystem services in response to the Nerbioi estuary ecosystem restoration. Estuar Coast Shelf Sci 208:96–106

Prayaga P, Rolfe J, Stoeckl N (2010) The value of recreational fishing in the great barrier reef, Australia: a pooled revealed preference and contingent behaviour model. Mar Policy 34(2): 244–251

Radinger J, Matern S, Klefoth T, Wolter C, Feldhege F, Monk CT, Arlinghaus R (2023) Ecosystem-based management outperforms species-focused stocking for enhancing fish populations. Science 379(6635):946–951

Rambonilaza T, Kerouaz F (2023) Valuing harvest regulation changes in recreational fisheries with a discrete choice experiment study: what can we learn from a synthetic review? Econ Anal Policy 79:40–54

Ryan KL, Desfosses CJ, Denham AM, Taylor SM, Jackson G (2021) Initial insights on the impact of COVID-19 on boat-based recreational fishing in Western Australia. Mar Policy 132:104646

Sawynok MA, Sawynok W (2021) Impact of Covid-19 on recreational fishing. Australian recreational fishing foundation., FRDC report 2019-213,. Deakin, Australian Capital Territory

Schuch E, Richter A (2022) Tracing intuitive judgement of experts in fish stock assessment data. Fish Fish 23(3):758–767

Silwal P, Poudyal NC, Engman A, Chen X, Cavasos K (2023) Catch orientation and public lake stocking preferences: a typology of freshwater anglers. Fish Res 257:106500

Skov C, Hyder K, Gundelund C, Ahvonen A, Baudrier J, Borch T, deCarvalho S, Erzini K, Ferter K, Grati F, van derHammen T, Hinriksson J, Houtman R, Kagervall A, Kapiris K, Karlsson M, Lejk AM, Lyle JM, Martinez-Escauriaza R, Moilanen P, Mugerza E, Olesen HJ, Papadopoulos A, Pita P, Pontes J, Radford Z, Radtke K, Rangel M, Sagué O, Sande HA, Strehlow HV, Tutiņš R, Veiga P, Verleye T, Vølstad JH, Watson JW, Weltersbach MS, Ustups D, Venturelli PA (2021) Expert opinion on using angler smartphone apps to inform marine fisheries management: status, prospects, and needs. ICES J Mar Sci 78(3):967–978

Smith VL (1982) Microeconomic systems as an experimental science. Am Econ Rev 72(5): 923–955

Smith LT (1999) Decolonizing methodologies: research and indigenous peoples. Zed Books, London

Steffe AS, Taylor SM, Blight SJ, Ryan KL, Desfosses CL, Tate AS, Smallwood CB, Lai EK, Trinnie FI, Wise BS (2017) Framework for integration of data from remotely operated cameras into recreational fishery assessments in Western Australia. Fisheries Research Report No. 286,. Department of Primary Industries and Regional Development, Western Australia

Stevens JP (2012) Applied multivariate statistics for the social sciences. Taylor & Francis

Stoop J, Noussair CN, van Soest D (2012) From the lab to the field: cooperation among fishermen. J Polit Econ 120(6):1027–1056

Tarrant MA, Manfredo MJ, Bayley PB, Hess R (1993) Effects of recall bias and nonresponse bias on self-report estimates of angling participation. N Am J Fish Manag 13(2):217–222

Thomas AS, Gavin MC, Milfont TL (2015) Estimating non-compliance among recreational fishers: insights into factors affecting the usefulness of the randomized response and item count techniques. Biol Conserv 189:24–32

Thomas AS, Milfont TL, Gavin MC (2016) A new approach to identifying the drivers of regulation compliance using multivariate behavioural models. PLoS One 11(10):e0163868

Trisos CH, Auerbach J, Katti M (2021) Decoloniality and anti-oppressive practices for a more ethical ecology. Nat Ecol Evol 5:1205–1212

Trudeau A, Beardmore B, Gerrish GA, Sass GG, Jensen OP (2022) Social fish-tancing in Wisconsin: the effects of the COVID -19 pandemic on statewide license sales and fishing effort in northern Inland Lakes. N Am J Fish Manag 42(6):1530–1540

van Poorten BT, Post JR (2005) Seasonal fishery dynamics of a previously unexploited rainbow trout population with contrasts to established fisheries. N Am J Fish Manag 25(1):329–345

van Soest DP (2013) Experiments on common property management. In: Shogren JF (ed) Encyclopedia of energy, natural resource and environmental economics. Elsevier, Amsterdam, pp 293–298

Venturelli PA, Hyder K, Skov C (2017) Angler apps as a source of recreational fisheries data: opportunities, challenges and proposed standards. Fish Fish 18(3):578–595

Viceisza ACG (2012) Treating the field as a lab: a basic guide to conducting economics experiments for policymaking. International Food Policy Research Institute, Washington, DC

Viceisza ACG (2016) Creating a lab in the field: economics experiments for policymaking. J Econ Surv 30(5):835–854

Walker JR, Foote L, Sullivan MG (2007) Effectiveness of enforcement to deter illegal angling harvest of northern pike in Alberta. N Am J Fish Manag 27(4):1369–1377

Walters C (1986) Adaptive management of renewable resources. Macmillan Publishing Company, New York

Ward HGM, Quinn MS, Post JR (2013) Angler characteristics and management implications in a large, multistock, spatially structured recreational fishery. N Am J Fish Manag 33(3):576–584

Wattage P, Mardle S, Pascoe S (2005) Evaluation of the importance of fisheries management objectives using choice-experiments. Ecol Econ 55(1):85–95

Webster M, Sell J (2014) Laboratory experiments in the social sciences. Academic Press, Amsterdam

Whitehead JC, Dumas CF, Landry CE, Herstine J (2011) Valuing bag limits in the North Carolina charter boat fishery with combined revealed and stated preference data. Mar Resour Econ 26(3): 233–241

Wilmer H, Meadow AM, Brymer AB, Carroll SR, Ferguson DB, Garba I, Greene C, Owen G, Peck DE (2021) Expanded ethical principles for research partnership and transdisciplinary natural resource management science. Environ Manag 68(4):453–467

Wilson KL, Foos A, Barker OE, Farineau A, De Gisi J, Post JR (2020) Social-ecological feedbacks drive spatial exploitation in a northern freshwater fishery: a halo of depletion. J Appl Ecol 57(2): 206–218

Young TB (2019) McNeil River state game sanctuary permit lottery applicant preferences and marginal willingness to pay for permit application: a best-worst discrete choice experiment. MSc. University of Alaska Fairbanks

Zhang W, Sohngen B (2017) Heterogeneous preferences for water quality attributes: a discrete choice experiment of Lake Erie recreational anglers. Center for Agricultural and Rural Development, Iowa State University, Ames

Digital Data Mining

Paul Venturelli, Christian Skov, Asta Audzijonyte, Jonathan Belmaker, Ana Sofia Cardoso, Layla Gordon, Casper Gundelund, Kieran Hyder, Ivan Jarić, Robert J. Lennox, Warren M. Potts, Kirsten Vacura, Ana Sofia Vaz, and Valerio Sbragaglia

Abstract Recreational fishers who interact with internet-connected and smart devices generate large volumes of digital data that can be mined to gain valuable and often unique insight into recreational fisher behaviours. Common sources of digital data include general online activity (e.g., searches, page views), social media (e.g., fishing forums, Facebook), smartphone applications, and connected devices (e.g., smartphones, fish finders). Research is stimulating a great deal of interest in

P. Venturelli (✉)
Department of Biology, Ball State University, Muncie, IN, USA
e-mail: paventurelli@bsu.edu

C. Skov · C. Gundelund
Section of Freshwater Fisheries and Ecology, Technical University of Denmark, Silkeborg, Denmark
e-mail: ck@aqua.dtu.dk; cgun@aqua.dtu.dk

A. Audzijonyte
Institute for Marine and Antarctic Studies, University of Tasmania & Centre for Marine Socioecology, Hobart, TAS, Australia
e-mail: asta.audzijonyte@utas.edu.au

J. Belmaker
School of Zoology, Faculty of Life Sciences and the Steinhardt Museum of Natural History, Tel Aviv University, Tel Aviv, Israel
e-mail: jbelmaker@tauex.tau.ac.il

A. S. Cardoso
CIBIO, Centro de Investigação em Biodiversidade e Recursos Genéticos, InBIO Laboratório Associado, Campus de Vairão, Universidade do Porto, Vairão, Portugal

Departamento de Biologia, Faculdade de Ciências, Universidade do Porto, Porto, Portugal

BIOPOLIS Program in Genomics, Biodiversity and Land Planning, CIBIO, Campus de Vairão, Vairão, Portugal
e-mail: sofia.cardoso@cibio.up.pt

L. Gordon
Albert Katz International School for Desert Studies, The Jacob Blaustein Institutes for Desert Research, Ben-Gurion University of the Negev, Midreshet Ben-Gurion, Israel
e-mail: gordonl@post.bgu.ac.il

K. L. Pope et al. (eds.), *Understanding Recreational Fishers*, Fish & Fisheries Series 45, https://doi.org/10.1007/978-3-031-99739-6_15

these data, but fisheries monitoring and management applications are rare. Our aim is to facilitate research and the appropriate adoption and integration of digital behaviour data within fisheries research, management, and governance. We begin with an inventory of available data sources, the types of fisher behaviours described, and methods for obtaining suitable and relevant datasets. We then identify barriers to the use of these data within a solutions-oriented framework that emphasises privacy and transparency, standardisation, non-probabilistic techniques, validation, and integration. Although digital data for describing fisher behaviours tend to accumulate passively, we describe digital platforms as a rare opportunity to engage large numbers of fishers in active data collection, and influence behaviours through education and outreach.

Keywords Big data · Connected devices · Fisher engagement · Online content · Smartphone apps · Social media

---

K. Hyder
Center for Environment, Fisheries and Aquaculture Science, Suffolk, UK

School of Environmental Sciences, University of East Anglia, Norfolk, UK
e-mail: kieran.hyder@cefas.gov.uk

I. Jarić
Université Paris-Saclay, CNRS, AgroParisTech, Ecologie Systématique Evolution, Orsay, France

Biology Centre of the Czech Academy of Sciences, Institute of Hydrobiology, České Budějovice, Czech Republic

University of South Bohemia, Faculty of Science, Department of Ecosystem Biology, České Budějovice, Czech Republic
e-mail: ivan.jaric@universite-paris-saclay.fr

R. J. Lennox
Norwegian Institute for Nature Research, Trondheim, Norway
e-mail: robert.lennox@nina.no

W. M. Potts
Department of Ichthyology and Fisheries Science, Rhodes University, Eastern Cape, South Africa
e-mail: w.potts@ru.ac.za

K. Vacura
Muckleshoot Indian Tribe, WA, USA
e-mail: vacur006@umn.edu

A. S. Vaz
CIBIO, Centro de Investigação em Biodiversidade e Recursos Genéticos, InBIO Laboratório Associado, Campus de Vairão, Universidade do Porto, Vairão, Portugal

NBI—Natural Business Intelligence, Vila Real, Portugal

V. Sbragaglia
Department of Marine Renewable Resources, Institute of Marine Sciences, Barcelona, Spain
e-mail: sbragaglia@icm.csic.es

1 Introduction

Digital data and computing power are fuelling a big data revolution in recreational fisheries science and most other sectors of society (Venturelli et al. 2017; Bradley et al. 2019; Cooke et al. 2022; Lennox et al. 2022). Digital data represent information that is stored electronically on a computer or server. Most fisheries scientists and managers are familiar with digital data in the form of digitised historical records, contemporary data that were entered manually into a database (e.g., from a netting survey), or auto-generated data (e.g., from a temperature logger). However, recreational fishers also generate large volumes of digital data using the internet, social media, and mobile and connected devices (Venturelli et al. 2017; Bradley et al. 2019; Vitale et al. 2021; Cooke et al. 2022; Lennox et al. 2022). These data are generated when fishers enter information, upload media, or use websites, smartphone applications (apps), or devices that track activity. Fisher-generated digital data are attractive because they are perceived to be a large and inexpensive source of valuable information that can always, and over large areas, provide unique insights.

Non-fisheries examples of digital data mining are common and diverse. Digital data mining is the process by which digital data are gathered, managed, and analysed to learn about the world or make predictions. For example, researchers have used internet page visits to explore the cultural importance of reptiles (Roll et al. 2016), estimate product preferences (Cirqueira et al. 2019), assess public mental health (Ford 2020), and gauge interest in public areas (Guedes-Santos et al. 2021). Internet search volume has also been used in a diversity of applications (Jun et al. 2018), including health (Ho et al. 2019), economics (Chancellor and Counts 2018), capital markets (Ye and Li 2017), conservation (Van Huynh 2023), and election polling (Prado-Román et al. 2021). Applications involving social media data and online forums are equally diverse (Verma et al. 2016). Biological or environmental examples include the use of Instagram and Flickr data to understand human preferences for natural areas (Keeler et al. 2015; Hausmann et al. 2018), YouTube data to estimate compliance with nature-based regulations (Otsuka and Yamakoshi 2020), X (formerly Twitter) data to study emotions related to alternative energy sources (Jain and Jain 2019), and Facebook data to track invasive species (Schifani and Paolinelli 2018). These and other topics are also being explored via app data. Notable examples include the use of data from eBird, a popular birding app, in hundreds of peer-reviewed publications (eBird 2022), and the use of data from Strava, a popular cycling app, to identify travel patterns and collision hotspots, estimate travel demand and air pollution exposure, and explain route choice (Lee and Sener 2021). Finally, researchers are also mining data from mobile and smart devices themselves (as opposed to the apps that they support). Examples include location and accelerometer data to estimate disease spread (Silva et al. 2021), quantify traffic to natural spaces (Monz et al. 2019), and detect events such as falls (Stampfler et al. 2022). Much of this mining is in support of the development of a variety of decision-support tools for individuals and organisations (Sunhare et al. 2022).

Fisher-generated digital data are increasingly being mined to quantify and predict a broad suite of recreational fishing metrics related to behaviour (Table 1). A pioneering study by Baker Jr and Oeschger (2009) recruited marine recreational anglers to report effort and catch data via a text messaging system. The field has since grown exponentially and expanded to include both active and passive data from apps, social media, online activity, and connected devices (Fig. 1a). Most studies have focused on mining digital sources of effort and harvest data (Fig. 1b) to determine the feasibility of complementing or even replacing traditional approaches such as surveys, diaries, and in situ observations (Pollock et al. 1994). However, novel applications of fisher-generated digital data include illegal fishing (Shiffman et al. 2017), invasive species distribution and spread (Banha et al. 2015; Weir et al. 2022; Fischer et al. 2023), fisher satisfaction (Gundelund et al. 2022; Johansen et al. 2022), and behavioural responses to the COVID-19 pandemic (Gundelund and Skov 2021; Britton et al. 2023; Audzijonyte et al. 2023). Although research is stimulating a great deal of interest and enthusiasm in fisher-generated digital data (e.g., Skov et al. 2021), applications within fisheries remain rare.

In this chapter, we describe the state of the art of digital data mining as it applies to fisher behaviour and offer practical guidance toward facilitating research and the appropriate adoption and integration of digital behaviour data within fisheries research, management, and governance. Our definition of fisher behaviour is intentionally broad to include individual actions, precursors to behaviour such as individual attitudes and perceptions (Hunt et al. 2013), and the collective behaviours of groups. Likewise, the data mining techniques and applications herein are likely to be applicable beyond behaviour and fisheries. We begin by describing common sources of fisher-generated digital data, and then outline common methods to obtain, manage, and analyse them. We also scrutinise past work and identify barriers to the use of fisher-generated digital data within a solutions-oriented framework that emphasises privacy and transparency, standardisation, non-probabilistic techniques, validation, and integration (Fig. 1).

2 Sources of Fisher-Generated Digital Data

Approaches to gathering fisher-generated digital data lie on a continuum between passive and active (Fig. 2). Passive approaches imply that recreational fishers are unaware (or do not care) that their data can provide insight into individual or group behaviour. Examples range from aggregated data such as general online activity (e.g., search trends or page views; Wilde and Pope 2013) to user-specific data from social media platforms and connected devices (Sbragaglia et al. 2020a; Lennox et al. 2022; Allison et al. 2023; Audzijonyte et al. 2023). In contrast, active approaches imply that fishers are aware of and choose to contribute their data to researchers or managers. Active approaches include the use of fishing apps or websites within a citizen science framework (Venturelli et al. 2017) and surveys and questionnaires that can be shared *en masse* through general and specialised social media (e.g., Shiffman et al. 2017; Sbragaglia et al. 2020b; Foster et al. 2023).

Table 1 Examples of peer-reviewed publications by source that have used digital data mining to quantify and predict various recreational fisher dimensions and behaviours (cells are shaded according to the number of studies therein)

Dimension or behaviour	**Online activity**	**Social media**	**Fishing app**	**Connected device**
Attitudes, perceptions, or sentiment		Shiffman et al. (2017), Maya-Jariego et al. (2021)*, Spragaglia et al. (2022), Allison et al. (2023), Vacura et al. (2023)	Gundelund et al. (2022), Hook et al. (2022), Johansen et al. (2022)	
Catch composition (size or species)	Winkler et al. (2022)	Shiffman et al. (2017), Giovos et al. (2018), Spragaglia et al. (2020, 2021, 2022), Maya-Jariego et al. (2021)*, Eryaşar and Saygu (2022), Hall et al. (2022), Martinazzo et al. (2022), Grabowski et al. (2023)	Gundelund et al. (2021), Johnston et al. (2021), Skov et al. (2022)	
Catch rate or total harvest	Carter et al. (2015)	Hall et al. (2022)	Liu et al. (2017), Gundelund et al. (2020, 2021), Midway et al. (2020), Gundelund and Skov (2021, 2023), Johnston et al. (2021), Hook et al. (2022)	

(continued)

Table 1 (continued)

Fisher demographics		Shiffman et al. (2017), Hoad and Wilson (2020), Maya-Jariego et al. (2021)*, Vitale et al. (2021)	Gundelund et al. (2020, 2021), Gundelund and Skov (2021), Britton et al. (2023)	
Fisher movement patterns			Papenfuss et al. (2015), Hook et al. (2022), Weir et al. (2022), Fischer et al. (2023)	Fricke et al. (2020)
Fishing effort in space or time	Belhabib et al. (2016)	Martin et al. (2014), Giovos et al. (2018), Monkman et al. (2018a,b), Spragaglia et al. (2020, 2022), Maya-Jariego et al. (2021)*, Eryaşar and Saygu (2022), Hall et al. (2022), Martinazzo et al. (2022), Grabowski et al. (2023)	Papenfuss et al. (2015), Jiorle et al. (2016), Gundelund and Skov (2021), Gundelund et al. (2021, 2023), Johnston et al. (2021), Hook et al. (2022), Britton et al. (2023), DePiper et al. (2023)	Fricke et al. (2020), Dainys et al. (2022), Audzijonyte et al. (2023)
Fishing trip duration		Monkman et al. (2018a)	Midway et al. (2020), Gundelund et al. (2021)	
Fishing trip number or frequency (activity)		Martin et al. (2014), Monkman et al. (2018b), Hall et al. (2022), Martinazzo et al. (2022)	Midway et al. (2020), Gundelund and Skov (2021), Gundelund et al. (2021, 2023), Hook et al. (2022)	Fricke et al. (2020)
Illegal fishing		Shiffman et al. (2017),		

		Maya-Jariego et al. (2021)*		
Interest in fishing	Martin et al. (2012), Wilde and Pope (2013)**, Winkler et al. (2022), Britton et al. (2023)	Winkler et al. (2022)		
Preferred species or fishing method		Shiffman et al. (2017), Giovos et al. (2018), Monkman et al. (2018b), Maya-Jariego et al. (2021), Eryaşar and Saygu (2022), Grabowski et al. (2023)	Gundelund et al. (2020), Skov et al. (2022) , DePiper et al. (2023)	
Release rate		Shiffman et al. (2017)	Gundelund and Skov (2021), Gundelund et al. (2021, 2023), Johansen et al. (2022), Skov et al. (2022)	

*Concern has been raised that this study violates several privacy rules (see Sbragaglia 2021b).

**But see Ficetola (2013) and McCallum and Bury (2014).

□ = 0, □ = 1 or 2, □ = 3-5, □ = 6-8, ■ = 9+

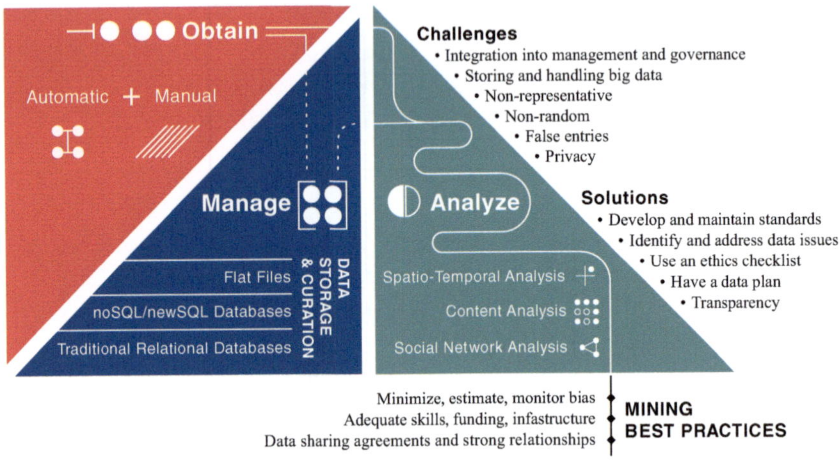

Fig. 1 A graphic summary of key points in the how to mine digital data, challenges, and solutions and best practices sections

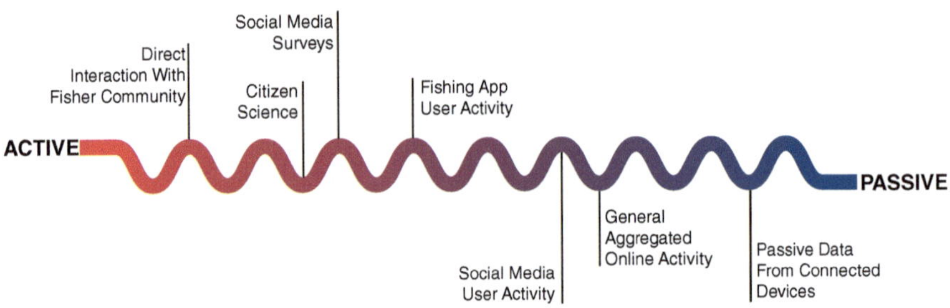

Fig. 2 The relative position of fisher-generated sources of digital data along a continuum of fisher awareness and engagement (data generation is most active when fishers are aware of and engaged in the data collection process, and entirely passive when they are not; positions are approximate and can overlap)

2.1 *Passive Approaches*

2.1.1 General Aggregated Online Activity

Aggregated data on public online activity are a promising source of information for recreational fisheries research, management, and governance (Jarić et al. 2020a; Lennox et al. 2022). Commonly used types of aggregated data for this purpose include web search volumes obtained from search engines and associated platforms such as Google Trends, and data on visitation rates for web pages (Correia et al. 2021). Visitation rates, or pageviews, can be obtained directly from open-access platforms such as Wikipedia, or through web analytics services such as Google

Analytics. Aggregated data can provide information on recreational fishing participation, as well as regional and global variation in public fishing over seasons or years (Martin et al. 2012; Wilde and Pope 2013; Carter et al. 2015). For example, Winkler et al. (2022) used online search volume to document the emergence of coastal drone fishing, which is the use of drones to scout, chum, or drop or retrieve a line. Although aggregate data are unlikely to describe specific or high-resolution aspects of recreational fisher behaviour, general reactions and sentiments can be used as indicators of, for example, the acceptance of management, conservation, and policy measures and interventions (Correia et al. 2021).

2.1.2 Social Media and Online Forums

Social media is an integral part of modern recreational fishing culture (Sbragaglia et al. 2020a). Although the proportion of recreational fishers who share their catches on social media may be low (e.g., 12–21% in Vitale et al. 2021), most fishers probably consume or interact with digital content. Popular platforms include Facebook, YouTube, and Instagram (Vitale et al. 2021), not to mention the scores of fisheries forums that are dedicated to specific fishing locations, species, topics, or issues. People not only use these platforms to connect with their peers and feel gratified (Ruggiero 2000) but also search for information on specific topics. Common topics among recreational fishers include techniques and opinions related to strategies, gear, technical support, fish behaviour, and other subjects such as environmental issues (Sbragaglia et al. 2020a). Therefore, social media can support the study of several attributes of recreational fisher behaviour, including demographics (Shiffman et al. 2017; Hoad and Wilson 2020; Vitale et al. 2021) and values and attitudes (Allison et al. 2023). For example, researchers used data from YouTube to characterise the behaviour of specialised grouper fishers that would have been difficult to obtain by traditional means (Sbragaglia et al. 2021a), or to quantify sentiment and behaviour related to invasive species (Sbragaglia et al. 2022). Online forum examples include an assessment of practices, perspectives, attitudes, and trust among shark fishers (Shiffman et al. 2017) and sentiment analysis to identify drivers of fisher happiness in a large region of the United States (Vacura et al. 2023).

2.1.3 Fishing Apps

Many recreational fishers use fishing apps to record, and share catch images and details (e.g., date and time, location, species, size), and as a source of information such as local regulations and where, when, and how to fish. Most of these specialised social media platforms also allow fishers to record non-traditional catch data such as the method of fishing, bait or tackle, water depth at which the fish was caught, and weather conditions. The first peer-reviewed analysis of passively collected app data demonstrated their potential to describe both fisher movements and regional or seasonal fishing effort that were in general agreement with conventional approaches (Papenfuss et al. 2015). Passively collected fishing app data have since been used to compare with data from mail, creel, and gillnet surveys (Jiorle et al. 2016; Johnston et al. 2022), estimate harvest (Liu et al. 2017), explain and predict the spread of invasive species (Weir et al. 2022), describe continent-scale changes in fishing effort in relation to the COVID-19 pandemic (Britton et al. 2023), and identify the potential impacts of offshore wind development on recreational fishing (DePiper et al. 2023). The emerging consensus is that passively collected app data (i) tend to produce reasonable results at low spatial and temporal resolutions, (ii) can produce reasonable results at higher resolutions if sample sizes are large, but (iii) may not track fisheries-independent estimates such as abundance and relative community composition. An interesting yet unexplored avenue of research is the nature of the social networks that can form within the fishing apps that have social media features, how information spreads through these networks, and how this information affects fisher behaviour.

2.1.4 Connected Devices

Passive data are also generated when fishers use devices that are connected to the internet and each other via wireless networks and protocols. These include smartphones, boat-based electronics such as navigational devices and fish finders, (underwater) cameras, and smart rods. Connected devices are underutilised in the context of understanding fisher behaviour (Fig. 1a) except for portable sonar devices that fishers use to survey aquatic habitat. Researchers have used location data from these sonar devices to estimate the amount of fishing activity (effort) in a single lake in comparison to visual surveys (Dainys et al. 2022), fishing effort changes in response to the COVID-19 pandemic in four European countries (Audzijonyte et al. 2023), and effort, fishing frequency, fishing site selection, and the movement behaviour of fishers over much of the continental United States (Fricke et al. 2020).

2.2 *Active Approaches*

Smartphone apps and social media provide enormous opportunities for fisheries scientists and managers to actively engage with recreational fishers through citizen science projects that use websites and apps, as well as to disseminate surveys to fishers who use apps and social media. An added advantage of active approaches is the potential for a two-way flow of information between fishers and decision makers that influences fisher behaviour directly and encourages their active participation in management and governance.

2.2.1 Fishing Apps

Citizen science projects that recruit and encourage fishers to report fishing activity and other behaviours via a fishing app are becoming increasingly popular among research institutions, government agencies, non-governmental organisations, and, to a lesser extent, private companies. The Fangstjournalen app is the most published example to date. It was launched in 2016 by the Technical University of Denmark to collect angler behaviour and catch data from recreational angling fisheries in Denmark. The app had more than 16,000 registered users (representing ~5% of all anglers in Denmark) and ~ 80,000 reported fishing trips by fall 2023. Data from the app have supported many publications (Gundelund et al. 2020, 2021, 2022, 2023; Gundelund and Skov 2021; Skov et al. 2022; Britton et al. 2023; Audzijonyte et al. 2023) on a variety of topics related to the behaviours of fishers (Table 1); including the only known app-based studies of fisher demographics (Gundelund et al. 2022). Other published examples include trip and harvest details via an app program that was piloted by the Louisiana Department of Wildlife and Fisheries (Midway et al. 2020); catch, effort, and trip details for sea fishing in England and Wales via the Substance and the Centre for Environment, Fisheries and Aquaculture Science's Sea Angler Diary program (Hook et al. 2022); and fisher satisfaction and catch-and-release behaviour along the eastern coast of Norway via the Institute of Marine Research's ToutXP app (Johansen et al. 2022). Researchers have used data from iNaturalist, a general biodiversity app that fishers often post their catches to, to estimate the prevalence and distribution of a fish pathogen (Happel 2019), but not fisher behaviour (Happel et al. 2020). The line between passive and active fishing apps will be blurred if active app use becomes so normalised that a large proportion of users are unaware that they are participating in a citizen science project, or a large proportion of passive app users become aware that their data are being used for research or management (e.g., as a result of press releases, publications, or invitations to provide data).

2.2.2 Engagement with Social Media Users

Directly interacting with fishers through social media platforms is an undeveloped method of generating digital fisher data. We are aware of two studies that disseminated surveys through social media: Foster et al. (2023) posted a survey on popular spearfishing Facebook and WhatsApp groups to collect information on spearfisher behaviour and ecological information, and (Hamelin et al. 2022) posted a survey to Facebook and Instagram as part of a wider dissemination effort to determine the cultural value of a Canadian recreational Atlantic mackerel (*Scomber scombrus*) fishery. It would be interesting to assess whether cultural values could be inferred directly from fisher-generated digital data.

Direct interaction and information sharing stand out among digital sources of fisher data because they have the potential to improve engagement in the management process, stakeholder trust, and fisher environmental behaviours (Allison et al.

2023). Several projects have strived to use digital platforms to engage recreational fishers around science-based advice. For example, the Keepemwet Fishing movement uses a variety of social media platforms to provide science-based best practice guidelines for catch-and-release fishing (Danylchuk et al. 2018). Although the digital dissemination of information may improve knowledge and empower recreational fishers, it has rarely been successful in improving environmental behaviour (Steg and Vlek 2009). In a recreational fisheries context, Mannheim et al. (2018) documented that the implementation of a pro-environmental strategy had much promise for improving recreational fishers' environmental behaviour, whereas Allison et al. (2023) made several suggestions on how pro-environmental strategies may be implemented using digital interactions with recreational fishers. They linked various pathways to environmental action (which stress personal, social, and contextual factors) to the main functions (informational, relational, and experiential) of social media and suggested that engaging recreational fishers through social media platforms had enormous potential for improving their environmental behaviour. Although they focused primarily on the potential of existing Facebook groups for implementing pro-environmental strategies, they made several recommendations: (i) carefully selecting a target social media group or platform; (ii) building relationships and formulating agreements that outline the roles and responsibilities with group administrators; (iii) identifying and partnering with appropriate influencers and role models; (iv) conducting baseline assessments of environmental attitudes and behaviour, and identifying appropriate techniques for intervention; and (v) continually evaluating the efficacy of the intervention (Allison et al. 2023). Whereas these strategies are broad and may need to be modified depending on the social platform selected, they provide some practical guidelines for developing holistic digital engagement projects that aim to improve the environmental behaviour of recreational fishers.

3 How to Mine Digital Data

This section reviews common approaches to obtaining, managing, and analysing fisher-generated digital data. These steps are integral to a wider data management life cycle (Krogman et al. 2024); therefore, they should be considered alongside planning, data cleansing, archiving, and sharing (Sievert et al. 2026). The information here should be relevant for a diversity of digital data sources and fisher behaviours. There is a large and evolving list of free software available to obtain, manage, and analyse digital data. The most common packages and libraries are supported by the R (R Core Team 2021) and Python (Van Rossum and Drake 19951995) programming languages. However, given that specific packages and libraries will become outdated, we encourage readers to identify appropriate software when designing mining strategies. Understanding and using this software requires good computer knowledge or a willingness to learn; therefore, it is important to either develop this knowledge (e.g., through training and resources), or identify skilled collaborators.

Finally, we encourage readers to obtain relevant training and approval early in the mining process to avoid or be prepared for any ethical or legal concerns that arise when working with human subjects' data (see Sect. 4.4).

3.1 Obtain Data

The first step in mining existing digital data is to obtain them (see Data Acquisition in Krogman et al. 2024). This step is trivial for projects that collect and control their own data (e.g., app-based citizen science projects, surveys administered through social media). Obtaining publicly available data often requires computer programming skills, whereas non-public data (e.g., from a commercial entity, government agency, or different research lab) often requires formal collaborations or data sharing, usage, or purchasing agreements. In all cases, care must be taken to protect sensitive personal information and comply with any terms of use to ensure ethical and responsible data acquisition practices (Jarić et al. 2020b, Zook et al. 2017, Section 5.4).

Large volumes of fisher-generated digital data are available publicly. Examples include Google search trends, YouTube posts, and iNaturalist logs. Approaches for obtaining such data range from manual to automatic (Toivonen et al. 2019). Manual approaches are appropriate when advanced programming and machine learning applications are not available or justified (e.g., small sample sizes or limited resources). However, manually searching through large volumes of data can be time-consuming at best or impossible at worst, monotonous, and prone to error. Alternatively, publicly open data can be obtained automatically through Application Programming Interfaces (APIs) and web scraping. APIs are software interfaces that combine several protocols and tools for managing and accessing digital data from online (usually social media) platforms (e.g., YouTube API and Google mobility and search APIs). These interfaces generally provide free access to researchers upon an authorisation request, although some may restrict the type and volume of data that can be freely accessed (Correia et al. 2021; Jarić et al. 2020a). There are two types of APIs: streaming APIs that automatically deliver data in real time (e.g., via X's API), and REST (Representational State Transfer) APIs that only deliver specific data when requested. Accessing digital data through APIs requires programming skills, but many platforms are becoming increasingly user-friendly. For example, some freely available platforms (e.g., InVEST) already integrate specific modules to obtain and analyse digital data from online sources (e.g., Flickr). Web scraping is another approach to obtaining publicly available fisher behaviour data. It uses automated scripts to download digital data from websites (Jarić et al. 2020a). However, whereas APIs and data software packages usually comply with platform privacy and security settings, scraping is often prohibited or even blocked because it violates the Terms of Use and Developer Policy conditions of platforms and data services (Correia et al. 2021; Toivonen et al. 2019). Thus, scraping is usually reserved for public web content such as fisher forums (e.g., Vacura et al. 2023).

Active collaboration and formal agreements are often required to obtain digital fisher data that belong to companies, agencies, or institutions. Data sharing and usage agreements allowed researchers to obtain data from the privately owned Fishbrain (Weir et al. 2022), iFish Alberta (Papenfuss et al. 2015), and MyCatch (Johnston et al. 2022) apps, as well as the Deeper sonar device (Dainys et al. 2022). These companies shared their data freely, but this is not always the case. For example, time-based location data from smartphones were purchased from the AirSage company to evaluate the relationship between water quality and traffic to coastal recreation areas (Furey et al. 2022), and from the StreetLight company to estimate trail use in a nature reserve (Creany et al. 2021). Data sharing and usage agreements can be a viable tactic for obtaining large volumes of data without the need for high-performance equipment or technical expertise. Most potential collaborators are open to data sharing as a source of revenue, means of demonstrating the value of their service or data, and opportunities to contribute to sustainable fishing. Agreements should be developed or reviewed with the help of a legal expert. Most agreements define parties and roles, describe data sharing and use, set expectations for confidentiality and security, and outline a process for resolving disputes.

3.2 Manage and Store Data

This subsection covers key themes within an overall data management cycle that is described in depth by Krogman et al. (2024). Primarily, researchers must responsibly, safely, and ethically manage, store, and disseminate fisher-generated digital data (Zook et al. 2017; Di Minin et al. 2021; Sbragaglia et al. 2021b). Some online platforms have specific rules regarding the storage and use of their data, which are displayed in the Terms and Conditions of each platform or service. These rules may hamper the adoption of data scraping procedures or limit access to specific data, and often imply the need for anonymisation or pseudo-anonymisation whenever sensitive or private information is collected (e.g., username or age). Disclosing this information can violate the rights of the users who created the data, especially children and vulnerable adults, which makes the responsible management and storage of digital data crucial. Besides the specific Terms and Conditions of each platform, managing digital information to protect user data should always respect the legislation and regulations applied to the different users and analysts; for instance, projects involving European data should adhere to the General Data Protection Regulation (GDPR) from EU Regulation 2016/679, which is currently the strictest personal data protection policy in the world. General ethical and responsible practices suggest that researchers should extract and store only the minimum amount of data that is needed, anonymise or pseudo-anonymise sensitive information, and implement data protection and security safeguards to avoid data leakage (Di Minin et al. 2021). Ideally, personal information should not even be extracted or stored, and whenever it is essential for research, the datasets with personal information must be protected via passwords and other security measures.

The most appropriate method for structuring and storing digital data depends on the type (e.g., textual content, videos, or images) and amount of information to be stored, the format (e.g., txt, xlsx, or jpg) and structure (e.g., array, stack, or tree) of the data, and the available memory resources (Toivonen et al. 2019; Brown et al. 2017). There are three categories of databases for managing digital data: flat files, relational databases, and noSQL or newSQL databases.

Flat files are two-dimensional databases that store data in an easy-to-access tabular format (e.g., Excel or Google spreadsheets and comma-separated values (CSV) files). Flat files are relatively easy to develop. The amount of information that can be stored in a flat file is limited by the management software that is being used (e.g., Microsoft Excel supports up to 1,048,576 rows and 16,384 columns in 2023) or system memory. A downside of flat files is that they require all information about a case to be stored on a single row, which means that larger databases contain redundant information. For example, a flat file of catches by a fisher would repeat the same fisher information for each catch and therefore make the database unnecessarily large, unwieldy, and inefficient. Relational databases solve this issue by allowing multiple tables to be linked through one or more relationships (Batrinca and Treleaven 2015). Relational databases are ideal for copious amounts of hierarchical data. In our catch example, catch and fisher information would be in separate tables that are linked through the fisher ID. If the catch table included a waterbody ID, then the ID could link to a table that contained information about each waterbody. Microsoft Access is a common relational database management system. Relational databases are easier to filter and modify than flat files but require more planning and maintenance. Finally, noSQL and newSQL are modern database management systems that allow users to reliably organise and query large volumes of growing and changing data in a manner that is both flexible and scalable (e.g., Apache CouchDB). This type of database is common among organisations that process large numbers of transactions, such as financial institutions, online retailers, healthcare providers, and governments. These noSQL and newSQL solutions are ideal for projects like fishing apps that are likely to evolve over time and have many users who supply data and query the database. Although noSQL and newSQL databases require advanced programming skills, they can be outsourced to collaborators, IT departments, or companies. See Sievert et al. (2026) for more information on different database structures.

One can store any type of database on a personal computer or local server if it is not too large. Larger databases require storage hardware (e.g., external hard drives, servers) or storage services (e.g., Amazon Web Services, Google Cloud Platform). In-house storage is recommended if the hardware is already in place or the storage will be long term; otherwise, cloud storage is both economical and convenient (Silva et al. 2022).

3.3 Analyse Data

In addition to common statistical techniques such as regression, there are three main groups of analysis that are commonly applied to fisher-generated digital data: spatiotemporal analysis, content analysis, and social network analysis. Comparative analyses are described in Sect. 5.3.1 *Identifying Data Quality Issues*. All analyses should be documented to be reproducible in response to new data, turnover in personnel, and external review (Krogman et al. 2024).

3.3.1 Spatiotemporal Analyses

These analyses take advantage of the geolocation (e.g., coordinates of data) and timestamp (e.g., date of data creation) information that is generated by mobile devices (Batrinca and Treleaven 2015); for example, when fishers record their catches or fishing trips via an app, use a fishfinder or other GPS-enabled device and applications, or enable location and activity tracking through passive means (e.g., photo metadata). Geolocation data can also be gathered from the content of social media posts (e.g., title and description of YouTube videos; Sbragaglia et al. 2020a). Spatiotemporal data can be relatively accurate and detailed, including the date, hour, minutes, and seconds for the timestamp, and the spatial coordinates or bounding boxes for the geolocation (Correia et al. 2021). Such data could be used to either assess fisher activity in a specific area or a range of locations (Yan et al. 2018) or follow individual or general movement trajectories to assess fisher behaviour and activity (Toivonen et al. 2019). The first type of analysis could generate a powerful understanding of fishing activity (i.e., effort) across space and time, either as an absolute or relative measure (see the references in the *Effort in space or time* row in Table 1). Alternatively, tracking individual fisher trajectories and movement could provide important information about the pathways along which fishers can spread invasive species or diseases (see references in the *Fisher movement patterns* row in Table 1), help managers to prioritise the locations of fisher access points and facilities (see Lee and Sener 2021 for a bicycling example), or track behavioural responses to regulatory changes or other factors that might affect fisher behaviour.

Spatiotemporal analyses are typically conducted via geospatial software. Esri's ArcGIS is arguably the most popular choice among fisheries professionals, but it is expensive without an institutional licence. Free options include open software such as R (Wikle et al. 2019) and Python (Alam et al. 2022). Integration across software is common (e.g., calling Python from within ArcGIS), and there is software that facilitates interoperability (e.g., Roberts et al. 2010; Alam et al. 2022).

3.3.2 Content Analysis

Content analysis is the study of the words, objects, or sounds that are found in fisher-generated text, photos, and videos. The content must be extracted before conducting qualitative or quantitative analyses (Kochalski et al. 2026). Manual extraction is feasible for small or straightforward datasets (e.g., McClenachan 2009; Shiffman et al. 2017), but it could be replaced by combining API and specific text analysis (Sbragaglia et al. 2020a, 2022), or even machine learning algorithms for datasets that are larger or more complex (Correia et al. 2021; Tuia et al. 2022). Deep learning is a type of machine learning algorithm that uses artificial neural networks to automatically identify high-dimensional nonlinear patterns and features in the data. These algorithms perform tasks with little to no human supervision. Computer vision and natural language processing are two high-performance deep learning methods that are of particular interest when analysing visual and textual content (Schulz and Behnke 2012).

Computer vision automatically extracts, processes, and analyses information and patterns from text, images, and videos. This allows for the detection, localisation, and classification of visual elements of interest. Computer vision is a rapidly developing field. Algorithms have been used widely in ecological and conservation research to identify species, classify animal behaviour, and estimate biodiversity in camera-trap images, audio recordings, and videos (Christin et al. 2019; Correia et al. 2021). Fisheries examples include monitoring commercial vessels and discards (Ferreira et al. 2017; French et al. 2020), and counting fish in aquaculture (Babu et al. 2023). Computer vision could be applied to photos or videos that were uploaded to social media or apps to study specific aspects of fisher behaviour, such as target species and lengths (Silva et al. 2022), handling practices, and even sentiment from facial expressions and audio. Table 2 contains various freely available tools for audio and video analysis.

Natural language processing combines knowledge from linguistics and computer science to process and analyse textual content. Applications include speech and entity recognition, sentiment analysis, and topic modelling (Schulz and Behnke 2012). To our knowledge, only sentiment analysis (i.e., the identification, extraction, and quantification of opinions, attitudes, and emotions) has been used to study recreational fishers. Sbragaglia et al. (2022) analysed comments associated with YouTube videos and determined that Italian spearfishers engaged more with fishing actions and expressed positive emotions, whereas anglers focused more on fishing strategy and gears and expressed more negative emotions. Both groups expressed negative and positive sentiments towards the invasiveness and aggressiveness of specific species. Vacura et al. (2023) analysed text within online fishing forums and determined that angler positivity increased with angler density, bag limit, and season length. Sentiment analysis may be particularly effective for evaluating the ethical dimension of fishing, particularly as it relates to animal rights (e.g., comments stemming from posts that depict regulation violations or poor handling practices). Table 2 contains various freely available tools for textual content analysis.

Table 2 Examples of analytical tools that are currently available for extracting information from digital data

Purpose	Tool or library name	Software platform	Notes
Audio processing, segmentation, classification, and detection	pyAudioAnalysis	Python	
	Wrassp	R	
Image and video processing, segmentation, classification, and detection	Tensorflow	Python	
	PyTorch	Python	
	Keras	Python	
	Rvision	R	
Textual content processing, tokenisation, word frequency, and entity recognition	spaCy	Python	
	NLTK	Python	
	Spacyr	R	
	Tm	R	
Textual content classification, speech recognition, topic modelling, and sentiment analysis	Tensorflow	Python	Incorporates pre-existing sentiment dictionaries and lexicons
	PyTorch	Python	
	pysentiment2	Python	Incorporates pre-existing sentiment dictionaries and lexicons
	koRpus	R	
	Syuzhet	R	
	SentimentAnalysis	R	
Social network analysis	Cytoscape	Stand-alone open source	Extendible platform used in biological and social network analysis
	Gephi	Stand-alone open source	Explores, analyses, and visualises network data
	Igraph	*R*	
	NodeXL	Excel add-in	
	Pajek	Stand-alone freely available program	Commonly used for large, complex networks
	UCINet	Stand-alone program for purchase	A comprehensive package that is popular among academics

These and other deep learning algorithms can be implemented using freely available and user-friendly interfaces. For example, the Google Colaboratory (Colab) currently provides a free processing service for a limited number of hours per day (Carneiro et al. 2018). The Colab provides an online working environment with many Python libraries already installed, which helps inexperienced users to start with simple models and tools. Free resources can be sufficient for small-scale analyses, thereby increasing the accessibility of machine learning methods. Silva et al. (2022) provide a concise overview of some of the tools for large data storage and analysis with machine learning methods.

3.3.3 Social Network Analysis

Social network analysis is used in social science and other disciplines to study the relationships and interactions among individuals, groups, or organisations (Wasserman and Faust 1994; Borgatti et al. 2018). The main goal of social network analysis is to provide quantitative indicators on the patterns of connections and the structure of relationships among social entities (e.g., people, organisations, or even concepts). Social entities are represented as nodes in a network, and the relationships or interactions between them are represented as edges. These networks can be analysed to uncover patterns, identify key players, understand information flow, and explore the overall structure and dynamics of social systems (Mitchell 1969).

Recreational fishers interact with each other, forming complex networks on digital platforms such as apps (e.g., Fishbrain; Weir et al. 2022) and especially social media (e.g., YouTube; Sbragaglia et al. 2020a). Most importantly, recreational fishers can use social relationships within a network to gather information that informs their behaviours. Therefore, social network analysis can elucidate the structure of the network and how information flows through it, and ultimately help to explain recreational fisher behaviours (Wasserman and Faust 1994; Freeman 2004). Although social network analysis has already been suggested as a key analytical process for social media data in the context of conservation and management (Toivonen et al. 2019), we are aware of only one study that has applied it to recreational fishing: Maya-Jariego et al. (2021) analysed the social network of a group of poachers, but Sbragaglia et al. (2021b) raised concerns about privacy violations, so we do not discuss it further.

Two studies involving small-scale commercial fisheries are examples of how social network analysis may be used in the context of recreational fisheries science and management. Turner et al. (2014) analysed networks of information flow among commercial fishers to identify individuals who had the most access to information about the fishing behaviour of others. Similarly, researchers who used social network analysis showed that the structure of information sharing among fishers depended on what fishing information was being shared (e.g., the structure for sharing fishing information differed from the network for sharing fishing regulations; Arlidge et al.

2021). The authors concluded that these networks implied differences in whether individual fishers were expected to receive information related to a planned conservation intervention.

There are several user-friendly software tools available for social network analysis that vary in their required skill level (Table 2). When choosing among social network analysis software, specific needs, the data size and complexity, and familiarity with network analysis concepts should be considered. Many of these tools offer tutorials and documentation to help users get started.

4 Challenges

Digital data have an enormous potential to contribute to characterising and understanding recreational fisher behaviour, but there are several challenges related to data accessibility and consistency, skills capacity and development, data quality issues, privacy and ethics concerns, and barriers to the integration of digital data into management and governance frameworks. We are intentionally critical, here; we encourage readers to also consider the solutions and best practices that are described in Sect. 5.

4.1 Access and Consistency

The potential benefits of fisher-generated digital data can only be realised if researchers and managers can reliably access these data, and the data are collected in a consistent manner. Access and consistency are usually concerns when working with data that are collected and owned by external entities such as collaborators or companies such as Google, YouTube, Fishbrain (www.fishbrain.com), and Deeper (www.deepersonar.com). Access can be obtained via an Application Programming Interface (API) that may require a one-time registration, or through a longer data sharing agreement process. In either case, access can change or be revoked for any number of reasons and with little warning. Thus, projects, research programs, and decision-support tools that use external data rely on the good will and longevity of the companies that control these data. Other potential threats to data access include file or server issues, lost passwords, and insufficient resources to pay for subscriptions or storage services.

Researchers who use external data to conduct analyses over space or time should be aware of potential changes to how the platform is used and by whom, or how the data are post-processed within the platform (Correia et al. 2021). Companies that produce these data regularly alter background algorithms and front-end user interfaces (e.g., by changing default settings, data fields, layouts, and product features). These changes can affect specific data fields, or the popularity of a product within a population or among specific market segments. Popularity can also change organically or because of targeted marketing campaigns. Researchers should also understand that they are unlikely to be consulted about, or informed of, changes that can add noise to the data, prevent simple comparisons, or lead to erroneous conclusions.

4.2 Data Science Competency and Capacity

Most current fisheries professionals lack the necessary skills and expertise for obtaining, storing, processing, analysing, and interpreting fisher-generated digital data. Working with such data tends to require some programming skills (e.g., understanding several coding languages), familiarity with machine learning and a diversity of software, a strong statistical background, experience with handling large datasets, and an awareness of privacy rules and legal requirements. These data science skills are rarely emphasised within traditional, undergraduate-level fisheries programs, and are only occasionally introduced at the graduate level. Professional development opportunities are available if individuals are interested and their organisation is supportive (e.g., conference workshops, free or low-cost online courses and tutorials). Finally, fisheries professionals who are trained to work with digital data cannot do so without access to appropriate software and hardware (see Sect. 3), which can be limited by a lack of existing infrastructure, limited budgets, or policies that limit external options.

4.3 Data Quality

Most concerns about the quality of fisher-generated digital data stem from issues that are associated with sampling bias (i.e., who is reporting the data) and reporting bias (i.e., how the data are reported). The nature and extent of these biases are likely to differ among digital platforms and vary in time and space.

4.3.1 Sampling Bias

Probabilistic sampling simply means that every fisher has an equal or known chance of contributing data. This is usually achieved via random or stratified random sampling and is one of the foundations of estimating statistics that are representative of a population (Vaske et al. 2026). A creel survey is an example of a probabilistic sampling approach when the sample frame is the total fishing activity in a fishery; mail surveys probabilistically sample all fishers, regardless of their fishing frequency (Pollock et al. 1994). On the other hand, applications that rely on fishers to self-report their data are non-probabilistic. Fisher diary programs, and all forms of digital data that are described in this chapter, are examples of non-probabilistic sampling schemes. Researchers and managers have little or no control over the fishers who contributed data via digital platforms, even when using a citizen science approach, so the segment of the fisher population that supplied that data is usually unknown. As a result of these issues, non-probabilistic sampling approaches violate the common statistical assumption of random selection (Brick et al. 2022). If this violation results in a biased sample of fishers (e.g., avid, urban, and young fishers; Jiorle et al. 2016, Gundelund et al. 2020, Vitale et al. 2021), then conclusions about fisher behaviour are likely to be biased (Lewin et al. 2023).

4.3.2 Reporting Bias and Error

Digital data can still be incomplete or incorrect even if sampling bias is not an issue. Fishers who share their data on social media, fishing apps, or connected devices may not log all their fishing trips. If a trip is reported, then it may lack complete information. For example, fishers may not report all catches during a trip, but instead focus on memorable catches of their target species. Similarly, an individual catch report may omit information, such as how the fish was captured and whether it was released. Finally, the information that is reported may be accidentally or intentionally incorrect (reviewed by Venturelli et al. 2017). These reporting issues can add noise to the data, or bias estimates of fisher behaviour (e.g., average number of fishing trips, catch rate per hour of fishing).

Bias and error likely result from both the fisher (e.g., their motivation to provide data) and the device or platform that is used to record the data. Fishers who are not sufficiently incentivised to interact with a device—be it to share through social media, maintain a personal catch log, enhance their fishing experience, or contribute to a citizen science project—may not be sufficiently motivated to minimise reporting bias and error (or report at all). Although motivation is well studied in the context of citizen science projects in general (e.g., Robinson et al. 2021) and angler diary programs in particular (Cooke et al. 2000), motivation to share data digitally remains poorly understood. This is especially the case for platforms that were not intended for fisheries research or management (i.e., passive approaches) because the motivation to share is likely confounded with the motivation to use the platform for its intended purpose (e.g., as a social media experience), which depends on the quality of the interface and a fisher's overall user experience. Finally, and importantly, platforms can introduce bias and error because of design features (Venturelli et al. 2017), bugs, conversion and configuration errors, compatibility issues, and network and database problems.

4.4 Ethics

It is essential to consider data privacy concerns and ethical principles associated with human-subjects research, even if the researcher is using digital data that are publicly available (Zimmer 2010, Monkman et al. 2018c, Di Minin et al. 2021). Both passive and active digital fisher data have the potential to include personal or otherwise sensitive information such as name, face, age, gender, race, marital status, and home location. Although the extent to which personal or potentially sensitive information is collected differs among digital data types and platforms, a lot of information can be inferred from photographs and videos. A researcher who has personal or sensitive information is ethically (and usually legally) required to ensure that it is stored securely and not revealed or misused during dissemination (Sbragaglia et al. 2021b).

Fisher-generated digital data can also contain information that leads to ethical dilemmas. For example, digital data may reveal illegal behaviours (Shiffman et al. 2017; Sbragaglia et al. 2021b). If this information is shared with or subpoenaed by authorities, then it could lead to the criminalisation of users and ultimately demotivate fishers to provide data. The information that fishers share could also be used by other fishers to target protected or threatened species. Researchers who publish this information need to pay specific attention to these ethical issues and meet reasonable expectations of privacy (Di Minin et al. 2021).

4.5 *Integration*

Examples of fisher-generated digital data being used in support of recreational fisheries management and governance are rare because the data are not trusted, and the means to incorporate them into the decision process have not been developed. If digital data quality is an issue, then it stands to reason statisticians, science advisors, and managers will critically view these data (e.g., Brick et al. 2022). Scepticism can also result from a lack of familiarity with or understanding of these data. In addition, methods for combining these results with more formal probabilistic methods for assessing specific fisher behaviours remain underdeveloped. Addressing these issues requires statistical and technical skills that are not currently widespread in fisheries research and management.

5 Solutions and Best Practices

This section offers solutions and best practices to address the challenges that were described in the previous section. Suggestions are based on common sense, our experiences working with these data, and the relevant literature. Fisher-generated digital data are diverse, so the information herein is not intended to be "one size fits all"; we encourage the reader to consider the applicability and feasibility of each solution in the context of their project. Common to all subsections was the need for projects to identify and secure dedicated sources of funding and other resources for maintaining projects; generating, managing, accessing, and analysing data; conducting necessary research; and forming strong ties with both fishers and decision-makers.

5.1 Access and Consistency

Access to data from companies and other external entities cannot be guaranteed, but projects can take steps to ensure that access is more likely. Primarily, projects should attempt to form strong working relationships or even collaborations with the companies that are supplying the data (e.g., Fricke et al. 2020, Dainys et al. 2022, Weir et al. 2022, DePiper et al. 2023, and Fischer et al. 2023 all have company personnel as co-authors). These relationships should be built upon mutual benefit and trust, and regular communication. How a company benefits from supplying the data should be clear. For example, the work might align with a company goal or vision, or perhaps executives are avid fishers or simply excited to see the data being used in interesting ways. In addition, just as projects need to trust that they will have access to the data, companies need to trust that their data will be managed and used appropriately. These and other technical aspects of a relationship should be described in data sharing agreements or contracts that are developed with legal counsel.

Projects that maintain a close working relationship with data suppliers may be more likely to be privy to, or even involved in, changes that can affect data volume and quality. Unfortunately, projects that access data through standard APIs that are offered by large companies such as Google and YouTube may struggle to build close working relationships. If access and consistency issues are untenable, then researchers and agencies can endeavour to develop their own digital solutions. These are likely to take the form of app- and web-based citizen science projects. Such projects should follow best practices for ensuring long-term data storage and access (see Krogman et al. 2024; Sievert et al. 2026).

5.2 Data Science Competency and Capacity

Digital data are becoming increasingly abundant and likely to play a pivotal role in recreational fisheries research and management (Cooke et al. 2022; Lennox et al. 2022; Skov et al. 2022). It is therefore incumbent upon research groups and organisations to hire, train, and provision personnel to ensure that such data sources can be effectively managed and used. Quick fixes include ad hoc training via the diversity of online courses and tools, hiring non-fisheries data scientists, and outsourcing to collaborators or industry. However, we also advocate for updates to university curricula that encourage the development of data science skills and competencies among future fisheries professionals.

5.3 Data Quality

This subsection focuses on major strategies for identifying and addressing the data quality issues that were described in Sect. 4.3. This section updates and expands on the data quality solutions in Venturelli et al. (2017), which were specific to fishing apps, and based on less experience and study. To the extent possible and as necessary, we recommend minimising the potential for error and bias, using comparative analysis to identify error and bias, and monitoring error and bias over time. These solutions should be applied iteratively so that issues are identified, actions are taken, outcomes are evaluated, and solutions are adjusted.

5.3.1 Identifying Data Quality Issues

Although it is easy to imagine (and in some cases document) the myriad ways in which fisher-generated digital data contain errors, it is more important to determine whether or how data quality issues affect estimates of fisher behaviour. Estimates of fisher behaviour derived from digital data should be validated via comparisons to estimates that were obtained by conventional (usually probabilistic) means (e.g., Pollock et al. 1994). Ideally, both estimates should be from the same spatiotemporal frame and resolution. Common statistical approaches include regression (e.g., Papenfuss et al. 2015; Johnston et al. 2022), comparing the mean of the raw, unweighted data from the non-probabilistic sample to the 95% confidence interval from the probabilistic sample (e.g., Bethlehem 2010, 2015), and comparing the ratios of the two means via a non-parametric bootstrapping approach (e.g., Gundelund et al. 2021 and references therein).

Comparative analyses are common in studies involving data from fishing apps. Research has validated app-based results against results from both on-site and off-site survey methods (Papenfuss et al. 2015; Jiorle et al. 2016; Gundelund et al. 2020, 2021, 2023; Johnston et al. 2022; Dainys et al. 2022). Generally speaking, results suggest that app data tend to produce reasonable results at low spatial and temporal resolutions and can perform well at higher resolutions if data collection is active (e.g., via citizen science) and sample sizes are large. However, until more comparative research has been conducted, we caution against assuming that specific results apply across fisheries given the high degree of variation among fisheries, platforms, and study scales and resolutions.

An important caveat when conducting comparative analyses is that conventional estimates are also subject to error. For example, on- and off-site survey methods can suffer from self-selection bias, avidity bias, non-response bias, recall bias, and reporting bias, to name a few (see Pollock et al. 1994; Venturelli et al. 2017). If a conventional estimate is also biassed, then this complicates validation checks. Detecting and correcting for errors in conventional approaches are beyond the scope of this chapter. Here, we simply caution the reader against assuming that digital data are responsible for any difference between estimates; it is entirely possible that digital data more closely approximate truth. This was the case in an analysis that compared multiple metrics from well-established recall surveys and a citizen science app for the recreational northern pike (*Esox lucius*) fishery in Denmark (Gundelund et al. 2023). The authors attributed disagreement to recall, non-response, and location identification issues within the established recall survey.

5.3.2 Addressing Data Quality Issues

If validation checks suggest that digital- and probability-based estimates of behaviour differ in predictable ways, and one is willing to assume that the latter is a close approximation of truth, then there is a compelling argument to be made for addressing data quality issues by applying a correction factor. For example, Papenfuss et al. (2015) and Gundelund et al. (2021, 2023) contain regression equations and correction factors for converting app data into conventional fisheries metrics. Correction factors may change over time because of changing fisheries, platforms, and users, so comparative studies should occur at regular intervals (Dainys et al. 2022). Gundelund et al. (2023) observed consistent bias in app-based estimates of catch and effort from three fisheries over 10 seasons.

Validation checks should not replace efforts to address data quality issues at their source. Strategies to motivate a diverse set of fishers to share complete and correct information are crucial (see also Venturelli et al. 2017). Motivation begins with platforms that are well-designed and pleasant to use (e.g., ease of data entry, appropriate gamification). In addition, platforms should educate fishers about the need for high-quality data. Value can also be encouraged through attractive features that rely on quality data (e.g., leaderboards, recommendations about where, how, or when to fish) and by providing regular updates. Finally, it is essential to reinforce and build fisher trust in how the data will be used and by whom. Fishers may be particularly concerned that their data will be "used against them," so they should be assured that sharing with law enforcement will only be through a subpoena, and that information that they deem to be sensitive—like secret spots—will not be revealed (e.g., by lagging data and presenting them at a low spatiotemporal resolution). Platforms should communicate clearly, honestly, and frequently about whether fisher-generated digital data could be used to tighten regulations. If the platform allows, it should encourage a two-way flow of information that can influence behaviour directly and encourage active participation in the management process.

Data quality issues can also be addressed post-collection via outlier checks (e.g., impossible fish sizes or locations) and mathematical adjustments. Weighting adjustment compares auxiliary information from the target population to those from the non-probability sample. If differences exist, then weights are calculated to correct for the under- or overrepresentation of certain groups. This technique is commonly used to correct for non-response in probability surveys (Särndal and Lundström 2005). Propensity weighting is another option (Rosenbaum and Rubin 1983, 1984). It relies on robust variables within a logistic modelling framework to predict model survey participation (yes, no) in different populations. It ultimately generates propensity scores for population stratification. Researchers have also explored Bayesian approaches that use informed priors to incorporate existing knowledge, and hybrid approaches that combine non-probability and probability methods (van der Hammen et al. 2016; Liu et al. 2017).

5.4 Ethics

Fisher-generated digital data must be obtained, stored, and used in a way that respects the privacy of the fishers who supplied them, adheres to the ethical principles that are associated with human-subject research, and does not violate existing terms of use or applicable laws. Ethical actions build trust, which can improve data volume and quality. As noted in Sect. 5.3, projects should clearly and formally state the purpose and intended use of fisher data beyond what might already be covered by existing terms and conditions. This includes setting aside resources for a long-term communication strategy that reinforces essential information, such as how the fisher-generated data will benefit the social, economic, and ecological sustainability of recreational fishing, and whether this could include the implementation of stricter regulations.

Ethical concerns with digital data also relate to privacy and data anonymisation (Monkman et al. 2018c; Di Minin et al. 2021). The following is a checklist for the responsible use of social media data (Sbragaglia et al. 2021b) that can also apply to projects that involve data from fishing forums, fishing apps, and connected devices.

- Read all applicable terms of use and be aware that they can change frequently.
- Conduct an in-depth ethics assessment (ideally in consultation with an expert in human-subjects research or privacy considerations).
- Only download necessary personal or sensitive information and then take steps to anonymise it.
- Comply with data storage and security requirements.
- Anonymise content (i.e., use paraphrased text).
- Remove sensitive geographical locations to protect fisher privacy and the locations of protected species.
- Interpret the results carefully to ensure that the above steps were effective.

5.5 Integration

Most of the publications that are cited in Table 1 are novel, stand-alone, proof-of-concept studies. To the best of our knowledge, these studies were not integrated into fisheries management; probably because managers did not trust the data, and a process for integrating the data had yet to be developed (Skov et al. 2021). Addressing these issues will require robust standards and a clear process for integrating trusted data and results.

Well-defined standards are essential for enhancing the quality, reliability, and integrity of fisher-generated digital data (Bonar et al. 2024). These standards should be based on evidence and best practices and encompass all aspects of the data management process (Krogman et al. 2024), including collection, storage, sharing, and analysis. Minimum data standards are particularly important because they not only define the essential data elements to be collected but also facilitate quality control and establish data sharing and reporting requirements. For example, Venturelli et al. (2017) proposed minimum standards for generating unbiased estimates of recreational effort and catch via fishing apps. Standards can also be adapted from related approaches, such as traditional logbooks and diaries (Cooke et al. 2000).

Standards should be embedded within a formal process for integrating the information that is obtained from fisher-generated digital data in the fisheries management decision-making process. Established processes include the National Oceanic and Atmospheric Administration's Recreational Fishing Survey and Data Standards, and the International Council for the Exploration of the Sea Workshop on Standards and Guidelines for Fisheries Dependent Data (ICES 2021). Plans include written documentation, survey design, data quality considerations, and transition plans. Plans should be developed collaboratively with all user groups, including fishers, scientists, managers, and the public.

6 Conclusion

Digital data mining has the potential to generate insights into recreational fisher behaviour, related aspects of human dimensions, and a diversity of other topics. A substantial proportion (0.12–0.21) of recreational fishers report sharing their catches through public digital platforms (Vitale et al. 2021), and more than 40 studies have used fisher-generated digital data in the context of fisher behaviour (Table 1, Fig. 3). The fact that almost 73% of these papers were published in 2020 or later suggests that this is an emerging research area within recreational fisheries science. We anticipate a great deal of research and application in the next decade (Skov et al. 2021; Cooke et al. 2022).

Research to date has used this new type of data to answer old questions (e.g., catch and effort; Table 1), and provide new insights that would have been difficult or impossible to obtain via other means. Examples of the latter include illegal fishing (Shiffman et al. 2017), invasive species distribution and spread (Banha et al. 2015; Weir et al. 2022; Fischer et al. 2023; Sbragaglia et al. 2022), reaction to distributional range shifts of species due to climate change (Sbragaglia et al. 2021a; Sbragaglia et al. 2024), fisher satisfaction and engagement (Sbragaglia et al. 2020a; Gundelund et al. 2022; Johansen et al. 2022), behavioural responses to the COVID-19 pandemic (Gundelund and Skov 2021; Britton et al. 2023; Audzijonyte et al. 2023), and the potential impacts of offshore wind development on recreational fishing (DePiper et al. 2023). Digital platforms also can provide an effective opportunity to engage large numbers of fishers in active data collection (e.g., via citizen science; Sbragaglia et al. 2020b), which can not only contribute to a reliable data stream (Link et al. 2021) and positively influence stewardship (Sbragaglia and Arlinghaus 2020; Shephard et al. 2023), but also increase communication, collaboration, trust, and compliance.

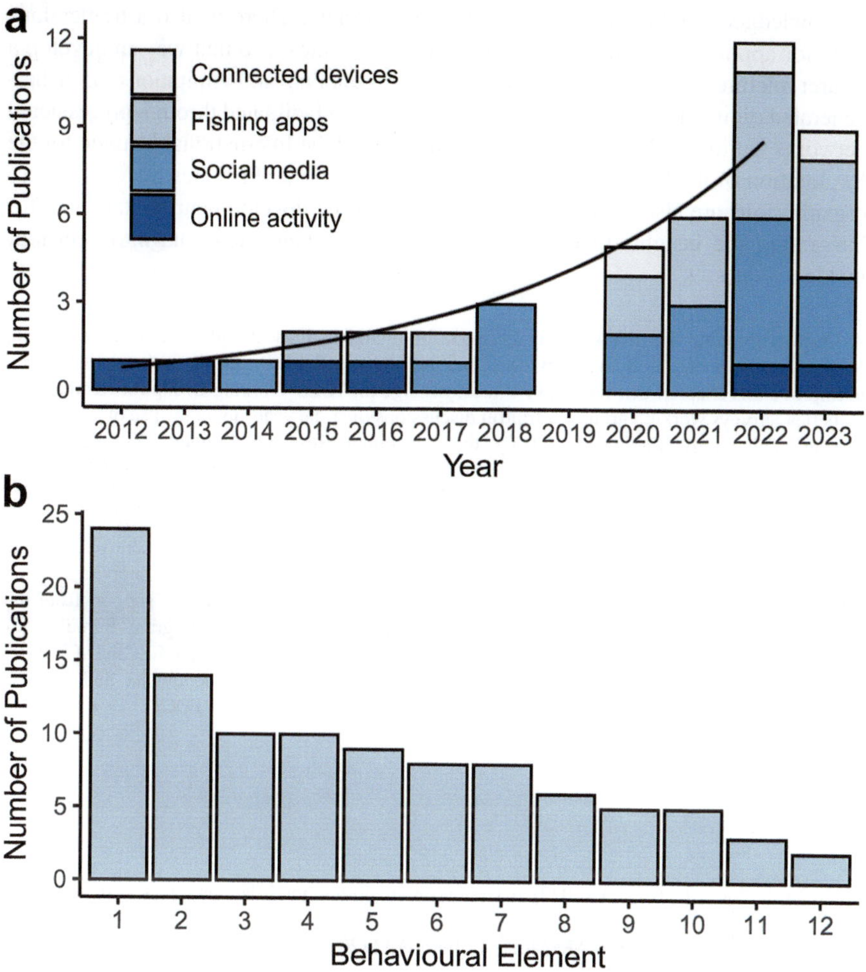

Fig. 3 Number of peer-reviewed publications by source (panel a) and topic (panel b) that have used digital data mining to quantify and predict recreational fisher dimensions and behaviours (1 = fishing effort in space or time, 2 = catch composition, 3 = catch rate or total harvest, 4 = fishing trip number or frequency, 5 = preferred species or method, 6 = fisher demographics, 7 = attitudes, perceptions, or sentiment, 8 = release rate, 9 = fisher movement patterns, 10 = interest in fishing, 11 = fishing trip duration, 12 = illegal fishing; see Table 1 for raw data; we excluded the year 2023 to the regression line [in panel a] [number = $0.6185e^{0.2391*\text{year}}$, $R^2 = 0.94$, $p = 3.09e{-06}$] because the publication list was finalised prior to the end of the year; publications in panel b [104] exceed publications in panel a [44] because most publications assessed multiple dimensions or behaviours)

The potential of fisher-generated digital data to shed light on the human dimensions of fisheries topics such as recreational fisher behaviour, and the degree to which it will be used within management, will likely be context-dependent. Fisheries-generated data have shortcomings just like any other method in fisheries science. Issues related to data access, management, quality, ethics, and integration must be acknowledged, studied, and addressed as appropriate. There is also a tremendous need for applied research in diverse fisheries and contexts so that we can develop a clearer picture of the strengths, weaknesses, benefits, and limitations of fisher-generated digital data. This research is underway and facilitated through professional networks within the American Fisheries Society and the International Council for the Exploration of the Sea's Working Group for Recreational Fisheries Surveys. These organisations are also well-positioned to assist in the development of standards for integrating the insights that are gained from fisher-generated data into decision-making processes.

Acknowledgments We thank Penny Forester for finalising Figs. 2 and 3. Rebecca Krogman, Michael Sullivan, one anonymous reviewer, and the editors provided valuable feedback that helped to improve this chapter. CS, KH, and PV acknowledge the ICES Working Group for Recreational Fisheries Surveys working group for fruitful discussions over the years about the use of digital data mining. This study was supported by the Australian Research Council Discovery grant (DP220102446) and 2020 Pew Fellowship in Marine Conservation awarded to AA. CS and CG received funding from the Danish Rod and Net Fish License funds (Project 39122). VS was supported by a Ramón y Cajal research fellowship (RYC2021-033065-I) granted by the Spanish Ministry of Science and Innovation and acknowledged by the Spanish government through the "Severo Ochoa Centre of Excellence" accreditation (CEX2019-000928-S). IJ was supported by the grant no. 23-07278S from the Czech Science Foundation. ASV acknowledges support from the FCT–Portuguese Foundation for Science and Technology through the program Stimulus for Scientific Employment–Individual Support [contract reference 2020.01175.CEECIND/CP1601/CT0009], and project ClimateMedia–Understanding climate change phenomena and impacts from digital technology and social media [contract reference 2022.06965.PTDC].

References

Alam MM, Torgo L, Bifet A (2022) A survey on spatio-temporal data analytics systems. ACM Comput Surv 54(10s):1–38

Allison C, Winkler AC, Childs AR, Muller C, Potts WM (2023) Can social media platforms be used to foster improved environmental behaviour in recreational fisheries? Fish Res 258:106544

Arlidge WN, Firth JA, Alfaro-Shigueto J, Ibanez B, Mangel JC, Squires D, Milner-Gulfland EJ (2021) Assessing information-sharing networks within small-scale fisheries and the implications for conservation interventions. R Soc Open Sci 8:211240

Audzijonyte A, Mateos-González F, Dainys J, Gundelund C, Skov C, Tyrell DeWeber J, Venturelli P, Vienožinskis V, Smith C (2023) High-resolution app data reveal sustained increases in recreational fishing effort in Europe during and after COVID-19 lockdowns. R Soc Open Sci 10:230408

Babu KM, Bentall D, Ashton DT, Puklowski M, Fantham W, Lin HT, Tuckey NPL, Wellenreuther M, Jesson LK (2023) Computer vision in aquaculture: a case study of juvenile fish counting. J R Soc N Z 53:52–68

Baker MS Jr, Oeschger I (2009) Description and initial evaluation of a text message based reporting method for marine recreational anglers. Mar Coastal Fish 1:143–154

Banha F, Ilhéu M, Anastácio PM (2015) Angling web forums as an additional tool for detection of new fish introductions: the first record of *Perca fluviatilis* in continental Portugal. Knowl Manag Aquat Ecosyst 416:3

Batrinca B, Treleaven PC (2015) Social media analytics: a survey of techniques, tools and platforms. AI Soc 30:89–116

Belhabib D, Campredon P, Lazar N, Sumaila RU, Baye BC, Kane EA, Pauly D (2016) Best for pleasure, not for business: evaluating recreational marine fisheries in West Africa using unconventional sources of data. Palgrave Commun 2:1–10

Bethlehem J (2010) Selection bias in web surveys. Int Stat Rev 78:161–188

Bethlehem J (2015) Essay: sunday shopping-the case of three surveys. Surv Res Methods-Ger 9(3): 221–230

Bonar SA, Mercado-Silva N, Pope KL (eds) (2024) Standard methods for sampling north American freshwater fishes, 2nd edn. American Fisheries Society, Bethesda

Borgatti SP, Everett MG, Johnson JC (2018) Analyzing social networks, 2nd edn. Sage, London

Bradley D, Merrifield M, Miller KM, Lomonico S, Wilson JR, Gleason MG (2019) Opportunities to improve fisheries management through innovative technology and advanced data systems. Fish Fish 20:564–583

Brick JM, Andrews WR, Foster J (2022) A review of nonprobability sampling using mobile apps for fishing effort and catch surveys. Trans Am Fish Soc 151:42–49

Britton JR, Pinder AC, Alós J, Arlinghaus R, Danylchuk AJ, Edwards W, Freire KMF, Gundelund C, Hyder K, Jarić I, Lennox R, Lewin W-C, Lynch AJ, Midway SR, Potts WM, Ryan KL, Skov C, Strehlow HV, Tracey SR, Tsuboi J-I, Venturelli PA, Weir JL, Weltersbach MS, Cooke SJ (2023) Global responses to the COVID-19 pandemic by recreational anglers: considerations for developing more resilient and sustainable fisheries. Rev Fish Biol Fish 33: 1095–1111

Brown DM, Soto-Corominas A, Surez JL, de la Rosa J (2017) Overview—the social media data processing pipeline. In: Sloan L, Quan-Haase A (eds) The SAGE handbook of social media research methods. Sage, London, pp 125–145

Carneiro T, Da Nóbrega RVM, Nepomuceno T, Bian G, de Albuquerque H, Filho P (2018) Performance analysis of Google Colaboratory as a tool for accelerating deep learning applications. IEEE Access 6:61677–61685

Carter DW, Crosson S, Liese C (2015) Nowcasting intraseasonal recreational fishing harvest with internet search volume. PLoS One 10:e0137752

Chancellor S, Counts S (2018) Measuring employment demand using internet search data. In: Mandryk R, Hancock M, Perry M, Cox A (eds) Proceedings of the 2018 CHI conference on human factors in computing systems, Montreal. Association for Computing Machinery, New York, pp 1–14

Christin S, Hervet É, Lecomte N (2019) Applications for deep learning in ecology. Methods Ecol Evol 10:1632–1644

Cirqueira D, Hofer M, Nedbal D, Helfert M, Bezbradica M (2019) Customer purchase behavior prediction in e-commerce: a conceptual framework and research agenda. In: Ceci M, Loglisci C, Manco G, Masciari E, Ras Z (eds) New frontiers in mining complex patterns. NFMCP 2019. Lecture notes in computer science, vol 11948. Springer, Cham, pp 119–136

Cooke SJ, Dunlop WI, Macclennan D, Power G (2000) Applications and characteristics of angler diary programmes in Ontario, Canada. Fish Manag Ecol 7:473–487

Cooke SJ, Docker MF, Mandrak NE, Young N, Heath DD, Jeffries KM, Howarth A, Brownscombe JW, Livernois J, Semeniuk CAD, Venturelli PA, Danylchuk AJ, Lennox RJ, Jarić L, Fisk AT, Vandergoot CS, Britton JR, Muir AM (2022) Technoscience and the modernization of freshwater fisheries assessment and management. Environ Technol Innov 28:102865

Correia RA, Ladle R, Jarić I, Malhado ACM, Mittermeier JC, Roll U, Soriano-Redondo A, Veríssimo D, Fink C, Hausmann A, Guedes-Santos J, Vardi R, Di Minin E (2021) Digital data sources and methods for conservation culturomics. Conserv Biol 35:398–411

Creany NE, Monz CA, D'Antonio A, Sisneros-Kidd A, Wilkins EJ, Nesbitt J, Mitrovich M (2021) Estimating trail use and visitor spatial distribution using mobile device data: an example from the nature Reserve of Orange County, California, USA. Environ Chall 4:100171

Dainys J, Gorfine H, Mateos-González F, Skov C, Urbanavičius R, Audzijonyte A (2022) Angling counts: harnessing the power of technological advances for recreational fishing surveys. Fish Res 254:106410

Danylchuk AJ, Danylchuk SC, Kosiarski A, Cooke SJ, Huskey B (2018) Keepemwet fishing-an emerging social brand for disseminating best practices for catch-and-release in recreational fisheries. Fish Res 205:52–56

DePiper G, Corvi D, Steinback S, Arrington DA, Blalock R, Roman N (2023) Leveraging data from a private recreational fishing application to begin to understand potential impacts from offshore wind development. ICES J Mar Sci 82(3):fsad154

Di Minin E, Fink C, Hausmann A, Kremer J, Kulkarni R (2021) How to address data privacy concerns when using social media data in conservation science. Conserv Biol 35:437–446

eBird (2022) Publications. https://science.ebird.org/en/research-and-conservation/publications. Accessed 3 Mar 2023

Eryaşar AR, Saygu İ (2022) Using social media to identify recreational bluefish angling in the Mediterranean and Black Sea. Mar Policy 135:104834

Ferreira JC, Branquinho J, Chavez PC, Piedade F (2017) Computer vision algorithms fishing vessel monitoring—identification of vessel plate number. In: De Paz J, Julián V, Villarrubia G, Marreiros G, Novais P (eds) Ambient intelligence – software and applications – 8th international symposium on ambient intelligence (ISAmI 2017). ISAmI 2017, Advances in Intelligent Systems and Computing, vol 615. Springer, Cham, pp 9–17

Ficetola GF (2013) Is interest toward the environment really declining? The complexity of analyzing trends using internet search data. Biodivers Conserv 22:2983–2988

Fischer SM, Ramazi P, Simmons S, Poesch MS, Lewis MA (2023) Boosting propagule transport models with individual-specific data from mobile apps. J Appl Ecol 60:934–949

Ford MT (2020) Internet search and page view behavior scores: validity and usefulness as indicators of psychological states. In: Woo SE, Tay L, Proctor RW (eds) Big data in psychological research. American Psychological Association, Washington p, pp 89–107

Foster RM, Childs AR, Mann BQ, Potts WM (2023) Specialist spearfishers' ecological knowledge provides accurate information that improves the description of the reproduction of a data-deficient species, *Oplegnathus conwayi* in South Africa. Fish Res 257:106513

Freeman L (2004) The development of social network analysis. Empirical Press, Vancouver

French G, Mackiewicz M, Fisher M, Holah H, Kilburn R, Campbell N, Needle C (2020) Deep neural networks for analysis of fisheries surveillance video and automated monitoring of fish discards. ICES J Mar Sci 77:1340–1353

Fricke RM, Wood SA, Martin DR, Olden JD (2020) A bobber's perspective on angler-driven vectors of invasive species transmission. NeoBiota 60:97–115

Furey RP, Merrill NH, Sawyer JP, Mulvaney KK, Mazzotta MJ (2022) Evaluating water quality impacts on visitation to coastal recreation areas using data derived from cell phone locations. PLoS One 17:e0263649

Giovos I, Keramidas I, Antoniou C, Deidun A (2018) Identifying recreational fisheries in the Mediterranean Sea through social media. Fish Manag Ecol 25(4):287–295

Grabowski T, Benedum ME, Curley A, Dill-De Sa C, Shuey M (2023) Pandemic-driven changes in the nearshore non-commercial fishery in Hawai'i: catch photos posted to social media capture changes in fisher behavior. PeerJ 11:e14994

Guedes-Santos J, Correia RA, Jepson P, Ladle RJ (2021) Evaluating public interest in protected areas using Wikipedia page views. J Nat Conserv 63:126040

Gundelund C, Skov C (2021) Changes in angler demography and angling patterns during the Covid-19 lockdown in spring 2020 measured through a citizen science platform. Mar Policy 131:104602

Gundelund C, Arlinghaus R, Baktoft H, Hyder K, Venturelli P, Skov C (2020) Insights into the users of a citizen science platform for collecting recreational fisheries data. Fish Res 229:105597

Gundelund C, Venturelli P, Hartill BW, Hyder K, Olesen HJ, Skov C (2021) Evaluation of a citizen science platform for collecting fisheries data from coastal sea trout anglers. Can J Fish Aquat Sci 78:1576–1585

Gundelund C, Arlinghaus R, Birdsong M, Flávio H, Skov C (2022) Investigating angler satisfaction: the relevance of catch, motives and contextual conditions. Fish Res 250:106294

Gundelund C, Venturelli P, Hartill BW, Hyder K, Olesen HJ, Skov C (2023) Spatiotemporal and multispecies comparisons between a citizen science platform and recall surveys in recreational fisheries. Mar Policy 155:105780

Hall QA, Coffey DM, Streich MK, Fisher MR, Stunz GW (2022) Social media shines light on the "hidden" impact of nighttime guided-gigging charters on Texas' southern flounder fishery: a stab in the dark. PLoS One 17:e0269397

Hamelin KM, MacNeil MA, Curran K, Bailey M (2022) "The people's fish": sociocultural dimensions of recreational fishing for Atlantic mackerel in Nova Scotia. Front Mar Sci 9:971262

Happel A (2019) A volunteer-populated online database provides evidence for a geographic pattern in symptoms of black spot infections. Int J Parasitol Parasites Wildl 10:156–163

Happel A, Murchie KJ, Willink PW, Knapp CR (2020) Great Lakes fish finder app; a tool for biologists, managers and education practitioners. J Great Lakes Res 46:230–236

Hausmann A, Toivonen T, Slotow R, Tenkanen H, Moilanen A, Heikinheimo V, Di Minin E (2018) Social media data can be used to understand tourists' preferences for nature-based experiences in protected areas. Conserv Lett 11:E12343

Ho AFW, To BZYS, Koh JM, Cheong KH (2019) Forecasting hospital emergency department patient volume using internet search data. IEEE Access 7:93387–93395

Hoad C, Wilson O (2020) More than 'clips of dead fish': gender, technologisation, and localisation in the amateur music videos of Wellington spearfishers. Green Lett 24:252–268

Hook SA, Brown A, Bell B, Kroese J, Radford Z, Hyder K (2022) The impact of COVID-19 on participation, effort, physical activity, and well-being of sea anglers in the UK. Front Mar Sci 9: 815617

Hunt LM, Sutton SG, Arlinghaus R (2013) Illustrating the critical role of human dimensions research for understanding and managing recreational fisheries within a social-ecological system framework. Fish Manag Ecol 20:111–124

ICES (2021) ICES workshop on standards and guidelines for fisheries dependent data (WKDSG; outputs from 2020 meeting). ICES Sci Rep 3:1–90

Jain A, Jain V (2019) Sentiment classification of twitter data belonging to renewable energy using machine learning. J Info Optim Sci 40:521–533

Jarić I, Correia RA, Brook BW, Buettel JC, Courchamp F, Di Minin E, Firth JA, Gaston KJ, Jepson P, Kalinkat G, Ladle R, Soriano-Redondo A, Souza AT, Roll U (2020a) iEcology: harnessing large online resources to generate ecological insights. Trends Ecol Evol 35(7): 630–639

Jarić I, Roll U, Arlinghaus R, Belmaker J, Chen Y, China V, Douda K, Essl F, Jähnig SC, Jeschke JM, Kalinkat G, Kalous L, Ladle R, Lennox RJ, Rosa R, Sbragaglia V, Sherren K, Šmejkal M, Soriano-Redondo A, Souza AT, Wolter C, Correia RA (2020b) Expanding conservation culturomics and iEcology from terrestrial to aquatic realms. PLoS Biol 18:e3000935

Jiorle RP, Ahrens RN, Allen MS (2016) Assessing the utility of a smartphone app for recreational fishery catch data. Fisheries 41:758–766

Johansen K, Olsen EM, Haraldstad T, Arlinghaus R, Höglund E (2022) Digital data help explain drivers of angler satisfaction: an example from southern Norway. N Am J Fish Manag 42:1165–1172

Johnston FD, Simmons S, Poorten BV, Venturelli P (2022) Comparative analyses with conventional surveys reveal the potential for an angler app to contribute to recreational fisheries monitoring. Can J Fish Aquat Sci 79:31–46

Jun SP, Yoo HS, Choi S (2018) Ten years of research change using Google trends: from the perspective of big data utilizations and applications. Technol Forecast Soc 130:69–87

Keeler BL, Wood SA, Polasky S, Kling C, Thomas CT, Downing J (2015) Recreational demand for clean water: evidence from geotagged photographs by visitors to lakes. Front Ecol Environ 13: 76–81

Kochalski S, Barz F, Pita P, Harrison HL (2026) Quantitative and qualitative content analysis of text and images. In: Pope KL, Arlinghaus R, Hunt LM, Lynch AJ, van Poorten BT (eds) Understanding recreational fishers: disciplinary and interdisciplinary approaches for fisheries management. Springer Nature, Cham, p 507–532

Krogman RM, Bayer JM, Cooper AR, Kopaska J, Leonard NJ, Pritt JJ, Roe C, Tracy EE, Venturelli P, Wieferich D, Infante DM (2024) Standard data management practices. In: Bonar SA, Mercado-Silva N, Pope KL (eds) Standard methods for sampling North American freshwater fishes, 2nd edn. American Fisheries Society, Bethesda

Lee K, Sener IN (2021) Strava metro data for bicycle monitoring: a literature review. Transp Rev 41:27–47

Lennox RJ, Sbragaglia V, Vollset KW, Sortland LK, McClenachan L, Jarić I, Guckian ML, Ferter K, Danylchuk AJ, Cooke SJ, Arlinghaus R, Twardek WM (2022) Digital fisheries data in the internet age: emerging tools for research and monitoring using online data in recreational fisheries. Fish Fish 23:926–940

Lewin WC, Weltersbach MS, Haase K, Riepe C, Strehlow HV (2023) Potential biases in angler diary data: the impact of the diarist recruitment process on participation rates, catch, harvest, and effort estimates. Fish Res 258:106551

Link JS, Werner FE, Werner K, Walter J, Strom M, Seki MP, Schwing F, Rusin J, Porch CE, Osgood K, Moline K, Methot RD, Lynch PD, Lipton D, Koch K, Howell EA, Hare JA, Foy RJ, Detlor D, Desfosse L, Crofts J, Cabana N (2021) A NOAA fisheries science perspective on the conditions during and after COVID-19: challenges, observations, and some possible solutions, or why the future is upon us. Can J Fish Aquat Sci 78:1–12

Liu B, Stokes L, Topping T, Stunz G (2017) Estimation of a total from a population of unknown size and application to estimating recreational red snapper catch in Texas. J Surv Stat Methodol 5:350–371

Mannheim SL, Childs AR, Butler EC, Winkler AC, Parkinson MC, Farthing MW, Zweig T, McCord M, Drobniewska N, Potts WM (2018) Working with, not against recreational anglers: evaluating a pro-environmental behavioural strategy for improving catch-and-release behaviour. Fish Res 206:44–56

Martin DR, Pracheil BM, DeBoer JA, Wilde GR, Pope KL (2012) Using the internet to understand angler behavior in the information age. Fisheries 37(10):458–463

Martin DR, Chizinski CJ, Eskridge KM, Eskridge KM, Pope KL (2014) Using posts to an online social network to assess fishing effort. Fish Res 157:24–27

Martinazzo GM, Giareta EP, Bornatowski H, Abilhoa V, Freitas M (2022) A look at the unknown: potential impact of marine recreational fishing on threatened species in the southern Atlantic Ocean. Ocean Coast Manag 218:106044

Maya-Jariego I, Martinez-Alba I, Alieva D (2021) "Plenty of black money": Netnography of illegal recreational underwater fishing in southern Spain. Mar Policy 126:104411

McCallum ML, Bury GW (2014) Public interest in the environment is falling: a response to Ficetola 2013. Biodivers Conserv 23:1057–1062

McClenachan L (2009) Documenting loss of large trophy fish from the Florida keys with historical photographs. Conserv Biol 23:636–643

Midway SR, Adriance J, Banks P, Haukebo S, Caffey R (2020) Electronic self-reporting: angler attitudes and behaviors in the recreational red snapper fishery. N Am J Fish Manag 40:1119–1132

Mitchell JC (ed) (1969) Social networks in urban situations. Manchester University Press, Manchester

Monkman G, Kaiser MJ, Hyder K (2018a) Heterogeneous public and local knowledge provides a qualitative indicator of coastal use by marine recreational fishers. J Environ Manag 228:495–505

Monkman G, Kaiser MJ, Hyder K (2018b) Text and data mining of social media to map wildlife recreation activity. Biol Conserv 228:89–99

Monkman G, Kaiser MJ, Hyder K (2018c) The ethics of using social media in fisheries research. Rev Fish Sci Aquac 26:235–242

Monz C, Mitrovich M, D'Antonio A, Sisneros-Kidd A (2019) Using mobile device data to estimate visitation in parks and protected areas: an example from the nature reserve of orange county, California. J Park Recreat Adm 37:92–109

Otsuka R, Yamakoshi G (2020) Analyzing the popularity of YouTube videos that violate mountain gorilla tourism regulations. PLoS One 15:e0232085

Papenfuss JT, Phelps N, Fulton D, Venturelli PA (2015) Smartphones reveal angler behavior: a case study of a popular mobile fishing application in Alberta, Canada. Fisheries 40(7):318–327

Pollock KH, Jones CM, Brown TL (1994) Angler survey methods and their applications in fisheries management. American Fisheries Society, Bethesda

Prado-Román C, Gómez-Martínez R, Orden-Cruz C (2021) Google trends as a predictor of presidential elections: the United States versus Canada. Am Behav Sci 65:666–680

R Core Team (2021) R: a language and environment for statistical computing. R Foundation for Statistical Computing, Vienna

Roberts JJ, Best BD, Dunn DC, Treml EA, Halpin PN (2010) Marine geospatial ecology tools: an integrated framework for ecological geoprocessing with ArcGIS, python, R, MATLAB, and C++. Environ Model Softw 25(10):1197–1207

Robinson JA, Kocman D, Speyer O, Gerasopoulos E (2021) Meeting volunteer expectations-a review of volunteer motivations in citizen science and best practices for their retention through implementation of functional features in CS tools. J Environ Plann Man 64(12):2089–2113

Roll U, Mittermeier JC, Diaz GI (2016) Using Wikipedia page views to explore the cultural importance of global reptiles. Biol Conserv 204:42–50

Rosenbaum PR, Rubin DB (1983) The central role of the propensity score in observational studies for causal effects. Biometrika 70:41–55

Rosenbaum PR, Rubin DB (1984) Reducing bias in observational studies using subclassification on the propensity score. J Am Stat Assoc 79:516–524

Ruggiero TE (2000) Uses and gratifications theory in the 21st century. Mass Commun Soc 3:3–37

Särndal CE, Lundström S (2005) Estimation in surveys with nonresponse. Wiley, Hoboken

Sbragaglia V, Arlinghaus R (2020) Conservation bottom-up initiatives in marine recreational spearfishing suggest the emergence of positive attitudes towards conservation. Sci Mar 84: 441–444

Sbragaglia V, Correia RA, Coco S, Arlinghaus R (2020a) Data mining on YouTube reveals fisher group-specific harvesting patterns and social engagement in recreational anglers and spearfishers. ICES J Mar Sci 77:2234–2244

Sbragaglia V, Cerri J, Bolognini L, Dragićević B, Dulćić J, Grati F, Azzurro E (2020b) Local ecological knowledge of recreational fishers reveals different meridionalization dynamics of two Mediterranean subregions. Mar Ecol Prog Ser 634:147–157

Sbragaglia V, Coco S, Correia RA, Coll M, Arlinghaus R (2021a) Analyzing publicly available videos about recreational fishing reveals key ecological and social insights: a case study about groupers in the Mediterranean Sea. Sci Total Environ 765:142672

Sbragaglia V, Correia RA, Di Minin E (2021b) Responsible use of social media data is needed: a reply to Maya-Jariego et al. "Plenty of black money: Netnography of illegal recreational underwater fishing in southern Spain.". Mar Policy 134:104780

Sbragaglia V, Espasandín L, Coco S, Felici A (2022) Recreational angling and spearfishing on social media: insights on harvesting patterns, social engagement and sentiments related to the distributional range shift of a marine invasive species. Rev Fish Biol Fish 32(2):687–700

Sbragaglia V, Espasandín L, Jarić VR I, Ramírez F, Coll M (2024) Tracking ongoing transboundary marine distributional range shifts in the digital era. Mar Ecol Prog Ser:SHIFTav3. 728:103–114 https://www.int-res.com/abstracts/meps/v728/meps14309

Schifani E, Paolinelli R (2018) Notas/notes forums and social media help to discover exotic species in Europe and monitor their spread: the case of *Exaireta spinigera* (Wiedemann, 1830)(*Diptera, Stratiomyidae*) in the Italian peninsula and Sicily. Graellsia 74:e079

Schulz H, Behnke S (2012) Deep learning. KI–Kunstl Intell 26:357–363

Shephard S, List CJ, Arlinghaus R (2023) Reviving the unique potential of recreational fishers as environmental stewards of aquatic ecosystems. Fish Fish 24:339–351

Shiffman DS, Macdonald C, Ganz HY, Hammerschlag N (2017) Fishing practices and representations of shark conservation issues among users of a land-based shark angling online forum. Fish Res 196:13–26

Sievert N, Krogman RM, Embke H (2026) Data standardisation and management to facilitate large-scale and interdisciplinary approaches. In: Pope KL, Arlinghaus R, Hunt LM, Lynch AJ, van Poorten BT (eds) Understanding recreational fishers: disciplinary and interdisciplinary approaches for fisheries management. Springer Nature, Cham, p 747–782

Silva JCS, de Lima Silva DF, Neto ADSD, Ferraz A, Melo JL, Júnior NRF, de Almeida Filho AT (2021) A city cluster risk-based approach for Sars-CoV-2 and isolation barriers based on anonymized mobile phone users' location data. Sustain Cities Soc 65:102574

Silva CN, Dainys J, Simmons S, Vienožinskis V, Audzijonyte A (2022) A scalable open-source framework for machine learning-based image collection, annotation and classification: a case study for automatic fish species identification. Sustainability 14:14324

Skov C, Hyder K, Gundelund C, Ahvonen A, Baudrier J, Borch T, deCarvalho S, Erzini K, Ferter K, Grati F, van derHammen T, Hinriksson J, Houtman R, Kagervall A, Kapiris K, Karlsson M, Lejk AM, Lyle JM, Martinez-Escauriaza R, Moilanen P, Mugerza E, Olesen HJ, Papadopoulos A, Pita P, Pontes J, Radford Z, Radtke K, Rangel M, Sagué O, Sande HA, Strehlow HV, Tutiņš R, Veiga P, Verleye T, Vølstad JH, Watson JW, Weltersbach MS, Ustups D, Venturelli PA (2021) Expert opinion on using angler smartphone apps to inform marine fisheries management: status, prospects, and needs. ICES J Mar Sci 78(3):967–978

Skov C, Gundelund C, Weltersbach MS, Ferter K, Bertelsen SK, Jepsen N (2022) Catch and release angling for sea trout explored by citizen science: angler behavior, hooking location and bleeding patterns. Fish Res 255:106451

Stampfler T, Elgendi M, Fletcher RR, Menon C (2022) Fall detection using accelerometer-based smartphones: where do we go from here? Front Public Health 10:996021

Steg L, Vlek C (2009) Encouraging pro-environmental behaviour: an integrative review and research agenda. J Environ Psychol 29:309–317

Sunhare P, Chowdhary RR, Chattopadhyay MK (2022) Internet of things and data mining: an application oriented survey. J King Saud Univ-Comput Inf Sci 34:3569–3590

Toivonen T, Heikinheimo V, Fink C, Hausmann A, Hiippala T, Järv O, Tenkanen H, Di Minin E (2019) Social media data for conservation science: a methodological overview. Biol Conserv 233:298–315

Tuia D, Kellenberger B, Beery S, Costelloe BR, Zuffi S, Risse B, Mathis A, Mathis MW, van Langevelde F, Burghardt T, Kays R, Klinck H, Wikelski M, Couzin ID, van Horn G, Crofoot MC, Stewart CV, Berger-Wolf T (2022) Perspectives in machine learning for wildlife conservation. Nat Commun 13:1–15

Turner RA, Polunin NV, Stead SM (2014) Social networks and fishers' behavior: exploring the links between information flow and fishing success in the Northumberland lobster fishery. Ecol Soc 19:38

Vacura K, Camp E, Venturelli P (2023) What makes anglers happy: a sentiment analysis of walleye angler forums in the United States. Trans Am Fish Soc 152:749–759

van der Hammen T, de Graaf M, Lyle JM (2016) Estimating catches of marine and freshwater recreational fisheries in The Netherlands using an online panel survey. ICES J Mar Sci 73:441–450

Van Huynh A (2023) Effect of IUCN red list category on public attention to mammals. Conserv Biol 37:e14050

Van Rossum G, Drake Jr FL (1995) Python reference manual. https://doi.org/10.1007/s10676-010-9227-5

Vaske JJ, Needham MD, Miller CA (2026) Fundamentals of sampling and implementing human dimensions of fishing surveys. In: Pope KL, Arlinghaus R, Hunt LM, Lynch AJ, van Poorten BT (eds) Understanding recreational fishers: disciplinary and interdisciplinary approaches for fisheries management. Springer Nature, Cham, p 327–353

Venturelli PA, Hyder K, Skov C (2017) Angler apps as a source of recreational fisheries data: opportunities, challenges and proposed standards. Fish Fish 18:578–595

Verma JP, Agrawal S, Patel B, Patel A (2016) Big data analytics: challenges and applications for text, audio, video, and social media data. Int J Soft Comput Art Int 5(1):41–51

Vitale G, Dedeu AL, Pujol M, Sbragaglia V (2021) Characterizing the profile of recreational fishers who share their catches on social media. Front Mar Sci 8:768047

Wasserman S, Faust K (1994) Social network analysis: methods and applications. Cambridge University Press, Cambridge

Weir JL, Vacura K, Bagga J, Berland A, Hyder K, Skov C, Attby J, Venturelli PA (2022) Big data from a popular app reveals that fishing creates superhighways for aquatic invaders. PNAS Nexus 1:pgac075

Wikle CK, Zammit-Mangion A, Cressie N (2019) Spatio-temporal statistics with R. Chapman & Hall/CRC, Boca Raton

Wilde GR, Pope KL (2013) Worldwide trends in fishing interest indicated by internet search volume. Fish Manag Ecol 20:211–222

Winkler AC, Butler EC, Attwood CG, Mann BQ, Potts WM (2022) The emergence of marine recreational drone fishing: regional trends and emerging concerns. Ambio 51:638–651

Yan Y, Kuo CL, Feng CC, Huang W, Fan H, Zipf A (2018) Coupling maximum entropy modeling with geotagged social media data to determine the geographic distribution of tourists. Int J Geogr Inf Sci 32:1699–1736

Ye, Li (2017) https://doi.org/10.1186/s40854-017-0056-y

Zook M, Barocas S, Boyd D, Crawford K, Keller E, Gangadharan SP, Goodman A, Hollander R, Koenig BA, Metcalf J, Narayanan A, Nelson A, Pasquale F (2017) Ten simple rules for responsible big data research. PloS Comput Biol 13:e1005399

Mixed Methods Research: Applications and Considerations for the Human Dimensions of Recreational Fisheries

Elizabeth Golebie, Carena J. van Riper, Sophia Kochalski, and Fiona Nunan

Abstract Mixed methods research combines qualitative and quantitative inquiry to understand complex problems. The human dimensions of recreational fisheries can benefit from mixed methods, given that they allow for generalizable claims about fisher behaviour while also providing rich information on individuals' experiences in the outdoors. In this chapter, we provide an overview of mixed methods, including guidelines for study design and applications for recreational-fisheries research. We begin by orienting mixed methods within scientific paradigms and explain the relevance of mixed methods in the context of interdisciplinary, multidisciplinary, and transdisciplinary research. We then offer guidance on how to select appropriate mixed methods designs (complementary, exploratory, and explanatory) and navigate their strengths and weaknesses. Finally, we present different methodologies and case examples from recreational fisheries research to learn from findings that have emerged from the combination of qualitative and quantitative methods, including mental maps, behavioural observations, participatory methods, focus groups, interviews, discourse analysis, and social network analysis. Corresponding with an increasing need for social science research in recreational-fisheries management and the importance of transformative outcomes, we contend that mixed methods research will become more widely used. This chapter, therefore, provides the overview needed for emerging recreational fisheries scientists to effectively implement mixed methods research.

E. Golebie (✉) · C. J. van Riper
Department of Natural Resources and Environmental Sciences, University of Illinois at Urbana-Champaign, Urbana, Illinois, United States of America
e-mail: golebie2@illinois.edu; cvanripe@illinois.edu

S. Kochalski
Center for Ocean and Society (CeOS), Kiel University, Kiel, Germany
e-mail: kochalski@ceos.uni-kiel.de

F. Nunan
International Development Department, University of Birmingham, Birmingham, United Kingdom
e-mail: f.s.nunan@bham.ac.uk

K. L. Pope et al. (eds.), *Understanding Recreational Fishers*, Fish & Fisheries Series 45, https://doi.org/10.1007/978-3-031-99739-6_16

Keywords Human dimensions · Interdisciplinary · Mixed methods · Qualitative · Quantitative · Recreational fishing

1 Introduction

Researchers within the human dimensions of recreational fisheries use different forms of knowledge to answer questions about aquatic social-ecological systems (Hunt et al. 2026). These researchers have predominantly employed quantitative methods (Weber et al. 2019) to collect descriptive information and test hypotheses, guided by principles of maintaining objectivity and minimizing bias (Creswell and Creswell 2017). In contrast, a minority of fisheries-related research has relied on qualitative methods to understand social phenomena underpinning human-environment interactions. These qualitative researchers tend to acknowledge subjectivity and build context-dependent knowledge (Barclay et al. 2017; Braun and Clarke 2013). Many researchers seek to integrate these paradigms and undertake both quantitative and qualitative inquiry. However, integration can cause tensions to arise due to differing goals and assumptions, as well as diverging attitudes, vocabularies, and research traditions (Johnson and Onwuegbuzie 2004). These tensions can be confronted through the conscious implementation of "mixed methods research" as an emerging strategy for knowledge production (Kinnebrew et al. 2021; Timans et al. 2019).

Local and regional fishery-related management problems have immense potential to be solved with mixed methods research guided by a philosophical foundation in pragmatism (Creswell and Creswell 2017; Dewey 1958). Pragmatism means that the strategy adopted in a research design depends on the problem at hand, thus facilitating alignment between decision-maker needs and research goals (Moon and Blackman 2014). Under this paradigm, mixed methods often use pragmatic logic that entails "studying facts and devising a theory to explain them" (Peirce 1934, p. 90). Many researchers unknowingly use mixed methods without describing it as such (Jefferson et al. 2021) and may benefit from a greater awareness of the context surrounding this paradigm and the challenges and opportunities it presents. Mixed methods underpinned by pragmatism can bridge disciplines and facilitate collaboration, though it may not fully resolve fundamental philosophical issues (Johnson and Onwuegbuzie 2004). Nevertheless, given the need for more human dimensions research supporting fisheries management (Dunham et al. 2018; Golebie et al. 2022; Hunt et al. 2013), especially in recreational fisheries, there is room for a broader application of a pragmatic research paradigm (Arlinghaus et al. 2017; Fenichel et al. 2013).

The use of multiple forms of knowledge inherent in mixed methods research opens possibilities for relational problem solving (West et al. 2020). Stakeholder involvement has been increasingly valued in recreational fisheries as managers seek to achieve complex social-ecological goals (Krogman et al. 2026; Fujitani et al. 2017; Ehrlich et al. 2023). Mixed methods research has the potential to enable stakeholder involvement, given that it may include multiple opportunities for participation at different time periods or through different platforms and methods. Further, when multiple knowledge systems are recognized, as often occurs with a pluralistic approach, recreational fishers are empowered to use their voice and actively participate in the research process (Burkett and Carter 2020). Consequently, the mixing of methods in human dimensions research may aid in understanding the diversity among recreational fishers in their behaviours, beliefs, and expectations for fisheries management (Magee et al. 2018).

2 Confluence of Integrated and Mixed Methods Research

The use of mixed methods in recreational fisheries research is particularly relevant to work that expands beyond a single discipline. This type of research can be characterized as one of three predominant types: multidisciplinary, interdisciplinary, and transdisciplinary research, with the latter two often collectively referred to as "integrated research" (Stock and Burton 2011; Szostak 2015; Tress et al. 2005). In integrated research, different disciplines, knowledge systems, methods, and research traditions can all be applied to solve problems more effectively. Such approaches are beneficial for tackling challenges tied to aquatic ecosystems, given their potential for understanding the dynamism in social-ecological systems (Arlinghaus et al. 2017; Kinnebrew et al. 2021; Lynch et al. 2026).

Multidisciplinary, interdisciplinary, and transdisciplinary research differ in terms of the degree of integration across disciplines and the application of knowledge for problem solving. First, "multidisciplinary" research is the least integrated form. This type of research pursuit is characterized by an overarching theme or purpose that is addressed by researchers from different disciplines who are working in parallel (Stock and Burton 2011; Tress et al. 2005). A more integrated form of research is "interdisciplinary," whereby members of a team represent different disciplines and bring those perspectives to develop a research approach and achieve applied outcomes. This cross-pollination occurs despite disciplinary boundaries, yet it requires greater collaboration and communication throughout a project (Stock and Burton 2011), particularly at the beginning of a collaboration (Bennett et al. 2017). Finally, transdisciplinary research aims to transcend disciplines and tackle real-world problems. One characteristic of transdisciplinary research is the close involvement of stakeholders and the value placed on problem-solving through the application of flexible methodologies. Transdisciplinary research thus requires co-designing and co-producing knowledge across researchers, stakeholders, and decision-makers (Mauser et al. 2023). Though difficult to implement, transdisciplinary research has been increasingly called for to address the challenges facing recreational fisheries management (Arlinghaus et al. 2014, 2017).

The approach taken to research—whether disciplinary, multidisciplinary, interdisciplinary, or transdisciplinary—has implications for how mixed methods research is designed and undertaken. That is, the degree of coordination in planning, data collection, and analysis will reflect the degree of integration in the broader research paradigm, with more integrated mixed methods found within interdisciplinary and transdisciplinary research (Kinnebrew et al. 2021). Transdisciplinary research is particularly likely to involve mixed methods as this research form requires flexible methodologies and a research approach that is oriented toward problem-solving (Stock and Burton 2011). The approaches taken in designing mixed methods research have implications for how a problem is conceptualized, a constituency engaged, and outcomes developed throughout the research process.

3 Mixed Methods Designs

Mixed methods may be implemented as a combination of different philosophical assumptions, research objectives, data sets, or methods of analysis (Johnson and Onwuegbuzie 2004), or even different forms of data dissemination. In this chapter, we focus on implementing mixed methods in data collection and analysis, which is characterized by the combination and integration of quantitative and qualitative inquiry and results in added value beyond the knowledge that could arise from just one form of inquiry. Embarking on mixed methods research first requires the identification of a research design. Mixed methods research designs can be categorized based on the ordering of quantitative and qualitative data collection and analysis as well as the purpose of each dataset and the relative emphasis placed on each (Creswell et al. 2003). The order in which data is collected is important because it determines how the different forms of knowledge build upon one another; thus, timeorder is the primary characteristic by which we define mixed methods research designs.

In this section, we explain five designs that are commonly used in mixed methods research and relevant to recreational fisheries work. The first three designs are drawn from the work of John W. Creswell and colleagues (Creswell and Creswell 2017; Creswell and Plano Clark 2018), which the reader may wish to reference for further information. The fourth and fifth designs are examples of more complex designs that are occasionally found in the recreational fisheries literature. To provide further examples of mixing methods, ten quantitative and qualitative methods commonly incorporated within these research designs are defined (Table 1). These lists are not exhaustive; many mixed methods research designs are possible (Creswell et al. 2003), and numerous research methods not discussed here may be useful for mixed methods.

3.1 Convergent Parallel Design

The convergent parallel design allows researchers to triangulate information on concepts of interest through phases of quantitative and qualitative data collection that occur at the same time. For instance, researchers might seek to understand

Table 1 Relevant quantitative and qualitative methods for mixed methods studies and examples of their use in recreational fisheries

Method	Explanation	Examples of integration in mixed methods
Mental models or cognitive maps	Concepts are assembled and their relationship to one another is represented by arrows showing direction (increase or decrease) and strength. Analysis of the relationships between the concepts and properties of the entire system, e.g. whether it changes continuously or eventually reaches a steady state. Data collection and analysis can be quantitative, qualitative, or mixed to collect causal knowledge, visualize agreement and difference, and increase awareness of mental models (see Jetter and Kok 2014 for guidance)	Conduct a survey to measure specialization and follow up with cognitive mapping to understand how recreational fishers at different specialization levels conceptualize the fishery (Gray et al. 2015) Capture recreational fishers' mental models of a fishery; use a qualitative approach to describe and define the components of the system, followed by a quantitative analysis of the relationships between the components (Aminpour et al. 2020)
Discourse analysis	A family of approaches that explore the meaning of language or text and how they are produced, consumed, and disseminated. Assumes that meaning is subjective, contested, and subject to change, and that context matters. Data analysis is mainly interpretive and qualitative, but quantitative frequency analysis methods are used to represent different forms of speech within or identify dominant discourses across texts (see Van Dijk 2011 and Boucquey 2017 for guidance)	Understand conflict among fishery user groups by synthesizing discourse analysis of social media comments and qualitative data generated from face-to-face interviews (Arlinghaus et al. 2022a, b). Analyse the causes and course of conflicts over salmon management by contrasting fishers' perceptions of fish stocks with historical catch numbers (Harrison et al. 2019b).
Social network analysis	Actors (e.g., fishers) are recorded and represented as nodes in a network, and their interactions as links between them. Possibility to identify particularly important actors, clusters of people who are connected, and characteristics of the entire community. Recent expansion to social-ecological networks that deal, for example, with the fit between used resources and their ecological interactions with social relationships and management structures. Generally, an approach with focus on quantitative aspects; mixed methods can enable understanding of the motivations of actors and the meaning of interactions (see Wasserman and Faust 1994 for guidance)	Examine fisher groups' interactions with other stakeholders. Complement social network analysis (quantitative) with expert interviews (quantitative) to understand *why* interactions took place and how they changed over time (Harrison et al. 2019b). Compare "whole networks" reflecting different fishing communities, using multiple methods including survey, social network analysis, interviews, and participant observation (Gorris et al. 2019).

(continued)

Table 1 (continued)

Method	Explanation	Examples of integration in mixed methods
Observations & behavioural mapping	Directly studying behaviour as it occurs at a location of interest. *Naturalistic observation* is an exploratory qualitative method where researchers observe what is naturally happening at the site and record as much detail as possible, without preconceived hypotheses. *Systematic observation* is a quantitative method in which specific behaviours are pre-identified, and the researcher counts or times the occurrence of behaviour within an observation period. *Behavioural mapping* measures spatial aspects of behaviour; researchers may record GPS coordinates of fisher locations and classify different areas of the study site according to utility. Observations may be quantitative or qualitative and can be integrated in mixed methods studies through convergent designs that compare observed and reported behaviour, or through designs where a qualitative phase of inquiry is used to investigate the meaning or rationale of observed behaviours (see Musante and DeWalt 2010 and Ciesielska et al. 2018)	Compare observed engagement of recreational fishers in illegal fishing activity with compliance behaviour reported in a survey (Bova et al. 2018). Observe recreational fisher catch-and-release behaviour to evaluate the efficacy of presentations designed to teach best handling practices for catch and release (Mannheim et al. 2018).
Participatory methods	Characterized by the involvement of "participants" in the research design, data collection, and analysis phases of research. Participatory research may involve the use of dedicated participatory methods, such as mapping fishing spots with a group of recreational fishers or may aim to solve an identified problem or challenge and be seen as "participatory action research." Participatory research could involve participatory video, where participants make their own videos as part of the research process, or the use of GIS to map participants' experiences. Participatory Mixed Methods Research (PMMR) reflects a specific approach that aims to bring Participatory Action Research (PAR) into the portfolio of	Synthesise and compare quantitative (PPGIS) and qualitative (interviews) measures of place-based marine and coastal values across multiple stakeholders, including recreational fishers (Brown et al. 2017). Identify the spatial values assigned by recreational fishers and SCUBA divers in a marine ecosystem and identify the reasons for these values using interviews (Seijo Núñez et al. 2021)

(continued)

Table 1 (continued)

Method	Explanation	Examples of integration in mixed methods
	approaches used within mixed methods research (see Burns et al. 2021 and Olson and Jason 2015 for guidance)	
Focus groups	A guided discussion with a small group of 5–15 people who share something in common. This shared identity, such as gender, age, residence, or occupation, makes for a comfortable environment in which participants feel able to share thoughts and experiences. Focus groups are facilitated by a moderator who aims to keep the discussion on track, encourage contributions from all participants, and respect confidentiality. Focus groups are often used to collect qualitative data, which may provide deeper insights into previous quantitative results, or inform the design of an upcoming quantitative phase of work (see Nyumba et al. 2018 for guidance)	After using surveys to understand fish consumption patterns among recreational fishers, use focus groups to explain survey results and gain feedback on risk communication messages (Lauber et al. 2017). Use focus groups prior to a stated choice experiment to select relevant attributes and design attributes levels that reflect a realistic range of characteristics of a fishing trip (Lew and Larson 2012). Validate classification of fishers identified through survey data and gain more information on attitudes towards recreational fisheries management within each group (Magee et al. 2018).
Interviews	A conversation with a single study participant guided by open-ended questions to understand the perspectives and experiences of the individual. Approaches range from informal, where no questions are decided in advance and the conversation is allowed to unfold naturally, to highly structured and standardized, where every question and statement posed by the interviewer is predetermined. Between those two extremes, researchers may use an interview guide to select and list the topics in advance, while allowing the wording and order of questions to flow with the conversation. Within mixed methods, interviews can be used to help understand and explain quantitative results, or to provide context needed to develop a quantitative phase of a study, such as a survey (see Young et al. 2018 and Dunn 2000 for guidance)	Conduct interviews after behavioural observations of recreational fishing waste disposal to understand the rationale behind observed behaviours (García et al. 2021). Understand recreational fisher perceptions of offshore wind turbines; use interview results to develop survey questions that will be most informative for the study context (Smythe et al. 2021).
Questionnaires	Sets of questions are asked of a representative population to quantify beliefs, behaviours, and characteristics	Use surveys to measure awareness of invasive species and familiarity with outreach campaigns; follow up with

(continued)

Table 1 (continued)

Method	Explanation	Examples of integration in mixed methods
	of the population of interest. Questionnaires are most commonly quantitative-focused and analysed statistically. Mixed methods researchers may include both quantitative and qualitative questions or may conduct qualitative work before or after survey administration (see Vaske et al. 2026a, b for guidance on questionnaire design and conducting surveys)	focus groups to gain insights on how campaigns directed at recreational fishers and boaters can be improved (Seekamp et al. 2016). Include open- and closed-ended questions in a survey of recreational fisheries experts to identify knowledge gaps and management priorities (Bower et al. 2020).
Experiments	Manipulating one or more independent variables to test hypotheses about the influence of independent variables on a dependent variable. Fisheries researchers may seek to understand the effect of varied environmental characteristics or diverse types of information on fisher behaviour. In a mixed methods study, experimental designs may be coupled with qualitative techniques such as interviews to provide in-depth detail on why a result may have been observed. Experiments may also be preceded by qualitative work that informs experimental design (see van Putten et al. 2026 for guidance on conducting experiments)	Conduct a stated choice experiment to understand fishing site preferences and complement with interviews to provide deeper insight on motivations (Golden et al. 2019). Test the effects of narrative messages regarding healthy fish consumption among recreational fishers, using messages informed by focus groups (Niederdeppe et al. 2019).
Content analysis	Analysis of text, images, or audio. Typically, an analysis of secondary data (e.g., newspapers, books, media) can provide quantitative (e.g., number of times a keyword appears) and qualitative (e.g., key themes present in the data) data. Mixed methods work may include both quantitative and qualitative content analysis of the same datasets. Additionally, content analysis can be integrated with other methods to inform or explain other stages of research (see Kochalski et al. 2026 for guidance on content analysis)	Use newspaper articles to determine historic trends in catch rate (quantitative) and the development (qualitative) of a recreational fishery (Chong-Montenegro et al. 2022). Analyse magazine articles to characterize changes in governance among recreational fishing organizations, blending qualitative (theme identification) and qualitative (theme frequency across time) methods (Daedlow et al. 2011)

recreational fisher catch and release behaviours by both directly observing the behaviours (quantitative), and by interviewing the fishers to ask them to describe their behaviours (qualitative). Likewise, a convergent parallel design may consist of a questionnaire in which fishers are asked to report their level of agreement or disagreement about the risks of invasive species (quantitative) and then to describe in their own words how they believe invasive species might impact their fishing experiences (qualitative). Following data collection, the two datasets are synthesized, and the results are compared.

Convergent parallel designs are best suited when the goal is to compare several types of data to look for similarities and differences, a process known as triangulation. For instance, in one study, researchers conducted behavioural observations of recreational fisher compliance with fishing regulations and followed up with surveys where fishers were asked to report whether they engaged in any illegal fishing activity (Bova et al. 2018). The authors then compared the observational and survey data to understand how willingly fishers might admit to performing undesirable behaviours. Thus, convergent parallel designs are helpful in identifying discrepancies in the data. Ideally, the same participants are included in both the quantitative and qualitative aspects of the study to increase the similarity of the two samples and ease comparisons.

One key consideration in selecting a convergent parallel design is time limitations and capacity on the part of the research team. If researchers can manage two data collection processes at the same time, this design is particularly efficient. However, if there is a small team or a single researcher, choosing a sequential design may be more manageable to avoid multiple data collection processes occurring simultaneously. Convergent designs may be considered an easier entry point to mixed methods than sequential designs because the data collection procedures and study design are not dependent on analysis of the first research phase; rather, all methods are predetermined and occur in parallel. In other words, the qualitative and quantitative datasets each have the same purpose. This design may thus be a helpful way for researchers looking to triangulate the results of quantitative work by incorporating qualitative work that poses the same research questions.

3.2 Exploratory Design

The exploratory sequential mixed methods design includes two phases of research: a qualitative phase occurs first and is followed by a quantitative phase. The most common application of this design involves using interviews or focus groups in an exploratory phase to help develop questionnaire items and content. After analysis of this first qualitative phase, a survey instrument is designed and then used for a second quantitative phase of data collection. Thus, the datasets are not compared or simultaneously analysed; rather, qualitative data are analysed first and inform the data-collection procedures for the quantitative data.

Preceding quantitative research with qualitative methods can uncover valuable insights that help to focus the direction of the study. Stakeholder analysis is also a useful precursor to research given its ability to identify relevant interest groups and partnerships that would increase the uptake of findings after a project (Pomeroy et al. 2016). For example, one study used mixed methods to understand how offshore wind farms affected recreational fishers using those locations (Smythe et al. 2021; Smith et al. 2018). Given previous research and discussion among managers about the potential conflict between wind turbines and fishers, the study initially sought to understand how to mitigate such conflicts. Before conducting a survey, researchers interviewed recreational fishers to gain a preliminary understanding of their perceptions of the wind turbines. Surprisingly, they found that recreational fishers enjoyed wind turbines, and felt they provided superior fish habitat, and indicated where to find the fish. The process of identifying relevant stakeholders and conducting interviews enabled the research team to develop a survey that accounted for the known viewpoints of anglers at this site. Based on the interviews, the researchers were able to design questions regarding the benefits provided by the turbines, rather than focusing only on potential negatives, which may have occurred in the absence of preliminary interview data.

There are several important considerations in using an exploratory sequential research design. This design is particularly useful when researchers are studying a topic or site at which there is minimal past research or information. Using interviews or focus groups to inform a survey may even be necessary if there is limited previous research on the concepts of interest. Research should be mindful; however, that the use of qualitative data to inform survey design does not preclude the need for rigorous assessment of measurement validity (Chapter "Measurement Instruments"), nor are qualitative methods meant to reflect an entire population. That is, qualitative work can be used as a basis for developing survey items, but these items still must be empirically evaluated prior to data analysis. Measurement analyses help to validate that the scale is measuring what the researcher intended; a guide to conducting such analyses can be found in Chapter "Measurement Instruments" (Kyle et al. 2026). Finally, participants who were involved in the qualitative phase should typically not be invited to take the survey during the quantitative phase; rather, a random sample of the target population should be identified and recruited for this phase of inquiry (Creswell et al. 2003).

3.3 Explanatory Design

The counterpart to exploratory designs is explanatory sequential mixed methods designs, whereby the collection and analysis of quantitative data are followed by a qualitative phase. A typical example of this is a quantitative, large-scale survey followed by in-depth interviews or focus groups with a subset of the original respondents. Within an explanatory design, the focus can be on either the quantitative or qualitative data, or they can be balanced; the research focus is not dictated by the design.

If an explanatory design is focused primarily on quantitative data, the qualitative work explains the causes or effects of research results. For example, Young et al. (2016) discovered that many people in a recreational fishing community reported having recently caught the biggest fish of their lives. Through open-ended questions, the researchers were able to understand why that finding emerged and conclude that the fishers were catching species that were remarkably similar in size. Therefore, it would not have been possible for the survey respondents to distinguish among the different-sized individuals across years. If the focus of an explanatory design is on the qualitative data, the quantitative part serves to pre-structure the data rather than closing knowledge gaps and uncovering new insights. For example, Magee et al. (2018) identified several types of recreational fishers using quantitative methods, but then used focus groups to explore the motivations and attitudes of the distinct groups, thereby uncovering patterns and formulating new hypotheses. The quantitative results were of interest themselves, but also enabled the group-specific qualitative analysis.

The explanatory design is ideal for allowing researchers to make generalizable statements but also provide deeper insights into such statements through rich, in-depth analysis. In contrast to the exploratory design, the participants in the quantitative and qualitative parts should be the same. The analytical challenge often lies in the smaller number of participants in qualitative compared to quantitative work. The solution is ensuring that a representative selection of participants is interviewed for the qualitative part, e.g., by quota sampling. For example, Magee et al. (2018) made certain that all fisher types identified in the quantitative analyses were subsequently represented in sufficient numbers in the qualitative focus group interviews. In this way, the qualitative findings providing deeper information from a selection of participants can be easily combined with the quantitative results when interpreting the data.

3.4 Embedded Designs

In embedded designs, quantitative or qualitative data is embedded within a larger design. Embedded designs may be considered variations of the convergent parallel, explanatory, or exploratory designs. The two types of data may be collected in the same phase of research, as in the convergent parallel design, or one may precede the other, as in the explanatory or exploratory designs. However, several characteristics distinguish embedded designs. Within embedded designs, either quantitative or qualitative data is prioritized, and fewer resources are dedicated to collecting a smaller amount of data for the other type. Further, the two data sets are not necessarily designed to converge but may be implemented to address separate research questions. The embedded data may be used to provide insights on results from the primary data, but they play a much smaller role than in the explanatory design.

One of the most common frameworks for an embedded design involves including some qualitative data collection during an experiment that is primarily collecting quantitative data and testing hypotheses. For example, Smyth et al. (2009) conducted a stated choice experiment to understand public preferences for lake management.

This experiment was quantitative and provided evidence that the safety of fish consumption was the most important factor in determining one's choice of fishing location. An open-ended (qualitative) question that was embedded in the survey after the experiment provided insights that aided in the researcher interpretation of the findings. Specifically, non-fisher participants in the experiment shared that they felt the safety of fish consumption was an indicator of the health of the lake, and thus an important factor in their decisions, though they were not fishing or consuming lake fish. In contrast to an explanatory design, the qualitative data collection was minimal (i.e., a single question, rather than a phase of interviews or focus groups) and was not developed based on previously analysed results but occurred within the same round of data collection. More robust qualitative data collection could provide the researchers with a greater understanding of fisher perceptions of the ecosystem, but given limited funding and time, an embedded design may still provide key insights.

There are several considerations in selecting an embedded design. Embedded designs allow for the collection of both quantitative and qualitative data in a less burdensome format, given that one type of data is prioritized. Thus, with limited time and resources, researchers may be able to add some qualitative data to a study. Researchers still must carefully consider the research objectives and purpose of adding qualitative data and developing an embedded design. The research questions should guide whether the quantitative data collection should be embedded before, during, or after the quantitative or experimental phase. It is less common to embed quantitative data collection within a predominantly qualitative research study. The lower sample sizes required for robust qualitative data collection may preclude statistical analysis from quantitative data derived from the same sample. Thus, researchers should carefully consider the research objectives and intended analysis when selecting the research design.

3.5 Three-Phase Design

Thus far, we have presented designs in which a quantitative and a qualitative phase are combined sequentially or simultaneously. Study designs may also include multiple quantitative and qualitative phases. Such multiphase designs belong to the category of complex designs and, in the case of three-step designs, can contain either a quantitative phase embedded in two qualitative phases, or a qualitative phase flanked by two quantitative phases.

A typical example of a quantitative => qualitative => quantitative three-phase design is when a quantitative social or biological survey can be followed by focus group discussions of the results, and based on that, a second, improved biological or social survey can be designed. The counterpart, which proceeds according to the qualitative => quantitative => qualitative model, can be chosen, for example, if a qualitative preliminary study is necessary before a quantitative survey in a relatively new research area, but the quantitative survey is also to be deepened later using qualitative methods. Both forms are common in research, though often not explicitly called a three-phase design. Compared to other forms of design, three-phase research designs have particular challenges in terms of time, cost, planning, and integration of the results, especially when dealing with large or complex data sets or projects where each individual research phase could already represent its project.

An example of a three-phase research design in recreational fisheries is a study by Cinner et al. (2016) that compared successful examples of coral reef management. The authors performed a quantitative analysis of 2514 coral reefs in 46 countries, linking fish biomass, an indicator of resource availability and ecosystem functioning, to 18 environmental and social variables. The authors identified reefs whose fish biomass was comparatively much higher than would be expected based on the environmental and social context. The reefs that defied expectations (positively or negatively) the most are what the authors called "bright spots" and "dark spots" (Cinner et al. 2016). In the second phase of the research, the researchers used qualitative methods to identify 10 factors that may explain why these places stood out positively or negatively from the global dataset. In the third phase, the researchers surveyed, again quantitatively, experts for the 15 bright and 35 dark spots identified in the first step (as well as for 14 average locations) to what extent the 10 characteristics identified in the second phase applied to their reef. Through the complex research design, they were able to identify key contextual factors and management interventions for improved management of coral reefs. Interestingly, the researchers themselves describe the second and third phases of the project as a "hypothesis-generating exploration" (Cinner et al. 2016). Although this case example did not exclusively involve recreational anglers, the authors demonstrated the benefits of a three-phase model while also suggesting that such a complex research approach may not be necessary.

4 Integrating Quantitative and Qualitative Methods and Data

In this section, we present three example studies that have engaged recreational fishers in human dimensions research focused on understanding behaviour and solving social–ecological problems. We highlight a range of geographic locales, including the Great Lakes Region, salmon fisheries in Europe, and the Great Barrier Reef, that embodied distinguishable challenges and called for different mixed methods designs, as well as select techniques (Table 1). That is, each example study was guided by a pragmatic understanding of knowledge production and the place-based needs for addressing resource management concerns.

4.1 Case Study #1: Mixed Methods Research to Address Aquatic Invasive Species in the Great Lakes Region

Aquatic invasive species are a key problem threatening freshwater fisheries, given the risk that they may substantially alter environmental conditions and impede recreational activities. Recreational fishers can help prevent the spread of invasive species by cleaning fishing equipment and boats before moving from one waterbody to another. Fisheries managers in the Great Lakes region wanted to know why fishers do or do not take these actions, and how outreach messages may be modified to best suit this audience. To address this issue, researchers adopted an interdisciplinary approach that integrated the fields of psychology and communication, with two research objectives: (i) determine the factors contributing to AIS-prevention behaviours; and (ii) evaluate the efficacy of outreach messages (Golebie et al. 2021a; Golebie and van Riper 2023).

To address the first objective, researchers chose an exploratory mixed methods design. Research began with a qualitative data collection phase in which focus groups of recreational fishers were conducted (Golebie et al. 2021a). Each focus group consisted of 2–5 recreational fishers. Researchers used a semi-structured guide to facilitate the discussion. During the focus groups, the research team learned about fishers' beliefs about aquatic invasive species. The researchers thematically analysed these results to understand how fishers were conceptualizing the benefits of acting and the barriers to doing so. These concepts were the foundation of questionnaire items for the second, quantitative, phase of the study. Drawing on established scales, researchers measured the perceived risks of aquatic invasive species (Golebie et al. 2021b), and perceived ability to act (Howell et al. 2015). These scales were complemented with what was learned about benefits and barriers during the focus group. Mixed methods were useful because it enabled researchers to gain a better understanding of the study population prior to disseminating a survey to a large group of fishers. Specifically, results showed that fishers perceived non-fishing boaters to be more responsible for problems associated with invasive species, leading researchers to segment into two groups reflecting these two audiences (Golebie et al. 2023).

To address the second objective, researchers used a convergent parallel design with an embedded experiment (Golebie and van Riper 2023). During the survey, fishers were asked to review one of three experimental outreach messages, and researchers measured their evaluations of the message and corresponding beliefs to evaluate the differences among the three messages. Both quantitative and qualitative responses were solicited regarding the perceptions of the outreach messages. The quantitative results indicated that the messages were perceived to be effective; however, the qualitative results revealed a wealth of information that fishers perceived to be missing from the messages. Fishers indicated that they needed more information about the species themselves, including a list of invasive species in Illinois, a map of where they are located, and a graphical guide to aid in species identification. Thus, mixed methods were useful because they highlighted a gap in current outreach initiatives that would have gone unnoticed with a sole reliance on quantitative message evaluations.

This study resulted in several applied outcomes for the management of fisheries in the Great Lakes Region. Having been guided by the qualitative phase of research to explore differences between boaters and fishers, the researchers learned that fishers did not feel cleaning fishing equipment made much of a difference in terms of AIS prevention, indicating that outreach messages should provide information on the importance of cleaning fishing equipment and the fact that every fisher can make a difference (Golebie et al. 2023). In contrast, boaters believed that cleaning their boat was important, but did not have high confidence in their ability to do so correctly (Golebie et al. 2023). Thus, outreach messages for boaters should focus on providing clear, accessible, information on the steps of remove-drain-dry and bolster self-efficacy. Results also demonstrated that the messages currently in use by Sea Grant were well-received but could benefit from additional information to help fishers identify invasive species and gain knowledge about which species were invasive in Illinois (Golebie and van Riper 2023).

4.2 Case Study #2: Participatory Mapping of Recreational Use in Australia

Recreation and resource management agencies are challenged to balance multiple and sometimes competing interests of user groups involved in consumptive versus non-consumptive activities. Building place-based knowledge of the values associated with places that support this range of activities is particularly important to (i) consider the potential for impacts incurred from overuse of protected land and seascapes, (ii) minimize social conflicts between fishers and other users, and (iii) support environmental planning decisions that are often made at spatially explicit scales. To address each of these priorities, the present case study examined the relative importance and spatial locations of "social values" (i.e., perceived benefits of places including material and nonmaterial qualities) that recreationists assigned to Hinchinbrook Island National Park within Australia's Great Barrier Reef Marine Park (Van Riper et al. 2012). This setting accommodated a range of recreational activities, including fishing in one of Queensland's most popular fishing destinations called the Hinchinbrook Channel that includes barramundi (*Lates calcarifer*) and mangrove jack (*Lutjanus argentimaculatus*). Hiking also occurs on the island, particularly along the long-distance (32-km) Thorsborne Trail.

In 2012, researchers conducted participatory mapping exercises with residents and tourists who participated in on-site and mailback surveys about their experience on Hinchinbrook Island ($n = 209$, response rate = 52%). These methods were employed to identify key recreational fishing destinations where overuse may become an issue in the future and to study preferences for social values assigned to places across user groups engaged in consumptive (e.g., fishing) and non-consumptive (e.g., hiking) activities. The spatially explicit data generated from survey respondents as part of their participatory involvement in this study were analysed in relation to GIS layers (e.g., elevation, slope, distance to key features) to understand how biophysical conditions on the island affected recreationists' expressed values of places (van Riper et al. 2012).

Results showed that a range of material and non-material social values were associated with places across the land and seascapes of Hinchinbrook Island, particularly aesthetics, recreation, and biological diversity. Fishers indicated that a broader array of places was aesthetically pleasing, associated with biological diversity, and useful for supporting their favourite recreational activities. Discernible trends emerged when comparing recreationists' social values of places with average levels of distance to waterbodies and the coastline, distance to the tail system, and the average slope of the landscape. These social-ecological patterns were in contrast to the results generated through engagement with recreationists who engaged in non-consumptive activities.

Findings from this case study were shared to support the decisions being made by a number of resource and recreation management agencies, such as the Queensland Parks and Wildlife Service, the Great Barrier Reef Marine Park Authority, and Fisheries Queensland (van Riper et al. 2016). Both hotspots (i.e., places of value abundance) and coldspots (i.e., undervalued places that are not recognized by users but are important to management agencies)—derived from participatory mapping processes—were identified as outputs from this research process. The intensity and types of social values associated with places can signal which areas are most likely to support high-quality visitor experiences, be associated with human use, or cause either social conflicts or environmental impacts. Alternatively, unrecognized resources like mangrove estuaries and sea grass beds in the context of this research created opportunities for agencies to increase awareness of unknown but important ecological features. In addition to generating spatially explicit knowledge to direct managerial attention to particular locales, the social-ecological framework adopted in this research proved to be a useful basis for discussing not only what recreationists valued but also the changes in landscape metrics that elicited such responses. Considering these patterns between consumptive and non-consumptive users was also a key benefit from the results generated as part of this case study.

4.3 Case Study #3: Mixed Methods Research to Understand Salmon Conservation in Europe

The protection of endangered fish species and the preservation or restoration of their stocks is a key concern of recreational fisheries management. In Europe, migratory fish, which spend some life stages in freshwater and others in the sea, are among the most threatened fish species. These include living fossils such as sturgeons and other historically important species such as European eel (*Anguilla anguilla*) and Atlantic salmon (*Salmo salar*). In the period 2013–2018, two transnational European projects pursued an interdisciplinary approach to the conservation of migratory fish species by not only bringing together natural and social sciences, but also different social science disciplines. By combining quantitative and qualitative methods, the researchers were able to simultaneously consider the social and political environment in which management of inland fisheries is embedded, and the motivations and actions of directly involved stakeholders such as fisheries managers, anglers, and hatchery operators, as both levels affect fish stocks and their management.

In the first project, expert interviews and qualitative content analysis (Table 1) of secondary data revealed the factors that prompted rapid policy change for native and non-native salmonids in Canada, Norway, and parts of Sweden, and small, incremental changes in Germany and France (Aas et al. 2018). For example, decision-making processes in Canada and the two Scandinavian countries are hierarchical, but the levels of governance are well linked (Aas et al. 2018). In contrast, in Germany and France, many decisions are made at the local level. Against this background, a second, quantitative study was conducted to better understand the motivation of local angling club managers in Germany and France (Riepe et al. 2017; Fujitani et al. 2020).

In the second project, a comparative, quantitative 4-country survey of the public on freshwater and migratory fish (Kochalski et al. 2019, 2022; Riepe et al. 2019, 2021) ran at the same time as qualitative interviews and participant observation of stakeholders engaged with salmon hatcheries (Harrison et al. 2018a, b, 2019a, b). Both approaches showed the importance of deeper values and worldview for salmon conservation (Harrison et al. 2019a; Riepe et al. 2021) and that restoring or maintaining declining salmon populations benefits people (Harrison et al. 2018a; Riepe et al. 2019; Kochalski et al. 2022). Although the quantitative studies revealed differences and similarities between the different European countries (Kochalski et al. 2019, 2022, Riepe et al. 2019, 2021), the researchers mostly found similarities between countries at the stakeholder level (Harrison et al. 2018a). The qualitative approach was also well suited to analyse conflicts and developments over time (Harrison et al. 2019b), similar to the governance analysis (Aas et al. 2018).

During both projects, the qualitative work informed the quantitative work (exploratory design), and vice versa (explanatory design). For example, recreational fishers gained psychological, social, and conservation benefits from their work with Atlantic salmon (Harrison et al. 2018a), which served as an inspiration for which ecosystem services were included in a second round of the quantitative surveys. As a result, a high appreciation of the public for aesthetic-educational ecosystem services was discovered (Kochalski et al. 2022) that would not have been queried to that extent based solely on the literature on fish ecosystem services (Holmlund and Hammer 1999). Mostly, however, the projects followed a parallel design in which the integration of qualitative and quantitative research took place at the level of the overall interpretation of the results. Thanks to mixed methods, distinct levels could be linked together (Creswell and Plano Clark 2018), e.g., the stakeholder level and the level of the public. This is particularly useful when you view fisheries as complex socio-ecological systems (McGinnis and Ostrom 2014; Arlinghaus et al. 2017). In our example, the combination of qualitative and quantitative studies showed that engaged and motivated stakeholders are essential to generate interest and enthusiasm for freshwater and migratory fish among the public.

5 Challenges in Mixed Methods Research

Conducting mixed methods research is not without challenges. Although all research involves anticipating and mitigating potential difficulties, mixed methods research presents additional challenges associated with the very nature of research design and implementation (Watkins and Gioia 2015). Four areas of challenges are identified here: (i) the existence of contrasting, and potentially conflicting, assumptions and philosophies that underpin different methods; (ii) the range and depth of skills and experience needed within individuals and teams; (iii) the time and resources needed to support complex research involving multiple methods; and (iv) integrating data generated by different methods.

5.1 Underpinning Assumptions and Philosophies

Mixed methods research is often assumed to involve both quantitative and qualitative research methods that each carry philosophical assumptions about the nature of reality (i.e., ontology) and knowledge (i.e., epistemology) (Hunt et al. 2026; Pearce 2015). For more details on the philosophies of research, see Moon and Blackman (2014). Traditionally, philosophies underpinning quantitative and qualitative research have been seen to conflict with each other. For instance, quantitative work is often paired with a post-positivist epistemology and seeks to be generalizable, whereas qualitative work embraces subjectivity and is necessarily context-dependent. Additionally, there is a clear gap between work that assumes one reality exists and work that assumes mental or social constructions are the source of meaning and knowledge. Post-positivist researchers may see value in social science research for understanding how well people's beliefs align with reality, whereas interpretive researchers may view people's diverse beliefs as realities in and of themselves that are worthy of study. Thus, fisheries research that seeks to identify logical truths may not seem immediately compatible with research that seeks to understand multiple socially constructed realities.

Despite these differences, quantitative and qualitative research guiding fisheries management do not need to be in conflict, provided researchers remain sensitive to the norms of disciplinary research (Patterson and Williams 1998). Many researchers conducting mixed methods research claim to take a pragmatic approach to solving problems because they do what "works best" and is most practical (Shan 2022). Pearce (2015) called for mixed methods research to go beyond the dichotomy of quantitative and qualitative techniques and recognize that these methods are not intrinsically aligned with objective and subjective philosophies or specific assumptions. Similarly, Shan (2022) provided a critical review of how mixed methods research should move beyond pragmatism towards transformative approaches and dialectical pluralism, which suggests that there are multiple realities and therefore reality can be conceptualized in separate ways. Adopting such a pluralist approach may be difficult but can be achieved with reflection and communication about assumptions of knowledge and reality that underpin different research traditions.

5.2 *Skills and Experience*

It is challenging for fishery researchers to build the skills and experience needed to use multiple methods. In a review of mixed methods research, Zhou and Wu (2022) found that researchers often indicated a lack of confidence in how they adopted and implemented their research designs. Thus, honestly evaluating gaps in one's experience is an important starting point in mixed methods work (Watkins and Gioia 2015). These gaps can be addressed at the onset of a research project by engaging collaborators who have expertise in each method (Martin 2020). With a collaborative approach, the team can collectively possess the skills and experience needed without each researcher independently requiring in-depth experience with all methods used. Creswell and Plano Clark (2018) advise that researchers should have their own, first-hand experience in designing and undertaking both quantitative and qualitative research separately before embarking on mixed methods research. Though it may be unrealistic for all team members to have this skillset, expertise in each method represented by at least one team member is key. For students or young researchers who need to use new and multiple methods for a fisheries project, additional sources of support may include training courses or mentorship with researchers who have experience with unfamiliar methods. This is particularly relevant for fisheries research, where it is common that researchers change (sub-) disciplines over the course of their career to solve environmental and social issues they encounter more effectively.

5.3 *Time and Resources*

Mixed methods research often requires a greater investment of time and resources than quantitative or qualitative research can accomplish alone, due to the collection of multiple types of data and the need to integrate and analyse data from multiple phases of work (Creswell and Plano Clark 2018). The collection of data sequentially or in parallel also impacts the time and resources needed. When conducting research with a parallel design, there may be opportunities for synergies that save time and money, such as by collecting quantitative and qualitative data while in the field. In the case of sequential designs, synergies are more difficult to obtain, but when the first phase of the study mainly serves to prepare the second, the additional time invested by adding a preliminary phase will pay off many times over in an improved design of the main study. Hesse-Biber (2010) cautioned that there may be significant gaps between phases of mixed methods research when findings generated from one phase are needed to inform the design and implementation of a subsequent phase. Such gaps in research add to the time burden, which may be infeasible if the research is part of a funded project or graduate student research. Without gaps and with a tightly integrated design, problems and delays in one phase will directly affect subsequent phases. The need for additional training or to bring in additional team members with the skills and experience in certain methods could bring additional demands on time and budget. Thus, we recommend beginning a mixed methods project with the expectation that additional time will be needed. Building in additional time from the start may help to mitigate time pressure as a barrier to data integration.

5.4 Integrating Data

Bringing together data and developing synthetic findings from multiple methods is integral to mixed methods research, yet challenging in practice (Maxwell et al. 2015). In the human dimensions of fisheries, the challenge is compounded by the need to integrate not only among the social sciences but between the social and natural sciences. In-depth discussions on the front-end of a project are therefore crucial to clarify the roles of collaborators and integration of ideas through a conceptual model (Arlinghaus et al. 2014). In a review of the experience of carrying out mixed methods research, Zhou and Wu (2022) were challenged to integrate data generated by different methods. These include a lack of uniformity in the use of measurements, making connections between datasets, merging data, and dealing with conflicting or contrasting findings generated by different methods (Zhou and Wu 2022). Maxwell et al. (2015) offer several lessons for integration, observing that integration within mixed methods research concerns not only the data but also the design, methods, and reporting of findings. To help overcome these challenges and enhance collaboration, all researchers in a team should be involved with multiple phases of the research process, especially the initial conceptualization of the study (Maxwell et al. 2015). Familiarization, active dialogue, and reflexivity facilitate meaningful integration of disciplines and enable fruitful multi-, inter-, and transdisciplinary research outcomes (Hakkarainen et al. 2020).

6 Conclusion

Demand for mixed methods research in the human dimensions of fisheries is likely to grow in response to current challenges in recreational fisheries management. Grounded in pragmatism (Creswell and Creswell 2017), mixed methods work is particularly suited to address applied problems from several perspectives. For instance, managers may address the decline in recreational fishing through work that integrates spatial data on fishing participation with qualitative interviews about the causes and effects of individual fishing experiments. Others may make decisions about stocking by holding focus groups to understand fisher preferences and a survey to identify the proportion of common beliefs among the broader fishing population (Arlinghaus et al. 2022a, b). Mixed methods work that incorporates both biological and social science (Murphy et al. 2021) may be particularly valuable as recreational fisheries managers seek to augment biological data collection efforts with local knowledge shared by fishers. This chapter provides an array of ideas for fisheries researchers to consider how to engage multiple data sources to better understand recreational fishers.

Acknowledgments We thank two anonymous reviewers for their constructive feedback on earlier versions of this chapter.

References

Aas Ø, Cucherousset J, Fleming IA, Wolter C, Höjesjö J, Buoro M, Santoul F, Johnsson JI, Hindar K, Arlinghaus R (2018) Salmonid stocking in five North Atlantic jurisdictions: identifying drivers and barriers to policy change. Aquat Conserv 28(6):1451–1464

Aminpour P, Gray SA, Jetter AJ, Introne JE, Singer A, Arlinghaus R (2020) Wisdom of stakeholder crowds in complex social-ecological systems. Nat Sustain 3(3):191–199

Arlinghaus R, Hunt LM, Post JR, Allen MS (2014) Not fish not meat: some guidance on how to study fisheries from an interdisciplinary perspective. Future of fisheries: perspectives for emerging professionals. American Fisheries Society, Bethesda, pp 223–230

Arlinghaus R, Alós J, Beardmore B, Daedlow K, Dorow M, Fujitani M, Hühn D, Haider W, Hunt LM, Johnson BM, Johnston F, Klefoth T, Matsumura S, Monk C, Pagel T, Post JR, Rapp T, Riepe C, Ward H, Wolter C (2017) Understanding and managing freshwater recreational fisheries as complex adaptive social-ecological systems. Rev Fish Sci Aquac 25(1):1–41

Arlinghaus R, Riepe C, Theis S, Pagel T, Fujitani M (2022a) Dysfunctional information feedbacks cause the emergence of management panaceas in social-ecological systems: the case of fish stocking in inland recreational fisheries. J Outdoor Recreat Tour 38:100475

Arlinghaus R, Vogt A, Kömle D, Niessner D, Ehrlich E, Rittweg T, Droll J (2022b) Ursachenanalyse von Berufsfischer-Angler Konflikten am Beispiel der Nutzung von Hechten (Esox lucius) in den Boddengewässern Mecklenburg-Vorpommerns. Zeitschrift für Fischerei 2:27–27

Barclay K, Voyer M, Mazur N, Payne AM, Mauli S, Kinch J, Fabinyi M, Smith G (2017) The importance of qualitative social research for effective fisheries management. Fish Res 186:426–438

Bennett NJ, Roth R, Klain SC, Chan K, Christie P, Clark DA, Cullman G, Curran D, Durbin TJ, Epstein G, Greenberg A, Nelson MP, Sandlos J, Stedman R, Teel TL, Thomas R, Veríssimo D, Wyborn C (2017) Conservation social science: understanding and integrating human dimensions to improve conservation. Biol Conserv 205:93–108

Boucquey N (2017) 'That's my livelihood, it's your fun': the conflicting moral economies of commercial and recreational fishing. J Rural Stud 54:138–150

Bova CS, Aswani S, Farthing MW, Potts WM (2018) Limitations of the random response technique and a call to implement the ballot box method for estimating recreational angler compliance using surveys. Fish Res 208:34–41

Bower SD, Aas Ø, Arlinghaus R, Douglas Beard T, Cowx IG, Danylchuk AJ, Freire KMF, Potts WM, Sutton SG, Cooke SJ (2020) Knowledge gaps and management priorities for recreational fisheries in the developing world. Rev Fish Sci Aquac 28(4):518–535

Braun V, Clarke V (2013) Successful qualitative research: a practical guide for beginners. Sage, London

Brown G, Strickland-Munro J, Kobryn H, Moore SA (2017) Mixed methods participatory GIS: an evaluation of the validity of qualitative and quantitative mapping methods. Appl Geogr 79:153–166

Burkett E, Carter A (2020) It's not about the fish: women's experiences in a gendered recreation landscape. Leis Sci 44(7):1013–1030

Burns S, Howard J, Ospina SM (eds) (2021) The sage handbook of participatory research and inquiry. Sage, London

Chong-Montenegro C, Thurstan RH, Campbell AB, Cunningham ET, Pandolfi JM (2022) Historical reconstruction and social context of recreational fisheries: The Australian East Coast Barramundi. Fish Manag Ecol 29(1):44–56

Ciesielska M, Boström KW, Öhlander M (2018) Observation methods. In: Ciesielska M, Jemielniak D (eds) Qualitative methodologies in organization studies, vol 2. Palgrave Macmillan, Cham, pp 33–52

Cinner JE, Huchery C, MacNeil MA, Graham NA, McClanahan TR, Maina J, Maire E, Kittinger JN, Hicks CC, Mora C, Allison EH, D'Agata S, Hoey A, Feary DA, Crowder L, Williams ID, Kulbicki M, Vigliola L, Wantiez L, Edgar G, Stuart-Smith RD, Sandin SA, Green AL, Hardt MJ, Beger M, Friedlander A, Campbell SJ, Holmes KE, Wilson SK, Brokovich E, Brooks AJ, Cruz-Motta JJ, Booth DJ, Chabanet P, Gough C, Tupper M, Ferse SCA, Sumaila UR, Mouillot D (2016) Bright spots among the world's coral reefs. Nature 535(7612):416–419

Creswell JW, Creswell JD (2017) Research design: qualitative, quantitative, and mixed methods approaches. Sage, Los Angeles

Creswell JW, Plano Clark VL (2018) Designing and conducting mixed methods research. Sage, Los Angeles

Creswell JW, Plano Clark VL, Gutmann ML, Hanson WE (2003) Advanced mixed methods research designs. In: Tashakkori A, Teddle C (eds) Handbook of mixed methods in social and behavioral research. Sage, Thousand Oaks, pp 209–240

Daedlow K, Beckmann V, Arlinghaus R (2011) Assessing an adaptive cycle in a social system under external pressure to change: the importance of intergroup relations in recreational fisheries governance. Ecol Soc 16(2):3

Dewey J (1958) Experience and nature. Dover Publications, New York

Dunham JB, Angermeier PL, Crausbay SD, Cravens AE, Gosnell H, McEvoy J, Moritz MA, Raheem N, Sanford T (2018) Rivers are social-ecological systems: time to integrate human dimensions into riverscape ecology and management. Wires Water 5(4):e1291

Dunn K (2000) Interviewing. In: Hay I (ed) Qualitative research methods in human geography. Oxford University Press, Oxford, pp 50–82

Ehrlich E, Bath AJ, Niessner D, Koemle D, Cyrus EM, Arlinghaus R (2023) Co-designing management recommendations with stakeholders: a case study about northern pike (Esox lucius) in the southern Baltic Sea. Fish Res 263:106687

Fenichel EP, Abbott JK, Huang B (2013) Modelling angler behaviour as a part of the management system: synthesizing a multi-disciplinary literature. Fish Fish 14(2):137–157

Fujitani M, McFall A, Randler C, Arlinghaus R (2017) Participatory adaptive management leads to environmental learning outcomes extending beyond the sphere of science. Sci Adv 3(6): e1602516

Fujitani ML, Riepe C, Pagel T, Buoro M, Santoul F, Lassus R, Cucherousset J, Arlinghaus R (2020) Ecological and social constraints are key for voluntary investments into renewable natural resources. Global Environ Chang 63:102125

García GO, Valenti AG, Zumpano F, Hernandez MM, Castano MV, Friedman I, Cabral VN, Favero M, Seco Pon JP (2021) Conservation approach in a coastal reserve in Argentina to promote the responsible disposal of litter derived from recreational fisheries. Ocean Coast Manage 214:105899

Golden AS, Free CM, Jensen OP (2019) Angler preferences and satisfaction in a high-threshold bucket-list recreational fishery. Fish Res 220:105364

Golebie EJ, van Riper CJ (2023) Enhancing aquatic invasive species outreach through values-framed messages. Environ Commun 17(1):67–86

Golebie E, Joffe-Nelson N, Siever A, Hitzroth G, Huegelmann A, van Riper CJ (2021a) Addressing barriers to aquatic invasive species prevention behaviors among Illinois recreational water users. Report, Illinois Department of Natural Resources, Urbana-Champaign

Golebie E, van Riper CJ, Suski C, Stedman R (2021b) Reducing invasive species transport among recreational anglers: the importance of values and risk perceptions. N Am J Fish Manag 41(6): 1812–1825

Golebie EJ, van Riper CJ, Arlinghaus R, Gaddy M, Jang S, Kochalski S, Lu Y, Olden JD, Stedman R, Suski C (2022) Words matter: a systematic review of communication in non-native aquatic species literature. Neobiota 74:1–28

Golebie EJ, van Riper CJ, Hitzroth G, Huegelmann A, Joffe-Nelson N (2023) Barriers to participation in aquatic invasive species prevention among Illinois, USA recreational water users. Biol Invasions 25:2549–2565

Gorris P, Glaser M, Idrus R, Yusuf A (2019) The role of social structure for governing natural resources in decentralized political systems: insights from governing a fishery in Indonesia. Public Adm 97(3):654–670

Gray S, Hilsberg J, McFall A, Arlinghaus R (2015) The structure and function of angler mental models about fish population ecology: the influence of specialization and target species. J Outdoor Recreat Tour 12:1–13

Hakkarainen V, Anderson CB, Eriksson M, van Riper CJ, Horcea-Milcu A, Raymond CM (2020) Grounding IPBES experts' views on the multiple values of nature in epistemology, knowledge and collaborative science. Environ Sci Pol 105:11–18

Harrison HL, Kochalski S, Arlinghaus R, Aas Ø (2018a) "Nature's little helpers": a benefits approach to voluntary cultivation of hatchery fish to support wild Atlantic salmon (Salmo salar) populations in Norway, Wales, and Germany. Fish Res 204:348–360

Harrison HL, Rybråten S, Aas Ø (2018b) Hatching knowledge: a case study on the hybridization of local ecological knowledge and scientific knowledge in small-scale Atlantic salmon (Salmo salar) cultivation in Norway. Hum Ecol 46(4):449–459

Harrison HL, Hauer J, Nielsen JØ, Aas Ø (2019a) Disputing nature in the Anthropocene: technology as friend and foe in the struggle to conserve wild Atlantic salmon (Salmo salar). Ecol Soc 24(3):13

Harrison HL, Kochalski S, Arlinghaus R, Aas Ø (2019b) Do you care about the river?' a critical discourse analysis and lessons for management of social conflict over Atlantic salmon (Salmo salar) conservation in the case of voluntary stocking in Wales. People Nat 1(4):507–523

Hesse-Biber SN (2010) Mixed methods research: merging theory with practice. Guilford Press, New York

Holmlund CM, Hammer M (1999) Ecosystem services generated by fish populations. Ecol Econ 29(2):253–268

Howell AP, Shaw BR, Alvarez G (2015) Bait shop owners as opinion leaders: a test of the theory of planned behavior to predict pro-environmental outreach behaviors and intentions. Environ Behav 47(10):1107–1126

Hunt LM, Sutton SG, Arlinghaus R (2013) Illustrating the critical role of human dimensions research for understanding and managing recreational fisheries within a social-ecological system framework. Fish Manag Ecol 20(2–3):111–124

Hunt LM, Boucquey N, Beardmore B, Christensen J, Fulton DC, Mackay M, Melstrom RT, Post JR, Schroeder SA, Thurstan RH, van Putten EI, Arlinghaus R (2026) Introducing disciplinary perspectives and interdisciplinary possibilities for understanding recreational fishers and fisheries. In: Pope KL, Arlinghaus R, Hunt LM, Lynch AJ, van Poorten BT (eds) Understanding recreational fishers: disciplinary and interdisciplinary approaches for fisheries management. Springer Nature, Cham, p 57–94

Jefferson R, McKinley E, Griffin H, Nimmo A, Fletcher S (2021) Public perceptions of the ocean: lessons for marine conservation from a global research review. Front Mar Sci 8:1705

Jetter AJ, Kok K (2014) Fuzzy cognitive maps for futures studies – a methodological assessment of concepts and methods. Futures 61:45–57

Johnson RB, Onwuegbuzie AJ (2004) Mixed methods research: a research paradigm whose time has come. Educ Res 33(7):14–26

Kinnebrew E, Shoffner E, Farah-Pérez A, Mills-Novoa M, Siegel K (2021) Approaches to interdisciplinary mixed methods research in land-change science and environmental management. Conserv Biol 35(1):130–141

Kochalski S, Riepe C, Fujitani M, Aas Ø, Arlinghaus R (2019) Public perception of river fish biodiversity in four European countries. Conserv Biol 33(1):164–175

Kochalski S, Riepe C, Arlinghaus R (2022) Perceived socio-cultural ecosystem services provided by wild Atlantic Salmon populations in four European countries. Aquat Ecosyst Health 25(3): 12–21

Kochalski S, Barz F, Pita P, Harrison HL (2026) Quantitative and qualitative content analysis of text and images. In: Pope KL, Arlinghaus R, Hunt LM, Lynch AJ, van Poorten BT (eds) Understanding recreational fishers: disciplinary and interdisciplinary approaches for fisheries management. Springer Nature, Cham, p 507–532

Krogman RM, Taylor SM, Adams AJ, Boucek RE, Lombardo SM, Crandall CA, Steffen SF, Acosta A, Boné Morón E, Haukebo S, Hyder K, Radford ZM, Potts WM (2026) Role of recreational fisher information in fisheries management. In: Pope KL, Arlinghaus R, Hunt LM, Lynch AJ, van Poorten BT (eds) Understanding recreational fishers: disciplinary and interdisciplinary approaches for fisheries management. Springer Nature, Cham, p 21–54

Kyle G, Landon AC, Pilgreen D, Golebie E (2026) Measurement instruments. In: Pope KL, Arlinghaus R, Hunt LM, Lynch AJ, van Poorten BT (eds) Understanding recreational fishers: disciplinary and interdisciplinary approaches for fisheries management. Springer Nature, Cham, p 355–385

Lauber TB, Connelly NA, Niederdeppe J, Knuth BA (2017) Urban anglers in the Great Lakes region: fish consumption patterns, influences, and responses to advisory messages. Sci Total Environ 590:495–501

Lew DK, Larson DM (2012) Economic values for saltwater sport fishing in Alaska: a stated preference analysis. N Am J Fish Manag 32(4):745–759

Lynch AJ, Hunt LM, Beardmore B, van Poorten BT, Pope KL, Arlinghaus R (2026) Complexity and integration of recreational fisheries. In: Pope KL, Arlinghaus R, Hunt LM, Lynch AJ, van Poorten BT (eds) Understanding recreational fishers: disciplinary and interdisciplinary approaches for fisheries management. Springer Nature, Cham, p 3–19

Magee C, Voyer M, McIlgorm A, Li O (2018) Chasing the thrill or just passing the time? Trialing a new mixed methods approach to understanding heterogeneity amongst recreational fishers based on motivations. Fish Res 199:107–118

Mannheim SL, Childs AR, Butler EC, Winkler AC, Parkinson MC, Farthing MW, Zweig T, McCord M, Drobniewska N, Potts WM (2018) Working with, not against recreational anglers: evaluating a pro-environmental behavioural strategy for improving catch-and-release behaviour. Fish Res 206:44–56

Martin VY (2020) Four common problems in environmental social research undertaken by natural scientists. Bioscience 70(1):13–16

Mauser W, Klepper G, Rice M, Schmalzbauer BS, Hackmann H, Leemans R, Moore H (2023) Transdisciplinary global change research: the co-creation of knowledge for sustainability. Curr Opin In Env Sustain 5(3–4):420–431

Maxwell J, Chmiel M, Rogers SE (2015) Designing integration in multimethod and mixed methods research. In: Hesse-Biber SN, Burke Johnson R (eds) The oxford handbook of multimethod and mixed methods research inquiry. Oxford University Press, Oxford, pp 223–239

McGinnis MD, Ostrom E (2014) Social-ecological system framework: initial changes and continuing challenges. Ecol Soc 19:2

Moon K, Blackman D (2014) A guide to understanding social science research for natural scientists. Conserv Biol 28(5):1167–1177

Murphy R, Cunningham C, Harris BP, Brown C (2021) Qualitative and quantitative fisher perceptions to complement natural science data for managing fisheries. Fisheries 46(5):209–219

Musante K, DeWalt BR (2010) Participant observation: a guide for fieldworkers. AltaMira Press, Plymouth

Niederdeppe J, Connelly NA, Lauber TB, Knuth BA (2019) Effects of a personal narrative in messages designed to promote healthy fish consumption among women of childbearing age. Health Commun 34(8):825–837

Nyumba T, Wilson K, Derrick CJ, Mukherjee N (2018) The use of focus group discussion methodology: insights from two decades of application in conservation. Methods Ecol Evol 9(1):20–32

Olson BD, Jason LA (2015) Participatory mixed methods research. In: Hesse-Biber SN, Burke Johnson R (eds) The oxford handbook of multimethod and mixed methods research inquiry. Oxford University Press, Oxford, pp 393–405

Patterson ME, Williams DR (1998) Paradigms and problems: the practice of social science in natural resource management. Soc Nat Resour 11:279–295

Pearce LD (2015) Thinking outside the Q boxes: further motivating a mixed research perspective. In: Hesse-Biber SN, Burke Johnson R (eds) The oxford handbook of multimethod and mixed methods research inquiry. Oxford University Press, Oxford, pp 42–56

Peirce CS (1934) Lectures on pragmatism: three kinds of goodness. In: Hartshorne C, Weiss P, Burks A (eds) Collected papers of Charles Sanders Peirce, vol 5. Harvard University Press, Cambridge

Pomeroy R, Parks J, Courtney K, Mattich N (2016) Improving marine fisheries management in Southeast Asia: results of a regional fisheries stakeholder analysis. Mar Policy 65:20–29

Riepe C, Fujitani M, Cucherousset J, Pagel T, Buoro M, Santoul F, Lassus R, Arlinghaus R (2017) What determines the behavioral intention of local-level fisheries managers to alter fish stocking practices in freshwater recreational fisheries of two European countries? Fish Res 194:173–187

Riepe C, Meyerhoff J, Fujitani M, Aas Ø, Radinger J, Kochalski S, Arlinghaus R (2019) Managing river fish biodiversity generates substantial economic benefits in four European countries. Environ Manag 63(6):759–776

Riepe C, Liebe U, Fujitani M, Kochalski S, Aas Ø, Arlinghaus R (2021) Values, beliefs, norms, and conservation-oriented behaviors toward native fish biodiversity in rivers: evidence from four European countries. Soc Nat Resour 34(6):703–724

Seekamp E, McCreary A, Mayer J, Zack S, Charlebois P, Pasternak L (2016) Exploring the efficacy of an aquatic invasive species prevention campaign among water recreationists. Biol Invasions 18(6):1745–1758

Seijo Núñez C, Calado H, McClintock W, Gil AJF, Fonseca C (2021) Mapping recreational ecosystem services from stakeholders' perspective in the Azores. One Ecosyst 6:1–17

Shan Y (2022) Philosophical foundations of mixed methods research. Philos Compass 17:e12804

Smith H, Smythe T, Moore A, Bidwell D, McCann J (2018) The social dynamics of turbine tourism and recreation: Introducing a mixed-method approach to the study of the first US offshore wind farm. Energy Res Soc Sci 45:307–317

Smyth RL, Watzin MC, Manning RE (2009) Investigating public preferences for managing Lake Champlain using a choice experiment. J Environ Manag 90(1):615–623

Smythe T, Bidwell D, Tyler G (2021) Optimistic with reservations: the impacts of the United States' first offshore wind farm on the recreational fishing experience. Mar Policy 127:104440

Stock P, Burton RJ (2011) Defining terms for integrated (multi-inter-trans-disciplinary) sustainability research. Sustainability 3(8):1090–1113

Szostak R (2015) Interdisciplinary and transdisciplinary multimethod and mixed methods research. In: Hesse-Biber SN, Burke Johnson R (eds) The oxford handbook of multimethod and mixed methods research inquiry. Oxford University Press, Oxford, pp 128–143

Timans R, Wouters P, Heilbron J (2019) Mixed methods research: what it is and what it could be. Theor Soc 48(2):193–216

Tress G, Tress B, Fry G (2005) Clarifying integrative research concepts in landscape ecology. Landsc Ecol 20(4):479–493

van Dijk TA (ed) (2011) Discourse studies. A multidisciplinary introduction. Sage, London

van Putten EI, Trudeau A, Mackay M, Obregón C, Post JR, Richter A, Schuch E, Solomon CT (2026) Economic experiments in recreational fishing studies. In: Pope KL, Arlinghaus R, Hunt LM, Lynch AJ, van Poorten BT (eds) Understanding recreational fishers: disciplinary and interdisciplinary approaches for fisheries management. Springer Nature, Cham, p 411–442

Van Riper CJ, Kyle GT, Sutton SG, Barnes M, Sherrouse BC (2012) Mapping outdoor recreationists' perceived social values for ecosystem services at Hinchinbrook Island National Park, Australia. Appl Geogr 35:164–173

Van Riper CJ, Sutton S, Kyle GT, Stewart W, Tobin RC (2016) Bridging managers' place meanings and environmental governance of the Great Barrier Reef Marine Park. Soc Nat Resour 29(11): 1342–1358

Vaske JJ, Needham MD, Miller CA (2026a) Fundamentals of sampling and implementing human dimensions of fishing surveys. In: Pope KL, Arlinghaus R, Hunt LM, Lynch AJ, van Poorten BT (eds) Understanding recreational fishers: disciplinary and interdisciplinary approaches for fisheries management. Springer Nature, Cham, p 327–353

Vaske JJ, Needham MD, Miller CM, Sponarski C (2026b) Guidelines for writing and constructing recreational fisher surveys. In: Pope KL, Arlinghaus R, Hunt LM, Lynch AJ, van Poorten BT (eds) Understanding recreational fishers: disciplinary and interdisciplinary approaches for fisheries management. Springer Nature, Cham, p 287–325

Wasserman S, Faust K (1994) Social network analysis: methods and applications. Cambridge University Press, Cambridge

Watkins DC, Gioia D (2015) Mixed methods research. Oxford University Press, New York

Weber CT, Borit M, Aschan M (2019) An interdisciplinary insight into the human dimension in fisheries models. A systematic literature review in a European Union context. Front Mar Sci 6: 369

West S, Haider LJ, Stålhammar S, Woroniecki S (2020) A relational turn for sustainability science? Relational thinking, leverage points and transformations. Ecosystems 16(1):304–325

Young MA, Foale S, Bellwood DR (2016) Why do fishers fish? A cross-cultural examination of the motivations for fishing. Mar Policy 66:114–123

Young JC, Rose DC, Mumby HS, Benitez-Capistros F, Derrick CJ, Finch T, Garcia C, Home C, Marwaha E, Morgans C, Parkinson S, Shah J, Wilson KA, Mukherjee N (2018) A methodological guide to using and reporting on interviews in conservation science research. Methods Ecol Evol 9(1):10–19

Zhou Y, Wu ML (2022) Reported methodological challenges in empirical mixed methods articles: a review on JMMR and IJMRA. J Mix Method Res 16(1):47–63

Quantitative and Qualitative Content Analysis of Text and Images

Sophia Kochalski, Fanny Barz, Pablo Pita, and Hannah L. Harrison

Abstract Content analysis is a research technique that allows recreational-fisheries researchers to draw conclusions about the world from images and textual data. In this chapter, we present detailed guidance for content analysis, ranging from formulating the research question to visualizing the results, and including quantitative and qualitative approaches to coding, analysis, and interpretation. Drawing from a rich pool of over 20 studies from recreational fishing research, we find that the method has many applications in human dimensions and beyond. It is used to describe and explain opinions, arguments, and the behaviour of recreational fishers. Because content analysis can be applied to a variety of different data sources, the approach is particularly suitable for interdisciplinary research that bridges disciplinary boundaries and enables a holistic view of complex systems and behaviours. Popular applications in recreational fisheries include comparative and longitudinal studies, using diverse data sources such as transcripts of interviews, policy documents, and media excerpts. While highlighting the method's strengths, we also illuminate instances where it may not be the optimal choice, elucidate challenges when coding certain material, such as images, and give an overview of helpful programming languages and software applications. Future developments promise exciting opportunities, propelled by technological advancements and evolving research paradigms.

Keywords Data coding · Data collection · Discourse analysis · Document analysis · Research paradigm · Social media · Text mining

---

S. Kochalski (✉)
Center for Ocean and Society (CeOS), Kiel University, Kiel, Germany
e-mail: kochalski@ceos.uni-kiel.de

F. Barz
Thünen Institute of Baltic Sea Fisheries, Rostock, Germany

P. Pita
CICA - Centro Interdisciplinar de Química e Bioloxía, Departamento de Bioloxía, Universidade da Coruña, A Coruña, Spain

H. L. Harrison
Marine Affairs Program, Dalhousie University, Halifax, NS, Canada

K. L. Pope et al. (eds.), *Understanding Recreational Fishers*, Fish & Fisheries Series 45, https://doi.org/10.1007/978-3-031-99739-6_17

1 Introduction

> Watch your thoughts; they become words.
> Watch your words; they become actions. […]
> Lao-Tze

Sustainable recreational fishing requires a comprehensive understanding of the people involved as well as biological, ecological, and governance aspects (Arlinghaus et al. 2017; Hunt et al. 2013; Lynch et al. 2026). To achieve this, extracting information from policy documents (Pita et al. 2018), online videos (Fennell 2017; Sbragaglia et al. 2020; Lennox et al. 2022), or fishing magazines (Daedlow et al. 2011; Mordue 2013) using content analysis can be helpful. Content analysis is a structured approach to organize, analyse, and interpret large volumes of images and textual data, employing a systematic classification process of coding to reduce complexity and uncover meaningful themes and patterns (Hsieh and Shannon 2005; Patton 2002).

Content analysis is widely applied in recreational fisheries research to extract biological, economic, and social information. For example, Sutton et al. (2009) used content analysis to examine why Australians stop or resume recreational fishing, whereas Kadagi et al. (2020) explored the perceptions and conflicts among artisanal and recreational fishers in Kenya. Thurstan et al. (2018) employed content analysis to illuminate the impact of technological, social, and regulatory changes on catch rate trends in a recreational snapper fishery (*Chrysophrys auratus*). Table 1 lists more examples showcasing the various applications in recreational fisheries research, along with references to foundational texts (Berelson 1952; Holsti 1968; Krippendorff 1989; Weber 1990). This chapter also cites instructional materials focused on the practical application of content analysis (Bell 2001; Neuendorf 2002; Früh 2007; Insch et al. 1997) and resources specifically addressing the coding process (Graneheim and Lundman 2004; Hsieh and Shannon 2005; Gibbs 2007; Elo and Kyngäs 2008).

As a versatile and interdisciplinary tool, content analysis is particularly relevant in studying complex social–ecological systems like recreational fishing (Arlinghaus et al. 2017). Beyond human dimension studies (Boucquey 2017; Schakner et al. 2019), it finds applications in resource monitoring and data acquisition (Young et al. 2015; Pita et al. 2020; Chong-Montenegro et al. 2022) and fisheries management research (Castro et al. 2018; Harrison et al. 2019). The method enables researchers to describe various aspects of recreational fishing and related activities, such as fly tying, camping, and cooking (e.g., Fennell 2017). Moreover, content analysis allows for the summary and contrast of different viewpoints and arguments regarding management measures (e.g., Brummer et al. 2017), as well as the analysis of factors explaining people's behaviour, opinions, and the resulting dynamics at the social–ecological system level, including responses to political and institutional shifts over time (e.g., Daedlow et al. 2011, 2013).

This chapter provides a detailed guide for implementing content analysis in recreational fisheries research, covering the entire research process from formulating

Table 1 Twenty Examples for Content Analysis in the Recreational Fisheries Peer-reviewed Scientific Literature

Reference (Country)	Studied behaviour	Data sources	Approach to content analysis	Use of mixed methods	Purpose of the content analysis
Aas et al. (2002) (Central and Northern Europe)	Perspectives on catch-and-release fishing	Scientific articles and texts from (non-) governmental agencies and organizations	Qualitative; Weber (1990)	No	Exploratory study; identify underlying meaning attributed to catch-and-release fishing
Bertella (2016) (Norway)	Tourists' experience of fishing and factors that shape this experience	Promotional material for tourists and TripAdvisor reviews	Qualitative; Mayring (2000)	No	Exploratory case study
Boucquey (2017) (USA)	Fishers' values, worldviews, fishing experiences and perceptions of others	Semi-structured interviews with fishers and pictures they took from their activity	Qualitative and visual; No reference to a specific school	Yes	Show what lies behind competing claims to fishing spaces and species
Brummer et al. (2017) (Germany)	Stakeholders' opinions and concerns about aquatic macrophyte de-weeding	Press articles (newspapers, online publications)	Qualitative; Mayring (2015)	Yes	Enrichment of other data sources
Castro et al. (2018) (Portugal)	Stakeholders' arguments against regulating a recreational fishery for the purpose if nature conservation	National and regional newspapers	Quantitative; No reference to a specific school	Yes	Analyse media data and overview of public debate; systematic approach that reveals surprising results
Chong-Montenegro et al. (2022) (Australia)	How did fishers' catch rates for barramundi change over time and why?	Historical newspaper articles (1869–1952)	Qualitative and quantitative; No reference to a specific school	Yes	Access historical data; make use of valuable data sources; enrich, triangulate, and explain historic catch data
Comby et al. (2019) (France)	Mobilization of recreational fishers around the topics of pollution and	Newspaper	Qualitative and quantitative; Berelson (1952)	Yes	Efficient approach for historical analysis of discursive trajectories and

(continued)

Table 1 (continued)

Reference (Country)	Studied behaviour	Data sources	Approach to content analysis	Use of mixed methods	Purpose of the content analysis
	dam building in its historic context				political strategies.
Daedlow et al. (2011) (Germany)	Support for changing the governance system after a disturbance	Angling magazines	Quantitative; Früh (2007)	Yes	Access historical data; test predictions of model in a social system
Fennell (2017) (Mostly Canada and United States)	Fishing behaviour and parallel behaviours (fly-tying, tent camping, consumption of alcohol)	Videos of fly-fishing posted by recreational fishers on Vimeo	Quantitative; Weber (1990)	No	Assess specialists' social world; possibility to consider parallel behaviours.
Harrison et al. (2019) (United Kingdom)	Stakeholders' beliefs, values, and arguments for and against salmon stocking	Interview data, online print media, social media, and policy documents	Qualitative; Corbin and Strauss (1990), Strauss and Corbin (1990)	Yes	Concrete work with text; open coding to develop concepts and categories
Kadagi et al. (2020) (Kenya)	Fishers' motivations, perceptions and opinions, and the uses and value of billfish	Semi-structured interviews with fishers	Qualitative; Graneheim and Lundman (2004) (among others); Method not explicitly named	No	Exploratory use; unexpected results through deductive coding; foundation for new conceptual model
Mordue (2013) (United Kingdom)	Identity of fly-fishers created and represented in angling magazines	Several editions of one preeminent game-angling periodical	Quantitative; No reference to a specific school	Yes	Triangulation; analyse angling magazines ("important cultural brokers")
Pita et al. (2018) (European Union, Spain, Portugal, United Kingdom)	Occurrence of the Ecosystem Approach to Fisheries (EAF) in regulations on marine recreational fisheries	Legal articles	Quantitative; No reference to a specific school	No	Analysis of large multi-country dataset; topic modelling
Sbragaglia et al. (2020) (Italy)	Recreational fishers' opinions, feelings	YouTube Videos	Quantitative; Elo and Kyngäs	No	Complementary data on

(continued)

Table 1 (continued)

Reference (Country)	Studied behaviour	Data sources	Approach to content analysis	Use of mixed methods	Purpose of the content analysis
	and priorities, and their conversations with each other		(2008), Holsti (1968), Mayring (2004)		recreational fisheries; test hypotheses
Schakner et al. (2019) (USA)	Recreational fishers' and the media's attitudes and perceptions of California sea lions	Newspapers	Quantitative; Krippendorff (1989)	Yes	Analyse media data ("critical role in framing environmental issues")
Shie (2020) (Taiwan)	People's values and motivations to self-organize in indigenous mountain villages	Media reports	Qualitative; Mayring (1983, cited in Flick 2002): 190–194	Yes	Acquire background knowledge about case studies; identify key actors; develop chronology; combine different sources
Shiffman et al. (2017) (USA)	Recreational fishing practices and fishers' views on shark conservation	Angling online forums	Quantitative, qualitative, and visual; No reference to a specific school	Yes	Access rich data source as "increasing number of social interactions occur online"; exploratory; non-intrusive; describe trends; detect cases of illegal fishing
Sutton et al. (2009) (Australia)	Factors why non-active fishers stopped fishing and how to encourage them to start again	Open-ended survey questions	Quantitative; Weber (1990)	Yes	Explorative
Thurstan et al. (2018) (Australia)	Trends and transitions in a recreational fishery across 140 years	Archival sources and fisher interviews	Qualitative; Braun and Clarke (2006), Hsieh and Shannon (2005)	No	Bridge disciplines; historic analysis; lack of data in recreational fisheries

a research question to presenting results. The guide encompasses both qualitative and quantitative approaches, addressing various data sources and challenges related to data collection, coding, and analysis. Though emphasizing text data, we also offer examples of content analysis applied to images (Merriam-Webster n.d.). Furthermore, we present quality criteria, explore capabilities and limitations, and suggest potential future applications in recreational fisheries research.

2 Preparatory Phase

Take Home Messages 1

- Qualitative content analysis examines opinions, behaviours, or phenomena, whereas quantitative analysis assesses their prevalence.
- Researchers should choose the most suitable method for analysing texts and images based on the research question, data type, and case study.
- Content analysis is applicable to both naturally occurring data and data generated for research purposes, each with its strengths and weaknesses.

2.1 Deciding on a Content Analysis

A content analysis begins by identifying the research problem, leading to the formulation of research questions and testable hypotheses (Mayring 2000; Insch et al. 1997). However, content analysis can also address open-ended inquiries in surveys (e.g., Arlinghaus and Mehner 2003; Sutton et al. 2009) and facilitate exploratory studies (e.g., Aas et al. 2002; Bertella 2016). Whether the study involves investigating a new topic or testing predetermined hypotheses influences the choice of coding method—whether codes are defined before or during data analysis (Hsieh and Shannon 2005; Elo and Kyngäs 2008). The key is to ensure that the selected methods and research design align with the research question, study case, and available data (Strübing et al. 2018).

In recreational fisheries research, content analysis adeptly summarizes arguments and opinions from interviews, public debates, and online comments, while also elucidating behavioural patterns of recreational fishers (Table 1). It is one of several methods for analysing text and images (Flick et al. 2004; Van Leeuwen and Jewitt 2001), and different questions may call for alternative methods, possibly building on content analysis but delving deeper. For instance, content analysis is often applied to historical and longitudinal data (Young et al. 2015; Thurstan et al. 2018; Chong-Montenegro et al. 2022), but if different phases and the chronology of events are important, narrative analysis could be a better option (Griffin and May 2012). Similarly, content analysis can reveal shared meaning within closely-knit groups (Fennell 2017), but conversation analysis examining the interactions between people offers additional insights into how this meaning is created (Sidnell and Stivers 2012). Qualitative content analysis is particularly similar to some forms of discourse analysis and the terms are sometimes used interchangeably (Hardy et al. 2004). However, critical discourse analysis, for example, considers not only the textual content but also the production, change, and negotiation of meanings from a power perspective (Fairclough 2013). Content analysis mostly excels in addressing comparative research questions (Bell 2001).

2.2 *Selecting a Research Design*

Content analysis encompasses qualitative and quantitative strands (Bos and Tarnai 1999). Qualitative content analysis reveals beliefs, norms, attitudes, and practices, whereas quantitative content analysis determines their prevalence in the study population, using a text corpus (Hunston 2022) as the sample. Despite distinct assumptions and theoretical perspectives (Flick et al. 2004; Moon and Blackman 2014), these content analysis types can be used concurrently or sequentially in the same study. Recreational fisheries research provides several examples of this mixed-method approach (Shiffman et al. 2017; Comby et al. 2019; Chong-Montenegro et al. 2022), presenting associated opportunities and challenges (Creswell and Clark 2018; Golebie et al. 2026).

Identifying key concepts and themes is essential in content analysis. When emphasizing the frequency of words, concepts, and phenomena, we term it quantitative content analysis. Whether examining people's assessment of recreational fishers (Sbragaglia et al. 2020), recreational fishing legislation (Pita et al. 2018), fish distribution patterns (Monkman et al. 2018), or the perceived state of the environment (Pita et al. 2020), all can be subjects of quantitative content analysis. This approach is particularly useful for handling large datasets, where modern data mining techniques offer cost-effective analysis options (Lennox et al. 2022).

Qualitative content analysis serves as a gateway to understanding people's perspectives and complex social phenomena (Flick et al. 2004). In recreational fisheries, it offers nuanced insights into values, attitudes, and experiences (Bertella 2016; Boucquey 2017; Kadagi et al. 2020), as well as the reasoning behind specific policy choices (Daedlow et al. 2013). Approaches within qualitative content analysis range from focusing on communication content (e.g., Berelson 1952; Neuendorf 2002) to interpretive, hermeneutic methods that unveil underlying meanings (Bos and Tarnai 1999). The method extends beyond summarizing texts, requiring immersion in and deep engagement with the data (Patton 2002; Forman and Damschroder 2007).

Content analysis also complements other research methods in recreational fisheries studies (Table 1). It can serve as an exploratory tool before the main study (e.g., Castro et al. 2018) or deepen and clarify findings (e.g., Sutton et al. 2009). Another option is to use it for triangulation, supplementing and verifying other methods. For instance, Daedlow et al. (2011) analysed fishing magazines' texts to corroborate statements from fishing association members whose recollection of events may be incomplete. Note that though this chapter focuses on content analysis as a primary method, the same considerations and steps generally apply when integrating it with other methods.

2.3 Choosing the Data Source

Content analysis is applied to artifacts (Finnegan 2014) representing, reproducing, or recording communication, behaviour, and events, thereby making fleeting or one-time occurrences accessible to research. Artefacts can be broadly categorized as those produced specifically for research purposes (e.g., participant photos, workshop results) and naturally occurring data (e.g., policy documents, online reviews, online videos). Common data sources for content analysis are (Table 1):

- Scientific studies
- Recorded verbal expressions such as interviews and open-ended survey questions
- Policy documents and historic texts
- Mass media
- Magazines and brochures
- Historical photographs
- Comments posted on social media and in online forums

Text is the prevalent data form in content analysis for recreational fisheries research, with less attention given to image analysis. Historical photos have been used to estimate fish lengths (Bonney et al. 2021), and studies have analysed videos and pictures shared on social media platforms to document recreational fishers' practices (Shiffman et al. 2017), along with visuals in angling magazines for understanding recreational fishers' self-perception (Mordue 2013). There remain untapped sources of visual data with the potential to generate novel insights for fisheries research, such as maps, documentaries, and street art, and photographs taken in research projects as part of the photovoice method (Hergenrather et al. 2009).

Different data types come with strengths and weaknesses, as well as ethical challenges. Data generated for research allows explicit participant consent and the exploration of new, previously unaddressed topics. Naturally occurring data offers the possibility of presenting historical perspectives and is non-intrusive (Smith and Emden 2004), thereby reducing the burden for the study subjects. Ethical concerns related to naturally occurring data, such as protecting personal data and dealing with observed illegal behaviour, can be mitigated by anonymizing online comments (Sbragaglia et al. 2020), blurring images (Shiffman et al. 2017), and aggregating data to prevent individual identification. For more details on ethical challenges for both types of data, please see Zimmer (2020), Monkman et al. (2018), and Minin et al. (2021), all of which are also discussed in relation to big data and data mining elsewhere in this book (Venturelli et al. 2026).

Data sources for content analysis exist in both the analogue or digital domains, each with its merits and potential biases. In media, constraints include the political stance of sources (Shie 2020) and different coverage based on the scope of the news outlets, spanning local, regional, national, or international media (Castro et al. 2018). Online perspectives may not be representative of the general population as not all members of a group may have equal access to or use of the "new media". For example, avid recreational fishers tend to be more active on social media than casual ones (Vitale et al. 2021; Venturelli et al. 2026). Concerns about their public self-image (Bertella 2016) may lead recreational fishers to report more successful catches than unsuccessful ones or express socially desired views and avoid discussions of controversial or illegal practices. Overall, it is good practice to consider the consequences and biases introduced by the choice of a data source and to present these considerations when reporting on the design and planning of a content analysis.

3 Empirical Phase

Take Home Messages 2

- The corpus is the collection of text or images to which codes can be applied.
- Coding is the process of systematically assigning meaning through labelling and organizing.
- The unit for coding can range from individual words to entire data records.
- Data records may require digitization and other preparations before the content analysis.

3.1 Collecting the Data

Data collection involves systematically gathering data from one or more sources aligned with the research objectives. This process may entail obtaining primary data, such as conducting surveys or verbal interviews with recreational fishers, or retrieving secondary data, like relevant records from a database using keywords. The outcome is a corpus (Hunston 2022), a collection of text or images to which codes can be applied.

The corpus is often simply the entirety of collected data. When data sources contain both text and images, they can be treated as separate or as a unified entity (Van Leeuwen and Jewitt 2001). The corpus is then composed of either text or images, or a combination of both. Examples of such image-text units include photographs accompanied by titles or other explanations, captions of articles, dialogues in films, or words and sentences integrated into works of art. The data used for the analysis may differ from the sampling unit. For example, Sbragaglia et al. (2020) sampled YouTube videos, though their data collection focused on the comments accompanying the videos. In this case, the corpus comprises the YouTube comments rather than the video content.

When creating the corpus, researchers need to consider the amount of data they can realistically work through. The analysis can be limited by refining the research question or by using selection criteria for quantitative or qualitative research. For instance, Fennell (2017) identified 780 relevant angler-generated videos but chose to analyse every third video based on a systematic random sampling procedure and established statistical criteria for sample size determination. In qualitative studies, theoretical saturation serves as a criterion for determining corpus size and composition. It indicates that the researcher no longer acquires significant new information by incorporating additional data, signalling that "the researcher becomes empirically confident that a category is saturated" (Glaser and Strauss 1967:61). To gain a comprehensive understanding using this criterion, it is beneficial to purposefully search for contrasting and complementary material and to draw on multiple data sources.

3.2 Deciding on the Coding Unit

Coding is the systematic process by which a researcher assigns meaning to words or other coding units through labelling and organization (Graneheim and Lundman 2004; Hsieh and Shannon 2005; Gibbs 2007; Elo and Kyngäs 2008). The part of a text or image to which the researcher attaches a category label is the coding unit, the smallest unit for analysis. Often, the coding has multiple levels, ranging from high-level aggregation (e.g., assessing recreational fishers' preferred management tools with a code like "angling regulation") to smaller units (e.g., specifying the exact proposed regulation). The following coding units could be used:

Single Words This method involves analysing individual words or visual elements (e.g., "fish" and "vessel"). In these studies, the focus is not typically on individual words themselves, but on grouping them into themes or categories for analysis (e.g., Pita et al. 2018). Complex terms like "catch per unit of effort," although made up of multiple words, are often treated as a single unit for counting, enhancing the comprehension of text information.

Sentences, Paragraphs, or Thematic Sections For example, Daedlow et al. (2011) chose paragraphs as coding units to encode recreational fishers' opinions into short statements. An equivalent for images is to treat entire frames from a video or panels from a comic as coding units.

The Entire Text, Image, Video, or Sampling Unit In this scenario, an entire newspaper article could serve as the coding unit, with the code then representing the main topic of the article (Castro et al. 2018; Comby et al. 2019; Schakner et al. 2019). This procedure can be complemented by finer-meshed analyses.

Image analysis poses unique challenges in determining the coding unit compared to textual analysis. The researcher must identify elements within an image worthy of coding and to which category they belong. These can be living beings, objects, and events, their occurrence and properties, or include formal attributes like colours and shapes. Although some variables might be straightforward, others can present complexities, especially with categories exhibiting numerous expressions. Precision, mutual exclusivity, and exhaustiveness are crucial for categories in content analysis (Bell 2001). There may be a temptation to delve into sub-elements, such as clothing or facial expressions, in the image of a recreational fisher holding up a catch. However, given the wealth of information that the human eye captures in visual data, it may be advisable to limit the analysis of images to the most important elements and features (e.g., Mordue 2013).

3.3 *Preparing the Data for Analysis*

Preparing audio and video for content analysis recordings typically requires conversion into transcripts (Kowal and O'Connell 2014) or a series of still images (Bell 2001). Transcription can be time-consuming and costly if using external service providers, but advancements in automatic text recognition and artificial intelligence are likely to speed up the process in the future. To be able to use software tools for qualitative and quantitative analysis, as detailed in Sect. 5, the corpus must be formatted in a readable and searchable manner. If the source content lacks these characteristics, it can be digitally encoded using special OCR (Optical Character Recognition) software available online.

Quantitative content analysis requires further data preparation to convert text into "bags of words". Text mining software accomplishes this through tokenization, a process involving breaking down the text into separable components (Ghavimi 2019). Tokenization may include converting letters to lowercase and removing stop words. Lists of stop words and topic-specific dictionaries in the correct language (e.g., Sbragaglia et al. 2022) are critical for producing an appropriate list of tokens (usually words). Despite careful preparation, comparing results with the original text is essential when using text mining for content analysis due to challenges such as different spellings and ambiguous words.

Two additional text normalization procedures, namely stemming and lemmatization, can improve analytical results after tokenization, especially when dealing with different grammatical forms of the same word or word families (Ghavimi 2019). Stemming identifies the root form of a family of related words, where the root itself may not have a meaningful interpretation (e.g., "stud" for "studies", "study", and "student"). In contrast, lemmatization yields a root form with a shared meaning of different inflected forms (e.g., "studies" as the lemma form for "studies" and "study"). Word frequencies can differ depending on whether stemming and lemmatization procedures have been performed. We show word frequencies for this chapter (Fig. 1a), illustrating that stemming and lemmatization emphasize more common words, resulting in changes to the content analysis results.

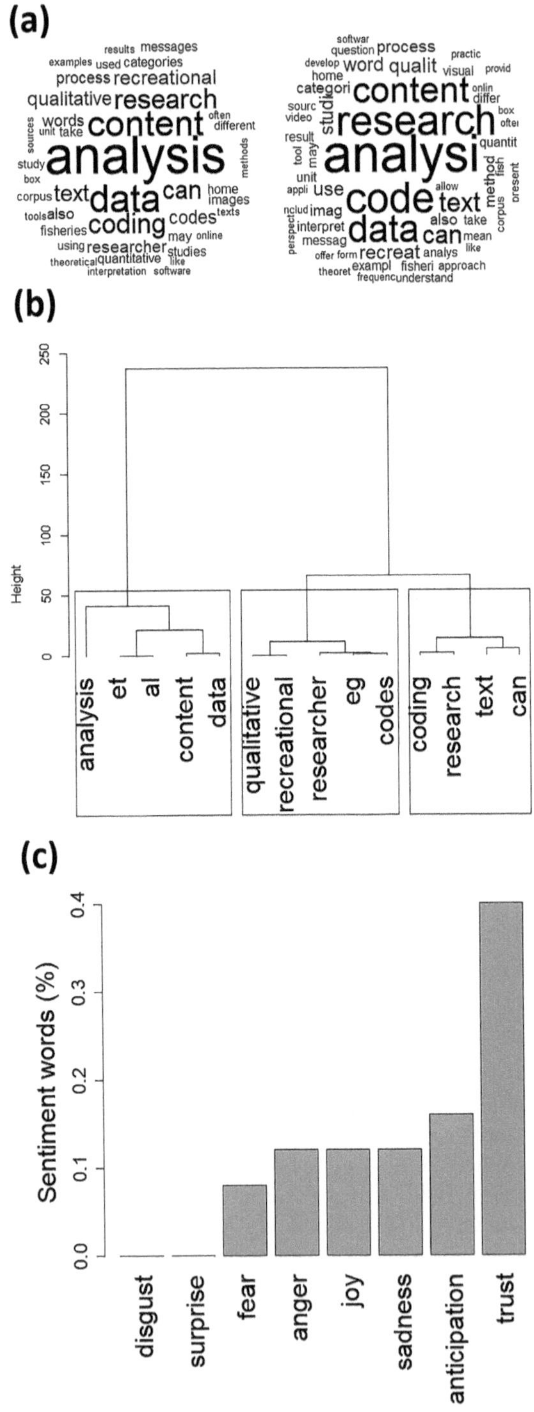

Fig. 1 Word Clouds (**a**) Created with the Text of This Chapter (left panel was created using the raw text, whereas stemming and lemmatization were performed in the case of the right panel), Cluster Dendrogram (**b**) Illustrating Relationships of the Most Frequent Words Used in This Chapter (boxes show the three clusters considered in this analysis), and Sentiment Analysis (**c**) Performed on the frequently occurring words in this chapter showing percentage of words associated with different emotions (Appendix 1 provides script code used to generate all panels)

4 Analytical Phase

Take Home Messages 3

- As part of a quantitative analysis, various methods can be used: frequency analysis, concordance analysis, sentiment analysis, and multidimensional analysis.
- In qualitative content analysis, a distinction is made between inductive and deductive coding.
- Software and online tools can support content analysis and visualization of both quantitative and qualitative data.

4.1 Quantitative Analysis

The following four popular methods are applicable for conducting quantitative content analysis. The underlying data can be "bags of words" created using computer-aided coding methods, but also codes that were created manually or qualitatively.

Frequency analysis means searching for specific words, topics, or categories and counting their occurrence. The more often words occur, the more significance they carry in the analysed data, assuming that the choice of words is rarely random (Archer 2016). Raw or relative frequencies can be measured to gain insights, particularly when comparing records or exploring variations over time and among different subpopulations (e.g., recreational fishers vs. other resource users).

Concordance analysis involves compiling a comprehensive list of each appearance of a designated word in the corpus, featuring all the words both preceding and following it. His helps discern meaningful patterns, such as whether fisheries should be "developed" or "stopped". Co-occurrence relationships, formed when words frequently appear together, enhance contextual understanding. Though less frequent, concordance analysis can also elucidate the sequence of topics in a text or measure spatial proximity between visual elements. It extends to associating distant terms and concepts with the original term, encompassing the entire data record. Using similarity and dissimilarity matrices and clustering (Moisl 2015), it is then possible to describe the subject of each text or identify themes resulting from frequently mentioned terms (Fig. 1b).

Sentiment Analysis refers to categorizing or tagging codes based on their emotional content. This method has gained popularity over alternative classification approaches because of the development of algorithms that automatically recognize lexical, morphological, or syntactic patterns to classify them as emotions and intentions (Fig. 1c). Classification models created by machine learning systems further contribute to sentiment analyses and intention detection (Liu 2020; Nandal et al. 2020; Płaza et al. 2022). Sentiment analysis enables the identification of concealed intentions and emotions in texts from sources like emails, interviews, fishing apps, or social media.

Multidimensional Analysis classifies coding units into multiple categories. For example, Sbraraglia et al. (2020) analysed comments on YouTube videos, coding them by topic. Additionally, each comment was classified by its communicative purpose (e.g., advice, opinion, general conversation) and the expressed feelings' valence (e.g., positive, negative, neutral). This approach allowed a multidimensional comparison between recreational spear fishers and anglers targeting the same fish species.

These analyses can be carried out through computer-assistance tools. Python and R, both open-source programming languages, offer libraries for text processing, analytical operations, and visualizations. Out-of-the-box tools include the web-based reading and analysis environment Voyant Tools, Google's Ngram Viewer for charting the frequencies of text fragments over time, and the network analysis program Gephi for visualizations. Qualitative software like Atlas.ti and MaxQDA can also determine word frequencies. For multi-word terms, text mining tools such as those offered online by the United Kingdom's National Centre for Text Mining (NaCTeM) provide additional support. Some of the tools also help with tokenization, stemming, and lemmatization.

4.2 *Qualitative Analysis*

Qualitative data analysis involves an iterative coding process concurrent with developing an understanding of the data. Coding, the process by which a researcher systematically assigns meaning to text through labelling and organizing, is a crucial analytical step in the qualitative paradigm, distinguishing it from mere data processing. This labour-intensive process requires reflexivity from the researcher and clear documentation of their thinking and reasoning (Gibbs 2007), particularly later in the coding process when researchers start to condense codes and form categories.

Coding for qualitative content benefits from comprehensive context knowledge of the subject matter, acquired through researching current literature, writing field notes, and transcribing audio material, as well as exploring related research (e.g., Boucquey 2017). This engagement with the case allows for free association and hypothesis development, inspiring later systematic coding (Forman and Damschroder 2007). Below, we outline steps to (i) inductive and (ii) deductive coding (Forman and Damschroder 2007; Gibbs 2007; Insch et al. 1997). We encourage the reader to treat the following descriptions as a starting point on which to build their own coding practice.

Inductive Coding, also known as conventional content analysis, open coding, or data-driven coding, is employed in qualitative content analysis to investigate phenomena through text, such as transcripts from qualitative interviews. In this approach, the researcher avoids using a pre-defined coding framework, allowing the research question and data to guide the understanding of the text. Codes are applied to the text as the researcher progresses, leading to the development of key ideas and understandings of the phenomena of interest. During analysis, the researcher usually sorts codes into categories based on their relationships within the text (Hsieh and Shannon 2005). These categories, considered emergent, help structure key ideas into meaningful groupings (Coffee and Atkinson 1996). Inductive coding is well-suited for explorative studies and any data sources that allow the interviewee to raise perspectives they find valuable and set their own relevancies (e.g., Brummer et al. 2017). Often, a subsample of interviews or text is used for developing the final coding scheme that is then applied to the rest of the sample (Insch et al. 1997).

Deductive Coding, also called directed content analysis or concept-driven coding, adopts a structured approach where codes are derived from a pre-existing conceptual or theoretical framework established before the text analysis begins. This framework may be shaped by existing literature, prior research, or other information sources, giving the researcher a mental structure for analysing their data. The researcher may start the process with some codes or thematic ideas in mind, then apply codes that match or are related to those pre-existing categories. This type of coding works well for answering specific questions about a chosen topic.

Qualitative content analysis often requires several rounds of coding. The first round focuses on "big-picture" codes, whereas subsequent rounds serve for understanding and fine-tuning these initial codes. Starting from the second round of coding, the researcher assigns codes to more abstract categories and gradually develops a code book. An effective code book not only lists codes but also indicates their categories, provides detailed code descriptions, and offers one or more examples (Forman and Damschroder 2007). While coding, the researcher should keep notes (e.g., memos) to define codes and explain the reasoning behind labelling and grouping. Modern software packages allow this process to be documented transparently. The final round of coding is dedicated to refining all codes, categories, and sub-categories to eliminate redundancy, ensure clear definitions, and capture all relevant nuances.

Coding may be conducted analogue or with software. The use of software for computer-assisted qualitative data analysis (CAQDAS is the commonly used acronym) has become increasingly common in qualitative research, with popular programs like MAXQDA, NVIVO, and Atlas.ti (Woods et al. 2016). Some CAQDAS tools also allow for the annotation and analysis of still images. R software also offers a CAQDAS package (Chandra and Shang 2019), facilitating the handling of large and complex data sets. These programs also support coding transparency within teams and allow the integration of text and non-text content (e.g., photos) into the same analysis space (Gibbs 2013).

Some criticize the use of software, raising concerns about the "tactile-digital divide" (i.e., a distance created between the researcher and data by the screen interface) and coding traps (i.e., the risk of creating so many codes that they lose meaning and become overwhelming) (Gilbert 2002). These critiques emphasize that CAQDAS software should be viewed as an organizing and visualization tool for researchers during content analysis. New users should familiarize themselves with the program and experiment with various tools to develop their own approach to creating codes, groupings, analytical notes or memos, and visualizing data sets effectively.

5 Data Interpretation and Visualization

Take Home Messages 4

- When interpreting data, it is necessary to step back from the codes.
- Good visualizations enhance transparency and reproducibility.

5.1 Data Interpretation

One risk in content analysis is staying too close to the codes and categories, potentially overlooking opportunities for further interpretation and synthesis. Good examples for an extensive interpretation in recreational fisheries research include studies by Boucquey (2017), Harrison et al. (2019), and Mordue (2013). Data interpretation can be a challenging step in the research process, and it requires the researcher to fully understand the research purpose and be cognizant of their own interpretive biases.

For inductive research, the goal is to interpret the data in a high-level manner so that the researcher can provide a descriptive answer to the research question while preserving key nuances of the original data. In other words, the researcher describes and explains the observed phenomenon based on the original data. In deductive research, this step requires explaining how the codes fit or do not fit into the pre-existing theoretical framework that guided the coding. Coding results that diverge from the original framework may shed light on ways to expand theoretical understanding of the content topic (e.g., Kadagi et al. 2020) or reveal flaws in the framework design, presenting an opportunity for learning.

All aspects of the coding process, including codes, categories, memos, quotes, classifications, and networks, contribute to data interpretation. Especially for codes and categories, it is often helpful to shift focus from their individual context to their applications and examples in the entire corpus. This reorganization of the data aids in interpretation, as does displaying key findings and writing short descriptive and interpretive summaries. The various forms of analysing and engaging with the data material and the holistic consideration thereof make up much of the art of content analysis (Forman and Damschroder 2007).

5.2 *Data Visualization*

After thoroughly analysing the results, clear and understandable communication becomes crucial. Visualizations play a vital role in facilitating this process. For illustration, we provide some examples of visualizations of quantitative data, based on the text of this chapter as input (Fig. 1). Frequencies can be presented through frequency tables, word clouds, or bar plots. For concordance of terms or categories, contingency tables and network representations are straightforward options. Visualizations for cluster solutions (e.g., dendrograms) and ordinations (e.g., plots for multidimensional scaling) can also be used. In these cases, applying a correlation coefficient threshold can help focus on the relevant information.

Visualizing qualitative analyses might be more difficult, but an accessible representation of the process and the results is still achievable. The researcher can provide details on codes, dimensions, and sub-categories in the text (Daedlow et al. 2011, 2013) or draw a flow diagram of the coding process (Brummer et al. 2017). Results can be shown by specifying codes and categories, either by showing the entire code book (Kadagai et al. 2020) or by giving examples (Bertella 2016). Finally, relationships among codes or categories can be visualized as networks (Mendoza et al. 2022). Incorporating quotes (Daedlow et al. 2011; Thurstan et al. 2018; Harrison et al. 2019) and other text segments (Chong-Montenegro et al. 2022) further corroborates and illustrates findings.

6 Quality Criteria

Take Home Messages 5

- Quality criteria differ for quantitative and qualitative content analysis because they usually follow different research paradigms.
- The important quality criteria for content analysis include reflexivity, positionality, validity, and reliability.

6.1 *Researcher Reflexivity and Positionality*

Reflexivity is the process of looking inward to identify and interrogate one's own judgements, beliefs, and practices as relevant to the collection, interpretation, and reporting of data. Particularly qualitative research acknowledges the benefits that come from the researcher's awareness and transparency regarding their subjectivity and biases. This can take different forms:

- personal reflexivity, where the researcher examines their own personal experiences to understand the impact on their research.
- functional reflexivity, where the researcher examines their role as a researcher; that is, their judgements and rationale behind research choices, and how their research values impact their work. Importantly, we describe reflexivity as a process, as it should be undertaken as regular practice in the course of a researcher's work.
- disciplinary reflexivity, which looks at the place and function of research in the grander scheme of the field or discipline (Gubrium et al. 2012; Wilkinson 1988).

Reflexivity is not a method itself but a way of thinking (May and Perry 2013) that allows a researcher to account for what they bring of themselves to their research. Personal biases can be reflected, for example, by working with a group or supervision, thus improving the validity claim of one's own work (Flick 2018, p. 171).

Another important process is identifying and clarifying one's positionality as a researcher. Positionality is the position that a researcher assumes within their research (Savin-Baden and Major 2013), both in relation to the subject of the study and the context in which the research takes place. Some parts of a researcher's positionality are fixed, such as race, gender identity, or country of origin. Others, like political or religious beliefs, are more subjective to the life experiences of the researcher (Darwin Holmes 2020). Acknowledging positionality allows researchers to investigate the ethical dilemmas and choices presented by qualitative inquiry, and to be further aware of the assumptions and preconceptions they bring to their work.

6.2 *Validity, Reliability, and Theoretical Penetration*

Two important standards for content analysis are validity and reliability (Neuendorf 2002; Kyle et al. 2026). Internal or construct validity refers to the credibility of any identified cause-effect relationships, and external validity means that the results of the study can be used to draw conclusions about other situations or the general study population (Neuendorf 2002). Therefore, the appropriate selection of texts for content analysis is essential, as well as consideration for the extent to which those texts reflect different groups.

Reliability refers to consistency and whether a result can be reproduced under the same conditions (Kyle et al. 2026). To reduce the possible bias of a single coder, multiple coders are often involved in content analysis (Mayring 2014). These coders need to be trained, and the inter-coder reliability, i.e., whether the same codes were given to the same text, can be calculated (Krippendorff 2009). In the case of disagreement, the decision can be made by a third coder, the codes can be removed, or consensus can be sought. Some studies opt against multiple coders to ensure consistency or costs safe. Regardless of a researcher's approach, it is critical to think about the reliability of content analysis and to transparently document and justify the decisions made.

The quality criteria mentioned cannot simply be transferred to qualitative methodological procedures, as these have some special features and are often based on different epistemological assumptions (Moon and Blackman 2014). Approaches advocating for the partial transfer of criteria like external validity and inter-coder reliability to qualitative content analysis recommend employing and coordinating multiple people during the coding process (Graneheim and Lundman 2004; Forman and Damschroder 2007) and adhering to pre-established analytical steps (Mayring 2014). Using multiple coders reduces bias in the coding but does not completely solve the subjectivity issues that is always present in qualitative content analysis. However, it is precisely this "subjectivity" and the ability to assign nuanced meaning to text that cannot possible to be inferred from mere quantitative surveys and that embeds the corpus in local cultures and practices (see Boucquey et al. 2026).

A frequently utilized quality criterion in qualitative research is theoretical penetration, signifying that empirical data becomes accessible through the application of theoretical perspectives. By employing theory and the researcher's perspective, insights are derived from the data that surpass a mere summary and rephrasing of the interpretations immanent to the field. The term "field-immanent," commonly used in research fields such as sociology, anthropology, and ethnography, pertains to interpretations or perspectives from within the context or setting being studied. This may include statements and interpretations found in analysed texts or self-descriptions provided by the interviewed individuals. Through the application of theoretical concepts to empirical data, the field can be opened up, revealing new aspects in the material. A theoretical perspective also translates the analysed cases into concepts, establishing connections to other cases and enhancing the benefits for subsequent studies (Strübing et al. 2018).

7 Final Reflections

Content analysis is a flexible research technique that can be used to explore, describe, explain, and compare recreational fishers' behaviours and test other hypotheses in recreational fisheries and its management and policymaking. In this chapter, we provided a detailed guide to content analysis, from framing research questions to delving into quantitative and qualitative coding, analysis, and interpretation. As the many examples from recreational fisheries science highlighted, content analysis does not simply summarize "what is said" (Krippendorff 1989), but it can lead to surprising and novel results that would not have been obtained from a superficial perusal of the texts or images (Patton 2002) or quantitative surveys with structured answers. Though we presented a unified guide, the diverse schools and versions of content analysis, especially in the qualitative realm (Table 1), invite readers to view this text as a foundation for building their coding practices and exploring varied analysis methods for text and images (Flick et al. 2004; Van Leeuwen and Jewitt 2001). For those new to coding, we encourage perseverance, as familiarity and ease in coding develop with practice, as well as—where possible—partnership with experienced coders.

Looking ahead, the future of content analysis in recreational fisheries research is poised for exciting developments that ride on technological advancements and evolving research approaches that are often built on digital data (Lennox et al. 2022). However, history also offers exciting opportunities with text dating back to the medieval ages (Thurstan et al. 2026) or analysis of ongoing policy challenges and associated text records. We encourage recreational-fishing researchers to seize the opportunities presented by these trends to fortify the depth and impact of their studies.

Appendix 1: Program R Code for Producing Panels in Fig. 1

Code script for Fig. 1a by using the libraries *tm* (Feinerer and Hornik 2014), and *wordcloud* (Fellows 2014).

```
#Needed libraries
library(tm)
library(wordcloud)
library(textstem)
#Text to be included. Problematic symbols like quotation marks and
brackets were manually removed
Text=c("copy and paste the text here")
Text_vector <- VectorSource(Text)
#Create corpus
Text_corpus <- (VCorpus(Text_vector, readerControl = list(reader =
readPlain, language = 'en')))
#Clean corpus
Text_corpus <- tm_map(Text_corpus, content_transformer(tolower))#
Convert to lowercase
Text_corpus <- tm_map(Text_corpus,removePunctuation)# Remove
punctuation marks
Text_corpus <- tm_map(Text_corpus, removeWords, stopwords("en"))#
Remove stop words
Text_corpus <- tm_map(Text_corpus, stripWhitespace)# Remove blank
spaces
Text_corpus <- tm_map(Text_corpus, removeNumbers)# Remove numbers
Text_corpus <- tm_map(Text_corpus, PlainTextDocument)# Ensure that it
is plain text
#Stemming
Text_corpus_st <- tm_map(Text_corpus, stemDocument, 'english')
Text_corpus_st <- tm_map(Text_corpus_st, PlainTextDocument)
```

```
#Lemmatization
Text_corpus_lem=lemmatize_words(Text_corpus_st, dictionary=
lexicon::hash_lemmas)
Text_corpus_lem <- tm_map(Text_corpus_lem, PlainTextDocument)
#Create term matrices
Text_corpus_matrix <- DocumentTermMatrix(Text_corpus)
Text_corpus_matrix_lem <- DocumentTermMatrix(Text_corpus_lem)
#Wordclouds
par(mfrow=c(1,2),mar=c(0,1,0,1))
#Original. No steming, nor lemmatization
set.seed(120)
palette <- brewer.pal(9,"Greys")[-(1:6)]
freq <- colSums(as.matrix(Text_corpus_matrix))
wordcloud(names(freq), freq, min.freq=10, rot.per=0.2,
colors=palette, random.order=F)
#After steming and lemmatization
set.seed(120)
palette <- brewer.pal(9,"Greys")[-(1:6)]
freq <- colSums(as.matrix(Text_corpus_matrix_lem))
wordcloud(names(freq), freq, min.freq=10, rot.per=0.2,
colors=palette, random.order=F)
```

Code script for Fig. 1b, following the previous example, by using the libraries *tm* (Feinerer and Hornik 2014), *cluster* (Maechler et al. 2022), and *quanteda* (Benoit et al. 2018):

```
#Needed libraries
library(tm)
library(cluster)
library(quanteda)
#Convert to quanteda corpus, clean token and term matrix
Text_corpus_q <- corpus(Text_corpus)
Text_corpus_q_matrix <- dfm(Text_corpus_q)
#Select frequent terms (>3)
Text_corpus_q_matrix_trim <- dfm_trim(Text_corpus_q_matrix,
min_termfreq = 4, verbose = TRUE)
#Create cluster dendogram with 3 groups and borders applying Ward's
minimum variance method
d <- dist(t(Text_corpus_q_matrix_trim), method='euclidian')
fit <- hclust(d=d, method='ward.D')
plot(fit, hang=-1,main='',sub='',xlab='')
groups <- cutree(fit, k=3)
rect.hclust(fit, k=3, border='black')
```

Code script for Fig. 1c, following the previous example, by using the library *syuzhet* (Jockers 2015):

```
#Needed library
library(syuzhet)
#Classify and count words into emotions: anger, anticipation, disgust,
fear, joy, sadness, surprise, and trust
Text_emot <- get_nrc_sentiment(Text)
#Plot sentiment words (%)
par(mar=c(10,6,2,0))
barplot(sort(colSums(prop.table(Text_emot[,1:8]))), horiz=FALSE,
cex.lab=2, cex.axis=1.5, cex.names=2, las=3, main='',
ylab='Sentiment words (%)')
```

References

Aas Ø, Thailing CE, Ditton RB (2002) Controversy over catch-and-release recreational fishing in Europe. In: Pitcher TJ, Hollingworth CE (eds) Recreational fisheries: ecological, economic and social evaluation. Blackwell Publishing Ltd, Oxford, pp 95–106

Archer D (2016) Does frequency really matter? In: Archer D (ed) What's in a word-list? Routledge, London, pp 1–15

Arlinghaus R, Mehner T (2003) Management preferences of urban anglers: habitat rehabilitation versus other options. Fisheries 28(6):10–17

Arlinghaus R, Alós J, Beardmore B, Daedlow K, Dorow M, Fujitani M, Hühn D, Haider W, Hunt LM, Johnson BM, Johnston F, Klefoth T, Matsumura S, Monk C, Pagel T, Post JR, Rapp T, Riepe C, Ward H, Wolter C (2017) Understanding and managing freshwater recreational fisheries as complex adaptive social-ecological systems. Rev Fish Sci Aquac 25(1):1–41

Bell P (2001) Content analysis of visual images. In: Van Leeuwen T, Jewitt C (eds) Handbook of visual analysis. SAGE Publications, London/Thousand Oaks, pp 10–34

Benoit K, Watanabe K, Wang H, Nulty P, Obeng A, Müller S et al (2018) quanteda: An R package for the quantitative analysis of textual data. J Open Source Softw 3:774

Berelson B (1952) Content analysis in communication research. The Free Press, Glencoe

Bertella G (2016) Experiencing nature in animal-based tourism. J Outdoor Recreat Tour 14:22–26

Bonney R, Byrd J, Carmichael JT, Cunningham L, Oremland L, Shirk J, Von Harten A (2021) Sea change: using citizen science to inform fisheries management. Bioscience 71(5):519–530

Bos W, Tarnai C (1999) Content analysis in empirical social research. Int J Educ Res 31(8): 659–671

Boucquey N (2017) 'That's my livelihood, it's your fun': the conflicting moral economies of commercial and recreational fishing. J Rural Stud 54:138–150

Boucquey N, Fly J, Boonstra WJ, Drakopulos L, Leong KM, Marjadi MN, Wise S, Fairbanks L (2026) Political ecology for understanding recreational fishers and fisheries. In: Pope KL, Arlinghaus R, Hunt LM, Lynch AJ, van Poorten BT (eds) Understanding recreational fishers: disciplinary and interdisciplinary approaches for fisheries management. Springer Nature, Cham, p 197–232

Braun V, Clarke V (2006) Using thematic analysis in psychology. Qual Res Psychol 3(2):77–101

Brummer V, Roth S, Röhl M, Herbes C (2017) Nature conservation against all? Aquatic macrophyte de-weeding–cut or conserve? A stakeholder analysis. Landsc Online 54:1–14

Castro P, Seixas E, Neca P, Bettencourt L (2018) Successfully contesting the policy sphere: examining through the press a case of local protests changing new ecological laws. Polit Psychol 39(1):107–123

Chandra Y, Shang L (2019) Qualitative research using R: a systematic approach. Springer, Singapore

Chong-Montenegro C, Thurstan RH, Campbell AB, Cunningham ET, Pandolfi JM (2022) Historical reconstruction and social context of recreational fisheries: the Australian East Coast Barramundi. Fish Manag Ecol 29(1):44–56

Coffee A, Atkinson P (1996) Making sense of qualitative data: complementary research strategies. Sage Publications, Thousand Oaks

Comby E, Le Lay YF, Piégay H (2019) Power and changing riverscapes: the socioecological fix and newspaper discourse concerning the Rhône River (France) since 1945. Ann Am Assoc Geogr 109(6):1671–1690

Corbin JM, Strauss A (1990) Grounded theory research: procedures, canons, and evaluative criteria. Qual Sociol 13(1):3–21

Creswell JW, Plano Clark VL (2018) Designing and conducting mixed methods research, 3rd edn. Sage Publications, Los Angeles

Daedlow K, Beckmann V, Arlinghaus R (2011) Assessing an adaptive cycle in a social system under external pressure to change: the importance of intergroup relations in recreational fisheries governance. Ecol Soc 16(2):3

Daedlow K, Beckmann V, Schlueter M, Arlinghaus R (2013) Explaining institutional persistence, adaptation, and transformation in East German recreational-fisheries governance after the German reunification in 1990. Ecol Econ 96:36–50

Darwin Holmes AG (2020) Researcher positionality-a consideration of its influence and place in qualitative research-a new researcher guide. Shanlax Int J Educ 8(4):1–10

Di Minin E, Fink C, Hausmann A, Kremer J, Kulkarni R (2021) How to address data privacy concerns when using social media data in conservation science. Conserv Biol 35(2):437–446

Elo S, Kyngäs H (2008) The qualitative content analysis process. J Adv Nurs 62(1):107–115

Fairclough N (2013) Critical discourse analysis. In: Handford M, Gee JP (eds) The Routledge handbook of discourse analysis. Routledge, London, pp 9–20

Feinerer I, Hornik K (2014) tm: Text mining package. https://cran.r-project.org/package=tm

Fellows I (2014) wordcloud: Word clouds. https://cran.r-project.org/package=wordcloud

Fennell D (2017) Fly-fishing on Vimeo: cas(t)ing the channel. Hum Dimens Wildl 22(1):18–29

Finnegan R (2014) Communicating: the multiple modes of human communication, 2nd edn. Routledge, London/New York

Flick U (2002) Qualitative research-state of the art. Soc Sci Inf 41(1):5–24

Flick U, Von Kardorff E, Steinke I (2004) What is qualitative research? An introduction to the field. In: Flick U, Von Kardorff E, Steinke I (eds) A companion to qualitative research. Sage Publications, Thousand Oaks, pp 3–11

Flick, U (2018) Designing qualitative research. Sage

Forman J, Damschroder L (2007) Qualitative content analysis. In: Jacoby L, Siminoff LA (eds) Empirical methods for bioethics: a primer. Adv Bioethics, vol 11. Emerald Group Publishing Limited, Leeds, pp 39–62

Früh W (2007) Inhaltsanalyse. Theorie und Praxis, 6th edn. UVK Verlagsgesellschaft mbH, Konstanz

Ghavimi P (2019) Big data analytics methods: analytics techniques in data mining, deep learning and natural language processing, 2nd edn. De Gruyter Inc, Boston/Berlin

Gibbs GR (2007) Thematic coding and categorization. In: Gibbs GR (ed) Analyzing qualitative data. SAGE Publications Ltd, London, pp 38–55

Gibbs GR (2013) Using software in qualitative analysis. In: Flick U (ed) The SAGE handbook of qualitative data analysis. SAGE Publishing Ltd, Los Angeles, pp 277–294

Gilbert LS (2002) Going the distance: 'Closeness' in qualitative data analysis software. Int J Soc Res Method 5(3):215–228

Glaser B, Strauss A (1967) Grounded theory: the discovery of grounded theory. Sociology 12(1): 27–49

Golebie E, van Riper CJ, Kochalski S, Nunan F (2026) Mixed methods research: applications and considerations for the human dimensions of recreational fisheries. In: Pope KL, Arlinghaus R, Hunt LM, Lynch AJ, van Poorten BT (eds) Understanding recreational fishers: disciplinary and interdisciplinary approaches for fisheries management. Springer Nature, Cham, p 481–506

Graneheim UH, Lundman B (2004) Qualitative content analysis in nursing research: concepts, procedures and measures to achieve trustworthiness. Nurs Educ Today 24(2):105–112

Griffin A, May V (2012) Narrative analysis and interpretative phenomenological analysis. In: Seale C (ed) Researching society and culture, 3rd edn. SAGE Publications, London, pp 441–458

Gubrium JF, Holstein JA, Marvasti AB, McKinney KD (eds) (2012) The SAGE handbook of interview research: the complexity of the craft, 2nd edn. SAGE Publications, Thousand Oaks

Hardy C, Harley B, Phillips N (2004) Discourse analysis and content analysis: two solitudes. Qual Multi-Method Res 2(1):19–22

Harrison HL, Kochalski S, Arlinghaus R, Aas Ø (2019) Do you care about the river? A critical discourse analysis and lessons for management of social conflict over Atlantic salmon (Salmo salar) conservation in the case of voluntary stocking in Wales. People Nat 1(4):507–523

Hergenrather KC, Rhodes SD, Cowan CA, Bardhoshi G, Pula S (2009) Photovoice as community-based participatory research: a qualitative review. Am J Health Behav 33(6):686–698

Holsti OR (1968) Content analysis. In: Lindzey G, Aronson E (eds) Handbook of social psychology, vol 2. Addison-Wesley, Reading, pp 596–692

Hsieh HF, Shannon SE (2005) Three approaches to qualitative content analysis. Qual Health Res 15(9):1277–1288

Hunston S (2022) Corpora in applied linguistics, 2nd edn. Cambridge University Press, Cambridge

Hunt LM, Sutton SG, Arlinghaus R (2013) Illustrating the critical role of human dimensions research for understanding and managing recreational fisheries within a social-ecological system framework. Fish Manag Ecol 20(2–3):111–124

Insch GS, Moore JE, Murphy LD (1997) Content analysis in leadership research: examples, procedures, and suggestions for future use. Leadersh Q 8(1):1–25

Jockers ML (2015) syuzhet: Extract Sentiment and Plot Arcs from Text. https://cran.rproject.org/package=syuzhet

Kadagi NI, Wambiji N, Swisher ME (2020) Potential for conflicts in recreational and artisanal billfish fisheries on the coast of Kenya. Mar Policy 117:103960

Kowal S, O'Connell DC (2014) Transcription as a crucial step of data analysis. In: Flick U (ed) The SAGE handbook of qualitative data analysis. SAGE Publishing Ltd, Los Angeles, pp 64–79

Krippendorff K (1989) Content analysis. In: Barnouw E, Gabner G, Schramm W, Worth TL, Gross L (eds) International encyclopedia of communication, vol 1. Oxford University Press, New York, pp 403–407

Krippendorff K (2009) Testing the reliability of content analysis data. In: Krippendorff K, Bock MA (eds) The content analysis reader. SAGE Publication, Thousand Oaks, pp 350–357

Kyle G, Landon AC, Pilgreen D, Golebie E (2026) Measurement instruments. In: Pope KL, Arlinghaus R, Hunt LM, Lynch AJ, van Poorten BT (eds) Understanding recreational fishers: disciplinary and interdisciplinary approaches for fisheries management. Springer Nature, Cham, p 355–385

Lennox RJ, Sbragaglia V, Vollset KW, Sortland LK, McClenachan L, Jarić I, Guckian ML, Ferter K, Danylchuk AJ, Cooke SJ, Arlinghaus R, Twardek WM (2022) Digital fisheries data in the Internet age: emerging tools for research and monitoring using online data in recreational fisheries. Fish Fish 23(4):926–940

Liu B (2020) Sentiment analysis: mining opinions, sentiments, and emotions, 2nd edn. Cambridge University Press, Cambridge

Lynch AJ, Hunt LM, Beardmore B, van Poorten BT, Pope KL, Arlinghaus R (2026) Complexity and integration of recreational fisheries. In: Pope KL, Arlinghaus R, Hunt LM, Lynch AJ, van Poorten BT (eds) Understanding recreational fishers: disciplinary and interdisciplinary approaches for fisheries management. Springer Nature, Cham, p 3–19

Maechler M, Rousseeuw P, Struyf A, Hubert M, Hornik K (2022) cluster: Cluster analysis basics and extensions. https://cran.r-project.org/package=cluster

May T, Perry B (2013) Reflexivity and the practice of qualitative research. In: Flick U (ed) The SAGE handbook of qualitative data analysis. SAGE Publishing Ltd, Los Angeles, pp 109–122

Mayring P (1983) Qualitative inhaltsanalyse. Grundlagen und Techniken. Beltz, Weinheim

Mayring P (2000) Qualitative content analysis. Forum Qual Soc Res 1(2):20

Mayring P (2004) Qualitative content analysis. In: Flick U, Von Kardorff E, Steinke I (eds) A companion to qualitative research. Sage Publications, Thousand Oaks, pp 266–269

Mayring P (2014) Qualitative content analysis: theoretical foundation, basic procedures and software solution. Beltz, Klagenfurt Austria

Mayring P (2015) Qualitative content analysis: theoretical background and procedures. In: Bikner-Ahsbahs A, Knipping C, Presmeg N (eds) Approaches to qualitative research in mathematics education: examples of methodology and methods. Springer, Dordrecht, pp 365–380

Mendoza JN, Prūse B, Mattalia G, Kochalski S, Ciriaco A, Sõukand R (2022) Fishers' perspectives: the drivers behind the decline in fish catch in Laguna Lake, Philippines. Marit Stud 21(4): 569–585

Merriam-Webster (n.d.) Image. https://www.merriam-webster.com/dictionary/image. Accessed 17 September 2023

Moisl H (2015) Cluster analysis for corpus linguistics, vol 66. Walter de Gruyter GmbH & Co KG, Berlin/Munich/Boston

Monkman GG, Kaiser M, Hyder K (2018) The ethics of using social media in fisheries research. Rev Fish Sci Aquac 26(2):235–242

Moon K, Blackman D (2014) A guide to understanding social science research for natural scientists. Conserv Biol 28(5):1167–1177

Mordue T (2013) Travels into nature and society with rod and line. Ann Tour Res 43:100–120

Nandal N, Tanwar R, Pruthi J (2020) Machine learning based aspect level sentiment analysis for Amazon products. Spat Inf Res 28(5):601–607

Neuendorf KA (2002) The content analysis guidebook. Sage Publications, Thousand Oaks

Patton MQ (2002) Two decades of developments in qualitative inquiry: a personal, experiential perspective. Qual Soc Work 1(3):261–283

Pita P, Villasante S, Arlinghaus R, Gomes P, Strehlow HV, Veiga P, Vingada J, Hyder K (2018) A matter of scales: does the management of marine recreational fisheries follow the ecosystem approach to fisheries in Europe? Mar Policy 97:61–71

Pita P, Antelo M, Hyder K, Vingada J, Villasante S (2020) The use of recreational fishers' ecological knowledge to assess the conservation status of marine ecosystems. Front Mar Sci 7:242

Płaza M, Trusz S, Kęczkowska J, Boksa E, Sadowski S, Koruba Z (2022) Machine learning algorithms for detection and classifications of emotions in contact center applications. Sensors 22(14):5311

Savin-Baden M, Major C (2013) Qualitative research: the essential guide to theory and practice. Routledge, London

Sbragaglia V, Correia RA, Coco S, Arlinghaus R (2020) Data mining on YouTube reveals fisher group-specific harvesting patterns and social engagement in recreational anglers and spearfishers. ICES J Mar Sci 77(6):2234–2244

Sbragaglia V, Espasandin L, Coco S, Felici A, Correia RA, Coll M, Arlinghaus R (2022) Recreational angling and spearfishing on social media: insights on harvesting patterns, social engagement and sentiments related to the distributional range shift of a marine invasive species. Rev Fish Biol Fisher 32(2):687–700

Schakner Z, Purdy C, Blumstein DT (2019) Contrasting attitudes and perceptions of California sea lions by recreational anglers and the media. Mar Policy 109:103710

Shie YJ (2020) Indigenous legacy for building resilience: a case study of Taiwanese mountain river ecotourism. Tour Manag Perspect 33:100612

Shiffman DS, Macdonald C, Ganz HY, Hammerschlag N (2017) Fishing practices and representations of shark conservation issues among users of a land-based shark angling online forum. Fish Res 196:13–26

Sidnell J, Stivers T (eds) (2012) The handbook of conversation analysis. Wiley, Chichester
Smith C, Emden C (2004) Non-intrusive research: ideas and guidelines for expedient thesis completion. Aust J Adv Nurs 22(1):37–43
Strauss A, Corbin J (1990) Basics of qualitative research: grounded theory procedures and techniques. Sage Publications, Los Angeles
Strübing J, Hirschauer S, Ayaß R, Krähnke U, Scheffer T (2018) Gütekriterien qualitativer Sozialforschung. Ein Diskussionsanstoß Zeitschrift für Soziologie 47(2):83–100
Sutton SG, Dew K, Higgs J (2009) Why do people drop out of recreational fishing? A study of lapsed fishers from Queensland, Australia. Fisheries 34(9):443–452
Thurstan RH, Buckley SM, Pandolfi JM (2018) Trends and transitions observed in an iconic recreational fishery across 140 years. Glob Environ Change 52:22–36
Thurstan RH, Chong-Montenegro C, Christensen J, Clark A, Gaynor A (2026) Environmental history and historical ecology for understanding recreational fishers and fisheries. In: Pope KL, Arlinghaus R, Hunt LM, Lynch AJ, van Poorten BT (eds) Understanding recreational fishers: disciplinary and interdisciplinary approaches for fisheries management. Springer Nature, Cham, p 233–259
Van Leeuwen T, Jewitt C (eds) (2001) The handbook of visual analysis. Sage, London
Venturelli P, Skov C, Audzijonyte A, Belmaker J, Ana Cardoso AS, Gordon L, Gundelund C, Hyder K, Jarić I, Lennox RJ, Potts WM, Vacura K, Vaz AS, Sbragaglia V (2026) Digital data mining. In: Pope KL, Arlinghaus R, Hunt LM, Lynch AJ, van Poorten BT (eds) Understanding recreational fishers: disciplinary and interdisciplinary approaches for fisheries management. Springer Nature, Cham, p 443–479
Vitale G, Dedeu AL, Pujol M, Sbragaglia V (2021) Characterizing the profile of recreational fishers who share their catches on social media. Front Mar Sci 8:768047
Weber R (1990) Basic content analysis, The Sage University Papers Series, vol 49, 2nd edn. Sage Publications, Newbury Park
Wilkinson S (1988) The role of reflexivity in feminist psychology. Women Stud Int Forum 11(5): 493–502
Woods M, Paulus T, Atkins DP, Macklin R (2016) Advancing qualitative research using qualitative data analysis software (QDAS)? Reviewing potential versus practice in published studies using ATLAS.ti and NVivo, 1994-2013. Soc Sci Comput Rev 34(5):59–617
Young MA, Foale S, Bellwood DR (2015) Dynamic catch trends in the history of recreational spearfishing in Australia. Conserv Biol 29(3):784–794
Zimmer M (2020) "But the data is already public": on the ethics of research in Facebook. In: Miller KW, Taddeo M (eds) The ethics of information technologies. Routledge, London, pp 229–241

Knowledge Co-production in Recreational Fisheries Science and Management

Robert Arlinghaus, Marie Fujitani, Elias Ehrlich, Monica T. Engel, and Steven J. Cooke

Abstract Recreational fisheries need flexible approaches to knowledge production and decision support that involve interaction with stakeholders of civil society (e.g., recreational fishers, guides, conservationists), governments, and other organised and non-organised actors. Co-production of knowledge grounds research in relevant societal challenges and yields outputs that can have a transformative impact on practice, management, and governance. The term co-production is variously defined and used in the literature. We consider co-production as a process where research questions are informed by practical problems or co-developed, studies are implemented, and findings are interpreted jointly by scientists and other actors in problem-oriented ways that meet their collective interests and needs, while bridging

R. Arlinghaus (✉)
Department of Fish Biology, Fisheries and Aquaculture, Leibniz Institute of Freshwater Ecology and Inland Fisheries, Berlin, Germany

Division of Integrative Fisheries Management, Humboldt-Universität zu Berlin, Berlin, Germany
e-mail: robert.arlinghaus@igb-berlin.de

M. Fujitani
Leibniz Centre for Tropical Marine Research, Bremen, Germany

University of Bremen, Bremen, Germany

E. Ehrlich
Department of Fish Biology, Fisheries and Aquaculture, Leibniz Institute of Freshwater Ecology and Inland Fisheries, Berlin, Germany

Institute of Biochemistry and Biology, University of Potsdam, Potsdam, Germany

M. T. Engel
Bath and Associates, Human Dimensions Consulting, Middle Cove, Newfoundland and Labrador, Canada

S. J. Cooke
Canadian Centre for Evidence-Based Conservation, Department of Biology and Institute of Environmental and Interdisciplinary Science, Carleton University, Ottawa, ON, Canada

K. L. Pope et al. (eds.), *Understanding Recreational Fishers*, Fish & Fisheries Series 45, https://doi.org/10.1007/978-3-031-99739-6_18

different knowledge domains and ways of knowing. The concept shares overlap with co-design, co-creation, mode-2 science, transdisciplinary science, action research, citizen and community science, co-learning, co-assessment, and other related "co-terms," but is not equivalent. In this chapter, we introduce and define co-production and place it in the context of research and management (and contrast it with co-management) related to recreational fisheries, laying out the aspirations, benefits, and challenges. A short narrative review of co-production work in recreational fisheries is provided. We write from a Western academic perspective on the key steps of co-production: identifying which stakeholders to include, exploring how to co-design research, determining data needs and roles, and suggesting one procedural approach. We also refer to leading methodological guidelines. We discuss the challenges and give guidance on weaving diverse visions, world views, and types and sources of knowledge and the importance of being aware of power asymmetries, roles, and (often hidden) norms and values in co-producing knowledge. The chapter concludes with three co-production case studies from recreational fisheries as examples.

Keywords Citizen science · Co-design · Rights holders · Stakeholders · Transdisciplinary

> Co-production is better for society. It also leads to better research — both technically, because it accounts for more factors, and ethically, because it's more equitable…what emerges will be more suitable for take-up Nature (2018)

1 Introduction

Managing recreational fisheries is a complex challenge at the interface between social, environmental, and cultural interactions (Arlinghaus et al. 2017). Recreational-fisheries management has added layers to consider beyond primarily valuing catch as with commercial fisheries. Instead, recreational fishers value the multifaceted fishing experience (Holland and Ditton 1992; Fedler and Ditton 1994; Hunt et al. 2019; Birdsong et al. 2021). Different recreational fishers value distinct aspects of the fishing experience including some that are completely independent of catching fish (e.g., the social or experiential aspect of the fishing trip). Individual recreational fishers are heterogeneous not only in their skill and their priorities, but also in how they behave and respond to changes in policy or the social-ecological environment (Murray-Jones and Steffe 2000; Lynch 2006; McCluskey and Lewison 2008; Johnston et al. 2010; Hunt et al. 2025). Recreational fishers come from diverse socio-cultural and demographic backgrounds, and the recreationally fished waters are often shared in use with other members of the public, as well as other fishery sectors, such as commercial or Indigenous fisheries. In some areas of the world, recreational fishing communities, through their organisations, are not only natural resource users, but they can have management rights and responsibilities for the waters they fish that are legally binding and secured (e.g., in German inland fisheries; Daedlow et al. 2011). The same holds for Indigenous communities who are rights holders about fisheries. Almost certainly, all recreational fishers have some level of personal interest or stakes in management issues. Conventional approaches to management decision-making and underlying knowledge generation to inform decisions tend to exclude diverse stakeholders. Also, the convention is that scientists should primarily define research questions, conduct research, and interpret findings that are then used, alongside other information sources in management decision-making. Such approaches fail to create pathways for engagement of scientists or managers with different stakeholders, and in doing so can lead to science-based knowledge that is less relevant and less likely to be embraced by decision makers, fishers, other participants, and stakeholders (Hakkarainen et al. 2022). Co-production of knowledge is an increasingly common approach of engaging in joint inquiry towards informing the best, scientifically sound management decisions; it is the topic of this chapter and one way to bridge, synthesise, and work together across historically separated dimensions (e.g., science and society).

Given the multi-faceted nature of recreational fisheries and the millions of people being affected by management decisions, pursuing triple-bottom-line sustainability, an ideal where ecological, social, and economic sustainability are jointly achieved at some compromise (Dichmont et al. 2021), requires agility. Management challenges in recreational fisheries can be considered in the context of complex adaptive systems (Arlinghaus et al. 2017) whereby the problems are often "wicked" (Rittel and Webber 1973). Addressing such problems requires diverse inputs from various actors, including individual stakeholders and organised interest groups as well as rights holders, to recognise and effectively consider the complexity of the system (Bammer 2019). One way by which stakeholder values and knowledge outside academia can be considered is by bringing interested people into dialogue and synthesising what they have to say and believe in a transparent and productive way. Co-production goes beyond the "simple" inclusion or consideration of stakeholders, participants, and other actors, by emphasising the joint production of new knowledge that is in turn used to address pertinent management problems (Hakkarainen et al. 2022). In this chapter, we present broad and flexible roadmaps to knowledge co-production that can fit the needs of a particular case in recreational fisheries. This chapter will acquaint readers with the state of co-production and provide guidelines for embracing a co-production approach in recreational fisheries. Whereas no general recipe for co-production exists, there are elements that will nearly always be important. We provide insights about those elements and conclude with case studies illustrating various approaches to co-production in recreational fisheries.

2 Defining Co-production of Knowledge for Sustainability of Natural Resource Use

Knowledge co-production is a loosely defined, constantly evolving term for a specific form of knowledge generation that goes beyond academia (i.e. beyond the domain of scientists and researchers). It can be associated with an overlapping series of related concepts that emphasise some form of collaboration among science and society, such as transdisciplinarity (Hirsch Hadorn et al. 2006), integrative research, (adaptive) co-management, co-creation, co-assessment, co-learning, action research, mode-2 science, or co-design (Hakkarainen et al. 2022). The differentiation among these concepts is not obvious (Norström et al. 2020), but we direct the reader to Hakkarainen et al. (2022) who does an excellent job in clarifying the various co-concepts and their relations (summarised in Fig. 1, see also Glossary).

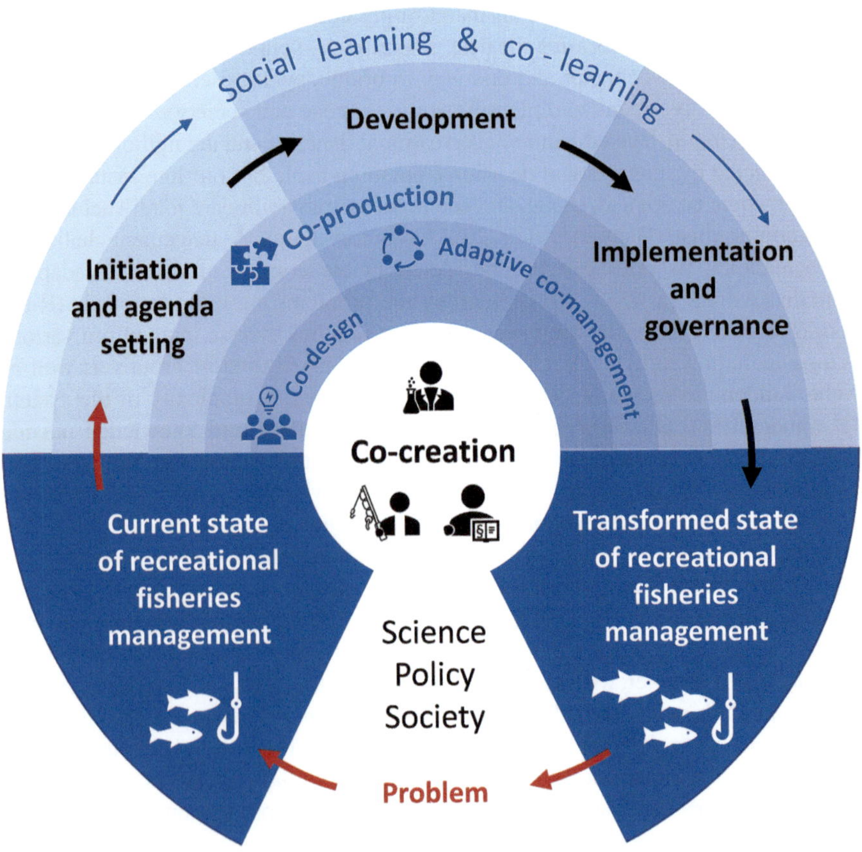

Fig. 1 An overview of the various co-concepts aimed at supporting collaboration among scientists, recreational fishers, and other actors, interest groups, and stakeholders. (Modified from Hakkarainen et al. 2022)

The roots of knowledge co-production are related to Latour (1988), who considered knowledge and action as reciprocal. Miller and Wydorn (2018) trace co-production to three main academic fields: public administration, science and technology studies, and sustainability science, which involves fisheries science. Within the latter tradition, Norström et al. (2020) defined co-production as "iterative and collaborative processes involving diverse types of expertise, knowledge, and actors to produce context-specific knowledge." Cooke et al. (2021) defined knowledge co-production specifically for fisheries as "the contribution of different knowledge sources, ways of knowing, and perspectives from different user groups with the goal of co-creating knowledge and information to inform fisheries management and conservation." A third definition that follows the first two but provides additional key areas of focus describes co-production as a "processes that iteratively unite ways of knowing and acting leading to mutual reinforcement and reciprocal transformation of societal outcomes" (Wyborn et al. 2019). This aspect of co-production is further elaborated by Moser (2016) in that a co-production process itself can be an instrument of transformation, changing not only the research projects envisioned but the people involved (Fujitani et al. 2017), while inviting critical reflection of the role of science in society, norms, and power asymmetries.

Though definitions vary and co-concepts overlap (Fig. 1), key to all definitions of co-production mentioned above is that both scientists and practitioners, stakeholders, and rights holders are involved in the knowledge generation and that four key procedural principles are met (Norström et al. 2020): (i) context-based, (ii) pluralistic, (iii) goal-oriented, and (iv) interactive (see Fig. 2 for details). If all four principles are met, the process will lead to co-production of knowledge among scientists and practitioners working towards solving a concrete problem in recreational fisheries (e.g., whether fish stocking produces positive outcomes or how to design best-practices in catch-and-release to limit impacts on a key species).

Hakkarainen et al. (2022) reviewed the aspiration of co-production in the context of natural resources management more generally, concluding that co-production projects exist in four variants with different outputs intended (Table 1): (i) outcome oriented, (ii) practical and pragmatic, (iii) empowering, and

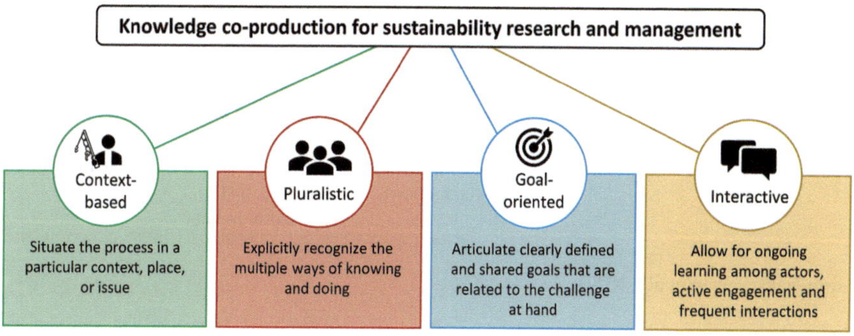

Fig. 2 Principles for knowledge co-production in sustainability sciences and specifically fisheries. (Modified from Norström et al. 2020)

Table 1 Variants of co-production of knowledge in natural resource management contexts

Variants	Outcomes oriented	Practical and pragmatic	Empowering	Transformative
Promises	Better understanding of data and monitoring	Increased relevance, scope, and usability of science, building of adaptive capacity	Creating partnerships that lead to better fisheries management	Better understanding and democratising management of fisheries by developing synergies across knowledge systems
Role of actors outside academia	Informants	Stakeholders	Partners	Co-researchers (often Indigenous or local people with rights on resources)
What kind of co-production?	From knowledge transfer to co-sharing. Actors are valued more as implementers of research outcomes (not necessarily because of their knowledge)	Co-production as a skill-dependent activity that enhances co-ownership of knowledge and builds capacity and stewardship	All are involved as equal partners and co-producers of knowledge. The process is more bound in time and scope than transformation	Appreciating different knowledge systems and embracing epistemological differences. Linking co-production with political issues (e.g., human rights, Indigenous rights)
Example from recreational fisheries	Recreational fishers sharing catch data in an online application that is then used in science for assessment	Conducting stocking experiments in collaboration with recreational fisher communities where the general design is developed by scientist without involvement of recreational fishers (see second case study in later section)	Working on improving catch-and-release practices in joint partnership of scientists and recreational fishers, where the goal is to identify best-practices that are encoded in local regulations or codes of conduct (see first case study in later section)	Co-development of visions, objectives, research approaches and data interpretation, with the goal to solve conflicts in co-exploited situations (see third case study in later section)

From Hakkarainen et al. (2022)

(iv) transformative. The promise of outcome-orientation is a better understanding of data and improved monitoring. Stakeholders can have distinct roles in this process, from merely becoming informants about fishery developments to becoming partners in the research process. If people outside academia stay as mere informants, these actors may be valued more as implementers of research outcomes than as knowledge holders. Practical and pragmatic approaches have the aspiration to increase the relevance, scope, and usability of science and to build adaptive capacity for management. Stakeholders involved become "knowledge stakeholders," and the co-production process is meant to lead to co-ownership of knowledge. The promise over empowerment is to create partnerships among scientists and non-scientists that lead to better fisheries management where all involved are co-producers and equal partners. Finally, the aspiration of transformative approaches is that actors outside academia become co-researchers to better understand and more democratically manage natural resource systems, where different knowledge sources are appreciated and embraced, and co-production of knowledge is linked with political dimensions of change and transformation.

3 Benefits of Co-production

Co-production can help to solve key scientific or practical issues in recreational fisheries as in other natural resource contexts. These solutions and issues range from questions stemming from a single scientific discipline (e.g., what is the impact of fisheries harvest on a local fish biomass and size structure), to basic research questions enmeshed in the complex social-ecological systems of a fishery (e.g., what are the principles of information spread in recreational fisher networks), to applied knowledge intended to directly inform management actions and policies (e.g., what are the social or economic factors determining the outcomes of a new harvest regulation).

The following recommendations are written from a Western perspective, envisioning someone with a science or resource management background in the primary role of designer or instigator of the process, with a normative approach of co-producing knowledge to achieve a shared goal with others (e.g., stakeholders, participants in the fishery, managers, rights holders). However, the frameworks discussed here are applicable in other situations. Our focus on knowledge produced rather than the policy outcomes as well as the explicit participation of people with stakes, rights, or influence, separates it from co-management, though the concepts may overlap (Fig. 1).

Knowledge co-production brings together different perspectives, ways of knowing and sensing, and epistemologies, and thus, draws on deliberately diverse sources of knowledge. It can thus provide a more complete, holistic, and importantly inclusive picture about a problem of relevance to recreational fisheries locally or regionally. In co-production, the lines of inquiry are typically developed collaboratively (Table 1 and Fig. 1), leading to research questions that may spin off in new directions, linking societal concern and political systems, fitting the local context, and responding to real needs of people "on the ground" (Reed et al. 2014; Meadow et al. 2015; Lemos et al. 2018). Co-production calls into question traditional roles of knowledge producers and consumers, "forcing" scientists, managers, rights holders, and all stakeholders to confront their own biases and consider other perspectives and ways of knowing. This aims to increase equity, buy in, and acceptance of the research process and results, which shall strengthen local ownership and build local management and evaluative capacity. In most co-production processes, local stakeholders and rights holders are constantly involved and informed throughout a transparent process that tends to build trust among all involved, which can be key to having new knowledge accepted and used by practitioners (Nguyen et al. 2019a, b). Collaboration in a knowledge-generating process with management implications will often also lead to a sense of ownership of and foster local stewardship of natural resources, and support for co-produced management outcomes. Co-production thus aims to bridge science-society-policy gaps, bridge the knowledge-action divide (Fazey et al. 2014; Cvitanovic et al. 2015, 2016; Metz et al. 2019), and facilitate transformative change through the interactions inherent in the co-creative research process to address specific sustainability issues in fisheries and beyond (Boaz et al. 2018; Bammer 2019). In summary, the key benefits of co-production to recreational fisheries are (adapted from Cooke et al. 2021):

- Integrating the diverse ecological and social knowledge held by people on a given problem.
- Increasing the acceptability and applicability of research results to solve sustainability issues.
- Increasing the relevance of science for social benefit.
- Increasing the uptake of research results to inform management and bridge the knowledge-action gap.
- Enhancing the human capacity to deal with problems in the future.
- Increasing interpersonal trust and building relationships among stakeholders, rights holders, and other actors and among science and society.
- Strengthening the research process in terms of relevance, credibility, saliency, and legitimacy.
- Confronting and mitigating biases of those involved in the process, including scientists.

4 Limitations and Challenges of Co-production

Achieving co-production is extremely challenging as working across scientific disciplines, regions, and societal groups requires new methods and concepts regarding communication and a willingness of all parties to engage and stay involved over time. Collaborative research often relies explicitly on "boundary objects" (Star and Griesemer 1989; Nel et al. 2016), which include conflicting narratives about an ongoing management problem and tools emerging from the realm of scientific investigation, such as data tables, logbooks, models, or maps, that can structure and share data between different user groups. These objects used for creating, storing, and sharing knowledge are understood and used differently by distinct groups, which cannot be taken for granted if they are to function well in co-producing knowledge. Researchers and non-research stakeholders need to reflect throughout the process. This is cognitively challenging, resource intensive, and very often time-consuming.

Co-production processes generally require skills that are outside the standard research toolkit of most fisheries biologists, including identifying, motivating, and involving relevant stakeholders, considering equity issues, addressing communication and language obstacles, and conflict resolution. Unfortunately, few formal opportunities exist for learning the suite of skills needed for co-production processes in fisheries science more generally (Cooke et al. 2021), and the idiosyncratic nature of each co-production relationship means that even with a strong background, one is always learning anew. Each co-production relationship fundamentally requires trust and relationship building, which necessitates large investments in face-to-face time and willingness of all people to understand and respect each other (Brinkerhoff 2002; Austin 2004).

Whereas one can design co-production very purposefully (examples later in this chapter), it can also develop more informally as relationships among science and society evolve and opportunities emerge. At times, it may involve just a few participants whereas in other instances co-produced projects become massive undertakings with hundreds of people directly or indirectly involved. Here, the timelines of researchers, management, and other stakeholders may not align with the speed and nature of research and outputs. Such issues are often the case for research or government institutional arrangements, considering constraints set by funding and associated budget and project timelines. Other limitations and challenges to co-production involve psychological, scientific, procedural, and financial aspects.

From a psychological perspective, delegating scientific inquiry into a co-design process risks losing scientific autonomy and having the research affected, if not driven and motivated by various norms and expectations of involved people outside academia. Whereas this is mainly an issue for scientists (Young et al. 2016), it may also lead to stakeholders and right holders avoiding engaging for assorted reasons including blue fear (i.e., the fear that outcomes of the research may have negative consequences such as motivating stricter regulatory control in the future). Stakeholders and other actors may also fear inferiority and the outcomes of research results affecting future decisions on which their own well-being is based. In particular, the results of co-production processes may find their way into management, which can be perceived as threatening to stakeholders and rights holders, or research results may disclose sensitive information (e.g., locations of fishing grounds) that people do not want to be revealed. It is extremely important to be open and transparent about how research results will be used, who owns them and what can or cannot follow for stakeholders and scientists to buy in. The necessary transparency can lead to processes breaking down in terms of functionality if the fear of consequences reduces the continuous willingness to engage.

A primary driver for some forms of co-production is the expectation, by researchers, managers, or funders, that it will increase the likelihood that the research output will influence policy and management (Table 1). However, this expectation itself, and the expectations of involved stakeholders, are just one of many challenges that must be navigated and managed through the process. One way these expectations may collide is when researchers are confronted by changes in the original research question and study design, especially when co-production is practised from the problem identification stage and stakeholders are involved in co-developing the research questions and approaches. These changes may bring the co-production process outside of the researchers' area of expertise, comfort zone or interest. Flexibility and constant modifications are part of the iterative and collaborative process, but the reality may be jarring for some. Thus, one must be realistic in the scientific team about how much room is available to change the research focus and whether and to what degree one is willing to engage in novel problem framing after engaging with stakeholders. If co-production begins in the problem framing phase, there must be room, in terms of capacity and flexibility, from oversight boards or funding institutions, to allow for stakeholder input to increase the relevance and applicability of the production process. Bringing stakeholders into a research program that is already predefined is co-production in a very contrived sense, and maybe better described as classical citizen science, where stakeholders are often mere informants in an already structured process. It is also possible to engage stakeholders after research findings are in hand to interpret and think about the application of knowledge, which is termed co-assessment (Sutherland et al. 2017). Researchers may engage in co-assessment but mistakenly consider it to be co-production (see glossary and Fig. 1).

From a scientific perspective, the risks include, especially if research questions are co-developed with stakeholders and data streams depend on stakeholder collaboration, that certain stakeholder norms and expectations can lead to strategic behaviour that undermines the research process in terms of data quality. For example, as mentioned before stakeholders may fear that the information they provide may later be used for more regulations, which can result in data (including verbal communication of observations) being withheld or shared in a misguided way. Seeking compromises in the research design phase may then not result asking the difficult, controversial questions, meaning those that instil high conflict potentials, which in turn can lead to projects that do not tackle the challenging issues. Also, some of the most controversial questions may simply not be answerable given the available resources and sites (e.g., due to lack of replication), which may create disappointment on the side of stakeholders and limit participation in future processes.

Procedurally, co-production entails a great ability of the leading team and all contributing people to compromise and to have the time and resources to engage. This particularly holds for people who invest free time into the process. During the process, interests, norms, and expectations may be strategically used, causing role conflicts on the side of scientists or managers. It is difficult for anyone in the process, including scientists, to escape normative expectations of stakeholders, which may perceive the scientist or other partners in the process to be biased. Situations like this can delay or break the process, leading to conflicts or even active resentment of the process itself due to non-compliance and other types of strategic behaviour to ensure one's own preferences are met. Effective co-production also requires a high level of skills, including facilitation and moderation, from those running the projects, which take time to develop and if poorly designed can quickly degrade a process. Scientists are usually working in project cycles with deadlines, which can be incompatible with a process that usually takes time to unfold positive relationships with a given stakeholder community (Cooke et al. 2021).

Perceived power imbalances can also occur in co-production processes. Typically, scientists with a high degree of formal training engage with stakeholders who often lack this training, which gives scientists an advantage in interpreting and understanding complex data. Scientists are also the people that typically guide publications and reports in writing, which means they have interpretational power about outcomes. When managers are involved and decision-makers do not delegate at least some responsibility for the ultimate decisions to be made for management away to the participating stakeholder group, the process might be perceived as fruitless as the ultimate power to install decisions out of the process may still rest with those that are not at the table.

One key aim of co-production is to bring together diverse perspectives, and though this diversity is often a strength, it can also be a source of conflict. Diversity in perspectives can mean diverging interests and values, disciplinary or cultural differences, and incompatible mental models. For this reason, finding a shared vision (Thoms 1997) in a co-production project is crucial, and this vision must be developed early in the process so that all follow a common framework and are felt as being heard.

Expectations must be carefully managed in any co-production relationship. There may be political or institutional barriers, as well as (hidden) power imbalances that need to be addressed for partnerships. Co-producing knowledge successfully depends on clear roles and responsibilities for all scientists and other people involved. This extends to transparency and early negotiation of which stakeholders are included at what stages of the process.

Finally, co-production demands time and money investments, resources that may not always be available. Successful co-production initiatives can easily take years to occur and must be built on trust, which again is time-consuming to develop. Relationship building is often disregarded by academics concerned solely on publication outputs. Developing relationships is ultimately about sharing time and own values, meeting face-to-face, having meals together and thereby building trust. Repeated interactions between groups are often necessary. In this context, formal workshops often need to be professionally moderated. If the resources are not available, the process can quickly break down and be disappointing to all. Then, instead of co-production, co-assessment of current knowledge bases (or new knowledge that has been generated) offers a suitable alternative and first step (Sutherland et al. 2017). In summary, co-designed research with an aspiration leading to co-production will always be more time and resource intensive and may be more unpredictable than "business as usual" science. Such research may be entirely incompatible with disciplinary expectations of many graduate programs.

Special considerations are needed when considering engaging in co-production with Indigenous communities and other rights holders. Indigenous community members are not stakeholders, but rights holders. The same may apply, for example, to angler communities in central Europe who are rights holders in local fisheries. In such a context, co-production involves a particularly salient ethical issue of who has the right to use the new knowledge and how it can be disseminated. Clearly, all research practices in co-production must obtain consent of those that retain the fishing rights, but even if consent is provided, a researcher can still violate morals and negatively affect local communities if, for example, a research paper is written that discloses specific fishing grounds or make statements about future management recommendations that could conflict with local norms. Tackling this issue well is difficult and necessitates a high degree of ethical standards in data use and reporting and continuous discussions with leaders of communities about research results and its use. The OCAP (Ownership, Control, Agency, and Possession) principles must be adhered to when engaging with Indigenous knowledge systems specifically (Schnarch 2004).

5 Types of Stakeholder Engagement in Co-production

There are several types of knowledge co-production, and various levels of involvement and intensity by stakeholders or Indigenous communities and other rights holders. There is no one right answer to the proper level of this involvement, as this is dictated by goals, needs, and resources of the project. However, wrong answers exist. Furthermore, the scientific or management stakeholders may not be the only ones who can dictate the needs and feasibility of a co-production process. All participants, including those initiating and guiding the co-production process, need to be willing to step out of their comfort zone, be challenged, and be open to changing their minds and deviating from their original ideas and biases.

That said, co-production should be seen as a spectrum, not a ladder where "more" is better, but a wheel (as in participation, Davidson 1998, Fig. 3) where the appropriate level for a given case is selected. Regardless of the level of involvement, the co-production process must be respectful and transparent about the aims and level of involvement sought and always be based on full consent by the affected parties and those involved in the process. A summary of different typologies and levels of stakeholder engagement is provided (Table 2) to show the diversity of stakeholder relationships possible. Many of them can be effective vehicles for co-production. Employing co-production need not be dogmatic as the needs of a particular case will dictate the specificities. However, most of the guidance in this

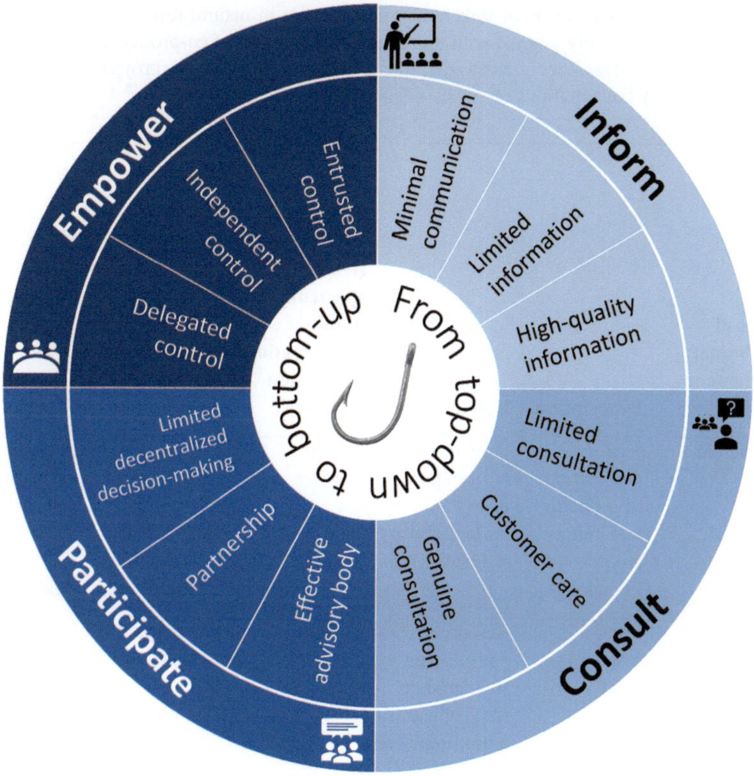

Fig. 3 The wheel of participation with stakeholders. (Modified from Dooris and Heritage 2011, after Davidson 1998)

chapter focuses on "collaborative" co-production (Table 2) as this yields the most benefits, including empowerment. Whereas this kind of co-production is time and resource-intensive, it is the most rewarding and shows the most promise to close the knowledge-action gap for using research results (Fujitani et al. 2017; Cooke et al. 2021).

6 Frameworks for Co-production Processes

A first step for researchers to begin an engaging co-production is co-production within the academic disciplines. This can be challenging enough, to move beyond the concepts of multidisciplinarity (scientists from multiple disciplines working alongside each other on the same question or facets of the same question, but remaining within their respective disciplines in terms of methods, theories and

Table 2 Typologies and levels of stakeholder engagement in natural resource management (the grey section is most closely aligned with this chapter's definition of co-production, allowing for a flexible continuum of science stakeholder interactions across a given project with respectful, inclusive, and early consultations of stakeholders outside academia)

Modes of stakeholder engagement (Biggs 1989)	Description	Relationship	International Association for Public Participation (IAP2) spectrum of public participation in decision making	Description	Relationship
Contractual	Test application of modern technology or knowledge	Flow of information from researchers to stakeholders	Inform	Provide objective and balanced information	Flow of information from researchers to stakeholders
Consultative	Use research to solve real-world problems	Researchers consult with stakeholders for problem definition, research design, and diffusion of findings	Consult	Obtain public feedback on analysis and decisions	Listen to concerns, provide feedback on public input influenced decisions
Collaborative	Learn from stakeholders to guide research	Partnerships with joint diagnosis of problem and ongoing collaboration	Involve	Work directly with public so that concerns reflected	Ensure input is considered in each step, providing feedback
			Collaborate	Partner with public in all aspects of the process	Work together to formulate solutions
Collegial	Strengthen local research capacity	Researchers encourage local research	Empower	Place decision in hands of the public	Implement what the public decides

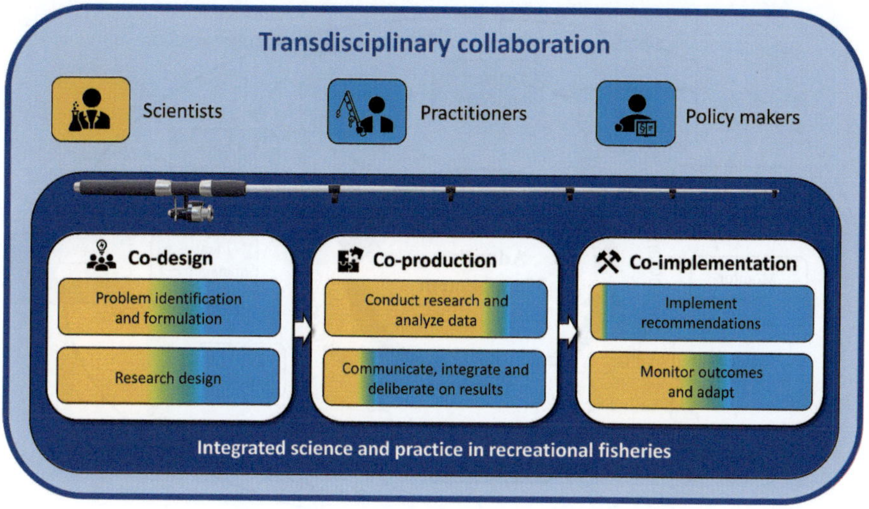

Fig. 4 Transdisciplinary science in recreational fisheries, including a co-production step. (Modified from Bennett et al. 2017)

knowledge production) to interdisciplinarity (topics, methods and concepts cross disciplinary boundaries and are jointly developed, knowledge is co-produced across disciplines) towards transdisciplinarity (crossing of disciplinary and academic boundaries, bridging science and society through integration of actors from multiple disciplines and beyond academia, joint goal-setting and development of integrated knowledge and theory). We draw from the literature on interdisciplinary and transdisciplinary knowledge integration to provide insights into knowledge co-production. The framework of Bennett et al. (2017) provides a useful and succinct roadmap for transdisciplinary conservation teams (Fig. 4). Note that in this framework, co-production is highlighted in the area about research and results, whereas in our definition, we include co-design and co-implementation as areas that may be flexibly part of the co-production process (see also Fig. 1).

The framework by Bennett et al. (2017) emphasises the circular approach to co-production, which runs through to the implementation phase of results and the monitoring of outcomes. This is closely related to what is known in environmental sciences as adaptive management with structured decision-making and stakeholder involvement (Fig. 5, see Robinson et al. 2026). Stakeholder involvement here, crucially, also means that ecological knowledge and perceptions of stakeholders are used to design objectives, synthesise what we know about system behaviour and the unknowns, and all further steps to fill the knowledge gaps and inform decisions to solve the fishery sustainability problem. The difference is that the adaptive management framework with stakeholder involvement emphasises strongly the testing of management options (e.g., does a harvest regulation work?), given what

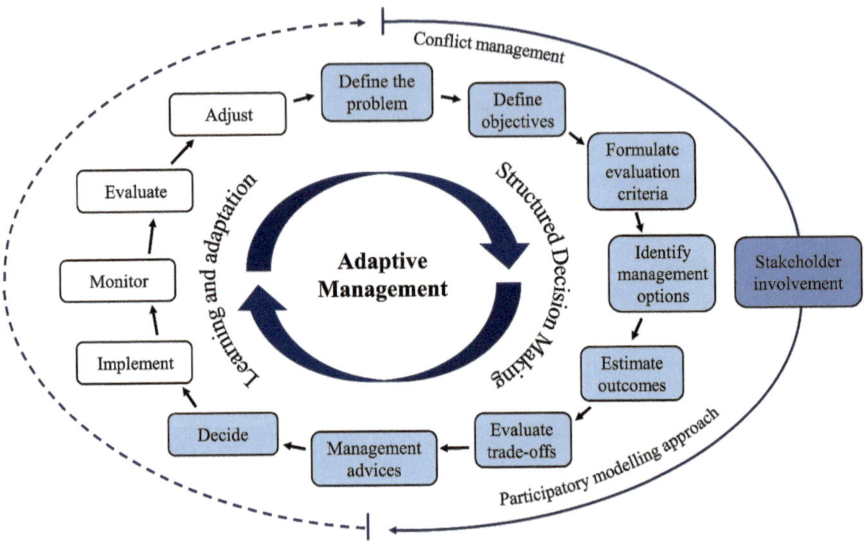

Fig. 5 Examples of adaptive management with structured decision making involving stakeholders. (From Ehrlich et al. 2023a)

is currently known or co-produced in the process, though other forms of co-production of knowledge unrelated to management are obviously possible and not part of the adaptive management process (e.g., what causes stakeholder conflicts?).

Both frameworks just mentioned share commonalities in terms of the key steps to collaboration of science and society within a co-production system. In the most basic case, the process consists of seven steps (see Box 1). They include identifying the issue and of stakeholders along with building relationships and trust. Often, the issues are brought to the researchers by the stakeholders. This is followed by generating a shared understanding about the issue at stake and clarifying the roles, power structure, and responsibilities of each partner in the process. This is further followed by the actual research process, ending with making the research available to others and applying the results. Later in this chapter, we propose a concrete four-step procedural approach that builds on these principles for effective co-production processes.

Box 1: Key Points to Consider in Knowledge Co-production Processes and Key References

Steps and recommendations	Key resources and references
Background reading on co-production	https://i2s.anu.edu.au/resources https://naturalsciences.ch/co-producing-knowledge-explained/methods/td-net_toolbox http://i2Insights.org https://www.futureearthcoasts.org/our-coastal-futures/ Reed and Abernathy (2018), Hickley et al. (2018), Bergman et al. (2012)
1. Identify the scope of the issue and relevant stakeholders, potential partners, and assess the state of existing knowledge	Reed et al. (2009), Beier et al. (2017), Lyon et al. (2020)
2. Build relationships and trust, reflect on hierarchies, power imbalances, establish conflict resolution mechanisms	Wilson (2008), Kovach (2010), Balser and McClusky (2005), Ostrom (2005), Chapman and Schott (2020)
3. Clarify roles, responsibilities, and expectations, develop an understanding of the critical unknowns and how to manage them	Hickley et al. (2018), Breckwoldt et al. (2021)
4. Generate a shared understanding, aim, scope, and vision of the goals of the collaboration as well as a model against which to judge each knowledge contribution	Mach et al. (2020), Cvitanovic et al. (2021), Thoms (1997), Baumgärtner et al. (2008)
5. Plan and implement the research, generate new knowledge, weave it with existing knowledge	Bammer (2013)
6. Make research available, transmitting and transferring the knowledge	Cooke et al. (2017a), Beck et al. (2020), Sopinka et al. (2020)
7. Apply knowledge, looking towards continuation in the medium and long-term, if applicable	Cooke et al. (2021)
Throughout: Adapt and learn from the process	Walters (1986), Fujitani et al. (2017), Ehrlich et al. (2023b)

Central to successful co-production is identifying, accounting for, and weaving diverse visions and worldviews, given power asymmetries and positionality in co-producing knowledge. Careful attention is therefore paid to potential disparities throughout all phases of a research project, from design to implementation, to foster strong working relationships and a partnership of equals. However, the specifics of each partnership and context are so diverse that no one-size-fits-all recommendation can be provided (Bammer 2013). One popular framework to find a workable consensus based on interests instead of personal positions and relationships is principled negotiation (Fisher et al. 2011). This is a structured and rational approach that is well-suited to some applications, but it is Western-centric and may not be effective in other cultural contexts. A more general recommendation is to smooth power asymmetries and flatten hierarchies, but this is also a Western academic perspective and it would be dishonest to think some entrenched imbalances can easily be smoothed. Even so, all involved in designing or running processes should be aware of hierarchies and power structures, how they impact relationships and information sharing, and how to handle them in a situationally appropriate way. A neutral facilitator can play a decisive role here, especially if conflicting situations are debated (e.g., impacts of fish-eating birds, recreational and commercial fisher conflicts).

An emerging paradigm is that of co-productive agility. This concept refers to the willingness and ability of diverse actors to iteratively engage in reflexive dialogues to grow shared ideas and actions that would not have been possible from the outset (Chambers et al. 2022). Co-productive agility will not emerge in all cases and presumably requires time to evolve. However, when it does, the potential for transformative change is strong (Chambers et al. 2022).

Transdisciplinary research with stakeholders usually aims at generating three types of knowledge related to: objectives and goals, how the system functions, and how to steer the systems in more desired directions (Hirsch Hadorn et al. 2006). Co-production can generate the same knowledge types. Methods must be devised on how to synthesise diverse knowledge sources (e.g., through modelling, environmental synthesis, Cooke et al. 2026, or causal mapping, van Maurik et al. 2023). Beyond a Western-centric approach, Indigenous frameworks exist that are wary of knowledge integration as assimilation, with a preference for terms such as "weaving" or "braiding" or "bridging" (Reid et al. 2019). Methods that bring together diverse ontologies and epistemological backgrounds can serve as further roadmaps to knowledge co-production that seeks to eliminate hierarchies and instead consider knowledge in parallel, such as the "Two-Eyed Seeing" in Canada (Reid et al. 2020; Bartlett et al. 2012) and the "Two-Ways" management in Australia (Hoffmann et al. 2012; Muller 2012).

7 Co-production in Recreational Fisheries and What It Is Not

Works on the co-production of knowledge in a fisheries context are increasing in number, including some applications within recreational fisheries. For example, Piczak et al. (2022) presented recommendations for Canada's fisheries to further highlight knowledge co-production, Gervasi et al. (2022) showed how angling guides can co-produce testable hypotheses in the crevalle jack (*Caranx hippos*) fishery in the Florida Keys, and Fujitani et al. (2017) presented quantitative data on how co-producing knowledge about fish stocking changes mental models, attitudes, and norms of the involved anglers and managers in Germany. Cooke et al. (2021) provided general guidelines for co-production in fisheries management, whereas Mills et al. (2022) described a global United Nations Ocean Decade programme to co-produce knowledge for climate-resilient fisheries. Studies of co-production with Indigenous and local knowledge are particularly rich in the literature on small-scale fisheries (Silvano et al. 2023). However, there are few works solely focusing on knowledge co-production and recreational fisheries (Cooke et al. 2021), although there are a few that have used participatory modelling of stakeholders and scientist to co-design management regulations that are accepted by all and where the computer model was the knowledge integrator (Wilberg et al. 2026). Later, we provide three case studies that followed co-production processes as idealised in this chapter, and more examples exist (summarised in Cooke et al. 2021).

One paper explicitly focusing on recreational fisheries and knowledge co-production highlighted the fact that in these arrangements, recreational fishers create data about themselves, but they also may have incentives for a fishery to be managed a certain way (Eden 2012). Eden's (2012) discussion of co-production reflects on a theme found in fisheries knowledge co-production research, that recreational fishers as a source of data are seen as collaborators in co-production. This is a narrow definition of citizen science, where fishers merely inform the process by generating information and data. Recreational fisher knowledge may be laid out to fill gaps, particularly in data-poor fisheries to run parallel to observed data and fill gaps in our scientific knowledge (Bevilacqua et al. 2016). They may contribute participatory maps, locate and report invasive species, and help tag and report marked fish or co-design testable hypotheses (Granek et al. 2008; Gervasi et al. 2022). Often, this knowledge is incorporated into existing fisheries models (Bentley et al. 2019; Ainsworth and Pitcher 2005; Bevilacqua et al. 2016). However, this contrived role in the research process is not what we emphasise as core to the co-production. In these situations, research stakeholders should be careful to differentiate recreational fishers as units of research observation (informational role), from fishers as consultants, and fishers as collaborators (Fig. 3). Though it was emphasised earlier that knowledge co-production can be a spectrum, fishers as study objects and token consultation of fishers do not fall within our scope of co-production. Researchers should also be aware of the implicit or explicit hierarchy that merely informational creates, with the Western scientific tradition producing knowledge that other ways of knowing are "validated" against or used as supplements due to data paucity. Model-based syntheses, though promising to integrate several types of knowledge, are nonetheless a managerial-technological process (that) automatically subjugates and silences forms of knowledge that it does not know how to validate, translate, and analyse (Said and Trouillet 2020). A solution might be to engage in participatory modelling, but here the recreational fishers are part of the model-building process (e.g., help building a model of cause-and-effect), which in turn is used to outline novel scenarios (Irwin et al. 2011; Ehrlich et al. 2023a). This is a vastly different role than "just" being informants to scientists. We urge readers not to confuse co-production with a simple observational role that is common in many citizen science applications in recreational fisheries.

Research on knowledge co-production in fisheries supports broader themes found in the co-production literature, with specifics for the fishery context. Young et al. (2016) described co-management of Pacific salmon fisheries in Canada's Fraser River and assessed how different stakeholders use knowledge. Stakeholders and participants (e.g., active recreational fishers) often draw on tacit, informal, and experiential knowledge above scientific knowledge (which can in fact, when diversity is accounted for, mirror scientific knowledge, Aminpour et al. 2020). Specifically, government employees were found to focus on the immediate applications of research to known problems, whereas other non-research and non-government stakeholders were interested in the political and direct implications (Young et al. 2016). Thus, communication must be tailored to the priorities of different stakeholder groups. Olson and Pinto de Silva (2019) catalogued the shift from basic and disciplinary research to applied, policy-driven, and participatory knowledge production through ninety oral histories of fisheries science practitioners. They found that this shift is expected, and even desired, and yet scientists struggle with the multiple new roles they must play as well as with the politicisation of their research. Other ways of engagement have rendered recreational fishers and managers of angling clubs as co-designers of the research and interpretation process (Fujitani et al. 2017; Ehrlich et al. 2023a, b) or as generators of testable hypotheses about fishery developments (Gervasi et al. 2022, see also the case studies below in this chapter) .

8 Ways of Doing Co-production in Recreational Fisheries

Co-production in recreational fisheries can come in four forms: (i) revealing and synthesising existing knowledge, (ii) generating new data and observations, (iii) identifying and shaping new research questions, and (iv) developing partnerships across the entire research process. We describe briefly what these areas are and suggest tools that have worked for us in our own co-production projects.

Synthesising knowledge occurs when scientific and other types of knowledge are brought together to understand the state of the art. Knowledge bases that are held outside academics are variously described as practitioner knowledge, traditional (ecological and social) knowledge, or "lay" ecologies. The word "lay" is meant to describe knowledge by non-scientists, and clearly, stakeholders can and do hold immense knowledge that is not codified in the literature. Whereas such knowledge bases often remain hidden to scientists, when properly collected and integrated, they can mirror the best scientific knowledge about fisheries and their dynamics (Aminpour et al. 2020). It is important to represent the full diversity of stakeholders and rights-based holder knowledge if one is to approach the best understanding, as Aminpour et al. (2020) showed in their study on ecological and social factors impacting northern pike (*Esox lucius*) ecology and exploitation. Diversity here means not only diverse people, but also people with diverse roles (e.g., recreational fishers, managers) because within any group (including scientists) there is a tendency to seek confirmatory knowledge and for biases to affect knowledge generation.

The wisdom of the crowd idea, where the diversity of individual knowledge well approximates truth, crucially depends on the principle of hearing diverse voices and considering the knowledge at the individual level and assessing it without the influence of others (Arlinghaus and Krause 2013). Examples of including non-academic knowledge bases revolve around identifying ecological and social phenomena (e.g., spawning sites or migratory routes of exploited fishes or the attitudes, mental models, and behaviours of recreational fishers). Focus groups are a standard means by which researchers capture the knowledge of users when designing research; capturing traditional knowledge can go far beyond this. For example, Roser et al. (2023) used traditional knowledge about past spawning places held by anglers, guides, fishers, water managers, residents, and farmers to reconstruct the level of anadromy exhibited by northern pike in a coastal fishery in Germany. Methods to synthesise existing knowledge range from mapping techniques (e.g., assessment of mental models (Gray et al. 2015) or participatory mapping by stakeholders (Schmitz Nunez et al. 2021)), verbal and digital reports of observations (e.g., capture locations of invasive species through digital apps (Venturelli et al. 2017)), to participatory synthesis exercises through verbal reports in association with peer-reviewed and other literature (e.g., systematic reviews (Cooke et al. 2026)). Synthesis is also possible through the building of participatory models that are used to examine the likely outcomes of different management regulations, as shown in several case studies involving anglers (Irwin et al. 2011; Ehrlich et al. 2023a).

Generating new data and observations are perhaps the most obvious way of co-production. Such approaches can generate a wealth of data that would otherwise not be available to researchers. The key challenge is to develop and maintain a two-way interaction among the research team and the individual stakeholders such that data generated and conclusions drawn are fed back from scientists to stakeholders. Digital applications (apps) are gaining momentum and can serve this role (Venturelli et al. 2026). In the case study presented later, the research team sent digital maps of recapture locations of marked fishes back to reporting anglers. Therefore, their knowledge was put into context, and they could learn from the movement distances of the fishes they reported. In workshop settings, data can be gathered through maps or mind maps (e.g., when identifying areas that are hotspots of the phenomenon of interest).

The most advanced process of knowledge co-production involves joint identification of issues, shaping and development of research questions, and co-design of research projects, with subsequent involvement of actors outside academia in data collection and analysis. This process is a typical situation when a management agency or a fishing club with its own fishing rights teams up with a research team to find answers to a question of common interest. One formalised approach to achieve outcomes is the four-step approach outlined below in paired workshop-field settings, where repeated workshops are used to: identify the problem to be solved, frame it, decide on a research path and data to be collected, distribute roles, and allow repeated interactions to discuss intermediate research outcomes and results. Although the formal analysis step of the raw data is typically delegated to the research team, stakeholders can still be involved in onsite data generation and should ideally always form part of result interpretation. One of the case studies shown below is an example of this approach, although many other examples exist with less formalised, yet still active involvement of recreational fishers in research design and interpretation (summarised in Cooke et al. 2021; see also Danylchuk et al. 2011; Griffin et al. 2023).

To maximise the benefit of workshop settings where co-production is developed, it is critical to develop trust beforehand, and one might need to engage a trained facilitator of the workshop. Multiple facilitation techniques exist to raise knowledge and break people's aversion to speaking up and engage, such as silent discussions or the use of the five-times-why question approach to unleash in-depth argumentative structures. Note that it is critical to carefully reflect on who owns the final research results, how to get written and moral consent on how to use the research results, and how to formally document each session and exchange mutually agreed protocols among all participants. All participants and perhaps even others should scrutinise sensitive research results before release, and researchers must pay attention to the possible misuse and misinterpretation of research findings (Cooke et al. 2017b). All results, even those that seem uncontroversial at first, may interfere with the morals, norms, and expectations of stakeholders. In co-production scenarios, paying due diligence to stakeholders is key throughout the entire process and in the aftermath, which is resource-intensive and demands critical people skills of all that are involved. Although one can never please all people, getting the process wrong will typically result in damage that will not be easily repaired and may completely break down relationships and potential future opportunities.

9 A Suggested Four-Step Approach to Co-production Processes in Recreational Fisheries

One way to think about co-production is a so-called four-step approach (Bath 2010). Although other approaches exist (see case studies) and many will not be as formalised, the four-step approach has worked for the authors in multiple cases, specifically in those conditions where conflicts are to be solved through collaboration. It starts with effectively listening to diverse groups, beginning with one-on-one interviews to identify the common issues and construct a common ground matrix. Next, a balancing of individual views and understanding the amount of consensus within each interest group is accomplished, possibly assisted through quantitative research of members of the group or the use of secondary data about the views held in the community. The third step involves the implementation of an applied, often facilitated workshop series whereby repeated interactions occur. Fourth, effective content evaluation of the process is undertaken that moves beyond the traditional objective and goal-oriented evaluations. This four-step approach has been used effectively in a variety of wildlife issues and parts of it within recreational fisheries issues (Ehrlich et al. 2023b). A brief description of the key steps of this process follows.

9.1 Step 1: Build Relationships with Diverse Actors

The first step for a researcher to design a co-production process is to thoroughly listen to the perspectives of different interest groups and individual people, often referred to as stakeholders but can also include rights holders, and to build trusted relationships. This typically involves face-to-face interviews or discussions with individual people that are relevant in each fisheries context. Whereas one could formally call this step a stakeholder analysis, it is, importantly, meant to develop an understanding of the issues and arguments present in the system and to build trust and relations. One-on-one meetings and discussions are key, which also serve the task of identifying key individuals and the issue at hand. The result of the first step of the process could be developing a common ground matrix that identifies, for each stakeholder group, areas where there is mutual understanding, worldviews, and values, and others where there is disagreement. The former serves for trust building and engagement further in the process (by focusing on those areas that are shared), whereas the latter serves to identify the new knowledge that is to be gathered or synthesised through the process or where mediation may be needed, which can be assisted by new knowledge jointly produced.

9.2 Step 2: Understanding the Amount of Consensus and Disagreement Present in a System

Many co-production processes involve some form of continued formal workshops or more informal conversations in which stakeholders and rights holders, scientists, and managers unite to develop a shared understanding, integrate knowledge, plan new knowledge generation, and discuss and evaluate results (see next step). The level of engagement and the need for formal synthesis of a given situation depend on the issue at hand, the scale of the issue, and the level of conflict present. If no conflict is present and the spatial scale of the issue is small, formal workshops may be unnecessary, and consensus can be reached merely through informal conversations of a few key people involved. If conflicts are present or the size of the community is large (e.g., thousands of users), formalised workshop settings with representatives of the various groups will be superior as beginning departure points for in-depth discussions.

Addressing conflicting situations in large populations of users (e.g., recreational fishers in conflict with commercial fishers, or fisheries with conservation) benefits from the availability of either secondary human-dimensions data or survey information that can tell the smaller workshop group about how prevalent certain views are in the wider community of users. This step will not be needed in a conflict-free situation where a small team of scientists and managers, and stakeholders unite to seek answers to a communal problem. However, objective human dimensions data on values, beliefs, and attitudes will help find compromises in more conflicting situations and for large populations. It is next necessary to choose the workshop team wisely (e.g., involve people that represent an underlying interest group and make these people accountable for informing their reference groups about ongoing processes and results). Subsequently, if resources are available and following step 1, quantitative surveys of stakeholders and rights holders (e.g., local recreational and other fisher populations) are useful to assess the current level of consensus on a given topic that is to be solved through co-production and to moderate workshop discussions. Alternatively, the team might have access to secondary data from already completed assessments about how the local stakeholders perceive a given problem. These data can be used to quantify the beliefs, values, norms, and attitudes of a large sample of users that are not in the room, assess the degree of polarisation on contentious topics (e.g., which management regulations to follow, the perceived underlying reasons for a fish population decline). Statistical tools (Vaske et al. 2010) might be applied to survey questions to be able to show in the subsequent workshops and meetings what the perception of the public and of key subgroups is on a given topic. Such data are relevant to moderate processes during workshop settings and to guide discussions. For example, vocal people in workshops may voice opinions that recreational fishers in the region have a certain attitude on a given topic. The perception may represent 1% or 90% of the fisher population or some value in between, but it cannot be validated without further data. Quantitative data following the principles outlined elsewhere in the book can provide such data, if they are currently not available, and thereby guide discussions.

9.3 *Step 3: Create Pathways for Collaboration and Engagement, e.g., Through Moderated Human Dimension Workshops*

The process continues by establishing a group of key stakeholders that are critical for understanding and solving the issue under investigation and by inviting the people to a series of workshops with a goal to synthesise knowledge and to reach consensus on the topic, and to document any other issues that do not reach consensus. Other collaboration formats are possible, but workshops seem like the most straightforward approach in large-scale conflicting situations. The number of workshops needed depends on the issue and the available information base and the level of disagreement, but the key is to invite key stakeholders and all rights holders, including those representing a larger group of stakeholders, whose task is to reveal what they know and do not know about a given topic, develop visions and objectives, and then engage in a continued process of issue identification and knowledge synthesis, revealing what is known from various perspectives, the open question to be solved, and favoured solutions (Ehrlich et al. 2023b).

The open questions typically identified in a series of workshops scattered throughout the project lifetime can then serve to nail down research questions and the process of co-generation of answers through novel research or clever weaving of existing knowledge bases. Roles and expectations about who is involved in the knowledge generation and how the knowledge is synthesised must be clearly clarified, and rules of efficient but equal communication must be agreed upon by all. Practitioners should ideally be involved as partners in the research that is then developed outside of the workshop settings by the research-practitioner team (e.g., joint marking and recapturing of fish, surveying people, collecting field observations, or implementing novel management interventions). The ongoing results are then repeatedly and iteratively brought back to the workshop settings, where they are jointly interpreted and evaluated, and next steps identified. It is in this step that co-production of new knowledge happens. If conflicts are severe, it is crucially important that the process is moderated by neutral people outside the core project team who are skilled at using moderation and facilitation techniques. The moderator should be broadly knowledgeable about the issues. In less conflicting situations, the communication pathway can be organised separately by those involved without external help.

9.4 Step 4: Documenting Results and Evaluation of Procedural Outcomes

The last step is a thorough reporting of outcomes and recommendations and evaluation of procedural outcomes (Table 1), which will typically generate results far beyond the solving of concrete fisheries management problems (e.g., new networks and new relationships or innovative approaches to management). The process shall evaluate all key principles of effective knowledge co-production that Norström et al. (2020) identifies: assessing the context-based principle (does the outcome produce results relevant to the context at stake?), assessing the pluralistic principle (have all voices been heard and included?), assessing the goal-oriented principle (did the process enable articulation, revision and achievement of desired goals, including reflexivity?) and assessing the interactive principle (what was the nature, frequency and quality of interactions among participants?).

Ideally, the outcome of joint knowledge production will result in a significant advance in the current understanding of the issue. The outcomes may also directly feed into management processes—an aspect that typically demands managers and policy makers to be part of the process, or the latter delegating responsibility for identification of future pathways to be delegated to the group at the onset of the process (see the case studies for examples). Of course, other formats of co-production of knowledge are possible (e.g., see Cooke et al. 2021).

10 Transitioning from Co-developed Knowledge to Co-implementation and Dissemination

The next phase after the co-producing knowledge on a given issue can be the design of co-implementation and dissemination processes (Fig. 1). Ideally, decision-makers are either part of the process of knowledge co-production or delegated responsibility for action away to the group. Then, the implementation of research results would be straightforward as long as legally and financially possible. Otherwise, the project team must identify leverage and entry points or wait for windows of opportunity to implement the research results. Some results may have implications for personal behaviours of actors (e.g., best practice of catch-and-release) in which case implementation is closely related to dissemination to the wider community so that a wide array of users is exposed to the novel findings. Influencers and other role models are key partners in this regard. Dissemination is often an integral component of co-production and can take many forms from multiplicative effects through stakeholder networks, capacity building to structured activities such as offering seminars, models, brochures and policy briefs, social media activity and podcasts, video or other movie formats, and presentations and publications, typically written in local non-scientific languages to facilitate knowledge uptake. Abundant resources and trained personnel (and other resources) are needed for effective science communication, but the benefits can be high. As a rule, a peer-reviewed publication will only reach scientists and some managers, whereas practitioners, most managers, and other stakeholders and rights holders need tailored approaches to dissemination via alternative channels.

11 Case Studies of Co-production in Recreational Fisheries

We finally present several case studies intended to demonstrate the diverse ways in which co-production can occur. It is important to note that although each case study followed the same general four-step process, there was significant variation in how the steps were implemented. This helps to emphasise that the steps can be scaled to a specific issue and modified to suit a given context, which is consistent with the advice from Djenontin and Meadow (2018). In some cases, where there is rapid and easy development of trusting relationships and agreement around issues of mutual interest, there may be less need for formal processes. Again, the larger the number of actors involved and the greater the conflict or lack of consensus, the greater the need for more formal, data-informed approaches such as social science surveys and structured facilitation that were previously discussed in the section above.

11.1 Co-Production Case Study: Muskellunge Science and Management

Muskellunge (*Esox masquinongy*), or musky, is an enigmatic freshwater fish species in the Midwest of North America. They grow to attain large sizes (over 1.3 m in length and 30 kg in mass) and are considered elusive, low-abundance, and difficult-to-catch fish for recreational anglers and spearers. In fact, they have been referred to as the "fish of 10,000 casts." Recreational anglers who target muskellunge tend to be highly specialised and are not motivated to harvest this species as release rates are >90% today (e.g., Fayram 2003). Specialised muskellunge anglers are often associated with fishing clubs (such as Muskies Canada or Muskies, Inc. in the United States of America) that bring together like-minded individuals interested in musky fishing and conservation.

Beginning in 2006, the Fish Ecology and Conservation Physiology Laboratory at Carleton University began to engage with the local chapter (Ottawa, Canada) of Muskies Canada. Initially, this involved providing a seminar and attending chapter meetings to develop relationships and understand the respective needs and abilities of the academic and Muskies Canada team (step 1 above). It soon became clear that there were mutual interests and strong consensus around several topics, including restoration and the associated habitat science (thus, allowing us to forgo step 2 in a formal way). That led to an interest in putting radio tags into muskellunge in Dow's Lake—a water body on the edge of Carleton's campus (Gillis et al. 2010). Muskies Canada representatives were concerned about the effects of water reductions during the fall as well as various development activities on muskellunge. Muskies Canada provided funding for the transmitters as well as several volunteer anglers to assist with fish capture. There were many opportunities for students to interact directly with the anglers, which created mutual learning opportunities (step 3—but doing so also connected back to step 1). As the relationship developed further, it became evident that there were concerns about the fate of muskellunge that were caught and released. A potential graduate student who was an avid muskellunge angler became engaged and eventually led to a graduate-level project on catch-and-release (Landsman et al. 2011) that involved working together with Muskies Canada to assist with tagging and tracking fish. Importantly, Muskies Canada assisted with sharing findings within and beyond their community through the development of educational materials. Besides doing science, the team from Carleton and Muskies Canada organised a symposium that brought together additional actors from across the Midwest, culminating in a conference report (Midwood et al. 2015). The relationship between the Carleton team and Muskies Canada now extends over

18 years and many other projects beyond the ones described here. Step 4 was never formally conducted, although both the academic team and Muskies Canada re-evaluated the relationship and project process along the way. The fact that the team is still co-producing knowledge almost two decades after when it began is telling.

There are a few lessons that arose from this work that contributed to the success. First and foremost, there was a sense of mutual respect where it was clear that Muskies Canada members were holders of a rich knowledge base on musky biology. This made it easy to embed Muskies Canada anglers in academic research and to embed students within the Muskies Canada organisation. Likewise, Muskies Canada was keen to work closely with academics and to bring science to the table. Second, through spending time together and developing strong personal relationships, it was easy to identify priority topics that were of mutual interest to the academics and Muskies Canada. Never has the academic team proposed a project for funding or partnership without those ideas emerging from discussions with Muskies Canada members and leadership. Third, there has been an opportunity to acknowledge the work that Muskies Canada has done in enabling co-production. For example, the academic team nominated Muskies Canada and Muskies, Inc. for a conservation award from the American Fisheries Society (which they won), and several Muskies Canada members are co-authors on peer-reviewed publications based on this work that further reflect their level of engagement.

11.2 Examining Fish Stocking Outcomes in Collaboration with Local Angling Clubs in Germany

"Hand in hand for sustainable stocking" was the subtitle of the Besatzfisch (German: stocked fish) project that ran from 2010 to 2016 in north-western Germany (Arlinghaus et al. 2015, 2022). Fish stocking plays a vital role in freshwater recreational fisheries management in Germany and elsewhere. Though the literature suggests that stocking can potentially have serious economic and ecological consequences if not conducted in a thoughtful way (e.g., Cowx et al. 2010; Laikre et al. 2010; Lorenzen et al. 2012; Johnston et al. 2018), understanding how stocking is practised in German recreational fisheries and identifying recommendations to move forward would be best achieved through knowledge co-production of anglers, angling clubs, and scientists. To that end, the Besatzfisch project team at Leibniz Institute of Freshwater Ecology and Inland Fisheries in Berlin Germany, designed a participatory active adaptive experiment in fish stocking in the locally managed water bodies that included volunteer angling clubs as fishing rights holders and partners in a co-production framework (summarised in Arlinghaus et al. 2015; Fujitani et al. 2017). Whereas subsequent projects focused on habitat enhancement in lakes (Radinger et al. 2023; Arlinghaus et al. 2023a), we discuss the key co-production principles based on the fish stocking project.

In the project Besatzfish, scientists and anglers in 17 angling clubs investigated together the ecological, economic, and social aspects of stocking and jointly conducted release experiments with marked model species (northern pike as a model of a recruiting and carp (*Cyprinus carpio*) as a model of a non-recruiting species) to co-produce knowledge about when stocking adds to the stock and when it fails. The research design was both interdisciplinary and transdisciplinary, with strong co-design aspects. In each angling club, representative groups of 10–30 club members were formed, which included the managers of the clubs and water bailiffs. Using collaborative and moderated workshops, they designed, conducted, and evaluated experiments to test stocking in their own club waters. Other clubs were passively involved (i.e., only received seminars about stocking) or served as placebo controls who were not exposed to training about stocking or the joint execution of release experiments, but where also seminars for joint knowledge exchange on other topics than stocking were held (Fujitani et al. 2017). During the process, which followed the four-step approach mentioned above (including an initial "listening phase" by the research team to understand key issues and standardised surveys, Arlinghaus et al. 2014), stakeholders were exposed to scientific knowledge about stocking through seminars and some clubs were additionally involved in designing and interpreting the outcome of the fish stocking experiments. Ordinary club members who were not part of the workshop settings were constantly informed about the project results through the club leadership, lectures, and a website, and all were invited to take part in regular fish assessment and participated in filling out logbooks, electrofishing, and tagging and releasing fish to both learn about the process and contribute own data. Besides the scientific outputs from the project, which showed conditions where stocking did and did not work to increase fisheries above natural recruitment (Arlinghaus et al. 2015; Johnston et al. 2018; Hühn et al. 2023), the project also demonstrated the impact of the participatory active adaptive management and knowledge co-production on learning, retention, and behavioural intentions of the involved anglers (Fujitani et al. 2017). The knowledge, attitudes, norms, and behavioural intentions of involved managers and anglers strongly changed as a result of the co-creation of stocking results, and the acceptability of stocking as a management tool strongly declined in response (Fujitani et al. 2017). To also disseminate the project results beyond the core participating angling club, the research team developed a nationwide seminar series and a handbook on sustainable fish stocking (Arlinghaus et al. 2015) as well as a fisheries management textbook in German (Arlinghaus et al. 2017), several brochures, comics, calendars, videos, podcasts, and a 60-minute documentary that was viewed by over 160,000 people in Germany alone (https://www.youtube.com/watch?v=27Ar-A5PLA0). As a result, stocking practices are beginning to change all over Germany. In 2014, the German UNESCO Commission honoured Besatzfisch for its education concept.

11.3 Co-designing Knowledge and Management Recommendation for Coastal Northern Pike Co-exploited by Anglers and Commercial Fishers

The northern pike stock in the brackish lagoons of the Baltic Sea in northeastern Germany, an area of large conservation interest, is under government-regulated open-access and co-exploited by commercial and recreational fishers (Arlinghaus et al. 2021; Koemle et al. 2021, 2022a, b, van Gemert et al. 2022; Arlinghaus et al. 2023b, c). Like muskie in the USA, pike is a popular angling fish that becomes exceptionally large in these brackish lagoons due to suitable food conditions and attracts not only local but also many tourist anglers (Arlinghaus et al. 2021; Koemle et al. 2021, 2022b). Whereas pike is a relevant economic factor for the region (Strehlow et al. 2025), catch rates have decreased in recent years. The pike stock is also targeted by a small-scale commercial fishery, and yields have also fallen over time (van Gemert et al. 2022; Olsson et al. 2023). In recent years, natural predators like cormorants (*Phalacrocorax carbo sinensis*) and grey seals (*Halichoerus grypus*) have risen and contribute to pre-recruit and recruited pike mortality (Arlinghaus et al. 2021). A recent study using catch-only stock assessment models documented that the pike stock is slightly growth- and certainly size-overfished (van Gemert et al. 2022). These results were supported by length-based assessments (Fitzgerald et al. 2023) and mirror stakeholder perspectives (van Gemert et al. 2022).

A collaborative BODDENHECHT (lagoon pike) project started in 2019 and had the overall goal to develop a management plan for Baltic pike that is widely accepted by all interest groups. It was funded by the Ministry for Agriculture, Environment and Fisheries at Mecklenburg-Western Pomerania (MWP) and EU funds tailored for structural aid to the fishery. The transdisciplinary project was designed in closely collaboration with angling guides, fishers, and other interested groups. It was coordinated by a scientist team at Leibniz Institute of Freshwater Ecology and Inland Fisheries in Berlin, Germany, that performed studies on important biological (e.g., spawning areas, migratory behaviour, meta-population structure, growth, and mortality) and socio-economic factors (e.g., economic importance and angler and fisher attitudes to management) being relevant for the understanding and predicting dynamics of the pike stock and its fishery. The entire project was done in collaboration with other academic and non-academic partners and strongly involved fishers, anglers, and guides (e.g., in mark-and-recapture experiments, social surveys, joint workshop for management plan development). The research was intended to provide the body of knowledge needed for effective management decisions. The project followed exactly the four-step approach introduced in this chapter because it was critical to include stakeholder knowledge and bring all actors together that were currently mired in conflict (Arlinghaus et al. 2022). A neutral facilitator was hired who managed the 11 workshop settings that were needed to reach consensus on 35 of 54 management options (Ehrlich et al. 2023a, b).

After an initial phase of stakeholder analysis and abundant face-to-face interactions for problem identification and trust building (step 1), step 2 generated abundant quantitative information on the attitudes and preferences of fishers (Koemle et al. 2022a) and anglers (Koemle et al. 2022b; Slaton et al. 2023). A stakeholder group with representatives from all interest groups was formed, which started in late 2019 and was facilitated by the scientist team and external neutral facilitators. The group engaged in a repeated workshop series (step 3) with 11 workshops being conducted over 3 years and a total of 60 hours in attendance. The Ministry for Agriculture, Environment and Fisheries (Ministerium für Klimaschutz, Landwirtschaft, Ländliche Räume und Umwelt in MWP) delegated responsibility to the stakeholder group to develop agreed-upon management recommendations that the ministry would then examine for implementation. This empowerment gave the stakeholder group a mandate and strong incentive to work cooperatively.

Over the course of the workshops, the stakeholders developed a shared vision, mutually agreed, quantifiable objectives, and identified and discussed the trade-offs involved in the management options. During the process, ecological and social knowledge was synthesised, and research gaps were identified, which the research team addressed throughout the process in close collaboration with recreational anglers, guides, and commercial fishers (e.g., assessing fishing mortality by sector and understanding movement ecology of the pike, inter lagoon exchange rates, genetic structure of the stock, and stakeholder views on management options). Quantitative social science data generated in step 2 were used in various workshops to outline representative views by fishers and anglers towards the acceptability of certain management tools that were being considered. Participatory models were designed and implemented during some of the workshops, based on cognitive maps about cause-and-effect expressed by the stakeholders, which served as a key knowledge integration mechanism (Ehrlich et al. 2023a). The cognitive maps were translated by the scientist team into mathematical models, which were used in further workshop rounds to assess the possible effectiveness of tools such as harvest limits, stocking, habitat improvement, and cormorant predation control. The models were all informed by ongoing research completed outside the workshop setting that was communicated to the stakeholders through further meetings. The stakeholder group thereby defined the scientific inputs they wanted and needed throughout the process, which involved repeated interactions with the team scientists to clarify important intermediate results that were stepping stones to evaluate management options considering new knowledge.

The results in step 4 were crafted into a management plan that was agreed and endorsed by the stakeholder group (Ehrlich et al. 2023b). This product was handed over to the ministry of fisheries in a public meeting that included a symposium, with a strong recommendation to implement the actions. Besides this output, the project generated new networks, trust, and understanding and resolved conflicts among the stakeholders, although contentious issues remained where interest groups clashed in perspectives (e.g., on the question of whether commercial fishing should be regulated by quotas in the future). The minister for fisheries repeatedly stressed that they intend to implement the recommendations (Backhaus 2021). At the time this book was printed, the first management measures were implemented, and more will likely follow through the renewal of the coastal fishery bylaws in 2026. The overall process was thoroughly evaluated, and research results were documented in an extensive 800-page German monograph and associated video material (Arlinghaus et al. 2023c). In a follow-up phase, the team engaged in various other forms of knowledge dissemination, which has generated abundant public attention. The project received an award for excellence in transdisciplinary research.

12 Conclusions

Co-production of knowledge among recreational fisheries scientists and various actors in the recreational fishing realm (e.g., recreational fishers, guides, outfitters, fisheries managers) is crucial to solve concrete problems locally, especially those where academic knowledge is insufficient and issues to be solved are uncertain or polarised. Co-production of new knowledge in recreational fisheries can generate important solutions, help overcome local issues and build bridges among conflicting parties that otherwise may not have been developed. The process, however, is often time and resource intensive, demands skilled staff and resources, and is replete with challenges of expectations, norms, and strategic behaviour that need to be navigated. This chapter suggests a four-step approach to effective knowledge co-production that has delivered positive results in multiple cases and is thus recommended as a roadmap for future application. The approach can be scaled to a given issue and can range from a very formal process (e.g., the Besatzfisch and Boddenhecht projects above) to one that is much more informal (muskellunge example above). Scientists are advised to pay particular attention to ensure they are an honest knowledge broker and if necessary to involve trained facilitators in the process, especially if the issue at hand is contentious. When done correctly, the benefits of co-production to science and management of recreational fisheries can be extensive.

Acknowledgments RA thanks the German Ministry for Education and Research, the EU (European Maritime Fisheries Fund), and the State of Mecklenburg-Vorpommern for funding of the projects Besatzfisch, Baggersee, and Boddenhecht, and the many actors, stakeholders, and managers involved in these projects over the years. We also thank Alistair Bath for being our mentor and help designing and running some of the participatory processes reported in this chapter, most notably also the Boddenhecht project. RA thanks Fritz Feldhege with help with literature formatting. SJC was supported by the Natural Sciences and Engineering Research Council of Canada and the Genome Canada GenFish Project.

References

Ainsworth CH, Pitcher TJ (2005) Using local ecological knowledge in ecosystem models. Fisheries assessment and management in data-limited situations. Alaska Sea Grant College Program, University of Alaska Fairbanks, Fairbanks, Alaska, USA, pp 289–304

Aminpour P, Gray SA, Jetter AJ, Introne JW, Singer A, Arlinghaus R (2020) Wisdom of stakeholder crowds in complex social-ecological systems. Nat Sustain 3:191–199

Arlinghaus R, Krause J (2013) Wisdom of the crowd and natural resource management. Trends Ecol Evol 28:8–11

Arlinghaus R, Beardmore B, Riepe C, Meyerhoff J, Pagel T (2014) Species-specific preferences of German recreational anglers for freshwater fishing experiences, with emphasis on the intrinsic utilities of fish stocking and wild fishes. J Fish Biol 85:1843–1867

Arlinghaus R, Cyrus E-M, Eschbach E, Fujitani M, Hühn D, Johnston F, Pagel T, Riepe C (2015) Hand in Hand für eine nachhaltige Angelfischerei: Ergebnisse und Empfehlungen aus fünf Jahren praxisorientierter Forschung zu Fischbesatz und seinen Alternativen. Berichte des IGB Band 28

Arlinghaus R, Alós J, Beardmore B, Daedlow K, Dorow M, Fujitani M, Hühn D, Haider W, Hunt LM, Johnson BM, Johnston F, Klefoth T, Matsumura S, Monk C, Pagel T, Post JR, Rapp T, Riepe C, Ward H, Wolter C (2017) Understanding and managing freshwater recreational fisheries as complex adaptive social-ecological systems. Rev Fish Sci Aquacult 25:1–41

Arlinghaus R, Lucas J, Weltersbach MS, Koemle D, Winkler HM, Riepe C, Kühn C, Strehlow HV (2021) Niche overlap among anglers, fishers and cormorants and their removals of fish biomass: a case from brackish lagoon ecosystems in the southern Baltic Sea. Fish Res 238:105894

Arlinghaus R, Vogt A, Koemle D, Niessner D, Ehrlich E, Rittweg T, Droll J (2022) Ursachenanalyse von Berufsfischer-Angler Konflikten am Beispiel der Nutzung von Hechten (Esox lucius) in den Boddengewässern Mecklenburg-Vorpommerns. FischZeit 2(2):1–27

Arlinghaus R, Klefoth T, Matern S, Radinger J, Nikolaus N, Meyerhoff J, Schafft M, Cyrus E-M, Emmrich M, Hering D, Wolter C (2023a) Biodiversität, Angeln und Gesellschaft: Wissensbasierte Empfehlungen für ein nachhaltiges Fischereimanagement an Baggerseen. Berichte des IGB 32

Arlinghaus R, Rittweg T, Dhellemmes F, Koemle D, van Gemert R, Schubert H, Niessner D, Möller S, Droll J, Friedland R, Lewin WC, Dorow D, Westphal L, Ehrlich E, Strehlow HW, Weltersbach MS, Roser P, Braun M, Feldhege F, Winkler H (2023b) A synthesis of a coastal northern pike (*Esox lucius*) fishery and its social-ecological environment in the southern Baltic Sea: implications for the management of mixed commercial-recreational fisheries. Fish Res 263:106663

Arlinghaus R, Braun M, Dhellemmes F, Ehrlich E, Feldhege FH, Koemle D, Niessner D, Palder J, Radinger J, Riepe C, Rittweg T, Roser P, Winkler H (2023c) Boddenhecht: Ökologie, Nutzung und Schutz von Hechten in den Küstengewässern Mecklenburg-Vorpommerns. Berichte des IGB 33

Austin DE (2004) Partnerships, not projects! Improving the environment through collaborative research and action. Hum Organ 63:419–430

Backhaus T (2021) MV bereitet Management für Hechtbestände in Küstengewässern vor. Pressemitteilung Nr. 020/2021 des Ministeriums für Landwirtschaft und Umwelt M-V. Available via https://www.regierung-mv.de/Landesregierung/lm/Aktuell/?id=167410&processor=processor.sa.pressemitteilung. Accessed 27 Dec 2023

Balser D, McClusky J (2005) Managing stakeholder relationships and nonprofit effectiveness. Nonprofit Manag Lead 15:295–315

Bammer G (2013) Disciplining interdisciplinarity: integration and implementation sciences for researching complex real-world problems. Australian National University Press, Canberra

Bammer G (2019) Key issues in co-creation with stakeholders when research problems are complex. Evid Policy 15:423–435

Bartlett C, Marshall M, Marshall A (2012) Two-eyed seeing and other lessons learned within a co-learning journey of bringing together indigenous and mainstream knowledges and ways of knowing. J Environ Stud Sci 2:331–340

Bath AJ (2010) Human Dimensions in Wolf Management in Savoie and Des Alpes Maritimes, France Results targeted toward designing a more effective communication campaign and building better public awareness materials. LIFE-Nature Project Le Retour Du Loup Dans Les Alpes Françaises and the Large Carnivore Initiative for Europe (LCI) 151 pp

Baumgärtner S, Becker C, Frank K, Müller B, Quaas M (2008) Relating the philosophy and practice of ecological economics: the role of concepts, models, and case studies in inter- and transdisciplinary sustainability research. Ecol Econ 67:384–393

Beck S, Bergenholtz C, Bogers M, Brasseur T-M, Conradsen ML, Di Marco D, Distel AP, Dobusch L, Dörler D, Effert A, Fecher B, Filiou D, Frederiksen L, Gillier T, Grimpe C, Gruber M, Haeussler C, Heigl F, Hoisl K, Hyslop K, Kokshagina O, LaFlamme M, Lawson C, Lifshitz-Assaf H, Lukas W, Nordberg M, Norn MT, Poetz M, Ponti M, Pruschak G, Priego LP, Radziwon A, Rafner J, Romanova G, Tucci CL, Tuertscher P, Vedel JB, Velden T, Verganti R, Wareham J, Wiggins A, Xu SM (2020) The open innovation in science research field: a collaborative conceptualisation approach. Ind Innov 29(2):136–185

Beier P, Hansen LJ, Helbrecht L, Behar D (2017) A how-to guide for coproduction of actionable science. Conserv Lett 10(3):288–296

Bennet NJ, Roth R, Klain SC, Chan KM, Clark DA, Cullman G, Epstein G, Nelson MP, Stedman R, Teel TL, Thomas RE, Wyborn C, Curran D, Greenberg A, Sandlos J, Verissimo D (2017) Mainstreaming the social sciences in conservation. Conserv Biol 31:56–66

Bentley JW, Serpetti N, Fox C, Heymans JJ, Reid DG, Coll M (2019) Fishers' knowledge improves the accuracy of food web model predictions. ICES J Mar Sci 76(4):897–912

Bergmann M, Jahn T, Knobloch T, Krohn W, Pohl C, Schramm E (2012) Methods for transdisciplinary research: a primer for practice. Campus Verlag, Frankfurt

Bevilacqua AHV, Carvalho AR, Angelini R, Christensen V (2016) More than anecdotes: fishers‘ ecological knowledge can fill gaps for ecosystem modeling. PLoS One 11(5):e0155655

Birdsong M, Hunt LM, Arlinghaus R (2021) Recreational angler satisfaction: what drives it? Fish Fish 22:682–706

Boaz A, Hanney S, Borst R, O‘Shea A, Kok M (2018) How to engage stakeholders in research: design principles to support improvement. Health Res Policy Syst 16:60

Breckwoldt A, Lopes PFM, Selim SA (2021) Look who's asking-reflections on participatory and transdisciplinary marine research approaches. Front Mar Sci 8:627502

Brinkerhoff JM (2002) Assessing and improving partnership relationships and outcomes: a proposed framework. Eval Program Plan 25:215–231

Chambers JM, Wyborn C, Klenk NL, Ryan M, Serban A, Bennett NJ, Brennan R, Charli-Joseph L, Fernández-Giménez ME, Galvin KA, Goldstein BE, Haller T, Hill R, Munera C, Nel JL, Österblom H, Reid RS, Riechers M, Spierenburg M, Tengö M, Bennett E, Brandeis A, Chatterton P, Cockburn JJ, Cvitanovic C, Dumrongrojwatthana P, Paz Durán A, Gerber J-D, Green JMH, Gruby R, Guerrero AM, Horcea-Milcu A-I, Montana J, Steyaert P, Zaehringer JG, Bednarek AT, Curran K, Fada SJ, Hutton J, Leimona B, Pickering T, Rondeau R (2022) Co-productive agility and four collaborative pathways to sustainability transformations. Glob Environ Chang 72:102422

Chapman JM, Schott S (2020) Knowledge co-evolution: generating new understanding through bridging and strengthening distinct knowledge systems and empowering local knowledge holders. Sustain Sci 15(3):931–943

Cooke SJ, Gallagher AJ, Sopinka NM, Nguyen VM, Skubel RA, Hammerschlag N, Boon S, Young N, Danylchuk AJ (2017a) Considerations for effective science communication. Facets 2:233–248

Cooke SJ, Nguyen VM, Kessel ST, Hussey NE, Young N, Ford AT (2017b) Troubling issues at the frontier of animal tracking for conservation and management. Conserv Biol 31(5):1205–1207

Cooke SJ, Nguyen VM, Chapman JM, Reid AJ, Landsman SJ, Young N, Hinch SG, Schott S, Mandrak NE, Semeniuk CAD (2021) Knowledge co-production: a pathway to effective fisheries management, conservation, and governance. Fisheries 46:89–97

Cooke SJ, Rytwinski T, Harper M, Howarth A, Nakagawa S, Hunt LM (2026) Evidence synthesis in recreational fisheries science and management. In: Pope KL, Arlinghaus R, Hunt LM, Lynch AJ, van Poorten BT (eds) Understanding recreational fishers: disciplinary and interdisciplinary approaches for fisheries management. Springer Nature, Cham, p 573–604

Cowx IG, Arlinghaus R, Cooke SJ (2010) Harmonizing recreational fisheries and conservation objectives for aquatic biodiversity in inland waters. J Fish Biol 76:2194–2215
Cvitanovic C, Hobday AJ, van Kerkhoff L, Wilson SK, Dobbs K, Marshall NA (2015) Improving knowledge exchange among scientists and decision-makers to facilitate the adaptive governance of marine resources: a review of knowledge and research needs. Ocean Coast Manag 112:25–35
Cvitanovic C, McDonald J, Hobday AJ (2016) From science to action: principles for undertaking environmental research that enables knowledge exchange and evidence-based decision-making. J Environ Manag 183:864–874
Cvitanovic C, Shellock RJ, Mackay M, van Putten EI, Karcher DB, Dickey-Collas M, Ballesteros M (2021) Strategies for building and managing 'trust' to enable knowledge exchange at the interface of environmental science and policy. Environ Sci Pol 123:179–189
Daedlow K, Beard TD Jr, Arlinghaus R (2011) A property rights-based view on management of inland recreational fisheries: contrasting common and public fishing rights regimes in Germany and the United States. Am Fish Soc Symp 75:13–38
Danylchuk AJ, Cooke SJ, Susky CD, Goldberg TL, Petersen D, Danylchuk SE (2011) Involving recreational anglers in developing best handling practices for catch-and-release fishing of bonefish: using citizen science to further stewardship. Am Fish Soc Sym 75:95–111
Davidson S (1998) Spinning the wheel of empowerment. Planning 1262:14–15
Dichmont CM, Dowling NA, Pascoe S, Cannard T, Pears RJ, Breen S, Roberts T, Leigh GM, Mangel M (2021) Operationalizing triple bottom line harvest strategies. ICES J Mar Sci 78:731–742
Djenontin INS, Meadow AM (2018) The art of co-production of knowledge in environmental sciences and management: lessons from international practice. Environ Manag 61:885–903
Dooris M, Heritage Z (2011) Healthy cities: facilitating the active participation and empowerment of local people. J Urban Health 90(suppl.1):74–91
Eden S (2012) Counting fish: performative data, anglers' knowledge-practices and environmental measurement. Geoforum 43:1014–1023
Ehrlich E, Bath AJ, Niessner D, Koemle D, Cyrus E-M, Arlinghaus R (2023a) Co-designing management recommendations with stakeholders: a case study about northern pike (*Esox lucius*) in the southern Baltic Sea. Fish Res 263:106687
Ehrlich E, Niessner D, Arlinghaus R (2023b) Empfehlungen für das künftige Management des Boddenhechts (Esox lucius) aus Sicht verschiedener Interessengruppen. Leibniz-Institut für Gewässerökologie und Binnenfischerei, Berlin
Fayram AM (2003) A comparison of regulatory and voluntary release of muskellunge and walleyes in Northern Wisconsin. N Am J Fish Manag 23:619–624
Fazey I, Bunse L, Msika J, Pinke M, Preedy K, Evely AC, Lambert E, Hastings E, Morris S, Reed MS (2014) Evaluating knowledge exchange in interdisciplinary and multi-stakeholder research. Glob Environ Chang 25:204–220
Fedler AJ, Ditton RB (1994) Understanding angler motivations in fisheries management. Fisheries 19:6–13
Fisher R, Ury WL, Patton B (2011) Getting to yes: negotiating agreement without giving in, New edn. Penguin Random House, New York
Fitzgerald CJ, Droll J, Shepard S, Monk CT, Rittweg T, Arlinghaus R (2023) Length-based assessment of an exploited coastal pike (*Esox lucius*) stock (Rügen, southern Baltic Sea) underscores the crucial relevance of growth and natural mortality for assessment outcomes. Fish Res 263:106667
Fujitani M, McFall A, Randler C, Arlinghaus R (2017) Participatory adaptive management leads to environmental learning outcomes extending beyond the sphere of science. Sci Adv 3:e1602516
Gervasi CL, Santos RO, Rezek RJ, James WR, Boucek RE, Bradshaw C, Kavanagh C, Osborne J, Rehage JS (2022) Bottom-up conservation: using translational ecology to inform conservation priorities for a recreational fishery. Can J Fish Aquat Sci 79:47–62

Gillis NC, Rapp T, Hasler CT, Wachelka H, Cooke SJ (2010) Spatial ecology of adult muskellunge (*Esox masquinongy*) in the urban Ottawa reach of the historic Rideau Canal, Canada. Aquat Living Resour 23:225–230

Granek EF, Madin EMP, Brown MA, Figueira WF, Cameron DS, Hogan Z, Kristianson G, de Villiers P, Williams JE, Post JR, Zahn S, Arlinghaus R (2008) Engaging recreational fishers in management and conservation: global case studies. Conserv Biol 22:1125–1134

Gray S, Hilsberg J, McFall A, Arlinghaus R (2015) The structure and function of angler mental models about fish population ecology: the influence of specialization and target species. J Outdoor Recreat Tour 12:1–13

Griffin LP, Casselberry GA, Markowitz EM, Brownscombe JW, Adams AJ, Horn B, Cooke SJ, Danylchuk AJ (2023) Angler and guide perceptions provide insights into the status and threats of the Atlantic tarpon (*Megalops atlanticus*) fishery. Mar Policy 151:105569

Hakkarainen V, Mäkinen-Rostedt K, Horcea-Milcu A, D'Amato D, Jämsä J, Soini K (2022) Transdisciplinary research in natural resources management: towards an integrative and transformative use of co-concepts. Sustain Dev 30:309–325

Hickey G, Brearley S, Coldham T, Denegri S, Green G, Staniszewska S, Tembo D, Torok K, Turner K (2018) Guidance on co-producing a research project. Southampton University. INVOLVE. Available via https://www.invo.org.uk/wp-content/uploads/2019/04/Copro_Guidance_Feb19.pdf. Accessed 28 Dec 2023

Hirsch Hadorn G, Bradley D, Pohl C, Rist S, Wiesmann U (2006) Implications of transdisciplinarity for sustainability research. Ecol Econ 60:119–128

Hoffmann BD, Roeger S, Wise P, Dermer J, Yunupingu B, Lacey D, Yunupingu D, Marika B, Marika M, Panton B (2012) Achieving highly successful multiple agency collaborations in a cross-cultural environment: experiences and lessons from Dhimurru Aboriginal Corporation and partners. Ecol Manage Restor 13:42–50

Holland SM, Ditton RB (1992) Fishing trip satisfaction: a typology of anglers. N Am J Fish Manag 12:28–33

Hühn D, Gwinn DC, Shaw SL, Alos J, Allen MS, Pagel T, Skov C, Arlinghaus R (2023) Density- and size-dependent mechanisms modulate the outcome of stocking in a naturally recruiting freshwater piscivore (northern pike, Esox lucius): a replicated whole-lake experiment. Fish Res 267:106799

Hunt LM, Camp E, Van Poorten B, Arlinghaus R (2019) Catch and non-catch-related determinants of where anglers fish: a review of three decades of site choice research in recreational fisheries. Rev Fish Sci Aquacult 27:261–286

Hunt LM, Arlinghaus R, Scott D, Kyle G (2025) Diversity of anglers: drivers and implications for fisheries management, Chapter 5. In: Neal JW, Lang TJ, Krogman RM, Kurzawski K, Hunt KM, Taylor J (eds) Angler recruitment, retention, and reactivation: the future of fisheries and aquatic conservation summary and synthesis. Am Fish Soc, pp 1–28

Irwin BJ, Wilberg MJ, Jones ML, Bence JR (2011) Applying structured decision making to recreational fisheries management. Fisheries 36(3):113–122

Johnston FD, Arlinghaus R, Dieckmann U (2010) Diversity and complexity of angler behaviour drive socially optimal input and output regulations in a bioeconomic recreational-fisheries model. Can J Fish Aquat Sci 67:1507–1531

Johnston FD, Allen MS, Beardmore B, Riepe C, Pagel T, Hühn D, Arlinghaus R (2018) How ecological processes shape the outcomes of stock enhancement and harvest regulations in recreational fisheries. Ecol Appl 28(8):2033–2054

Koemle D, Beardmore B, Dorow M, Arlinghaus R (2021) The human dimensions of recreational anglers targeting freshwater species in coastal ecosystems, with implications for management. N Am J Fish Manag 41:1572–1590

Koemle D, Niessner D, Arlinghaus R (2022a) Gesellschaftliche Bedeutung der Erwerbsfischerei und Einflussfaktoren auf die Küstenfischerei Mecklenburg-Vorpommerns aus Sicht von Berufsfischern. Fischerei & Fischmarkt in M-V 4:42–48

Koemle D, Meyerhoff J, Arlinghaus R (2022b) How catch uncertainty and harvest regulations drive anglers' choice for pike (Esox lucius) fishing in the Baltic Sea. Fish Res 256:106480

Kovach M (2010) Indigenous methodologies: characteristics, conversations, and contexts. University of Toronto Press, Toronto

Laikre L, Schwartz MK, Waple RS, Ryman N, The GeM Working Group (2010) Compromising genetic diversity in the wild: unmonitored largescale release of plants and animals. Trends Ecol Evol 25:520–529

Landsman SJ, Wachelka HJ, Suski CD, Cooke SJ (2011) Evaluation of the physiology, and survival of adult muskellunge (Esox masquinongy) captured and released by specialized anglers. Fish Res 110:377–386

Latour B (1988) The pasteurization of France. Harvard University Press, Harvard

Lemos MC, Arnott JC, Ardoin NM, Baja K, Bednarek AT, Dewulf A, Fieseler C, Goodrich KA, Jagannathan K, Klenk N (2018) To co-produce or not to co-produce. Nat Sustain 1:722–724

Lorenzen K, Beveridge MCM, Mangel M (2012) Cultured fish: integrative biology and management of domestication and interactions with wild fish. Biol Rev 87:639–660

Lynch TP (2006) Incorporation of recreational fishing effort into design of marine protected areas. Conserv Biol 20:1466–1476

Lyon C, Cordell D, Jacobs B, Martin-Ortega J, Marshall R, Camargo-Valero MA, Sherry E (2020) Five pillars for stakeholder analyses in sustainability transformations: the global case of phosphorus. Environ Sci Pol 107:80–89

Mach KJ, Lemos MC, Am M, Wyborn C, Klenk N, Arnott JC, Ardoin NM, Fieseler C, Moss RH, Nichols L, Stults M, Vaughan C, Wong-Parodi G (2020) Actionable knowledge and the art of engagement. Curr Opin Environ Sustain 42:30–37

McClusky SM, Lewison RL (2008) Quantifying fishing effort: a synthesis of current methods and their applications. Fish Fish 9:188–200

Meadow AM, Ferguson DB, Guido Z, Horangic A, Owen G, Wall T (2015) Moving toward the deliberate coproduction of climate science knowledge. Weather Clim Soc 7:179–191

Metz A, Boaz A, Robert G (2019) Co-creative approaches to knowledge production: what next for bridging the research to practice gap? Evid Policy 15:331–337

Midwood JD, Kerr SJ, Levick P, Cooke SJ (2015) Muskellunge science and management: progress through partnerships. Environ Biol Fish 98:2031–2035

Miller CA, Wyborn C (2018) Co-production in global sustainability: histories and theories. Environ Sci Pol 113:88–965

Mills KE, Armitage D, Eurich JG, Kleisner KM, Pecl GT, Tokunaga K, Browman H (2022) Co-production of knowledge and strategies to support climate resilient fisheries. ICES J Mar Sci 80:358–361

Moser SC (2016) Can science on transformation transform science? Lessons from co-design. Curr Opin Environ Sustain 20:106–115

Muller S (2012) Two ways': bringing indigenous and nonindigenous knowledges together. In: Weir JK (ed) Country, native title and ecology. Australian University Press, Canberra, pp 59–79

Murray-Jones S, Steffe AS (2000) A comparison between the commercial and recreational fisheries of the surf clam, *Donax deltoides*. Fish Res 44:219–233

Nature (2018) The best research is produced when researchers and communities work together. Nature 562:7

Nel JL, Roux DJ, Driver A, Hill L, Maherry AC, Snaddon K, Petersen CR, Smith-Adao LB, Van Deventer H, Reyers B (2016) Knowledge co-production and boundary work to promote implementation of conservation plans. Conserv Biol 30:176–188

Nguyen VM, Young N, Corriveau M, Hinch SG, Cooke SJ (2019a) What is "usable" knowledge? Perceived barriers for integrating new knowledge into management of an iconic Canadian fishery. Can J Fish Aquat Sci 76:463–474

Nguyen VM, Young N, Brownscombe JW, Cooke SJ (2019b) Collaboration and engagement produce more actionable science: quantitatively analyzing uptake of fish tracking studies. Ecol Appl 29(6):e01943

Norström AV, Cvitanovic C, Löf MF, West S, Wyborn C, Balvanera P, Bednarek AT, Bennett EM, Biggs R, de Bremond A, Campbell BM, Canadell JG, Carpenter SR, Folke C, Fulton EA, Gaffney O, Gelcich S, Jouffray J-B, Leach M, Le Tissier M, Martín-López B, Louder E, Loutre M-F, Meadow AM, Nagendra H, Payne D, Peterson GD, Reyers B, Scholes R, Speranza CI, Spierenburg M, Stafford-Smith M, Tengö M, van der Hel S, van Putten I, Österblom H (2020) Principles for knowledge co-production in sustainability research. Nat Sustain 3:182–190

Olson J, Pinto da Silva P (2019) Knowledge production at the science-policy interface: lessons from fisheries scientists. Sci Public Policy 47(1):47–55

Olsson J, Andersson ML, Bergström U, Arlinghaus R, Audzijonyte A, Berg S, Briekmane L, Dainys J, Dalby Ravn H, Droll J, Dziemian Ł, Fey DP, van Gemert R, Greszkiewicz M, Grochowski A, Jakubaviciute E, Lozys L, Lejk AM, Mustamäki N, Naddafi A, Olin M, Saks L, Skov C, Smolinski S, Svirgsden S, Tiainen J, Östman Ö (2023) A pan-Baltic assessment of temporal trends in coastal pike populations. Fish Res 260:106594

Ostrom E (2005) Understanding institutional diversity. Princeton University Press, Princeton

Piczak ML, Brooks JL, Bard B, Bihun CJ, Howarth A, Jeanson AL, LaRochelle L, Bennett JR, Lapointe NWR, Mandrak NE, Cooke SJ, Ferrari M (2022) Revisiting the challenge: perspectives on Canada's freshwater fisheries policies three decades after the Pearse Report. Facets 7: 912–935

Radinger J, Matern S, Klefoth T, Wolter C, Feldhege F, Monk CT, Arlinghaus R (2023) Ecosystem-based management outperforms species-focused stocking for enhancing fish populations. Science 379(6635):946–951

Reed MG, Abernethy P (2018) Facilitating co-production of transdisciplinary knowledge for sustainability: working with Canadian Biosphere Reserve practitioners. Soc Nat Resour 31: 39–56

Reed MS, Graves A, Dandy N, Posthumus H, Hubacek K, Morris J, Prell C, Quinn CH, Stringer LC (2009) Who's in and why? A typology of stakeholder analysis methods for natural resource management. J Environ Manag 90:1933–1949

Reed MG, Godmaire H, Abernethy P, Guertin M-A (2014) Building a community of practice for sustainability: strengthening learning and collective action of Canadian biosphere reserves through a national partnership. J Environ Manag 145:230–239

Reid AJ, Carlson AK, Creed IF, Eliason EJ, Gell PA, Johnson PT, Kidd KA, MacCormack TJ, Olden JD, Ormerod SJ, Smol JP, Taylor WW, Tockner K, Vermaire JC, Dudgeon D, Cooke SJ (2019) Emerging threats and persistent conservation challenges for freshwater biodiversity. Biol Rev 94:849–873

Reid AJ, Eckert LE, Lane JF, Young N, Hinch SG, Darimont CT, Cooke SJ, Ban NC, Marshall A (2020) "Two-Eyed Seeing": an indigenous framework to transform fisheries research and management. Fish Fish 22:243–261

Rittel HW, Webber MM (1973) Dilemmas in a general theory of planning. Policy Sci 4:155–169

Robinson KF, Arlinghaus R, Camp EV, Post JR, Stedman R, Sutton SG (2026) Adaptive management for recreational fisheries decisions in the face of uncertainty. In: Pope KL, Arlinghaus R, Hunt LM, Lynch AJ, van Poorten BT (eds) Understanding recreational fishers: disciplinary and interdisciplinary approaches for fisheries management. Springer Nature, Cham, p 719–746

Roser P, Dhellemmes F, Rittweg T, Möller S, Winkler H, Lukyanova O, Niessner D, Schütt J, Kühn C, Dennenmoser S, Nolte AW, Radinger J, Koemle D, Arlinghaus R (2023) Synthesizing historic and current evidence for anadromy in a northern pike (*Esox lucius* L.) meta-population inhabiting brackish lagoons of the southern Baltic Sea, with implications for management. Fish Res 263:106670

Said A, Trouillet B (2020) Bringing 'deep knowledge' of fisheries into marine spatial planning. Marit Stud 19:347–357

Schmitz Nunes MU, Cardoso OR, Silvano RAM, Fávaro LF (2021) Participatory mapping and fishers' knowledge about fish and shrimp migration in a subtropical coastal ecosystem. Estuar Coast Shelf Sci 258:107412

Schnarch B (2004) Ownership, control, access, and possession (OCAP) or self-determination applied to research: a critical analysis of contemporary First Nations research and some options for First Nations communities. Int J Indig Health 1(1):80–95

Silvano RAM, Baird IG, Begoossi A, Hallwass G, Huntington HR, Lopes PFM, Parlee B, Berkes F (2023) Fishers' multidimensional knowledge advances fisheries and aquatic science. Trends Ecol Evol 38:8–12

Slaton C, Koemle D, Birdsong M, Arlinghaus R (2023) Explaining attitudes to management actions and beliefs about other user groups and conservation with angler characteristics: a case study in a coastal pike (*Esox lucius*) fishery in the southern Baltic Sea, Germany. Fish Res 263:106669

Sopinka NM, Coristine LE, DeRosa MC, Rochman CM (2020) Envisioning the scientific paper of the future. FACETS 5(1):1–16

Star SL, Griesemer JR (1989) Institutional ecology, translations' and boundary objects: amateurs and professionals in Berkeley's Museum of Vertebrate Zoology, 1907-39. Soc Stud Sci 19:387–420

Strehlow HV, Korzhenevych A, Lucas J, Lewin WC, Weltersbach MS, Riepe C, Arlinghaus R (2025) Economic impact of resident and nonresident marine anglers to the local economy in Mecklenburg-Western Pomerania, Germany. Fish Manage Ecol 32, e12664

Sutherland WJ, Shackelford G, Rose DC (2017) Collaborating with communities: co-production or co-assessment? Oryx 51:569–570

Thoms P (1997) Creating a shared vision with a project team. J Proj Manag 11(1):33–35

van Gemert R, Koemle D, Winkler H, Arlinghaus R (2022) Data-poor stock assessment of fish stocks co-exploited by commercial and recreational fisheries: applications to pike *Esox lucius* in the western Baltic Sea. Fish Manag Ecol 29:16–28

van Maurik Matuk FA, Verschuuren B, Morseletto P, Krause T, Ludwig D, Cooke SJ, Haverroth M, Maeesters M, Mattijssen TJ, Keßler S, Lanza TR, Milberg E, Ming LC, Hernández-Vélez CA, da Silva KMT, Souza MPV, Souza VO, Fernandes JW, dos Reis Carvalho BL (2023) Advancing co-production for transformative change by synthesizing guidance from case studies on the sustainable management and governance of natural resources. Environ Sci Pol 149:103574

Vaske JJ, Beaman J, Barreto H, Shelby LB (2010) An extension and further validation of the potential for conflict index. Leis Sci 32:240–254

Venturelli PA, Hyder K, Skov C (2017) Angler apps as a source of recreational fisheries data: opportunities, challenges and proposed standards. Fish Fish 18:578–595

Venturelli P, Skov C, Audzijonyte A, Belmaker J, Ana Cardoso AS, Gordon L, Gundelund C, Hyder K, Jarić I, Lennox RJ, Potts WM, Vacura K, Vaz AS, Sbragaglia V (2026) Digital data mining. In: Pope KL, Arlinghaus R, Hunt LM, Lynch AJ, van Poorten BT (eds) Understanding recreational fishers: disciplinary and interdisciplinary approaches for fisheries management. Springer Nature, Cham, p 443–479

Walters CJ (1986) Adaptive management of renewable resources. Collier MacMillan, Basingstoke

Wilberg MJ, Irwin BJ, Jones ML (2026) Participatory modelling to support evaluation of management actions for recreational fisheries. In: Pope KL, Arlinghaus R, Hunt LM, Lynch AJ, van Poorten BT (eds) Understanding recreational fishers: disciplinary and interdisciplinary approaches for fisheries management. Springer Nature, Cham, p 693–718

Wilson S (2008) Research is ceremony: indigenous research methods. Fernwood Publishing, Winnipeg

Wyborn C, Datta A, Montana J, Ryan M, Leith P, Chaffin B, Miller C, van Kerkhoff L (2019) Co-producing sustainability: reordering the governance of science, policy, and practice. Annu Rev Environ Resour 44:319–346

Young N, Nguyen VM, Corriveau M, Cooke SJ, Hinch SG (2016) Knowledge users' perspectives and advice on how to improve knowledge exchange and mobilization in the case of a co-managed fishery. Environ Sci Pol 66:170–178

Evidence Synthesis in Recreational Fisheries Science and Management

Steven J. Cooke, Trina Rytwinski, Meagan Harper, Andrew Howarth, Shinichi Nakagawa, and Len M. Hunt

Abstract Researchers have studied the behaviours of recreational fishers and conducted science to support recreational fisheries management for over 50 years, resulting in a large volume of literature. Yet, evidence users (e.g., practitioners and fishery managers) rarely have the time to sift through and reconcile the literature, leading to biased approaches from selective use of evidence. To help evidence users make sense of these studies in a manner that minimizes bias, various evidence synthesis methods can be used. However, these methods are not all robust, can themselves be subject to bias, and their uncritical use could lead to poor management decisions and outcomes. Evidence synthesis has been applied in recreational fisheries management in several contexts (e.g., to understand how social–ecological systems theory applies to recreational fisheries; to evaluate the effectiveness of angler gear choice on biological outcomes for fish or the impact of recreation angling on other taxa than fish), yet is often conducted using less rigorous, informal synthesis methodologies. By contrast, meta-analysis to support stock assessment is reasonably common, but these approaches are uncommon in recreational fisheries. Here, we briefly review the suite of evidence-synthesis methods available, consider their strengths and weaknesses, and outline key steps involved in their conduct. We explore synthesis methods including traditional narrative literature reviews, systematic maps, rapid evidence syntheses, (quantitative) meta-analysis, and systematic reviews (which often include meta-analysis). In doing so, we also briefly review past evidence syntheses on this topic and summarize ways in which evidence synthesis can be best used to support recreational fisheries management in the future.

S. J. Cooke (✉) · T. Rytwinski · M. Harper · A. Howarth
Canadian Centre for Evidence-Informed Conservation, Department of Biology and Institute of Environmental and Interdisciplinary Science, Carleton University, Ottawa, ON, Canada
e-mail: StevenCooke@cunet.carleton.ca

S. Nakagawa
Evolution and Ecology Research Centre, School of Biological and Environmental Sciences, University of New South Wales, Sydney, NSW, Australia

L. M. Hunt
Ontario Ministry of Natural Resources, Thunder Bay, ON, Canada

K. L. Pope et al. (eds.), *Understanding Recreational Fishers*, Fish & Fisheries Series 45, https://doi.org/10.1007/978-3-031-99739-6_19

Keywords Decisions · Evidence · Fisheries management · Literature review · Meta-analysis · Systematic review

1 Introduction

Empirical research is very much the foundation of knowledge generation, whether it be in the health sciences (Bowling 2014), education (Drew et al. 2007), or the environmental sciences (Montello and Sutton 2012). Such studies—whether observational or experimental—generate important understanding of how the world works, ideally when informed by underlying theory. Yet, we also know that a similar study conducted on a different system or in a different context could result in vastly different findings. One empirical study is but one perspective on how the world works (Yanow and Schwartz-Shea 2015). There are many opportunities for bias or error to be introduced into empirical research (Sica 2006; Smith and Noble 2014). Moreover, the ecological and social world is complicated and that is particularly the case when studying distinct species, assemblages, ecosystems, and social systems as well as their interactions, as is so common in fisheries science. The resulting complexities represent routine struggles of practitioners and decision makers. In the context of this chapter, practitioners and decision makers are those involved in the governance and management of recreational fisheries, often but not always as part of government agencies. They tend to operate at local scales (e.g., waterbodies, basins) but may be involved in more regional, state, or provincial level, national, or even international activities. Although decision makers and practitioners tend to be tasked with making decisions and implementing various interventions (e.g., harvest regulations, restoration), they may also be involved in fisheries assessment. A practitioner or decision maker that relies on a single empirical study could be misguided: e.g., one may inadvertently or intentionally select the one study that reveals that a given intervention is beneficial for a given issue, yet ignore the many other studies that were equivocal. It is now well understood that greater understanding and potential to minimize bias come from synthesizing the evidence on a given topic through some form of evidence synthesis of the published record (Briner and Denyer 2012). We acknowledge that decisions involve considering various outcomes and always involve uncertainty, but evidence synthesis has the potential to reduce such uncertainty (Downey et al. 2022b).

Evidence syntheses come in many forms. From mini-reviews conducted in a matter of days and that are subject to many biases to the "gold standard" systematic reviews that use rigorous, repeatable methods in an attempt to minimize bias and often include meta-analysis (when the evidence base is amenable to it—e.g., quantitative rather than qualitative data), with many other forms in between. Some reviews are entirely narrative, whereas others use meta-analytical approaches to analyse existing data to determine the collective effect of a given impact or outcome. The relative strengths and weaknesses of each form of evidence synthesis are varied and need to be considered carefully when deciding which evidence synthesis method to use, or for interpreting the relative bias associated with existing evidence syntheses. Assuming evidence synthesis is being done in the service of decision makers, involving them in the process from the beginning to help determine which type of evidence synthesis is necessary is an important first step. Social science research focused on environmental decision makers (including fisheries managers) has revealed that they value evidence syntheses, understand that evidence syntheses are not all created equal, and would use systematic reviews if they were available on topics relevant to the decisions they make (Thomas-Walters et al. 2021). In other disciplines such as education and healthcare, robust evidence syntheses have become the foundation of decision-making (e.g., Pullin and Knight 2001), with this concept still novel for other realms, such as those related to environmental management (Sutherland et al. 2004) or recreational fisheries. What is clear is that for a variety of reasons (expansion of evidence base, recognition of the value of evidence synthesis), evidence syntheses are being increasingly recognized as important vehicles for improving decisions.

Recreational fishing is a popular fisheries sector. It is defined by the United Nations (UN) Food and Agriculture Organization (FAO) as "fishing of aquatic animals (mainly fish) that do not constitute the individual's primary resource to meet basic nutritional needs and are not generally sold or otherwise traded on export, domestic or black markets" (FAO 2012). Although recreational fishing varies in popularity among regions, in developed nations it is often the dominant sector targeting fish in inland and some nearshore coastal waters (Arlinghaus et al. 2002). Recreational fisheries are considered good examples of complex social–ecological systems (Arlinghaus et al. 2017) whereby recreational fishers influence aquatic ecosystems while aquatic ecosystems influence people engaged in fishing. In the context of recreational fisheries science and management, there is growing recognition that effective management requires adoption of social–ecological systems frameworks and thinking, given the inherent interconnections between the environment (including fish) and people (i.e., the recreational fishers), and the reality that many systems are constantly changing (Arlinghaus et al. 2016). As such, there is a need to conduct science of all forms (i.e., spanning the natural and social sciences) and generate evidence to inform decision-making (FAO 2012), to ensure that recreational fisheries are sustainable, as well as beneficial to people and ecosystems today and in the future (Elmer et al. 2017). Given the complexities of managing recreational fisheries, evidence synthesis is a potentially powerful tool for understanding disparate evidence and identifying what it means when aggregated. Evidence synthesis also serves to determine the generalizability of observations and understand the contexts in which evidence from one context can be applied to another. More broadly, evidence synthesis (in particular meta-analysis) has been embraced in other domains of fisheries science and management (reviewed in Hilborn and Liermann 1998; Myers 2001; Thorson et al. 2015), suggesting that evidence synthesis could easily be mainstreamed for recreational fisheries and the human dimensions of this sector.

The objective of this chapter is to summarize the suite of evidence synthesis methods available, consider their strengths and weaknesses (Table 1 and 2), and outline key steps involved in their conduct with a particular focus on recreational fisheries science and management. We explore synthesis methods including

Table 1 Features of different evidence synthesis (components that are specific strengths of each type are highlighted in bold, and weaknesses are underlined; based on summaries by Dicks et al. 2014b; Cook et al. 2017; Dicks et al. 2017; CEE 2018)

	Traditional Narrative Review	Meta-analysis	Systematic Review	Rapid Review	Systematic Map	Evidence Synopses
Description	**Exploratory syntheses used in cases where specific/testable questions may be lacking** (i.e., historical contexts, theory building) or where literature is too diverse for other synthesis techniques.	Aggregative syntheses combining **comparable studies statistically** to test a hypothesis	Aggregative evidence syntheses **answering defined questions using transparent, repeatable methods, in an unbiased way** through narrative and potentially quantitative or qualitative synthesis (may or may not include meta-analysis).	An accelerated evidence synthesis similar to a systematic review, but on a **short time scale** to meet emergency decision-making requirements.	Exploratory syntheses that **collate and describe the state of knowledge** and available literature on a topic.	**Summarizes** existing evidence syntheses **for managers and policy makers** (i.e., short summaries or briefs)
Scope	Broad	Specific	Specific	Specific	Broad	Variable[a]
Defined methodology applied	Rarely	Sometimes	**Always**	Sometimes	**Always**	Variable
A priori protocol published	Rarely	Rarely	**Always**	**Always**	**Always**	Variable
Systematic literature search used	Rarely	Sometimes	**Always**	Sometimes	**Always**	Variable
Specific inclusion	Sometimes	**Always**	**Always**	**Always**	**Always**	Variable

criteria applied						
Critical appraisal of study validity conducted	Rarely	Rarely	**Always**	Sometime	Sometimes	Variable
Quantitative synthesis conducted	Never	**Always**	Sometimes	Sometimes	Not applicable	Variable
Qualitative synthesis conducted	Never	Not applicable	**Sometimes**	Sometimes	Not applicable	Variable
Narrative synthesis conducted	**Always**	Sometimes	Always	**Always**	**Always**	Variable
Resources (time, money, expertise)	**Low**	Moderate	Very high	Moderate	Moderate/High	**Low**
Type of outcomes	Narrative synthesis[b]	May include brief description of literature included in meta-analysis **Measures of effect**[b]	Literature database Narrative synthesis Meta-analysis with measure of effect **Measures of quality of the available literature base**	Preliminary literature database Preliminary narrative or quantitative synthesis May include measure of overall effect or quality of the literature base	**Extensive literature database** Narrative synthesis	**Easy-to-read, quick summary** of outcomes of a synthesis May include specific recommendations for managers and decision makers
Risk of bias	High	Moderate/High	**Low**	Moderate	**Low**	Variable
Objectivity	Low	Moderate/Low	**High**	**High**	**High**	Variable
Transparency	Low	Moderate/Low	**High**	**High**	**High**	Variable

(continued)

Table 1 (continued)

	Traditional Narrative Review	Meta-analysis	Systematic Review	Rapid Review	Systematic Map	Evidence Synopses
Replicability	Low	Moderate/Low	**High**	**High**	**High**	Low
Example questions	How did X develop in a specific topic?	What is the effectiveness of an intervention? What is the effect of X on Y? What is the prevalence of a phenomenon? How reliable is a specific method?	What is the effectiveness of an intervention? What is the effect of X on Y? What is the prevalence of a phenomenon? How reliable is a specific method?	What is the effectiveness of an intervention? What is the effect of X on Y? What is the prevalence of a phenomenon? How reliable is a specific method?	What evidence currently exists? How much evidence is available? What interventions and outcomes are studied most frequently? What methods are used?	Dependent on the synthesis being summarized

[a]Dependent on the type of synthesis being summarized
[b]In some instances, authors use "vote-counting" (e.g., counting studies with significant vs non-significant results) to compare the number of positive or negative effects across studies. This should be avoided whenever possible because results can often be misleading. Vote-counting is highly subjective due to the difficulty of defining positive and negative outcomes and it does not provide a measure of magnitude of effect or take study validity and power into account (CEE 2018)

Table 2 Further information for the features of different evidence synthesis presented in Table 1 (components that are specific strengths of each type are highlighted in bold, and weaknesses are underlined (based on summaries by Dicks et al. 2014b; Cook et al. 2017; Dicks et al. 2017; CEE 2018))

	Traditional Narrative Review	Meta-analysis	Systematic Review	Rapid Review	Systematic Map	Evidence Synopses
Scope: Describes whether questions are normally formatted to determine an effect (specific) or are more flexible and can be used to explore various aspects of the literature base (broad) beyond testing a hypothesis or determination of an effect.						
	Broad	Specific	Specific	Specific	Broad	Variable[a]
Components of evidence synthesis: The inclusion of different components of evidence syntheses are determined by the type of evidence syntheses [i.e., systematic reviews need not always include meta-analysis), but where possible, the inclusion of these components can improve the relatability, repeatability, and transparency of the synthesis (Cook et al. 2017; CEE 2018)]. Here, we highlight whether different evidence syntheses commonly include particular components, or never include them, and identify whether this inclusion or exclusion of components is a strength or weakness of a synthesis. Note that some evidence syntheses may claim to meet the standards of different types, but in some instances may fall short. Tools such as the CEE's CEEDER (https://environmentalevidence.org/ceeder/; Pullin et al. 2022) can help decision makers determine if the evidence they are using meets high standards in conduct and reporting.						
Defined methodology applied: Defined methodologies assist authors in following a consistent approach that improves transparency and reproducibility. They can also decrease bias. Defined methodologies may be provided by organizations such as the Collaboration for Environmental Evidence (CEE 2018), or the Cochran Collaboration (Higgins et al. 2019).						
	Rarely	Sometimes	**Always**	Sometimes	**Always**	Variable
A priori protocol published: A priori protocols act as a guide for the conduct of an evidence synthesis from question formulation to analysis, decreases bias and decreases the risk of mission creep.						
	Rarely	Rarely	**Always**	**Always**	**Always**	Variable
Systematic literature search used: Systematic methods of literature searching from multiple sources (i.e., databases, search engines) help to ensure that the database created during synthesis is comprehensive and contains as much relevant literature as possible. If strategies are not comprehensive, bias may be introduced when relevant evidence is omitted. To be systematic, there should be a documented search strategy with clear, repeatable search strings that includes searches for grey literature.						
	Rarely	Sometimes	**Always**	Sometimes	**Always**	Variable
Specific inclusion criteria applied: Pre-defined inclusion criteria that avoid ambiguous terms help decrease reviewer bias and ensure articles are included in a transparent manner during the review process. Inclusion criteria ensure that subjectivity during screening is decreased, especially if multiple individuals are involved. Screening criteria should be tested through consistency checking among reviewers to avoid reviewer bias.						

(continued)

Table 2 (continued)

	Traditional Narrative Review	Meta-analysis	Systematic Review	Rapid Review	Systematic Map	Evidence Synopses
	Sometimes	**Always**	**Always**	**Always**	**Always**	Variable
Critical appraisal of study validity conducted: Enables weighting of evidence included in a synthesis based on the reliability of a study. Helps decrease the impact of bias within primary literature itself, while also ensuring that the generalizability of studies, as they relate to the synthesis question, is considered.						
	Rarely	Rarely	**Always**	Sometimes	Sometimes	Variable
Quantitative synthesis conducted: If sufficient information is available in the published literature, quantitative analysis (i.e., meta-analysis) may be conducted to determine an overall effect of an intervention.						
	Never	**Always**	Sometimes	Sometimes	Not applicable	Variable
Qualitative synthesis conducted: Syntheses of primary literature using qualitative techniques require different methods of synthesis than those reporting quantitative results. These are newer techniques in evidences synthesis and are normally found only in systematic reviews.						
	Never	Not applicable	**Sometimes**	Sometimes	Not applicable	Variable
Narrative synthesis conducted: Describes the literature base to provide context for further analysis (if possible).						
	Always	Sometimes	Always	Always	Always	Variable
Resources (time, money, expertise): Evidence syntheses take variable amounts of time and effort and require different expertise depending on the types of analyses conducted (e.g., to conduct a full systematic review requires both topic and methodological expertise) and vary significantly in person hours depending on resources available and the features of the evidence syntheses being considered (Haddaway and Westgate 2019). When technical expertise is moderate (i.e., systematic maps or meta-analysis), the size of the evidence base can strongly influence the resources needed (Cook et al. 2017).						
	Low	Moderate	Very high	Moderate	Moderate/High	**Low**
Type of outcomes:						
	Narrative synthesis[b]	May include brief description of literature included in meta-analysis **Measures of effect**[b]	Literature database Narrative synthesis Meta-analysis with measure of effect **Measures of quality of the available literature base**	Preliminary literature database Preliminary narrative or quantitative synthesis May include measure of overall effect or quality of the literature base	**Extensive literature database** Narrative synthesis	**Easy-to-read, quick summary of outcomes of a synthesis** May include specific recommendations for managers and decision makers

Risk of Bias, Objectivity, Transparency, and Replicability: Methods used should always aim to minimize risk of bias through inclusion of checks and balances (i.e., consistency checks, inclusion of critical appraisal etc.), while maximizing objectivity (CEE 2018). Transparency and replicability through robust, well-documented methods ensure that readers fully understand all critical steps used during the synthesis process and that other reviewers can repeat the synthesis when new evidence becomes available (Cook et al. 2017).						
Risk of bias:						
	High	Moderate/High	**Low**	Moderate	**Low**	Variable
Objectivity:						
	Low	Moderate/Low	**High**	**High**	**High**	Variable
Transparency:						
	Low	Moderate/Low	**High**	**High**	**High**	Variable
Replicability:						
	Low	Moderate/Low	**High**	**High**	**High**	Low
Example questions	How did X develop in a specific topic?	What is the effectiveness of an intervention? What is the effect of X on Y? What is the prevalence of a phenomenon? How reliable is a specific method?	What is the effectiveness of an intervention? What is the effect of X on Y? What is the prevalence of a phenomenon? How reliable is a specific method?	What is the effectiveness of an intervention? What is the effect of X on Y? What is the prevalence of a phenomenon? How reliable is a specific method?	What evidence currently exists? How much evidence is available? What interventions and outcomes are studied most frequently? What methods are used?	Dependent on the synthesis being summarized

[a]Dependent on the type of synthesis being summarized

[b]In some instances, authors use "vote-counting" (e.g., counting studies with significant vs non-significant results) to compare the number of positive or negative effects across studies. This should be avoided whenever possible because results can often be misleading. Vote-counting is highly subjective due to the difficulty of defining positive and negative outcomes and it does not provide a measure of magnitude of effect or take study validity and power into account (CEE 2018)

traditional narrative literature reviews, systematic maps, rapid evidence syntheses, meta-analysis, and systematic reviews (which often include meta-analysis). In doing so, we also briefly review past evidence syntheses that deal with recreational fisheries and summarize ways in which evidence synthesis can be best used to support recreational fisheries management today and in the future. In many ways, this chapter serves as a primer given that there are lengthy guidance documents (e.g., the Collaboration for Environmental Evidence guidelines for conducting systematic reviews) and entire textbooks on topics like meta-analysis.

2 Why Is Evidence Synthesis Relevant to Recreational Fisheries Science and Management?

2.1 Fisheries Managers and Other Practitioners Lack Time to Keep Up with the Entirety of an Evolving Evidence Base

The evidence base is continually expanding. This is, of course, beneficial but makes it more challenging for evidence users to keep up with new knowledge. Fisheries managers and other practitioners rarely have time to devote to reading new material, given that time itself is a precious resource (Pullin et al. 2004). To be clear, this is not about laziness or lack of commitment. It would take several days per week to try and keep on top of the evolving recreational fisheries literature, which is simply not reasonable. Evidence syntheses serve as collations of evidence to help practitioners focus their limited time on key resources.

2.2 Fisheries Managers and Other Practitioners Often Lack Access to the Entirety of an Evidence Base

Even if a fisheries manager has time to keep up with new research, that assumes that they have access to the entirety of the evidence base. Yet, it is well known that most of the evidence base is in the form of peer-reviewed papers that require subscription access. Some managers embedded within larger agencies have access to library support, but that is not always the case (e.g., small angling club context in central Europe, Daedlow et al. 2011). Although open access publishing is transforming access (albeit slowly; Willinsky 2006), the reality is that journal pay walls still limit access. Social science surveys of fisheries managers in North America have explicitly identified access as a major barrier (Nguyen et al. 2019). Without access to the entirety of the evidence base, key information can be missed. Evidence syntheses bring together evidence that can otherwise be difficult to access. Moreover, evidence syntheses are often published as open access so they can be accessible to evidence users.

2.3 Fisheries Managers and Other Practitioners Tend to Rely on Experience and May Ignore New Evidence

For a variety of complex reasons, it is common for fisheries managers and other environmental practitioners to rely on experience rather than using the entirety of the evidence base—especially new information (Pullin et al. 2004; Kadykalo et al. 2021). There are many barriers to including new evidence in decision-making that extend from the individual to the institution (for a summary in stocking, see Arlinghaus et al. 2022). This is problematic in that objective perspectives can be constrained, given that tradition and experience become codified as norms in decision-making. Evidence syntheses can illuminate alternative ways of doing things and provide the rationale for doing so, even if it means change.

2.4 Individual Empirical Studies Vary in Quality and Thus Can Be Biased in Various Ways

Research is not easy, and for assorted reasons (e.g., poor experimental design, errors, low sample sizes, other contextual influences), each study varies in quality, potentially introducing bias (Delgado-Rodriguez and Llorca 2004). Moreover, authors may take liberties and draw conclusions that are not well-supported by their findings. Although peer review attempts to improve the quality of research, it too is imperfect (Smith 2006). Indeed, criticisms can be found with every study; therefore, replication of individual study questions is important. If one were to read just one paper and uncritically accept that the author's conclusions are valid, there is potential for major bias that could lead to managers making poor (or even incorrect or counterproductive) decisions. Some evidence synthesis methods (e.g., systematic reviews) include quality appraisal phases intended to assess the potential for bias. Moreover, adoption of best practices for evidence syntheses can reduce biases related to searches and inclusion.

2.5 Fish and Fisheries Are Diverse and There is Need to Understand the Context in Which Different Pressures or Interventions Result in Various Outcomes

Fish are diverse, with ~30,000 known species (Helfman et al. 2009). There is also great variation in environments, ranging from small, inland waters to estuaries and the high seas. The fisheries sectors also vary dramatically (e.g., subsistence, commercial, recreational), and there is also much variation within sectors related to regional, cultural, geopolitical, or socio-economic factors. As such, it is necessary to determine the contexts in which various pressures (e.g., habitat loss, water quality, exploitation) and interventions (e.g., harvest controls, effort controls, habitat restoration) are effective. Evidence synthesis can be done in ways that consider how context influences impacts and outcomes, thus helping managers to determine when findings from one system can be applied to another. Meta-analytical approaches can be used to assess stock status, which has become common in the marine commercial sector (Hilborn and Liermann 1998; Myers 2001) and is increasingly being used in the recreational sector (Thorson et al. 2015; Cahill et al. 2022). Meta-analysis can also be used to synthesize classical human dimension context, such as studying the drivers of angler satisfaction (Birdsong et al. 2021) or preferences (Hunt et al. 2019).

2.6 Decisions Made by Fisheries Managers Are Often Controversial Such That There Is Need for Robust Evidentiary Support

Conflict and controversy are common when dealing with natural resources and environmental issues (Lewicki et al. 2003), including those relevant to recreational fisheries management (Arlinghaus 2005). Indeed, some recreational fisheries issues are so challenging that they are deemed to be "wicked problems" (Bower et al. 2017). It is common for recreational fishers that occur on a given waterbody to have various segments that may target distinct species or derive satisfaction in diverse ways that lead to conflict over fisheries management decisions (Hunt et al. 2023). Sometimes recreational fisheries management actions conflict with biodiversity conservation goals (Cowx et al. 2010). The more controversy or conflict, the greater the need for robust evidentiary support to inform management. Evidence syntheses, especially when conducted using robust methods, can provide fisheries managers with the evidence they need to make tough decisions—even if those decisions may not be popular among stakeholder groups (Cooke et al. 2017a).

2.7 Evidence Synthesis Reveals Gaps in the Evidence Base That Can Be Addressed Through Targeted Empirical Research

The evidence landscape is highly variable, with some topics being very well studied with an abundance of high-quality studies, whereas for other topics, there may be nothing. In the recreational fisheries realm, this is common, with much known about a few species (i.e., the most popular species in developed countries or research emphasizing specific countries and cultures, Birdsong et al. 2021). Although evidence synthesis is often done to identify what we do know about a given topic, it is equally useful for identifying what we do not know. These evidence gaps can serve to inform future research that could be conducted by (or funded by) the management agencies in need of such evidence. It is also a signal to the research community where there may be profitable partnerships and research topics when engaging in applied research (Cooke et al. 2020).

2.8 *Evidence Synthesis Can Identify Deficiencies in How Empirical Research Is Conducted That Can Yield Direction for Future Researchers to Develop a Robust Evidence Base*

Evidence syntheses that include aspects of critical appraisal can be useful for identifying the characteristics of research that are important for generating a robust evidence base (Sutton et al. 2009). Critical appraisal requires careful thought on aspects of experimental design that represent the "gold standard" for a given discipline or question while also recognizing inherent challenges that are common with some topics, including many relevant to environmental science and recreational fisheries (Cooke et al. 2017b). It is not uncommon for a systematic review to conclude that the overall evidence base for a given topic is weak (e.g., as Taylor et al. (2019) report for use of habitat restoration to improve substrate for spawning fish), and in doing so, set the bar for what researchers should strive for. In the context of human dimensions of recreational fisheries, for which there have been relatively few evidence syntheses (but see Johnston et al. 2006; Hunt et al. 2019; Birdsong et al. 2021; Schafft et al. 2021), this represents an important opportunity for enhancing the evidence base so that the evidence that is generated is of high quality and thus can be used in future evidence syntheses.

2.9 *Evidence Syntheses Provide the "Big Picture" on What Works and What Doesn't to Guide Evidence-Based or Evidence-Informed Decision-Making*

Evidence-based decision making involves embracing evidence fully, whereas evidence-informed decision-making tends to acknowledge that other factors (e.g., politics, socio-economic reasons) may influence decisions. It is easy to cherry-pick findings or ignore vast swaths of the evidence base. Yet, what we need to keep our eye on is the "big picture" of what works and what does not. Almost always, one can find a study to refute or support one's worldview. Yet, a robust evidence synthesis can bring together diverse sources of evidence to yield insight into what works and what does not—and in which contexts. There are always exceptions with science, and uncertainty is common in environmental decision-making (Ricci et al. 2003). Yet, it is those characteristics of science that reinforce the importance of evidence syntheses that bring together diverse studies and use robust methods (e.g., meta-analysis) to identify the primary signals in the evidence base (Sutton et al. 2009; Bartolucci and Hillegass 2010). In the context of recreational fisheries, it is equally easy to cherry-pick, given a reasonably substantial evidence base. Moreover, many recreational fisheries issues are emotionally charged, emphasizing the importance of objective, synthetic guidance (Elmer et al. 2017).

3 Types of Evidence Syntheses and an Overview of Their Conduct

Here, we provide an overview of the primary types of evidence synthesis and summaries of their conduct. The order in which they are presented does not reflect their relative rigor, but rather the concepts that are foundational to different forms of review. For example, rapid reviews are presented after systematic reviews because they involve a subset of steps found in other approaches, so it is sensible to first present the full systematic review model. We also wish to be clear that meta-analysis is a form of analysis that can be conducted with evidence and can be incorporated into other evidence synthesis methods (e.g., systematic review with meta-analysis).

3.1 Traditional Narrative Reviews

Narrative or traditional reviews are largely unstructured ways to synthesize literature. These reviews are generally comprehensive, cover a variety of topics, but often lack clarity of methods, such as how literature searches are conducted and how the relevance and validity of reviewed materials are assessed (Collins and Fauser 2005). The lack of structure when conducting narrative reviews results in criticisms of unreliability of results (Baethge et al. 2019) and concerns about subjectivity and bias affecting the validity of conclusions (e.g., Ferrari 2015; Siddaway et al. 2019).

We use a simple approach to differentiate narrative and systematic reviews (described in detail below). Systematic reviews focus on both structured ways that literature is searched and compiled and the presence of clearly identified and testable research questions. Narrative studies can include instances where researchers use structured methods for searching for literature but lack specific and testable research questions. Our distinction here is not universally accepted. For example, Green et al. (2006) differentiate qualitative systematic reviews from narrative reviews, whereas Siddaway et al. (2019) include narrative reviews as a potential type of systematic review. Our general definition is, however, consistent with calls by researchers to add structure when conducting narrative reviews (e.g., Ferrari 2015; Baethge et al. 2019).

Systematic reviews offer some advantages over narrative reviews in terms of transparency, reproducibility, explicitness, and clarity (Collins and Fauser 2005). Researchers conducting systematic reviews are explicit and transparent in their disclosure of methods and their purpose, thus making them reproducible (Siddaway et al. 2019). By contrast, researchers conducting narrative reviews seldom divulge (and often do not include) methods, heightening concerns about subjectivity, bias, and lack of reproducibility. Despite these advantages of systematic over narrative reviews, we believe that a few cases exist where narrative reviews are warranted.

The first case where narrative reviews offer an advantage over systematic reviews is when the literature review focuses on describing a field of study. Providing a historical context of literature is one such case where narrative reviews are commonly applied (Baumeister and Leary 1997). Systematic reviews are ill-conditioned in this case, as their limited scope and reliance on highly prescribed methods (Collins and Fauser 2005; Ferrari 2015) are at odds with the more exploratory focus of a historical context review. Arlinghaus et al. (2007) provide one such example of a narrative-based, historical context review from recreational fisheries through their comprehensive review of catch-and-release fishing, which included a human dimensions component about the drivers of catch-and-release behaviour of anglers.

The second case involves reviews that rely on inductive, rather than deductive reasoning. Deductive reasoning is synonymous with inference, whereby one moves from the general to the specific. For example, the science-based process of hypothesis testing starts with a general hypothesis that is evaluated using inferences drawn from specific methods and observations. In inductive reasoning, the researcher starts from the specific and works their way to the general. A literature review focused on developing theory is an example of inductive reasoning, whereby the observations (reviewed literature) are used to develop general theory or hypotheses. Narrative reviews are a valuable technique for theory building (Baumeister and Leary 1997). An example of this type of narrative review that is applied to recreation fisheries is from Hunt (2005), who reviewed literature from resource economics to theorize how recreational fishers choose fishing sites.

The third case for using narrative reviews is when results from the literature are based on different methods or conceptualization (e.g., Baumeister and Leary 1997; Siddaway et al. 2019). Quantitative systematic reviews fail in this case because of a lack of comparability among the reviewed studies. This failure can be overcome with narrative reviews that permit researchers to increase the scope of the review and to rely on less prescribed methods. Hyder et al. (2020) provide an example with their general review of marine recreational fisheries using literature and researchers' expressions of interest for contributing to a special issue to identify key themes and knowledge gaps.

Although narrative reviews are suitable in a few cases, researchers strongly recommend that additional structure be added to these reviews to address concerns with subjectivity and bias (e.g., Green et al. 2006; Ferrari 2015). Inconsistent guidance for conducting these reviews yields inconsistent methods. It is possible to produce reviews based on expert opinion rather than evidence from the literature (Wright et al. 2007) but that also comes with biases. Consequently, researchers conducting narrative reviews should better conduct (and communicate) their methods in ways that are reproducible and follow the growing advice on how to conduct these reviews (e.g., Ferrari 2015; Siddaway et al. 2019).

3.2 Meta-Analysis

A meta-analysis statistically aggregates effect sizes from studies investigating the same or related topics. Effect sizes represent the degree of evidence and are often statistics measuring relationships between two variables (e.g., correlation and its transformation, Fisher's Zr) and differences between two groups in continuous measurements (e.g., response ratio and standardized mean difference [SMD], also known as Cohen's *d* or Hedges' *d*) or in mortality or survival (e.g., odds ratio or risk ratio; White et al. 2021). Most (non-medical) meta-analyses use "unitless" statistics (e.g., all the effect size statistics mentioned above) so that meta-analysts can synthesize results measured in different units from different studies in a statistical model (note that one's research question dictates what effect size one chooses for their meta-analysis; see Nakagawa et al. 2017).

A formal meta-analysis weighs each effect size by the inverse of its corresponding sampling variance. How sampling variance is calculated differs among effect size statistics (e.g., Zr, SMD, and response ratio), but sampling variance always comprises sample size. Perhaps, it is intuitive that effect sizes (i.e., evidence) need weighting according to their sample sizes. Some meta-analyses are unweighted, but such meta-analyses have two shortfalls. First, unweighted meta-analyses are less precise because they give every effect size the same weight, rather than giving more weight to more precisely estimated effect sizes. Therefore, unweighted meta-analysis is more susceptible to publication bias. Second, unweighted meta-analysis cannot estimate heterogeneity among studies, which is due to differences in sample size (more on each concept [i.e., publication bias and heterogeneity] below; Nakagawa et al. 2023a, b).

Meta-analysis has become an essential part of the scientific process across different fields, especially in medical, social, and biological sciences (including in stock assessment for commercial fisheries (Hilborn and Liermann 1998) and even in recreational fisheries (Thorson et al. 2015; Cahill et al. 2022)). Although approximately a quarter of a million meta-analyses exist, the main aims of meta-analyses should usually fall into two categories: (1) estimating the overall effect size (evidence) and its heterogeneity, and (2) explaining the observed heterogeneity (Gurevitch et al. 2018). We can quantify an overall mean and heterogeneity among effect sizes using the "random-effects" model (another traditionally used model is called the "fixed-effect" model, but this model is now rarely used because it assumes zero heterogeneity). We may realize that an overall estimate itself may mean extraordinarily little without a measure of heterogeneity. This is because zero overall means with low heterogeneity likely represent the lack of an effect, whereas zero mean with high heterogeneity could mean true positive and negative effects depending on the contexts. An example of meta-analysis in the human dimensions of recreational fisheries is the meta-analysis by Birdsong et al. (2021), who looked at all the available evidence on drivers of angler satisfaction across the globe. A more ecological example is Schafft et al. (2021), taking a recreation ecology lens to look at the ecological impacts of recreational fishing and other water-based recreation on freshwater taxa. A final example is Hunt et al. (2019), who summarized all studies on angler preferences, but the way preferences are assessed in utility models prevented the use of formal meta-analytical approaches based on effect sizes.

There is a need for a meta-regression model to explain heterogeneity among effect sizes. A meta-analysis can be thought of as a model with just an intercept (i.e., overall mean), whereas a meta-regression is a linear model with predictors. In meta-regression, predictors are often referred to as "moderators." Moderators can be any variables contributing to variation among effect sizes (e.g., separate locations, methods, and subject characteristics). A meta-regression can reveal patterns called "review-generated" evidence to contrast with "study-generated" evidence. For example, meta-regression can identify consistent differences between different methods or taxa or among different angler groups fishing different environments (Birdsong et al. 2021). Of relevance, Nakagawa and Santos (2012) stated that there are three types of moderators: biological, methodological, and contextual; the last kind of moderator relates to publication bias (see below). In a similar vein, Thorson et al. (2015) discussed three sources of heterogeneity that meta-regression could deal with: experimental variability (sampling differences), parametric (contextual differences), and functional variability (assumed underlying social or biological mechanistic differences). By identifying appropriate moderators, the importance of these different variabilities can be quantified and used to make context-dependent and species-specific predictions (Spake et al. 2022; for an example in human dimensions of angling, see Birdsong et al. (2021)).

Though often overlooked (Nakagawa et al. 2022), all meta-analyses (including a systematic review with meta-analysis) should have the third aim of quantifying publication bias in the dataset. Publication bias is almost certainly present in the data set (sample of studies); it comes in different forms, but the most famous one is due to the selective publication of statistically significant results. Such selective publication skews meta-analytic results, leading to biased conclusions. In a meta-analysis, this kind of publication bias manifests as the "small-study" effect, wherein studies with small sample sizes often have large effect sizes (Sterne et al. 2011). Another common type of publication bias is known as the "decline" effect, where the magnitude of the effect declines over time (Koricheva and Kulinskaya 2019). Recently, Nakagawa et al. (2022) reviewed methods for detecting and assessing the impacts of publication biases. Also, they proposed that in multilevel or hierarchical meta-regression models, one could detect and correct both the small-study and the decline effects by fitting the inverse of "effective" sample sizes and publication year, respectively, as moderators (Yang et al. 2023; cf. Andrews and Kasy 2019). Furthermore, they have noted that the most sophisticated, complex class of the methods are "selection models," which considers the underlying process of how publication and selection bias occur, although implementations for relevant models for ecological and environmental research remain a challenge. Therefore, it is a common misunderstanding that publication bias renders meta-analytic results invalid. Notably, testing for publication bias allows one to quantify bias across studies, whereas critical appraisal attempts to estimate the risk of bias for each study. It is possible to include some elements of critical appraisal about each study in a meta-regression analysis as moderators (e.g., whether studies use randomization or blinding).

Finally, for a meta-analysis, or more accurately, a systematic review with meta-analysis, there is a well-known reporting guideline: PRISMA (Preferred Reporting Items for Systematic Reviews and Meta-Analyses; Moher et al. 2009), or its recent update PRISMA 2020 (Page et al. 2021). However, PRISMA focuses on medical studies, so more relevant reporting guidelines for ecological and environmental studies are PRISMA-EcoEvo—a PRISMA extension for ecology and evolution (O'Dea et al. 2021). Another useful framework is ROSES (RepOrting standards for Systematic Evidence Syntheses), which is tailored reporting guidelines for systematic reviews and maps in environmental research, but it is equally relevant to meta-analysis (Haddaway et al. 2018).

3.3 Systematic Reviews (With or Without Meta-Analysis)

In the evidence synthesis world, systematic reviews are considered the gold standard for reliability (Haddaway et al. 2015). This is because they are a highly structured form of evidence synthesis where methods are detailed a priori in a protocol that has been externally peer-reviewed and published. The protocol is essential to minimize bias introduced from those conducting the review (e.g., resulting from ad-hoc decisions made or "mission creep" during the synthesis process). The common types of questions amenable to systematic reviews are: What is the effectiveness of an intervention? What is the effect of X on Y? What is the prevalence of a phenomenon? How reliable is a specific method?

The overarching goal of a systematic review is to answer a well-defined question as precisely and transparently as possible in an unbiased way. To accomplish this, the process includes the following key stages (detailed further in CEE 2018):

1. Creation of an advisory team with relevant stakeholders (including evidence users such as recreational fisheries managers).
2. Comprehensive searches for all relevant evidence on the specific topic from a variety of sources, including commercially published scientific literature and grey literature (that may or not be peer-reviewed).
3. Careful screening of all evidence according to predetermined inclusion criteria.
4. Critical appraisal of study validity (i.e., internal validity: a detailed assessment of the risk of bias by examining aspects of the methodology used to generate the study data [e.g., representative sampling, appropriate methods of measurement, and robust statistical analyses]; as well as external validity: an assessment of the generalisability of each study as it relates to the review question).

5. Transparent coding of study characteristics and extraction of study findings.
6. Well-documented and comprehensive synthesis of the qualitative or quantitative study findings.
7. Transparent reporting (and communication) of the review results with use of extensive supplementary information.

One of the components that often distinguishes systematic review from other forms of evidence synthesis is the inclusion of the critical appraisal of study validity (fourth key stage). When this component is not included in a synthesis, reviewers are essentially treating each study equally with respect to its reliability. If bias is present in primary studies, their results will be incorrect. Subsequently, if an evidence synthesis is based on incorrect evidence, synthesis results will also be incorrect, resulting in misleading conclusions (Boutron et al. 2019; Rytwinski et al. 2021). By following the above (stepwise and rigorous) methodology (including critical appraisal of study validity), systematic reviews identify risks of bias in the evidence itself, minimize bias in the way that evidence is identified, selected, and discussed, allow for repeatability, and thus provide reliable findings that could inform decision-making (Pullin et al. 2016; CEE 2018). These factors are what make systematic reviews more comprehensive and less open to potential bias than other review formats that summarize the literature in an unstructured way (e.g., traditional reviews and stand-alone meta-analyses) (Roberts et al. 2006; Koricheva and Gurevitch 2014; Pullin et al. 2020). However, systematic reviews are, as a result, more resource-intensive and easily take one year or more to complete (Haddaway and Westgate 2019).

Different forms of synthesis are possible with systematic review: narrative, quantitative, and qualitative. The type of synthesis used depends on the specific review question, as well as the available evidence base to address that question with respect to the amount and reliability of evidence, and the type of study data (i.e., qualitative, quantitative, or mixed study findings) (CEE 2018).

3.3.1 Narrative Synthesis

All systematic reviews include some form of narrative synthesis whereby findings of individual primary studies are synthesized with tables and figures (often with descriptive statistics) and supporting text to explain the context (e.g., study and population characteristics, study design, relevant outcomes). Included in the narrative synthesis should be a statement on the assessment of study validity. Sometimes, narrative synthesis is all that is feasible in terms of drawing study findings together (e.g., when faced with a pool of disparate studies of relatively high susceptibility to bias). Other times, quantitative synthesis (meta-analysis) or qualitative synthesis is also possible and appropriate.

3.3.2 Quantitative Synthesis

When attempting to measure the effect of an intervention or exposure, a quantitative synthesis is sought to provide a combined effect and a measure of its variance within and between studies, as well as to explore potential sources of heterogeneity (see Sect. 3.2 above). When the evidence base of a systematic review is comprised of studies with both higher and lower risks of bias, as noted previously, it is important to understand the potential impact of this bias on review results. This can be accomplished with meta-analysis, for instance, by means of developing study weights that account for study quality (Schafft et al. 2021) or sensitivity analysis (i.e., comparing results of models that include all studies, studies at lower risk of bias only, and studies at higher risk of bias only; see Rytwinski et al. 2021). To date, most systematic reviews in environmental management are syntheses of quantitative study findings (i.e., those with the intention of doing a meta-analysis but that was not possible given the available evidence base or those that were able to include a meta-analysis).

3.3.3 Qualitative Synthesis

In the environmental and conservation social science realm, it is common to use qualitative methods (e.g., interviews, observations of behaviour) to capture data (usually textual) on the views or reactions of individual people in relation to a question. Here, the goal of the qualitative synthesis is to describe the potential range of views to increase understanding of the context of some policy- and practice-relevant issue or issues (CEE 2018). A synthesis of evidence from qualitative research can add value to research and decision-making by exploring various questions. For example: how might a management strategy be best implemented? What are people's beliefs and attitudes towards an intervention? What is the extent of adoption of a conservation intervention? What are the barriers and facilitators to acceptability for an intervention? What are the priorities and challenges for local communities? (CEE 2018; Macura et al. 2019). Qualitative synthesis may stand alone or complement quantitative synthesis and contribute to a mixed methods approach that brings these evidence types together. Qualitative research related to the human dimensions of conservation and environment, though growing in popularity, has rarely been subject to evidence synthesis (CEE 2018). This is largely due to limited methodological guidance for environmental researchers to undertake such exercises, though progress is being made here (see Macura et al. 2019).

There are various resources available to support researchers undertaking a systematic review. For instance, the Collaboration for Environmental Evidence (CEE; www.environmentalevidence.org) provides step-by-step methodological guidelines (CEE 2018), a critical appraisal tool for evaluating risk of bias (prototype: https://environmentalevidence.org/cee-critical-appraisal-tool/), a set of reporting standards of review conduct (Haddaway et al. 2018), and an online tool for supporting conduct of evidence syntheses to follow the established CEE standards (Kohl et al. 2018).

We failed to find any examples of systematic reviews done to a high standard related to recreational fisheries management or the human dimensions within it. There are some that deal with other aspects of fisheries management, such as habitat enhancement and restoration (e.g., Taylor et al. 2019), but none that consider the tools typically used to understand recreational fishing impacts or interventions (e.g., harvest regulations, recreational fisher education).

3.4 Rapid Reviews

Rapid reviews, as their name suggests, are intended to be done with sufficient speed that they can be used to support decision-making when there is insufficient time (or funding) to enable a more comprehensive, systematic review. To save time, rapid reviews involve taking various "shortcuts" that represent the usual steps in a systematic review. For example, the scope of a search can be limited by species (e.g., only marine fish or only fish from the family Centrarchidae) or geography (e.g., only reservoirs in Europe or only estuaries in Australia). Similarly, the search string can be focused on a reduced subset of more focused search terms. Efforts to search websites or incorporate grey literature may also be omitted. All the aforementioned actions will yield a smaller suite of papers and, subsequently, reduce the time needed for screening and analysis. Yet, using a more limited evidence base also has consequences that could lead to bias. Other ways to expedite the review process include dropping the critical appraisal component of reviews. Doing so means that all evidence is treated equally, which fails to acknowledge the range of study "quality." Not surprisingly, this also introduces bias and makes it difficult to identify what aspects of study design are particularly important for subsequent studies that aim to improve or increase the evidence base.

Because of the shortcuts and associated biases that arise in rapid reviews (relative to full, systematic reviews), this approach receives much criticism, based largely on comparative studies between these and full, systematic reviews. Gannan et al. (2010) conducted a systematic review on the use of rapid reviews and revealed that they took from three weeks to six months to conduct (compared to systematic reviews, which tend to require 1 or more years, depending on resources). The authors determined that the most common methods for accelerating the review process were largely related to search limitations (e.g., only using certain years, focusing on a single database, using only a single language [usually English], and only using electronic sources). However, a variety of other methods have been used, including using a single reviewer for activities such as title and abstract screening or critical appraisal, or not preparing a formal protocol prior to conducting the review. The authors concluded that biases may be introduced due to shortened timeframes for literature searching, article retrieval, and appraisal, with the potential to impact outcomes.

For recreational fisheries topics, the evidence base is generally small, and there tends to be equally large quantities of studies published as technical reports. For that reason, limiting searches to "academic" databases or intentionally excluding technical reports could lead to substantial bias. Nonetheless, rapid reviews are still preferable compared to basing decisions on cherry-picked empirical studies. Decisions that need to be made quickly may not have the benefit of waiting for extended periods for a full review. The key to rapid review is that end-users of that review are aware of biases and limitations therein. We are unaware of any rapid reviews specific to recreational fisheries, and the nearest example is a rapid review of the recreational, cultural, and environmental meanings of water resources for Australian river communities (Downey et al. 2022a).

3.5 Systematic Maps

Systematic maps are a form of evidence synthesis used to describe published literature on a topic of interest. They use structured methodologies similar to those of systematic reviews to collate and describe (i.e., map) the evidence available, identifying where, what type, and how much evidence is present in the literature (James et al. 2016; CEE 2018). Systematic maps do not attempt to measure an effect (e.g., the impact of X on Y) or answer a specific research question, but instead work to illustrate the state of the literature in a comprehensive, objective, and transparent manner. Questions are often broad and can include several elements. For example, what evidence currently exists? How much evidence is available? What interventions or outcomes are studied most frequently? What methods are used? Systematic maps may also be used to address questions of impact where systematic reviews are not feasible (Pullin et al. 2016). Systematic maps consider both the breadth and depth of available evidence (i.e., peer-reviewed, grey literature, primary, secondary, quantitative, or qualitative) and are frequently a preliminary step to assess evidence prior to initiating more effect-specific or hypothesis-driven reviews (Pullin et al. 2016).

The stages of a systematic map follow closely those of a systematic review (see Sect. 3.3 above) with some modification (CEE 2018). After team development (including methodological and topical experts, as well as stakeholders) and question formulation, an a priori protocol detailing the steps of the mapping exercise and steps to decrease bias during the process is developed (Haddaway et al. 2016). Protocols are peer-reviewed to ensure high quality methodologies are used during the mapping process and registered (i.e., through registries such as CEE's PROCEED https://environmentalevidence.org/proceed/), which reduces the risk of duplicate maps from different teams and allows the research community to provide comments prior to the completion of the synthesis (Haddaway et al. 2016). The mapping process generally follows five steps (for methodological guidelines, see CEE 2018), although a sixth step (critical appraisal) may be added in some cases:

1. Comprehensive searches for relevant evidence on specific topics, based on explicit, scoped search strings formulated during the protocol, and considering multiple sources of commercially published and grey literature.
2. Screening to ensure studies relate directly to the mapping question, based on predetermined inclusion criteria.
3. Metadata coding to describe each study in relation to the mapping question, based on predetermined coding criteria updated iteratively during the process (not including extraction of outcome measures such as effect sizes).
4. Exploratory data synthesis (normally narrative-based with descriptive statistics, well documented and comprehensive, and covering all components of the mapping question).
5. Transparent reporting and communication, especially in relation to any changes in the mapping process as they relate to the published protocol, with extensive use of supplementary information (including the full systematic map database) to maximize transparency.

Unlike systematic reviews, where critical appraisal of study validity (e.g., measuring internal validity [risk of bias within a study] or external validity [generalizability of a study to the review question]) is required, critical appraisal is optional for systematic maps (CEE 2018). Some validity indicators may be captured during metadata coding (see Rytwinski et al. 2020), but the decision to include critical appraisal in systematic mapping does not influence the mapping process. Although systematic maps only include metadata coding rather than extracted data for assessment of effects (James et al. 2016), systematic maps often have larger scopes and literature bases than the more specific questions asked during systematic reviews and can have moderate to high resource requirements (Cook et al. 2017). Careful scoping during protocol development can keep the map size manageable. Systematic maps and reviews are often up to date for just a brief period after the initial search, so map updates may be required periodically, especially if the primary research continues to expand rapidly (Bayliss et al. 2016).

The outcomes of a systematic map include a descriptive, exploratory synthesis of available evidence, highlighted knowledge clusters (i.e., subsets of the evidence base that are sufficiently covered to allow for further, secondary research such as systematic reviewing), or knowledge gaps (i.e., areas that are underrepresented in the current literature that require further primary research before secondary research can occur) (Haddaway et al. 2016). These syntheses include tables and figures (often with descriptive statistics) to contextualize the available literature. This output can aid decision makers in determining where there is evidence for further systematic review that would aid in current management decisions, or where future efforts should be targeted (i.e., knowledge gaps). In addition, most systematic maps provide a searchable database that can be used by decision makers looking for specific supporting evidence within their topic of interest.

3.6 Evidence Synopses

Evidence synopses take evidence aggregation a step further by condensing evidence syntheses in any of the above-mentioned forms into comparatively short summaries. The purpose of this is to provide management or policy briefings to front-line decision makers, for whom sifting through long reviews is impractical (e.g., due to time constraints). Recreational fishery managers tend to fit this description and already face numerous challenges related to recreational users (e.g., conflicts, non-compliance with regulations; Arlinghaus and Cooke 2009). These tall tasks make staying up to date on fisheries science exceedingly difficult and create a demand for more "rapidly digestible" evidence synopses.

Demands for rapidly digestible evidence syntheses have been realized and met outside of natural resource management and conservation. Healthcare, for example, now benefits from improved evidence synthesis and synopsis services in combination with more and higher quality clinical research (Haynes 2001). Put simply, evidence synopsis allows the benefits of rapidly accumulating clinical research to be realized more fully. Evidence synopses in clinical science are numerous, and centre on wide-ranging topics such as the effectiveness of early versus late tracheostomy in critically ill patients (Keeping 2016), the effect of general health checkups on total mortality (Sox 2013), and treatment of the common cold using zinc (Hemilä 2015). In 2013, the Journal of the American Medical Association introduced Clinical Evidence Synopsis as a new type of article to be published monthly. The purpose of this was to make accessible the most important information from high-quality systematic reviews for clinicians in their day-to-day practice (McDermott and Livingston 2013). Here, as elsewhere, the target audience is front-line decision makers (i.e., practicing clinicians) whose decisions must be, more than anything else, timely.

From healthcare to recreational fishery management, evidence synopses contain the most essential, practicable information gleaned from the existing research on a topic. As such, evidence synopses omit much of the basic information underlying that which is most relevant to end-users (i.e., practitioners). Good synopses, however, contain clear links to foundational information should the user require them. The key quality and advantage of these synopses is that they are "fast and frugal," i.e., fit for use "on the run" (Haynes 2001). Evidence synopses can be created for various purposes and at various scales (e.g., for specific actions, for single species or other taxa). Conservation Evidence—a free information source and authority on evidence synopsis in conservation—provides evidence synopses for specific actions within a given context (e.g., conservation of bats; Conservation Evidence 2020). Here, existing scientific information is aggregated and summarized for actions such as water-level management in freshwater wetlands. As in clinical science, examples of evidence synopsis in natural resource management and conservation are numerous: terrestrial mammal conservation (Littlewood et al. 2020) and carnivore conservation (Lockhart and Staurt 2013) are just two of the many topics for which evidence synopsis has proven valuable.

Beyond accessibility, evidence synopsis can facilitate more transparent evidence-based decision-making, provided that the trail of evidence and uncertainty around a topic are documented clearly (Dicks et al. 2014a). Research by Walsh et al. (2015) shows that the impact of scientific research on conservation may depend on its accessibility to end-users as much as or more than the innate qualities of said research—a result which strongly supports evidence synopsis as a means to evidence-based conservation. So long as there are demands for fast and frugal decision-making tools in recreational fisheries management, evidence synopsis will prove valuable to front-line practitioners.

4 Mainstreaming Evidence Synthesis in Recreational Fisheries Science and Management

In general, evidence syntheses have been poorly embraced in the recreational fisheries science and management communities, despite their many benefits, particularly in the human dimensions of the sector (with some notable exceptions, e.g., Hunt et al. 2019; Birdsong et al. 2021). This needs to be overcome to harness the inherent power that arises from bringing together disparate empirical studies. There is not one method that should be applied to all scenarios. As outlined here, there are a variety of tools, each with its strengths and weaknesses. In science, no matter the method used, there will always be limitations. So, the key to evidence synthesis is knowing the extent to which those limitations result in biases that could impact outcomes. One way to approach this is to consider the costs of "getting it wrong." What if that management approach that is deemed effective by a rapid review is in fact ineffective? What if a decision is delayed, waiting for a systematic review with meta-analysis to be conducted? The reality is that because recreational fisheries systems are complex and dynamic, fisheries decision makers are used to dealing with uncertainty (Cabanellas-Reboredo et al. 2017) and information-poor scenarios (Post et al. 2002), which is partly responsible for the reliance on adaptive management approaches (Arlinghaus et al. 2017). Evidence syntheses can be strengthened and their relevance ensured by including evidence users (e.g., recreational fisheries managers) in the process. Ideally, this would begin at the question formulation stage and while determining which evidence synthesis method is most appropriate.

Given that recreational fisheries are spatially, biologically, and socially diverse, managers have no alternative but to adopt an evidence-synthesis approach. Resources are sufficiently limited that it is not possible to "research" every issue in every context or to adaptively manage one's way through every recreational fishery management decision. Social science studies have revealed that environmental decision makers tend to be aware of the range of evidence synthesis tools and will use them if available (Thomas-Walters et al. 2021). Nonetheless, there is room for professional development training opportunities for both evidence users (i.e., fisheries managers) as well researchers that would be most likely to do such syntheses.

There is more work to do before evidence synthesis will be "mainstreamed," including changing institutional practices for how decisions are made and creating a culture of evidence use. There are opportunities to interface evidence synthesis with decision support tools such as adaptive management frameworks (see Robinson et al. 2026). Because of a tendency for fisheries management decisions to be based on tradition or rely upon input from colleagues rather than use the entire evidence base (Young et al. 2016), it is almost certain that some of the decisions being made are contrary to the evidence base. The best approach to increasing uptake of evidence synthesis in fisheries management will be through case studies that demonstrate how this has been applied and the benefits of doing so. Unfortunately, such cases remain elusive in the recreational fisheries sphere, despite being common in other domains such as health and education.

5 Conclusion

Recreational fisheries are complex social–ecological systems that are challenging to understand and manage (Arlinghaus et al. 2017; Elmer et al. 2017). Also notable is the fact that recreational fisheries managers are often tasked with making decisions that involve trade-offs, controversy, or even what are considered "wicked problems." Hence, decisions that are made need to consider the best available evidence to guide them. Recreational fisheries management has long been approached from a trial-and-error perspective, i.e., slowly learning what works and then using that experience to shape individual and institutional norms. Only recently have rigorous active adaptive experiments that also track the response and learning outcome of practitioners been developed (Fujitani et al. 2017). Social science surveys reveal that fisheries managers are failing to make use of the broader evidence base and tend to rely on tradition (Young et al. 2016). We submit that recreational fisheries management could be improved if it were to adopt more of an evidence-based approach that was partially supported by evidence synthesis. As outlined above, evidence syntheses come in many forms, each with their own strengths and weaknesses. Each of these methods has a place, but we wish to emphasize the particular importance of systematic reviews that incorporate meta-analysis. Systematic reviews (ideally with meta-analysis, although this may not always be possible given characteristics of the evidence base) are considered the gold standard of evidence synthesis, as they are rigorous, repeatable, and focus strongly on reducing bias. When quantitative data can be extracted, it is possible to conduct powerful meta-analyses that can provide managers with important insight into how the world works. But also, qualitative inferences are hugely relevant if they follow a rigorous search and synthesis approach. We hope that our paper will help to equip recreational fisheries managers and scientists with sufficient information to make decisions about why evidence syntheses could be of use, how to select and even conduct the appropriate form of synthesis, and to interpret the findings arising from them.

Acknowledgments We thank several anonymous referees as well as the book editorial team for their thoughtful comments on this manuscript. No funding sources are declared.

References

Andrews I, Kasy M (2019) Identification of and correction for publication bias. Am Econ Rev 109(8):2766–2794

Arlinghaus R (2005) A conceptual framework to identify and understand conflicts in recreational fisheries systems, with implications for sustainable management. Aquat Resour Cult Dev 1(2): 145–174

Arlinghaus R, Cooke SJ (2009) Recreational fisheries: socioeconomic importance, conservation issues and management challenges. In: Dickson B, Hutton J, Adams WM (eds) Recreational hunting, conservation and rural livelihoods: science and practice. Wiley-Blackwell, United Kingdom, pp 39–58

Arlinghaus R, Mehner T, Cowx IG (2002) Reconciling traditional inland fisheries management and sustainability in industrialized countries, with emphasis on Europe. Fish Fish 3(4):261–316

Arlinghaus R, Cooke SJ, Lyman J, Policansky D, Schwab A, Suski C, Sutton SG, Thorstad EB (2007) Understanding the complexity of catch-and-release in recreational fishing: an integrative synthesis of global knowledge from historical, ethical, social, and biological perspectives. Rev Fish Sci 15(1–2):75–167

Arlinghaus R, Cooke SJ, Sutton SG, Danylchuk AJ, Potts W, Freire KDM, Van Anrooy R (2016) Recommendations for the future of recreational fisheries to prepare the social-ecological system to cope with change. Fish Manag Ecol 23(3–4):177–186

Arlinghaus R, Alós J, Beardmore B, Daedlow K, Dorow M, Fujitani M, Hühn D, Haider W, Hunt L, Johnson BM, Johnston FD, Klefoth T, Matsumura S, Monk CT, Pagel T, Post JR, Rapp T, Riepe C, Ward H, Wolter C (2017) Understanding and managing freshwater recreational fisheries as complex adaptive social-ecological systems. Rev Fish Sci Aquac 25(1):1–41

Arlinghaus R, Riepe C, Theis S, Pagel T, Fujitani M (2022) Dysfunctional information feedbacks cause the emergence of management panaceas in social-ecological systems: the case of fish stocking in inland recreational fisheries. J Outdoor Recreat Tour 38:100475

Baethge C, Goldbeck-Wood S, Mertens S (2019) SANRA-a scale for the quality assessment of narrative review articles. Res Integrity Peer Rev 4:1

Bartolucci AA, Hillegass WB (2010) Overview, strengths, and limitations of systematic reviews and meta-analyses. In: Chiappelli F, Caldeira Brant XM, Neagos N, Oluwadara OO, Ramchandani MH (eds) Evidence-based practice: toward optimizing clinical outcomes. Springer, Berlin/Heidelberg, pp 17–33

Baumeister R, Leary MR (1997) Writing narrative literature reviews. Rev Gen Psychol 1(3): 311–320

Bayliss HR, Haddaway NR, Eales J, Frampton GK, James KL (2016) Updating and amending systematic reviews and systematic maps in environmental management. Environ Evid 5(1):20

Birdsong M, Hunt LM, Arlinghaus R (2021) Recreational angler satisfaction: what drives it? Fish Fish 22(4):682–706

Boutron I, Page MJ, Higgins JPT, Altman DG, Lundh A, Hróbjartsson A (2019) Considering bias and conflicts of interest among the included studies. In: Higgins JPT, Thomas J, Chandler J, Cumpston M, Li T, Page MJ, Welch VA (eds) Cochrane handbook for systematic reviews of interventions. Chichester, UK: John Wiley & Sons

Bower SD, Danylchuk AJ, Raghavan R, Danylchuk SC, Pinder AC, Alter AM, Cooke SJ (2017) Involving recreational fisheries stakeholders in development of research and conservation priorities for mahseer (Tor spp.) of India through collaborative workshops. Fish Res 186:665–671

Bowling A (2014) Research methods in health: investigating health and health services. McGraw-Hill Education, London, UK

Briner RB, Denyer D (2012) Systematic review and evidence synthesis as a practice and scholarship tool. In: Rousseau D (ed) Oxford handbook of evidence-based management: companies, classrooms and research. Oxford University Press, New York, pp 112–129

Cabanellas-Reboredo M, Palmer M, Alos J, Morales-Nin B (2017) Estimating harvest and its uncertainty in heterogeneous recreational fisheries. Fish Res 188:100–111

Cahill CL, Walters CJ, Paul AJ, Sullivan MG, Post JR (2022) Unveiling the recovery dynamics of walleye after the invisible collapse. Can J Fish Aquat Sci 79(5):708–723

CEE (Collaboration for Environmental Evidence) (2018) Guidelines and standards for evidence synthesis in environmental management, ver. 5.0. http://www.environmentalevidence.org/information-for-authors. Accessed 15 Oct 2023

Conservation Evidence (2020, Sept 12) Synopses. https://www.conservationevidence.com/synopsis/index#what-is-a-synopsis. Accessed 15 Oct 2023

Cook CN, Nichols SJ, Webb JA, Fuller RA, Richards RM (2017) Simplifying the selection of evidence synthesis methods to inform environmental decisions: a guide for decision makers and scientists. Biol Conserv 213:135–145

Cooke SJ, Birnie-Gauvin K, Lennox RJ, Taylor JJ, Rytwinski T, Rummer JL, Haddaway NR (2017a) How experimental biology and ecology can support evidence-based decision-making in conservation: avoiding pitfalls and enabling application. Conserv Physiol 5:1

Cooke SJ, Wesch S, Donaldson LA, Wilson AD, Haddaway NR (2017b) A call for evidence-based conservation and management of fisheries and aquatic resources. Fisheries 42(3):143–149

Cooke SJ, Rytwinski T, Taylor JJ, Nyboer EA, Nguyen VM, Bennett JR, Smol JP (2020) On "success" in applied environmental research-what is it, how can it be achieved, and how does one know when it has been achieved? Environ Rev 28(4):357–372

Collins JA, Fauser BC (2005) Balancing the strengths of systematic and narrative reviews. Human reproduction update 11(2), pp.103–104

Cowx IG, Arlinghaus R, Cooke SJ (2010) Harmonizing recreational fisheries and conservation objectives for aquatic biodiversity in inland waters. J Fish Biol 76(9):2194–2215

Daedlow KA, Beard TD, Arlinghaus R (2011) A property rights-based view on management of inland. American Fisheries Society Symposium 75:13–38

Delgado-Rodriguez M, Llorca J (2004) Bias. J Epidemiol Commun H 58(8):635–641

Dicks LV, Hodge I, Randall NP, Scharlemann JP, Siriwardena GM, Smith HG, Sutherland WJ (2014a) A transparent process for "evidence-informed" policy making. Conserv Lett 7(2): 119–125

Dicks LV, Walsh JC, Sutherland WJ (2014b) Organising evidence for environmental management decisions: a '4S' hierarchy. Trends Ecol Evol 29(11):607–613. https://doi.org/10.1016/j.tree.2014.09.004

Dicks LV, Haddaway NR, Hernandez-Morcillo M, Mattson B, Randall N, Failler P, Wittmer H (2017) Knowledge synthesis for environmental decisions: an evaluation of existing methods, and guidance for their selection, use and development: a report from the EKLIPSE project. European Union Funding for Research and Innovation, p 81

Downey H, Spelten E, Holmes K, Van Vuuren J (2022a) A rapid review of recreational, cultural, and environmental meanings of water for Australian river communities. Soc Nat Resour 35(1): 1–19

Downey H, Bretagnolle V, Brick C, Bulman CR, Cooke SJ, Dean M et al. (2022b) Principles for the production of evidence-based guidance for conservation actions. Conservation Science and Practice 4(5):e12663

Drew CJ, Hardman ML, Hosp JL (2007) Designing and conducting research in education. Sage Publications, Thousand Oaks

Elmer LK, Kelly LA, Rivest S, Steell SC, Twardek WM, Danylchuk AJ, Arlinghaus R, Bennett JR, Cooke SJ (2017) Angling into the future: ten commandments for recreational fisheries science, management, and stewardship in a good Anthropocene. Environ Manag 60(2):165–175

FAO (United Nations Food and Agriculture Organization) (2012) Recreational fisheries. United Nations Food and Agriculture Organization, Technical Guidelines for Responsible Fisheries, no. 13, Rome, Italy

Ferrari R (2015) Writing narrative style literature reviews. Med Writ 24(4):230–235

Fujitani M, McFall A, Randler C, Arlinghaus R (2017) Participatory adaptive management leads to environmental learning outcomes extending beyond the sphere of science. Science Advances. Jun 14;3(6):e1602516

Ganann R, Ciliska D, Thomas H (2010) Expediting systematic reviews: methods and implications of rapid reviews. Implement Sci 5(1):1–10

Green BN, Johnson CD, Adams A (2006) Writing narrative literature reviews for peer-reviewed journals: secrets of the trade. J Chiropr Med 5(3):101–117

Gurevitch J, Koricheva J, Nakagawa S, Stewart G (2018) Meta-analysis and the science of research synthesis. Nature 555:175–182

Haddaway NR, Westgate MJ (2019) Predicting the time needed for environmental systematic reviews and systematic maps. Conserv Biol 33:434–443

Haddaway NR, Woodcock P, Macura B, Collins A (2015) Making literature reviews more reliable through application of lessons from systematic reviews. Conserv Biol 29(6):1596–1605

Haddaway NR, Bernes C, Jonsson BG, Hedlund K (2016) The benefits of systematic mapping to evidence-based environmental management. Ambio 45(5):613–620

Haddaway NR, Macura B, Whaley P, Pullin AS (2018) ROSES reporting standards for systematic evidence syntheses: pro forma, flow-diagram and descriptive summary of the plan and conduct of environmental systematic reviews and systematic maps. Environ Evid 7(1):1–8

Haynes RB (2001) Of studies, syntheses, synopses, and systems: the "4S" evolution of services for finding current best evidence. BMJ Evid Base Med 6(2):36–38

Helfman G, Collette BB, Facey DE, Bowen BW (2009) The diversity of fishes: biology, evolution, and ecology. Wiley, Malden

Hemilä H (2015) Common cold treatment using zinc. JAMA J Am Med Assoc 314(7):730–730

Higgins JP, Thomas J, Chandler J, Cumpston M, Li T, Page MJ, Welch VA (2019) Cochrane handbook for systematic reviews of interventions (Version 6.0). Cochrane, UK

Hilborn R, Liermann M (1998) Standing on the shoulders of giants: learning from experience in fisheries. Rev Fish Biol Fisher 8:273–283

Hunt LM (2005) Recreational fishing site choice models: insights and future opportunities. Hum Dimens Wildl 10(3):153–172

Hunt LM, Camp E, van Poorten B, Arlinghaus R (2019) Catch and non-catchrelated determinants of where anglers fish: a review of three decades of site choice research in recreational fisheries. Reviews in Fisheries Science & Aquaculture. Jul 3;27(3):261–86

Hunt LM, Arlinghaus, R., Scott, D. & Kyle G (2023) Diversity of anglers: drivers and implications for fisheries management. In: Neal JW, Lang TJ, Krogman RM, Kurzawski KF, Taylor JB & Hunt KM (Eds.) Angler recruitment, retention, and reactivation: the future of fisheries and aquatic conservation. Bethesda, Maryland: American Fisheries Society, pp. 113–140. Available from: https://doi.org/10.47886/9781934874738.ch5

Hyder K, Maravelias CD, Kraan M, Radford Z, Prellezo R (2020) Marine recreational fisheries-current state and future opportunities. ICES J Mar Sci 77(6):2171–2180

James KL, Randall NP, Haddaway NR (2016) A methodology for systematic mapping in environmental sciences. Environ Evid 5(1):7

Johnston RJ, Ranson MH, Besedin EY, Helm EC (2006) What determines willingness to pay per fish? A meta-analysis of recreational fishing values. Mar Resour Econ 21(1):1–32

Kadykalo AN, Cooke SJ, Young N (2021) The role of western-based scientific, indigenous and local knowledge in wildlife management and conservation. People Nat 3(3):610–626

Keeping A (2016) Early versus late tracheostomy for critically ill patients: a clinical evidence synopsis of a recent Cochrane review. Can J Respir Ther 52(1):27

Kohl C, McIntosh EJ, Unger S, Haddaway NR, Kecke S, Schiemann J, Wilhelm R (2018) Online tools supporting the conduct and reporting of systematic reviews and systematic maps: a case study on CADIMA and review of existing tools. Environ Evid 7(1):1–17

Koricheva J, Gurevitch J (2014) Uses and misuses of meta-analysis in plant ecology. J Ecol 102(4): 828–844

Koricheva J, Kulinskaya E (2019) Temporal instability of evidence base: a threat to policy making? Trends Ecol Evol 34:895–902

Lewicki R, Gray B, Elliott M (eds) (2003) Making sense of intractable environmental conflicts: concepts and cases. Washington, DC, USA: Island Press

Littlewood NA, Rocha R, Smith RK, Martin PK, Lockhart SL, Schoonover RF, Wilman E, Bladon AJ, Sainsbury KA, Pimm S, Sutherland WJ (2020) Terrestrial mammal conservation: global evidence for the effects of interventions for terrestrial mammals excluding bats and primates. Open Book Publishers, Cambridge, UK

Lockhart S, Stuart P (2013) Carnivore conservation evidence: framework and connections. Digital Depositories at Duke, Durham

Macura B, Sŭskevĭcs M, Garside R, Hannes K, Rees R, Rodela R (2019) Systematic reviews of qualitative evidence for environmental policy and management: an overview of different methodological options. Environ Evid 8(1):1–11

McDermott MM, Livingston EH (2013) Introducing JAMA clinical evidence synopsis: from systematic reviews to clinical practice. JAMA J Am Med Assoc 309(1):89–89

Moher D, Liberati A, Tetzlaff J, Altman DG, Grp P (2009) Preferred reporting items for systematic reviews and meta-analyses: the PRISMA statement. PLoS Med 6(7):e1000097

Montello D, Sutton P (2012) An introduction to scientific research methods in geography and environmental studies, vol 1. Sage Publications, London

Myers RA (2001) Stock and recruitment: generalizations about maximum reproductive rate, density dependence, and variability using meta-analytic approaches. ICES J Mar Sci 58(5):937–951

Nakagawa S, Santos ES (2012) Methodological issues and advances in biological meta-analysis. Evolutionary Ecology Sep;26(5):1253–74

Nakagawa S, Noble DWA, Senior AM, Lagisz M (2017) Meta-evaluation of meta-analysis: ten appraisal questions for biologists. BMC Biol 15(1):18

Nakagawa S, Lagisz M, Jennions MD, Koricheva J, Noble DWA, Parker TH, Sánchez-Tójar A, Yang Y, O'Dea RE (2022) Methods for testing publication bias in ecological and evolutionary meta-analyses. Methods Ecol Evol 13:4–21

Nakagawa S, Yang Y, Macartney EL, Spake R, Lagisz M (2023a) Quantitative evidence synthesis: a practical guide on meta-analysis, meta-regression, and publication bias tests for environmental sciences. Environ Evid 12(1):1–9

Nakagawa S, Noble DW, Lagisz M, Spake R, Viechtbauer W, Senior AM (2023b) A robust and readily implementable method for the meta-analysis of response ratios with and without missing standard deviations. Ecol Lett 26(2):232–244

Nguyen VM, Young N, Corriveau M, Hinch SG, Cooke SJ (2019) What is "usable" knowledge? Perceived barriers for integrating new knowledge into management of an iconic Canadian fishery. Can J Fish Aquat Sci 76(3):463–474

O'Dea RE, Lagisz M, Jennions MD, Koricheva J, Noble DWA, Parker TH, Gurevitch J, Page MJ, Stewart G, Moher D, Nakagawa S (2021) Preferred reporting items for systematic reviews and meta-analyses in ecology and evolutionary biology: a PRISMA extension. Biol Rev 96:1695–1722

Page MJ, McKenzie JE, Bossuyt PM, Boutron I, Hoffmann TC, Mulrow CD, Shamseer L, Tetzlaff JM, Akl EA, Brennan SE, Chou R, Glanville J, Grimshaw JM, Hróbjartsson A, Lalu MM, Li T, Loder EW, Mayo-Wilson E, McDonald S, McGuinness LA, Stewart LA, Thomas J, Tricco AC, Welch VA, Whiting P, Moher D (2021) The PRISMA 2020 statement: an updated guideline for reporting systematic reviews. PLoS Med 18:e1003583

Post JR, Sullivan M, Cox S, Lester NP, Walters CJ, Parkinson EA, Paul AJ, Jackson LJ, Shuter BJ (2002) Canada's recreational fisheries: the invisible collapse? Fisheries 27(1):6–17

Pullin AS, Knight TM (2001) Effectiveness in conservation practice: pointers from medicine and public health. Conserv Biol 15(1):50–54

Pullin AS, Knight TM, Stone DA, Charman K (2004) Do conservation managers use scientific evidence to support their decision-making? Biol Conserv 119(2):245–252

Pullin AS, Frampton G, Jong R, Kohl C, Livoreil B, Lux A, Pataki G, Petrokofsky G, Podhora A, Saarikoski H, Santamaria L, Schindler S, Sousa-Pinta I, Vandewalle M, Wittmer H (2016) Selecting appropriate methods of knowledge synthesis to inform biodiversity policy. Biodivers Conserv 25(7):1285–1300

Pullin AS, Cheng SH, Cooke SJ, Haddaway NR, Macura B, McKinnon MC, Taylor JJ (2020) Informing conservation decisions through evidence synthesis and communication. In: Sutherland WJ, Brotherton PNM, Davies ZG, Ockendon N, Pettorelli N, Vickery JA (eds) Conservation research, policy and practice. Cambridge University Press, Cambridge, UK, pp 114–128

Pullin AS, Cheng SH, Jackson JD, Eales J, Envall I, Fada SJ, Frampton GK, Harper M, Kadykalo AN, Kohl C, Konno K, Livoreil B, Ouedraogo D, O'Leary BC, Pullin G, Randall N, Rees R, Smith A, Sordello R, Sterling EJ, Twardek WM, Woodcock P (2022) Standards of conduct and reporting in evidence syntheses that could inform environmental policy and management decisions. Environ Evid 11(1):16

Ricci PF, Rice D, Ziagos J, Cox LA Jr (2003) Precaution, uncertainty and causation in environmental decisions. Environ Int 29(1):1–19

Roberts PD, Stewart GB, Pullin AS (2006) Are review articles a reliable source of evidence to support conservation and environmental management? A comparison with medicine. Biol Conserv 132:409–423

Robinson KF, Arlinghaus R, Camp EV, Post JR, Stedman R, Sutton SG (2026) Adaptive management for recreational fisheries decisions in the face of uncertainty. In: Pope KL, Arlinghaus R, Hunt LM, Lynch AJ, van Poorten BT (eds) Understanding recreational fishers: disciplinary and interdisciplinary approaches for fisheries management. Springer Nature, Cham, p 719–746

Rytwinski T, Harper TJJ, Bennett JR, Donaldson LA, Smokorowski KE, Clarke K, Bradford MJ, Ghamry H, Olden JD, Boisclair D, Cooke SJ (2020) What are the effects of flow-regime changes on fish productivity in temperate regions? A systematic map. Environ Evid 9(1):7

Rytwinski T, Cooke SJ, Taylor JJ, Roche DG, Smith PA, Mitchell GW, Smokorowsk KE, Prior KA, Bennett JR (2021) Acting in the face of evidentiary ambiguity, bias, and absence arising from systematic reviews in applied environmental science. Sci Total Environ 775:145122

Schafft M, Wegner B, Meyer N, Wolter C, Arlinghaus R (2021) Ecological impacts of water-based recreational activities on freshwater ecosystems: a global meta-analysis. P Roy Soc B-Biol Sci 288(1959):20211623

Sica GT (2006) Bias in research studies. Radiology 238(3):780–789

Siddaway AP, Wood AM, Hedges LV (2019) How to do a systematic review: a best practice guide for conducting and reporting narrative reviews, meta-analyses, and meta-syntheses. Annu Rev Psychol 70(1):747–770

Smith R (2006) Peer review: a flawed process at the heart of science and journals. J Roy Soc Med 99(4):178–182

Smith J, Noble H (2014) Bias in research. Evid Based Nurs 17(4):100–101

Sox HC (2013) The health checkup: was it ever effective? Could it be effective? JAMA J Am Med Assoc 309(23):2496–2497

Spake R, O'Dea RE, Nakagawa S, Doncaster CP, Ryo M, Callaghan CT, Bullock JM (2022) Improving quantitative synthesis to achieve generality in ecology. Nat Ecol Evol 6(12): 1818–1828

Sterne JAC, Sutton AJ, Ioannidis JPA, Terrin N, Jones D, Lau J, Carpenter J, Rücker G, Harbord RM, Schmid CH, Tetzlaff J, Deeks JJ, Peters J, Macaskill P, Schwarzer G, Duval S, Altman DG, Moher D, Higgins JP (2011) Recommendations for examining and interpreting funnel plot asymmetry in meta-analyses of randomised controlled trials. Br Med J 343:d4002

Sutherland WJ, Pullin AS, Dolman PM, Knight TM (2004) The need for evidence-based conservation. Trends Ecol Evol 19(6):305–308

Sutton AJ, Cooper NJ, Jones DR (2009) Evidence synthesis as the key to more coherent and efficient research. BMC Med Res Methodol 9(1):1–9

Taylor JJ, Rytwinski T, Bennett JR, Smokorowski KE, Lapointe NW, Janusz R, Clarke KD, Tonn B, Walsh JC, Cooke SJ (2019) The effectiveness of spawning habitat creation or enhancement for substrate-spawning temperate fish: a systematic review. Environ Evid 8(1): 1–31

Thomas-Walters L, Nyboer EA, Taylor JJ, Rytwinski T, Lane JF, Young N, Bennett JR, Nguyen VM, Harron N, Altken SM, Auld G, Browne DR, Jacob AL, Prior K, Smith PA, Smokorowski KE, Alexander SM, Cooke SJ (2021) An optimistic outlook on the use of evidence syntheses to inform environmental decision-making. Conserv Sci Pract 3(6):e426

Thorson JT, Cope JM, Kleisner KM, Samhouri JF, Shelton AO, Ward EJ (2015) Giants' shoulders 15 years later: lessons, challenges and guidelines in fisheries meta-analysis. Fish Fish 16(2): 342–361

Walsh JC, Dicks LV, Sutherland WJ (2015) The effect of scientific evidence on conservation practitioners' management decisions. Conserv Biol 29(1):88–98

White IR, Schmid CH, Stijnen T (2021) Choice of effect measure and issues in extracting outcome data. In: Schmid CH, Stijnen T, White IR (eds) Handbook of meta-analysis. CRC Press, Boca Raton

Willinsky J (2006) The access principle: the case for open access to research and scholarship, vol 20. MIT Press, Cambridge, MA

Wright RW, Brand RA, Dunn W, Spindler KP (2007) How to write a systematic review. Clin Orthop Relat R 455:23–29

Yang Y, Sánchez-Tójar A, O'Dea RE, Noble DW, Koricheva J, Jennions MD, Parker TH, Lagisz M, Nakagawa S (2023) Publication bias impacts on effect size, statistical power, and magnitude (Type M) and sign (Type S) errors in ecology and evolutionary biology. BMC Biol 21(1):1–20

Yanow D, Schwartz-Shea P (2015) Interpretation and method: empirical research methods and the interpretive turn. Routledge, New York

Young N, Corriveau M, Nguyen VM, Cooke SJ, Hinch SG (2016) How do potential knowledge users evaluate new claims about a contested resource? Problems of power and politics in knowledge exchange and mobilization. J Environ Manag 184:380–388

Part IV
Interdisciplinary Integration for Recreational Fisheries Management

Integrating Diverse Stakeholder Knowledge in Recreational Fisheries Decision Making

Edward V. Camp, Brett T. van Poorten, Kelly F. Robinson, Hannah L. Harrison, Vivian M. Nguyen, Chelsey A. Crandall, Kai Lorenzen, William R. Casola, Michel Dedual, and Kotryna Klizentyte

Abstract Scientists and policymakers increasingly recognize that recreational fisheries decision making should in some way involve the knowledge of affected stakeholders. There are myriad studies about integrating stakeholder knowledge, but they span broad disciplines and topics. Lacking is (i) simple taxonomies that organize general approaches for integrating knowledge, and (ii) a review and examples of their application to specifically recreational fisheries and specifically diverse stakeholders. This chapter first defines the terms "stakeholders" and "recreational" and recognizes that globally, the decision making for non-commercial fisheries often occurs outside of the North American concept of management agencies and even purely leisure-based fisheries. Most of the chapter reviews common approaches for integrating stakeholder knowledge, from near-complete external governance with minimal input from stakeholders (top-down,

E. V. Camp (✉) · K. Lorenzen · W. R. Casola · K. Klizentyte
School of Forest, Fisheries, and Geomatics Sciences, University of Florida, Gainesville, FL, USA
e-mail: edvcamp@ufl.edu

B. T. van Poorten
School of Resource and Environmental Management, Simon Fraser University, Burnaby, BC, Canada

K. F. Robinson
U.S. Geological Survey, Georgia Cooperative Fish and Wildlife Research Unit, Warnell School of Forestry and Natural Resources, University of Georgia, Athens, GA, USA

H. L. Harrison
Marine Affairs Program, Dalhousie University, Halifax, NS, Canada

V. M. Nguyen
Department of Biology, Institute of Environmental and Interdisciplinary Science, Carleton University, Ottawa, ON, Ottawa, ON, Canada

C. A. Crandall
Fish and Wildlife Research Institute, Florida Fish and Wildlife Conservation Commission, Gainesville, FL, USA

M. Dedual
Department of Conservation, Taupo Fishery, Turangi, New Zealand

K. L. Pope et al. (eds.), *Understanding Recreational Fishers*, Fish & Fisheries Series 45, https://doi.org/10.1007/978-3-031-99739-6_20

command-and-control) to self-governance by stakeholders. Organizationally, we adapt the existing cooperative management continuum to first separate non-participatory from participatory approaches and then separate participatory approaches into those aimed at creating knowledge (to eventually inform future decisions) and those focused on decision making itself. The approaches we consider include observing stakeholders (e.g., via survey), incorporating multiple types of knowing (e.g., local traditional knowledge), collaborative data collection (e.g., citizen science), collaborative modelling (e.g., mental models and quantitative modelling), participatory decision making (co-management proper and the special cases of structured decision making and adaptive management), and finally self-governance (including private management of recreational fisheries). We define each according to a broader literature beyond recreational fisheries, describe the recognized advantages and disadvantages, describe or summarize applications to recreational fisheries, and finally mention relevance for integrating specifically *diverse* stakeholder knowledge. We then provide advice for how to select different approaches, including situations in which certain approaches are more likely to produce intended consequences. We conclude by describing potential challenges to the more useful integration of diverse stakeholder knowledge and needs in recreational fisheries decision making.

Keywords Collaborative data collection · Collaborative modelling · Local traditional knowledge · Participatory decision making · Self-governance by stakeholders · Top-down governance

1 Introduction

Sustaining recreational fisheries generally requires decision- making. For example, decisions are often made to prevent overharvest or to protect important habitat or water quality. Balancing greater short-term use of recreational fisheries with long-term conservation drives the need for decision making—whether by fishers and other affected human communities (stakeholders; though the term "stakeholders" is often used broadly in fisheries science, in some contexts Indigenous peoples are more appropriately referred to as rights-holders, reflecting their relationship and various treaty or other legal claims to lands and waters), or external governance entities like management agencies. Effective recreational fishery decisions should be generally (i) scientifically sound, and (ii) accepted by stakeholders. Achieving both usually requires involving stakeholders—and their knowledge—in the decision-making process to some degree (Craik 1943; Pinkerton 1994; Berkes et al. 2000; Berkes 2007; Arlinghaus et al. 2016). Critically, our premise that stakeholder knowledge should ideally be integrated together into decision making should not imply that either stakeholder or scientific or government knowledge is more "important"—such determinations are very context specific and may ultimately be less important than recognizing knowledge from both scientists and non-scientists (see Sects. 2.2.2.1 and 2.2.2.2 for further detail).

One of the benefits and challenges of involving stakeholders in decision making is their diversity and the diversity of their knowledge. People engage with recreational fisheries to satisfy a variety of psychological (e.g., enjoyment) and physical (e.g., catch, harvest) outcomes (Arlinghaus 2006b). Diverse motivations lead to diverse ways of perceiving utility (Hunt et al. 2019), and diverse behaviours, experiences, and ultimately knowledge. We consider knowledge the sum of fisher cognitive processes that incorporate positive (what is) and normative (what should be) components. Though western scientific worldviews have typically focused on positive (e.g., what the "science" is) approaches for informing management, ignoring normative (what people believe should occur) is not possible nor advisable (Fenichel et al. 2013). Stakeholders differing in the centrality of fishing to their lifestyle, as well as their experiences (related and unrelated to demographics) and character traits, possess different knowledge about the fishery system (positive), which may translate to diverse knowledge about how fisheries decisions should be made (normative). Integrating this stakeholder's positive knowledge about the system (e.g., local ecological knowledge, LEK) can bolster what scientists know and improve decision-making quality, and integrating normative knowledge about how stakeholders want the system to operate can increase stakeholder acceptance of decisions and overall satisfaction (Berkes 2007; Fenichel et al. 2013). Integrating stakeholder knowledge, regardless of type, lends legitimacy to the decision -making process and broader governance (Pinkerton 1994). Integrating knowledge that is diverse—in terms of the content of knowledge and cognitive processes behind knowledge, as well as the demographics of humans from which it comes—is therefore important for providing greater governance legitimacy among more stakeholders.

Integrating diverse types of knowledge and knowledge from stakeholders who are inherently diverse are both challenging. One reason is that recreational (broadly defined) fisheries decision-makers may be less diverse in their knowledge, values, and demographics than their stakeholders (Bess 2001; Floyd et al. 2006; Morales et al. 2020). This may be because the decision-makers in any governance system cannot represent the full spectrum of values and knowledge of stakeholders, as well as historic societal inequity (Fabinyi et al. 2015; Arismendi and Penaluna 2016; Gustavsson et al. 2021). The challenge of integrating diverse stakeholder knowledge is often compounded by institutional barriers (Furman et al. 2023), perhaps the most pervasive and robust of which is historical precedent that may have hardened into institutional culture that diverse stakeholder knowledge is not needed or is not especially valuable (Reid et al. 2021). Such institutional cultures may spawn further practical barriers like a lack of personnel trained in gathering knowledge, a lack of funding and tasks devoted to such, and dominance of decision-making processes ill-suited for incorporating stakeholder knowledge (Silver et al. 2022). Such barriers can persist even after a change in institutional culture (Reid et al. 2021).

This chapter describes some of the recognized approaches for integrating diverse knowledge in terms of their process, requirements, and constraints, and then offers perspectives on how approaches might be selected. We first define the terms to set bounds for what the chapter covers—i.e., defining "recreation," "stakeholders." The bulk of the chapter reviews approaches for integrating diverse stakeholder knowledge. We define and describe each approach in terms of its broader (i.e., including beyond recreational fisheries) advantages and disadvantages and then summarize applications to specifically recreational fisheries and relevance for specifically *diverse* stakeholder knowledge—which again includes both knowledge from diverse stakeholders as well as diverse knowledge from any stakeholders. All approaches are summarized (Table 1). We then provide some advice on selecting an approach to implement. The chapter concludes with perspectives on the challenges and opportunities of the subject, including where the field will need to grow to do a better job of integrating diverse knowledge in decision making.

1.1 Disambiguation–Definitions of Terms and Bounds of Chapter

We consider recreational fisheries to be those where a primary motivation for fishing is for leisure and enjoyment, even if consumption or livelihood is another important motivation (Boucquey 2017; Nyboer et al. 2022). It is important to recognize that recreational fisheries may also include those with food as a primary motivation because the concept of "recreational fishing" or fishing solely for leisure does not translate as well outside of Europe and European-colonized countries (Cooke et al. 2018). Therefore, we include information about recreational as well as more small-scale, subsistence, and artisanal fisheries. We continue to use the term "recreational" but what we mean is fisheries that are not strictly commercial in their gear and economic characteristics, and from which there is some psychological benefit or motivation beyond survival or livelihood alone.

Within this chapter, we use the imperfect term stakeholders to describe people affected by the outcome of any decision. We recognize the term "stakeholder" is imperfect because it has connections to colonial historical—and ongoing—injustice and is considered offensive to some groups, especially Indigenous communities, who may use the term "rightsholders" instead. At the same time, the term stakeholder remains ubiquitous within fisheries literature. Therefore, when we say stakeholders, we mean stakeholders and rightsholders. We consider stakeholders to include anyone who values the resource (fish, water, ecosystem) or derives value from it, i.e., fishers and non-fishers. Most fisheries decisions are made with current and future fishers in mind, but decisions that benefit fishers may hurt or otherwise affect non-fishers. This is because of trade-offs between different sectors (e.g., water for agriculture vs. maintaining fishery-suitable levels, Lynch et al. 2017), as well as opportunity costs of monitoring and managing fisheries. Thinking broadly about who to include in the decision process will affect the perspectives considered and the acceptance of, and compliance with, the final decision (Gregory et al. 2012).

Knowledge includes goals and objectives of the system, opinions and preferences on different actions, or an innate or learned understanding of aspects of the social or

Table 1 Summary of the advantages and disadvantages of each approach this chapter covers, along with notes on application to recreational fisheries and diverse knowledge

Approaches		Advantages	Disadvantages	Application to recreational-specific fisheries	Application to diverse knowledge
2.1 Non-participatory approaches—Top-down governance		Inexpensive, simple, sometimes appropriate	No participatory benefits	Common in North American freshwater recreational fisheries	Occasionally diverse demographics targeted with surveys
2.2 Participatory approaches	2.2.2.1 Incorporating multiple ways of knowing	Better decisions, relationships, deeper understanding of fish system	Lengthy process (years) that challenges ideas of western science	Sparsely applied to purely leisure recreational fisheries; more common in artisanal and small-scale	Often applied to minority and less privileged groups
	2.2.2.2 Collaborative data collection and use	More and sometimes better data for sometimes less money; trust	Still has costs, which often are borne by volunteers	Usually used to get trip and catch information or biological samples	Volunteers are usually non-random subset of society and fishery stakeholders
	2.2.2.3 Collaborative modelling	See how parts of fishery fit in system; build trust	Time and permission mean non-random sample	Understanding differences among groups of fishers (within and outside recreational)	Difficult to stratify to get *diverse* mental models, also non-random
	2.2.3.1 Decision science approaches	Best approach for complex, multi-stakeholder issue with competing objectives	Requires substantial support (time, facilitation, amenable government structure)	Less commonly applied because hard, usually used for intra sectoral conflict (rec. vs. comm.)	Has been applied to issues involving Indigenous peoples, but rarely
	2.2.3.2 Co-management	All the participatory benefits—better outcomes, more trust	Usually does not work because it is difficult	Usually only occurs when required, some exceptions	Little work done on *which* stakeholders share power
	2.3 Self-governance	Fewer bureaucratic issues, possibly avoid Tragedy of the Commons	Requires trust in fishers by government *and* public	Rare in North America, common in Europe, elsewhere	Less work done on which stakeholders hold power

ecological subsystems being considered. Knowledge may also be based on a technical understanding of the ecological, social, or governance structures that are embedded in the system. Everyone has their own unique views and opinions; understanding the full range of views will help build a better model of the system.

Discussing the integration of diverse knowledge can be difficult, particularly with respect to how we integrate diverse knowledge *systems*. For example, western fisheries governance entities have usually operated within a western knowledge system and have not often incorporated or acknowledged different systems of knowledge, especially those belonging to Indigenous nations (Reid et al. 2021). Further, as working with Indigenous nations is increasingly important in some countries like the USA and Canada, this work has often meant taking the information needed by decision makers (who often are not Indigenous) and ignoring the broader worldview from which they were taken (Toki 2010). Western science is starting to recognize that there are multiple worldviews that all have equal value and should be considered side-by-side, rather than either-or (Reid et al. 2021). This implies that knowledge considered important under different worldviews should be considered when rendering decisions that apply to people of multiple worldviews.

2 A Brief Overview and Taxonomy of Approaches for Integrating Diverse Knowledge

Understanding how diverse stakeholder knowledge can be integrated into recreational fisheries decision making requires recognizing the broader concepts of how stakeholders are involved in the governance of fisheries or other natural resources. This is essentially cooperative management (co-management). It is critical to understand that though co-management, especially in North American fisheries management, is often described as the specific situation where managers and stakeholders share decision-making power, technically, co-management is not a single homogeneous approach but a continuum. The co-management continuum is described by Sen and Nielsen (1996) and is based on earlier studies (McCay 1993; Berkes 1994). On one end of the spectrum (Fig. 1) decision-making power rests fully with the government, which executes top-down, command-and-control management of fisheries. Self-governance is on the other end, in which stakeholders themselves hold full power to make fisheries decisions. Stakeholder participation increases left-to-right, moving from full government control to self-governance. The "stops," or categories, along this continuum describe the way government and stakeholders interact. For example, the *Instructive* and *Consultative* approaches describe how stakeholders might be communicated with (e.g., by making announcements; *Instructive*) or provide information (e.g., allow public comment, or survey stakeholders; *Consultative*).

It is critical to recognize that categories of the co-management continuum do not align one-to-one with the approaches discussed in this chapter. For example, almost

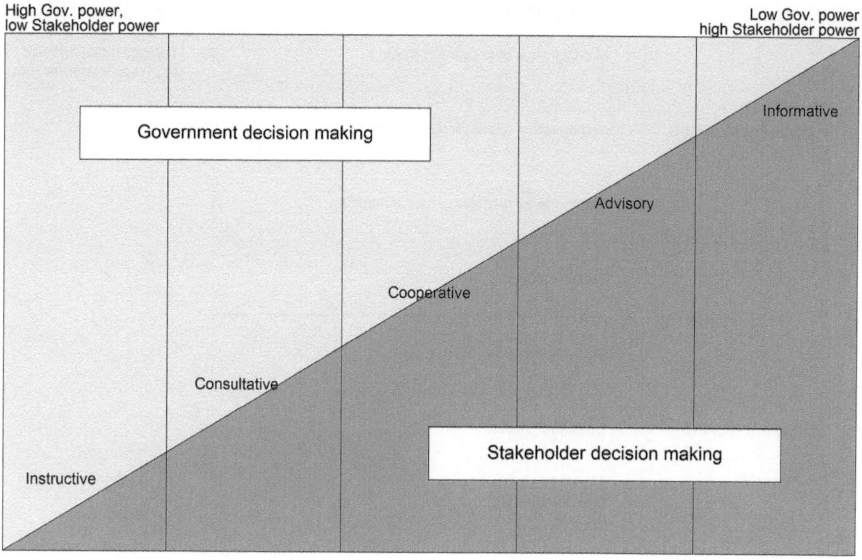

Fig. 1 The Cooperative Management Continuum. (Adapted from Sen and Nielsen 1996, which itself was adapted from McCay 1993 and Berkes 1994)

all the approaches we describe (excluding top-down and perhaps self-governance) *could* be conducted in the *Cooperative* co-management category. Therefore, we organize the approaches for integrating stakeholder knowledge into decision making according to two rubrics (Fig. 2).

We first separate approaches that are non-participatory from participatory. Non-participatory is essentially solely top-down governance, applied in an instructive or consultative manner, in which information flows "one-way" (e.g., either governments telling stakeholders what will happen, or governments receiving information from stakeholders), with no expectation of two-way communication. The rest of the approaches we cover are participatory in that there is at least two-way communication. We divide participatory approaches into two broad categories: knowledge creation and decision making. Knowledge creation focuses on collaborative development of knowledge (two-way learning) that can directly or indirectly affect decision making. We consider the participatory knowledge creation approaches to include multiple ways of knowing (e.g., Local Traditional Knowledge; LTK), collaborative research (including citizen science, co-production), and collaborative modelling (e.g., mental models). Participatory decision making involves stakeholders having a role in the actual decision making. The key distinction between knowledge creation and decision making is that though knowledge creation *can* lead to affecting decisions, it is not required to, whereas decision making implies some decision-making power being allocated to stakeholders. Participatory decision-making approaches include decision science (e.g., structured decision making; SDM), co-management "proper" (in which both managers and

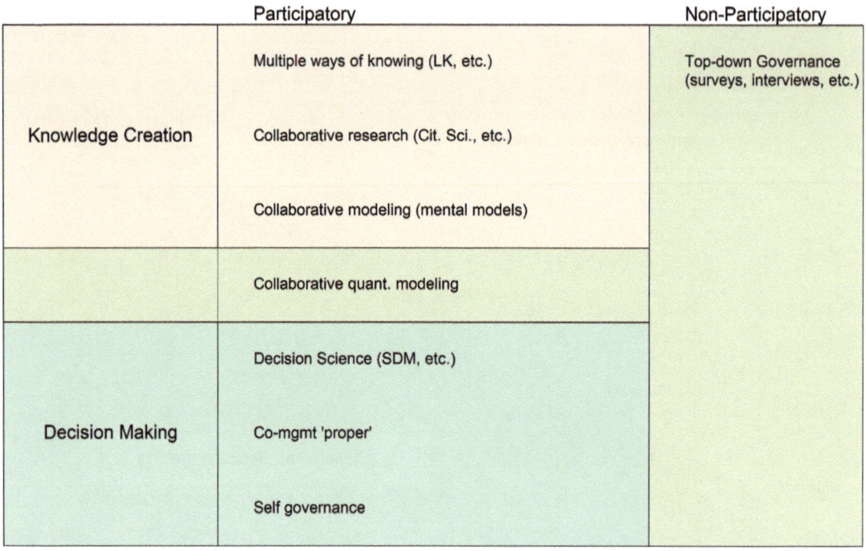

Fig. 2 Diagram of the approaches covered in this chapter, organized into participatory (with the participatory approaches further separated into knowledge creation and decision making) and non-participatory Options (non-participatory, top-down governance spans both knowledge creation and decision making, whereas collaborative modeling occupies the overlap between participatory knowledge creation and participatory decision making)

stakeholders have seats at the decision-making table), and self-governance. We note that knowledge creation and decision making can overlap, especially with participatory quantitative modelling, in which stakeholders are involved in developing models usually used to not only learn about the system but also prioritize management decisions.

2.1 Non-Participatory Approaches–Top-Down Governance

2.1.1 Overview

The top-down governance approach involves government decision-makers (e.g., management agencies) holding essentially all decision-making power, and using information of, or from stakeholders without directly engaging with them. The simplest example of this would be a management agency making a recreational fishery decision, such as what species of fish to stock, based on either assumptions or surveys about what stakeholders would want. In contrast to participatory approaches, this is typically a one-way approach. Stakeholders may be observed, measured, and possibly even understood, but stakeholders are not actively involved in how their knowledge is understood or used. This deprives stakeholders of power, invites mistakes when integrating stakeholder knowledge, and provides little opportunity to remedy such mistakes. The non-participatory, top-down approach has probably been the "status quo" for recreational fisheries decision making, at least for non-federally managed species in North America. Top-down governance subsumes both knowledge creation and decision making (Fig. 2), and usually operates via Informative or Consultative categories of the co-management continuum (Fig. 1).

2.1.2 Advantages and Disadvantages

The primary advantage of non-participatory approaches is their ease. If this approach is the status quo, no changes to decision making are required to continue it. If stakeholder knowledge is desired, it can be gathered via a survey, which, in comparison to participatory approaches, is very inexpensive and easy to implement (although careful construction is still important to avoid bias; Vaske et al. 2026; Kyle et al. 2026). Another advantage is that concentrating decision-making power in the government may facilitate making unpopular decisions. This may be more common when management agencies act on behalf of the broader or future public (such as is the case with the Public Trust Doctrine that guides and governs most state agencies in the USA) (Johnson and Galloway 1994; Henquinet and Dobson 2005), or where the decision-space is defined by higher government levels. An example might be where agencies are protecting native but non-sportfish species that may be threatened by non-native species popular with anglers (Camp et al. 2020), as has occurred with the federally protected Humpback chub (*Gila cypha*) in the Colorado River in the U.S. (Coggins Jr and Yard 2010) and whitebait species (*Galaxiidae*) in New Zealand (Lintermans 2000). If fisheries decision-makers face legal restrictions regarding choices, there may be little room for decisions to be altered, regardless of what stakeholders may want. In such cases, a non-participatory approach may reach the inevitable answer faster.

Of course, it is possible for government entities, like management agencies, to make scientifically sound decisions even when not following participatory approaches. These should be most likely when stakeholders and resources are more homogenous (e.g., a single species fishery, in which the recreational sector is the only fishery the recreational fishers are mostly homogeneous with respect to preferences, and there are minimal non-fishery objectives such as endangered species or water conservation) rather than cases with diverse species, waters, and stakeholders (van Poorten and Camp 2019). Still, even in these cases, agencies would likely benefit from well-designed surveys that provide them with accurate representations of what stakeholders want.

The downside of forgoing participatory approaches is, obviously, that there is no participation of stakeholders in the decision-making process. This has several likely effects. Decision-makers are almost certainly less likely to understand what stakeholders know and want, and thus are less likely to take actions that best serve stakeholders, or that are based on the best understanding of the resource. But at a more fundamental level, non-participatory government forgoes the potential benefits of greater trust, shared understanding, and improved communication that are all described as potential benefits of participatory processes (Gregory et al. 2012). A key disadvantage, as described below, is that non-participatory processes will usually make it much more difficult to specifically understand diverse knowledge.

2.1.3 Application to Recreational-Specific Fisheries

We do not go into detail on the application of non-participatory processes to recreational fisheries, mostly because it is commonly the status quo and because it should be well-covered in other chapters of the book that focus specifically on survey approaches (see Vaske et al. 2026). There is a myriad of examples of surveys used to understand recreational fishers knowledge. Some surveys focus on very applied questions, such as stakeholder preferences for certain management actions, whereas others focus on *why* stakeholders hold certain opinions and preferences, typically connecting responses to broader behavioural and psychological theories.

2.1.4 Application to Specifically Diverse Knowledge

One of the key disadvantages of non-participatory approaches is that it can be especially difficult to access specifically diverse knowledge, or conversely, especially easy to omit or ignore diverse perspectives. This may result in decision making that is driven by, and perhaps best serves, the most powerful or most vocal stakeholder groups. Non-participatory approaches like surveys can be used to target (via strata) knowledge from more diverse stakeholders, but doing so can be practically difficult. Most often, it requires purposefully stratifying surveys to focus on non-majority angler groups (e.g., Morales et al. 2020). However, there are still some short-falls, including (i) stratifying surveys can only address diversity via demographics that are recorded and available, such as sex, age, race, and ethnicity, and these not likely sufficient to represent diversity in motivations, behaviour, or management preferences, (ii) the availability of demographic information is usually collected via fishing licenses, which varies by state and may miss large sections of populations that are not required to purchase licenses (e.g., in some cases shore-based or younger or older fishers), and finally (iii) targeted surveys will probably be impractical in cases where fishing licenses are not required or databases are not maintained, which may in fact be many areas where non-commercial fishing occurs. Surveys that are not specifically targeted towards diverse segments of the population can still be used to understand differences among stakeholder groups (e.g., by asking respondents to self-identify). This can be useful for looking for a more complex diversity than demographics (e.g., motivations, typologies). The downside with such an approach, of course, is that without a priori knowledge of the distribution of anglers, it is difficult to determine a sufficient sample size from less-represented groups, or whether response rates from these groups will be sufficient if they feel less empowered and entitled to have their opinions heard.

2.2 *Participatory Approaches*

2.2.1 Overview of Participatory Approaches

2.2.1.1 Definitions and Background

There are many participatory approaches; here we provide an explanation for how we organize them. Participatory approaches are generally defined as involving non-scientists in research or decision making (Reed 2008; Lopes et al. 2013). Participatory approaches emphasize two-way learning–both decision-makers and stakeholders learn about the fisheries systems (including about each other) or have some amount of capacity or power in decision making. Born largely out of (international economic) development work in the 1980s and 1990s (Chambers 1992; Pretty et al. 1995), participatory approaches are a large and growing area of research that crosses many disciplines and involves a few similar ideas with sometimes overlapping definitions (Roque et al. 2022). For example, we consider participatory approaches to include co-management in its different forms (adaptive, place-based), citizen science (and related ideas like co-creation and participatory science), and uses of stakeholder knowledge such as local ecological knowledge, local traditional knowledge, and Indigenous knowledge. The point is that the "participatory approach bin" is large and its taxonomy can be confusing (Voinov and Bousquet 2010; Schwermer et al. 2020). Regardless, there are increasing calls for the application of participatory approaches across many natural resource management systems (Lauber et al. 2012), including fisheries (Jentoft and McCay 1995; Lopes et al. 2013). The majority of literature on fisheries participatory approaches focuses on small-scale commercial or artisanal fisheries, rather than strictly recreational fisheries (see Dedual et al. 2013). This may be changing as participatory approaches are now required to be used for some federally managed fish species (recreational, commercial, or both) in the USA and Canada (Wiber et al. 2009), and North American recreational fisheries managers increasingly describe their approaches as participatory (Crandall et al. 2019).

2.2.1.2 Advantages and Disadvantages

There are some universally recognized advantages and challenges that apply to many of the different participatory approaches that will be discussed (Pinkerton 1994; Berkes 2007; Dedual et al. 2013). When appropriately implemented, participatory approaches can:

- Lead to greater stakeholder investment, involvement, and trust in the decision-making process.
- Increase acceptance of decisions made.
- Increase the quantity and quality of data or information available for use in decision making.

- Increase shared fishery understanding among stakeholders, scientists, and decision-makers.
- Introduce ideas for making better decisions that are new, at least to decision-makers.
- Democratize the decision-making process, which may increase fairness and equity.

Participatory approaches can also have some disadvantages. Some processes can require long time frames and dedicated personnel to appropriately implement. Virtually none of the participatory approaches are "no risk"—if they are performed poorly, or the information gathered from participation is not actually used, they may erode stakeholder trust and lead to greater contention than had the participatory approach not been attempted (Crandall et al. 2019). Importantly, most participatory approaches will cost additional resources, such as experts to appropriately lead or facilitate them, and incur changes to the decision-making process (which take energy and time) to ensure the information is integrated.

We consider participatory approaches to both creating knowledge and to making decisions (Fig. 2). Within participatory approaches to creating knowledge, we first describe approaches for incorporating multiple ways of knowing, which focuses on local, traditional, and Indigenous knowledge. We then describe collaborative data collection approaches, which are often called citizen science or co-creation. Finally, we briefly describe collaborative modelling, separating conceptual modelling (e.g., cognitive fuzzy mapping) and quantitative modelling. Although virtually all participatory decision-making approaches can be considered co-management, we separate specifically named decision-science approaches to co-management, such as structured decision making, from general co-management. However, both would comprise the *Cooperative* category of the co-management continuum (Fig. 1). We consider self-governance as a participatory approach in practice because in almost all cases there is some at least upper-level governmental oversight or authority, though this approach aligns with the *Advisory* or *Informative* categories of the co-management continuum.

2.2.2 Participatory Science and the Cooperative Creation of Knowledge

2.2.2.1 Incorporating Multiple Ways of Knowing

2.2.2.1.1 Overview

Several related and sometimes overlapping terms are used to describe knowledge obtained and held by stakeholders that is generally not learned from formal, academic education, but from sociocultural heritage, tradition, and lived experiences engaging with or immersed in a resource (Berkes et al. 2000; Thornton and Scheer 2012). These terms include local ecological knowledge (LEK), traditional ecological knowledge (TEK), Indigenous knowledge (IK), local traditional knowledge (LTK), community knowledge (CK), as well as translational ecology (TE). We will use the term local knowledge (LK) as shorthand to encompass these multiple ideas, but readers should be aware that in some cases IK may be more appropriate (Mauro and Hardison 2000), that TEK was and is often used by international organizations such as the International Union for Conservation of Nature (IUCN; Johannes 1989), and that LEK is one of the most widely used terms, especially in fisheries (Beaudreau and Levin 2014). LK is regularly studied in commercial, artisanal, and non-commercial fisheries, but these constitute applications of a broader topic to a specific field, and so some brief background on LK is necessary before discussing how it may be used to integrate diverse stakeholder knowledge into decision making.

LK functionally describes a system of knowledge that includes not only biological information and relational information describing ecological and environmental linkages but also a more holistic understanding of resources and systems that includes experiences, observations, narratives, and skills passed along over generations (Berkes et al. 2000; Thornton and Scheer 2012). LK, a form of tacit knowledge, is increasingly being made explicit—transcribed from its owners into academic formats via interviews or other methods (Beaudreau and Levin 2014), though this process has important ethical considerations regarding ownership, moral considerations regarding whether and why LK must be translated into reductionist approaches to be valued (Mauro and Hardison 2000), and critiques that question whether such translation furthers colonial approaches in fisheries science (Silver et al. 2022). Nonetheless, broad categories of LK exist in primary literature today, including descriptions of LK, assessments of relationships between LK and outcomes at the individual level (e.g., how does an individual's knowledge affect their fishing success), the role of LK in policy making, appropriate ethical methods for conducting LK gathering and research, and comparison of LK to knowledge generated through western science traditions (Thornton and Scheer 2012).

Incorporating multiple ways of knowing generally corresponds to the *Cooperative* category of the co-management continuum (Fig. 1) and fits firmly within the knowledge creation category of participatory approaches (Fig. 2).

2.2.2.1.2 *Advantages and Disadvantages*

There are two general types of benefits expected from appropriately employing multiple ways of knowing in fisheries decision making: more legitimate and acceptable decision-making outcomes and better relationships between stakeholders and decision-makers (Cash et al. 2002). The broader literature describes the benefit of improved quality of management decisions and better conservation outcomes by employing LK approaches (Johannes 1989; Moller et al. 2004). This is especially true where there is little scientifically derived information or where existing data time-series are short (Beaudreau and Levin 2014). Most broader studies (e.g., Gadgil et al. 2003; Folke 2004) focus on the benefits LK provides in terms of creating better linkages, shared understanding, respect, and trust among decision-makers and stakeholders or rightsholders. This is seen as an intrinsic benefit on its own, but it can also foment more cooperatively managed systems that are recognized as benefiting socioecological resilience (Mackinson and Nottestad 1998; Olsson et al. 2004;

Camp et al. 2020). Thornton and Scheer (2012) provide an extensive review documenting that LK approaches are often used in marine resource settings to (i) augment a knowledge base to improve management, (ii) assess, develop, or improve management techniques, and (iii) to inform and improve conservation strategies.

One challenge in incorporating LK is that decision-makers may not be prepared to understand its underlying ontologies, and therefore may struggle to appreciate its value or how to incorporate these ways of knowing into decision-making processes. An important starting point is to understand that LK (and other ways of knowing outside of Western science) holds inherent value and does not require validation by Western science to be applied. Rather, Western science and LK should be viewed as different epistemological approaches to understanding the world (see Moon and Blackman 2014). Specific to Indigenous knowledge, Reid et al. (2021) offer several mental models of Two-Eyed Seeing, a method by which Western science and IK are used together to draw from the strengths of both.

Time investments can be a practical challenge to managers who seek to solicit and meaningfully integrate multiple ways of knowing into decision-making processes. In their review, Thornton and Scheer (2012) highlight studies advising against more rapid transmission (e.g., one-off interviews) and instead emphasize the need for deeper ethnographic and participatory approaches (e.g., embedding within culture) that can and usually should span years. This is especially the case where there are greater cultural and power differences between current decision-makers and rightsholders or stakeholders, as is often the case with IK. Using inappropriate processes to obtain LK may be insulting or offensive, potentially erasing whatever benefits they otherwise would have brought.

2.2.2.1.3 Application to Recreational-Specific Fisheries

Within fisheries, LK has commonly been used to describe traditional and artisanal fisheries that are not strictly recreational, such as small-scale commercial or artisanal and subsistence fisheries, and especially those without extensive data, or where harvesters have been practicing for prolonged periods (i.e., multiple generations). This is especially the case when focusing on First Nations and Indigenous LK or IK, with several examples of researchers developing a better and more holistic understanding of local fisheries via engaging with Indigenous knowledge and knowledge holders (Deepananda et al. 2015; Raju et al. 2016; Ban et al. 2017).

In one example, Beaudreau and Levin (2014) interviewed commercial and recreational fishers, including Indigenous rightsholders around Puget Sound, WA, USA about the status of several species of rockfish (e.g., *Sebastes* spp.), including both commercially and recreationally exploited species, and used their results to demonstrate declines among species for which formal assessments were lacking. Though this study was not specific to recreational fishing applications, the findings exemplify how such information can improve decision making where a single fishery fills multiple roles (leisure, food, revenue), or where multiple distinct fisheries target the same species or species complexes.

More recently LK has been applied to fully leisure-based recreational fisheries and brings to light how additional inclusion of LK or IK could begin to more directly address issues of inequity within recreational fisheries; e.g., in fisheries historically used by Indigenous peoples but that are now typically dominated by non-indigenous fishers. For example, Gervasi et al. (2022a) used translational ecology and specifically LEK approaches to assess the stock status and research priorities for a recreationally targeted species crevalle jack (*Caranx hippos*), and Boucek et al. (2022) described reviewing tournament records of bonefish (*Albula vulpes*), though it is not clear whether their interviews captured the diversity of stakeholders in these systems or focused more on a specific sector of recreational fishers.

Finally, an example of how different knowledge types can be hybridized through recreational fishing-related activities can be found in Harrison et al.'s (2018) work with voluntarily operated Atlantic salmon (*Salmo salar*) hatcheries in Norway. This study found that recreational anglers who were operating small-scale conservation-oriented salmon hatcheries combined their LEK with scientific knowledge gleaned from interactions with researchers and policymakers. The study identified several reasons why hatchery operators were motivated to combine multiple types of knowledge; perhaps more important was as a means of coping with changing hatchery regulations (Harrison et al. 2018). For example, changing regulations required hatchery volunteers to collect tissue samples for genetic analysis to guide appropriate broodstock pair selection, a task that anglers would have previously completed using their LEK of fish size, spawning areas, and run timing. Similarly, this study also identified knowledge hybridization as a means by which hatchery groups sought to maintain their relevance and involvement in Norwegian salmon management decisions, particularly as they pertained to the use of hatcheries and stocking programs (Harrison et al. 2018).

From these examples, it is clear that researchers, policymakers, managers, *and* stake-holders benefit from the consideration and incorporation of multiple ways of knowing in fisheries management. How to incorporate such ways of knowing remains an evolving question, with multiple frameworks available to guide decision-makers (see Reid et al. 2021). Bridging ways of knowing, including the epistemologies that underlie them, remains an ongoing opportunity for learning and development in twenty-first-century fisheries management and thinking.

2.2.2.1.4 Application to Specifically Diverse Knowledge

LK approaches, at least within the global north countries, often focus on the perspectives of diverse participants. Often (and especially when considering IK), First Nations, Indigenous, or Native peoples are involved with fisheries LK approaches (e.g., Thornton et al. 2010). To the extent that these or any other participants may not have much power within current fisheries decision-making processes, greater recognition of their knowledge can represent a diversification of the knowledge used in decision-making processes (Reid et al. 2021; Silver et al. 2022). However, as seen by recent studies by Boucek et al. (2022) and Gervasi et al. (2022b), LK approaches are also being applied to the stakeholders or participants most commonly engaged with by recreational fisheries decision-makers, at least in North American fisheries–leisure fishers. Thus, LK cannot be assumed to, by definition, provide a diversity of views. It may be especially important for researchers or decision-makers to consider how they design LK approaches to ensure knowledge from distinct groups is included.

2.2.2.2 Collaborative Research

2.2.2.2.1 Overview

Another approach to integrating stakeholders is to involve them directly in the scientific collection of data. Traditionally, within fisheries and related fields (ecology, natural resource conservation), this is often called "citizen science"—when a non-scientist volunteer collects data or is similarly involved as part of a scientific project or process (Silvertown 2009). Citizen science is also called community science (Bellquist et al. 2022), collaborative research (Hoefnagel et al. 2006), co-creation (Wood et al. 2015; Crossin et al. 2017), and community-based monitoring (Sharpe and Conrad 2006). For the purposes of this chapter, we refer to all these terms as "collaborative research." Collaborative research studies can greatly differ in terms of volunteer roles. For example, collaborative monitoring projects range from volunteers collecting data that is analysed, interpreted, and provided to decision-makers by professionals, to cases where volunteer stakeholders are involved through the entire process (Danielsen et al. 2009). However, most studies of collaborative fisheries research focus on the collection of data by volunteer stakeholders according to monitoring plans designed by scientists, who often analyse and interpret the data (Lewandowski and Specht 2015; Venturelli et al. 2017). This is essentially "crowdsourcing" data to get better information quality or quantity and promote better decision making (Fairclough et al. 2014; Duncan et al. 2017). But the process of democratizing science via collaborative research can also result in greater dialogue between scientists and the public that can result in better accountability and community empowerment (Dedual et al. 2013; Druschke and Seltzer 2012; Charles et al. 2020). Despite collaborative research dating back centuries (multiple papers point out that there was a point in time at which many if not most scientists were actually employed as something else and volunteered to collect information (Duncan et al. 2017)), it is still likely underutilized in recreational fisheries (Arlinghaus et al. 2016), especially with newer technologies such as apps and smartphones (Venturelli et al. 2017; Mazumdar et al. 2018).

Collaborative research generally corresponds to the *Cooperative* category of the co-management continuum (Fig. 1), and usually would fall within the knowledge creation category of participatory approaches (Fig. 2).

2.2.2.2.2 *Advantages and Disadvantages*

Most studies describing collaborative research emphasize the benefits of the approach to data collection—typically hailing collaborative research to acquire greater quantity or quality of data that could be transformed into more-informed and conceivably better decisions. A broad, not fisheries-specific, meta-analysis by Lewandowski and Specht (2015) demonstrated that studies focusing on stakeholders tended to be more methodological, such as predicting the quality of data collection or comparing stakeholder-collected data to those obtained by experts. This is similar within fisheries. Fairclough et al. (2014) pointed to the benefits of reducing monitoring costs by having fishers donate fish parts (e.g., skeletons) to supplant more costly biological sampling necessary for stock assessments, Hyder et al. (2014) and Duncan et al. (2017) highlighted similar benefits of increasing the quantity of data using collaborative techniques, and Lamine et al. (2018) specifically described the benefits of obtaining information from less-studied locations and more rare species. Such cost-effective data collection may be further enhanced with the advent of app-based collaborative research (Venturelli et al. 2017; Gundelund et al. 2020). Beyond more and better data for less, acquiring data more quickly (which is especially possible with online or app-based approaches) can potentially provide an "early warning system" of change, providing notice sooner than traditional scientific monitoring (Sharpe and Conrad 2006).

Collaborative research can also benefit decision-making systems beyond providing data. Bellquist et al. (2022) summarizes some of the non-data benefits of citizen science, including greater community ownership of science and potentially, investment in process and outcomes, and community cohesiveness. Hoefnagel et al. (2006) add to this the benefits of shared understanding and improved communication among scientists and stakeholders. Støttrup et al. (2018) described how recreational fishers collaborated with scientists to develop sampling methodology for building useful time series for European flounder, and how this resulted in participants functioning as "ambassadors" for science within their communities.

There are a few primary disadvantages to collaborative research described in broad (beyond fisheries) overviews (Silvertown 2009; Danielsen et al. 2009). One is that the data collected by non-scientists is usually subjected to at least initial validation by traditional scientist-collected data. This can require local and external (to the stakeholders) expertise for training and validation (Danielsen et al. 2009), but a governance system can be well-designed to accept and integrate non-expert-collected information, especially into goals beyond just better science (e.g., community improvement, environmental sustainability, better policies) (Vann-Sander et al. 2016). Such requirements can eat into the potential cost saving of collaborative research, and if done poorly, could diminish other benefits (e.g., trust among scientists and stakeholders).

Another disadvantage is that volunteers will bear some of the costs saved by researchers and managers. This may be minimized when collaborative research can be integrated into "regular" planned recreational fishing trips (e.g., app-based data collection), but even here, fishers will spend time entering information that they might otherwise spend fishing. For example, many anglers do not measure each fish they catch—even those harvested. Measuring and recording that information may be seen as a nuisance, though length-based composition data are often critical for assessment methods, and difficult to obtain in some systems. A related challenge with collaborative research that Silvertown (2009) and others (e.g., Vann-Sander et al. 2016; Støttrup et al. 2018) point out is that volunteers must receive some reward for their contributing energy and time for collaborative research to return positive social capital. The reward may be in the form of feedback (e.g., recognition, small tokens of appreciation like t-shirts or hats). But often a more meaningful reward is the inclusion of the stakeholders in the actual interpretation and evaluation of the data, leading to actual participation in the decision making (Danielsen et al. 2009). This may mean moving beyond a "science-centric" approach, where the focus on collaborative research is the generation of data useful to scientists, to other systems, such as management-centric (in which the focus is on providing better management and policy decisions), eco-centric (focus on better resource outcomes), or citizen-centric (focus on meeting personal interests and community goals) (Vann-Sander et al. 2016).

2.2.2.2.3 Application to Recreational-Specific Fisheries

There are many ways that collaborative research can be applied to recreational fisheries. The two most common applications of collaborative research to recreational fisheries decision making are (i) collection of biological samples, and (ii) reporting of fishing trips and catches. Both can affect fisheries management decisions, often through traditional scientific approaches such as stock assessments. For example, Fairclough et al. (2014) described a collaborative research approach where skeletons of harvested fish were collected from fishers, from which scientists could extract hard structures (otoliths, spines) used to age fish. Age information can directly alter known maximum ages from which age- or size-based natural mortality estimates are often derived (Amy et al. 2018), and age of catches can be useful for understanding selectivity (and with great enough sample sizes, catch composition) that are both critical to modern stock assessment.

Recreational fishing records can provide information on catches that can be used to develop indices of relative abundance, occurrence, and possible fisher behavioural shifts, such as targeting. These record types include recreational fishing trips and catches, whether app-based (Venturelli et al. 2017; Gundelund et al. 2020) or logbooks (Bellquist et al. 2022; Harris et al. 2021), or observation or survey-based methods (Støttrup et al. 2018; Wambiji et al. 2022). Some of these studies demonstrate how the information gathered from collaborative research has been translated to affect fisheries management decisions. Monk and He (2019) showed how a collaborative research program was directly used in the assessment and management of rockfishes in the US Pacific. Bellquist et al. (2022) demonstrated that for-hire fishing logbooks could be used to assess the effects of fisheries regulations on popular *Paralabrax* spp. (e.g., Kelp Bass, Sand Bass), and Bellquist et al. (2022)

documented the use of collaboratively developed management strategy evaluations to affect recreational abalone fisheries.

2.2.2.2.4 Application to Specifically Diverse Knowledge

Collaborative research is inherently volunteer-oriented; thus, the people who are engaged are not likely to be random. For example, Gundelund et al. (2020) showed that contributors to an app-based collaborative research project were younger, more avid, and psychologically involved anglers, and Støttrup et al. (2018) detail how the volunteers in longer-term studies were often the most highly motivated. The potential challenge of this is that more specialized or motivated anglers, or anglers to whom fishing is particularly central to their lifestyle, may already think much more like scientists and managers than other fishers (Gray et al. 2015a). This suggests that even collaborative research that goes beyond data collection to engaging volunteers with interpretation or decision making may do so with only a subset of the fishing population, and one that may be already more like the scientists and managers.

Though concerns over self-selecting bias in collaborative research volunteers might be minimized if the collaborative research program focuses on the volunteer collection of purely biological data (e.g., water sample, environmental quality; Poisson et al. 2020), self-selecting bias may need to be addressed in projects aimed at understanding either anglers or fishery related aspects (e.g., catch rates, catchability, utility functions, preferences for management actions). In addition to a specialization bias, participants may also skew towards wealthier individuals who can afford the time to volunteer. If a greater diversity of stakeholders, or chronically underrepresented groups, are to be involved in collaborative research, steps may need to be taken to specifically engage these stakeholders (e.g., developing "strata" or sub-groups from which to elicit volunteers). For example, Dobbs et al. (2016) demonstrated how specifically focusing on Indigenous people in Western Australia led to greater collective understanding of aquatic ecosystem values, threats, and management priorities.

On the other hand, the quality and quantity of collaborative research data are also critical to project success and management decisions (Gundelund et al. 2020), and the quality and quantity of data tend to be improved by greater stakeholder retention and thus experience (Edgar and Stuart-Smith 2009) in collaborative research projects (Sharpe and Conrad 2006). Therefore, researchers may need to consider offsetting costs or otherwise facilitating the engagement of diverse volunteers.

2.2.2.3 Collaborative Modelling

2.2.2.3.1 Overview

Collaborative modelling provides a framework for all parties involved in a decision to contribute their beliefs and knowledge to the decision process. Some collaborative modelling refers to qualitative and semi-quantitative mental modelling (Gray et al.

2014; Murphy Jr et al. 2021), such as cognitive fuzzy mapping (Gray et al. 2015b; Li et al. 2016), whereas other collaborative modelling describes the development and use of quantitative models (usually population dynamics, assessment, and or fisheries management models) in a collaborative setting between scientists and stakeholders (Miller et al. 2010). The latter is dealt with in Wilberg et al. (2026) and so is only briefly mentioned here. Instead, we focus on mental models.

People use mental models that combine information with reasoning to frame and guide behaviour when they make decisions, talk to others, read, or otherwise receive information (Craik 1943; Carley and Palmquist 1992; Johnson-Laird 2001). Mental models are defined as internal (to the stakeholder) cognitive representations of how the "world," or a specific system, works (Biggs et al. 2011) and are generally qualitative in nature (Jones et al. 2011). The concept of mental models was theorized mostly at the individual level (Johnson-Laird 2001), i.e., individual rightsholders or stakeholders are believed to possess their own mental models (Biggs et al. 2011). Thus, mental model research is usually collected via direct interviews, such as semi-structured interviews, though some group approaches like the Q-method can also be used (Bueno and Schiavetti 2019).

Mental models include not just what a person believes about an issue (i.e., a preference), but how their beliefs or knowledge of different components of a system are related. For example, a mental model could describe not only what environmental and fishery factors a person believes mst drive fish populations, but also how those factors are affected by management actions, decision-makers, and scientists. Mental models are understood to encompass the cognitive aspects related to decision making–beliefs, values, attitudes, remembered experiences–and are believed to be helpful in influencing future behaviour, including related to resources like fisheries (Biggs et al. 2011; Gray et al. 2012a, b, 2015b). This makes understanding mental models valuable not just for better understanding stakeholders, but also potentially for guiding predictions of stakeholder responses to different management or governance changes (Gray et al. 2012a, b; Murphy Jr et al. 2021).

If mental models are internal to the individual, researchers need a way to extract them so that they can be described, assessed, and used (Carley and Palmquist 1992). Mental model approaches use different specific tools, such as cognitive mapping (sometimes fuzzy cognitive mapping), a semi-quantitative method that typically visually describes (usually lines on two-dimensional planes) how stakeholders connect ideas, such as weather, fish abundance, and management actions (Murphy Jr et al. 2021). For example, Gray et al. (2015a) used cognitive mapping techniques to demonstrate that more specialized German northern pike anglers' mental models were more like those of managers, compared to less specialized anglers, suggesting lesser potential for conflict between these anglers and managers.

Collaborative modelling, like previous approaches, generally corresponds to the i category of the co-management continuum (Fig. 1). In most cases it will fall within the knowledge creation category of participatory approaches (Fig. 2). However, there may be some cases in which the collaborative mental models developed are directly used to inform decision making, though this is most likely if other approaches (such as co-management proper) are already in place.

2.2.2.3.2 *Advantages and Disadvantages*

The obvious advantage of mental models is that they specifically focus on the linkages among components of a system and can be used to infer the perceived greatest or strongest linkages and determinants. Most other approaches either focus on finer-scale connections or preferences or do not explicitly elicit information about how aspects are connected.

The related main advantage of mental model approaches is that they can provide a more complete or "deeper" description of not just what, but how stakeholders think. Most approaches used in fisheries for assessing stakeholder perceptions focus on asking specific questions about specific issues. This usually explicitly or implicitly assumes or imposes conceptual, if not structural, linkages between factors, such as fish populations and environmental drivers (Brewer 2013). For example, survey questions that ask for stakeholder preferences for fisheries regulations to restore a collapsed stock would seem to imply that fishing was related to the decline, or at least that moderation of fishing could help restore the stock, whereas stakeholders may believe the fish population changes were driven by non-fishing factors such as environmental fluctuations. Alternatively, using mental models specifically encourages broader and more encompassing feedback that can help researchers recognize factors they had perhaps not considered but that might be especially important to stakeholders (Horowitz et al. 2018).

Using mental model approaches to understand and integrate stakeholder knowledge into decision making has disadvantages. The semi-quantitative nature of the specific methods used to assess mental models (e.g., cognitive fuzzy mapping) reveals the challenge of translating words from different individual stakeholders into a common framework that generalizes the population studied. For example, alternative terms and names may be used to describe the same concept and need to be understood well enough to correctly aggregate them. The time-consuming nature of mental model work also rarely allows for probabilistic sampling, typically because it relies on volunteers willing to be interviewed. This means sample sizes are often small (usually $<$100 individuals sampled) and moreover, not necessarily representative of the fishing population. This may be less of a concern when focusing on fisheries with comparatively few participants, such as for-hire fishers, captains, and guides (Gervasi et al. 2022a, b) or small-scale fisheries that operate in an extremely specific location (Brown et al. 2021). If the mental models are not elicited from stakeholders following a probabilistic sampling approach, and if sample sizes are substantially lower than the population, it may be challenging to ascribe the mental model results to the broader population. For this reason, mental models may be paired with probabilistic stated preference surveys in which mental models are used to ensure the survey is relatable to stakeholders' worldviews.

2.2.2.3.3 *Application to Recreational-Specific Fisheries*

There are three main ways that mental models have been used in fisheries, including recreational fisheries. First, models have been developed to simply better understand stakeholders because this understanding is considered important for stakeholder relationships and ultimately better management decision making. An example could include Gervasi et al. (2022b), who developed a conceptual model of a red snapper (*Lutjanus campechanus*) fishery based on interviews with for-hire fishing captains. Second, mental models can be applied to management decision making. Most mental models published in fisheries and related journals will describe the model results and apply them to inform decision making (Murphy Jr et al. 2021). Third, much of the work using mental models is comparative and tries to assess how similar are mental models of distinct groups or assess the correlation between mental models and other factors. Work by Gray et al. (2015a) exemplifies this, noting correlations between mental models and angler specialization, such that the mental models of more specialized anglers were more similar to those of managers. Comparing mental models among researchers, fishers, and decision-makers (e.g., Brown et al. 2021) is a comparatively common topic within fisheries, and other studies have focused on how mental models do or do not change after some management alteration (Wade and Biedenweg 2019).

2.2.2.3.4 Application to Specifically Diverse Knowledge

Like previous approaches that rely on volunteers, care must be taken to ensure that mental models are elicited from a diverse group of stakeholders. This may be addressed by stratification–asking for similar numbers of volunteer participants from different groups of people. The challenge with this is that mental models are based on internal cognitive processes which tend to not be outwardly visible, while people are often grouped by externally or more identifiable characteristics, such as belonging to a fisher, scientist, or decision-maker group (Wade and Biedenweg 2019; Brown et al. 2021). But as Gray et al. (2015a, b) showed, the mental models of some different groups may be similar (in this case, specialized anglers and managers), so even such stratification does not guarantee a diversity of mental models will be collected. This is especially true if diversity of mental models relates to some other factors (e.g., cultural worldview).

One approach for assessing how different mental models are is to use them to investigate social-ecological knowledge diversity, which examines the coupled human-environment interactions and acknowledges differences in the knowledge, value, and beliefs of various stakeholders or stakeholder groups (Schwermer et al. 2021). This approach may be especially useful for describing the diversity within the sampled mental models, and potentially, for understanding how individuals within the same population may have diverging opinions and preferences regarding the same management options, for example, native vs. stocked fish.

2.2.2.3.5 Brief Description of Quantitative Participatory Modelling

Quantitative participatory modelling can use existing frameworks that formally integrate knowledge or beliefs (such as from stakeholders) into an estimation or simulation framework (e.g., Bayesian Belief Networks; Röckmann et al. 2012;

Haapasaari et al. 2013; Ramos et al. 2017) or informally into fisheries quantitative models. Quantitative participatory modelling generally involves scientists developing quantitative models as would routinely be applied to inform management, but using a process that involves stakeholders in the actual model development. The stakeholder involvement usually includes allowing stakeholders to influence or determine aspects of the model or modelling, such as altering model structures, shifting parameter values, determining which uncertainty analyses to perform, and what management options are considered (Miller et al. 2010). For stakeholders to be able to provide this input, quantitative participatory modelling usually involves some formal or informal education of stakeholders regarding quantitative modelling. In turn, participatory modelling may involve stakeholders providing information that modellers lack, especially regarding LTK (e.g., spatial dynamics of fish or fisheries, fisher behaviours and their drivers). For additional detail on participatory modelling, see Wilberg et al. (2026).

Quantitative participatory modelling would almost always be within the *Cooperative* category of the co-management continuum (Fig. 1). We see this approach as often occupying the grey area boundary between knowledge creation and decision making (Fig. 2), because quantitative models are so often used to inform, prioritize, or even define management actions.

2.2.3 Participatory Decision Making

2.2.3.1 Definitions and Background

Participatory decision making is defined as the involvement of users or stakeholders in management decisions of common-pool resources (Pinkerton 1994; Linke and Bruckmeier 2015). Participatory decision making involves governance decision makers (e.g., managers, management agencies, government) ceding some of their decision-making power to stakeholders so that stakeholders and governance personnel share power (McCay and Jentoft 1996; NOAA 2015). Participatory decision making is differentiated from participatory science in that the latter does not provide stakeholders with any direct decision-making power. And participatory decision making is also notably different from self-governance or community-based resource management, in which users are wholly responsible for making decisions (Sen and Nielsen 1996; Symes 2006; Linke and Bruckmeier 2015). Instead, participatory decision making occupies the wide "middle ground" between these two extremes or ends, with decisions being made cooperatively by governance agencies, user groups, and sometimes research institutions (Jentoft 1989). Accordingly, participatory decision making is today mostly synonymous with cooperative management or co-management (Jentoft et al. 1998).

Co-management is a confusing term, even as used within this chapter, because the same term can refer to the broad continuum from top-down government control to self-governance (e.g., the co-management continuum, Fig. 1), the general category of participatory decision making (green rectangle, Fig. 2), and finally a specific but broad arrangement when both managers and stakeholders share power—such as both having a seat at the decision-making table. Even with this last, most specific definition, there is variation. Co-management has been divided into several sub-categories, like place-based co-management and adaptive co-management. Additionally, other decision-analytic approaches that often include stakeholders in the decision-making process essentially are co-management even if they are not explicitly called that. Within natural resources, especially in North America, Structured Decision Making (SDM) and sometimes Adaptive Management (AM) are examples of this and are described in detail in Robinson et al. (2026). For this chapter, we will split out decision-analytic approaches like SDM separately from co-management "proper," in which we include different "sub-species" of co-management, like place-based co-management or adaptive co-management.

2.2.3.2 Decision Science Approaches

2.2.3.2.1 Overview

Decision science provides a suite of methods that can be useful for making difficult decisions for recreational fisheries management and can be used to integrate stakeholder knowledge into these decisions. Decision analysis is the umbrella term that describes methods that can be used to quantitatively evaluate potential actions (Peterman and Peters 1998), having first been developed for the realm of decision making for business (Raiffa 1968). Many methods fall into the category of decision analysis, but three methods have been used more frequently for natural resources decision making: multi-criteria decision analysis (MCDA), structured decision making (SDM), and adaptive management. These frameworks provide methods to account for the common aspects that make natural resources decisions, including decisions for recreational fisheries, difficult, such as multiple, competing interests and objectives, as well as uncertainties that could affect the decision itself.

The three methods—multi-criteria decision analysis, structured decision making, and adaptive management—overlap substantially in approach and outcome. Multi-criteria decision analysis methods were developed to accommodate decisions that must account for multiple objectives, or criteria, each of which may be measured on a different scale (Converse 2020). These methods require clear articulation of the criteria upon which the alternatives will be evaluated, as well as the potential alternatives that could be implemented (Converse 2020). Therefore, MCDA applications in natural resources management can fit naturally into a larger decision-analytic process, like SDM or adaptive management. Structured decision making provides a series of steps in which decision makers and stakeholders can work together to define the problem, clearly articulate the objectives to be achieved, identify a set of alternative actions that could be implemented to achieve the objectives, predict the consequences of each alternative on each objective, and make trade-offs among objectives (Hammond et al. 1999). The SDM process works not only to make decisions but is a semi-structured process for helping diverse stakeholders to understand one another's perspectives and constraints. Although this process may often involve quantitative prediction of how different alternatives may impact important objectives, this is not necessary if all participants agree on the decision to be made. The MCDA process provides a quantitative method for making trade-offs among objectives in the SDM process when quantitative predictions are necessary. For this chapter, we will describe applications of MCDA as a tool within a larger SDM context. Note, however, that MCDA has been applied without the overarching problem framing steps of SDM. Finally, adaptive management is often viewed as an extension of SDM for iterative decisions (McFadden et al. 2011; Runge 2011). The goal of an adaptive management process is to make use of repeated decisions to reduce uncertainty, learning while doing by treating the decision as a treatment to be experimentally assessed (Walters and Hilborn 1978; Williams et al. 2007; Robinson et al. 2026).

Recreational fisheries decision-makers and stakeholders often grapple with how to make decisions that account for multiple ecological, economic, and social values (McDaniels et al. 2006), and often must do so while faced with a seemingly overwhelming amount of uncertainty. Decision analytic frameworks (i.e., MCDA, SDM, adaptive management) are most useful for integrating ecological and social science to evaluate actions for multiple objectives (Robinson et al. 2019), and SDM and adaptive management can be used to account for and potentially reduce uncertainties (Gregory et al. 2012; Walters and Hilborn 1978). Structured decision making has been implemented for a wide range of natural resources problems, including decisions for endangered species protection and recovery (Johnson et al. 2011; Cohen et al. 2016; Cummings et al. 2020), harvest of wildlife species (Peterson and Evans 2003; Peters et al. 2001; Robinson et al. 2016), and invasive species response actions (Blomquist et al. 2010; Moore and Runge 2012; Robinson et al. 2021a). Adaptive management, in which repeated decisions are used to reduce uncertainty that hinders the choice of alternative, has been applied less frequently, but has been used for decisions such as waterfowl harvest management (Williams and Johnson 1995; Johnson et al. 2015), freshwater fish stocking (Fujitani et al. 2017), water use and fish conservation (Melis et al. 2015), and commercial harvest and endangered species interactions (McGowan et al. 2015). Although many of the above decision problems were largely focused on ecological concerns, like endangered species conservation or uncertainties about population dynamics of a harvested or threatened species, each involved some need to consider social objectives, although the handling of these social concerns ranged widely in terms of the consideration received. The fields of SDM and adaptive management for natural resources have focused mainly on those predictions of how management actions will affect ecological objectives, without consideration for how to integrate social science and other diverse ways of knowing into the problem (Robinson et al. 2019). That said, there are examples in the realm of recreational fisheries management that provide insights into how this integration can be more fully realized. We focus here on a review of case studies of SDM and refer the reader to Chapter "Participatory Modelling to Support Evaluation of Management Actions for Recreational Fisheries" in this volume for a more fulsome discussion of the application of adaptive management for recreational fisheries (Robinson et al. 2026).

Decision science approaches almost entirely fall within the *Cooperative* category of the co-management continuum (Fig. 1), and the decision-making section of participatory approaches (Fig. 2).

2.2.3.2.2 Advantages and Disadvantages

Decision analytic frameworks like SDM (including MCDA) and adaptive management are especially useful in several natural resources management contexts. The scope of the decision will dictate the complexity of the decision analytic process, and the entire decision-making framework may not always be needed (Keeney 2004). Sometimes, taking the time to frame the problem by identifying the decision at hand, carefully considering the scope of the decision, and articulating the objectives and actions, can provide all the information necessary to make the decision (Keeney 2004; Gregory et al. 2012). However, if a decision is complex, if there are multiple values and objectives at play, and if there are no actions that would satisfy all the objectives, a more fully quantitative decision analysis may be necessary. In these cases, there may be constraints on the application of this process. Decision analytic processes that fully engage stakeholders and make the best use of ecological and social science can take a great deal of time to complete, on the order of months to years, especially when implementing an adaptive management framework. They can also require the expertise of external facilitators and quantitative analysts, which can lead in part to a need for more funding to complete this process. Ultimately, if the decision is large and is difficult because of competing values or uncertainty, the process to resolve the decision well will take time.

Although decision science processes can be time-intensive, they provide a framework for integrating stakeholders more fully into decisions for natural resources management. For example, decisions for stocking salmonine species into Lake Michigan, part of the Laurentian Great Lakes, have been made through an application of decision analysis that involved agency decision makers as well as recreational and commercial fishers (Jones and Bence 2009). The values-focused nature of decision analysis, in which objectives are identified and used as the basis for creating alternatives, gives stakeholders a means to voice their desires for the resource (Keeney 1992), such as species-specific preferences for salmonine harvest in Lake Michigan (Jones and Bence 2009). Decision analysis in natural resources management has largely tended to focus more on ecological objectives, with the creation of predictive models for fish and wildlife populations, and less on social objectives. However, the integration of ecological and social science can be a strong point of decision analysis, assuming that social scientists are involved in the process, leading to potentially more acceptable decisions (Robinson et al. 2019, 2021b). This strength of decision analytic frameworks is not, however, a substitute for conflict resolution. If groups are in conflict or if stakeholders or decision-makers are unwilling to be transparent about their objectives, then processes like SDM and adaptive management cannot be applied. In addition, if the decision-maker for a problem is unwilling to be part of a decision analytic process, the results of this process likely will not be implemented and will lead to frustration from those involved. Consequently, care must be taken to ensure that groups and decision-makers are willing and able to commit time, resources, and transparency to the process.

2.2.3.2.3 *Application to Recreational-Specific Fisheries*

Decisions for recreational fisheries management involve multiple stakeholders and decision-maker concerns, each of which should be accounted for to make a transparent and objective decision. These concerns could vary among recreational anglers themselves, such as different values placed on the catch or harvest of distinct species. Often, though, these concerns among recreational anglers are only part of the picture, in which economic concerns of commercial anglers and the businesses that support the fishing industry, conservation concerns, and cost constraints must also be accounted for. For example, in Queensland, Australia, coral reef finfish are managed through harvest quotas, spatial zoning, and seasonal closures. Use of this resource includes extraction by groups like recreational, commercial, charter, and Indigenous anglers, and non-extractive actions such as tourism and conservation (Walshe and Slade 2020). Decision-makers revisit decisions for seasonal closures periodically, which can lead to disagreements among stakeholders about the effectiveness and fairness of these closures. In a recent iteration of evaluating seasonal closures, a group of scientists and stakeholders that represented not only recreational interests, but also other extractive and non-extractive interests, were convened for an SDM process. Participants were able to work together to identify the suite of objectives and a set of seasonal closure scenarios that could feasibly be implemented. The group then focused on trade-offs among objectives, based on the value judgements of the various stakeholder groups, and worked to enhance transparency in how the ultimate seasonal closure was chosen. In this case, the group used the limited time that was afforded to them to build an understanding of how particular actions could affect the objectives of the various stakeholder groups at the table, thereby allowing for an open evaluation and discussion of trade-offs (Walshe and Slade 2020).

In some cases, more fully quantitative SDM processes can be implemented, which can lead to a more robust decision, but which can take a great deal of time and resources. For example, Lake Erie, which is one of the Laurentian Great Lakes, is the source of two of the largest freshwater fisheries in North America–walleye (*Sander vitreus*) and yellow perch (*Perca flavescens*). In particular, the walleye fishery is divided into a commercial segment, which is almost entirely implemented on the Canadian side of the lake, and a recreational segment, which is most heavily utilized in the United States (Berger et al. 2012). There have been conflicting values and mistrust between the two fishing sectors, as well as among stakeholders and management agencies, since the 1970s (Jones et al. 2016). As a result of these ongoing concerns, the agencies that have management authority on Lake Erie convened an SDM process among themselves to identify a harvest policy for walleye. Ultimately, the agency managers identified an ecologically sound decision, but they neglected to involve stakeholders in the decision-making process. As a result, this harvest policy was rejected by the angling community (Jones et al. 2016). Therefore, in 2009, the Lake Erie fisheries managers developed a new SDM process for walleye harvest management that included a group of recreational, commercial, and charter anglers, known as the Lake Erie Percid Management Group. This group worked together to identify objectives, which included recreational harvest desires, commercial catch needs, and sustainability of the walleye population, and develop a set of formal harvest control rules that incorporated fishing mortality and biomass considerations (Jones et al. 2016; Robinson et al. 2021b). They then refined the existing stock assessment model for Lake Erie walleye and developed a formal management strategy evaluation (Punt et al. 2016) in a participatory modelling framework to predict the effects of the various harvest control rules on the stakeholders' objectives (see above and Wilberg et al. 2026 for more information on participatory modelling). Ultimately, this SDM process allowed for the different angling groups to understand how different management actions would affect both their own objectives and the objectives of the other group, which led to a decision that was less conservative for recreational anglers but provided a large gain for commercial anglers. This result was likely only possible through this stakeholder-driven participatory process, which built trust among all parties involved (Jones et al. 2016; Robinson et al. 2021b).

2.2.3.2.4 Application to Specifically Diverse Knowledge

Although decision science is becoming more commonly used for natural resources management problems, including recreational fisheries, the purposeful use of decision science for diverse knowledge bases is seemingly less common. However, decision science can provide a framework for bringing this diversity together for recreational fishing decisions. For example, the decision-making process for non-native fish (rainbow trout, *Oncorhynchus mykiss*; brown trout, *Salmo trutta*) removal below Glen Canyon Dam on the Colorado River, the USA, included concerns from recreational fishers about continued sportfish harvest, as well as from Tribal representatives about their spiritual values regarding taking of life unnecessarily (Runge et al. 2011). One reason for the lack of examples with diverse knowledge bases may stem from the aforementioned time required for implementing participatory SDM. Much like in applications of knowledge co-creation described above, stakeholders who can devote the time, and often money for travel to these processes, likely have similar demographics and experiences. In addition, at least in the United States, fishing license sales fund the management agencies who make decisions, which can lead to a desire on the part of decision-makers to include those who "pay the bills." Research has shown, though, that including a diverse group of resource users in decisions for complex socio-ecological systems like recreational fisheries improves the resulting models and understanding of the system (Aminpour et al. 2020). Inclusion of diverse knowledge bases in SDM processes can lead to greater procedural justice (i.e., fairness in decision making), and, therefore, satisfaction with resulting decisions (Ruano-Chamorro et al. 2022).

2.2.3.3 Co-management "Proper"

2.2.3.3.1 Overview

All co-management systems involve the sharing of decision-making power, but co-management can look quite different depending on the system (Pinkerton 1994). Some co-management systems will have only two main parties, a centralized governance component and a single user group. Others may have multiple governance institutions (e.g., at national, regional, and local scales) and multiple user-groups (Pinkerton 1994) and could potentially include research or third-party groups. Some co-management is formal, with legally binding agreements about how decisions are made, whereas others are informal (Acheson 1975). Other factors, such as spatial scale of resources (Wilson et al. 2006) and historic dynamics of politics and power (e.g., an arch from self-governance pre-European colonization to centralized governance and sometimes to some cooperative management) can also influence the appearance of co-management (Hauck and Sowman 2001; Ayers and Kittinger 2014).

Co-management systems can be described by two axes. One axis is the sub-types of co-management, such as adaptive co-management. Each can be thought of as co-management plus an additional specific qualifier. Place-based co-management specifies the spatial scale considered by focusing on sharing power with regional or local stakeholders to make management decisions at finer scales, and which sometimes emphasizes the connection to place (Young et al. 2007). Adaptive co-management specifies that the management objectives should include learning, as it blends adaptive management (see Robinson et al. 2026) and co-management into an approach that accounts for uncertainty in systems processes and emphasizes active learning by treating management decisions as experiments, but making those decisions is a cooperative framework (Armitage et al. 2009). The two qualifiers could also be combined for adaptive place-based co-management.

The other axis useful for describing co-management details how power is shared in decision making because not all co-management is equal with respect to this (NOAA 2015). Power sharing could be described in terms of the process (*how* is power shared) and the degree (*how much* is power shared), but these are usually correlated, and so researchers have tended to describe both within the same scale, of which there are several. In this chapter we have used the co-management continuum by Sen and Nielsen (1996), based on earlier studies (McCay 1993; Berkes 1994) (Box 1, Fig. 1). An alternative is Bens (2005), which describes four distinct levels of stakeholder power in a co-management system:

1. agency has authority, makes decision, informs stakeholders (agency controlled)
2. agency gathers input from stakeholders before deciding
3. stakeholders decide and recommend action for agency to take
4. stakeholders decide to act and then implement (stakeholder controlled)

Box 1: Description of Sen and Nielsen (1996) co-management continuum categories

Sen and Nielsen (1996), based on earlier studies (McCay 1993; Berkes 1994) defined each of the "stops" along the co-management continuum. We base our definitions on theirs and explain what each might resemble in recreational fisheries.

Instructive–minimal power sharing because there is the opportunity for exchange of information between government and users, but not the expectation that such communication will occur or that it will influence decision making. This might look like a management agency allowing for public comment on pending decisions.

Consultative–government does in fact consult with users but has complete authority and power to make decisions without consideration of the consultation. This might look like a management agency having public meetings that ensure some consultation with the stakeholders but does not guarantee that this consultation will influence decision making.

Cooperative–government and users are seen as equal partners in making decisions, and decisions are jointly made with near-equal power sharing. An example of this might be US federal regional fisheries management councils, in which a panel of stakeholders representing different user groups vote on management actions.

Advisory–users advise government on what decisions should be made, with the formal or informal expectation that the government will follow this advice and take the advised actions. An example might be management agencies determining enhancement actions like fish stocking or habitat restoration at behest of organized fisher groups.

Informative–government delegates the authority to make management decisions to the user groups, who make them and then inform the government. Often user groups will also be partially responsible for implementing the actions.

Despite the diversity of approaches, most co-management systems should share some common attributes. Some attributes expected in the co-management of federally managed resources in the United States are taken from Dalton (2005) (NOAA 2015), and include (i) active participant involvement, (ii) decisions based on complete information, (iii) fair decision making, (iv) efficient decision making, and (v) positive participant interaction. These mostly correspond to earlier-described requirements, e.g., Pinkerton (1994).

Co-management "proper" approaches fall mostly within the *Cooperative* category of the co-management continuum (Fig. 1), and the participatory decision-making category of this chapter (Fig. 2).

2.2.3.3.2 Advantages and Disadvantages

The benefits of co-management largely overlap with those attributed in general to participatory processes (see Sect. 2.2.1). Most of the examples describe a spectrum of potential benefits, from better communication, shared understanding, and trust between government or management agencies and stakeholders, to better decisions and improved resources (NOAA 2015). Perhaps obviously, stakeholders receiving some power in the decision-making process feel more empowered and potentially invested in resource management (Ostrom et al. 1999; Gelcich et al. 2009). The power sharing can also translate to greater trust in management agencies (Gray et al. 2012a, b) and can potentially help reduce conflict among stakeholder groups (Wilson et al. 2003; Kearney 2002). Finally, stakeholders who have some power in the decision-making process are expected to be more satisfied with the decisions (Crandall et al. 2019).

One disadvantage of co-management is that it is exceedingly difficult. Sen and Nielsen (1996) reviewed 22 studies and concluded that none had conclusively been successful. Though there have been some noted successes (e.g., examples described above in Kearney (2002) and Plummer and Hashimoto (2011)), there are probably more accounts where co-management has achieved some but not all its objectives. For example, Obiero et al. (2015) noted that co-management on Lake Victoria in Kenya reduced conflict among users, but also failed to achieve the broader goals of meaningfully decreasing non-compliance and leading to more sustainable fisheries. Also, the initial benefits may not last, as DeVos and van Tatenhove (2011) suggest that trust initially gained with co-management may erode over time.

A risk of attempting co-management is that it may leave stakeholders disappointed and disillusioned if it fails to provide the expected shared power, better decision making, and improved outcomes (Chuenpagdee and Jentoft 2007). Co-management, like most participatory approaches, that does not live up to its expectations may in fact leave the system in a worse-off state than before its initiation (Crandall et al. 2019). An unfortunate but perhaps not uncommon case of this may occur if a government (e.g., a fisheries management agency) pursues co-management for the potential benefits of improved communication with and trust from stakeholders, but does not actually wish to cede any decision-making power (Conrad et al. 2011).

Co-management is often attempted but less commonly successful; thus, it is worth examining some recognized reasons for failure. One is history. Histories of mistrust or disenfranchisement may limit the willingness of certain groups to participate in co-management (Finkbeiner and Basurto 2015), as well as potentially affecting how quickly they become disenfranchised with co-management if it struggles or is slow to produce benefits. Another reason co-management fails is that insufficient time or resources are allocated. Sometimes co-management emerges in times of crisis, such as overexploitation of fish stocks (Sen and Nielsen 1996). In these cases, there may be hopes for quick fixes and a sense of urgency associated with outcomes that is inconsistent with the time investment needed for most co-management processes, and with the fact that shifts to co-management systems are not and should not be expected to always be smooth (Jentoft et al. 1998). Greater time and resources are needed to apply co-management to complex governance systems. Co-management struggles in systems with complex vertical and horizontal governance linkages (Armitage et al. 2009)—such as multiple actors (e.g., stakeholders, agencies, government) at each of multiple hierarchical levels of power (Pinkerton 1994; Begossi and Brown 2003). In such systems, decision authority is split among too many different entities, such that government consensus of how to cede power is difficult to obtain, and hard for stakeholders to track (Armitage 2005). A final but related challenge is that the "fit" (described as the connection of biophysical, governance, and socioeconomic systems) will almost always differ between co-management applications (Berkes 2007). This means that the specific approaches to co-management are not often transportable, but rather would need to be "tailor-made" to fit specific cases. Avoiding these pitfalls requires substantial commitments and coordination from government, stakeholders, and perhaps researchers. Professional, third-party (i.e., neither government nor stakeholder) facilitators are recommended at a minimum. Co-management has a lot of potential, which is why it is so commonly recommended, but successfully implementing co-management is exceedingly difficult.

2.2.3.3.3 Application to Recreational-Specific Fisheries

Much of the literature on co-management focuses on self-organizing, community-based systems, often occurring in less economically developed countries and in systems where, stakeholders are closer geographically to both the resource and to the government decision-makers (Armitage et al. 2009). Such systems may not well-match recreational fisheries management, especially in North America). Still, even in North American recreational fisheries systems, there are ample examples of co-management.

One of the most notable co-management applications is the management of federal fisheries in the United States by the National Marine Fisheries Service (NMFS, also known as NOAA Fisheries). Federal legislation (the Magnuson Stevens Act) mandates a co-management process for both commercial and recreational fisheries, which is achieved via regional stakeholder-led (recreational and commercial) fisheries management councils (May 2008). Although each council operates differently and independently, regulatory decisions require multiple levels of review (Sutinen and Johnston 2003). One of the main applications of the co-management process within NMFS are the decisions that allocate harvest of fish among recreational and commercial sectors (Abbott 2015; Abbott et al. 2018), however the councils also make decisions about harvest and when applicable, stock rebuilding (which determines recreational harvest season, size, and bag limits) (van Poorten et al. 2013; Lee et al. 2017), recreational fishing gear and release practices (Flaherty-Walia et al. 2016; Runde et al. 2021), and sector separation, such as among for-hire and private recreational fisheries (Doerpinghaus et al. 2014). The inclusion of the recreational sector in this co-management system provides substantial power to recreational fishery stakeholders and helps to legitimize their preferences, opinions, and information—i.e., their knowledge (MacKenzie and Cox 2013). The governance of these systems is also contentious, with conflict between recreational fishers, commercial fishers, and federal management agencies (Abbott et al. 2022). Even reasonably well-implemented co-management should not be assumed to prevent conflict.

Addressing allocation issues via co-management is not isolated to the United States. Kearney (2002) described the process by which conflict over allocation between recreational and commercial fishers targeting mostly inshore species in Victoria, Australia, was addressed via a co-management process. A co-management council was developed, consisting of government and commercial stakeholders, as well as recreational, conservation, and Aboriginal groups, with the goal of achieving some consensus or at least improved communication between resource users (or conservation groups) and government. Interestingly, the major recommendations of the council were in fact implemented, and apparently considered by both recreational and commercial fisheries as a major and mostly positive achievement. Kearney (2002) considered the positive outcome due to the following steps taken in the co-management process: (i) identifying the sources of conflict, (ii) gathering sufficient data to assess principal perceptions, (iii) having independent assessment of resource use and public debate of these assessments and their likely implications, and (iv) having broader interests than just the primary antagonists involved in the assessment process and advisory processes.

An example of adaptive co-management that took place in Ontario, Canada, is described in Plummer and Hashimoto (2011). The Grand River watershed was a historically rich coldwater fishery that suffered substantial degradation post-colonial settlement. Even though some water-quality-based improvements in fish populations were made in the 1980s, a formal co-management system was not initiated until 1995 by the creation of the Grand River Fisheries Management Planning Process. This process was a partnership between a local management entity, the Grand River Conservation Authority, and the provincial management agency, the Ontario Ministry and Natural Resources. What made the Planning Process co-management was that the steering committees, in which decision-making power was vested in governmental and non-governmental individuals (including individuals belonging to angling groups). The committees formed a plan for various management actions that a subsequent (but similarly mixed government and non-governmental individual) implementation committee translated into 42 specific projects, which were approved by both the government and the public. Two important observations that Plummer and Hashimoto (2011) highlight are that (i) this formal power-sharing partnership was a paradigm shift in which power had to be relinquished from the government and even the steering committees and given to the public, and (ii) that the seeds of this project's success were sewn by previous governmental agreements that included entrusting watershed-level regional management agencies with decision making, and some recognition of First Nations.

Co-management of recreational fisheries may be described less frequently in literature than expected, given the ubiquity of recreational fisheries and the prominence of co-management as a concept in fisheries governance. There are several possible reasons for this. One is that much what is technically co-management is not recognized as such. Many management agencies, especially state or regional (i.e., at more local levels than federal), engage in at least consultative co-management, via informal (phone calls, other feedback) or formal (public meetings, public comment) processes. This does not seem to be often described as co-management, at least in North America. A related reason co-management of recreational fisheries may receive less attention is that much of the co-management literature focuses on more stakeholder-led co-management, at the opposite end of the Sen and Nielsen (1996) spectrum (advisory or informative), such as community-based systems in which stakeholders are closer to both the resource and to the decision-makers (Armitage et al. 2009). Many of these coincide with commercial (Yandle 2003) or artisanal, small-scale, and subsistence fisheries (Maccord et al. 2007; Guidetti et al. 2010; Trimble and Berkes 2013), though there also exist leisure recreational fisheries that are mostly privately managed, especially in Europe (Arlinghaus 2006a). Finally, it appears that co-management is quickly becoming something that is recommended more than it is in fact achieved or at least assessed, given several broader papers describing the state and future of recreational fisheries governance (e.g., Pita and Villasante 2019; Brownscombe et al. 2019; Hyder et al. 2020; Potts et al. 2020). Though such work (correctly) does not describe co-management as a panacea for recreational fisheries, the apparent disparity between the frequency of its recommendation and its successful application highlights the practical challenges of effectively developing and evaluating co-management systems.

2.2.3.3.4 Application to Specifically Diverse Knowledge

Co-management that successfully achieves some transfer of power from government is to some stakeholders may or may not be perceived as fair or equitable among stakeholders. As with participatory approaches in general, it is possible, if not likely, that the most privileged and historically powerful stakeholders will have disproportionately greater power in a co-management system, and thus the knowledge of other, less powerful stakeholders may be ignored or downweighed. To avoid this, Smith (2012) outlines how important it is to acknowledge and address any existing power differentials to ensure the inclusion and empowerment of historically marginalized groups. This highlights a gap in the current co-management literature. Though most of the research has focused on theory and case studies of co-management (these focusing mostly on the successes, failures, and causes), there has been much less written about the practical aspects of co-management, such as determining how tasks should be divided among government and stakeholders, if and how any volunteerism or stakeholder work should be compensated, developing formal, legal institutional arrangements for power sharing, linking different levels of governance organizations (Carlsson and Berkes 2005; Jentoft 2005).

2.3 *Self-governance*

2.3.1 Overview

Self-governance describes stakeholders holding actual or effective complete power for governing their own fisheries, and management agencies playing a minimal role, if any. Self-governance would generally exist in the upper-right-most categories of the co-management continuum, *Advisory* and *Informative*. Self-governance is sometimes grouped within the broader category of co-management, as the latter is sometimes referred to as anything that is not a centralized, top-down, command and control system (Townsend and Shotton 2008), and thus we have considered it a participatory decision-making approach. However, technically, actual self-governance differs from co-management by having the fishers make their own decisions absent another government entity (Townsend and Shotton 2008). Self-governance has mostly been described for commercial fisheries rather than recreational fisheries. Further, self-governance is beyond much of the scope of this chapter, which has as its premise some separation between stakeholders and decision-makers. For these reasons, a full description of the use in recreational fisheries and application to diverse knowledge is not included. Instead, we describe some key points about self-governance and simple advantages and disadvantages.

2.3.1.1 Self-governance of Fisheries That are Neither Strictly Commercial nor Leisure Based

Though self-governance of leisure-based recreational fisheries has not been well-studied, self-governance of less than industrial commercial fisheries is common. Because so many fisheries throughout the world exist as the nexus of enjoyment, nutrition, and livelihood (Cooke et al. 2018), and because the concept of recreational fisheries does not always translate across cultures (Boucquey 2017; Butterly 2018; Nyboer et al. 2022), it is necessary to consider that self-governance may well apply to some of what we, in this chapter, term non-commercial fisheries.

2.3.1.2 Corporate Versus Cooperative Visions for Self-governance

What self-governance "looks like" can vary substantially depending on different philosophical and disciplinary views of the world. Ostrom's famous work on the commons and self-governance, as well as related work that it catalysed (e.g., Ostrom 1990, 2010; Ostrom et al. 1992; Dietz et al. 2003), demonstrated the potential for self-governance to be cooperative if not democratic, and certainly polycentric in a way that challenges the prior understanding that self-governance could not essentially avoid the tragedy of the commons and thus was untenable for common-pool resources such as fisheries. This challenge brought greater weight to earlier work, such as Johannes (1981) book, *Words of the Lagoon,* describing essentially effective self-governance of common pool fisheries, as well as Acheson's decades of work (Acheson 1975, 1988, 2003; Acheson and Gardner 2004) on commercial lobster fisheries in Maine. What has emerged is a valid theory that cooperative, decentralized governance of fisheries is possible, especially in settings with strong leadership, closer proximity to and dependency on resources (leading to stronger incentives), and sufficient social capital (Gutiérrez et al. 2011). Thus, one approach to understanding the potential for self-governance of fisheries is via cooperative, and typically community, self-organization. But it is not the only approach.

Townsend (1995) challenged the association of self-governance of fisheries with the concepts of democratic and cooperative decision making, suggesting that a corporate organizational structure might offer theoretical and practical advantages. The idea that fisheries could essentially be privately managed was not new, with work describing the potential for this sole ownership of fishing rights originating at least a decade earlier (Jones et al. 1980; Scott 1989). As the idea has moved forward, it has focused mostly on the potential industry ownership of fishing rights to eventually obviate the need for external management (Townsend and Shotton 2008; Yang et al. 2010). In nearly all the case studies described, however, the focus is on strictly non-leisure fisheries, and specifically those managed within some form of an individual transferable quota (ITQ), rights-based structure. The prospect of rights-based recreational fisheries has been considered (Abbott and Willard 2017; Abbott et al. 2018), and the notion of applying ITQ or similar systems to recreational fisheries is not new (Sutinen and Johnston 2003; Johnston et al. 2007) but has not seemed to become common.

2.3.2 Advantages and Disadvantages

Self-governance, if it is successful, can cut out massive amounts of bureaucratic procedures and potentially substantially streamline decision making. On the other hand, self-governance could lack broader oversight to ensure long-term sustainability, though at least there is demonstrated potential for both community and corporate organizations to avoid this. The different approaches (community vs. corporate) themselves pose different advantages. Community-based approaches generally retain access to the fishery for at least the public within the community, though they seem unlikely to maximize economic value. Corporate organization of fisheries self-governance (e.g., ITQ's) have long been lauded for their economically efficient solutions to fisheries sustainability issues and have been critiqued for their concentration of the "right to fish" from many to a few (Grafton 1996; Grafton et al. 2007; Sumaila 2010), if not their potential lack of distributional equity and fairness. Community-based approaches would generally be expected to reflect whatever value of diversity the community holds (which could be high or low), whereas the concentration of power with corporate-based approaches is likely to, if anything, decrease the diversity of knowledge in decision making–though this may or may not lead to worse economic and ecological outcomes. Finally, most unlike full government control, true self-governance is certainly rare within recreational fisheries and is probably most common in those peripheral "recreational" fisheries where fishing is also part of subsistence or livelihood.

3 Advice for Selecting an Approach for Integrating Diverse Knowledge

3.1 Considerations Before Choosing an Approach

We first describe four "take home" messages that emerge from the previous summaries and which we think should be considered before attempting to implement any of the aforementioned approaches. These are neither exhaustive nor definitive.

3.1.1 There Are Downsides to Getting It Wrong

There are substantial downsides to a government saying it will acknowledge, respect, and integrate stakeholder knowledge and then failing to do so—or being perceived to failing to do so. This can happen if (i) a poor job is done of working with stakeholders to exchange and understand knowledge (e.g., poor organization of surveys, citizen science, or co-management that makes it hard for stakeholders to understand the process, or if the process is not organized equitably), or (ii) the information is well received and understood, but is not actually used for management decision making. The fact that there was engagement with the stakeholders alone does not guarantee that it will lead to positive effects. For example, if stakeholders' expectations are raised that decision-makers will employ their knowledge, but then this fails to happen (either from i or ii or both, above), stakeholder satisfaction may decline because their expectations were raised and not met. Stakeholders may also become frustrated when decision-makers do not seem to understand them, or stakeholders may feel devalued if insufficient resources are put forth to do so. The most important implication of this is that whatever approaches are used, they should be sufficiently planned and supported (time, logistics, financial) so that they have a reasonable chance of succeeding. This emphasizes the importance of choosing the approaches not only based on what might work best or be best in an idealized world, but also considering practical aspects like what can be accomplished reasonably well.

3.1.2 It's Going to Be Difficult to Get Diverse Knowledge with Any Approach

It may be hard to appropriately use diverse knowledge because diverse knowledge is almost tautologically less readily available within the current governance system. This may be because diverse knowledge is in fact more rare in the population (e.g., there may be fewer stakeholders with certain, more diverse lived experiences and/or preferences, but it also may be because even if the knowledge itself is common but resides in people who have less decision making power. Either way, the knowledge is less easily available in the current system, regardless of the approaches used. For example, even with LK approaches that are often used specifically to better understand and share more diverse knowledge, historic injustice and power imbalances may retard the development of trust between researchers or governments, and stakeholders. Other approaches that rely on stakeholders having the financial freedom to volunteer to participate, such as citizen science or often even co-management approaches (that often hold meetings with stakeholders during work hours when scientists and managers can attend, but many working people cannot), may make it even more difficult to integrate diverse knowledge. The main implication of this is that decision-makers should explicitly plan for ways to reach diverse stakeholders and get diverse knowledge if that is what they are interested in. A second important implication is that it is likely to take effort just to understand the landscape of stakeholder knowledge—and specifically which stakeholders and knowledge are more diverse. For example, management agencies (or other governance entities) may need to apply some of the approaches here (e.g., surveys, interviews, public comment) to begin to better understand who they are not hearing from or disproportionately hearing from less.

3.1.3 More Powerful Approaches Will Be More Difficult

None of the approaches described is likely to be easy to implement well, but the ones that provide the clearest path for translating diverse knowledge into decision making power are also likely to be the hardest to accomplish. The co-management continuum provided by Sen and Nielsen (1996) describes the transition between decision-making power being held by, on one end, the government or non-stakeholders, and on the other, the stakeholders. This continuum may also map the relative challenge of implementing the different approaches well. This is because transitioning power from government to stakeholders is difficult logistically and practically, even if there is the social or political will and desire to do so. For example, transferring power from government to stakeholders may require legal changes that take time to accomplish. Legal changes are unlikely to be necessary with other approaches not involving a change in power (e.g., surveys, LK, citizen science, collaborative modelling). The first implication of this is that although care should be taken for all approaches (to avoid disappointed and dissatisfied stakeholders), government needs to budget additional resources (financial, time, personnel) when considering changes in the actual decision-making power structure.

This is especially true when decision-making power is altered in systems more complex in terms of stakeholders or governance. Greater complexity is characterized in terms of multiple vertical and horizontal entities. For example, a more complex stakeholder situation might occur if there are not only multiple (and potentially competing) stakeholder groups, but especially if there are multiple groups even for the same category but that may differ in power (e.g., multiple recreational fisheries stakeholders, such as tournament fishers, harvest oriented fishing clubs, and shore-based anglers lacking formally recognized organizations). Governance structures that feature multiple vertical and horizontal “layers” would have multiple entities, e.g., agencies, within various levels of government. For example, many states in the USA may have both a fish and wildlife agency but also a separate agency charged with environmental protection, species conservation, water management, or aquaculture. Each may be beholden to multi-state or more likely federal agencies and laws that affect different waters or species. Some combination of local, state, regional, and federal agencies may be affected by or need to be involved with changing the way that fisheries decisions are made. If the main implication is that shifting decision-making power is difficult, a sub-point is that it is likely to be especially difficult in systems with more complex stakeholder and government situations.

3.1.4 Approaches are Not Mutually Exclusive

It should be obvious that many of the approaches described are not mutually exclusive and may add substantial benefit if they are combined sequentially or subsequently. For example, surveys are one of the tools likely to be used when the government retains complete power and attempts to obtain stakeholder information via a one-way process—but surveys or other processes, like interviews and public comment, can also be used to augment other processes, such as citizen science, LK, collaborative modelling. Not only (as previously mentioned) may these approaches be necessary for identifying diverse voices and knowledge to follow up with or on, but they may function as an initial step in integrating diverse knowledge, and be strengthened by subsequent future actions (e.g., citizen science, collaborative modelling).

3.2 A General Path for Choosing an Approach

It is difficult and perhaps not possible to provide a complete recipe or set of instructions for determining how best to integrate diverse stakeholder knowledge into recreational fisheries decision making. We describe a simple, three-step framework which can and should be adapted depending on the situation details. One of the most crucial details is who currently has (more) decision-making power in the system—government or stakeholders. The framework described may most closely resemble a system in which the government (e.g., a fisheries management agency) has most of the decision-making power. However, this approach may be useful even in decentralized self-governance systems because it is possible (if not likely) that there will be variance in the individual powers held by different stakeholders, and that some stakeholders will at least have the power and capacity to gauge and potentially reallocate resources and decision-making processes.

3.2.1 Understand the System

Prior to deciding, the system should be understood at some basic level. We emphasize understanding the system in five terms.

1. *Consider what the constraints are that may affect what types of approaches are viable*. This is arguably the most crucial step because constraints tautologically control what can be changed, and ignoring them may render useless any of the steps that follow. Identifying constraints should include assessing the legal, social, and political systems in which the government is nested, to reveal what may be required to make a change in how decisions are made. For example, changing the management decision-making process for a recreational fishery targeting anadromous salmonids in western North America may require considering multiple sovereign entities (e.g., different Tribes, Canada, USA) depending on the fish stock and its migratory patterns in the sea, as well as federal and state laws determining the harvest of salmonids within state waters. Considering these constraints should result in an understanding of which decision processes could be changed, i.e., which decisions are worth integrating greater diversity of stakeholder knowledge into. For example, it may be possible to alter gear restrictions, access even if changes in size, bag, and season limits cannot be made without engaging in a much broader process. Note that consideration of constraints can and should include "soft" constraints, such as where social or political will limits certain types of actions that would otherwise be logistically or legally available.
2. *Assess understanding and prioritization of current system's diversity—how much is known about "how diverse" the stakeholder knowledge is*. It may be useful to recognize the broader (societal and political) context with respect to diversity, especially as it relates to injustice, prejudice, and power differentials between government and stakeholders, as well as among stakeholders.
3. *Identify where the system currently sits on the co-management continuum*—that is, how is the knowledge currently integrated into the decision-making process, which level of "co-management" according to Sen and Nielsen (1996) is the system operating at.
4. *Recognize social norms and expectations of the current and future governance systems*. This could involve describing how different stakeholders expect the government to act with respect to integrating stakeholders' knowledge, as well as if these stakeholders desire or feel entitled to an engagement differently than the current system. Lacking an understanding of stakeholder expectations can lead to those expectations being unmet and subsequent disillusionment with the decision process.
5. *Describe the resources available to facilitate the change*, including time, money, personnel, as well as any existing relationships or partnerships that can provide social capital or other leverage.

3.2.2 Decide on a Broad Category of Approach to Implement

The decision of what approach to use for integrating diverse knowledge is probably critically important but may not be obvious. We embed a smaller "recipe" for selecting a choice within this broader advice.

1. *Summarize the system in terms of terms listed in Understand the System* (Sect. 3.2.1). This summary can be used to (i) narrow down approaches that are viable within system constraints, (ii) decide if the approach should emphasize specifically diverse knowledge based on how much that is prioritized in the management system and how much is known about knowledge diversity in the system, (iii) assess what types of changes would be desired, given where the system currently is on the co-management continuum, and what stakeholders' expectations for changes are, and (iv) determine which of the remaining options can be accomplished within the resource constraints. This process should yield some "list" of viable approaches.
2. *Differentiate between the list of viable approaches.* This may include employing some quantitative or ordered ranking based on a single factor (e.g., cost) or multiple factors (e.g., cost, stakeholder preferences, probability of success), or describing options more qualitatively in terms of expected advantages and disadvantages.
3. *Make the selection.* The selection should ideally follow some objective and sound decision science approach, which could also include some of the principles (e.g., stakeholder involvement, looking to explicitly involve diverse stakeholders) that form the premise of this chapter.

3.2.3 Identify and Contact Stakeholders

Stakeholders will need to be informed about the potential changes in the approaches towards decision making, and subsequently involved to whatever degree the selected approach warrants (see Robert et al. 2026 to consider appropriate ways for communicating with stakeholders). Note that if an emphasis is to be placed on specifically diversity of knowledge, it may be necessary to stratify or otherwise affect the selection of stakeholders to increase the chances of achieving this goal. Doing this may require additional resources, such as surveys or interviews that are designed from a sampling perspective to reach and identify separate groups of stakeholders. Information from step 1 (understand the system, points 2 and 4) may be especially useful for this.

3.3 Preventative Actions After Choosing an Approach

There are several practical and shared challenges in participatory approaches that we briefly mention. These are likely to apply to all the approaches but are mostly different from the disadvantages recognized in the literature for each. Most of these challenges are especially applicable to participatory decision-making processes, i.e., co-management and potentially participatory modelling.

3.3.1 Avoid Overpromising

The excitement of describing potential benefits can seduce practitioners into describing outcomes that either cannot be reasonably assured (e.g., a decision will be made that stakeholders like in some number of months), or processes that are incompatible with constraints or resources (e.g., suggesting stakeholders will be able to make fisheries decisions when they may only be able to describe their preferences). Overpromising leads to unrealistically elevating expectations and consequently, disappointment when reality falls short. In addition to more episodic dissatisfaction of stakeholders, it can cause or increase distrust in decision-makers, which may exacerbate conflict. If stakeholders lose (more) trust in decision-makers, it will likely decrease the efficacy of future attempts at integrating stakeholders' knowledge, even if those future attempts are more accurately and modestly proposed.

3.3.2 Cultivate Appropriate Expectations

It may not be enough to simply not overpromise; it may be necessary to actively work to develop more accurate expectations for how the process of integrating diverse stakeholder knowledge will unfold. This is especially important for the timeframe involved. Most of the approaches described emphasize that substantial time (usually years) is necessary for appropriate implementation. Stakeholder expectations can be managed better by explaining early and often how long the process may take, as well as its current progress.

3.3.3 Discourage "Elite Capture"

Elite capture can occur when a cooperative stakeholder process falls under the control of one or more powerful stakeholders and therefore ceases to become more representative of the stakeholders at large. It is especially a problem with approaches that employ some "subgroup" of stakeholders who meet regularly, such as most proper co-management or participatory modelling approaches (e.g., Miller et al. 2010). If elite capture occurs, it can be difficult to "reclaim" the group, and sometimes results in entire processes fizzling out as stakeholders realize they do not have much say, or for multiple competing stakeholder processes to arise. Elite capture can be common in recreational fisheries, especially in more economically developed countries or where there are greater financial or time investments for fishing (e.g., big game, tourism-based, trophy, tournament). Discouraging elite capture is especially important to avoid if there is an emphasis placed on integrating diverse stakeholder knowledge. Recognizing the risks of elite capture is important to preventing it. Prevention methods include funding professional facilitation (as opposed to relying on scientists or managers who usually do not have such training), carefully selecting stakeholders to engage with to ensure a diversity of stakeholder types (e.g., through some sort of stakeholder analysis (Conroy and Peterson 2013)) and using processes for communication that allow people with differing communications preferences (e.g., written vs. spoken).

3.3.4 Be Wary of End-arounds

End-arounds occur when a process for integrating diverse stakeholder knowledge is subverted near the actual decision, often by disgruntled stakeholders applying pressure on decision-makers to discount or discard whatever approach has been followed. End-arounds typically take place in co-management approaches but can occur whenever actual decisions are being made based, to some extent, on stakeholder participation in an approach. An end-around can be devastating because it can kill otherwise successful approaches at the final step. Prevention and planning are key, especially at the system identification. For example, it is critical to recognize who will have the decision-making power (e.g., is it an agency itself or is it an elected official who is not required to agree with agency recommendations or is there actual legal standing to support collective stakeholder decision?). It is similarly important to ensure that a diversity of stakeholders is involved in the process—even if there are some individuals who may be disgruntled or difficult to engage. If they are already so, it may be much better to have them part of the group than to allow them to criticize it as outsiders, again depending on the decision-making power structure. The goal should not be to eliminate differences in opinions at all, but to ensure that diversity is represented in the approach followed. Finally, it is important to keep participating stakeholders informed about the potential for end-arounds, which can mediate expectations and help identify potential challenges that can be addressed before becoming more problematic.

3.4 Opportunities

More and more management agencies seem to be espousing participatory processes, stakeholder involvement or engagement, and sometimes explicit co-management. This likely increases the opportunities and needs for integrating diverse stakeholder knowledge into recreational fisheries decision making. Additionally, this process is completely compatible with some of the most dominant trends in fisheries decision making and management. Integrating stakeholder knowledge is one of the most common components of structured decision making (SDM), which includes a specific description of how stakeholders should be involved (Robinson et al. 2026). It is also compatible with adaptive management (see references on adaptive co-management). Finally, it works well with the trend towards managing at finer scales such as place-based management.

Acknowledgments We thank A. Lynch, B. Weidel, and two anonymous reviewers for helpful feedback on this chapter. Any use of trade, firm, or product names is for descriptive purposes only and does not imply endorsement by the U.S. Government.

References

Abbott JK (2015) Fighting over a red herring: the role of economics in recreational-commercial allocation disputes. Mar Resour Econ 30(1):1–20

Abbott JK, Lloyd-Smith P, Willard D, Adamowicz W (2018) Status-quo management of marine recreational fisheries undermines angler welfare. P Natl Acad Sci USA 115(36):8948–8953

Abbott JK, Lew DK, Whitehead JC, Woodward RT (2022) The future of fishing for fun: the economics and sustainable management of recreational fisheries. Rev Environ Econ Policy 16(2):262–281

Acheson JM (1975) The lobster fiefs: economic and ecological effects of territoriality in the Maine lobster industry. Hum Ecol 3(3):183–207

Acheson JM (1988) The lobster gangs of Maine. University Press of New England, Hanover

Acheson JM (2003) Capturing the commons: devising institutions to manage the Maine lobster industry. University Press of New England

Acheson JM, Gardner RJ (2004) Strategies, conflict, and the emergence of territoriality: the case of the Maine lobster industry. Am Anthropol 106(2):296–307

Aminpour P, Gray SA, Jetter AJ, Introne JE, Singer A, Arlinghaus R (2020) Wisdom of stakeholder crowds in complex social-ecological systems. Nat Sustain 3(3):191–199

Amy Y, Then John M, Hoenig Norman G, Hall David A, Hewitt (2018) Evaluating the predictive performance of empirical estimators of natural mortality rate using information on over 200 fish species ICES J of Marine Science 75(4):1509–1509. https://doi.org/10.1093/icesjms/fsx199

Arismendi I, Penaluna BE (2016) Examining diversity inequities in fisheries science: a call to action. Bioscience 66(7):584–591

Arlinghaus R (2006a) Overcoming human obstacles to conservation of recreational fishery resources, with emphasis on central Europe. Environ Conserv 33(1):46–59

Arlinghaus R (2006b) On the apparently striking disconnect between motivation and satisfaction in recreational fishing: the case of catch orientation of German anglers. N Am J Fish Manag 26(3): 592–605

Arlinghaus R, Cooke SJ, Sutton SG, Danylchuk AJ, Potts W, de Freire KMF, Alós J, da Silva ET, Cowx IG, Van Anrooy R (2016) Recommendations for the future of recreational fisheries to prepare the social-ecological system to cope with change. Fish Manag Ecol 23(3–4):177–186

Armitage D (2005) Adaptive capacity and community-based natural resource management. Environ Manag 35:703–715

Armitage DR, Plummer R, Berkes F, Arthur RI, Charles AT, Davidson-Hunt IJ, Diduck A, Doubleday N, Johnson DS, Marschke M, McConney P, Wollenberg EK (2009) Adaptive co-management for social-ecological complexity. Front Ecol Environ 7(2):95–102

Ayers AL, Kittinger JN (2014) Emergence of co-management governance for Hawai ‘i coral reef fisheries. Glob Environ Change 28(1):251–262

Ban NC, Eckert L, McGreer M, Frid A (2017) Indigenous knowledge as data for modern fishery management: a case study of Dungeness crab in Pacific Canada. Ecosyst Health Sust 3(8): 1379887

Beaudreau AH, Levin PS (2014) Advancing the use of local ecological knowledge for assessing data-poor species in coastal ecosystems. Ecol Appl 24(2):244–256

Begossi A, Brown D (2003) Experiences with fisheries co-management in Latin America and the Caribbean. In: Wilson DC, Nielsen JR, Degnbol P (eds) The fisheries co-management experience. Kluwer Academic Publishers, Dordrecht, pp 135–152

Bellquist L, Harford WJ, Hurd F, Jackson A, Prince JD, Freiwald J, Neumann A, Likins J, Wilson JR (2022) Use of management strategy evaluation to understand the value of citizen science in managing an iconic California recreational fishery. Estuar Coast Shelf S 278:108112

Bens I (2005) Facilitating with Ease! Jossey-Bass, San Francisco

Berger AM, Jones ML, Zhao Y, Bence JR (2012) Accounting for spatial population structure at scales relevant to life history improves stock assessment: the case for Lake Erie walleye Sander vitreus. Fish Res 115(116):44–59

Berkes F (1994) Co-management: bridging the two solitudes. North Perspect 22(2):18–20

Berkes F (2007) Adaptive co-management and complexity: exploring the many faces of co-management. In: Armitage D, Berkes F, Doubleday N (eds) Adaptive co-management: collaboration, learning and multi-level governance. UBC Press, Vancouver, pp 19–38

Berkes F, Colding J, Folke C (2000) Rediscovery of traditional ecological knowledge as adaptive management. Ecol Appl 10(5):1251–1262

Bess R (2001) New Zealand's indigenous people and their claims to fisheries resources. Mar Policy 25(1):23–32

Biggs D, Abel N, Knight AT, Leitch A, Langston A, Ban NC (2011) The implementation crisis in conservation planning: could "mental models" help? Conserv Lett 4(3):169–183

Blomquist SM, Johnson TD, Smith DR, Call GP, Miller BN, Thurman WM, McFadden JE, Parkin MJ, Boomer GS (2010) Structured decision making and rapid prototyping to plan a management response to an invasive species. J Fish Wildl Manag 1(1):19–32

Boucek RE, Rehage JS, Castillo NA, Dwoskin E, Lombardo SM, Santos R, Navarre C, Larkin M, Adams AJ (2022) Using recreational tournament records to construct a 53-year time series of the Florida Keys recreational Bonefish fishery. Environ Biol Fish 106(3):1–13

Boucquey N (2017) 'That's my livelihood, it's your fun': the conflicting moral economies of commercial and recreational fishing. J Rural Stud 54:138–150

Brewer TD (2013) Dominant discourses, among fishers and middlemen, of the factors affecting coral reef fish distributions in Solomon Islands. Mar Policy 37(C):245–253

Brown HO, Jacobson SK, Adams AE (2021) Defining success for a declining resource: mental models of oyster management in Apalachicola. Florida Environ Manag 68(3):377–392

Brownscombe JW, Hyder K, Potts W, Wilson KL, Pope KL, Danylchuk AJ, Cooke SJ, Clarke A, Arlinghous R, Post JR (2019) The future of recreational fisheries: advances in science, monitoring, management, and practice. Fish Res 211:247–255

Bueno P, Schiavetti A (2019) The influence of fisherman scale in the resilience of socio-ecological systems: an analysis using Q methodology. Ocean Coast Manag 169:214–224

Butterly L (2018) Fishing for rights: the water-food nexus and indigenous fishing in Australia's Northern Territory. Jurimetrics 59:43

Camp EV, Kaemingk MA, Ahrens RN, Potts WM, Pine WE III, Weyl OL, Pope KL (2020) Resilience management for conservation of inland recreational fisheries. Front Ecol Evol 7:498

Carley K, Palmquist M (1992) Extracting, representing, and analyzing mental models. Soc Forces 70(3):601–636

Carlsson L, Berkes F (2005) Co-management: concepts and methodological implications. J Environ Manag 75(1):65–76

Cash D, Clark WC, Alcock F, Dickson NM, Eckley N, Jäger J (2002) Salience, credibility, legitimacy and boundaries: linking research, assessment and decision making. John F. Kennedy School of Government, Faculty Research Working Paper Series RWP02–046, Harvard University, Cambridge, Massachusetts

Chambers R (1992) Rural appraisal: rapid, relaxed and participatory. IDS Discussion Paper No. 311, Institute of Development Studies, Sussex, UK

Charles A, Loucks L, Berkes F, Armitage D (2020) Community science: a typology and its implications for governance of social-ecological systems. Environ Sci Pol 106:77–86

Chuenpagdee R, Jentoft S (2007) Step zero for fisheries co-management: what precedes implementation. Mar Policy 31:657–668

Coggins LG Jr, Yard MD (2010) Mechanical removal of nonnative fish in the Colorado River within Grand Canyon. In: Melis T, Hamill JF, Bennett GE, Coggins LG, Grams PE, Kennedy TA, Kubly DM, Ralston BE (eds) Proceedings of the Colorado River basin science and resource management symposium, US Geological Survey Scientific Investigations, Report 5135, pp 227–234

Cohen JB, Hecht A, Robinson KF, Osnas EE, Tyre AJ, Davis C, Kocek A, Maslo B, Melvin SM (2016) To exclose nests or not: structured decision making for the conservation of a threatened species. Ecosphere 7:e01499

Conrad E, Cassar LF, Christie M, Fazey I (2011) Hearing but not listening? A participatory assessment of public participation in planning. Environ Plann C 29(5):761–782

Conroy MJ, Peterson JT (2013) Decision making in natural research management: a structured, adaptive approach. Wiley, West Sussex

Converse SJ (2020) Introduction to multi-criteria decision analysis. In: Runge MC, Converse SJ, Lyons JE, Smith DR (eds) Structured decision making: case studies in natural resource management. Johns Hopkins University Press, Baltimore, pp 51–61

Cooke SJ, Twardek WM, Lennox RJ, Zolderdo AJ, Bower SD, Gutowsky LF, Danylchuk AJ, Arlinghaus R, Beard D (2018) The nexus of fun and nutrition: recreational fishing is also about food. Fish Fish 19(2):201–224

Craik KJW (1943) The nature of explanation. Cambridge University Press, Cambridge, UK

Crandall CA, Monroe M, Dutka-Gianelli J, Lorenzen K (2019) Meaningful action gives satisfaction: stakeholder perspectives on participation in the management of marine recreational fisheries. Ocean Coast Manag 179:104872

Crossin GT, Heupel MR, Holbrook CM, Hussey NE, Lowerre-Barbieri SK, Nguyen VM, Raby GD, Cooke SJ (2017) Acoustic telemetry and fisheries management. Ecol Appl 27(4): 1031–1049

Cummings JW, Parkin M, Zelenak J, Bell H, Broderdorp K, Holt B, McCollough M, Smith T (2020) Applying expert elicitation of viability and persistence to a lynx species status assessment. Conserv Sci Pract 2(11):1–14

Dalton TM (2005) Beyond biogeography: a framework for involving the public in planning of U.S. Marine protected areas. Conserv Biol 19:1392–1401

Danielsen F, Burgess ND, Balmford A, Donald PF, Funder M, Jones JP, Alviola P, Balete DS, Blomley T, Brashares J, Child B, Enghoff M, Fjeldså J, Holt S, Hübertz H, Jensen AE, Jensen PM, Massao J, Mendoza MM, Ngaga Y, Poulsen MK, Rueda R, Sam M, Skielboe T, Stuart-Hill G, Topp-Jørgensen E, Yonten D (2009) Local participation in natural resource monitoring: a characterization of approaches. Conserv Biol 23(1):31–42

David, Symes (2006) Fisheries governance: A coming of age for fisheries social science? Fisheries Research 81(2-3):113–117. https://doi.org/10.1016/j.fishres.2006.06.015

Dedual M, Sague Pla O, Arlinghaus R, Clarke A, Ferter K, Geertz Hansen P, Gerdeaux D, Hames F, Kennelly SJ, Kleiven AR, Meraner A, Ueberschär B (2013) Communication between scientists, fishery managers and recreational fishers: lessons learned from a comparative analysis of international case studies. Fish Manag Ecol 20(2–3):234–246

Deepananda KA, Amarasinghe US, Jayasinghe-Mudalige UK (2015) Indigenous knowledge in the beach seine fisheries in Sri Lanka: an indispensable factor in community-based fisheries management. Mar Policy 57:69–77

DeVos BI, van Tatenhove JPM (2011) Trust relationships between fishers and government: new challenges for the co-management arrangements in the Dutch flatfish industry. Mar Policy 35: 218–225

Dietz T, Ostrom E, Stern PC (2003) The struggle to govern the commons. Science 302(5652): 1907–1912

Dobbs RJ, Davies CL, Walker ML, Pettit NE, Pusey BJ, Close PG, Akune Y, Walsham N, Smith B, Wiggan A, Cox P, Ward DP, Tingle F, Kennett R, Jackson M, Davies PM (2016) Collaborative research partnerships inform monitoring and management of aquatic ecosystems by Indigenous rangers. Rev Fish Biol Fisher 26(4):711–725

Doerpinghaus J, Hentrich K, Troup M, Stavrinaky A, Anderson S (2014) An assessment of sector separation on the Gulf of Mexico recreational red snapper fishery. Mar Policy 50:309–317
Douglas Clyde, Wilson Mahfuzuddin, Ahmed Susanna V, Siar Usha, Kanagaratnam (2006) Cross-scale linkages and adaptive management: Fisheries co-management in Asia Marine Policy 30(5):523–533 https://doi.org/10.1016/j.marpol.2005.07.001
Druschke CG, Seltzer CE (2012) Failures of engagement: lessons learned from a citizen science pilot study. Appl Environ Educ Commun 11(3–4):178–188
Duncan C, McKinley Abe J, Miller-Rushing Heidi L, Ballard Rick, Bonney Hutch, Brown Susan C, Cook-Patton Daniel M, Evans Rebecca A, French Julia K, Parrish Tina B, Phillips Sean F, Ryan Lea A, Shanley Jennifer L, Shirk Kristine F, Stepenuck Jake F, Weltzin Andrea, Wiggins Owen D, Boyle Russell D, Briggs Stuart F, Chapin David A, Hewitt Peter W, Preuss Michael A, Soukup (2017) Citizen science can improve conservation science natural resource management and environmental protection Biological Conservation 20815–28. https://doi.org/10.1016/j.biocon.2016.05.015
Edgar GJ, Stuart-Smith RD (2009) Ecological effects of marine protected areas on rocky reef communities-a continental-scale analysis. Mar Ecol-Prog Ser 388:51–62
Elinor, Ostrom Joanna, Burger Christopher B, Field Richard B, Norgaard David, Policansky (1999) Revisiting the Commons: Local Lessons Global Challenges Science 284(5412):278-282 https://doi.org/10.1126/science.284.5412.278
Fabinyi M, Foale S, Macintyre M (2015) Managing inequality or managing stocks? An ethnographic perspective on the governance of small-scale fisheries. Fish Fish 16(3):471–485
Fairclough DV, Brown JI, Carlish BJ, Crisafulli BM, Keay IS (2014) Breathing life into fisheries stock assessments with citizen science. Sci Rep 4(1):1–10
Fenichel EP, Gentner B, Arlinghaus R (2013) Normative considerations for recreational fishery management: a bioeconomic framework for linking positive science and normative fisheries policy decisions. Fish Manag Ecol 20(2–3):223–233
Finkbeiner EM, Basurto X (2015) Re-defining co-management to facilitate smallscale fisheries reform: an illustration from northwestern Mexico. Mar Policy 51:433–441
Flaherty-Walia KE, Winner BL, Tyler-Jedlund AJ, Davis JP (2016) Short-term discard mortality estimates for Gray Snapper in a West-Central Florida estuary and adjacent nearshore Gulf of Mexico waters. N Am J Fish Manag 36(2):329–340
Floyd MF, Nicholas L, Lee I, Lee JH, Scott D (2006) Social stratification in recreational fishing participation: research and policy implications. Leisure Sci 28(4):351–368
Folke C (2004) Traditional knowledge in social–ecological systems. Ecology and society, 9(3)
Fujitani M, McFall A, Randler C, Arlinghaus R (2017) Participatory adaptive management leads to environmental learning outcomes extending beyond the sphere of science. Sci Adv 3(6):1–12
Furman KL, Harlan SL, Barbieri L, Scyphers SB (2023) Social equity in shore-based fisheries: identifying and understanding barriers to access. Mar Policy 148:105355
Gadgil M, Olsson P, Berkes F, Folke C (2003) Exploring the role of local ecological knowledge in ecosystem management: three case studies. Navigating social-ecological systems: building resilience for complexity and change, 189:209
Gelcich S, Godoy N, Castilla JC (2009) Artisanal fishers' perceptions regarding coastal co-management policies in Chile and their potentials to scale-up marine biodiversity conservation. Ocean Coast Manag 52(8):424–432
Gervasi C, McPherson M, Karnauskas M (2022a) System dynamics of red snapper populations in the Gulf of Mexico to support ecosystem considerations in the assessment and management process. SouthEast Data, Assessment, and Review, SEDAR74-DW-16, North Charleston, South Carolina
Gervasi CL, Santos RO, Rezek RJ, James WR, Boucek RE, Bradshaw C, Kavanagh C, Osborne J, Rehage JS (2022b) Bottom-up conservation: using translational ecology to inform conservation priorities for a recreational fishery. Can J Fish Aquat Sci 79(1):47–62
Grafton RQ (1996) Individual transferable quotas: theory and practice. Rev Fish Biol Fish 6:5–20

Grafton RQ, Kompas T, Hilborn RW (2007) Economics of overexploitation revisited. Science 318(5856):1601

Gray S, Shwom R, Jordan R (2012a) Understanding factors that influence stakeholder trust of natural resource science and institutions. Environ Manag 49(3):663–674

Gray S, Chan A, Clark D, Jordan R (2012b) Modeling the integration of stakeholder knowledge in social-ecological decision making: benefits and limitations to knowledge diversity. Ecol Model 229:88–96

Gray SA, Zanre E, Gray SR (2014) Fuzzy cognitive maps as representations of mental models and group beliefs. In: Fuzzy cognitive maps for applied sciences and engineering. Springer, Berlin/Heidelberg, pp 29–48

Gray S, Hilsberg J, McFall A, Arlinghaus R (2015a) The structure and function of angler mental models about fish population ecology: the influence of specialization and target species. J Outdoor Recreat Tour 12:1–13

Gray SA, Gray S, De Kok JL, Helfgott AE, O'Dwyer B, Jordan R, Nyaki A (2015b) Using fuzzy cognitive mapping as a participatory approach to analyze change, preferred states, and perceived resilience of social-ecological systems. Ecol Soc 20(2):11

Gregory R, Failing L, Harstone M, Long G, McDaniels T, Ohlson D (2012) Structured decision making: a practical guide to environmental management choices. Wiley, UK

Guidetti P, Bussotti S, Pizzolante F, Ciccolella A (2010) Assessing the potential of an artisanal fishing co-management in the Marine Protected Area of Torre Guaceto (southern Adriatic Sea, SE Italy). Fish Res 101(3):180–187

Gundelund C, Arlinghaus R, Baktoft H, Hyder K, Venturelli P, Skov C (2020) Insights into the users of a citizen science platform for collecting recreational fisheries data. Fish Res 229:105597

Gustavsson M, Frangoudes K, Lindström L, Burgos MCÁ, de la Torre-Castro M (2021) Gender and Blue Justice in small-scale fisheries governance. Mar Policy 133:104743

Gutiérrez NL, Hilborn R, Defeo O (2011) Leadership, social capital and incentives promote successful fisheries. Nature 470(7334):386–389

Haapasaari P, Mäntyniemi S, Kuikka S (2013) Involving stakeholders in building integrated fisheries models using Bayesian methods. Environ Manag 51(6):1247–1261

Hammond JS, Keeney RL, Raiffa H (1999) Smart choices: a practical guide to making better life decisions. Broadway Books, New York

Harris D, Johnston D, Yeoh D (2021) More for less: citizen science supporting the management of small-scale recreational fisheries. Reg Stud Mar Sci 48:102047

Harrison HL, Rybråten S, Aas Ø (2018) Hatching knowledge: a case study on the hybridization of local ecological knowledge and scientific knowledge in small-scale Atlantic salmon (*Salmo salar*) cultivation in Norway. Hum Ecol 46(4):449–459

Hauck M, Sowman M (2001) Coastal and fisheries co-management in South Africa: an overview and analysis. Mar Policy 25(3):173–185

Henquinet JW, Dobson T (2005) The public trust doctrine and sustainable ecosystems: a Great Lakes fisheries case study. NYU Envtl LJ 14:322

Hoefnagel E, Burnett A, Wilson DC (2006) The knowledge base of co-management. In: Motos L, Wilson DC (eds) The knowledge base for fisheries management. Elsevier, Amersterdam, pp 85–108

Horowitz J, Pressey RL, Gurney GG, Wenger AS, Pahang KA (2018) Investigating stakeholder perceptions of fish decline: making sense of multiple mental models. Sustainability 10(4):1222

Hunt LM, Camp E, van Poorten B, Arlinghaus R (2019) Catch and non-catch-related determinants of where anglers fish: a review of three decades of site choice research in recreational fisheries. Rev Fish Sci Aquac 27(3):261–286

Hyder K, Armstrong M, Ferter K, Strehlow HV (2014) Recreational sea fishing–the high value forgotten catch. ICES insight 51:8–15

Hyder K, Maravelias CD, Kraan M, Radford Z, Prellezo R (2020) Marine recreational fisheries-current state and future opportunities. ICES J Mar Sci 77(6):2171–2180

Jentoft S (1989) Fisheries co-management: delegating government responsibility to fishermen's organizations. Mar Policy 13(2):137–154

Jentoft S (2005) Fisheries co-management as empowerment. Mar Policy 29(1):1–7

Jentoft S, McCay B (1995) User participation in fisheries management: lessons drawn from international experiences. Mar Policy 19(3):227–246

Jentoft S, McCay BJ, Wilson DC (1998) Social theory and fisheries co-management. Mar Policy 22(4–5):423–436

Johannes RE (1981) Words of the lagoon: fishing and marine lore in the Palau district of Micronesia. University of California Press

Johannes RE (ed) (1989) Traditional ecological knowledge: a collection of essays. IUCN, Cambridge, UK

Johnson RW, Galloway WC (1994) Protection of biodiversity under the public trust doctrine. Tul Envtl L J 8(1):21–32

Johnson FA, Breininger DR, Duncan BW, Nichols JD, Runge MC, Williams BK (2011) A Markov decision process for managing habitat for Florida scrub-jays. J Fish Wildl Manag 2(2):234–246

Johnson FA, Boomer GS, Williams BK, Nichols JD, Case DJ (2015) Multilevel learning in the adaptive management of waterfowl harvests: 20 years and counting. Wildlife Soc B 39(1):9–19

Johnson-Laird PN (2001) Mental models and deduction. Trends Cogn Sci 5(10):434–442

Johnston RJ, Holland DS, Maharaj V, Campson TW (2007) Fish harvest tags: an alternative management approach for recreational fisheries in the US Gulf of Mexico. Mar Policy 31(4): 505–516

Jones M, Bence J (2009) Uncertainty and fishery management in the North American Great Lakes: lessons from applications of decision analysis. In: Krueger C, Zimmerman C (eds) Pacific salmon: ecology and management of western Alaska's populations Symposium. American Fisheries Society, Bethesda, pp 1059–1081

Jones RA, Pearse PH, Scott AD (1980) Conditions for cooperation on joint projects by independent jurisdictions. Can J Econ 13(2):231–249

Jones NA, Ross H, Lynam T, Perez P, Leitch A (2011) Mental models: an interdisciplinary synthesis of theory and methods. Ecol Soc 16(1):13

Jones ML, Catalano MJ, Peterson LK, Berger AM (2016) Stakeholder-centered development of a harvest control rule for Lake Erie walleye Sander vitreus. In: Edwards C, Dankel D (eds) Management science in fisheries: an introduction to simulation-based methods. Routledge, London, pp 163–183

Joshua K, Abbott Daniel, Willard (2017) Rights-based management for recreational for-hire fisheries: Evidence from a policy trial Fisheries Research 196106-116 https://doi.org/10.1016/j.fishres.2017.08.014

Kearney RE (2002) Co-management: the resolution of conflict between commercial and recreational fishers in Victoria, Australia. Ocean Coast Manag 45(4–5):201–214

Keeney RL (1992) Value-focused thinking: a path to creative decision making. Harvard University Press, Cambridge, MA

Keeney RL (2004) Making better decision makers. Decis Anal 1(4):193–204

Kyle G, Landon A, Pilgreen D, Golebie E (2026) Measurement Instruments. In: Pope KL, Arlinghaus R, Hunt LM, Lynch AJ, van Poorten BT (eds) Understanding recreational fishers: disciplinary and interdisciplinary approaches for fisheries management. Springer Nature, Cham, p 355–385

Lamine EB, Di Franco A, Romdhane MS, Francour P (2018) Can citizen science contribute to fish assemblages monitoring in understudied areas? The case study of Tunisian marine protected areas. Estuar Coast Shelf S 200:420–427

Lauber TB, Decker DJ, Leong KM, Chase LC, Schusler TM (2012) Stakeholder engagement in wildlife management. Hum Dimens Wildl:139–156

Lee MY, Steinback S, Wallmo K (2017) Applying a bioeconomic model to recreational fisheries management: groundfish in the northeast United States. Mar Resour Econ 32(2):191–216

Lewandowski E, Specht H (2015) Influence of volunteer and project characteristics on data quality of biological surveys. Conserv Biol 29(3):713–723

Li O, Gray SA, Sutton SG (2016) Mapping recreational fishers' informal learning of scientific information using a fuzzy cognitive mapping approach to mental modelling. Fisheries Manag Ecol 23(3–4):315–329

Linke S, Bruckmeier K (2015) Co-management in fisheries-experiences and changing approaches in Europe. Ocean Coast Manag 104:170–181

Lintermans M (2000) Recolonization by the mountain galaxias *Galaxias olidus* of a montane stream after the eradication of rainbow trout *Oncorhynchus mykiss*. Mar Freshw Res 51(8):799–804

Lopes PF, Rosa EM, Salyvonchyk S, Nora V, Begossi A (2013) Suggestions for fixing top-down coastal fisheries management through participatory approaches. Mar Policy 40:100–110

Lyall, Bellquist Brice, Semmens Stephen, Stohs Alayna, Siddall (2017) Impacts of recently implemented recreational fisheries regulations on the Commercial Passenger Fishing Vessel fishery for Paralabrax sp. in California Marine Policy 86:134–143.https://doi.org/10.1016/j.marpol.2017.09.017

Lynch AJ, Cooke SJ, Beard TD Jr, Kao YC, Lorenzen K, Song AM, Allen MS, Basher Z, Bunnell DB, Camp E, Cowx IG, Freedman JA, Nguyen VM, Nohner J, Rogers MW, Siders ZA, Taylor WW, Youn SJ (2017) Grand challenges in the management and conservation of North American inland fishes and fisheries. Fisheries 42(2):115–124

Maccord PF, Silvano RA, Ramires MS, Clauzet M, Begossi A (2007) Dynamics of artisanal fisheries in two Brazilian Amazonian reserves: implications to co-management. Hydrobiologia 583(1):365–376

MacKenzie CJ, Cox SP (2013) Building legitimacy of the recreational fishing sector in mixed commercial-recreational fisheries. Ocean Coast Manag 75:11–19

Mackinson S, Nottestad L (1998) Points of view: combining local and scientific knowledge. Rev Fish Biol Fisher 8:481–490

Mauro F, Hardison PD (2000) Traditional knowledge of indigenous and local communities: international debate and policy initiatives. Ecol Appl 10(5):1263–1269

May CK (2008) Achieving sustainability in US fisheries: community engagement in co-management. Sustain Dev 16(6):390–400

Mazumdar S, Ceccaroni L, Piera J, Hölker F, Berre A, Arlinghaus R, Bowser A (2018) Citizen science technologies and new opportunities for participation. In: Hecker S, Haklay M, Bowser A, Makuch Z, Bogel H, Bonn A (eds) Citizen science: innovation in open science, society and policy. UCL Press, London, pp 303–320

McCay BJ (1993) Management regimes, property rights and the performance of natural resource systems. The Beijer International Institute of Ecological Economics Performance of Natural Resource Systems, Stockholm

McCay BJ, Jentoft S (1996) From the bottom up: participatory issues in fisheries management. Society & Natural Resources, 9(3):237–250

McDaniels T, Longstaff H, Dowlatabadi H (2006) A value-based framework for risk management decisions involving multiple scales: a salmon aquaculture example. Environ Sci Pol 9(5): 423–438

McFadden JE, Hiller TL, Tyre AJ (2011) Evaluating the efficacy of adaptive management approaches: is there a formula for success? J Environ Manag 92(5):1354–1359

McGowan CP, Smith DR, Nichols JD, Lyons JE, Sweka J, Kalasz K, Niles LJ, Wong R, Brust J, Davis M, Spear B (2015) Implementation of a framework for multi-species, multi-objective adaptive management in Delaware Bay. Biol Conserv 191:759–769

Melis TS, Walters CJ, Korman J (2015) Surprise and opportunity for learning in Grand Canyon: the Glen Canyon dam adaptive management program. Ecol Soc 20(3):33

Miller TJ, Blair JA, Ihde TF, Jones RM, Secor DH, Wilberg MJ (2010) FishSmart: an innovative role for science in stakeholder-centered approaches to fisheries management. Fisheries 35(9): 424–433

Moller H, Berkes F, Lyver POB, Kislalioglu M (2004) Combining science and traditional ecological knowledge: monitoring populations for co-management. Ecology and society, 9(3)

Monk MH, He X (2019) The combined status of Gopher, *Sebastes carnatus*, and Black-and-Yellow, *Sebastes chrysomelas*, Rockfishes in U.S. waters off California in 2019. Pacific Fishery Management Council, Portland

Moon K, Blackman D (2014) A guide to understanding social science research for natural scientists. Conserv Biol 28(5):1167–1177

Moore JL, Runge MC (2012) Combining structured decision making and value-of-information analyses to identify robust management strategies. Conserv Biol 26(5):810–820

Morales N, Camp E, Cleary M (2020) Modeling target species selection behavior among diverse freshwater anglers in Florida, USA. Fish Res 231:105693

Murphy R Jr, Estabrooks A, Gauvin J, Gray S, Kroska AC, Wolf N, Harris BP (2021) Using mental models to quantify linear and non-linear relationships in complex fishery systems. Mar Policy 132:104695

NOAA (2015) Introduction to stakeholder participation. Office for coastal management. Available at: https://coast.noaa.gov/data/digitalcoast/pdf/stakeholder-participation.pdf. Accessed 18 July 2018

Nyboer EA, Embke HS, Robertson AM, Arlinghaus R, Bower S, Baigun C, Beard D, Cooke SJ, Cowx IG, Koehn JD, Lyach R, Milardi M, Potts W, Lynch AJ (2022) Overturning stereotypes: the fuzzy boundary between recreational and subsistence inland fisheries. Fish Fish 23(6): 1282–1298

Obiero KO, Abila RO, Njiru MJ, Raburu PO, Achieng AO, Kundu R, Ogello EO, Munguti JM, Lawrence T (2015) The challenges of management: recent experiences in implementing fisheries co-management in Lake Victoria, Kenya. Lakes Reserv Res Manag 20(3):139–154

Olsson P, Folke C, Berkes F (2004) Adaptive co-management for building resilience in social-ecological systems. Environ Manag 34:75–90

Ostrom E (1990) Governing the commons: the evolution of institutions for collective action. Cambridge University Press, Cambridge, UK

Ostrom E (2010) The challenge of self-governance in complex contemporary environments. J Specul Philos 24(4):316–332

Ostrom E, Walker J, Gardner R (1992) Covenants with and without a sword: self-governance is possible. Am Polit Sci Rev 86(2):404–417

Peterman R, Peters CN (1998) Decision analysis: taking uncertainties into account in forest resource management. In: Sit V, Taylor B (eds) Statistical methods for adaptive management studies. Land management handbook. BC Ministry of Forests, Victoria, pp 105–128

Peters CN, Marmorek DR, Deriso RB (2001) Application of decision analysis to evaluate recovery actions for threatened Snake River fall Chinook salmon (Oncorhynchus tshawytscha). Can J Fish Aquat Sci 58(12):2447–2458

Peterson JT, Evans JW (2003) Quantitative decision analysis for sport fisheries management. Fisheries 28(1):10–21

Pinkerton E (1994) Local fisheries co-management: a review of inter-national experiences and their implications for salmon management in British Columbia. Can J Fish Aquat Sci 5511:2363–2378

Pita P, Villasante S (2019) The building of a management system for marine recreational fisheries in Galicia (NW Spain). Ocean Coast Manag 169:191–200

Plummer R, Hashimoto A (2011) Adaptive co-management and the need for situated thinking in collaborative conservation. Hum Dimens Wildl 16(4):222–235

Poisson AC, McCullough IM, Cheruvelil KS, Elliott KC, Latimore JA, Soranno PA (2020) Quantifying the contribution of citizen science to broad-scale ecological databases. Front Ecol Environ 18(1):19–26

Potts WM, Downey-Breedt N, Obregon P, Hyder K, Bealey R, Sauer WH (2020) What constitutes effective governance of recreational fisheries?-A global review. Fish Fish 21(1):91–103

Pretty JN, Guijt I, Thompson J, Scoones I (1995) Participatory learning and action: a trainers guide. International Institute for Environment and Development, London

Punt AE, Butterworth DS, de Moor CL, De Oliveira JAA, Haddon M (2016) Management strategy evaluation: best practices. Fish Fish 17(2):303–334

Raiffa H (1968) Decision analysis: introductory lectures on choices under uncertainty, vol 19. Random House, New York, p 180

Raju CS, Rao JCS, Rao KG, Simhachalam G (2016) Fishing methods, use of indigenous knowledge and traditional practices in fisheries management of Lake Kolleru. J Entomol Zool Stud 4(5): 37–44

Ralph E, Townsend (1995) Fisheries self-governance: corporate or cooperative structures? Marine Policy 19(1):39–45 https://doi.org/10.1016/0308-597X(95)92571-N

Ramos J, Caetano M, Himes-Cornell A, dos Santos MN (2017) Stakeholders' conceptualization of offshore aquaculture and small-scale fisheries interactions using a Bayesian approach. Ocean Coast Manag 138:70–82

Reed MS (2008) Stakeholder participation for environmental management: a literature review. Biol Conserv 141(10):2417–2431

Reid AJ, Eckert LE, Lane JF, Young N, Hinch SG, Darimont CT, Cooke SJ, Ban NC, Marshall A (2021) "Two-Eyed Seeing": an Indigenous framework to transform fisheries research and management. Fish Fish 22(2):243–261

Robert Arlinghaus, Marie Fujitani, Elias Ehrlich, Monica Engel, Steven J. Cooke (2026) Knowledge Co-production in Recreational Fisheries Science and Management. In: Pope KL, Arlinghaus R, Hunt LM, Lynch AJ, van Poorten BT (eds) Understanding recreational fishers: disciplinary and interdisciplinary approaches for fisheries management. Springer Nature, Cham, p 533–572

Robinson KF, Fuller AK, Hurst JE, Swift BL, Kirsch A, Farquhar J, Decker DJ, Siemer WF (2016) Structured decision making as a framework for large-scale wildlife harvest management decisions. Ecosphere 7(12):e01613

Robinson KF, Fuller AK, Stedman RC, Siemer WF, Decker DJ (2019) Integration of social and ecological sciences for natural resource decision making: challenges and opportunities. Environ Manag 63:565–573

Robinson KF, DuFour M, Jones M, Herbst S, Newcomb T, Boase J, Brenden T, Chapman D, Dettmers J, Francis J, Hartman T, Kočovský P, Locke B, Mayer C, Tyson J (2021a) Using decision analysis to collaboratively respond to invasive species threats: a case study of Lake Erie grass carp (Ctenopharyngodon idella). J Great Lakes Res 47:108–119

Robinson KF, Fuller AK, Jones ML (2021b) Using structured decision making to incorporate ecological and social values into harvest decisions: case studies of white-tailed deer and walleye. In: Pope KL, Powell LA (eds) Harvest of fish and wildlife: new paradigms for sustainable management. CRC Press, Boca Raton, pp 131–146

Robinson KF, Arlinghaus R, Camp EV, Post JR, Stedman R, Sutton SG (2026) Adaptive management for recreational fisheries decisions in the face of uncertainty. In: Pope KL, Arlinghaus R, Hunt LM, Lynch AJ, van Poorten BT (eds) Understanding recreational fishers: disciplinary and interdisciplinary approaches for fisheries management. Springer Nature, Cham, p 719–746

Röckmann C, Ulrich C, Dreyer M, Bell E, Borodzicz E, Haapasaari P, Hauge KH, Howell D, Mäntyniemi S, Miller D, Tserpes G, Pastoors M (2012) The added value of participatory modelling in fisheries management-what has been learnt? Mar Policy 36(5):1072–1085

Roque A, Wutich A, Quimby B, Porter S, Zheng M, Hossain MJ, Brewis A (2022) Participatory approaches in water research: a review. WIRES Water 9(2):e1577

Ruano-Chamorro C, Gurney GG, Cinner JE (2022) Advancing procedural justice in conservation. Conserv Lett 15:1–12

Runde BJ, Bacheler NM, Shertzer KW, Rudershausen PJ, Sauls B, Buckel JA (2021) Discard mortality of red snapper released with descender devices in the US South Atlantic. Mar Coast Fish 13(5):478–495

Runge MC (2011) An introduction to adaptive management for threatened and endangered species. J Fish Wildl Manag 2(2):220–233
Runge MC, Bean E, Smith DR, Kokos S (2011) Non-Native fish control below glen canyon dam-report from a structured decision making project. U.S. Geological Survey, Reston
Schwermer H, Barz F, Zablotski Y (2020) A literature review on stakeholder participation in coastal and marine fisheries. In: Jungblut S, Liebich V, Bode-Dalby M (eds) YOUMARES 9-the oceans: our research, our future. Springer Nature, Cham, pp 21–43
Schwermer H, Aminpour P, Reza C, Funk S, Möllmann C, Gray S (2021) Modeling and understanding social-ecological knowledge diversity. Conserv Sci Pract 3(5):e396
Scott AD (1989) Conceptual origins of rights based fishing. In: Neher PA, Arnason R, Mollett N (eds) Rights based fishing, vol 169. Springer Link, Dordrecht, pp 11–38
Sen S, Nielsen JR (1996) Fisheries co-management: a comparative analysis. Mar Policy 20(5): 405–418
Sharpe A, Conrad C (2006) Community based ecological monitoring in Nova Scotia: challenges and opportunities. Environ Monit Assess 113(1):395–409
Silver JJ, Okamoto DK, Armitage D, Alexander SM, Atleo C, Burt JM, Jingwas RJN, Lee LC, Muhl E-K, Salomon AK, Stoll JS (2022) Fish, people, and systems of power: understanding and disrupting feedback between colonialism and fisheries science. Am Nat 200(1):168–180
Silvertown J (2009) A new dawn for citizen science. Trends Ecol Evol 24(9):467–471
Smith SL (2012) Toward inclusive co-management: factors influencing stakeholder participation. Coast Manag 40:327–357
Støttrup JG, Kokkalis A, Brown EJ, Olsen J, Andersen SK, Pedersen EM (2018) Harvesting geo-spatial data on coastal fish assemblages through coordinated citizen science. Fish Res 208:86–96
Sumaila UR (2010) A cautionary note on individual transferable quotas. Ecol Soc 15(3):8
Sutinen JG, Johnston RJ (2003) Angling management organizations: integrating the recreational sector into fishery management. Mar Policy 27(6):471–487
Thornton TF, Scheer AM (2012) Collaborative engagement of local and traditional knowledge and science in marine environments: a review. Ecol Soc 17(3):25
Thornton TF, Moss ML, Butler VL, Hebert J, Funk F (2010) Local and traditional knowledge and the historical ecology of Pacific Herring in Alaska. J Ecol Anthropol 14(1):81–88
Toki V (2010) Adopting a Maori property rights approach to fisheries. NZJ Envtl L 14:197
Townsend R, Shotton R (2008) Fisheries self-governance: new directions in fisheries management. In: Townsend R, Shotton R, Uchida H (eds) Case studies in fisheries self-governance. FAO Fisheries Technical Paper 504, Rome, Italy, pp 1–19
Trimble M, Berkes F (2013) Participatory research towards co-management: lessons from artisanal fisheries in coastal Uruguay. J Environ Manag 128:768–778
van Poorten BT, Camp EV (2019) Addressing challenges common to modern recreational fisheries with a buffet-style landscape management approach. Rev Fish Sci Aquac 27(4):393–416
Van Poorten BT, Cox SP, Cooper AB (2013) Efficacy of harvest and minimum size limit regulations for controlling short-term harvest in recreational fisheries. Fisheries Manag Ecol 20(2–3):258–267
Vann-Sander S, Clifton J, Harvey E (2016) Can citizen science work? Perceptions of the role and utility of citizen science in a marine policy and management context. Mar Policy 72:82–93
Vaske JJ, Needham MD, Miller CA (2026) Fundamentals of sampling and implementing human dimensions of fishing surveys. In: Pope KL, Arlinghaus R, Hunt LM, Lynch AJ, van Poorten BT (eds) Understanding recreational fishers: disciplinary and interdisciplinary approaches for fisheries management. Springer Nature, Cham, p 327–353
Venturelli PA, Hyder K, Skov C (2017) Angler apps as a source of recreational fisheries data: opportunities, challenges and proposed standards. Fish Fish 18(3):578–595
Voinov A, Bousquet F (2010) Modelling with stakeholders. Environ Model Softw 25:1268–1281
Wade E, Biedenweg K (2019) Exploring the diversity of mental models associated with Belize's Managed Access Fisheries Policy. Ocean Coast Manag 178:104868

Walshe TV, Slade S (2020) Spawning closures for coral reef fin fish. In: Runge MC, Converse SJ, Lyons JE, Smith DR (eds) Structured decision making: case studies in natural resource management. Johns Hopkins University Press, Baltimore, pp 72–82

Walters CJ, Hilborn R (1978) Ecological optimization and adaptive management. Annu Rev Ecol Syst 9:157–188

Wambiji N, Kadagi NI, Everett BI, Temple AJ, Kiszka JJ, Kimani E, Berggren P (2022) Integrating long-term citizen science data and contemporary artisanal fishery survey data to investigate recreational and small-scale shark fisheries in Kenya. Aquat Conserv 32(8):1306–1322

Wiber M, Charles A, Kearney J, Berkes F (2009) Enhancing community empowerment through participatory. Fish Res Mar Policy 33(1):172–179

Wilberg MJ, Irwin BJ, Jones ML (2026) Participatory modelling to support evaluation of management actions for recreational fisheries. In: Pope KL, Arlinghaus R, Hunt LM, Lynch AJ, van Poorten BT (eds) Understanding recreational fishers: disciplinary and interdisciplinary approaches for fisheries management. Springer Nature, Cham, p 693–718

Williams BK, Johnson FA (1995) Adaptive management and the regulation of waterfowl harvests. Wildlife Soc B 23(3):430–436

Williams BK, Szaro RC, Shapiro CD (2007) Adaptive management: the U.S. Department of the Interior technical guide. U.S. Department of the Interior, Washington, D.C.

Wilson DC, Nielsen JR, Degnbol P (eds) (2003) The fisheries co-management experience: accomplishments, challenges and prospects. Springer-Science+Business Media, B.V

Wood KA, Stillman RA, Goss-Custard JD (2015) Co-creation of individual-based models by practitioners and modellers to inform environmental decision making. J Appl Ecol 52(4): 810–815

Yandle T (2003) The challenge of building successful stakeholder organizations: New Zealand's experience in developing a fisheries co-management regime. Mar Policy 27(2):179–192

Yang YW, Frazer A, Rees E (2010) Self-governance within a QMS framework-the evolution of self-governance in the New Zealand Bluff oyster fishery. Mar Policy 34(2):261–267

Young OR, Osherenko G, Ekstrom J, Crowder LB, Ogden J, Wilson JA, Day JC, Douvere F, Ehler CN, McLeod KL, Halpern B, Peach R (2007) Solving the crisis in ocean governance: place-based management of marine ecosystems. Environ Sci Pol 49(4):20–32

Integrated Models of the Social–Ecological Dynamics of Recreational Fisheries

Olaf P. Jensen, Robert Arlinghaus, Abigail S. Golden, Marco A. Janssen, Christopher T. Solomon, and Brett T. van Poorten

Abstract Integrated models of recreational fisheries as social–ecological systems contain—at a minimum—two dynamically linked components: a fishing effort or harvest dynamics sub-model (representing recreational fisher behaviour) and a population dynamics sub-model (representing fish dynamics in response to exploitation). Here, we review and categorize the use of integrated models, provide a set of general instructions for building them, and identify gaps and opportunities for further development. The structure of coupled social–ecological models diverges along two major paths: agent-based models that follow the behaviour of individual autonomous agents (generally fishers or fish) and models that track the aggregate dynamics of a fish population and the fishing effort exerted on it by fishers. Most integrated models published so far are lopsided in their development. That is, they often contain one sub-model (harvest or population dynamics) that is detailed and grounded in empirical data, whereas the other sub-model is a more generic representation of that process. The future of integrated models depends on increased

---

O. P. Jensen (✉)
Center for Limnology, University of Wisconsin-Madison, Madison, WI, USA

R. Arlinghaus
Department of Fish Biology, Fisheries and Aquaculture, Leibniz Institute of Freshwater Ecology and Inland Fisheries, Berlin, Germany

Division of Integrative Fisheries Management, Humboldt-Universität zu Berlin, Berlin, Germany

A. S. Golden
School of Aquatic and Fishery Sciences, University of Washington, Seattle, WA, USA

M. A. Janssen
School of Sustainability, Arizona State University, Tempe, AZ, USA

C. T. Solomon
Cary Institute of Ecosystem Studies, Millbrook, NY, USA

B. T. van Poorten
School of Resource and Environmental Management, Simon Fraser University, Burnaby, BC, Canada

K. L. Pope et al. (eds.), *Understanding Recreational Fishers*, Fish & Fisheries Series 45, https://doi.org/10.1007/978-3-031-99739-6_21

collaboration between the social and ecological sciences, a rigorous quantification of both fish and fisher behavioural patterns, and a confrontation of model predictions with actual system behaviour.

Keywords Coupled human and natural system · Hyperdepletion · Hyperstability · Recreational fisheries · Social-ecological system

1 Introduction

Integrated mathematical models of recreational fisheries social–ecological systems (hereafter "integrated models") explicitly represent the dynamics of fish populations and the behaviour of humans, i.e., of recreational fishers and managers, along with the dynamic links between the two (McConnell and Sutinen 1979; Johnston et al. 2010; Fenichel et al. 2013). One of the most basic behavioural reactions of fishers to ecological or social changes in the fisheries social–ecological system (SES) is to change where and how much they fish. This is why site choice and the effort of fishers as harvesters of a resource (hereafter "effort") are key behaviours that, in many integrated models, provide the link between fish and people. In integrated models, human behaviours, such as site choice and fishing effort, are a dynamic function of changes in the fish population, fishing regulations, or general social and economic characteristics (e.g., cost of fishing), including fish population abundance, catch rate, or the size of fish in the catch (Hunt et al. 2019). Fish population characteristics, and some other aspects of a fishery to which fishers respond (e.g., crowding), are in turn a function of local fishing effort. A simple non-mathematical representation of these feedbacks (Fig. 1) depicts stabilizing feedbacks between fish abundance and fishing effort leading to an equilibrium even in the absence of management. That this depiction of a recreational fisheries SES leaves out many important processes and is thus only useful as a "straw man" starting point should be readily apparent. Much of the work of developing integrated models lies in elaborating, expanding, and formalizing the dynamic relationships between fish populations and fishers shown in this conceptual model.

In its most basic mathematical form, the link between fishing effort and population dynamics sub-models may be represented through the widely used catch equation:

$$C_t = q\, E_t\, N_t \tag{1}$$

where C_t is catch at time t, E_t is the fishing effort expended at time t, N_t is the fish population abundance at time t, and q is the portion of the stock captured by one unit of effort, i.e., the "catchability coefficient." The catch equation (Eq. 1) should be seen only as a mathematical "cartoon" of the complex interactions of fishers and ecological systems; it describes a very simplified version of reality, but one that might in some cases be sufficient to capture key dynamics if parameterized well. Clearly, the equation ignores many salient features of real recreational fisheries, e.g.,

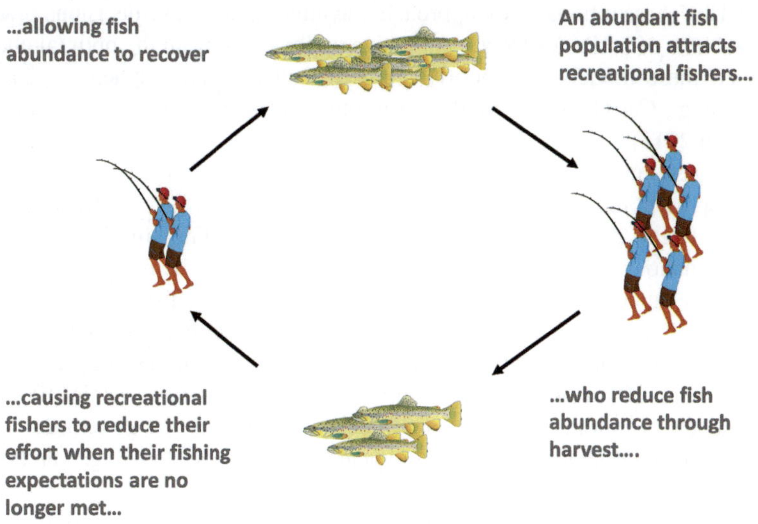

Fig. 1 Conceptual Model of Stabilizing Feedbacks Between a Harvested Fish Population and Fishing Effort. (Images courtesy of Integration and Application Network; ian.umces.edu/media-library)

spatial heterogeneity in fish aggregation or habitat association, vulnerability of only a fraction of the fish population to angling, and non-random fishing patterns over space and time.

Mathematical models representing the dynamics of harvested fish populations (the N component of Eq. 1) have a long history (often dated back to Baranov 1918) and are treated in detail in numerous textbooks (e.g., Ricker 1975; Hilborn and Walters 1992; Walters and Martell 2004). Representations of the E component of the catch equation, i.e., how effort changes through time, are much less standardized and have historically received less attention, especially in the fisheries ecology literature. Often, the product of q and E_t has been simplified to represent the fishing mortality rate F_t exogenously acting on an exploited system, rather than directly modelling dynamic human behavioural processes that lead to changes in effort, and perhaps in catchability. Explicit and implicit sub-models of effort dynamics in *commercial* fisheries have long been a critical component of fishery bioeconomic models (Gordon 1954; Schaefer 1957; Clark 1973; Smith and Wilen 2003) and have received renewed interest in recent years, in part because of their potential value for extracting stock status information from catch time series (Thorson et al. 2013). Typically, effort dynamics of commercial fishers have been estimated with variants of economic representations of fleet behaviour following random utility theory and a fully rational actor model of human behaviour (e.g., Smith and Wilen 2003, see Melstrom et al. (2026) for more information about random utility models). Similar utility-based, rational actor approaches have been used to parametrize the basic bioeconomic model of recreational fisheries (McConnell and Sutinen 1979) in several recent applications (Massey et al. 2006; Johnston et al. 2010, 2015; Lee et al. 2017). These rational actor approaches assume agents make decisions based on a known suite of attributes for all alternative options and make the optimal choices. However, these models may poorly represent actual behaviour across numerous fisheries (e.g., Carrella et al. 2020) and in other decision contexts in general (e.g., Kahneman 2011).

Agent-based models (Schulze et al. 2017) are a separate framework that has been used to represent the integrated social–ecological dynamics of recreational fisheries (Gao and Hailu 2013; Ayllón et al. 2018; Cenek and Franklin 2017). In these models, equations for population and effort dynamics may be replaced by a set of decision rules that guide the behaviour of individual agents (fishers and sometimes fish). "Agent" is a generic term that may encompass explicitly modelling individuals or a collection of individuals representing some "type" of resource user (as in Johnston et al. 2010) that may be informed by quantitative, deterministic utility-based models of rational actors or other decision heuristics that govern the human response to changes in the social or ecological fabric of the system (e.g., van Poorten et al. 2011).

Integrated models that include more complexity and "variation" are often a place to challenge those assumptions by enhancing the realism of fish behaviour (e.g., Cox and Walters 2002; Camp et al. 2015) or representing variation in fisher behaviour rather than aggregating across the often-heterogeneous responses of fishers to ecological and management changes in the fishery (e.g., Johnston et al. 2010, 2013). In this sense, integrated models are a key step in interdisciplinary research and management—one which forces the researcher to develop a system perspective and an explicit view of how ecological and social systems interact. Such models have also been key to integrating knowledge derived from the natural and social sciences and may thus serve as a knowledge incubator (Arlinghaus et al. 2014) by forcing teams of researchers to bring the different models of how the ecological or social subsystems respond to change.

Management strategy evaluation is a related and widely used tool in marine (and generally, commercial) fishery management (Smith 1994; Punt et al. 2016) that overlaps with the integrated models discussed here, but it differs in at least two key respects. First, Management Strategy Evaluations frequently, though not always (e.g., Mapstone et al. 2008), contain simpler effort or harvest dynamics rules—often a harvest quota set by managers within a Management Strategy Evaluation is assumed to be equivalent to the actual harvest, perhaps with some random implementation error. Second, Management Strategy Evaluations often explicitly represent the stock assessment process as part of a closed-loop simulation. This addition allows the information available to managers to differ in critical ways from the underlying true values in a simulated fish population. Because management strategy evaluations are an established and widely studied tool in fisheries science, and an excellent guide for their use already exists (Punt et al. 2016), we do not cover them here.

This chapter focuses on the use of quantitative integrated models for the study and management of *recreational* fisheries. Qualitative and semi-quantitative models of recreational fisheries SES can be a valuable tool, but one that is beyond the scope of this chapter. These quantitative integrated models represent a more recent development in fisheries science, have only recently been broadly reviewed (Solomon et al. 2020), and are not covered in many fisheries science textbooks. They are also substantially different from the much more established bioeconomic models of commercial fisheries in that the motivations and behaviours of recreational fishers are much more complex and heterogenous than the profit-seeking motivation assumed for commercial fishers. Here, we review the types and uses of integrated models, present a guide to developing them, and identify gaps and opportunities for their further development.

2 Review of Integrated Models

2.1 *What Integrated Models Exist?*

Integrated models of recreational fisheries dynamics are still relatively rare, although they are becoming more common. Fenichel et al. (2013) were the first to review how fisher behaviour was represented in integrated models, outlining that approaches mainly followed a utility-based tradition and cross-referencing even among the social sciences was rare (e.g., recreational fisheries economists rarely cited other social sciences and vice versa). One reason for the lack of cross-referencing might be that economists were often interested in using fisher samples to develop methods for valuing environmental quality, rather than applying the resulting behavioural sub-models to represent fisher behaviours explicitly. Cole and Ward (1994) may be the first explicit quantitative representation of fish-fisher interactions, closely followed by Johnson and Carpenter (1994), although using completely different modelling frameworks. A more recent review of peer-reviewed dynamic models of recreational fisheries identified 49 papers that considered social–ecological processes, out of a total of 179 modelling studies of recreational fisheries published between 1975 and 2019 (Solomon et al. 2020). All but three of these 49 studies were published after 2000, and two-thirds were published after 2010. These models considered a wide range of ecosystem types from lakes and rivers to coastal marine habitats, yet many focused on one of just a few well-studied fisheries systems, such as a marine park in western Australia and a landscape of lakes in western Canada.

2.2 *What Are the Applications for These Models?*

Integrated models have been used for a wide range of purposes, from tactical to strategic and from theoretical to heuristic to applied. Most commonly, these models have been used to understand how to set fishery regulations to maximize benefits such as catch or catch rate, fisher satisfaction, or ecological fishery sustainability. For instance, Post et al. (2003) considered how different length limits influenced catch, harvest, fish size, and sustainability for bull trout (*Salvelinus confluentus*), and Johnston et al. (2015) used a coupled model to examine how illegal harvest and hooking mortality undermined the suitability of harvest regulations. One of the key social processes in these studies and many others that followed was the response of fishing effort to changes in fishing quality; Post and colleagues saw that allowing effort to freely respond to fishing quality compromised the effectiveness of length limits as a tool for ensuring fishery sustainability.

A wide range of other uses of integrated models exist and include elaborations of different sub-components of the model to focus on outcomes or processes. For example, some studies have considered different benefits ranging from profitability to the biomass of non-target species (Olaussen and Skonhoft 2008; Gao and Hailu 2011). Other studies elaborated on the social, ecological, and social–ecological features and processes such as fisher heterogeneity, regulation compliance, hooking mortality, maternal effects, and vulnerable pool dynamics (Arlinghaus et al. 2010; Johnston et al. 2010, 2015; Camp et al. 2015; Fujiwara et al. 2019). Integrated models have also been used to evaluate diverse ways of regulating or managing the fishery, including stocking and time-area closures (Rogers et al. 2010; van Poorten et al. 2011; Gao and Hailu 2011; Johnston et al. 2018). One recent study used a social–ecological model in a decision analysis framework to identify management options that maximized benefits in the face of uncertainty about current conditions and future social and ecological changes (van Poorten and MacKenzie 2020).

If fishing effort responds to fishing quality at a given fishing location, then the distribution of effort across multiple fishing locations may be quite dynamic. Several studies have implicitly or explicitly modelled spatial dynamics in recreational fisheries to explore this idea and its implications for sustainable fisheries landscapes. One important early example modelled fisher decisions at multiple spatial and temporal scales to build insight about managing a landscape of lakes for ecological resilience, social resilience, and social welfare (Carpenter and Brock 2004). This study emphasized the risk of serial collapses of neighbouring fisheries and the importance of diversifying fisheries regulations across the landscape. Recent work in this vein includes both spatially abstracted models focused on general principles (Matsumura et al. 2019; van Poorten and Camp 2019) and geographically specific ones focused on evaluating management alternatives in a particular fisheries landscape (Hunt et al. 2011; Carruthers et al. 2019; Wilson et al. 2020).

Another recent trend in integrated modelling of recreational fisheries has focused on drivers of system dynamics that can be difficult to detect or control, because they change slowly or occur at a spatial scale much larger than that at which the fishery is managed. Climate change is one classic example of such a driver, which Carpenter et al. (2017) consider, along with other slow or big drivers like habitat change and the value of fishing to fishers, to illustrate the concept of managing fisheries within a safe operating space (Scheffer et al. 2015). Similarly, Nieman and Solomon (2022) use the bioeconomic model of Stoeven (2014) to suggest that "slow social change" in transportation infrastructure, fishing technology, and the relative value to fishers of catch versus effort might all lead to erosion of the open-access equilibrium in recreational fisheries. Further consideration of slow ecological and social changes, such as trait evolution, shifting social norms, changing fisher preferences, and network properties among fishers all seem likely to be interesting and productive.

This is just a sampling of the ways in which integrated models of recreational fisheries have been used. They have also been used to explore the emergence and implications of stocking and its interactions with human modification of fish habitat (van Poorten et al. 2011; Camp et al. 2017; Ziegler et al. 2017; Johnston et al. 2018); to estimate the costs and benefits of environmental interventions that impact fisheries (Hickey and Diaz 1999; Massey et al. 2006); to show how the interaction of social and ecological processes might drive evolution of fish traits (Arlinghaus et al. 2009) or alter the sustainability of open-access, unmanaged fisheries (Golden et al. 2022); to guide resilient management of fisheries in the face of regime shifts between different stable states (Biggs et al. 2009; Horan et al. 2011); and for many other purposes (Solomon et al. 2020).

Despite this diversity, in general, all these efforts to integrate social and ecological perspectives in models of recreational fisheries share one overarching goal: to help us to think insightfully about the processes that drive the dynamics of these systems, so that we can better understand and manage them.

2.3 *What Do These Models Look Like?*

The structure of integrated models has varied widely, according to the questions that researchers are interested in and the processes that they know or speculate to be essential for understanding those questions. We summarize some key features of the integrated model structure here, drawing heavily on the models reviewed by Solomon et al. (2020).

An integrated model of a recreational fishery includes both ecological and social processes. Typically, the ecological component is a model of fish population dynamics; usually the focus has been on a single species in a single ecosystem, but, in some cases, authors have considered interacting species, metapopulation dynamics, or multiple ecosystems linked by fishers who act like mobile predators (Carpenter and Brock 2004; Mapstone et al. 2008; Post et al. 2026). Most studies to date have used stage- or age-structured population models as a way to accommodate features like size-dependent vulnerability, regulations, or fecundity. Others have used a wide variety of other ecological models, from simple Schaefer biomass-dynamic models through agent-based models and complex food web models such as Ecopath with Ecosim (e.g., Alós et al. 2012; Townsend 2013; Thébaud et al. 2014; Ayllón et al. 2018).

The core social component of an integrated recreational fisheries model is often a model of fisher behavioural response that is used to represent changes in fishing effort. Models of fisher behavioural responses have followed a few main approaches, although there are examples of a broader diversity of modelling perspectives. The simplest and most common treatment of effort dynamics is an informal representation of utility maximization by fishers, assuming that aggregate effort responds in some straightforward way to a metric of fishing quality (e.g., Cox et al. 2003; Camp et al. 2015). Many other studies use a more formal utility model, which often incorporates additional catch- and noncatch-related determinants of effort such as travel costs, crowding, or regulation preferences (e.g., Massey et al. 2006; Johnston et al. 2010; Matsumura et al. 2019; Carruthers et al. 2019).

These approaches all rely heavily on a rational actor model and parameterize models of fisher behaviour with choice models fitted to stated or revealed preference data (Hunt et al. 2011). However, it must be clearly stated that this decision may simply be caused by the fact that economists—seeking a simple quantitative expression of a complex behaviour (fishing activity/site choice)—favour bioeconomic, utility and welfare-based modelling frameworks and such frameworks have dominated the quantitative representation of fisher behaviour so far. By no means are bioeconomic frameworks the only or a superior way of representing fisher behaviour as outlined by Schlüter et al. (2012). In fact, the response of humans to system changes may be represented in agent-based models by any reasonable set of decision rules that guide the behaviour of individual agents or a collective of actors and may also represent the interactions of actors (e.g., networks) and how users affect managers. Simple decision rules can sometimes reproduce actual fishing behaviour more accurately than a rational actor model (see Carrella et al. 2020 for an example from a commercial fishery). Müller-Hansen et al. (2017) review the various ways in how human behaviours can be represented in coupled social–ecological models, Schlüter et al. (2017) summarize the theories that are suitable to also represent boundedly rational actors in integrated models, and Wijermans et al. (2020) show novel applications to represent heterogenous fishers and their behaviour in models.

Heterogeneity among fishers in terms of preferences and behaviours has been an important elaboration of early human dimensions research (e.g., Fedler and Ditton 1994). In integrated models, this heterogeneity is typically represented by classes of fishers. For example, in agent-based models, super agents (where agents of the same type share the same preferences, but agents of other types differ in their preferences) are often used (Johnston et al. 2010, 2013, 2015; Matsumura et al. 2019), although in some agent-based models individual fishers all vary in their preferences (Innes-Gold et al. 2021). Some models incorporating heterogeneity in fisher preferences suggest that it may reduce fishery resilience by promoting higher effort across a landscape of diverse fishing opportunities (Johnston et al. 2010; Hunt et al. 2011; van Poorten and Camp 2019; Matsumura et al. 2019). This result runs counter to typical thinking about the role of diversity in social–ecological resilience (Solomon et al. 2020). However, though heterogeneity of fishers renders the total exploitation more effective (as even remote fisheries attract effort by some fisher types), it can put at jeopardy specific high-use fisheries. Integrated models that account for both ecological and social heterogeneity can examine such outcomes and generate testable hypotheses for coupled landscapes (e.g., Matsumura et al. 2019). Heterogeneity is not only needed to better understand the ecological impacts of fishing, but also how to optimize benefits to fishers by allowing fisher-type specific regulations or optimal compromise regulations to be revealed (Johnston et al. 2010). It is of specific importance to note that the same rule that covers fisher behaviour, e.g., choosing sites or opportunities that provide maximum benefit, may be used to study welfare impacts of policies, and thereby complement or even substitute traditional objectives such as maximum sustainable yield.

Other aspects of the structure of integrated models—such as their consideration of density-dependent processes, their assumptions about the information available to fishers, and their treatment of decision aids like heuristics and social norms—have received an interesting range of treatments, or in some cases an interesting lack of treatment (Solomon et al. 2020). For instance, most studies to date have implicitly represented fisheries governance as controlled by a centralized, rational planner, even though the actual process of governance can be more complex and involve two-way interactions between fisher and manager behaviour (Ward et al. 2016). In one of several exceptions to that generalization, van Poorten et al. (2011) considered the role of fisher (dis-)satisfaction in pressuring managers to stock fish, leading to a stocking-dependent fishery and the loss of wild stocks. The result suggests that remembering the past may predispose the SES toward stocking as a panacea (Arlinghaus et al. 2022).

A key message that emerges from the diversity of recreational effort models is that there is no one recipe to build an integrated model of a recreational fishery; instead, model structure must reflect the way the system works and the questions that the researcher seeks to understand.

3 Building Integrated Models

Based on our experience with distinct types of projects, we provide a set of steps in the modelling process that could be followed to develop an integrated model (see also Van Delden et al. 2011; Schlüter et al. 2014; Railsback and Grimm 2019). Although the list is presented here in a linear fashion, the actual modelling process is interactive and will likely involve looping between distinct stages (Fig. 2). It is important to remember that exciting and surprising outcomes often come at the intersection of multiple disciplines. Building integrated models is an obvious opportunity to enhance our understanding of SESs by merging key ideas, theories, and epistemologies within an interdisciplinary setting. When possible, representing human decisions, constraints, and values with tools and ideas from the appropriate disciplines will be key to creating current ideas and theories. An example of the application of these eight steps is presented (Box 1).

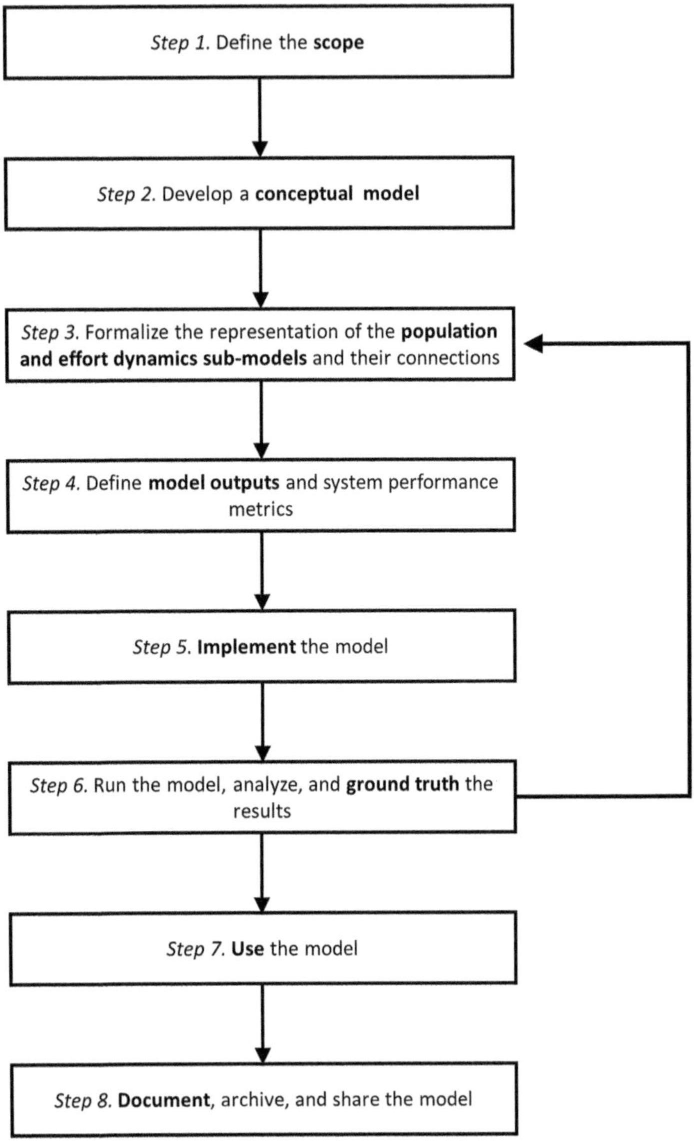

Fig. 2 Steps in the Development of an Integrated Model of a Recreational Fishery Social-ecological System

Box 1: Example Application of the Eight-Step Process of Integrated Model Development

This case study is derived from the research process that led to the journal article by Johnston et al. (2018), which is an integrated social-ecological model of a single-species fishery that is stock enhanced and can also be regulated by harvest regulations.

(continued)

Box 1 (continued)

Step 1: The model was developed to assist in generating and testing hypotheses about the outcomes of fish stocking in freshwater systems using a variety of recruiting and non-recruiting model species and compare outcomes to alternative management interventions, such as size-limits. The model was initially designed to help facilitate stakeholder involvement in workshop settings in a transdisciplinary project (Besatzfisch project, Arlinghaus et al. 2015; Fujitani et al. 2017), but was further developed as a stand-along simulation study (Johnston et al. 2018) and formed the basis of a planning software that was released online in German and English for use in planning fish stocking measures relative to harvest constraints in a coupled system of recreational fisheries (https://www.ifishman.de/praktikerinfo/software-hegeplanung/).

Step 2: The simulation model was first conceived as a conceptual model that outlines the interactions and "binding blocks" of the model (Fig. 3). The multidimensional utility experienced by recreational fishers served to operationalize fisheries quality, which affected effort decisions across the simulation. Management interventions were evaluated experimentally in silico, while accounting for dynamic feedback of the fish stock and the recreational fisher populations, moderated through utility dynamics.

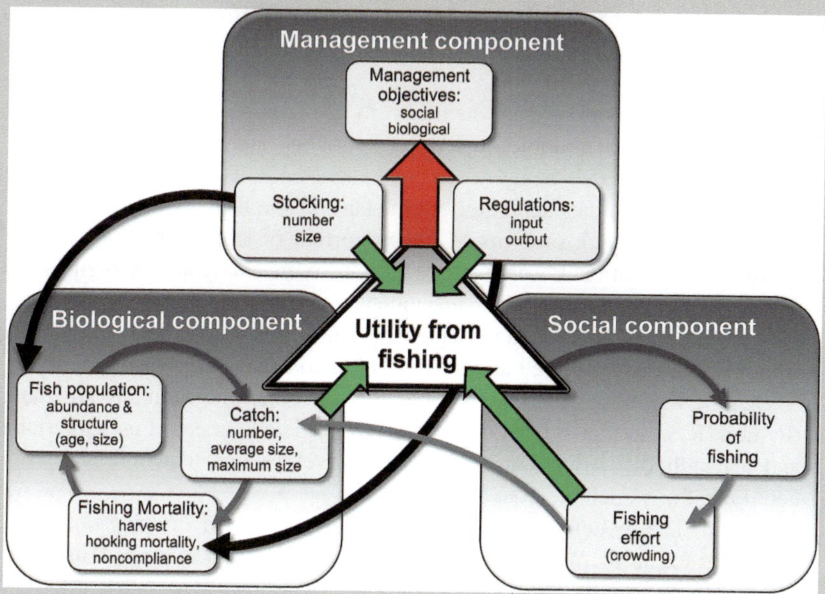

Fig. 3 Conceptual Model. (Inspired by Johnston et al. 2018)

(continued)

Box 1 (continued)

Step 3: The fish population model was conceptualized as an age and size-structured single species model with multiple forms of density-dependent feedback, e.g., a stock-recruitment function predicting abundance of age 1 fish (and hence controlling first year morality) and density-dependent growth. Parameters for the model were calibrated to a range of fish species, most notably pike (*Esox lucius*) and common carp (*Cyprinus carpio*), serving as examples of a recruiting (pike) and not naturally recruiting species. In the final application several more species were modelled, e.g., perch (*Perca fluviatilis*), trout (*Salmo trutta, Oncorhynchus mykiss*), zander (*Sander lucioperca*) and eel (*Anguilla anguilla*). Parameters came from literature searches or were estimated for pike and carp from local studies. Preference was given to generic relationships estimated previously in the literature across a wide family of species, e.g., the Lorenzen size-dependent mortality model (Lorenzen 2000). The recreational fisher behavioural model was informed by a dedicated stated preference survey that was administered to the region where the participatory stakeholder process took place (Arlinghaus et al. 2014). For the application to the coupled model, parameters were re-estimated. In the paper by Johnston et al. (2018) one representative recreational fisher type was simulated, but the available software allows multiple recreational fisher typologies (estimated based on latent class modelling based on the stated preference data, Arlinghaus et al. 2020). The link among fishing effort and impacts on the fish stock was conceptualized through catchability and the size-selectivity of the angling gear.

Step 4: The model output was meant to provide information for management and represent multiple objectives, fish stock and conservation related, recreational fisher well-being related and economics. To that end, the model tracked stock dynamics and equilibrium states of the fish stock, the composition of the final stock in terms of the proportion of stocked fish in the final population, recreational fisher welfare (as approximated by net willingness-to-pay as a welfare measure) and management costs to produce a given outcome. The output metrics were kept separately, but could have been integrated, which would have required assigning weights to each objective.

Step 5: The model was coded in C+, but any other programming language (Python, R) could have been used. In a further development, a user interface and a handbook (Arlinghaus 2017) were developed so that ordinary users could use the model.

Step 6: The model was calibrated to empirical data on stocking outcomes generated for northern pike (Hühn et al. 2023) and common carp (Arlinghaus et al. 2017b) and then used for simulation experiments of many different scenarios, such as variation in size limits, stocking rates, size of fishes stocked, stocking in environmental degraded and healthy conditions and for recreational fisher populations that are homogenous and heterogenous in preferences. Results were tabulated and visualized.

(continued)

Box 1 (continued)

Step 7: The model was first and foremost used in participatory workshop settings with managers and members of angling clubs to generate predictions about which outcomes to expect from different management interventions (Arlinghaus et al. 2015; Fujitani et al. 2017). The model was further developed into a free software designed for fishery managers in central Europe to examine stocking outcomes, and the model was further developed into a peer-reviewed paper which examined the relative functionality of stocking vs. harvest regulations in model species. The sensitivities of model outputs were examined across simulation scenarios and individual variation of parameters of the model.

Step 8: The model was first and foremost presented in the scientific literature, but also in a handbook for German fisheries managers. The software is available in both German and English and can be used to check the relative performance of stocking and harvest regulations such as size limits (minimum size of fish, harvest slots) and different stocking scenarios (varying stocking rates and sizes of fish stocks, all normalized to the same investment of resources). The user can track different performance metrics and compare outcomes visually.

3.1 Step 1: Define the Scope

The first task is to define the scope of the model. What resources (human and financial) are available to develop the model? What is the intended purpose of the model? Will the goal be to use the model to address a scientific question, to provide support for the decision-making of a practical problem, or to facilitate training on understanding the integrated system? Depending on the goal of the model, different follow-up steps are needed. If the focus is to address a scientific question, what are the hypotheses you want to test and what is the relevant academic literature to this question? If the focus is to support real-world decision-making, you may engage with various stakeholders to define the scope. If the focus is to use the model as an educational tool, you will need to define the learning outcomes.

Because a model is a simplification of reality, defining the scope of the model is critical to make the simplifying assumptions. What are the boundaries of the system? What are the main components? And what are the temporal and spatial dimensions? Without a clear intended purpose for the model, one cannot start the modelling process in an effective way.

3.2 Step 2: Develop a Conceptual Model

Once the scope of the model is defined, a creative process of conceptualizing the model starts. Building on models with similar scope, you will have to define the main state variables of the model, their interactions, the spatial and temporal dynamics, and the outcome variables of interest. In larger projects, the conceptual model development can often be a vehicle for communication and collaboration in the project. Project members will have to be explicit about assumptions and definitions to craft the conceptual model, and this benefits a project with members from diverse disciplines and backgrounds. Also, the model will "force" team members to contribute representations of key processes, and later, specific values for parameters, throughout the project to make the model operational. This process fosters interdisciplinary cooperation where the model is a tool to synthesize disparate disciplinary knowledge (Arlinghaus et al. 2014).

So far, we have not discussed the mathematical expression of the model. That was done purposely because the selection of the modelling technique should be informed by the purpose of the model (Schlüter et al. 2012, 2014, 2017). There are two general approaches that have commonly been used for integrated models of recreational fisheries. One is to construct a set of continuous (e.g., differential equations) or discrete-time representations of the changes in stocks and flows of the system. The other approach is to develop a set of algorithmic statements to define the decisions of individual actors (i.e., human or non-human). Both approaches could include spatial representations. The types of mechanisms one wishes to represent in the model will often determine whether a system dynamics or agent-based modelling approach is used.

Especially for integrated models, balancing the different knowledge domains can be a challenge. It is common that one likes to add more detail to those aspects of the model that one is more knowledgeable about. However, this could limit the benefit of developing an integrated model of social and ecological dynamics. To avoid this bias, it could be beneficial to use frameworks for building models of SESs (Binder et al. 2013; Schlüter et al. 2014; Hinkel et al. 2014) to organize the conceptual model.

At the most basic level, an integrated model of a recreational fishery requires two sub-models: (i) a model of the population dynamics of some biological resource, usually finfish or shellfish, and (ii) some rule or set of rules for how humans interact with this resource through harvest. Optionally, one can also add additional environmental, governance, or relationship processes that alter either of these two sub-models, such as nutrient loading, temperature shifts, regulatory options, or the subsidization of fish populations through stocking. Explicitly representing more of these processes does not necessarily lead to a better model. These choices should be guided by the intended use of the model, i.e., its motivating questions, as well as the available information for the system under study. To start defining the mechanisms of processes you would like to include within the model, one may have to rely on theories from the social and ecological sciences, mental models from stakeholders (Vasslides and Jensen 2015; Wilberg et al. 2026), or findings from empirical analysis from your case study that informs your model.

3.3 *Step 3: Formalize the Representation of the Population Dynamics and Fisher Behaviour Sub-Models and Their Connections*

For the population dynamics sub-model, the major decision points are whether to include size or age structure, spatial structure, and environmental effects. One of the simplest approaches is a surplus production model (Hilborn 2001), also known as a biomass dynamics model, which is based on the familiar logistic population growth equation in ecology with the addition of harvest:

$$B_t = B_{t-1} + \lambda\, B_{t-1}\,(1 - B_{t-1}/K) - C \tag{2}$$

where B_t is the biomass at time t, λ is the intrinsic rate of increase, K is the population carrying capacity, and C is catch (e.g., in a coupled SES, see Hunt et al. 2011). If the size structure of the population is important (e.g., management uses size limits to constrain harvest or recreational fishers disproportionately value large trophy fish), consider using an age-structured model (Deriso 1980; for an example in a coupled SES, see Johnston et al. 2010). An important downside of moving from an age-aggregated model, such as the surplus production model, to an age-structured model is the added complexity and increasing number of parameters that must be somehow estimated or provided. In particular, age-structured models typically require an additional equation representing the relationship between the abundance or biomass of mature individuals in the population (i.e., the "spawning stock") and the resulting offspring that survive to the first age class (the "recruits"). These stock-recruitment relationships typically involve the addition of at least two unknown parameters that are notoriously difficult to estimate from noisy stock-recruitment data (Myers and Barrowman 1996). Other factors like temperature-dependent population growth, immigration and emigration, stocking, interspecific interactions (predator-prey or competition), and spatial dynamics can all be added as well, depending on the research questions. In general, it is the research questions themselves that should guide the degree of abstraction or specificity in defining the equations that inform these relationships. Parameter values for the biological model can be based on theoretical reasoning, field studies of the specific species or fishery in question, or meta-analysis (Myers et al. 1999; Thorson et al. 2014). When in doubt, err on the side of simplicity and abstraction, especially for models that are not intended for use in tactical management (e.g., to set length or bag limits) of specific fisheries.

For the fisher behaviour sub-model, the simplest approach is constant effort or harvest (two vastly different assumptions) through time. This is highly unrealistic, especially in recreational fisheries that rarely include direct management controls on effort (Johnston et al. 2010) and it has a limited place in truly integrated modelling except when used in comparison with the other approaches in this list (see e.g., Post et al. 2003). There are many ways to model fisher behaviour so that effort or harvest varies dynamically in response to an output of the population dynamics sub-model.

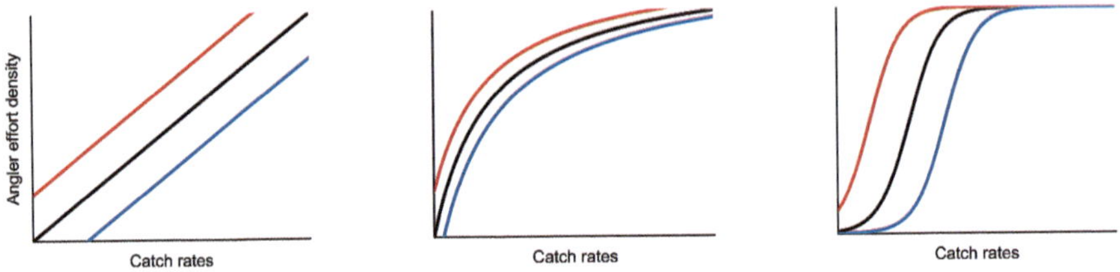

Fig. 4 Fishers' Numerical Effort Response to Catch May Be Linear, Asymptotic, or Logistic and May Have a Positive Y-intercept (destabilizing; red) or Negative Y-intercept (stabilizing; blue). (Adapted from Post 2013)

Ecologically informed models of fishing effort assume a Type II (asymptotic) or Type III (logistic) numerical or functional response (Post 2013) of effort to either population abundance or catch rates, although linear functions can also be used (Post et al. 2026; Fig. 4). Depending on the research or management problems being addressed, the response of fishing effort to changing abundance can be parameterized from empirical relationships estimated from stated or revealed preference data (Hunt et al. 2011; Matsumura et al. 2019; Golden et al. 2022) or can be defined with reasonable abstract functions that link effort directly to fishery characteristics (e.g., Camp et al. 2015; Cox et al. 2003; Post and Parkinson 2012).

When developing a fisher behaviour sub-model, a critical question to answer is: to what aspects of the biological system are fishers responding? Fishers do not have perfect knowledge of population dynamics; their knowledge is mediated by catch rates and the catchability of fish, regulations, awareness, and information networks. Many models treat fisher behaviour as a function of fish abundance as a substitute for a direct factor that fishers experience, such as catch per unit of effort. The unstated assumption here is that fishers integrate information across time, other fishing locations, and the experience of other fishers to intuitively predict abundance and allocate effort accordingly (Cox et al. 2003; Walters and Martell 2004). The (also generally unstated) basis for the substitution of abundance for catch rate is that, by rearranging Eq. 1, catch rates should be proportional to abundance: $C/E = qN$. Such implicit substitutions should be avoided as they can obscure mechanisms and assumptions—for example, the dubious but common assumption that catchability (q) is constant (Harley et al. 2001; Burgess et al. 2017; Feiner et al. 2020).

Other characteristics of the population besides abundance could also motivate fishing and therefore could be included as drivers of effort in this sub-model. For example, fishers may be motivated by catching trophy-size or consumption-size fish, in which case they should respond to changes in average fish size or catch rates of a particular size fraction of the fish population or harvest regulations. Recreational fishers are distinct from commercial fishers and many other resource users in that they value the act of fishing as well as the ability to catch fish (Stoeven 2014). Therefore, it is worth carefully considering how to model the behaviour of fishers when catch rates (or other biological metrics) are near or at zero. On the one hand, fishers who value the act of fishing for its own sake or continue fishing for other co-existing species may continue fishing even if they catch no fish (a positive y-intercept on the catch/effort relationship). On the other hand, fishers who care strongly about catching many fish (e.g., tournament fishers) may stop fishing at low but still non-zero catch rates (a negative y-intercept) (Fig. 4). These two behaviour types have critically important implications for a model's stability (Golden et al. 2022) and should be considered carefully.

Fishers also respond to non-catch-related aspects of each fishery, which may motivate them to target fish populations even though they will result in poor catch outcomes. In a meta-analysis of choice studies by recreational fishers, Hunt et al. (2019) identified a variety of factors, such as cost in time or money, regulations, and congestion, that affect site choice, some of which are often even more important than catch. However, the importance of these factors is often context-specific (Beardmore et al. 2015) and may even vary depending on how it is measured (Hunt et al. 2019). Ignoring non-catch-related drivers of fisher behaviour may lead to poor predictions of fishing satisfaction and impact on the resource.

A final consideration for the development of the effort sub-model is that fishers can vary in their behaviour and preferences even within a given fishery. Statistical tools are becoming more sophisticated in their ability to measure and evaluate these differences (e.g., latent class analysis using choice data). However, it is important to remember that qualitative knowledge can also inform the depiction of fishing effort vs. fish abundance relationships or (in agent-based models) associated decision rules. Such relationships developed from intimate qualitative knowledge of a fishery might be superior to utility-based approaches that must often assume fixed preferences, even though fisher preferences (e.g., for consumption vs. catch-and-release) are likely to change over time. Statistical approaches to understanding fisher behaviour are also based on a limited and frequently biased sample of fishers. Ultimately, it is an empirical question whether the assumptions of fisher behaviour in the model replicate observations. Especially when model predictions cannot be independently validated, they should be treated as thought experiments and not as perfect realizations of reality (Burgess et al. 2020; Carella et al. 2020; Madsen et al. 2021).

3.4 Step 4: Define Model Outputs and System Performance Metrics

Before implementing the model, it is important to define the output variables through which model behaviour will be evaluated. These outputs will again depend on the specific research questions. For integrated models, researchers will probably want to select variables from both the population dynamics and human behaviour sub-models that are informative about both the biological and social performance of the fishery. In commercial fisheries, biological performance is often evaluated using maximum sustainable yield or the presence of overfishing, which is defined in terms of the fishing mortality rate associated with maximum sustainable yield. These metrics are often easily calculated from common fish population dynamics models. For example, the maximum sustainable yield for the surplus production model in Eq. 2 has a simple analytical solution: $\lambda K/4$. However, maximum sustainable yield-based performance metrics are rarely useful for models of recreational fisheries (Johnston et al. 2010), which typically are not managed to maximize harvest but rather to balance fisher well-being with biological sustainability. Instead, optimal social yield (OSY) has been proposed as an alternative metric (Malvestuto and Hudgins 1996). Optimal social yield integrates "social, economic and biological considerations into a single measure of the utility (in terms of benefits, satisfaction or social welfare) a recreational fishery provides to society" (Johnston et al. 2010, 2013). Due to its complexity, OSY is still in the process of being operationalized, but the principle of integrating social, economic, and biological outcomes should be considered when defining model outputs. Model outputs representing social outcomes could include catch, effort, and integrated utility-based metrics of fisher welfare (Johnston et al. 2010). Where the governance component of the system is modelled, other metrics might prove useful, such as resilience or stability of the system or speed of implementing decisions or the general conflict proneness of the system.

3.5 Step 5: Implement the Model

There are many software platforms available to implement computational models, and we do not provide recommendations for a specific platform. Some may use modelling approach-specific platforms like Vensim (system dynamics) or NetLogo (agent-based modelling), whereas others may rely on more general-purpose platforms like R and Python. What might impact the choice of the platform are the practices within the community (to build on existing models and exchange the model with others), the programming experience within the team, and the development time available. In our experience, it is important that knowledge domain experts are involved during this part of the modelling cycle too. Do not assume that the implementation tasks can be outsourced to a computer programmer. During implementation, decisions will have to be made on how to translate the often-incomplete descriptions of the conceptual model into mathematical or algorithmic statements. Often, these decisions represent implicit assumptions about how the system works and should be made with care by those most familiar with the system.

3.6 Step 6: Run the Model, Analyse, and Ground Truth the Results

Once a version of the model has been implemented, one can engage in a series of activities to analyse the model. In the beginning, the analysis is focused on verification, testing whether the model is implemented correctly. At this stage, the model outputs are compared to simple logical predictions (e.g., as harvest increases, does the fish population decline?). Next, the model outputs can be compared to observations of the system in a process of "ground truthing," i.e., asking whether the model outputs or predictions match empirical observations of the system. More formally, to configure the model to an empirical case, one can use direct observations to inform parameter values or use calibration or statistical estimation to determine the parameter values that lead to the best fit with the data (e.g., Carruthers et al. 2019). With a more complex model, it may not be obvious what is a good fit because there are various spatial and temporal patterns in the data that may need to be explored, or processes may have been abstracted in the model to the point that they no longer have direct analogues in empirical data for comparison. Pattern-oriented modelling is an approach that aims to capture this complexity through replication of general empirical patterns rather than individual observations to gain more confidence in the structural integrity of the model (Grimm et al. 2005).

If the model is intended to be empirically based, multiple data sources can be used to evaluate the performance of the model, but care should be taken to make sure that these sources are independent of those used to parameterize the model. Key outputs of integrated models often include time series of fish population size, catch or landings, and fishing effort. These can be compared to fish survey catch-per-unit effort time series and creel survey estimates of catch and effort. Often, the researcher is faced with a choice: either use these data sources to help parameterize the model or save them to validate model performance. Though the same observation should not be used for both purposes, the time series can be split into training and test portions. Frequently, a small percentage (10–20%) of the observations—either randomly selected or from the end of the time series—are retained for comparison to model output in a process called cross-validation. Parameters of an integrated model could, in theory, be estimated from fitting the model to time series data (a process sometimes referred to as "inverse modelling") as is commonly done in stock assessment (see Walters and Martell 2004 Chap. 5 for a brief description). However, in practice, the necessary time series of recreational fishing metrics to fit the model are often lacking. If the performance of the integrated model is deemed to be sufficient, it is time to move to the next step. If performance is poor, however, it is better to return to *Step 3. Formalize the representation of the population and effort dynamics sub-models and their connections*. Poor cross-validation performance often indicates either that the value of a critical model parameter is wrong or that there are important processes that have not been considered.

3.7 *Step 7: Use the Model*

Once the model performance is within acceptable bounds (defined by the needs of the project and the data sources available for cross-validation), a sensitivity analysis can be performed where the researcher systematically varies the values of one or more parameters and notes the impact of this change on model outputs. Sensitivity analysis can also be incorporated in the previous step, if cross-validation performance is unacceptable, to identify parameters that might be inaccurate and the direction that the parameter must change to improve model performance. Parameters in the model that represent management options, such as the minimum length of fish retained by fishers or the season length, are prime candidates for sensitivity analysis as the results highlight management measures that might provide the most "bang for the buck."

Scenario testing is practically like sensitivity analysis, but it involves running the model with different discrete sets of parameter values that typically represent different management options or different potential future states (e.g., warmer vs. cooler climate conditions or restored vs. unrestored habitat). The development of scenarios for testing is best done in collaboration with managers, fishers, and other stakeholders or rights-based holders. Researchers are rarely in the best position to know which management actions are most feasible or acceptable to stakeholders, including political decision-makers, or to have an accurate sense of the costs of different alternative actions. Co-development of scenarios with stakeholders also increases interest and "buy-in" among those who may be the target audience for the knowledge gained from the integrated model. Ideally, stakeholder involvement begins significantly earlier in the process than Step 7 (see Camp et al. (2026) for more detailed information on this topic). A well-calibrated model may also be used to derive hypotheses and inform experiments of adaptive management, where model outputs are then confronted with actual system responses (Walters 1986). Responses can then be used to change the model. Often, integrated models will be used as a mechanistic tool to assess scenarios of possible futures, as full experimentation may not be possible or desired and is rarely conducted in practice (Walters 2007). In this sense, the model itself may be used as the subject of experimentation.

3.8 Step 8: Document, Archive, and Share the Model

In addition to publication in a relevant journal and communication of model results to relevant stakeholders, it is increasingly expected by the community, journals, and research sponsors that you make your model and associated documentation available in a public archive (Barton et al. 2022). Many of the benefits of public *data* archiving (Hampton et al. 2013; Roche et al. 2015), e.g., facilitating replication, exchange of knowledge, and transparency, also apply to archiving of *models*. Probably the simplest solution to model archiving is to maintain your code in a public GitHub repository and to provide a link to this repository in the journal article or technical report that describes the model. Model-specific archives also exist for some of the sub-models that might be components of an integrated model (e.g., the EcoBase archive for Ecopath models; Colléter et al. 2013). As part of this process, it is also useful to select a license for your model code that defines how others could build on your code, distribute the code, and whether commercial use is allowed. A software license protects contributors and users; an overview of different open-source software licenses is available at https://choosealicense.com/.

4 Conceptual Gaps, Challenges, and Opportunities for Integrated Models

Integrated models dealing with the interaction between social and ecological systems are becoming increasingly common; this is especially true within recreational fisheries (Solomon et al. 2020). Several papers have discussed frontiers for future research, either to develop a fuller understanding of interactions between SES subsystems (i.e., fish, fishers, and management) or to enhance the management value of these models (Schlüter et al. 2012; Fenichel et al. 2013; Ward et al. 2016; Arlinghaus et al. 2017a; Solomon et al. 2020). In this section, we suggest current gaps, challenges, and opportunities for integrated models to move forward. This discussion is informed and organized into the different disciplinary perspectives highlighted in earlier chapters of this book.

4.1 Gaps

Many integrated models do not consider long-term changes in the fishery. Historical ecology and environmental history are important, and often underappreciated, disciplines that can help contextualize any fishery and inform the development of integrated models (Thurstan et al. 2026). The most obvious application is for historical ecology, where this research can help to describe the development of a fishery from the perspective of the fished population, documenting changes in abundance and size-structure. In the marine realm, the most rapid declines in population biomass generally occurred early in a fishery's development, often before fishery-independent monitoring systems were in place (Worm et al. 2009; Hutchings et al. 2010). Relatively unexploited fish populations often exhibit an abundance of large, old individuals who help suppress recruitment of younger fish, much like the structure of an old-growth forest (Johnson 1976, 1994). Understanding the relative change in size structure and catch rates (and perhaps even relative abundance) helps contextualize the magnitude of change possible between current conditions and near-unfished conditions (e.g., Thurstan et al. 2018). This is important for parameterizing the population dynamics sub-model and in understanding how catch-related attributes of fishing utility and satisfaction may potentially change under different management actions.

Another gap in integrated models of recreational fisheries is that they generally either ignore social norms or assume they are constant over time. For example, many models ignore catch-and-release fishing altogether. In their literature review, Solomon et al. (2020) found only 10 of 49 integrated models incorporated voluntary catch and release; importantly, none of them suggested that voluntary catch and release may change over years and decades. This gap persists despite a growing literature documenting changes in release rates across several fisheries (Arlinghaus et al. 2007; Myers et al. 2008), a behaviour related to changes in biophysical and social contexts (Stensland et al. 2013). Many social norms change through individual interactions among fishers and may have profound impacts on long-term fishery performance, yet these changing rates, like other aspects of the networked social dynamics among fishers, are rarely considered in integrated models. Just as a broader historical perspective can improve the population dynamics sub-model, so too will the social components of integrated models be improved by a better understanding of long-term trends in fisher behaviours and motivations.

The ecological components of SES models are often especially detailed, particularly within recreational fisheries (Ward et al. 2016). Yet, there are still areas that could be more fully explored to help understand emergent properties in these systems and even help improve their value for management. For example, there is growing evidence that fish are not always available to be captured, either because they are spatially unavailable, behaviourally unwilling to attack bait, or are recovering from recent capture events (Cox et al. 2002; Askey et al. 2006; Matthias et al. 2014; Alós et al. 2015). Additionally, fish re-forming aggregations during stock decline can mean fish are always available in high densities, despite declining abundance. These behaviours, and others, cause catch rates to not reflect abundance, which affects the catch rate signals that both managers and fishers may use to alert them to stock decline but also affects the catch-based utility that motivates fishers to return to fish. Both mechanisms can lead to collapse, particularly when combined with other destabilizing mechanisms (Golden et al. 2022). Some SES models include a function allowing catchability to vary with abundance. Most commonly, this takes the form of an exponent, β, added to the catch equation (Eq. 1):

$$C_t = q\, E_t\, N_t^{\beta}; \tag{3}$$

with values of $\beta < 1$ indicating hyperstability and values of $\beta > 1$ indicating hyperdepletion (Fig. 5). To date, only a handful of empirical studies have estimated β for recreational fisheries (reviewed in Golden et al. 2022). Though more estimates of β would be useful, explicit inclusion of the *mechanisms,* both ecological (e.g., space-use, aggregation) and fisher-related (e.g., technological innovations in fish finders, social information networks), driving density-dependent catchability would permit more accurate long-term predictions of impacts on all nodes of the system (Ward et al. 2016; Solomon et al. 2020; Post et al. 2026).

Behaviour of fishers is also well-represented in SES models, though often with simplistic social models (see Melstrom et al. 2026). For example, the spatial distribution of fishers across patchy landscapes of fishing opportunities is a component of many SES models. Several SES models (e.g., Johnston et al. 2010; van

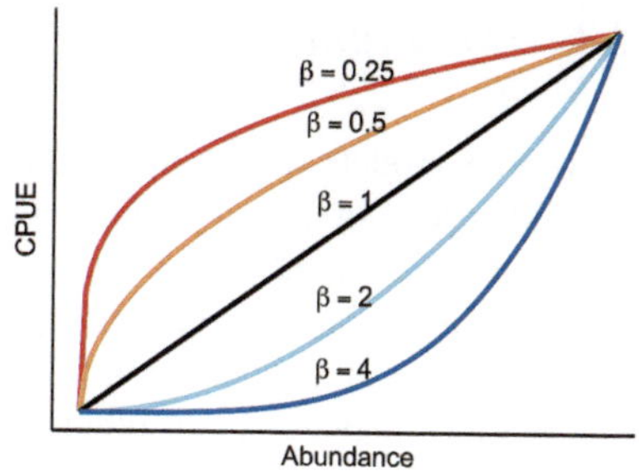

Fig. 5 Relationship Between Catch-per-unit Effort and Abundance Based on Different Values of the Shape Parameter β. (Reprinted from Golden et al. 2022)

Poorten and Camp 2019) have included heterogeneity in fisher motivation and skill, which would lead to effort sorting (less skilled fishers dropping out of declining fisheries, leaving a more highly skilled pool of fishers, thus maintaining high catch rates; Ward et al. 2013; van Poorten et al. 2016), yet the outcome of this process is rarely explored in integrated models. Other mechanisms leading to hyperstability in catch rates include both human and fish behaviour components. For example, fishers often fish preferentially in visible high-quality fish habitat that fish continually recolonize as density in these habitats is depleted through fishing (Dassow et al. 2020).

4.2 *Challenges*

The greatest challenge in building integrated models is appropriately representing the different systems from an interdisciplinary perspective. This is important because though many social processes can be approximated by ecological models (Post et al. 2026), or economic models (Melstrom et al. 2026), they are merely approximations and do not necessarily reflect processes that drive outcomes. Having a clear understanding of the goals of the project should help determine the appropriate disciplines that can or should be included.

Importantly, different disciplines may view variables differently, which can make the integration of different disciplinary perspectives difficult if project participants are not in constant communication and thinking ahead to analysis. For example, a resource economist may measure how fishers would respond to several fishery attributes, including crowding. However, an ecologist would model effort as the number of people on a water body (or people per area). Does this "effort" parameter adequately represent crowding as perceived by an individual fisher? If fishers crowd in an area and time, there may be extreme crowding that would not be evident from "effort." These considerations should be considered in the initial stages of development and represent the growing pains of interdisciplinary research.

4.3 Opportunities

Early SES models were built around random utility models (RUMs) and a rational actor idea derived from resource economics and random utility theory (Schlüter et al. 2012). As SES models are becoming more common and diverse in their applications, an important consideration will be the inclusion of the cognitive processes that lead to boundedly rational behaviour, including heuristics, habitats, emotions, norms and information networks; in other words, a move towards concepts central to behavioural economics (see MacKay et al. (2026) for a discussion of behavioural economics). To date, only a small fraction of the rich science of human decision-making has been included in integrated models. There is a need for modelers of fisheries SES to engage a broader swath of the social sciences in a conversation about how to best represent human behaviour in fisheries.

Information flow among fishers and networks of "like-minded" fisher types may be particularly important for SESs and outcomes (Solomon et al. 2020). Many SES models assume information is available across all fishers; few studies have explored how different fisher types access and apply information when making decisions related to fishing (Hunt et al. 2013). However, heterogeneity in information can influence ecosystem outcomes due to agent decisions (Carpenter et al. 1999). Moreover, access to information affects fishers' expectations and, therefore, satisfaction from a trip (Schramm et al. 1998). Understanding these dynamics on their own and applying these concepts within integrated models could potentially yield interesting and unexpected outcomes.

Similarly, information use is important for management. Although there are some well-monitored recreational fisheries, the most common scenario is that there is little or no information on either the biological or social components of the fishery on which management decisions might be based. Understanding the impacts of management decisions based on incomplete information is a mainstay of Management Strategy Evaluations (Punt et al. 2016), but few integrated models of recreational fisheries have explored how incomplete information affects management outcomes. The exceptions suggest that a lack of information by managers may contribute to eventual collapse (van Poorten et al. 2011).

Importantly, most SES models focus on the interactions between fishers and fish, and they ignore any dynamic feedback between these nodes and management. This is largely a by-product of the propensity for open-access recreational fisheries in countries where integrated models are created (Solomon et al. 2020). However, management is often responsive to changes in fishing conditions, even if this response is lagged, and this can influence long-term social–ecological dynamics (van Poorten et al. 2011; Ziegler et al. 2017). Determining where dynamic management feedbacks exist could dramatically influence social–ecological outcomes and emergent behaviours of these systems (Ward et al. 2016). Moreover, applying these decision rules could facilitate the creation of recreational fisheries closed-loop simulations (essentially a Management Strategy Evaluation), where the entire management loop—including information gathering, decision making, policy change, fisher response, fish population impact, and eventual updated information gathering—is modelled to explore long-term impacts on management outcomes. To date, closed-loop simulation is largely absent from integrated models (Schlüter et al. 2012), yet this could dramatically improve management outcomes and acceptance and management application of these models.

There is an obvious interaction among individuals and management organizations with distinct levels of agency and authority. How power struggles (Boucquey et al. 2026) impact management decisions, and therefore affect the attractiveness of different fisheries, is a phenomenon that has received little attention to date (Schlüter et al. 2012). These dynamics may be particularly important, especially as shared (Ziegler et al. 2021) or strongly influenced (van Poorten et al. 2011) decision-making by different agencies and stakeholders becomes increasingly common.

5 Conclusion

Integrated models are a powerful tool for understanding the dynamic interactions between fishers and the fish populations that they target. These interactions can run in both directions with mortality from fishers reducing fish populations and reduced catch rates or average fish size driving a decline in fishing effort (Fig. 1). It is a truism that "managing fisheries is managing people," and this applies especially to recreational fisheries for which harvest is rarely directly controlled by fishing quotas set by managers. Yet, recreational fishery management has often focused disproportionately on the biological dimensions of fisheries, carefully monitoring (where monitoring is done at all) the abundance, size structure, and recruitment of fish populations. Recent research (Golden et al. 2022) suggests that this focus may be exactly the inverse of the knowledge needed to understand sustainability of recreational fisheries, which depends on the responses of fishers to changes in fish populations. Though stabilizing feedbacks, such as those presented (Fig. 1), are plausible and perhaps even common, they are not inevitable. Understanding the social–ecological dynamics of recreational fisheries is necessary if we are to consistently achieve management goals. Integrated models are a crucial tool for gaining such understanding.

Acknowledgments This work was supported by a grant from the U.S. National Science Foundation (CNH #1716066) to CTS, MAJ, and OPJ. We thank Len Hunt, Abigail Lynch, Kevin Pope, and two anonymous reviewers for insightful comments on an earlier draft of this manuscript.

References

Alós J, Palmer M, Arlinghaus R (2012) Consistent selection towards low activity phenotypes when catchability depends on encounters among human predators and fish. PLoS One 7(10):e48030

Alós J, Palmer M, Trías P, Díaz-Gil C, Arlinghaus R (2014) Recreational angling intensity correlates with alteration of vulnerability to fishing in a carnivorous coastal fish species. Can J Fish Aquat Sci 72(2):217

Arlinghaus R (2017) Nachhaltiges Management von Angelgewässern: Ein Praxisleitfaden. Berichte des IGB, Heft, p 30

Arlinghaus R, Cooke SJ, Lyman J, Policansky D, Schwab A, Suski C, Sutton SG, Thorstad EB (2007) Understanding the complexity of catch-and-release in recreational fishing: an integrative synthesis of global knowledge from historical, ethical, social, and biological perspectives. Rev Fish Sci 15(1–2):75

Arlinghaus R, Matsumura S, Dieckmann U (2009) Quantifying selection differentials caused by recreational fishing: development of modeling framework and application to reproductive investment in pike (*Esox lucius*). Evol Appl 2(3):335–355

Arlinghaus R, Matsumura S, Dieckmann U (2010) The conservation and fishery benefits of protecting large pike (*Esox lucius* L.) by harvest regulations in recreational fishing. Biol Conserv 143(6):1444–1459

Arlinghaus R, Beardmore B, Riepe C, Meyerhoff J, Pagel T (2014) Species-specific preferences of German recreational anglers for freshwater fishing experiences, with emphasis on the intrinsic utilities of fish stocking and wild fishes. J Fish Biol 85:1843–1867

Arlinghaus R, Cyrus E-M, Eschbach E, Fujitani M, Hühn D, Johnston F, Pagel T, Riepe C (2015) Hand in Hand für eine nachhaltige Angelfischerei: Ergebnisse und Empfehlungen aus fünf Jahren praxisorientierter Forschung zu Fischbesatz und seinen Alternativen. Berichte des IGB Heft 28(2015):204

Arlinghaus R, Alós J, Beardmore B, Daedlow K, Dorow M, Fujitani M, Hühn D, Haider W, Hunt LM, Johnson BM, Johnston F, Klefoth T, Matsumura S, Monk C, Pagel T, Post JR, Rapp T, Riepe C, Ward H, Wolter C (2017a) Understanding and managing freshwater recreational fisheries as complex adaptive social-ecological systems. Rev Fish Sci Aquac 25(1):1

Arlinghaus R, Hühn D, Pagel T, Beck M, Rapp T, Wolter C (2017b) Fischereilicher Nutzen und gewässerökologische Auswirkungen des Besatzes mit Karpfen (*Cyprinus carpio*) in stehenden Gewässern: Ergebnisse und Schlussfolgerungen aktueller Ganzseeexperimente und Meta-Analysen. Fisch Fischmark Mecklenburg-Vorpommern 1:36–46

Arlinghaus R, Beardmore B, Riepe C, Pagel T (2020) Species-specific preference heterogeneity in German freshwater anglers, with implications for management. J Outdoor Recreat Tour 32: 100216

Arlinghaus R, Riepe C, Theis S, Pagel T, Fujitani M (2022) Dysfunctional information feedbacks cause the emergence of management panaceas in social-ecological systems: The case of fish stocking in inland recreational fisheries. Journal of Outdoor Recreation and Tourism. 38:100475

Askey PJ, Richards SA, Post JR, Parkinson EA (2006) Linking angling catch rates and fish learning under catch-and-release regulations. N Am J Fish Manag 26(4):1020

Ayllón D, Railsback SF, Almodóvar A, Nicola GG, Vincenzi S, Elvira B, Grimm V (2018) Eco-evolutionary responses to recreational fishing under different harvest regulations. Ecol Evol 8(19):9600–9613

Baranov FI (1918) On the question of the biological basis of fisheries. In: Bulletin of the Department of Fish Farming and Fishery Research (Nauchnyi Issledovatelskii Ikhtiologicheskii Institut Isvestia), pp 81–128

Barton CM, Lee A, Janssen MA, van der Leeuw S, Tucker GE, Porter C, Greenberg J, Swantek L, Frank K, Chen M, Jagers HRA (2022) How to make models more useful. P Natl Acad Sci USA 119(35):e2202112119

Beardmore B, Hunt LM, Haider W, Dorow M, Arlinghaus R (2015) Effectively managing angler satisfaction in recreational fisheries requires understanding the fish species and the anglers. Can J Fish Aquat Sci 72(4):500–513

Biggs R, Carpenter SR, Brock WA (2009) Turning back from the brink: detecting an impending regime shift in time to avert it. P Natl Acad Sci USA 106(3):826–831. https://doi.org/10.1073/pnas.0811729106

Binder CR, Hinkel J, Bots PWG, Pahl C, Binder CR, Hinkel J, Bots PWG, Pahl-Wostl C (2013) Research, part of a special feature on a framework for analyzing, comparing, and diagnosing social-ecological systems comparison of frameworks for analyzing social-ecological systems. Ecol Soc 18(4):19

Boucquey N, Fly J, Boonstra WJ, Drakopulos L, Leong KM, Marjadi MN, Wise S, Fairbanks L (2026) Political ecology for understanding recreational fishers and fisheries. In: Pope KL, Arlinghaus R, Hunt LM, Lynch AJ, and van Poorten BT (eds) Understanding recreational fishers: disciplinary and interdisciplinary approaches for fisheries management. Springer Nature, Cham, Switzerland, pp 197–232

Burgess MG, Costello C, Fredston-Hermann A, Pinsky ML, Gaines SD, Tilman D, Polasky S (2017) Range contraction enables harvesting to extinction. P Natl Acad Sci USA 114(15): 3945–3950

Burgess MG, Carrella E, Drexler M, Axtell RL, Bailey RM, Watson JR, Cabral RB, Clemence M, Costello C, Dorsett C, Gaines SD, Klein ES, Koralus P, Leonard G, Levin SA, Little LR, Lynham J, Madsen JK, Merkl A, Owashi B, Saul SE, Putten IE, Wilcox S (2020) Opportunities for agent-based modelling in human dimensions of fisheries. Fish Fish 21(3):570–587

Camp EV, van Poorten BT, Walters CJ (2015) Evaluating short openings as a management tool to maximize catch-related utility in catch-and-release fisheries. N Am J Fish Manag 35(6):1106

Camp EV, Larkin SL, Ahrens RNM, Lorenzen K (2017) Trade-offs between socioeconomic and conservation management objectives in stock enhancement of marine recreational fisheries. Fish Res 186:446–459

Camp EV, van Poorten BT, Robinson KF, Harrison HL, NguyenV , Crandall CA, Lorenzen K, Casola W, Dedual M, Klizentyte K (2026) Integrating diverse stakeholder knowledge in fisheries decision making. In: Pope KL, Arlinghaus R, Hunt LM, Lynch AJ, and van Poorten BT (eds) Understanding recreational fishers: disciplinary and interdisciplinary approaches for fisheries management. Springer Nature, Cham, Switzerland, pp 607–660

Carpenter SR, Brock WA (2004) Spatial complexity, resilience, and policy diversity: fishing on lake-rich landscapes. Ecol Soc 9(1):8

Carpenter S, Brock W, Hanson P (1999) Ecological and social dynamics in simple models of ecosystem management. Ecol Soc 3(2):4

Carpenter SR, Brock WA, Hansen GJA, Hansen JF, Hennessy JM, Isermann DA, Pedersen EJ, Perales KM, Rypel AL, Sass GG (2017) Defining a safe operating space for inland recreational fisheries. Fish Fish 18(6):1150–1160

Carrella E, Saul S, Marshall K, Burgess MG, Cabral RB, Bailey RM, Dorsett C, Drexler M, Madsen JK, Merkl A (2020) Simple adaptive rules describe fishing behavior better than perfect rationality in the US West Coast Groundfish Fishery. Ecol Econ 169:106449

Carruthers TR, Dabrowska K, Haider W, Parkinson EA, Varkey DA, Ward H, McAllister MK, Godin T, Van Poorten B, Askey PJ, Wilson KL, Hunt LM, Clarke A, Newton E, Walters C, Post JR (2019) Landscape-scale social and ecological outcomes of dynamic angler and fish behaviors: processes, data, and patterns. Can J Fish Aquat Sci 76(6):970

Cenek M, Franklin M (2017) An adaptable agent-based model for guiding multi-species Pacific salmon fisheries management within a SES framework. Ecol Model 360:132–149

Clark CW (1973) The Economics of Overexploitation: Severe depletion of renewable resources may result from high discount rates used by private exploiters. Science 181(4100): 630–634

Cole RA, Ward FA (1994) Optimum fisheries management policy: angler opportunity versus angler benefit. N Am J Fish Manag 14(1):22–33

Colléter M, Valls AE, Guitton J, Morissette L, Arreguín-Sánchez FF, Christensen V, Pauly D (2013) EcoBase: a repository solution to gather and communicate information from EwE models. Fish Cent Res Rep 21(1)

Cox SP, Walters C (2002) Modeling exploitation in recreational fisheries and implications for effort management on British Columbia rainbow trout lakes. North American Journal of Fisheries Management 22(1): 21–34

Cox SP, Walters CJ, Post JR (2003) A model-based evaluation of active management of recreational fishing effort. N Am J Fish Manag 23(4):1294

Dassow CJ, Ross AJ, Jensen OP, Sass GG, van Poorten BT, Solomon CT, Jones SE (2020) Experimental demonstration of catch hyperstability from habitat aggregation, not effort sorting, in a recreational fishery. Can J Fish Aquat Sci 77(4):762

Deriso RB (1980) Harvesting strategies and parameter estimation for an age-structured model. Can J Fish Aquat Sci 37(2):268

Fedler AJ, Ditton RB (1994) Understanding angler motivations in fisheries management. Fisheries 19(4):6–13

Feiner ZS, Wolter MH, Latzka AW (2020) "I will look for you, I will find you, and I will [harvest] you": persistent hyperstability in Wisconsin's recreational fishery. Fish Res 230:105679

Fenichel EP, Abbott JK, Huang B (2013) Modelling angler behavior as a part of the management system: synthesizing a multi-disciplinary literature. Fish Fish 14(2):137–157

Fujitani M, McFall A, Randler C, Arlinghaus R (2017) Participatory adaptive management leads to environmental learning outcomes extending beyond the sphere of science. Sci Adv 3:e1602516

Fujiwara M, Martinez-Andrade F, Wells RJD, Fisher M, Pawluk M, Livernois MC (2019) Climate-related factors cause changes in the diversity of fish and invertebrates in subtropical coast of the Gulf of Mexico. Commun Biol 2(1):403

Gao L, Hailu A (2011) Evaluating the effects of area closure for recreational fishing in a coral reef ecosystem: the benefits of an integrated economic and biophysical modeling. Ecol Econ 70(10): 1735–1745

Gao L, Hailu A (2013) Identifying preferred management options: An integrated agent-based recreational fishing simulation model with an AHP-TOPSIS evaluation method. Ecol Model 249:75–83

Golden AS, van Poorten B, Jensen OP (2022) Focusing on what matters most: evaluating multiple challenges to stability in recreational fisheries. Fish Fish 23(6):1418–1438

Gordon HS (1954) The economic theory of a common-property resource: the fishery. J Polit Econ 62(2):124–142

Grimm V, Revilla E, Berger U, Jeltsch F, Mooij WM, Railsback SF, Thulke HJ, Weiner J, Wiegand T, DL DA (2005) Pattern-oriented modeling of agent-based complex systems: lessons from ecology. Science 310(5750):987–991

Hampton SE, Strasser CA, Tewksbury JJ, Gram WK, Budden AE, Batcheller AL, Porter JH (2013) Big data and the future of ecology. Front Ecol Environ 11(3):156–162

Harley SJ, Myers RA, Dunn A (2001) Is catch-per-unit-effort proportional to abundance? Can J Fish Aquat Sci 58(9):1760–1772

Hickey JT, Diaz GE (1999) From flow to fish to dollars: an integrated approach to water allocation. J Am Water Resour As 35(5):1053–1067

Hilborn R (2001) Calculation of biomass trend, exploitation rate, and surplus production from survey and catch data. Can J Fish Aquat Sci 58(3):579

Hilborn R, Walters CJ (1992) Stock and recruitment. Quantitative fisheries stock assessment: choice, dynamics and uncertainty. Springer Link, New York, pp 241–296

Hinkel J, Bots PWG, Schlüter M (2014) Enhancing the Ostrom social-ecological system framework through formalization. Ecol Soc 19(3):art51

Horan RD, Fenichel EP, Drury KLS, Lodge DM (2011) Managing ecological thresholds in coupled environmental-human systems. P Natl Acad Sci USA 108(18):7333–7338

Hühn D, Gwinn DC, Shaw SL, Alos J, Allen MS, Pagel T, Skov C, Arlinghaus R (2023) Density- and size-dependent mechanisms modulate the outcome of stocking in a naturally recruiting freshwater piscivore (northern pike, *Esox lucius*): a replicated whole-lake experiment. Fish Res 267:106799

Hunt LM, Arlinghaus R, Lester N, Kushneriuk R (2011) The effects of regional angling effort, angler behavior, and harvesting efficiency on landscape patterns of overfishing. Ecol Appl 21(7):2555–2575

Hunt LM, Sutton SG, Arlinghaus R (2013) Illustrating the critical role of human dimensions research for understanding and managing recreational fisheries within a social-ecological system framework. Fish Manag Ecol 20(2–3):111

Hunt LM, Camp E, van Poorten B, Arlinghaus R (2019) Catch and non-catch-related determinants of where anglers fish: a review of three decades of site choice research in recreational fisheries. Rev Fish Sci Aquac 27(3):261–286

Hutchings JA, Minto C, Ricard D, Baum JK, Jensen OP (2010) Trends in the abundance of marine fishes. Can J Fish Aquat Sci 67:1205–1210

Innes-Gold A, Pavlowich T, Heinichen M, McManus MC, McNamee J, Collie J, Humphries A (2021) Exploring social-ecological trade-offs in fisheries using a coupled food web and human behavior model. Ecol Soc 26(2):40

Johnson L (1976) Ecology of arctic populations of lake trout, *Salvelinus namaycush*, Lake Whitefish, *Coregonus clupeaformis*, Arctic Char, *S. alpinus*, and associated species in unexploited lakes of the Canadian Northwest Territories. J Fish Res Board Can 33(11):2459–2488

Johnson L (1994) Long-term experiments on the stability of two fish populations in previously unexploited Arctic lakes. Can J Fish Aquat Sci 51(1):209

Johnson BM, Carpenter SR (1994) Functional and numerical responses: a framework for fish-angler interactions? Ecol Appl 4(4): 808–821

Johnston FD, Arlinghaus R, Dieckmann U (2010) Diversity and complexity of angler behavior drive socially optimal input and output regulations in a bioeconomic recreational-fisheries model. Can J Fish Aquat Sci 67(9):1507–1531

Johnston FD, Arlinghaus R, Dieckmann U (2013) Fish life history, angler behavior and optimal management of recreational fisheries. Fish Fish 14(4):554–579

Johnston FD, Beardmore B, Arlinghaus R (2014) Optimal management of recreational fisheries in the presence of hooking mortality and noncompliance—predictions from a bioeconomic model incorporating a mechanistic model of angler behavior. Can J Fish Aquat Sci 72(1):37–53

Johnston FD, Allen MS, Beardmore B, Riepe C, Pagel T, Hühn D, Arlinghaus R (2018) How ecological processes shape the outcomes of stock enhancement and harvest regulations in recreational fisheries. Ecol Appl 28(8):2033–2054

Kahneman D (2011) Thinking fast and slow. Farrar, Straus, and Giroux. American Psychological Association, Washington, DC

Lee M-Y, Steinback S, Wallmo K (2017) Applying a bioeconomic model to recreational fisheries management: groundfish in the Northeast United States. Mar Resour Econ 32(2):191–216

Lorenzen K (2000) Allometry of natural mortality as a basis for assessing optimal release size in fish-stocking programmes. Can J Fish Aquat Sci 57:2374–2381

Mackay M, van Putten EI, Richter A, Abatayo AL, Chuah S-H, Dankel DJ, Schadeberg A, Snyder H (2026) Behavioural economics for understanding recreational fishers and fisheries. In: Pope KL, Arlinghaus R, Hunt LM, Lynch AJ, and van Poorten BT (eds) Understanding recreational fishers: disciplinary and interdisciplinary approaches for fisheries management. Springer Nature, Cham, Switzerland, pp 261–284

Madsen JK, Bailey R, Carrella E, Koralus P (2021) From reactive towards anticipatory fishing agents. J Simul 15(1–2):23–37

Malvestuto SP, Hudgins MD (1996) Optimum yield for recreational fisheries management. Fisheries 21(6):6–17

Mapstone BD, Little LR, Punt AE, Davies CR, Smith ADM, Pantus F, McDonald AD, Williams AJ, Jones A (2008) Management strategy evaluation for line fishing in the Great Barrier Reef: balancing conservation and multi-sector fishery objectives. Fish Res 94(3):315–329

Massey DM, Newbold SC, Gentner B (2006) Valuing water quality changes using a bioeconomic model of a coastal recreational fishery. J Environ Econ Manag 52(1):482–500

Matsumura S, Beardmore B, Haider W, Dieckmann U, Arlinghaus R (2019) Ecological, angler, and spatial heterogeneity drive social and ecological outcomes in an integrated landscape model of freshwater recreational fisheries. Rev Fish Sci Aquac 27(2):1–28

Matthias BG, Allen MS, Ahrens RNM, Beard TD, Kerns JA (2014) Hide and seek: interplay of fish and anglers influences spatial fisheries management. Fisheries 39(6):261–269

McConnell KE, Sutinen JG (1979) Bioeconomic models of marine recreational fishing. J Environ Econ Manag 6(2):127–139

Melstrom RT, Jungers B, Bi X, Abbott JK (2026) Resource economics for understanding recreational fishers and fisheries. In: Pope KL, Arlinghaus R, Hunt LM, Lynch AJ, van Poorten BT (eds) Understanding recreational fishers: disciplinary and interdisciplinary approaches for fisheries management. Springer Nature, Cham, p 123–156

Müller-Hansen F, Schlüter M, Mäs M, Donges JF, Kolb JJ, Thonicke K, Heitzig J (2017) Towards representing human behavior and decision making in Earth system models—an overview of techniques and approaches. Earth Syst Dynam 8(4):977–1007

Myers RA, Barrowman NJ (1996) Is fish recruitment related to spawner abundance? Fish B-Noaa 94:707–724

Myers RA, Bowen KG, Barrowman NJ (1999) Maximum reproductive rate of fish at low population sizes. Can J Fish Aquat Sci 56(12):2404–2419

Myers R, Taylor J, Allen M, Bonvechio TF (2008) Temporal trends in voluntary release of largemouth bass. N Am J Fish Manag 28(2):428–433

Nieman CL, Solomon CT (2022) Slow social change: implications for open access recreational fisheries. Fish Fish 23:195–201

Olaussen JO, Skonhoft A (2008) A bioeconomic analysis of a wild Atlantic salmon (Salmo salar) recreational fishery. Mar Resour Econ 23(3):273–293

Post JR (2013) Resilient recreational fisheries or prone to collapse? A decade of research on the science and management of recreational fisheries. Fish Manag Ecol 20(2–3):99

Post JR, Parkinson EA (2012) Temporal and spatial patterns of angler effort across lake districts and policy options to sustain recreational fisheries. Can J Fish Aquat Sci 69(2):321

Post JR, Mushens C, Paul A, Sullivan M (2003) Assessment of alternative harvest regulations for sustaining recreational fisheries: model development and application to bull trout. N Am J Fish Manag 23(1):22–34

Post JR, Wilson KL, Johnston FD, Sass GG, Askey PJ, Ward HGM, Allen MS (2026) Ecology for understanding recreational fishers and fisheries. In: Pope KL, Arlinghaus R, Hunt LM, Lynch AJ, van Poorten BT (eds) Understanding recreational fishers: disciplinary and interdisciplinary approaches for fisheries management. Springer Nature, Cham, p 95–121

Punt AE, Butterworth DS, de Moor CL, De Oliveira JA, Haddon M (2016) Management strategy evaluation: best practices. Fish Fish 17(2):303–334

Railsback SF, Grimm V (2019) Agent-based and individual-based modeling: a practical introduction. Princeton University Press, Princeton

Ricker WE (1975) Computation and interpretation of biological statistics of fish populations. Bulletin of the Fisheries Research Board of Canada, Bulletin 191, Ottawa

Roche DG, Kruuk LE, Lanfear R, Binning SA (2015) Public data archiving in ecology and evolution: how well are we doing? PLoS Biol 13(11):e1002295

Rogers MW, Allen MS, Brown P, Hunt T, Fulton W, Ingram BA (2010) A simulation model to explore the relative value of stock enhancement versus harvest regulations for fishery sustainability. Ecol Model 221(6):919–926

Schaefer MB (1957) Some considerations of population dynamics and economics in relation to the management of the commercial marine fisheries. J Fish Res Board Can 14(5):669–681

Scheffer M, Barrett S, Carpenter R, Folke C, Green AJ, Holmgren M, Hughes TP, Kosten S, van de Leemput IA, Nepstad DC, van Nes EH, Peeters ETHM, Walker B (2015) Creating a safe operating space for iconic ecosystems. Science 347(6228):1317–1319

Schlüter M, McAllister RRJ, Arlinghaus R, Bunnefeld N, Eisenack K, Hölker F, Milner-Gulland EJ, Müller B, Nicholson E, Quaas M, Stöven M (2012) New horizons for managing the environment: a review of coupled social-ecological systems modeling. Nat Resour Model 25(1):219–272

Schlüter M, Hinkel J, Bots PWG, Arlinghaus R (2014) Application of the SES framework for model-based analysis of the dynamics of social-ecological systems. Ecol Soc 19(1):14

Schlüter M, Baeza A, Dressler G, Frank K, Groeneveld J, Jager W, Janssen MA, McAllister RRJ, Müller B, Orach K, Schwarz N, Wijermans N (2017) A framework for mapping and comparing behavioral theories in models of social-ecological systems. Ecol Econ 131:21–35

Schramm HL, Arey SD, Miko DA, Gerard PD (1998) Angler perceptions of fishing success and the effect of on-site catch rate information. Hum Dimens Wildl 3(3):1–10

Schulze J, Müller B, Groeneveld J, Grimm V (2017) Agent-based modelling of social-ecological systems: achievements, challenges, and a way forward. J Artif Soc S 20(2):8

Smith ADM (1994) Management strategy evaluation: the light on the hill. In: Hancock DA (ed) Population dynamics for fisheries management, pp 249–253

Smith MD, Wilen JE (2003) Economic impacts of marine reserves: the importance of spatial behavior. J Environ Econ Manag 46(2):183–206

Solomon CT, Dassow CJ, Iwicki CM, Jensen OP, Jones SE, Sass GG, Trudeau A, van Poorten BT, Whittaker D (2020) Frontiers in modelling social-ecological dynamics of recreational fisheries: a review and synthesis. Fish Fish 21(5):973–991

Stensland S, Aas Ø, Mehmetoglu M (2013) The influence of norms and consequences on voluntary catch and release angling behavior. Hum Dimens Wildl 18(5):373

Stoeven MT (2014) Enjoying catch and fishing effort: the effort effect in recreational fisheries. Environ Resour Econ 57(3):393–404

Thébaud O, Ellis N, Little LR, Doyen L, Marriott RJ (2014) Viability trade-offs in the evaluation of strategies to manage recreational fishing in a marine park. Ecol Indic 46:59–69

Thorson JT, Minto C, Minte-Vera CV, Kleisner KM, Longo C (2013) A new role for effort dynamics in the theory of harvested populations and data-poor stock assessment. Can J Fish Aquat Sci 70(12):1829–1844

Thorson JT, Jensen OP, Zipkin E (2014) How variable is recruitment for exploited marine fishes? A hierarchical model for testing life history theory. Can J Fish Aquat Sci 71:973–983

Thurstan RH, Buckley SM, Pandolfi JM (2018) Trends and transitions observed in an iconic recreational fishery across 140 years. Glob Environ Chang 52(2018):22–36

Thurstan RH, Chong-Montenegro C, Christensen J, Clark A, Gaynor A (2026) Environmental history and historical ecology for understanding recreational fishers and fisheries. In: Pope KL, Arlinghaus R, Hunt LM, Lynch AJ, and van Poorten BT (eds) Understanding recreational fishers: disciplinary and interdisciplinary approaches for fisheries management. Springer Nature, Cham, Switzerland, pp 233–259

Townsend H (2013) Comparing and coupling a water quality and a fisheries ecosystem model of the Chesapeake Bay for the exploratory assessment of resource management strategies. ICES J Mar Sci 71(3):703–712

Van Delden H, Seppelt R, White R, Jakeman AJ (2011) A methodology for the design and development of integrated models for policy support. Environ Model Softw 26(3):266–279

van Poorten BT, Camp EV (2019) Addressing challenges common to modern recreational fisheries with a buffet-style landscape management approach. Rev Fish Sci Aquac 27(2):1–24

van Poorten BT, MacKenzie CJA (2020) Using decision analysis to balance angler utility and conservation in a recreational fishery. N Am J Fish Manag 40(1):29–47

van Poorten BT, Arlinghaus R, Daedlow K, Haertel-Borer SS (2011) Social-ecological interactions, management panaceas, and the future of wild fish populations. P Natl Acad Sci USA 108(30): 12554–12559 van Poorten BT, Walters CJ, Ward HGM (2016) Predicting changes in the catchability coefficient through effort sorting as less skilled fishers exit the fishery during stock declines. Fish Res 183: 379–384

Vasslides J, Jensen OP (2015) Fuzzy cognitive mapping in support of integrated ecosystem assessments: developing a shared conceptual model among stakeholders. J Environ Manag 166(2016):348–356

Walters CJ (1986) Adaptive management of renewable resources. Macmillan Publishers Ltd, Basingstoke

Walters CJ (2007) Is adaptive management helping to solve fisheries problems? J Hum Environ 36(4):304–307

Walters CJ, Martell SJ (2004) Fisheries ecology and management. Princeton University Press, Princeton

Ward HGM, Quinn MS, Post JR (2013) Angler characteristics and management implications in a large, multistock, spatially structured recreational fishery. N Am J Fish Manag 33(3):576

Ward HGM, Allen MS, Camp EV, Cole N, Hunt LM, Matthias B, Post JR, Wilson K, Arlinghaus R (2016) Understanding and managing social-ecological feedbacks in spatially structured recreational fisheries: the overlooked behavioral dimension. Fisheries 41(9):524

Wijermans N, Boonstra WJ, Orach K, Hentati-Sundberg J, Schlüter M (2020) Behavioral diversity in fishing-towards a next generation of fishery models. Fish Fish 21(5):872–890

Wilberg MJ, Irwin BJ, Jones ML (2026) Participatory modelling to support evaluation of management actions for recreational fisheries. In: Pope KL, Arlinghaus R, Hunt LM, Lynch AJ, van Poorten BT (eds) Understanding recreational fishers: disciplinary and interdisciplinary approaches for fisheries management. Springer Nature, Cham, p 693–718

Wilson KL, Foos A, Barker OE, Farineau A, De Gisi J, Post JR (2020) Social-ecological feedbacks drive spatial exploitation in a northern freshwater fishery: a halo of depletion. J Appl Ecol 57(2): 206–218

Worm B, Hilborn R, Baum JK, Branch TA, Collie JS, Costello C, Fogarty MJ, Fulton EA, Hutchings JA, Jennings S, Jensen OP, Lotze HK, Mace PM, McClanahan TR, Minto C, Palumbi SR, Parma AM, Ricard D, Rosenberg AA, Watson R, Zeller D (2009) Rebuilding global fisheries. Science 325:578–585

Ziegler JP, Golebie EJ, Jones SE, Weidel BC, Solomon CT (2017) Social-ecological outcomes in recreational fisheries: the interaction of lakeshore development and stocking. Ecol Appl 27(1): 56–65

Ziegler JP, Jardine SL, Jones SE, van Poorten BT, Janssen MA, Solomon CT (2021) Investing in the commons: transient welfare creates incentives despite open access. Ecol Soc 26(2):16

Participatory Modelling to Support Evaluation of Management Actions for Recreational Fisheries

Michael J. Wilberg , Brian J. Irwin , and Michael L. Jones

Abstract Participatory modelling (PM) is a collaborative problem-solving approach that involves shared learning through the development of a model or multiple models with a group of participants and a modelling team. PM has the potential to be used more frequently to address recreational fisheries problems. We focus on our experience working with small groups (usually <40 people) to address natural resource issues. This chapter focuses on building a PM team and the communications that occur before, during, and after a PM process. The participants usually include people like recreational and commercial fishers, members of environmental non-governmental organizations, and fishery managers or other decision makers. Ground rules, expectations, and end goals should be established early in the PM process. We emphasize that benefits arise from including the participants in the development and validation of the models in a process that includes feedback between the modellers and participants. Although participants are ideally involved in all stages, their inclusion during model development and testing can be critical for buy-in within the group. When the modellers listen carefully to participant concerns and respond to those concerns during model development, we have experienced favourable process outcomes such as improved participant engagement and building of trust.

M. J. Wilberg (✉)
Chesapeake Biological Laboratory, University of Maryland Center for Environmental Science, Solomons, MD, USA
e-mail: wilberg@umces.edu

B. J. Irwin
U.S. Geological Survey--Georgia Cooperative Fish and Wildlife Research Unit, Athens, GA, USA

Warnell School of Forestry and Natural Resources, University of Georgia, Athens, GA, USA

M. L. Jones
Quantitative Fisheries Center, Department of Fisheries and Wildlife, Michigan State University, East Lansing, MI, USA

K. L. Pope et al. (eds.), *Understanding Recreational Fishers*, Fish & Fisheries Series 45, https://doi.org/10.1007/978-3-031-99739-6_22

Keywords Co-management · Coproduction of knowledge · Participatory processes

1 Introduction

Participatory modelling (PM) is a collaborative problem-solving approach that involves shared learning with a group of participants and a modelling team. The goal of PM is typically to organize what is known and unknown in support of management-level decision making. Thus, PM involves coproducing knowledge to evaluate potential options for addressing the problem (Moallemi et al. 2021). One aspect of the knowledge coproduction is the development of a model that helps participants better understand the problem they are trying to solve and the potential effectiveness of alternative suites of options. PM processes have the potential to increase the effectiveness and legitimacy of management measures because participants have helped develop, vet, and choose preferred options (Hare 2011; Robinson and Fuller 2017). Given these potential benefits, PM can aid in developing solutions to challenges of recreational fishery management (Arlinghaus et al. 2019; Röckman et al. 2012). PM can be used as a process to identify a range of goals, to co-develop knowledge among the participants, and as a "flight simulator" to test potential fishery management options before they are implemented.

PM has been used to address a wide range of problems in natural resources and sustainability science (Sterling et al. 2019). Many forms of PM have been described, and most involve establishing a shared representation of the problem through the development of a model that accounts for goals and objectives, participant knowledge, and consideration of key uncertainties (Hare 2011; Hovmand 2014; Robinson and Fuller 2017; Herrera de Leon and Kopainsky 2019). The PM process often involves iterative consideration of the objectives, options, and a model that describes the potential for options to achieve the objectives, often described at the "consequences" stage of a decision-making process. Other descriptions of PM exist (Hovmand 2014; Robinson and Fuller 2017; Moallemi et al. 2021), and they often share traits with processes like Adaptive Management (Crevier and Parrott 2019), Structured Decision Making (Irwin et al. 2011), and management strategy evaluation (Punt et al. 2016), the latter of which has many commercial fisheries applications (e.g. Smith et al. 1999). Though some of these processes may end with the development of conceptual or qualitative models, our own experience is primarily with extending conceptual models into quantitatively explicit numerical simulation models developed iteratively during and in between a series of participatory workshops. We refer to such stochastic scenario-forecasting tools as "operating models".

Coproduction of information is not unique to PM; in this chapter, we focus on PM as a way to inclusively conduct a model-based evaluation of the potential performance of alternative management actions for recreational fisheries. Here, we share some insights gained from different decision problems where a common thread was that the shared development of an operating model with a group of participants was instrumental for providing relevant decision-support information. For recreational fisheries, the "problem" or focus of these decisions can often be the assessment of different management options, including stocking decisions, or evaluation of regulations or harvest policies. In our experience, recreational fishery objectives typically relate to improving catch or harvest opportunities, maintaining an adequate fish population for future fishing opportunities, as well as experiential aspects like improving access and environmental quality, which could influence fisher recruitment and retention. As part of PM, a team can work together to define performance measures that will be used to assess how different management options are expected to achieve objectives. In addition to refining expectations for potential policy performance, the PM experience can also help reveal or diagnose various decision-relevant uncertainties and sometimes help assess risk. A major recurring concern or challenge is a real or perceived lack of data to guide the development and use of models (Fuller et al. 2020). Even so, we consider that in many cases, including those where data are deemed to be lacking and uncertainty is high, the PM approach can yield worthwhile benefits by collecting new data, uncovering existing data, and determining if new data are indeed needed. In subsequent sections, we further describe some of the more recognizable benefits and challenges of successfully implementing PM.

The goal of this chapter is to provide a roadmap for conducting a PM process and to provide guidance on familiar challenges in PM for recreational fisheries rather than an overview of PM broadly. Additionally, our focus in this chapter is on conducting PM processes to help solve recreational fishery problems or conflicts. PM has developed through convergent evolution from many fields, and thus, the literature on approaches is vast. Our processes stem from adaptive resource assessment (Walters 1986), but they share key aspects with other versions of PM (Moallemi et al. 2021). Collectively, we have several decades of experience working in the field of PM in a range of contexts, including recreational fisheries. These experiences include working with participants to develop quantitative models to help address fisheries problems. Therefore, this chapter will emphasize PM that involves quantitative, rather than qualitative or conceptual, models despite some PM processes that have a greater emphasis on conceptual models (e.g. Hovmand 2014). This chapter is particularly aimed at those responsible for leading a PM process, but other participants in PM processes may find value as well. We include examples from our experience and break down PM into the stages of planning, conducting the project, and completion. The ethics of PM processes, though generally being the same as scientific research (Weinbaum et al. 2019), include some aspects that fishery scientists may not have had to consider in other parts of their research. Therefore, we highlight ethical considerations that are particularly important for PM.

2 Leadership and Planning

Leadership and planning are critical components of successful PM processes (Hämäläinen et al. 2020). In our experience, establishing the initial leadership of and planning for PM involves dialogue between fishery managers and fishery scientists. In some cases, fishers or representatives of fishing groups may be among the first to recognize a fishery "problem" or be at the forefront of requesting the formation of a study to address potential fishery performance under a previous, current, or anticipated threat (e.g. invasive species, contaminants, overfishing, poor recruitment, unfavourable flow or temperature regimes). At this stage, key discussions include problem scoping, identifying individuals or agencies involved with monitoring or managing the system, anticipating decision timelines (e.g. for setting harvest regulations or assessing species status), and determining the funding or planning needs for the PM itself. Here and throughout the rest of the chapter, we use "system" to refer to all the interrelated aspects of the problem that the PM process is attempting to address.

2.1 Leadership

A committed leadership team is essential to guide the entire process and should include both experienced modellers and members with technical facilitation expertise. The leadership team should also include representatives of the "client" for the issue, typically a management agency or other decision-making body. Whenever possible, the modellers and facilitators should be primarily composed of people who are not perceived as having a personal stake in the outcome, so they may be considered more objective. However, a PM process can also be successfully conducted in cases where the modelling team has had a long-term engagement with the focal issue. For example, in a PM process for eastern oyster (*Crassostrea virginica*) restoration in the Maryland portion of the Chesapeake Bay, the modelling team had prior involvement with a range of management recommendations prior to the initiation of the PM process (Wilberg et al. 2011, 2013; Maryland DNR 2021). However, conflicts of interest should be avoided where a member of the modelling team has a personal stake in the outcome of the process. For example, if the perceived problem is post-release mortality, where a caught fish is subsequently released but may suffer unintended mortality from injury or stress, and a member of the modelling team has developed a commercially available de-hooking device that they have proposed as a solution for the problem, then their participation in the process would likely constitute a conflict of interest. Conversely, we do not view the successful completion of a PM process or the distribution of the information it produces to be a conflict of interest that would prevent scientists, researchers, modellers, or managers from engaging with the process or formally being part of a PM team. Note that a typical goal of a PM process would be to have a participant group make recommendations to help solve a pressing recreational fisheries problem. A misuse of PM would be for leadership to attempt to convince participants to accept their solution.

Early in the process, it is vital to clearly define and achieve consensus on the scope of the problem to be addressed. For example, in a PM process to consider recreational fishery management opportunities for king mackerel (*Scomberomorus cavalla*) in the South Atlantic region of the United States of America, some participants also wanted to consider the population and fisheries in the Gulf of Mexico. However, that geographic area was outside the scope of the defined project (Miller et al. 2010; Ihde et al. 2011). Similarly, in a PM project centred on oyster management in the Chesapeake Bay, considering land use effects on oysters would require a vastly different participant group and process than the one that had been assembled, and was therefore "outside" the scope of the process.

The leadership team should define their expected scope prior to engagement with participants, but they should also allow for flexibility so that participants can have input on scope and thereby begin to develop buy-in for the analysis. It can be helpful from a process standpoint to document the diversity of views on scope, even if the focus is narrower than some participants would have preferred. This sends the message that participants were listened to, and their ideas recorded, even if some components were not explicitly modelled. Further, these suggestions may serve future endeavours if subsequent iterations of PM can be expanded in scope or scale.

Another critical aspect of planning is to define the relationship between the PM process and the decision-making process it is intended to inform. In some cases, the PM process may be the decision-making process (i.e. the recommendations that are developed during PM will be automatically implemented). However, in many cases, PM is used to inform decision makers who may or may not have been part of the PM process. Therefore, it is critical to consider how information from the PM process will support the decision-making process. Ideally, decision makers should be aware of and supportive of PM and understand what to expect from the PM process so that they can inform their decision making. For instance, if decision or policy makers ignore the participants' recommendations, then it can make interactions with user groups worse. Avoiding this outcome was a key element of discussions at the initial stages of the Lake Erie walleye (*Sander vitreus*) PM process (Jones et al. 2016). The group established formal Terms of Reference for the process that included an explicit definition of the relationship between the PM process and decision making—namely that the recommendations emerging from the process would be advisory to decision makers, but not binding. This enabled the decision makers to participate in the PM process without fear of being bound by the consequences, which in turn facilitated the coproduction of solutions that were in fact acceptable to decision makers.

PM processes require a substantial amount of work for all involved, and they are also usually costly in terms of time and money. Therefore, PM processes should only be used in cases where the information generated during the process is likely to be used. For example, in the process of management of oysters in Maryland, the decision makers were the state government executive and legislators, and the participants were selected by the executive branch to represent key groups that were identified in state law. At the end of this process, the PM group made a suite of recommendations to the Maryland state government, and many of these recommendations were implemented in law the next year. One thing that facilitated this process was that members of the executive and legislative branches of state government were non-voting members of the participant group. Thus, the involvement of decision makers in the PM process can be extremely helpful in signalling the importance of the PM process to the participants and implementing the recommendations of the process. However, not all PM processes will result in changes in management or practices. For example, despite enthusiasm by the participants in a PM process for king mackerel, their management recommendations were not adopted (Ihde et al. 2011).

2.2 Participants

The participants of an inclusive decision-making process may represent a broad range of interests, and they may vary in how or why they are invested in the managed resource. Interested and affected parties represent the part of society affected by a potential decision—either directly or indirectly (Stern 2021). For example, in many fisheries in the United States of America, the goal is to manage a renewable resource for the long-term benefit of the nation or state. This creates an exceptionally large pool of people who could be perceived as having a personal stake in the resource, even including those from future generations. Participants in PM, by contrast, are the subset of those people who directly and fully engage in a PM process. Participants should include individuals from groups that are affected by the decisions or have an interest in the decision. In addition to direct users of the resource (e.g. recreational fishers or commercial fishers), a PM group might include participants who are not actively fishing. For instance, academics or agency biologists may have useful knowledge about the managed species or system and adjacent business owners (e.g. tackle shop owners) or landowners might have opinions about resource access and potential management actions that are preferential to different user groups (e.g. through gear, season, or access limits). Our processes have also commonly included representatives from environmental non-governmental organizations.

Not everyone with an interest in the issue can be involved in a PM project. The PM projects we have conducted usually range from groups of 12–30 participants, but some efforts have included larger groups (e.g. Feeney et al. 2019). Therefore, it is often necessary to select participants so that logistics and expenses are manageable with the resources available to the PM team. Generally, there are two approaches to selecting participants. In some cases, for example, working with an existing commission, the participants will be determined in a manner that is outside the influence of the PM leadership team. However, in most of the processes that we have been involved in, the client determines the participants, ideally in consultation with the leadership team. In other cases, the participants will be selected as part of the PM process by the leadership team through a scoping process (e.g. Miller et al. 2010).

Mixes of these two endpoints can also occur. More contentious cases may warrant the use of more formal criteria for selecting experts or balancing representation across interests.

Qualities for effective participants in a PM process include (1) the ability to commit the time, (2) being representative of a broader constituency, (3) being knowledgeable about the issue, and (4) good attitude about participatory work. PM processes often involve many meetings over the course of multiple years; thus, participants must be sufficiently engaged such that they continue through the entire process. Ideally, participants should be involved from the beginning and repeatedly engaged throughout the project. Therefore, one aspect of participant selection is the ability of each participant to engage in as much of the process as possible, ideally, the entire process. However, it is likely that some participants will leave during the process, and plans should be made at the beginning of the PM process for whether those participants will be replaced, and if so, how that will be done. Participants who have experience communicating for a constituency within their community (e.g. an officer of a fishing club or a regional biologist who has a history of interacting with constituents) may be more likely to be able to represent the broader interests or share the knowledge base of their group. Also, including these people as part of the PM group may lead to a higher degree of buy-in from that group. For the model to be useful, it needs to represent key aspects of the recreational fishery system (e.g. population dynamics, management, fisher behaviour). These key aspects will be different for each problem, but well-informed, knowledgeable participants will be able to help guide the development of the model to ensure that the key considerations are included in its development. Lastly, PM involves working within a group setting. Participants who are able to work within a group will be most effective.

Large groups of people rarely share the same worldviews, which consist of a range of perspectives, goals, values, and desirable outcomes (Lempert and Turner 2021). Including underrepresented and historically marginalized groups in PM is a topic of increasing importance for the legitimacy of PM processes (Hämäläinen et al. 2020). It is also important to ensure that a wide range of worldviews are present among the participants so that as wide a range as possible of potential problems and solutions is considered. However, engaging groups that have been historically excluded can be challenging because they may not have sufficient trust in the leadership team to participate in the process. Therefore, additional outreach may be necessary to recruit members from historically excluded groups.

Selecting individual participants will differ depending on the problem of interest. One useful technique to select participants that we have used on multiple occasions when the leadership team is tasked with choosing the participants is to conduct successive interviews. During these interviews, the leadership team should attempt to learn about pressing concerns, desired outcomes, as well as who else could well represent varied interests within a PM process. These kinds of interviews should start with a core group of people knowledgeable about the problem. For instance, a question asked of each interviewee could be: To whom do you recommend we talk? If the leadership team feels that certain groups are not represented in the interview process, they should strive to contact people with closer connections to those groups who may be able to open avenues of communication. In our experience, a concise list of names of people who would likely be effective participants in a PM process will materialize. Note that new constituencies may come to the attention of the leadership team during these interviews, and the composition of the participants may need to be reconsidered (e.g. adding more participants or changing the number of participants from distinct groups).

Lastly, determining how decision makers will engage with the PM project is important. Some options include having the decision makers as full members of the participant group, having the decision makers as ex-officio or non-voting members of the group, or not including decision makers in the group. We have had success with all three approaches. Not including decision makers in the group can run the risk of them being unprepared to act on the recommendations that come out of the PM process.

2.3 Planning the Process

A PM process should be designed to achieve specified goals, fit within an agreed-upon timeline, and accommodate the constraints of participants and the modelling team. To be successful, a PM process must develop and foster an environment of mutual respect; doing so should result in a range of benefits (Box 1). In particular, these process questions are worth addressing at the planning stage:

- *Who can contribute and at what stages?* Will only invited participants be allowed to contribute, or will there be avenues for broader engagement (e.g. public forums, presentations at commission meetings)?
- *Are there barriers to participation for some groups or individuals?* Can these barriers be addressed by altering meeting times or locations? Will some people need to take time off from their normal jobs to participate? If so, can or should they be compensated?
- *Can the meetings be virtual, or hybrid?* If virtual, do all the participants have the skills, appropriate time and space, and access to technology such that they can meaningfully participate? We note that in our experience, in-person engagement is better, particularly for fostering mutual respect, but that virtual processes can be effective for at least some tasks, such as initial inquiries about available data and sharing preliminary findings. Although a virtual process can sometimes achieve selected goals, the participants and PM leadership team need to be prepared to commit extra efforts to making it work, as the virtual environment may not ensure equitable access.
- *How will the leadership team interact with the participants?* Successful PM processes empower the participants as collaborators, helping to solve a problem. The modelling team will often have more power than many of the participants in the process due to their technical knowledge and experience with PM processes

(Hämäläinen et al. 2020). Additionally, there will likely be power imbalances among the participants. Therefore, the process should be designed to work around these differences in power to effectively engage all participants. An experienced facilitation team is vital to address these issues. All participants will have something to contribute, and some will be more vocal than others; missing the perspectives of less vocal participants will generally mean a less successful process. Facilitators may want to carefully plan how they will elicit information from participants. Options include working with individuals or small or large groups, as well as determining if content will be shared anonymously or be identified to individual participants. These elicitation choices can affect whose voice is magnified and to what extent participants are able to reflect on other opinions and potentially adjust following discussions (Runge et al. 2011; Morrison et al. 2016).

- *What is the endpoint of the process or phase?* A key part of the planning is deciding how the group will make decisions, which may include defining how "consensus" will be identified (e.g. Miller et al. 2010; Robinson et al. 2021). Though achieving unanimity among the participants is ideal, it may not be possible. For example, in a project on king mackerel in the south Atlantic region of the United States of America, we defined consensus as a 75% supermajority (Miller et al. 2010). In practice, this supermajority means that a wide range of interests need to come together to put forward any group decision. Because the king mackerel participants developed a strong working relationship, the final recommendations packet was unanimously adopted (Miller et al. 2010). Consensus is important because the goal of the process is to find options with which everyone can live. If a simple majority decision-making process is used for the PM process, many interests will likely not be met.
- *What will the final product be and how will it be produced?* Most commonly in our experience, the final product was a written report to the decision makers that described the participants, process, and the recommendations. The reports were mainly written by the leadership team and were usually vetted by all the participants before being finalized. Sometimes the document is co-produced by the modellers, facilitators, and the client (decision maker). It is wise to agree upon the responsibility for reporting at the onset of the project and to determine how the participants' efforts will be recognized.

Box 1: Potential Benefits and Challenges of Participatory Modelling

Benefits

- Buy-in
  - People want their concerns to be heard
  - Sense of ownership in process could lead to stewardship

(continued)

Box 1 (continued)

- Improve data organization, familiarity, use
  - Identify and add data that were not previously being used by management
- Evaluation of options as a shared experience
  - Adds process transparency, documented record, defensible
  - Able to identify impediments to implementation
- Prioritizing information needs
  - Identifying key uncertainties that influence policy performance
- Education
  - For example, increase awareness of risks of moving species among systems
  - Benefits to other projects (through building trust)

Challenges

- Time (including scheduling, event preparation, logistics)
- Cost
- Expertise
- Technical jargon
- Status quo inertia (people/authority/standards) or resistance to relinquishing authority
- Setting realistic expectations
- Quantity and interpretability of potential model outputs (not just producing a "basket of consequences")
- Shifting composition of participants (retirements, attrition, or recruitment of new participants by current participants—"stacking the deck")
- Assessing social learning and how participants determine value from participatory modelling

PM processes can range in duration from months to years, and it is important to be realistic from the outset about how long the process will last. In some cases, either the amount of funding available or the time available prior to a policy change can drive the timeline for PM. To the extent possible, the timeline for PM should align with the timing of the desired outcome of the process. It is worth bearing in mind that the degree of conflict among participants can influence the time needed for the PM group to effectively begin working together. For Lake Erie walleye fisheries, where there was a long history of conflict between recreational and commercial fishers, the PM process that resulted in a recommended harvest control rule took four years and twelve meetings (Jones et al. 2016). Trying to rush a PM process where there is substantial conflict can backfire. Once the focus, logistics, and potential membership of the process have been planned, then PM shifts to an implementation phase where facilitation, communication, and technical skills become more critical for fostering an environment of mutual respect and trust (Feeney et al. 2019).

3 Facilitation and Communication

Facilitation and communication are critical because (i) the participants need to feel that their point of view is being acknowledged and reflected in the analysis, and (ii) the practical objective of the process is very technical and challenging: to construct a quantitative model that represents the collective understanding of the system of interest. Quantitative modelling is a highly specialized skill, and not one that most participants will possess; however, a beneficial aspect of PM is that it can make assumptions more explicit when the process is a shared experience among the participants. For instance, if a participant has a hypothesis that can be explored with a model (e.g. is recreational fishing intensity responsive to sport fish density?), then that participant may be more likely to help modellers and managers think about how the system could operate at other steps of the effort. Therefore, effective facilitation and communication are keys to a successful PM process that achieves its goals (Table 1).

An environment of mutual respect is developed through transparency, clear communication, and respectful interactions. Having trained facilitators as part of the leadership team is vital. Ideally, the facilitators and the modellers have clear and continued communication—this ensures that the respect for the process cultivated by good facilitation helps to engender trust in the modellers as well, so that expectations and what can actually be modelled are not on vastly different tracks. If the modellers and the facilitators are separate members of the leadership team, it is crucial that they are comfortable working together and are respectful of the importance of both of their contributions. We encourage facilitators and modellers to gain some experience working together prior to engaging with the leadership and participants, even if only through the initial problem scoping. A related potential issue is that the modellers may already have substantial experience on the problem being addressed through PM, and they may already have developed some models or made management recommendations. In such situations, having the facilitators be neutral parties that are not perceived to have a stake in the outcome of the PM process can be beneficial.

Effective technical facilitation is a skill and an art. We emphasize "technical facilitation" here because the facilitators for a PM process will need to have a reasonable understanding of the technical issues surrounding the development of a quantitative model, even if they will not be responsible for the construction of the actual model (Fig. 1). Otherwise, they will not know the right questions to ask and will be too dependent on the modellers for guiding the conversation. So, a key skill for a PM process facilitator is a working knowledge of simulation modelling, and preferably also of the ecological system being modelled. For recreational fisheries,

Table 1 Summary of participatory modelling (coproducing models for scenario forecasting can improve the decision-making process and lead to indirect benefits, but several considerations are worthwhile at the start)

Questions	Considerations
What?	Participatory modelling is the coproduction of scenario forecasting tools that organize what is known and unknown for a decision process in order to predict potential outcomes of management alternatives to facilitate an examination of the degree to which objectives may be met, the severity of trade-offs, and the decision sensitivity to information quantity and quality.
Why?	Identify implementable options, potential magnitude of uncertainty. (a) Improve decision making (i.e. management outcomes) (b) Educate and engage participants such that people feel heard and want to remain engaged. (c) Identify future applied research needs.
How?	Traditionally in-person workshop-style meetings, however we expect virtual meetings and online surveys to become more routinely used. A facilitated participatory modelling process can engage diverse constituents but requires skilled personnel in both leading groups and technical applications. Participants may have personal preferences for format and anonymous contributions.
When?	When identifying objectives and corresponding performance metrics (i.e. model outputs)—objectives drive the remainder of the process. When identifying management alternatives to evaluate with the model—user groups can help identify options that will be difficult to implement. When identifying model assumptions—models help reveal assumptions so that stakeholders will share a common set of assumptions. When confronting anticipated trade-offs—co-evaluation of trade-offs can foster discussions about *acceptable* performance.
Whom?	Deciding whom to include is one of the most difficult areas to offer specific transferable guidance. Generally speaking for recreational fisheries management, interested participants are likely to include those individuals who are directly catching or harvesting fish. However, those individuals could be represented by spokespersons, club officials, or fish biologists and managers. Likewise, other potential participants, such as academic researchers, marina operators, or property owners (e.g. granting access rights) may have influential preferences or knowledge to be shared. Further, other participants with more indirect connections to a recreational fishery may be able to inform discussions about potential scenarios to explore or when assumptions are necessary. For instance, questions about other sources of predation (e.g. cormorant consumption of fishery target species as an unknown source of mortality), barriers to fish migration (e.g. potential consequences of dam removal), or operation of other fishery sectors (e.g. historical allocation of quota between recreational and commercial fishers) could help identify important information gaps.
Where?	Proximity to participants Neutral site; accessibility (space and time); cost; sometimes separate meetings? Space for "large" group and break-out groups Space conducive to discussions and collecting information (e.g. flip charts, projectors)

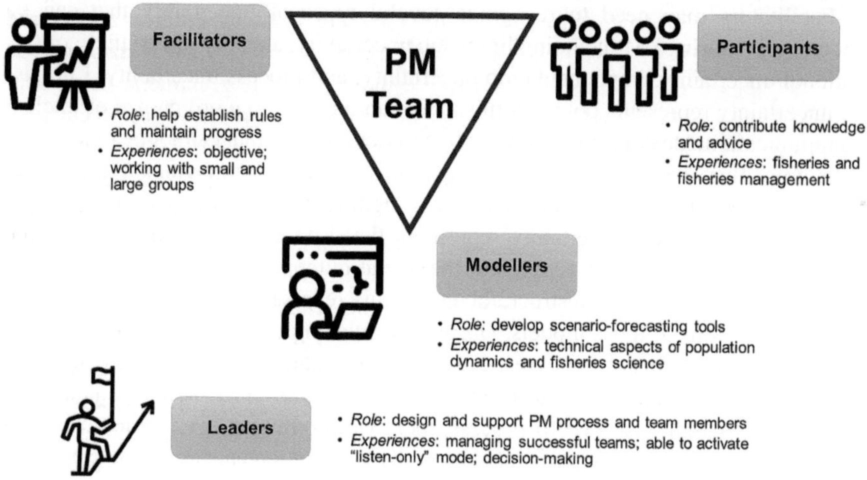

Fig. 1 Diagram of interactions for effective participatory modelling (pm) processes for recreational fisheries

that ecological knowledge base might include understanding the dynamics of fish recruitment, growth, and maturity, and fishing gear selectivity. Whether facilitators are from an outside group or members of the modelling team, their duty is to remain loyal to the process rather than to a particular interest group or favoured policy option, and they should also be perceived as independent of the outcome of the process (i.e. they do not have a stake in the ultimate decisions). In the case of Lake Erie walleye, facilitation and modelling were led by the same group, from the Quantitative Fisheries Center at Michigan State University. Institutions like the Quantitative Fisheries Center are well suited to this role, having been established to provide independent advice on fisheries science management at arm's length from Great Lakes fishery agencies. The Quantitative Fisheries Center facilitators and modellers worked closely together, with overlap in personnel for these two roles, which ensured facilitation focused on the technical needs of the PM process.

Good facilitators also need to have a communication style that draws out and organizes ideas from the participants. This means they need to be effective, active listeners and be able to translate sometimes arcane thoughts from participants into something that reflects what the participant was trying to say, to reframe it in a way that is accessible to the rest of the group. For example, a scientist participant may ask a technical question that is well beyond the comprehension of many other participants, and the facilitators need to "translate" the question into one that is more accessible for the whole group. On the other hand, a fisher may comment on some aspect of her fishing experience that is only peripherally relevant to the focus of discussion, and a good facilitator will find a way to make the comment seem relevant, thereby demonstrating respect for the participant's point of view.

Facilitators may need to recognize several types of uncertainty that can be encountered during PM including linguistic uncertainty, aleatory uncertainty, observational uncertainty, implementation uncertainty, and process uncertainty. Linguistic uncertainty represents potential misunderstanding due to word choice or unclear communication; aleatory uncertainty refers to randomness that cannot be reduced by collecting more data, observational uncertainty represents uncertainty due to our inability to exactly measure the state of a system (e.g. number of fish in a population); implementation uncertainty represents the difference between the intended outcome of a management action and the actual outcome; and process uncertainty includes alternative forms of structural uncertainty represented as different models or parameterizations of those models. During PM meetings, facilitators may especially need to attempt to resolve linguistic uncertainty where different participants may have alternative views of what a phrase means (Regan et al. 2002; Kujala et al. 2012). For example, in one PM process, there was confusion over what was meant by "fishery closure". Some participants initially thought it meant complete closure of the fishery, though others thought it meant closure after the catch quota had been reached (Feeney et al. 2019). Fortunately, linguistic uncertainty is generally reducible during PM via recognition and additional communication. Unfortunately, linguistic uncertainty is only one of several forms of uncertainty that may be encountered during a PM process (Runge et al. 2011; Irwin and Conroy 2013; Rounsevell et al. 2021). Other forms of uncertainty can more directly be incorporated into the operating model (Fig. 2).

Like any engagement process, it is important to distribute a written record of each meeting that describes the issues discussed and, most importantly, documents any decisions. Meeting minutes can help show participants that their ideas have been heard and can be used to show that a topic has already been covered when a participant tries to re-introduce a specific issue that has already been discussed.

In summary, effective facilitation is critical to a successful PM process. Facilitators bear a burden to keep PM moving forward so that participants can see that progress is being made. At the same time, effective facilitation should help participants to better appreciate contrasting points of view and keep them focused on salient matters. Good facilitation can help navigate controversial and complex issues and build consensus, even when it might seem elusive at the start of the process.

4 Collaborative Model Development

In the processes we have led, the goal has been to develop a model that allows the evaluation of the consequences of different management options in terms of their effect on meeting fishery objectives (Fig. 2). The model should be designed to accommodate the primary management actions the participants want to consider, the outcomes that are important to them, and incorporate the available information to the degree possible. The common components of our models have included population dynamics, fishery dynamics, and effects of management on the fishers and the

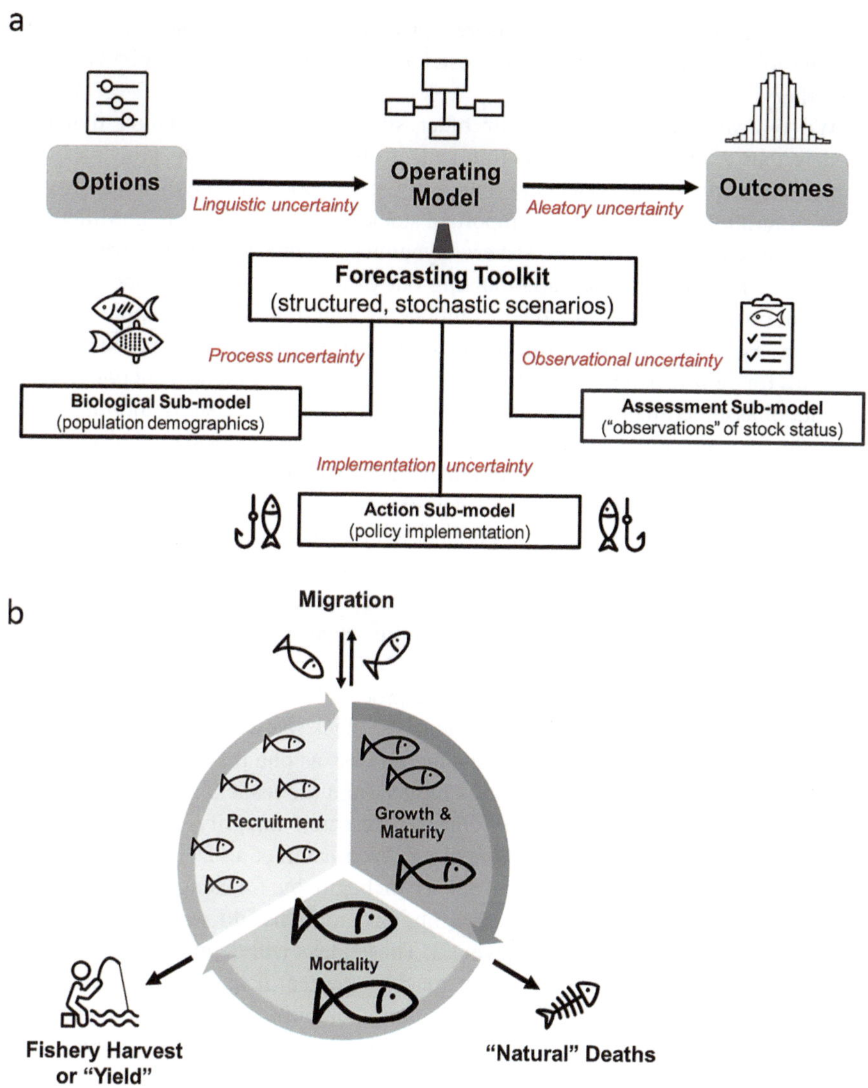

Fig. 2 Conceptual model (**a**) of the interactions among the population, fishery, and management submodels, and an example conceptual model (**b**) showing the population dynamics

stock. The population dynamics submodel describes how the abundance and biomass of the stock change over time and are affected by fishing and other drivers. The fishery dynamics submodel describes how fishers respond to management and the resource. The management submodel can include things like data collection, assessment, and how those activities affect decisions (if the management model includes decisions that change based on an observed state of the resource). The operating model produces performance measures that are used to compare management options to determine which are the best for achieving the suite of objectives for the fishery.

We typically use size-, age-, or other stage-structured population dynamics models that explicitly include recruitment, growth, and mortality processes with fisheries as explicit components to represent the fish population dynamics. In some cases, migration has also been an important stock dynamics process (e.g. Miller et al. 2010; Jones et al. 2016). The fishery dynamics submodel and the management submodels have been quite varied among our examples to meet the specific goals of each project. These submodel components can range from simple to quite complex. For example, in a project we conducted on recreational and commercial fisheries for yellow perch (*Perca flavescens*) in Lake Michigan, the fishery dynamics submodel assumed that the fishery would achieve the target fishing mortality rates from the management submodel (Irwin et al. 2008). The model should also include how the management options affect either the population (e.g. stocking or habitat enhancement) or fishery dynamics (e.g. fisher participation, and fish retention probabilities). We use population models over agent-based models due to their lower computational requirements and the historical use of population models in fisheries management. However, in cases where interactions among individuals or groups are thought to be an important feature, agent-based models may be more appropriate (e.g. Carr-Harris 2021).

For PM to be effective and impactful, the process for model development must be seen as collaborative, including frequent feedback between the modellers and participants. All participants must view the process as one they are contributing to, even when their modelling expertise is limited, and the leadership team must show integrity in working with the participants. Research integrity involves honesty, truthfulness, and transparency in modelling (Weinbaum et al. 2019). In our experience, an effective way to achieve this is to begin by focusing on management objectives and alternative management options (i.e. what do we want to achieve, and how do we want to try to achieve it?). These ideas will define the bounds of the model and can be articulated without much expertise in quantitative methods or fishery dynamic processes. Thus, a discussion of objectives and options can be inclusive, which is key to creating a collaborative environment. Often these discussions generate sets of objectives and options that include items that are unlikely to be modelled, but they capture the range of values held by the participants and identify management choices that are at the front of mind for at least some participants, even if the option may be unlikely to address the issue at hand. It can also be helpful to remind participants that the role of the model is to help describe plausible outcomes that relate to their stated objectives.

Objectives are often quite difficult for the participants to clearly define, but clearly defined objectives are key to constructing a useful model. One way to help participants think of objectives is by asking how they would tell if fishery management were successful (i.e. what would success look like?). This can be a useful starting point for developing objectives. It can also be helpful to start with a set of example objectives. A quite common trade-off in recreational fisheries is to want elevated levels of catch and an abundant fish stock. When exploring objectives, one can start with these concepts and then branch into other potential objectives. For example, are fishers interested in high catch rates rather than the retention of fish? Is the catch of trophy-sized fish an important aspect of the fisher experience? Considerations such as these will help guide model development because they point to performance measures that the model must be able to produce to speak to these objectives.

Identifying the options that the participants want to consider is also crucial to determine the model that needs to be built. For example, if participants want to evaluate spatial management measures (e.g. Miller et al. 2010), then a model with some spatial structure will likely be required. Similarly, if participants wish to evaluate harvest policies that depend on some aspect of the state of the fishery, the model needs to be able to quantify this state (e.g. a model that can represent a state-dependent biomass control rule) within the management submodel.

Once there is agreement on the options and objectives, the process of constructing the model can begin. In essence, the task is to construct a model that allows forecasting of the consequences of each management option, in terms of its effect on achieving, or not, the stated objectives. It is important that the participants believe they are working with the leadership team to develop the model de novo, which facilitates buy-in to the process and its consequences. Modellers will have some ideas for the model a priori, but in our experience, it is best if a preliminary model does not already exist when the process starts. This allows the process to determine the necessary components and how they are structured. Such an approach has also been successful in other contexts (Stern 2021). To develop these kinds of models requires a modelling team that has experience developing models for the specific PM problem and a desire to work with the participants to find creative ways to include new elements.

For some issues, models already exist that have been used to inform management, and for practical reasons, it might be preferable not to have to "reinvent the wheel". This was the case for Lake Erie walleye, where the model used to evaluate future management options was strongly informed by stock assessment models that existed prior to the start of the PM process. Several important refinements to this model were made during the PM process, which helped build trust, and a group of third-party experts was invited to review the revised models and then report on their findings to the process participants. This group of third-party experts comprised individuals recommended by the process participants themselves, which further facilitated trust building in the model coproduction process.

As noted earlier, PM processes usually require months or years to complete, although specific timelines may be driven by problem complexity and familiarity of the participants with the process and each other. Generally, once the conceptual model has been defined, it takes several meetings to construct a working numerical model that generates the forecasts the participants are seeking. At these meetings, ideas for modelling approaches are discussed, knowledge and uncertainty about the processes to be modelled are identified, and alternative modelling strategies are considered. This feedback between the modellers and participants is critical to ensuring that the participants' ideas are being represented in a meaningful way in the model. Between meetings, the modellers do the technical work of coding the model, debugging it, developing tools for processing the outputs to arrive at visualizations that will be informative to the participants, and so on. This stage of the PM process is analogous to the Analytical-Deliberative process described by Dietz and Stern (2008).

The deliberative part of this process is where the "participatory" in PM happens, and below we describe common approaches to encourage discussion between the modellers and participants. There is no common "formula" to describe how this process should occur, but across a range of experiences with PM, we have identified the following ideas, or tips, for ways to increase the likelihood of a successful outcome:

- *Try to use accessible language for the discussions.* The modelling team should commit to open communication to help the participants understand the model. The participants' understanding will vary, but descriptions of modelling decisions should be made in such a way that all participants are aware of the decision being considered, what evidence there is for a specific implementation, and the potential for alternative solutions. Clear communication also benefits from using consistent language when possible and using units familiar to the participants. For example, in our work in the United States of America, we commonly represent fish length in inches and weight in pounds, which are the metrics that the non-scientist participants are usually most familiar with. As with any technical discipline, modelling is full of jargon, and modellers need to be reminded of when they are using jargon excessively or unnecessarily. Misunderstandings among the participants and modelling team are common, and working through these misunderstandings helps to develop knowledge for the whole group.
- *Develop visualizations that help participants understand how the model works and how it uses available data.* Two tools we commonly use in our processes are to develop flow charts of processes in the model to help the participants understand its inner workings and provide plots of model components with the available data alongside them (e.g. Fig. 3). Exploring the data with the participants is an effective way to generate discussions about appropriate use of the data to inform the model and appropriate model structures. One way this can be done is to show the model and data together by process, for example, graph the model's representation of a process to show how it compares with the data.
- *Be honest about the limitations of the model.* At least some of the participants will rely on the modelling team's knowledge of the field for deciding whether the model is trustworthy in the sense of being useful for evaluating the relevant management conditions. We strive to be clear about what we consider best practices for modelling and for talking through the strengths and weaknesses of alternative approaches that we may take. Similarly, it can help to review other studies that help inform components of the model being developed to review their strengths and weaknesses. A common weakness of relying on studies from other recreational fisheries is that they were conducted in different systems, were relatively small-scale, or occurred a long time ago. The consequences of these

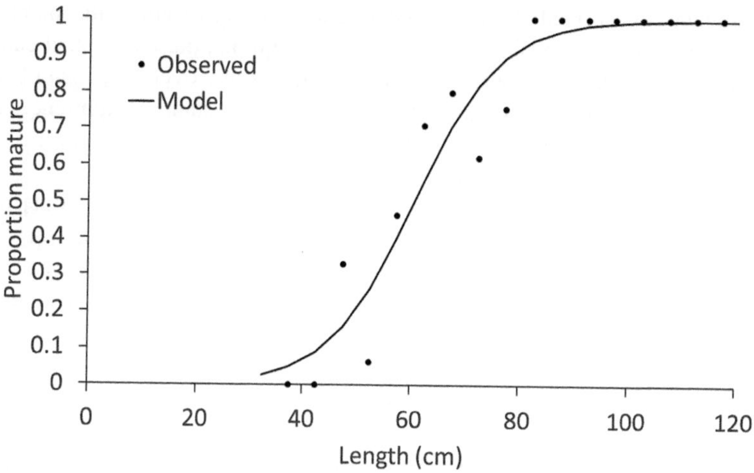

Fig. 3 Example plot showing maturity as a function of size (points) for king mackerel in Northwest Florida with a logistic curve (line) that was used in the model

issues and potential ways to proceed should be discussed with the participants. A good approach is to communicate the modellers' concerns about data and models with the participants to set up an opportunity for discussion. Many participants will find it surprising when a scientist admits openly that they do not know something, and these open admissions can be an opportunity to build trust.

- *Be pragmatic about the trade-off between realism and parsimony.* Participants will often want more realism in the model than the modelling team would choose. Sometimes trust in the model can be engendered by adopting a level of complexity that may not be, strictly speaking, necessary, but will signal to the participants that their ideas are being taken seriously. One example from the king mackerel study (Miller et al. 2010; Ihde et al. 2011) is that the participants were interested in trophy-sized fish as those above a specific weight. They also knew that fish condition varied, such that there was considerable variability in weight-at-length. Our initial model only included a mean length-weight relationship, but we expanded the model to include a distribution of weight-at-length to satisfy the request of the participants. In some cases, the modelling team can also be wrong about the need for additional complexity, which might provide new insights (Walters 1986).
- *Whenever possible, use participant information and insight to inform the model.* Often, participants will question decisions in the model. For example, the effect of tournament fishing was a major uncertainty in the PM project on king mackerel (Miller et al. 2010; Ihde et al. 2011). The PM team used expert judgement of several participants who ran tournaments and their data on tournament participation and weigh-in numbers to reduce the uncertainty on the magnitude of tournament catches, fishing mortality rates, and size selectivity of fishers participating in the tournaments. Similarly, in a PM project on oysters in the Chesapeake Bay,

participants questioned the assumptions that were made about the number of oysters per bushel (1 bushel ≈ 46 L). It turned out that the fishery had changed the use of containers that were used to hold the oysters over time, such that the original numbers were outdated. Participants with substantial experience in the fishery were able to help the PM team develop more appropriate relationships for the number of oysters per bushel used to track the fishery.

- *Acknowledge that some things cannot be included in the model.* Some of the participants' ideas will not be able to be incorporated into the model. Sometimes ideas voiced by the participants are valid, but they cannot be explicitly incorporated into the model due to time, modelling team expertise, data availability, or other reasons. In these cases, the participants can be reminded of which ideas were not fully explored due to procedural or technical constraints. However, if analyses can be done to explicitly consider a participant's concern, that will likely be the best option.
- *Consider the model's sensitivity to uncertain assumptions.* One secret weapon in the modelling toolbox is the use of scenarios to explore critical uncertainties. Even in cases where there are no data, models can be developed to represent hypotheses put forward by the participants (i.e. what if scenarios). These explorations can focus on the extent to which an option's performance differs across hypotheses. If performance is sensitive to assumptions, then it may be an area to target for more research. If the relative performance of options is not different across these scenarios, then this result will help the discussion shift to those uncertainties that do matter.

One common challenge is that the participants will often have specific options that they would like to consider that may be difficult to model. For example, participants may want to consider spatially varying size and bag limits that change during the course of a year. To explicitly represent these types of regulations would require a spatial model with a subannual time step that also includes how size limits and bag limits affect the fishery. Data may not be available to support this structure, and such a model would be much more complicated to construct, which would preclude an easy inclusion of this type of option. The modelling team is then left with the conundrum of how to include this type of option. One approach we have used is bounding experiments. If two options can be developed that serve as endpoints of the options the participants want to consider, then the option the participants want to consider should be between the two endpoints. Therefore, if both endpoints produce satisfactory performance, then the original complex option should also be satisfactory. For example, a project on recreational limits for summer flounder (*Paralichthys dentatus*) involved a proposed regulation that was a slot size limit with the allowance for retention of one fish larger than the upper limit of the slot. The accepted solution was to develop two management models as bounding cases: the first model included a slot limit without retention of fish larger than the upper limit, and the second model only included a minimum size limit.

Our PM efforts have avoided developing utility functions to attempt to summarize the collective objectives of the participants (Irwin et al. 2011). We have done this for two reasons: (1) not all aspects of the decision that are important to the participants can be quantified by the model, and (2) the relative importance of some objectives may be contingent on achieving other objectives. Therefore, a utility function approach may miss important considerations in the decision-making process. Similarly, the focus of our PM projects has been to find options that are *acceptable* to all participants involved in the process, rather than *optimal,* because defining "optimal" can cause the process to become stuck. PM processes, by their nature, involve negotiations among the participants about how to balance achieving a wide range of objectives. Sometimes, explicitly ranking the importance of participants' objectives can become counterproductive in practice even if the approach is theoretically sound.

4.1 Inclusion of Uncertainty

Formally recognizing uncertainty in the model is extremely important for both scientific rigour and fruitful engagement with the participants. Three common types of uncertainty in natural systems include parameter uncertainty, natural variability (i.e. process error), and structural uncertainty, and formally including these types of uncertainty requires different approaches. Parameter uncertainty represents that the model parameters are not exactly known. Parameter uncertainty can be included in models by randomly drawing sets of parameters from a multivariate distribution. Ideally, a multivariate distribution is used because parameters are often correlated. One way to estimate the joint distribution of the parameters is to fit the operating model to data (Punt et al. 2016). However, developing an operating model that also estimates the parameters before conducting simulations may require more time than is available in some PM processes. In these cases, distributions are often chosen or estimated for each parameter individually, which assumes that each parameter is independent. A potential problem with this approach is that some combinations of parameters may produce model dynamics that differ substantially from what has been observed. Natural variability represents the idea that population and human processes are subject to random variation over time. A common example of this type of variability is fish recruitment. Natural variability is typically included by having an additive or multiplicative process random error as part of the dynamic equations of the model. Structural uncertainty represents that there are often multiple hypotheses that can explain observations of the system. Structural uncertainty is most often included in PM by constructing multiple operating models, each of which represents a hypothesis.

The PM processes should always recognize how these uncertainties can affect the judgment of management options. Participants often find discussions of uncertainty vaguely dissatisfying; they seek clear and unequivocal answers to the questions the model is designed to address, and do not like the notion that "given the large uncertainty that exists, anything is possible". To facilitate a discussion of uncertainty that feels more meaningful to participants, it is useful to frame the discussion in terms of risk. In other words, encourage the definition of performance measures in terms like “the level of catch rates below which I would no longer go fishing,” and then present model results such as “the probability, given uncertainty, of catches falling below that level, given the management option being considered”. For the Lake Erie walleye fishery, the key modelling result that enabled consensus on a management recommendation was a plot that showed the trade-off between the risk of an unacceptable outcome for the recreational fishery and one for the commercial fishery (Jones et al. 2016).

4.2 Model Assessment and Validation

A critical aspect of PM is demonstrating that the model represents reality. The model must be vetted by the participants for them to have any faith in its output. Several approaches can be considered, and multiple approaches may be useful in any given circumstance. Model assessment and validation must, however, be handled with humility. By their very nature, the models we build in PM processes are not amenable to validation in the traditional statistical sense because we rarely know the true state of the system to which we could compare the model predictions.

The first approach to model validation is to use the model to describe a time series of the past (i.e. hindcasting). If the model cannot replicate the past, there may be concerns about using it to project the future. Hindcasting can be used for long time series or to describe specific events in the past. In some cases, participants may identify key events in the history of their fishery. If the model can accurately replicate those key events, it will bolster the confidence of the participants to use the model. It can also be helpful to use the model to make short-term forecasts to compare with anticipated new data. For example, because a PM process often extends over several years, new data will become available during the process. One approach we have used is to forecast the next year’s harvest data and compare that to newly acquired information. Lastly, a model can sometimes be fitted to time series of data or other estimates (Punt et al. 2016) to gauge if parameters appear realistic.

Even being able to reconstruct past dynamics is a weak validation approach because, presumably, the goal will usually be to forecast consequences of management options that have not been used in the past, that is, extrapolation. The modellers need not be apologetic about this challenge. The forecasts are best guesses, given our knowledge and uncertainty. Thus, it is important to recognize that the absolute outcome may have substantial uncertainty, but the focus of the modelling efforts is to evaluate whether one option is likely to be better than another for achieving a specific objective. Comparisons to hindcasts are useful but should not be oversold. Forecasting ecological and fishery dynamics is challenging under the best circumstances. Being open and honest about model limitations will help to develop trust with the participants. Specifically, the model is likely to better represent the relative performance of options than to be able to predict specific outcomes.

Also, the modellers should listen closely to concerns that are raised by the participants during this stage. If the model conflicts with the participants' experience and knowledge, it will be difficult for the group to use the model when making recommendations. When participants raise concerns of this nature, the modellers need to think creatively about how participant knowledge might be better integrated into the model or if that knowledge can be used as a part of the validation. These iterative adjustments can be of great benefit to PM, but often, the time required to make these adjustments can be difficult to anticipate. If critical uncertainties are identified that cannot be included in the validation, developing alternative scenarios to test if the performance of potential options depends on those uncertainties is a reasonable way forward.

Lastly, external review can be an important part of building trust in the model (Robinson et al. 2021). As noted earlier, in the PM process for Lake Erie walleye, the participants selected external reviewers to review the model during a multiday workshop in which the participants also participated. This process was helpful for the participants to get a second set of opinions about the model that had been produced during the PM process.

5 Communicating Findings and Completion of the Project

The PM process usually concludes with a set of recommendations that are made to the decision makers. As noted earlier, usually, members of the leadership team are the primary authors of the PM report. However, the report should ideally be vetted by the participant group. Similarly, the final report can include information on the participants and process in a way that acknowledges the contributions of the participants. Standards are beginning to be developed about how to acknowledge work conducted with communities (e.g. Canadian Science Publishing 2022). It is important to work out the details of how the efforts of participants will be acknowledged in reports, presentations, papers, and other communications early in the process.

6 Conclusions

PM can be a powerful tool for addressing recreational fisheries management issues because it is a process that engages with the people affected by management decisions and includes them in the development of potential solutions (e.g. Miller et al. 2010; Ihde et al. 2011; Robinson et al. 2021). Recreational fisheries often involve substantial uncertainty about the current state of the system, appropriate goals, and the effectiveness of potential options for achieving those goals. These large uncertainties can lead to conflict among participants and a sense of frustration for managers. PM is a promising approach for addressing challenges in recreational fisheries because of its potential to develop solutions that have been evaluated thoroughly and have a prominent level of participant support. Care must be taken to engage with the participants, however, for a successful effort. This includes careful planning and execution of the project, and genuine efforts to collaborate with the participants to find solutions to the problems. As described by Sterling et al. (2019), "Finally, knowing when to use PM and when not to, managing expectations, and effectively and equitably addressing conflicts is essential". Although data generating and technical modelling capabilities are rapidly developing and data analysts are working with increasing amounts of data, in our experience, the process by which these data and models are used to support decision making is still critically important for producing useful information and setting the stage for reducing uncertainty for future iterations of the decision.

Acknowledgements We thank the many colleagues, participants, decision makers, and funding agencies who have collaborated with us on participatory modelling projects. This chapter was substantially improved by helpful suggestions from A. Fuller, A. Lynch, K. Pope, and two anonymous reviewers. Any use of trade, firm, or product names is for descriptive purposes only and does not imply endorsement by the U.S. Government. This is contribution #2024-01 from the Quantitative Fisheries Center, Michigan State University.

References

Arlinghaus R, Abbott JK, Fenichel EP, Carpenter SR, Hunt LM, Alós J, Klefoth T, Cooke SJ, Hilborn R, Jensen OP, Wilberg MJ, Post JR, Manfredo MJ (2019) Governing the recreational dimension of global fisheries. P Natl Acad Sci USA 116:5209–5213

Canadian Science Publishing (2022) Author guidelines: How to report community-engaged research. https://cdnsciencepub.com/authors-and-reviewers/community-engaged-research. Accessed 7/28/2025

Carr-Harris A (2021) Recreational fluke MSE economic modeling overview. Mid-Atlantic fishery management council report. https://static1.squarespace.com/static/511cdc7fe4b00307a2628ac6/t/613a20985bf43548f0b82803/1631199385068/Economic+modeling+overview+2.pdf. Accessed 25 Jun 2023

Crevier LP, Parrott L (2019) Synergy between adaptive management and participatory modelling: the two processes as interconnected spirals. Ecol Inform 53:100982

Dietz T, Stern PC (eds) (2008) Public participation in environmental assessment and decision making. Panel on public participation in environmental assessment and decision making. The National Academies Press, Washington DC

Feeney RG, Boelke DV, Deroba JJ, Gaichas S, Irwin BJ, Lee M (2019) Integrating management strategy evaluation into fisheries management: advancing best practices for stakeholder inclusion based on an MSE for Northeast US Atlantic herring. Can J Fish Aquat Sci 76(7):1103–1111

Fuller AK, Decker DJ, Schiavone MV, Forstchen AB (2020) Ratcheting up rigor in wildlife management decision making. Wildlife Soc B 44(1):29–41

Hämäläinen RP, Miliszewska I, Voinov A (2020) Leadership in participatory modelling-is there a need for it? Environ Model Softw 133:104834

Hare M (2011) Forms of participatory modelling and its potential for widespread adoption in the water sector. Environ Policy Gov 21:386–402

Herrera de Leon HJ, Kopainsky B (2019) Do you bend or break? System dynamics in resilience planning for food security. Syst Dynam Rev 35:283–379

Hovmand JF (2014) Community based system dynamics. Springer, New York

Ihde TF, Wilberg MJ, Secor DH, Miller TJ (2011) FishSmart: harnessing the knowledge of stakeholders to enhance U.S. marine recreational fisheries with application to the Atlantic king mackerel fishery. In: Beard TD, Arlinghaus R, Sutton SG (eds) The angler in the environment: social, economic, biological, and ethical dimensions. American Fisheries Society, symposium 75, Bethesda, Maryland, pp 75–93

Irwin BJ, Conroy MJ (2013) Consideration of reference points for the management of renewable resources under an adaptive management paradigm. Environ Conserv 40(4):302–309

Irwin BJ, Wilberg MJ, Bence JR, Jones ML (2008) Evaluating alternative harvest policies for yellow perch in southern Lake Michigan. Fish Res 94:267–281

Irwin BJ, Wilberg MJ, Jones ML, Bence JR (2011) Applying structured decision making to recreational fisheries management. Fisheries 36(3):113–122

Jones ML, Catalano MJ, Peterson LK (2016) Stakeholder-centered development of a harvest control rule for Lake Erie walleye *Sander vitreus*. In: Edwards CTT, Dankel DJ (eds) Management science in fisheries. Routledge, Oxford, UK, pp 163–183

Kujala H, Burgman MA, Moilanen A (2012) Treatment of uncertainty in conservation under climate change. Conserv Lett 6(2):73–85

Lempert RJ, Turner S (2021) Engaging multiple worldviews with quantitative decision support: a robust decision-making demonstration using the lake model. Risk Anal 41:845–865

Maryland Department of Natural Resources (DNR) (2021) Final report: Oyster Advisory Commission Consensus Recommendations on Oyster Management. https://dnr.maryland.gov/fisheries/Documents/Final_Report_OAC_SB0808_HB0911.pdf. Accessed 19 Jan 2023

Miller TJ, Blair JA, Ihde TF, Jones RM, Secor DH, Wilberg MJ (2010) FishSmart: an innovative role for science in stakeholder-centered approaches to fisheries management. Fisheries 35:424–433

Moallemi EA, de Haan FJ, Hadjikakou M, Khatami S, Malekpour S, Smajgl A, Smith MS, Voinov A, Bandari R, Lamichhane P, Miller KK, Nicholson E, Novalia W, Ritchie EG, Rojas AM, Shaikh MA, Szetey K, Bryan BA (2021) Evaluating participatory modeling methods for co-creating pathways to sustainability. Earth's Future 9:e2020EF001843 https://doi.org/10.1029/2020EF001843

Morrison WE, Nelson MW, Griffis RB, Hare JA (2016) Methodology for assessing the vulnerability of marine and anadromous fish stocks in a changing climate. Fisheries 41:407–409

Punt AE, Butterworth DS, de Moor CL, De Oliveira JA, Haddon M (2016) Management strategy evaluation: best practices. Fish Fish 17(2):303–334

Regan HM, Colyvan M, Burgman MA (2002) A taxonomy and treatment of uncertainty for ecology and conservation biology. Ecol Appl 12(2):618–628

Robinson KF, Fuller AK (2017) Participatory modeling and structured decision making. In: Gray S, Paolisso M, Jordan R, Gray S (eds) Environmental modeling with stakeholders: theory, methods, and applications. Springer International Publishing, Cham, pp 83–101

Robinson KF, Fuller AK, Jones ML (2021) Using structured decision making to incorporate ecological and social values into harvest decisions: case studies of white-tailed deer and walleye. In: Pope KL, Powell LA (eds) Harvest of fish and wildlife: new paradigms for sustainable management. CRC Press, Boca Raton, pp 131–146

Röckmann C, Ulrich C, Dreyer M, Bell E, Borodzicz E, Haapasaari P, Hauge KH, Howell D, Mäntyniemi S, Miller D, Tserpes G, Pastoors M (2012) The added value of participatory modelling in fisheries management—What has been learnt? Marine Policy 36:1072–85

Rounsevell MD, Arneth A, Brown C, Cheung WW, Gimenez O, Holman I, Leadley P, Luján C, Mahevas S, Marechaux I, Pélissier R, Verburg PH, Vieilledent G, Wintle BA, Shin Y-J (2021) Identifying uncertainties in scenarios and models of socio-ecological systems in support of decision-making. One Earth 4(7):967–985

Runge MC, Converse SJ, Lyons JE (2011) Which uncertainty? Using expert elicitation and expected value of information to design an adaptive program. Biol Conserv 144(4):1214–1223

Smith ADM, Sainsbury KJ, Stevens RA (1999) Implementing effective fisheries-management systems-management strategy evaluation and the Australian partnership approach. ICES J Mar Sci 56:967–979

Sterling EJ, Zellner M, Jenni KE, Leong K, Glynn PD, BenDor TK, Bommel P, Hubacek K, Jetter AJ, Jordan R, Olabisi LS, Paolisso M, Gray S (2019) Try, try again: lessons learned from success and failure in participatory modeling. Elem Sci Anth 7:9

Stern PC (2021) On modeling worldviews in quantitative decision support. Risk Anal 41:870–873

Walters CJ (1986) Adaptive management of renewable resources. MacMillan Pub. Co, New York

Weinbaum C, Landree E, Blumenthal MS, Piquado T, Gutierrez CI (2019) Ethics in scientific research: an examination of ethical principles and emerging topics. RAND Corporation, Santa Monica

Wilberg MJ, Livings ME, Barkman JS, Morris BT, Robinson JM (2011) Overfishing, disease, habitat loss, and potential extirpation of oysters in upper Chesapeake Bay. Mar Ecol Prog Ser 436:131–144

Wilberg MJ, Wiedenmann JR, Robinson JM (2013) Sustainable exploitation and management of autogenic ecosystem engineers: application to oysters in Chesapeake Bay. Ecol Appl 23(4): 766–776

Adaptive Management for Recreational Fisheries Decisions in the Face of Uncertainty

Kelly F. Robinson, Robert Arlinghaus, Edward V. Camp, John R. Post, Richard Stedman, and Steven G. Sutton

Abstract Recreational fisheries management requires making decisions that consider social, economic, and ecological dimensions. The feedbacks among ecological changes and social responses by fishers create complexity and uncertainty. However, uncertainty can impede the integration of ecological and social dimensions, particularly when considering human behavioural responses to management decisions and ecological changes. One of the few ways to reduce these uncertainties is via experimentation at the system level. Adaptive management (and the larger umbrella of decision analysis) provides a framework to implement purposeful management experiments in a structured manner to learn through an iterative decision process, thereby allowing for the reduction in social and ecological uncertainties in response to planned policy interventions. We discuss the theory of adaptive management and conditions for which it is appropriate, as well as the benefits and costs of applying active versus passive adaptive management to reduce social and ecological

K. F. Robinson (✉)
U.S. Geological Survey, Georgia Cooperative Fish and Wildlife Research Unit, Warnell School of Forestry and Natural Resources, University of Georgia, Athens, GA, USA
e-mail: kfrobinson@uga.edu

R. Arlinghaus
Department of Fish Biology, Fisheries and Aquaculture, Leibniz Institute of Freshwater Ecology and Inland Fisheries, Berlin, Germany

Division of Integrative Fisheries Management, Humboldt-Universität zu Berlin, Berlin, Germany

E. V. Camp
School of Forest, Fisheries, and Geomatics Sciences, University of Florida, Gainesville, FL, USA

J. R. Post
Department of Biological Sciences, University of Calgary, Calgary, AB, Canada

R. Stedman
Center for Conservation Social Sciences, Department of Natural Resources and the Environment, Cornell University, Ithaca, NY, USA

S. G. Sutton
Community Outreach and Engagement, Atlantic Salmon Federation, Calais, ME, USA

K. L. Pope et al. (eds.), *Understanding Recreational Fishers*, Fish & Fisheries Series 45, https://doi.org/10.1007/978-3-031-99739-6_23

uncertainties for recreational fisheries. We then provide key examples from previous studies on decisions for zoning in the Great Barrier Reef, Australia, stocking decisions that include stakeholder learning in Germany, and decisions for stocking and harvest management in British Columbia, that demonstrate best practices for implementation of adaptive management to reduce uncertainties surrounding human behaviour and other social and ecological objectives related to recreational fisheries.

Keywords Active adaptive management · Decision analysis · Human behaviour · Institutional learning · Management · Social science · Uncertainty

1 Introduction

Making decisions in recreational fisheries management, like most natural resource decisions, requires understanding the potential effects of management actions on the ecosystem and the recreational fishing community alike (Bennett et al. 2017; Robinson et al. 2019). As stakeholders and rightsholders have diverse values and preferences, and there are often multiple ecological concerns in these decisions, decision makers also must account for multiple objectives. Therefore, integration of social and ecological science is important, but this integration remains difficult and is often inadequately executed in natural resource management decisions (Decker et al. 1992; Robinson et al. 2019). In natural resources management, including recreational fisheries management, any multi-objective decision should ideally account for ecological, economic, and social dimensions, as well as associated uncertainties that stem from difficult-to-predict feedbacks among ecological and human systems (Fulton et al. 2011). However, individuals affected by recreational fisheries decisions, and in some cases also policy makers and decision makers, have diverse values, or general preferred end states that influence behaviour (Rokeach 1973, applied to fisheries by van Riper and colleagues, for example, Golebie et al. 2021; Shin et al. 2022; Landon et al. 2026), which often are not clearly articulated, and may conflict with each other (Keeney 2004; Gregory et al. 2012). Stakeholders and rightsholders may also have different understandings of the system (Tyre and Michaels 2011; Williams and Brown 2018; Aminpour et al. 2020), and potentially mistrust of fisheries managers (Dedual et al. 2013; Robinson et al. 2021b), which can affect their responses to management interventions in unforeseen ways (Hunt et al. 2013). Throughout this chapter, we use the term "stakeholder" in the broadest sense to describe interested or affected parties, and in the specific case of recreational fishing, we refer to those who are part of the recreational fishing community. Please see Camp et al. (2026) for a broader discussion of recreational fishing stakeholders.

Decision analysis can provide a framework for structuring natural resources decisions, including differing values, and accounting for uncertainties inherent in human-nature interactions. The field of decision analysis is broad, bringing together concepts of formal model building and facilitation techniques in the service of making a decision (von Winterfeldt and Edwards 1986), and is rooted in the disciplines of economics, psychology, operations research, mathematics, and others

(Runge and Bean 2020). In the realm of natural resources management, structured decision making and adaptive management are the main decision analytic frameworks. In particular, when social or ecological uncertainty hinders the decision choice, one can apply adaptive management to reduce that uncertainty by implementing management strategies at the system level, while learning from responses to management as experiments (Walters and Hilborn 1978; Williams et al. 2007). Applications of adaptive management in which stakeholders are included in the decision-making process allow for incorporating and clarifying multiple objectives of those stakeholders (Keeney 2004; Gregory et al. 2012), identifying a more creative and complete list of management actions that could be implemented (Walters and Holling 1990; Gregory and Keeney 2002), conducting participatory modelling to understand the consequences of potential management actions on stakeholder values and evaluate trade-offs (Robinson and Fuller 2017; Camp et al. 2026; Wilberg et al. 2026), and providing an opportunity to bring together all decision makers in instances where multiple jurisdictions have management authority (Tyre and Michaels 2011; Robinson et al. 2021a, b).

2 Uncertainty and Its Effects on Recreational Fisheries Decisions

Social and ecological uncertainties can impede decisions for recreational fisheries, leading to suboptimal experiences and frustration by all involved. However, decisions for fisheries management inherently involve uncertainty—there will never be complete certainty about an ecological system (Walters 1986; Walters and Holling 1990). Multiple researchers have provided typologies of uncertainty that pertain to decision making (Williams 1997; Peterman and Peters 1998; Regan et al. 2002). Walters (1986) describes three types of uncertainty: noise, parametric uncertainty, and structural uncertainty, suggesting that the latter two are "measurable" and that all three manifest in social-ecological systems. The choice of tools available for decision makers to account for and potentially reduce uncertainty depends on the type of uncertainty that is hindering the decision at hand.

Noise is stochasticity that is irreducible. In recreational fisheries management decisions, examples of noise include environmental variation in dynamic systems, such as random variation in weather or unpredictable human effects on a system (Williams 1997). This can also be driven by non-stationary processes (Milly et al. 2008; Nichols et al. 2011; Schindler and Hilborn 2015). For example, climate change has led to non-stationarity—we cannot use past system states, such as temperature or rainfall regimes, to predict future states (Milly et al. 2008; Tyre and Michaels 2011). Non-stationarity can also manifest socially as changing values and objectives over time (Tyre and Michaels 2011).

Parametric and structural uncertainty *are* reducible through research. They hinder our ability to fully describe a system. These two types of measurable uncertainty are commonly considered in ecological decisions, such as uncertainty about the mortality or production rates of a fish stock (parametric) or the stock–recruitment relationship for a harvested fish population (structural; Smith and Walters 1981). Structural uncertainty can affect social components of fishery management decisions through partial controllability, which includes uncertainty in how humans will behave under particular management actions, sometimes referred to as implementation uncertainty in response to policies (Fulton et al. 2011).

Decision making under uncertainty requires understanding whether uncertainty is reducible through experimentation (i.e. amenable to adaptive management) or irreducible and therefore must simply be accounted for in the decision or through institutional learning (see below). Decision analysis (i.e. structured decision making and adaptive management) provides a framework and associated tools to account for and potentially reduce uncertainties that may hinder these decisions (FAO 2012). In many cases, accounting for uncertainty through sensitivity analyses and analyses of the risks associated with making decisions under different ranges of uncertainty can help clarify the decision (Keeney 2004; Gregory et al. 2012). However, in instances in which an uncertainty changes the preferred decision and decisions are iterative (i.e. can be made more than once), adaptive management can help reduce structural uncertainties and account for noise and non-stationarity (Runge et al. 2011; Tyre and Michaels 2011). A key justification for experimenting at the scale of the entire fishery is that this scale is necessary for understanding how complex adaptive systems respond ecologically and socially (Walters 1986; Arlinghaus et al. 2017).

3 What Is Adaptive Management?

Adaptive management is a focused method to reduce uncertainty inherent in natural resources management decisions. However, the term "adaptive management" has been used in many ways, leading to confusion about the process and what it means in a decision-making context. At its heart, adaptive management uses the iterative nature of a management process to learn and adjust future management actions (Fig. 1; Williams et al. 2007), while also recognizing that in many (most) systems, the best management action cannot be "fixed a priori but must instead be established through sequential reassessment of system states and dynamic relationships" (Walters and Hilborn 1978). However, adaptive management is more than trial and error or monitoring: it is a structured process that requires predictions of outcomes from management actions, experimentation, and application of learning to improve predictions and, therefore, the decision (Allen and Gunderson 2011; Bouwes et al. 2016; Pine et al. 2022). In complex social-ecological systems like recreational fisheries, decision makers can make use of the iterative nature of adaptive management to improve long-term goals for the system, both ecological and social (Martin and Pope 2011; Tyre and Michaels 2011; FAO 2012; Arlinghaus et al. 2017). Freshwater systems are especially good model environments for adaptive management because different regulations can be applied to different

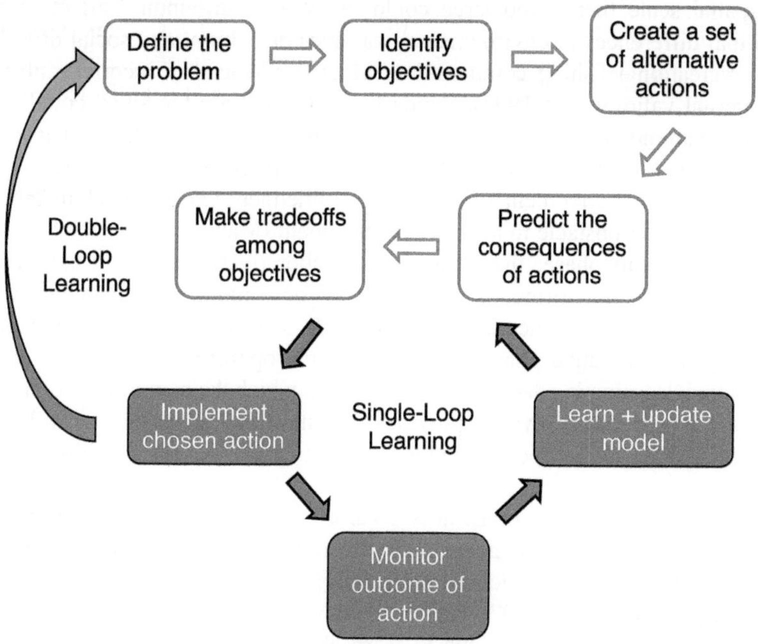

Fig. 1 The double-loop learning process of adaptive management (white boxes with grey borders depict the "set up" phase of adaptive management in which the problem is framed, and models are devised to examine outcomes of management actions; grey boxes are the "iterative" or "single-loop" phase of adaptive management in which the action is implemented, learning occurs and is used to update predictive models, the actions are evaluated, trade-offs are made, and an updated action is implemented). (Modified from Williams et al. 2007)

systems, such as lakes, streams, or watersheds, providing a set of systems in which replication can be achieved.

Adaptive management allows decision makers and included stakeholders to identify the problem's essential elements (e.g. the decision to be made, the values and objectives to be achieved, any constraints to the system, models and data to predict outcomes of actions, uncertainties that should be targeted). For example, structured decision making is often the framework for this setup phase for applications of adaptive management (Fig. 1; Runge 2011; Månsson et al. 2022). Structured decision making is what is known as a "values-based" process (Keeney 1992). In this setup phase, the decision makers and stakeholders first describe any constraints limiting what decisions can be made (e.g. social, political, legal, or financial limits) and articulate their values and preferences related to the decision. Those values are used to craft a set of objectives and performance measures, and actions identified are designed to achieve those objectives (Keeney 1992). In addition, the setup phase requires that the choice of the spatial and temporal scale be agreed upon early in the adaptive management process. This choice determines the scale of experimentation, which could have real effects on the choice of management action. Experimentation at a spatial scale that is too large could introduce confounding effects such as latitudinal differences in environmental variables or cultural and social differences in the recreational fishing communities, which could lead to incorrect inferences (i.e. internal validity, Lee 1993). Similarly, scales that are too small could lead to results that are not applicable beyond the experimental system (i.e. external validity, Lee 1993). Failure to properly address this setup phase of adaptive management, including lack of agreement on objectives, timelines, experimental methods, or alternatives, is an obstacle to adaptive management (Månsson et al. 2022).

After the setup phase, the group can collaboratively design an experimental framework with monitoring of the outcomes, or a monitoring plan that can be implemented to reduce uncertainty and provide data to update predictive models iteratively (i.e. "iterative phase"). In addition, the opportunity for double-loop, or institutional, learning in adaptive management, in which the components of the setup phase are revisited and revised, provides a method for addressing what Tyre and Michaels (2011) term "socially induced uncertainty", a form of non-stationarity (Fig. 1).

Adaptive management comes in two variants—active and passive. Both active and passive adaptive management intend to reduce uncertainty about the system through model creation, prediction, monitoring, and confrontation of the model with new data (Williams et al. 2007). However, active adaptive management implements experimental management actions to learn about which of a suite of alternative hypotheses of the system is correct (e.g. different fish stocking rates or different size limits applied to different lakes with responses of fishers observed; Fig. 2 and Table 1; for examples see Fujitani et al. 2017; Cahill et al. 2022). Passive adaptive

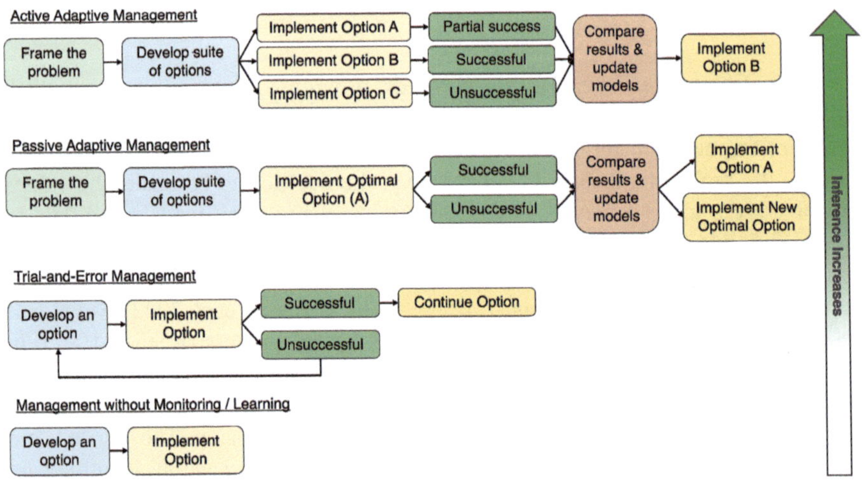

Fig. 2 Approaches to recreational fisheries management and the inference and learning possible with each of the approaches, ranging from no learning with no monitoring to a greater degree of learning with passive and active adaptive management (colours are used to delineate similar steps among the different approaches). (Modified from Allen et al. 2011)

Table 1 Pros and cons of active and passive adaptive management (am), as well as some relevant examples from recreational fisheries management

	Pros	Cons	Relevant examples from recreational fisheries
Active AM	Faster reduction in uncertainty Increased inference space Increased social-ecological realism Can be participatory throughout	High planning and implementation costs Difficult to secure long-term monitoring Replication difficult Potential for risk to humans or sensitive species via experimentation	McCook et al. (2010) Fujitani et al. (2017) Post et al. (2026) Mee et al. (2016) Cahill et al. (2022)
Passive AM	Improvement in parameter estimates for models Useful when there are institutional limitations on experimentation Less potential for risk to environment or humans	Slower reduction in uncertainty Can take more time than active AM Care must be taken to update models and adjust management action—can easily turn in to "trial and error" May confound effects of environment and management (Walters and Holling 1990)	

management seeks to understand the effect of a management action on the system by implementing the model-predicted best option and monitoring to evaluate the underlying hypotheses (Fig. 2 and Table 1; Walters and Hilborn 1978; Smith and Walters 1981; Walters and Holling 1990; Gregory et al. 2006; Allen and Gunderson 2011).

Active adaptive management can lead to faster reduction of uncertainty than passive adaptive management because designed manipulations usually allow for greater contrast in treatments over a shorter time. This experimentation can also potentially incur greater costs to humans or risks to sensitive species or ecosystems (Gregory et al. 2006). Therefore, active adaptive management is appropriate when a resource, such as a recreational fishery, needs management action, and action can be taken, but there is uncertainty about how the resource will respond to the management intervention (Allen and Gunderson 2011; Mee et al. 2016; Cahill et al. 2022). Passive adaptive management can be appropriate if there is interest in improving parameter estimates for the structural model that is used for prediction (e.g. the parameters of a Ricker stock-recruitment model (Walters and Hilborn 1976) or the knowledge of the response degree of a fisher population to a more restrictive harvest limit), or if there are institutional or regulatory limitations or constraints that preclude experimentation (Gregory et al. 2006). No matter which method of adaptive management is preferred or possible, there are also multiple reasons why adaptive management may not be useful or feasible.

4 When Should Adaptive Management Be Applied?

Although the framework of adaptive management provides a clear and structured guide to reducing uncertainty in coupled social-ecological systems, it has been applied infrequently in natural resources management, and even less often is it applied appropriately or well (Walters 1997, 2007; Runge 2020). Gregory et al. (2006) argue that the problems that are often ascribed to adaptive management are based on its being applied indiscriminately. If there is no uncertainty that impedes the decision to be made, or if there is no opportunity for iteration, adaptive management to reduce uncertainty is neither needed nor possible (Runge 2011). Other limitations are the willingness to invest in long-term monitoring of outcomes and the ability to track both ecological and social response variables (Walters and Holling 1990; Månsson et al. 2022). Williams et al. (2007) describe criteria for the application of adaptive management, including a management choice, uncertainty that confounds the decision, an ability to adapt management actions in response to learning, and inclusion of stakeholders in the process (Fig. 3).

Stakeholder inclusion is complex, as decision analytic processes require involving a diversity of stakeholders to represent the breadth of values and concerns while ensuring that the number of participants is not too large for effective group communication and trust (Williams and Brown 2018), in addition to individuals' willingness to commit extended time and resources to the process (Williams and Brown 2018; Pine et al. 2022). Importantly, stakeholder involvement can take many forms, ranging from survey work designed to elicit stakeholder values to full engagement throughout the process (Fujitani et al. 2017; Gregory 2017; Robinson et al. 2019, case studies below). For more description of the forms of stakeholder involvement, please see Camp et al. (2026). Natural resource management decisions that do not fully consider stakeholder values, both conflicting and shared, and needs in the process are less likely to be viewed as acceptable, which can lead to frustration among managers and stakeholders, as well as an inability to implement the desired management actions (Jones et al. 2016; van Riper et al. 2018; Robinson et al. 2021b).

Although adaptive management is intuitively compelling for reducing uncertainties that arise in decision-making processes, decision makers must first assess whether conducting adaptive management is necessary, and whether it is worth the potential costs and risks. As has already been articulated, adaptive management requires that there be some key uncertainty that affects the decision that would be made (Runge et al. 2011). For example, if there is uncertainty about how fish stocking works in the field and how fishers respond to it, only experimenting with stocking and observing the response of fishers can show whether the practice has the intended benefits. However, as described, application of adaptive management can be costly in terms of funding, time, and effort. Therefore, the results of an adaptive management process should be worth the effort, such that decision makers feel as though allocating funds and staff time that likely are scarce towards the adaptive management process will result in greater achievement of their objectives (Gregory et al. 2006).

Fig. 3 Key for determining if adaptive management (AM) is the appropriate framework for the decision at hand. (Modified from Williams et al. 2007)

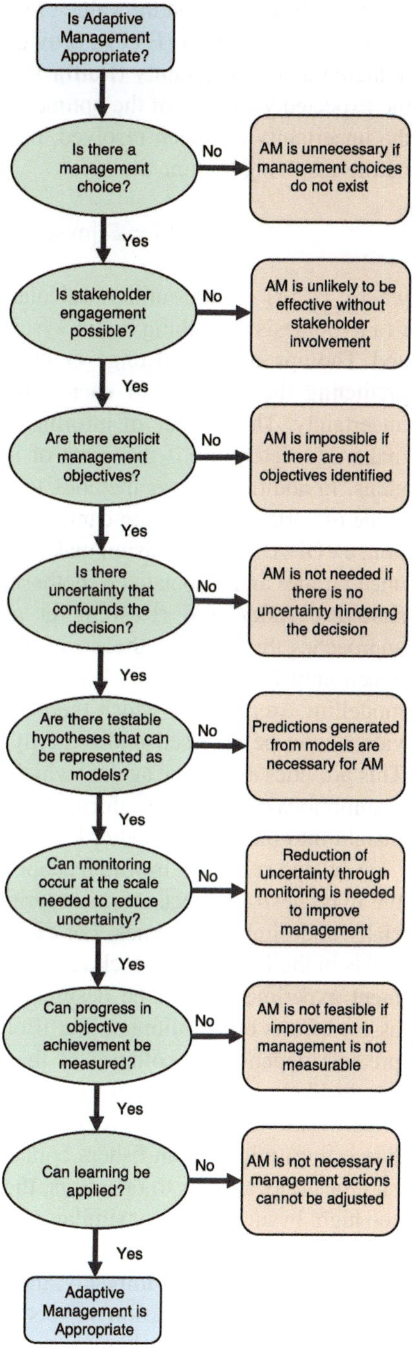

Methods to calculate the value of gathering new data, such as the expected value of perfect information (EVPI), provide a way to quantify the gains to be made by reducing a key uncertainty (Raiffa 1968; Runge et al. 2011). EVPI is calculated as the expected value (*E*) of the optimal action (from a suite of identified actions) after the uncertainty has been resolved, minus the expected value of the optimal action before resolving the uncertainty:

$$\text{EVPI} = E_s[\max{}_a U(a, s)] - \max{}_a E_s[U(a, s)]$$

where *U*(*a,s*) is the utility calculated when action *a* is taken under model *s* (a hypothesis describing how a system works; Raiffa and Schlaiffer 1961; Yokota and Thompson 2004; Runge et al. 2011). The utility scores are calculated by predicting the outcome of each action on each objective under each model of uncertainty. These value of information calculations can then be used by decision makers to determine if the costs of adaptive management outweigh the potential gains. In addition, newer methods to evaluate value of information when expected value of perfect information cannot be calculated (i.e. constructed value of information; cVOI), can help decision makers and stakeholders determine which of a suite of uncertainties and associated hypotheses that appear to be hindering decision making should be addressed (Rushing et al. 2020; Runge et al. 2023). Less formal approaches that are nonetheless useful for assessing the relative value of reducing uncertainty include "omniscient manager" (Walters and Martell 2004) simulation modelling exercises, in which theoretical optimal policy (and its resulting objective values) can be assessed under the hypothetical assumption of perfect knowledge. This provides a baseline against which objective values realized under more realistic assumptions of imperfect information can be compared, with the comparison giving insight into the value of better knowledge.

A related condition in which adaptive management may be inappropriate arises if the hypothesized effects of an experimental management action may include especially undesirable outcomes. For example, imagine a scenario in which uncertainty exists in the form of the stock-recruitment relationship. An active adaptive management experiment could be designed that would more severely harvest stocks to assess effects on recruitment, but if one of the hypothesized models of recruitment predicts depensatory collapse of the fishery at low stock levels, this experiment would likely be too risky for implementation (Walters 1986). Additionally, if the experiment (e.g. strong reduction in fish abundance through experimental harvesting) might mean fishers could leave the fishery to a large degree, incurring significant local costs to operators, the risk of conducting the experiments might be too high. In such a case (or similar, such as involving rare or endangered species), the potential benefits of reducing uncertainty may not be worth the risk of damaging the population or fishery. Ultimately, the choice to implement an adaptive management framework requires decision makers to weigh the risks and costs, but when the value of information is great, adaptive management provides a rigorous means to improve decision making for recreational fisheries and other natural resources management issues.

5 Adaptive Management for Recreational Fisheries

Ecological uncertainty can have both direct and indirect effects on recreational fisheries management objectives. Ecological objectives, such as maintaining population persistence, can be directly affected by ecological uncertainty about the system, whereas social objectives, such as recreational fisher satisfaction with expected harvest, can be indirectly affected by this ecological uncertainty. For example, when making decisions for managing a double-crested cormorant (*Nannopterum auritum*) colony to minimize effects on recreationally important fish like smallmouth bass (*Micropterus dolomieu*) and yellow perch (*Perca flavescens*) in Lakes Huron and Michigan, the effects of uncertainty about the spatial distribution of fish affected objectives for recreational fishing satisfaction (indirectly through fish abundance) and bird population abundance (Tsehaye et al. 2015). Ultimately, although the outcome of this process indicated that this uncertainty could influence decisions for cormorant management, the adaptive management process did not move forward because the order allowing for depredation of cormorants in the United States of America was vacated through litigation. This adaptive management framework is currently the basis for an ongoing decision-making project that addresses decisions for allocations of fish productivity to cormorants and fishers.

In addition to ecological uncertainties, social uncertainty in recreational fisheries management often manifests as uncertainty in recreational fisher behaviour and what drives it (e.g. values). Predicting human behaviour is exceedingly difficult (Heberlein 2012). Humans do not always behave according to their attitudes and beliefs (Stern 2000), which can lead to differences in planned outcomes from management actions, also known as partial controllability (Williams 1997) or implementation uncertainty (Fulton et al. 2011). Rejecting the idea that integration of human concerns *necessarily* introduces "surprise" into the analysis of social-ecological systems is crucially important (Stedman 2016), because accounting for human behavioural responses and learning how they respond ultimately helps us improve system understanding *and reduce* uncertainty (Arlinghaus et al. 2017). The human behavioural kind of uncertainty is often overlooked in decision-making processes. The integration of social science fully into decision analytic processes like adaptive management is rare (for an exception see Fujitani et al. 2017), and if models of behaviour are employed, they typically focus on predicting behavioural intention (e.g. Johnston et al. 2010, 2013, 2015), rather than the behaviour itself (Robinson et al. 2019).

Uncertainties in recreational fisher behaviour can, in turn, be reduced through adaptive management. For example, Martin and Pope (2011) describe a potential adaptive management framework for anglers that could be implemented to understand and better manage angler use of multiple waterbodies in a region. In this suggested application of adaptive management, the authors describe experimental fishing regulations on different waterbodies to discern between two hypotheses of angler behaviour: (i) anglers will not fish, and rather participate in another activity, if their preferred waterbody is unavailable for fishing (i.e. activity substitution; Shelby and Vaske 1991), or (ii) anglers who cannot fish on their preferred waterbody will choose to fish on another nearby waterbody (i.e. resource substitution; Shelby and Vaske 1991). This type of active adaptive management, in which competing hypotheses of angler action with implementation of a waterbody closure are evaluated through active experimentation, can be especially useful for managers who are interested in ensuring that certain waterbodies are not over- or underused with respect to the fishery resource (Martin and Pope 2011). Behavioural uncertainty that leads to partial controllability is not novel to recreational fishing, but few examples exist of implementing adaptive management in such a way to better understand resource user behaviour. Ultimately, reducing structural uncertainties—be they social or ecological—through adaptive management, particularly in a collaborative and participatory context with stakeholders, can lead to identifying management actions that are progressively better at achieving conservation, management, and social objectives. Further examples of adaptive management in recreational fisheries are provided below.

5.1 Institutional Learning (Double-Loop Learning) and Stakeholder Values

Adaptive management can be useful in accounting for uncertainties associated with non-stationarity, as well. This kind of institutional, or double-loop, learning differs from the technical learning that is accomplished in the iterative phase of adaptive management (Fig. 1). Whereas technical learning in the iterative phase is focused on reducible uncertainties (i.e. parametric and structural), in double-loop learning, the elements and structure of the decision itself are revisited to evaluate changes that may have arisen (Williams and Brown 2018). Socially induced uncertainties have an inherently irreducible component owing to the indeterminism of human choice (i.e. humans do not always act rationally with respect to their actual behaviours or with respect to their acceptance of evidence, and at times lack an ability to behave according to their wishes (Heberlein 2012)), and this cannot be wholly accounted for with probability (Tyre and Michaels 2011). Changes in objectives represent non-stationarity in beliefs and attitudes, whether this is at the level of the individuals engaged in the decision process (including changes in objective weights) or at the societal level. Accounting for changes in knowledge, beliefs, norms, and attitudes, as well as ultimately behaviour, through double-loop learning can lessen the possibility of surprises for managers (Tyre and Michaels 2011; Fujitani et al. 2017), and in a participatory framework, can engender trust from stakeholders (Williams and Brown 2018).

Through the double-loop learning process, decision makers and stakeholders can revisit and reevaluate the objectives, the suite of proposed actions, modelling framework, objective weights, and the monitoring procedures (Williams and Brown 2018). Examples of applications of double-loop learning are rare in natural resources management, and we were unable to find any examples in recreational fisheries. However, the double-loop process for adaptive management of commercial horseshoe crab (*Limulus polyphemus*) harvest in Delaware Bay, USA, provides an example of a program that has been long-lived and at times controversial. In this example, the decision for the appropriate level of horseshoe crab harvest depended on the availability of horseshoe crab eggs for consumption by endangered red knot birds (*Calidris canutus rufa*) as they migrate from South America to the Arctic. This adaptive management process was designed to include a double-loop process every four years, so that contentious issues could be revisited (McGowan et al. 2020). As the group learned more through the adaptive framework, they were able to step back to reevaluate the modelling framework for birds and crabs, the harvest management alternatives, the objective function, and the inclusion of sensitive biomedical data in the decision process (McGowan et al. 2020).

The design of a systematic method for double-loop learning is more challenging than the approaches applied for technical learning in the iterative phase (Williams and Brown 2018). The methods to measure and account for institutional uncertainty are less clear than for structural (i.e. reducible) uncertainty, and care must be taken to ensure that the process is accomplished correctly and for the right reasons. For example, premature calls for double-loop learning could be the result of stakeholders or decision makers attempting to manipulate the decision after finding that the results of the adaptive management process are not leading towards their preferred action (McGowan et al. 2020). Proper facilitation techniques, working agreements or terms of reference for the working group, and a commitment to transparency can promote trust among group members and lead to better outcomes for the adaptive management process when working through highly contentious problems (Williams and Brown 2018; McGowan et al. 2020).

6 Case Studies of Adaptive Management for Recreational Fisheries

Although there have been relatively few cases of the implementation of adaptive management in the conservation and management of natural resources, case studies in recreational fisheries management can provide guidance, serving as a roadmap for the reduction of uncertainty in these complex social-ecological systems. These case studies describe both applications of adaptive management with recreational fishers, as well as cases in which decision making could have been improved by the application of adaptive management (Table 2).

Table 2 Description of how the three case studies addressed the key ideas of adaptive management for recreational fisheries described in this chapter

Key ideas of adaptive management	Great barrier reef	Fish stocking in Germany	Rainbow trout stocking in British Columbia
Inclusion of recreational fishing community?	Only through public comment	Yes, through transdisciplinary science and management	Yes, through surveys and collaboration with social scientists
Social and ecological objectives	*Social*: None articulated *Ecological:* Protect biodiversity Maintain ecosystem health and resilience	*Social*: Maintain or enhance the recreational fishery *Ecological*: Enhance fish populations through stocking	*Social*: Providing quality fishing opportunities for citizens and visitors Maximizing fisher satisfaction Optimizing stocking rates across 800 lakes given fiscal investments *Ecological*: Diverting fishing effort from wild stocks with conservation concerns to the stocked fisheries
Ecological uncertainties	Effects of no-take zones on biodiversity and health of the ecosystem	Effects of distinct levels of stocking on fish population dynamics	Density dependent growth and mortality of rainbow trout in stocked systems
Social uncertainties	None evaluated	Effects of fish stocking on recreational fisher satisfaction Effects of transdisciplinary science and collaboration on recreational fisher knowledge, attitudes, and beliefs about stocking	Preferences and behaviours of fishers Effectiveness of fishing regulations
Models used	Simulation models of fish stocks (see McCook et al. 2010)	Bio-economic modelling of the systems (see Johnston et al. 2018)	Social–ecological systems model of the recreational fishery (see Carruthers et al. 2019)
Monitoring methods	Before and after monitoring of no-take and control zones Monitoring of compliance	Monitoring of fish populations in experimentally stocked ponds Surveys of recreational fishers	Surveys of fishers Monitoring of fisher responses to stocking experiments Monitoring of fish growth and survival
Active or passive adaptive management?	Active	Active	Active

6.1 *Case Study: Recreational Fishing and Rezoning the Great Barrier Reef Marine Park*

The Great Barrier Reef Marine Park extends approximately 2300 km along the northeast coast of Queensland, Australia. Since 1981, the park has been adaptively managed primarily through zoning plans that dictate where various activities can and cannot occur (Day et al. 2019). In 2004, a new zoning plan increased no-fishing zones from 4.5% to 33% of the total park area. The aim was to increase the protection of biodiversity in the park and maintain the health and resilience of the ecosystem by ensuring that representative areas of each of the 70 major bioregions within the park were included in a network of highly protected areas (Fernandes et al. 2005).

An extensive public education and consultation process was implemented prior to the rezoning to inform the public about the need for the rezoning, facilitate public input into the size and location of new zones, and provide opportunities for public comment on draft zoning plans. Through that process the Great Barrier Reef Marine Park Authority held approximately 600 public meetings and received over 31,000 written submissions from a wide range of stakeholders (Jago et al. 2007). Although the recreational fishing community was arguably the largest group (in terms of both number of people and impact on fisheries resources in the Park; Fig. 4) affected by the rezoning, managers resisted attempts to understand more deeply the perceptions, attitudes, and activities of recreational fishers regarding management of the Park and the proposed rezoning.

Fig. 4 Recreational fishers in the Great Barrier Reef, Australia

Viewed through the lens of conservation and protection of marine resources, the 2004 rezoning of the park is considered a successful application of adaptive management (McCook et al. 2010). However, viewed through the lens of managing recreational fishers' access to and impacts on marine resources, there is little evidence of the application of the principles and methods of adaptive management. There was little a priori effort to understand the objectives, values, and behaviours of the recreational fishing community; no effort to model or predict the consequences of new no-fishing zones on recreational fisher activities or values; and no long-term monitoring plan to understand the outcomes of the rezoning process for recreational fishers.

In the years following the 2004 rezoning, several independent studies investigated recreational fishers' responses to the increase in no-fishing areas within the park (Sutton and Tobin 2009; De Freitas et al. 2013). A majority (56%) of fishers surveyed reported losing at least one of their regular fishing locations under the new zoning plan. The most common compensation strategies used by affected fishers were: (i) fishing at locations within the Park where they had not fished previously; (ii) fishing more at locations within the Park where they had previously fished; and (iii) fishing at new locations outside the Park (e.g. creeks and estuaries that flow into the Park). On average, fishers' substitute locations were 27% closer to their boat ramp departure points compared with "lost" locations, resulting in a general shift in recreational fishing effort towards inshore areas with a concomitant shift in target species away from offshore species (e.g. coral trout, *Plectropomus* spp.) to inshore species (e.g. barramundi, *Lates calcarifer*) (De Freitas et al. 2013).

Recreational fishers' behavioural responses to the zoning changes have ecological implications (e.g. increased fishing pressure in areas that remain open to fishing or on some species) that could undermine the ecological objectives of the rezoning. Fishers' behavioural responses also have socio-economic implications (e.g. increased crowding and reduced satisfaction in areas left open to fishing) that could erode trust in the management agency and reduce public support for future conservation initiatives. Ongoing conservation and management of the Great Barrier Reef Marine Park that affects recreational fishing would benefit from a more integrated approach to adaptive management, which aims to understand and minimize uncertainty in both the ecological and social components of the system.

6.2 *Case Study: Active Adaptive Management for Sustainable Fish Stocking in Germany (Modified from Arlinghaus et al. 2017)*

Some of the authors were involved in designing and running a five-year inter- and transdisciplinary research program called Besatzfisch (stocked fish, www.besatzfisch.de) in Germany. Transdisciplinary research on stocking was conceptualized as a recursive process of adaptive knowledge production with deep involvement of practitioners in terms of problem framing, knowledge integration, and co-production of new knowledge (Gibbons et al., 1994; Klein et al., 2001; Hirsch Hadorn et al. 2008). The goal was to produce scientifically and socially relevant outcomes (Ruppert-Winkel et al. 2015) in relation to the fish stocking issue and to do so by engaging in a robust active adaptive stocking experiment to reduce the uncertainties related to outcomes of fish stocking with several species of fish in lakes. The key stakeholders involved were anglers and managers in organized angling clubs. Angling clubs are fishing rights holders in Germany, and any changes to the governance and management of local fisheries depend on active decisions by local angling clubs. Stocking is a contested issue, whose success or failure depends on a range of social, ecological, and evolutionary factors (Lorenzen et al. 2012; Hühn et al. 2014; Johnston et al. 2018). To learn about successful and unsuccessful stocking practices, as well as associated genetic and other ecological risks, a team of researchers partnered with 18 angling clubs in Lower Saxony, Germany. In close collaboration with the anglers and managers in the angling clubs, the research team developed an active adaptive stocking experiment involving radical stocking density treatments of northern pike (*Esox lucius*) and common carp (*Cyprinus carpio*) into a set of angler-managed gravel pits (in total 22 lakes were manipulated, with some serving as unstocked controls). Pike served as an example of a naturally recruiting species (Hühn et al. 2014), whereas carp served as an example of a non-naturally recruiting but highly desired and often-stocked species (Arlinghaus et al. 2017).

Workshops were used for developing goals and objectives as well as hypotheses about stocking outcomes and allocation of treatments to study lakes. Structured surveys were used to understand angler behaviours and perspectives towards stocking at the representative level of the associated angling clubs. Models and stakeholder knowledge were used to inform hypotheses during the workshops (Johnston et al. 2018) and to plan the details of stocking interventions at the local angling club lakes (Fig. 5). The stocking outcomes were monitored jointly through a series of workshops, creating opportunities for learning and reflexive processes. Anglers participated in fish surveys and completed angling diaries for monitoring carp catches, which are exceedingly difficult to enumerate with other gears than angling. Results were regularly discussed in workshop settings, where joint interpretation of findings and adaptation of future club policies occurred. Large-scale surveys of club anglers and workshop attendees were used to understand attitudes, norms, and other human dimensions related to stocking and to parameterize utility functions driving angler behaviours (Arlinghaus et al. 2014; Gray et al. 2015; Fujitani et al. 2016). These functions were incorporated into bioeconomic models designed as a learning tool for examining stocking policies (Johnston et al. 2018).

Moreover, the researchers evaluated different variants of pedagogical interventions, comparing the transdisciplinary cooperation with standard teaching of scientific content through a PowerPoint-based seminar with placebo controls where participants also received scientific knowledge, but the seminar did not include information about fish stocking. The actual stocking experiments only ran in the transdisciplinary angling clubs. Ecological results showed that stocking failed to have additive stock effects when the fish stocks naturally reproduce (exemplified by pike; Hühn et al. 2014; Arlinghaus et al. 2015). Social results showed that the

Fig. 5 Recreational fishers and scientists worked together for developing objectives and hypotheses for active adaptive management of fish stocking, as well as to implement the experimental stocking efforts. (Photo credit: Besatzfisch project, IGB)

integration of anglers in the jointly conducted active adaptive stocking experiments was instrumental to changing ecological knowledge, beliefs, and behavioural intentions about stocking, with effects retained when measured 10 months later (Fujitani et al. 2017). Merely passive consumption of technical information also affected knowledge, beliefs, and attitudes, but results were forgotten over time (Fujitani et al. 2016, 2017). Only transdisciplinary cooperation where adaptive management was combined with structured decision making and active stakeholder involvement was able to shift the norms of anglers to engage in alternatives to stocking and to rely less on stocking in the future management of the clubs. Details about this project can be found in Arlinghaus et al. (2015, 2022) and a documentary film (www.youtube.com/watch?v=sFMvsz4YufY).

6.3 Case Study: Landscape Scale Active Adaptive Management to Optimize the Rainbow Trout Fishery of British Columbia, Canada

The rainbow trout (*Oncorhynchus mykiss*) fishery of British Columbia, Canada, is large and complex, covering about 500,000 km^2, including about 3500 freshwater lakes (Cantin et al. 2021). Each year, about 300,000 licensed recreational fishers spend about $500 million (CAN), and about half of this is directed at the approximately 800 lakes stocked with rainbow trout with the goal of maximizing fishing opportunities (Post and Parkinson 2012; Askey et al. 2013; Freshwater Fisheries Society of British Columbia 2023). Rainbow trout are raised from wild gametes in an extensive hatchery system with funding that originates from freshwater recreational fishing license fees. Management goals for this stocked fishery include: (i) providing quality fishing opportunities for citizens and visitors, (ii) diverting fishing effort from wild stocks with conservation concerns to the stocked fisheries, (iii) maximizing fisher satisfaction, and (iv) optimizing stocking rates across 800 lakes given fiscal investments.

Optimizing stocking rates is a complex problem with many uncertainties involving fisher behaviour, with fishers originating from urban and rural areas in British Columbia, across Canada, and international tourists, all with diverse expectations and behaviours. Within British Columbia, fisher residences and fishing opportunities are spread across many hundreds of kilometres connected by a complex network of highways, secondary roads, gravel forest access roads, and trails (Parkinson et al. 2004; Post et al. 2008). In addition, fishers are a heterogenous group, differing substantially in willingness to travel and with a diversity of preferences for fishing quality and fishing site characteristics.

Key uncertainties in optimizing the provision of opportunities for fishers are how stocking rates and fishing regulations should be distributed across this landscape to provide fishing opportunities and satisfy the preferences of this diversity of fishers. At the core of these uncertainties are ecological questions (i.e. density-dependent growth and survival of rainbow trout), social questions (i.e. preferences and behaviours of a heterogeneous fisher population), and management questions (i.e. effective fishing regulations and provision of site amenities). Likely the largest source of uncertainty is the dynamic interactions among the social, ecological, and management processes across the large landscape (Ward et al. 2016). Therefore, a group of academic, government, and non-governmental organizations (NGO), fishery biologists, and human dimension specialists undertook a landscape-scale active adaptive management experiment to reduce uncertainty in how to optimize fishing outcomes for the diverse fisher population.

The approach is multi-faceted including analyses of fisher characteristics and behaviour, experimental analyses of stocking rate optimization, and experimental assessments of fishing regulations, all distributed across the landscape with strong contrasts in access travel costs to the broadly distributed recreational fishery (Fig. 6). Often adaptive management involves coordinated stakeholder input in determining policy options for evaluation. In this active adaptive management experiment, however, a key uncertainty was how the population of fishers may respond to alternate policy options across a landscape of opportunities, such that the fisher behavioural response was a key component of the social-ecological dynamic outcome and therefore key to optimizing the fishery. So, the experimental program was developed jointly by fishery and human dimensions specialists.

Fisher segmentation was assessed with on-site creel and stratified random mail and internet surveys (Ward et al. 2013; Dabrowksa et al. 2017). Five groups of fishers were identified that differed consistently in willingness to travel, preference

Fig. 6 Adaptive management efforts involving fishers in British Columbia, Canada

for fishing quality (size and catch rate), and other site amenities, and their relative proportion differed spatially. Fisher's preference for fish size and catch rates, and the underlying density-dependent growth and survival of targeted fish, resulted in an apparent size-CPUE (catch per unit effort) trade-off of equally perceived fishing quality within regions of equivalent travel distance to access fishing opportunities (Parkinson et al. 2004; Wilson et al. 2016; Post et al. 2026). This bivariate emergent measure of fishing quality shifted along travel distance axes across the province, with the lowest overall quality near large human population centres and increasing while radiating outward towards less populated areas.

Experimental changes in the status quo stocking rate (increases and reductions) were used in experimental lakes to identify optimum stocking policy (i.e. highest effort density per stocked fish) (Mee et al. 2016). Stocking at rates above the optimum resulted in reduced fish growth and survival, and therefore reduced fishing quality and reduced fishing effort. Lower than optimum stocking rates resulted in low catch rates and low overall fishing quality, resulting in lower than maximum fishing effort. This optimum was strongly driven by fisher demand—in a high-demand region closer to large cities, the stocking rate optimum was higher than in low-demand regions.

The research group enacted experimental changes in fishing regulations from relatively liberal to trophy (i.e. high daily limit changed to lower daily limit and minimum size to harvest) with control fisheries with no changes. Fisher responses to these regulation changes were site-specific, with both positive and negative effort responses driven at least in part by local fishing quality responses and the overall effort demand of the region where the experiments were implemented. Some responses were rapid, within a fishing season, and others were longer term. Responses appeared to be at least partly related to other fishing opportunities within the region and likely resulted in some local-scale effort redistribution. Several

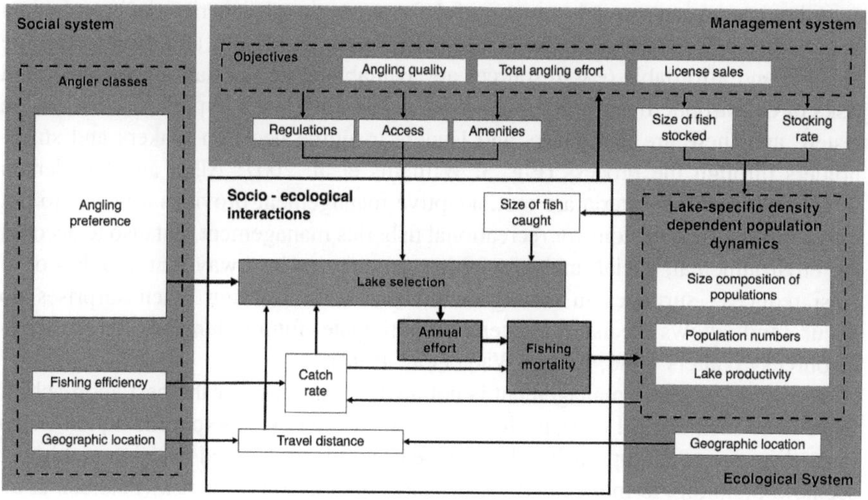

Fig. 7 Spatial social, ecological and management processes, and their dynamic interactions, modelled for an adaptive experimental management program designed to optimize the rainbow trout fishery of British Columbia, Canada. (Adapted from Carruthers et al. 2019)

models of this social-ecological system have been built that vary in complexity in both fisher behaviour and ecological dynamics to predict fishing outcomes like fisher effort, stocking efficiency, and utility (Parkinson et al. 2004; Post et al. 2008; Askey et al. 2013; Carruthers et al. 2019). The model analysis by Carruthers et al. (2019) is the most complete of these, with a spatial analysis of the interactions among ecological, social and management systems that captures the dynamics of this system across this large-scale spatial fishery (Fig. 7). These models are being used to predict fishing outcomes across a range of management policies for use in the next iteration of adaptive management of the fishery. The goal of the program continues to be to reduce uncertainty in the optimization of the fishery for the provision of fishing opportunities for local, regional, and tourist fishers.

7 Discussion

Adaptive management, through experimentation, institutional learning, and participatory modelling, provides a framework for reducing social and ecological uncertainties that hinder management. Although the complexity of implementing an adaptive management plan can seem daunting, the core concepts of the framework are fundamental elements of human behaviour (Possingham 2000). Humans learn by gathering evidence through experiences—conducting research, evaluating, and refining their models of how a system works (Possingham 2000), and adaptive management simply provides the framework to gather this evidence and conduct these evaluations more systematically at a spatial and temporal scale that is relevant to natural resources decisions (Walters and Hilborn 1978).

Adaptive management is most useful for recreational fisheries decisions when uncertainty affects the decision to be made, experimentation and monitoring are feasible and affordable (e.g. resources are available, outcomes are measurable, and risks of catastrophe from experimentation are low), there are hypotheses that can be tested, and there are facilitators and leaders to guide decision makers and stakeholders through the process (Fig. 3; Williams et al. 2007; Allen and Gunderson 2011). When these criteria are met, adaptive management provides a means to not only make better decisions for recreational fisheries management, but also to account for environmental, social, and economic uncertainties in a way that avoids—or at least reduces—surprises in management, allows for learning when surprises do occur, and allows decision makers to anticipate future demands on fisheries resources (Walters 1986; Endter-Wada et al. 1998).

However, adaptive management is not always necessary or the best approach. If there is no institutional capacity for change or if the risk of experimentation is too high, there are other approaches that can be used, like structured decision making or scenario planning, to make one-time decisions that account for risk (Peterson et al. 2003; Allen and Gunderson 2011). The key for all these decision-making approaches is to use values-focused methods to design a process that evaluates potential management actions based on their ability to achieve decision maker and stakeholder objectives.

In situations for which adaptive management is appropriate, decision makers must understand the need for applying the framework, leadership is paramount to engage managers and stakeholders in the complex process, and funding is necessary to monitor the ecological, economic, and social outcomes of experimental actions (Walters 2007). The case studies provided in this chapter describe applications of adaptive management that demonstrated strong leadership and engagement with stakeholders, whether that challenge was met (i.e. fish stocking in Germany) or left lacking (i.e. lessons from the Great Barrier Reef), as well as strong teams of social and ecological scientists to build complex models of social-ecological systems to predict outcomes of fishery management actions at the landscape scale (i.e. rainbow trout in British Columbia).

These case studies also provided ample evidence for the inclusion of stakeholder values and needs in management decisions. When the full suite of values is left unaccounted for in recreational fishing decisions, decision makers may lose support for actions that are otherwise scientifically feasible and reasonable, as well as the ability to better understand ecological and social implications of management decisions on humans and the ecosystem prior to implementation, as was shown in the Great Barrier Reef case. In contrast, when stakeholders are included as partners in adaptive management experiments, all parties learn together about the effects of management actions like fish stocking on population dynamics and fishery returns. Finally, when adaptive management is applied at a large scale to reduce social, as well as ecological, uncertainties, decision makers gain the opportunity to better understand fisher behaviours and needs and update management actions with improved learning, as was seen in the rainbow trout example.

Although adaptive management has most often been implemented to reduce uncertainty through the iterative phase of the process (Fig. 1; Williams et al. 2007), complex social–ecological systems often require engaging the potential for institutional change as well. Institutional change can emerge from shifting beliefs, norms, knowledge, and attitudes, changes in the capacity of stakeholders to act according to their wishes, or changes in decision-making authority. Adaptive management provides a unique opportunity for decision makers to understand and distinguish between ecologically and socially induced uncertainty, as described by Tyre and Michaels (2011). By parsing the sources of uncertainty, institutional, or double-loop, learning can be used to update the decision-making process, including the objectives and alternatives, for socially induced uncertainties (Tyre and Michaels 2011; FAO 2012; Williams and Brown 2018). In addition, institutional learning provides a mechanism for stakeholders and managers alike to ensure that their values are represented, to understand how the ecological and social components of the decision are linked, and to recognize the full suite of objectives and trade-offs necessary for complex systems (Gregory et al. 2006).

Acknowledgements We thank M. DuFour, N. Johnson, and A. Lynch, as well as two anonymous reviewers, for helpful comments on this chapter. Any use of trade, firm, or product names is for descriptive purposes only and does not imply endorsement by the U.S. Government.

References

Allen CR, Gunderson LH (2011) Pathology and failure in the design and implementation of adaptive management. J Environ Manag 92:1379–1384

Allen CR, Fontaine JJ, Pope KL, Garmestani AS (2011) Adaptive management for a turbulent future. J Environ Manag 92:1339–1345

Aminpour P, Gray SA, Jetter AJ, Introne JE, Singer A, Arlinghaus R (2020) Wisdom of stakeholder crowds in complex social-ecological systems. Nat Sustain 3:191–199

Arlinghaus R, Hunt LM, Post JR, Allen MS (2014) Not fish, not meat: some guidance on how to study fisheries from an interdisciplinary perspective. In: Taylor WW, Lynch A, Leonard N (eds) Future of fisheries: perspectives for the next generation of fisheries professionals. American Fisheries Society, Bethesda, pp 223–230

Arlinghaus R, Cyrus E-M, Eschbach E, Fujitani M, Hühn D, Johnston F, Pagel T, Riepe C (2015) Hand in Hand für eine nachhaltige Angelfischerei: Ergebnisse und Empfehlungen aus fünf Jahren praxisorientierter Forschung zu Fischbesatz und seinen Alternativen. Berichte des IGB Heft 28/2015

Arlinghaus R, Alós J, Beardmore B, Daedlow K, Dorow M, Fujitani M, Hühn D, Haider W, Hunt LM, Johnson BM, Johnston F, Klefoth T, Matsumura S, Monk C, Pagel T, Post JR, Rapp T, Riepe C, Ward H, Wolter C (2017) Understanding and managing freshwater recreational fisheries as complex adaptive social-ecological systems. Rev Fish Sci Aquac 25:1–41

Arlinghaus R, Riepe C, Theis S, Pagel T, Fujitani M (2022) Dysfunctional information feedbacks cause the emergence of management panaceas in social-ecological systems: the case of fish stocking in inland recreational fisheries. J Outdoor Recreat Tour 38:100475

Askey PJ, Parkinson EA, Post JR (2013) Linking fish and angler dynamics to assess stocking strategies for hatchery-dependent, open-access recreational fisheries. North Am J Fish Manag 33:557–568

Bennett NJ, Roth R, Klain SC, Chan K, Christie P, Clark DA, Cullman G, Curran D, Durbin TJ, Epstein G, Greenberg A, Nelson MP, Sandlos J, Stedman R, Teel TL, Thomas R, Veríssimo D, Wyborn C (2017) Conservation social science: understanding and integrating human dimensions to improve conservation. Biol Conserv 205:93–108

Bouwes N, Bennett S, Wheaton J (2016) Adapting adaptive management for testing the effectiveness of stream restoration: an intensively monitored watershed example. Fisheries 41:84–91

Cahill CL, Walters CJ, Paul AJ, Sullivan MG, Post JR (2022) Unveiling the recovery dynamics of walleye after the invisible collapse. Can J Fish Aquat Sci 79:708–723

Camp EV, van Poorten BT, Robinson KF, Harrison HL, NguyenV CCA, Lorenzen K, Casola W, Dedual M, Klizentyte K (2026) Integrating diverse stakeholder knowledge in fisheries decision making. In: Pope KL, Arlinghaus R, Hunt LM, Lynch AJ, van Poorten BT (eds) Understanding recreational fishers: disciplinary and interdisciplinary approaches for fisheries management. Springer Nature, Cham, p 607–660

Cantin A, Farineau A, Bender DJ, Post JR (2021) Predicting distribution, production, recreational fishing demand, and conservation concern of indigenous rainbow trout populations based on landscape characteristics. Can J Fish Aquat Sci 78:1096–1108

Carruthers TR, Dabrowska K, Haider W, Parkinson EA, Varkey DA, Ward H, McAllister MK, Godin T, Van Poorten B, Askey PJ, Wilson KL, Hunt LM, Clarke A, Newton E, Walters C, Post JR (2019) Landscape-scale social and ecological outcomes of dynamic angler and fish behaviours: processes, data, and patterns. Can J Fish Aquat Sci 76:970–988

Dabrowksa K, Hunt LM, Haider W (2017) Understanding how angler characteristics and context influence angler preferences for fishing sites. North Am J Fish Manag 37:1350–1361

Day JC, Kenchington RA, Tanzer JM, Cameron DS (2019) Marine zoning revisited: how decades of zoning the Great Barrier Reef has evolved as an effective spatial planning approach for marine ecosystem-based management. Aquat Conserv Mar Freshw Ecosyst 29:9–32

De Freitas DM, Sutton SG, Moloney JM, Lédée EJ, Tobin RC (2013) Spatial substitution strategies of recreational fishers in response to zoning changes in the Great Barrier Reef Marine Park. Mar Policy 40:145–153

Decker DJ, Brown TL, Connelly NA, Enck JW, Pomerantz GA, Purdy KG, Siemer WF (1992) Toward a comprehensive paradigm of wildlife management: integrating the human and biological dimensions. In: Mangun WR (ed) American fish and wildlife policy: the human dimension. Southern Illinois University Press, Carbondale, pp 33–54

Dedual M, Sague Pla O, Arlinghaus R, Clarke A, Ferter K, Geertz Hansen P, Gerdeaux D, Hames F, Kennelly SJ, Kleiven AR, Meraner A, Ueberschär B (2013) Communication between scientists, fishery managers and recreational fishers: lessons learned from a comparative analysis of international case studies. Fish Manag Ecol 20:234–246

Endter-Wada J, Blahna D, Krannich R, Brunson M (1998) A framework for understanding social science contributions to ecosystem management. Ecol Appl 8:891–904

FAO (2012) Recreational fisheries: FAO technical guidelines for responsible fisheries. FAO, Rome

Fernandes L, Day JO, Lewis A, Slegers S, Kerrigan B, Breen DA, Cameron D, Jago B, Hall J, Lowe D, Innes J, Tanzer J, Chadwick V, Thompson L, Gorman K, Simmons M, Barnett B, Sampson K, De'ath G, Mapstone B, Marsh H, Possingham H, Ball I, Ward T, Dobbs K, Aumend J, Slater D, Stapleton K (2005) Establishing representative no-take areas in the great barrier reef: large-scale implementation of theory on marine protected areas. Conserv Biol 19: 1733–1744

Freshwater Fisheries Society of British Columbia (2023) Media kit. https://www.gofishbc.com/PDFs/Footer/FreshwaterFisheries_MediaKit.aspx. Accessed 16 June 2023

Fujitani ML, McFall A, Randler C, Arlinghaus R (2016) Efficacy of lecture-based environmental education for biodiversity conservation: a robust controlled field experiment with recreational anglers engaged in self-organized fish stocking. J Appl Ecol 53:25–33

Fujitani M, McFall A, Randler C, Arlinghaus R (2017) Participatory adaptive management leads to environmental learning outcomes extending beyond the sphere of science. Sci Adv 3:1–12

Fulton EA, Smith ADM, Smith DC, Van Putten IE (2011) Human behaviour: the key source of uncertainty in fisheries management. Fish Fish 12:2–17

Golebie E, van Riper CJ, Suski C, Stedman R (2021) Reducing invasive species transport among recreational anglers: the importance of values and risk perceptions. North Am J Fish Manag 41: 1812–1825

Gray S, Hilsberg J, McFall A, Arlinghaus R (2015) The structure and function of angler mental models about fish population ecology: the influence of specialization and target species. J Outdoor Recreat Tour 12:1–13

Gregory RS (2017) The troubling logic of inclusivity in environmental consultations. Sci Technol Human Values 42:144–165

Gregory RS, Keeney RL (2002) Making smarter environmental management decisions. J Am Water Resour Assoc 38:1601–1612

Gregory R, Failing L, Higgins P (2006) Adaptive management and environmental decision making: a case study application to water use planning. Ecol Econ 58:434–447

Gregory R, Failing L, Harstone M, Long G, McDaniels T, Ohlson D (2012) Structured decision making: a practical guide to environmental management choices. Wiley-Blackwell, West Sussex

Hadorn GH, Biber-Klemm S, Grossenbacher-Mansuy W, Hoffmann-Riem H, Joye D, Pohl C, Wiesmann U, Zemp E (2008) The emergence of transdisciplinarity as a form of research. In: Hadorn GH, Hoffmann-Riem H, Biber-Klemm S, Grossenbacher-Mansuy W, Joye D, Pohl C, Wiesmann U, Zemp E (eds) Handbook of transdisciplinary research. Springer Science + Business Media B.V, pp 19–42

Heberlein TA (2012) Navigating environmental attitudes. Oxford University Press, Oxford

Hühn D, Lübke K, Skov C, Arlinghaus R (2014) Natural recruitment, density-dependent juvenile survival, and the potential for additive effects of stock enhancement: an experimental evaluation of stocking northern pike (*Esox lucius*) fry. Can J Fish Aquat Sci 71:1508–1519

Hunt LM, Sutton SG, Arlinghaus R (2013) Illustrating the critical role of human dimensions research for understanding and managing recreational fisheries within a social-ecological system framework. Fish Manag Ecol 20:111–124

Jago B, Day J, Fernandes L, Thompson L, Hall J, Sampson K, Cameron D, Russell M, Lowe D (2007) Involving the community early. The variable role that communities can play in MPA management: experiences and lessons learned on the Great Barrier Reef. In: Day J, Senior J, Monk S, Neal N (eds) First international marine protected areas congress conference proceedings, vol 2005. IMPAC1, Geelong, Australia, pp 353–355

Johnston FD, Arlinghaus R, Dieckmann U (2010) Diversity and complexity of angler behaviour drive socially optimal input and output regulations in a bioeconomic recreational-fisheries model. Can J Fish Aquat Sci 67:1507–1531

Johnston FD, Arlinghaus R, Dieckmann U (2013) Fish life history, angler behaviour and optimal management of recreational fisheries. Fish Fish 14:554–579

Johnston FD, Beardmore B, Arlinghaus R (2015) Optimal management of recreational fisheries in the presence of hooking mortality and noncompliance — predictions from a bioeconomic model incorporating a mechanistic model of angler behavior. Can J Fish Aquat Sci 72:37–53

Johnston FD, Allen MS, Beardmore B, Riepe C, Pagel T, Hühn D, Arlinghaus R (2018) How ecological processes shape the outcomes of stock enhancement and harvest regulations in recreational fisheries. Ecol Appl 28:2033–2054

Jones ML, Catalano MJ, Peterson LK, Berger AM (2016) Stakeholder-centered development of a harvest control rule for Lake Erie walleye. In: Edwards CT, Dankel DJ (eds) Management science in fisheries: an introduction to simulation-based methods. Routledge, London, pp 163–183

Keeney RL (2004) Framing public policy decisions. Int J Technol Policy Manag 4:95–115

Landon AC, Schroeder SA, Stefen SF, Arlinghaus R (2026) Social psychology for understanding recreational fishers and fisheries. In: Pope KL, Arlinghaus R, Hunt LM, Lynch AJ, van Poorten

BT (eds) Understanding recreational fishers: disciplinary and interdisciplinary approaches for fisheries management. Springer Nature, Cham, pp 157–196

Lee KN (1993) Compass and gyroscope: integrating science and politics for the environment. Island Press, Washington, DC

Lorenzen K, Beveridge MCM, Mangel M (2012) Cultured fish: integrative biology and management of domestication and interactions with wild fish. Biol Rev 87:639–660

Månsson J, Eriksson L, Hodgson I, Elmberg J, Bunnefeld N, Hessel R, Johansson M, Liljebäck N, Nilsson L, Olsson C, Pärt T, Sandström C, Tombre I, Redpath SM (2022) Understanding and overcoming obstacles in adaptive management. Trends Ecol Evol 38:55–71

Martin DR, Pope KL (2011) Luring anglers to enhance fisheries. J Environ Manag 92:1409–1413

McCook LJ, Ayling T, Cappo M, Choat JH, Evans RD, De Freitas DM, Heupel M, Hughes TP, Jones GP, Mapstone B, Marsh H, Mills M, Molloy FJ, Pitcher CR, Pressey RL, Russ GR, Sutton S, Sweatman H, Tobin R, Wachenfeld DR, Williamson DH (2010) Adaptive management of the Great Barrier Reef: a globally significant demonstration of the benefits of networks of marine reserves. Proc Natl Acad Sci USA 107:18278–18285

McGowan CP, Lyons JE, Smith DR (2020) Decision implementation and the double-loop process in adaptive management of horseshoe crab harvest in Delaware Bay. In: Runge MC, Converse SJ, Lyons JE, Smith DR (eds) Structured decision making: case studies in natural resource management. Johns Hopkins University Press, Baltimore, pp 258–268

Mee JA, Post JR, Ward H, Wilson KL, Newton E, Cantin A (2016) Interaction of ecological and angler processes: experimental stocking in an open access, spatially structured fishery. Ecol Appl 26:1693–1707

Milly PC, Betancourt J, Falkenmark M, Hirsch RM, Kundzewicz ZW, Lettenmaier DP, Stouffer RJ (2008) Stationarity is dead: whither water management? Science 319:573–574

Nichols JD, Koneff MD, Heglund PJ, Knutson MG, Seamans ME, Lyons JE, Morton JM, Jones MT, Boomer GS, Williams BK (2011) Climate change, uncertainty, and natural resource management. J Wildl Manag 75:6–18

Parkinson EA, Post JR, Cox SP (2004) Linking the dynamics of harvest effort to recruitment dynamics in a multistock, spatially structured fishery. Can J Fish Aquat Sci 61:1658–1670

Peterman R, Peters CNC (1998) Decision analysis: taking uncertainties into account in forest resource management. In: Sit V, Taylor B (eds) Statistical methods for adaptive management studies. Land management handbook no. 42, Victoria, BC, pp 105–127

Peterson GD, Cumming GS, Carpenter SR (2003) Scenario planning: a tool for conservation in an uncertain world. Conserv Biol 17:358–366

Pine WE, Johnson FA, Frederick PC, Coggins LG (2022) Adaptive management in practice and the problem of application at multiple scales—insights from oyster reef restoration on Florida's Gulf coast. Mar Coast Fish 14:1–13

Possingham HP (2000) Is environmental research a waste of time? Ecol Manag Restor 1:81–82

Post JR, Persson L, Parkinson EA, Van Kooten T (2008) Angler numerical response across landscapes and the collapse of freshwater fisheries. Ecol Appl 18:1038–1049

Post JR, Wilson KL, Johnston FD, Sass GG, Askey PJ, Ward HGM, Allen MS (2026) Ecology for understanding recreational fishers and fisheries. In: Pope KL, Arlinghaus R, Hunt LM, Lynch AJ, van Poorten BT (eds) Understanding recreational fishers: disciplinary and interdisciplinary approaches for fisheries management. Springer Nature, Cham, p 95–121

Raiffa H (1968) Decision analysis: introductory lectures on choices under uncertainty, vol 19. Random House, New York

Raiffa H, Schlaiffer R (1961) Applied statistical decision theory. Harvard University Press, Cambridge, MA

Regan HM, Colyvan M, Burgman MA (2002) A taxonomy and treatment of uncertainty for ecology and conservation biology. Ecol Appl 12:618–628

Robinson KF, Fuller AK (2017) Participatory modeling and structured decision making. In: Gray S, Paolisso M, Jordan R, Gray S (eds) Environmental modeling with stakeholders: theory, methods, and applications. Springer International, Cham, pp 83–101

Robinson KF, Fuller AK, Stedman RC, Siemer WF, Decker DJ (2019) Integration of social and ecological sciences for natural resource decision making: challenges and opportunities. Environ Manag 63:565–573

Robinson KF, DuFour M, Jones M, Herbst S, Newcomb T, Boase J, Brenden T, Chapman D, Dettmers J, Francis J, Hartman T, Kočovský P, Locke B, Mayer C, Tyson J (2021a) Using decision analysis to collaboratively respond to invasive species threats: a case study of Lake Erie grass carp (*Ctenopharyngodon idella*). J Great Lakes Res 47:108–119

Robinson KF, Fuller AK, Jones ML (2021b) Using structured decision making to incorporate ecological and social values into harvest decisions: case studies of white-tailed deer and walleye. In: Pope KL, Powell LA (eds) Harvest of fish and wildlife: new paradigms for sustainable management. CRC Press, Boca Raton, pp 131–146

Rokeach M (1973) The nature of human values. Free Press

Runge MC (2011) An introduction to adaptive management for threatened and endangered species. J Fish Wildl Manag 2:220–233

Runge MC (2020) Introduction to linked and dynamic decisions. In: Runge MC, Converse SJ, Lyons JE, Smith DR (eds) Structured decision making: case studies in natural resource management. Johns Hopkins University Press, Baltimore, pp 227–233

Runge MC, Bean E (2020) Decision analysis for managing public natural resources. In: Runge MC, Converse SJ, Lyons JE, Smith DR (eds) Structured decision making: case studies in natural resource management. Johns Hopkins University Press, Baltimore, pp 3–14

Runge MC, Converse SJ, Lyons JE (2011) Which uncertainty? Using expert elicitation and expected value of information to design an adaptive program. Biol Conserv 144:1214–1223

Runge MC, Rushing CS, Lyons JE, Rubenstein MA (2023) A simplified method for value of information using constructed scales. Decis Anal 20:187–241

Ruppert-Winkel C, Arlinghaus R, Deppisch S, Eisenack K, Gottschlich D, Hirschl B, Matzdorf B, Mölders T, Padmanabhan M, Selbmann K, Ziegler R, Plieninger T (2015) Characteristics, emerging needs, and challenges of transdisciplinary sustainability science: experiences from the German social-ecological research program. Ecol Soc 20(3):13

Rushing CS, Rubenstein M, Lyons JE, Runge MC (2020) Using value of information to prioritize research needs for migratory bird management under climate change: a case study using federal land acquisition in the United States. Biol Rev 95:1109–1130

Schindler DE, Hilborn R (2015) Prediction, precaution, and policy under global change. Science 347:953–954

Shelby B, Vaske JJ (1991) Resource and activity substitutes for recreational salmon fishing in New Zealand. Leis Sci 13:21–32

Shin S, van Riper CJ, Stedman RC, Suski CD (2022) The value of eudaimonia for understanding relationships among values and pro-environmental behavior. J Environ Psychol 80:101778

Smith ADM, Walters CJ (1981) Adaptive management of stock-recruitment systems. Can J Fish Aquat Sci 38:690–703

Stedman RC (2016) Subjectivity and social-ecological systems: a rigidity trap (and sense of place as a way out). Sustain Sci 11:891–901

Stern PC (2000) Toward a coherent theory of environmentally significant behavior. J Soc Issues 56: 407–424

Sutton SG, Tobin RC (2009) Recreational fishers' attitudes towards the 2004 rezoning of the Great Barrier Reef Marine Park. Environ Conserv 36:245–252

Tsehaye I, Jones ML, Irwin BJ, Fielder DG, Breck JE, Luukkonen DR (2015) A predictive model to inform adaptive management of double-crested cormorants and fisheries in Michigan. Nat Resour Model 28:348–377

Tyre AJ, Michaels S (2011) Confronting socially generated uncertainty in adaptive management. J Environ Manag 92:1365–1370

Van Riper CJ, Thiel A, Penker M, Braito M, Landon AC, Thomsen JM, Tucker CM (2018) Incorporating multilevel values into the social-ecological systems framework. Ecol Soc 23(3): 25

von Winterfeldt D, Edwards W (1986) Decision analysis and behavioral research. Cambridge University Press, Cambridge, UK

Walters CJ (1986) Adaptive management of renewable resources. MacMillan Publishing Company, New York

Walters CJ (1997) Challenges in adaptive management of riparian and coastal ecosystems. Ecol Soc 1(2):1

Walters CJ (2007) Is adaptive management helping to solve fisheries problems? AMBIO A J Hum Environ 36:304–307

Walters CJ, Hilborn R (1976) Adaptive control of fishing systems. J Fish Res Board Canada 33: 145–159

Walters CJ, Hilborn R (1978) Ecological optimization and adaptive management. Annu Rev Ecol Syst 9:157–188

Walters CJ, Holling CS (1990) Large-scale management experiments and learning by doing. Ecology 71:2060–2068

Walters CJ, Martell SJ (2004) Fisheries ecology and management. Princeton University Press, Princeton

Ward HGM, Quinn MS, Post JR (2013) Angler characteristics and management implications in a large, multistock, spatially structured recreational fishery. North Am J Fish Manag 33:576–584

Ward HG, Allen MS, Camp EV, Cole N, Hunt LM, Matthias B, Post JR, Wilson K, Arlinghaus R (2016) Understanding and managing social-ecological feedbacks in spatially structured recreational fisheries: the overlooked behavioral dimension. Fisheries 41:524–535

Wilberg MJ, Irwin BJ, Jones ML (2026) Participatory modelling to support evaluation of management actions for recreational fisheries. In: Pope KL, Arlinghaus R, Hunt LM, Lynch AJ, van Poorten BT (eds) Understanding recreational fishers: disciplinary and interdisciplinary approaches for fisheries management. Springer Nature, Cham, p 693–718

Williams BK (1997) Approaches to the management of waterfowl under uncertainty. Wildl Soc Bull 25:714–720

Williams BK, Brown ED (2018) Environmental management double-loop learning in adaptive management: the need, the challenge, and the opportunity. Environ Manag 62:995–1006

Williams BK, Szaro RC, Shapiro CD (2007) Adaptive management: the U.S. Department of the Interior technical guide. Adaptive Management Working Group, U.S. Department of the Interior, Washington, DC

Wilson KL, Cantin A, Ward HG, Newton ER, Mee JA, Varkey DA, Parkinson EA, Post JR (2016) Supply-demand equilibria and the size-number trade-off in spatially structured recreational fisheries. Ecol Appl 26:1086–1097

Yokota F, Thompson KM (2004) Value of information literature analysis: a review of applications in health risk management. Med Decis Mak 24:287–298

Data Standardisation and Management to Facilitate Large-Scale and Interdisciplinary Approaches Access

Nicholas Sievert, Rebecca M. Krogman, and Holly Embke

Abstract Bringing data related to recreational fishers and fisheries together across large scales can provide tremendous insight. Methods for collecting, analysing, and storing data can vary dramatically, which can have significant implications for the use of these data. Efforts to standardise data within organisations often increase the ability to compare datasets from different areas, monitor changes over time, and increase the utility of the data for management and research. Though employing standardised methodology and data architecture results in the most straightforward and robust opportunities for data integration, doing so may not be possible due to variation in data collection objectives, continuity with historical programs, and resource limitations. Additionally, managing data according to FAIR principles (findability, accessibility, interoperability, and reusability) helps support data sharing and large-scale research efforts. Key elements of integrable data include appropriate data structures, adequate documentation of methodology, interpretable and complete metadata, and accessible storage formats. We document the potential benefits of standardising data and offer example approaches. We also explore best practices regarding the formatting, storage, and transmission of recreational fisher data. Efforts to increase the level of standardisation and integrability of recreational fisher data can create opportunities to better understand fisher behaviour, needs, and fulfilment.

Keywords Data management · Integration · Metadata · Recreational fisher data · Standardisation

N. Sievert (✉)
Missouri Department of Conservation, Columbia, MO, USA
e-mail: nicholas.sievert@mdc.mo.gov

R. M. Krogman
Fisheries Research Section, Iowa Department of Natural Resources, Chariton, IA, USA

H. Embke
U.S. Geological Survey, Midwest Climate Adaptation Science Center, Saint Paul, MN, USA

K. L. Pope et al. (eds.), *Understanding Recreational Fishers*, Fish & Fisheries Series 45, https://doi.org/10.1007/978-3-031-99739-6_24

1 Small-Scale Data for Large-Scale Inference

The study of recreational fishers is important to achieving fisheries management goals (Krogman et al. 2026). Managers and researchers use a variety of approaches in their assessments, including creel surveys (de Kerckhove et al. 2026), experiments (van Putten et al. 2026), digital data mining (Venturelli et al. 2026), mixed methods (Golebie et al. 2026), and analysis of text and images (Kochalski et al. 2026). These approaches can vary in scale and inference and are suited to a variety of applications. Though many methods exist, in this chapter we will be focusing on creel surveys, specifically fisher intercept surveys; even so, the ideas shared in this chapter are applicable to a variety of approaches.

Creel surveys are typically aimed at capturing information about recreational fishers and the catch and harvest of fish. A variety of approaches, such as fisher intercept surveys, fisher diaries, mail or phone, or online surveys, as well as boat or trailer, or fisher counts, are considered creel surveys (Nieman et al. 2021). Fisher intercept surveys are often conducted by natural resource agencies and typically involve collecting information about recreational fishing (catch and harvest of fish, as well as fisher effort, preferences, and characteristics) via on-site interviews with fishers, and fishing pressure through counts (de Kerckhove et al. 2026). Data collected via fisher interviews can then be expanded based on the count data to generate overall estimates for a waterbody over the survey period (Pollock et al. 1994). In many cases, studies and surveys are conducted at small spatial scales and are often focused on individual waterbodies over relatively brief time spans (1 year or less). Creel surveys are often employed to collect information that can be used to guide management of a waterbody. Creel surveys can be used for evaluating angling regulations (harvest and size limits), deciding which species or how many fish to stock, and what types of habitat improvement efforts to undertake (Fayram et al. 2006; Hall and Douglas 2002; Krogman 2021). Understanding harvest by recreational fishers is also a critical component of fisheries management, and the information acquired through creel surveys, such as harvest estimates and the size structure of harvested fish, may be used in these analyses (Embke et al. 2020). Information from creel surveys can also inform investments in fisheries infrastructure, such as boat landings and aquaculture facilities. Although the primary objectives of many of these studies may be small-scale, the data being collected and the results of the analyses may have utility for large-scale efforts (Lynch et al. 2021).

There is a growing line of thought that the conservation and management of recreational fisheries can benefit from viewing fisheries as large-scale social–ecological systems (Hunt et al. 2013). In some systems, fish stocks are not confined to a unique location, such as salmon populations that are harvested along hundreds of miles of river across multiple states and provinces, and therefore need to be managed across broad spatial scales (Smith 1994). Additionally, recreational fishers allocate effort across the landscape based on a variety of factors including taxonomic and size preferences, travel distance, accessibility, facilities and infrastructure, and characteristics of the waterbody and surrounding landscape (Hunt et al. 2019). Recreational fisheries are often in a state of flux; therefore, considering forward-looking management may be crucial to long-term success. Climate change is causing shifts in species populations with substantial effects for recreational fishing (e.g. Braynen et al. 2022). In many areas, fisher demographics, preferences, and values are also shifting, which can have profound impacts on recreational fisheries (Thunberg and Fulcher 2006; van der Hammen and Chen 2020). Fisher preferences also often differ in target species given varying motivations, including those for "action", "quality", and "trophy" fisheries (Krogman and Stubbs 2021). Further, the growth of a catch-and-release ethos in recreational fisheries has affected where fishers choose to fish, what they fish for, and the amount of harvest that occurs (Sass and Shaw 2020; Arlinghaus et al. 2021). All these factors may have implications for how fisheries are managed for both individual waterbodies and among regional networks. Being able to evaluate patterns in fisher use, composition of catch and harvest, and other metrics related to recreational fishing across time and space is critical to our understanding of the effects of these shifts. The ability to leverage existing knowledge to address issues is also a potential benefit of compiling large-scale datasets. In addition to utilising these datasets to conduct large-scale assessments, researchers can make targeted comparisons. Information on how an existing regulation impacts fisheries in a neighbouring region might help predict what impact a similar change would have locally, or understanding angling dynamics in a nearby urban area may help predict changes likely to occur as human populations expand in other areas. The concept of "anglersheds", or social–ecological catchments, which visually depicts areas from which fishers travel to visit a waterbody, is growing in use and represents a major step forward in thinking of waterbodies as components of larger socio-ecological systems (Martin et al. 2015; Kaemingk et al. 2021; Nieman et al. 2021). In many cases, the ability to address large-scale issues in recreational fisheries will depend upon bringing fine-scale data sources together for large-scale analyses.

Management of information collected as part of small-scale investigations, such as fisher surveys, is often focused on meeting the specific and immediate needs of the research project without consideration of the broader value and utility of the data being collected. This can have major implications for where the dataset is stored, what form it takes, and what documentation exists, all of which impact the ability of that dataset to be utilised for large-scale management and research efforts. Often, data are stored in files that are not integrated into easily accessible systems. Commonly, the results of surveys and investigations are stored in local files on the computer of a manager or biologist and in some cases, only products such as reports are shared beyond those who initially collected and analysed the data. In other cases, data and results may be stored on a server where they can be accessed and shared among staff and users beyond those who initially collected the data. Within many management agencies, there are growing efforts to compile data in repositories and databases to make data accessible to agency staff and sometimes even the public.

Data can be presented in a variety of forms, including values reported in the text of a report, static tables and figures, spreadsheets, and relational databases. Additionally, the presence and quality of documentation describing the dataset (metadata) can vary dramatically. Descriptions of the characteristics of the study, units of measurement, and details of calculations (e.g. Harvest Rate = Harvest/Effort Hours) are all critical to the accurate interpretation of data. The accessibility of data (where they are stored), the format of data (how they are presented), and the presence of appropriate metadata (what they mean) play a key role in the ability of data to be shared and utilised for large-scale data-compilation efforts. To efficiently bring together small-scale data for large-scale analyses, data should be findable, accessible, interoperable, and reusable (FAIR; Wilkinson et al. 2016).

2 Standardisation and Integrable Data Management Practices (What They Are and Why They Matter)

The levels of standardisation and integrability that exist among datasets are critical in determining whether multiple datasets can be leveraged to address large-scale issues. Making comparisons among data collected, analysed, or reported in inconsistent ways can be problematic, as the inferences that can be drawn from one dataset may not match those that can be drawn from another. Data standardisation and integrable data management both deal with the relationship among multiple surveys or data collection events. Put simply, *standardisation* is a process by which various elements are made to be consistent among surveys, whereas *integrable data management* refers to the management of data in a way that facilitates the compilation of data across multiple datasets. Standardisation typically consists of setting standards or specific ways in which data are to be collected, analysed, and reported, to achieve desired levels of consistency among datasets. Integrable data management is focused on the ways data can be formatted, organised, documented, and stored to promote linkages among data sources. The *FAIR principles* (findability, accessibility, interoperability, and reusability) offer useful guidance in how data can be managed in ways that facilitate future use and integration of a dataset (Wilkinson et al. 2016). Standardising aspects of the data collection, analysis, and management process can aid in achieving the levels of consistency necessary for making comparisons or completing analyses across multiple datasets. The ability to bring data together from various sources for scientific analyses is greatly influenced by how the data are managed. Both processes are essential to expand the utility of creel surveys for fisheries science in general (Nieman et al. 2021). In this chapter, we will discuss key considerations related to both data standardisation and interoperability with the goal of providing guidance that we hope can be employed to facilitate the sharing of recreational fisher data to facilitate large-scale and interdisciplinary approaches.

3 Standardisation

3.1 Standardisation Benefits and Opportunity Loss

Standardisation of recreational fisher creel data within a management entity (e.g. a state natural resource agency or regional collaborative group) can yield important benefits. Like many fishery management activities, collecting recreational fisher data requires extensive time and financial resources. For instance, a single open-water intercept survey with one staff person may require $14,400 USD for a single season (one full-time person working for $15/hour for six months; R. M. Krogman, Iowa Department of Natural Resources (IA DNR), unpublished data). Therefore, inability to effectively use data derived from such a survey is a substantial opportunity loss and waste of limited resources (Sutinen 1989; Hall-Arber et al. 2009; Kolb et al. 2013; Bartley et al. 2015). Lack of standardisation and poor data management thus hinder reproducibility and accumulation of scientific evidence (NASEM 2016). This is apparent in the realm of recreational creel surveys, where a combination of datasets collected by many disparate agents has only recently been achieved (e.g. Lynch et al. 2021) but is extremely powerful when done well and applied to decision-making (e.g. Hyder et al. 2017).

Leveraging recreational fisher data over longer time and larger spatial scales has clear benefits (Nieman et al. 2021). For instance, long-term creel datasets may be available for a single location for many decades, facilitating the understanding of trends and fishery change over time (Cass-Calay and Schmidt 2009; Cowie 2021), but more often, numerous creel surveys have been completed in many locations over time within a management agency's purview. Trends and valuable lessons may still be learned if those data can be combined in a meaningful manner, and ideally standardised (Cook and Younk 1998; Hurley and Jackson 2002; Chizinski et al. 2014; Embke et al. 2020). Fisher surveys are the most common and possibly only source of data connecting the ecological and social components of a fishery for many locations, especially the metrics of catch, effort, and satisfaction (Nieman et al. 2021). The spatial and temporal scope of information derived from creel surveys can be expanded with standardisation, which allows comparisons to be made more widely across surveys. Commonalities amongst managed waterbodies may facilitate predictions for data-poor locations in time and space. For instance, the value of a potential fishery being considered for acquisition was recently estimated by comparing it to other waterbodies of similar distance from urban centres and surface area, yielding an approximation of direct annual expenditures likely to be seen from recreational fishers (R. M. Krogman, IA DNR, unpublished data). Creel data were utilised from standardised surveys across the state of Iowa to estimate that the new publicly accessible fishery would likely be worth at least $1 M USD per annum. As another example, many creel surveys were cancelled in Iowa during 2020 due to staffing health safety issues related to COVID-19, but one survey that was completed revealed a significant increase in recreational fishing pressure during the pandemic (Krogman 2021). That increase provided a quantitative sentinel record that aligned with qualitative observations of fishing across the state for which no creel data were available. This pandemic effect was documented through an email survey of anglers across the United States of America (Midway et al. 2021).

3.2 Example of the Standardisation Process

Moving to a standardised system of creel data management can be challenging and time-consuming, but it is feasible. We present a recent example from Iowa, where the Iowa Department of Natural Resources recently underwent a standardisation process for all creel surveys conducted on lentic and lotic inland waterbodies (R. M. Krogman, IA DNR, unpublished data).

In 2014, a new creel survey was initiated at Easter Lake, Iowa, to gather recreational fishing and general recreational information prior to an extensive park and reservoir renovation. Digital data collection was desired to reduce time dedicated to data entry for the staff, but several other creel surveys were ongoing or planned in other parts of the state. Thus, before designing a data acquisition tool, a thorough evaluation was completed of all historical creel survey study designs and question sets to identify commonalities, differences, and potential areas of compromise. Although some long-term creel surveys used methodology suggested by Rose (1956) whereas others used methodology suggested by McWilliams (1984), it was later found that McWilliams (1984) cited Rose (1956) and that all basic creel study stratification factors, catch and harvest metrics, and expansion analyses were identical. However, some questions differed slightly in wording and measurement units. Some surveys measured satisfaction by asking, "Are you satisfied with your fishing trip today? Yes or No″ whereas others measured satisfaction by asking, "Rate your satisfaction with your fishing trip today on a scale of 1 to 5, where 1 is very dissatisfied and 5 is very satisfied". All variations were compiled, and a meeting with all fisheries management biologists was conducted during which variations and their merits were discussed, and a standard wording was chosen. Recreational fisher types (boat, shore, icehouse, open ice, and wading fishers) were also standardised to ensure expansion of each group would be possible, and alternate fisher types were strongly discouraged. Some questions were identified as required for all creel surveys (e.g. fisher type, fishing party size, trip duration, zip code), whereas others were identified as optional (e.g. trip spending, desired amenities).

After standard questions, their precise wording, and their measurement units were finalised, a database was designed to hold all desired fields in a set of hierarchical tables, in which a survey encompassed many pressure counts and many interviews (Fig. 1). An interview encompassed zero to many fish catch records and numerous human dimensions fields such as spending and travel behaviour. This data architecture was echoed in the electronic data acquisition tool, an app for mobile tablets, which worked offline and synchronised with a single, centralised database on the cloud. The app form and associated database enforced data quality by requiring requisite fields, enforcing field formats, and warning of possible missing entries using various quality checks.

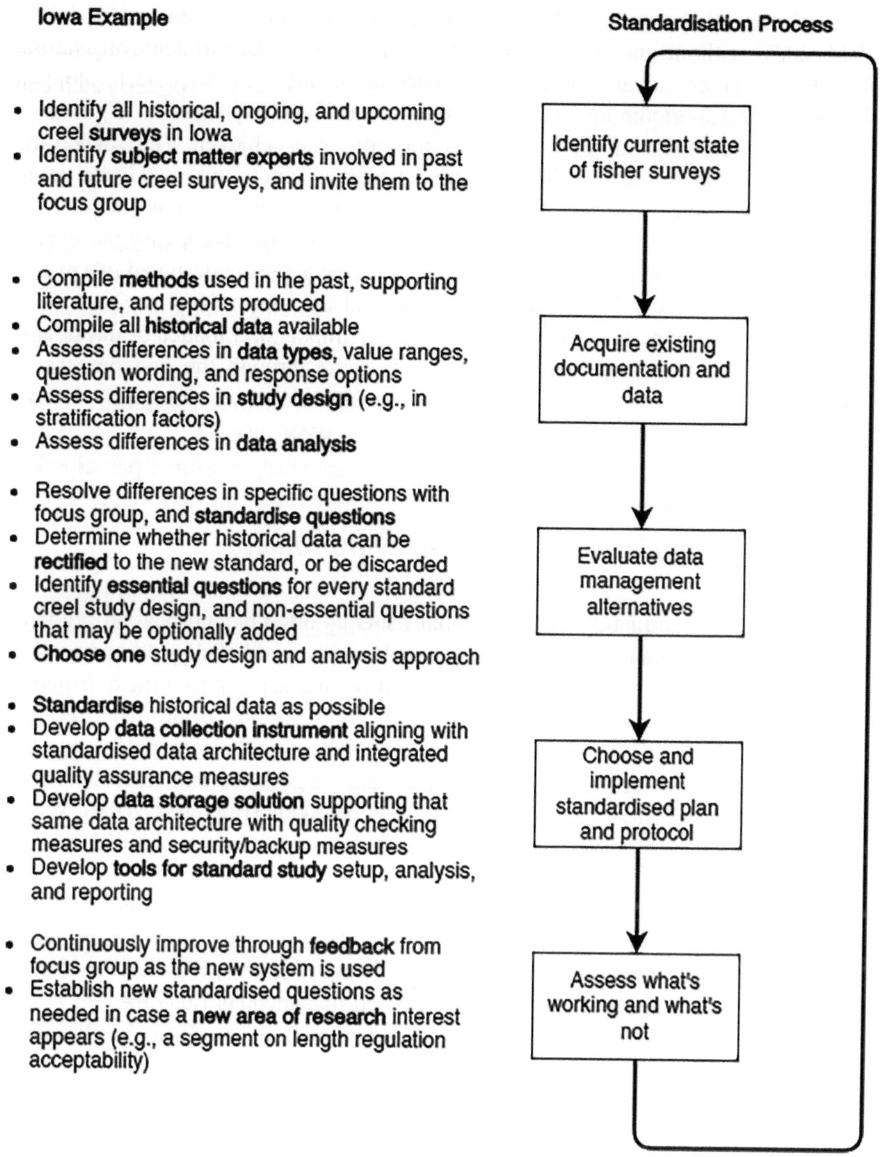

Fig. 1 Standardisation process for aligning fisher surveys, with an example from Iowa where all fisher surveys were standardised

All creel survey data fed into the same database, thus subsequent data analyses, including expansions, were able to follow a standard calculation process. Several calculations were built into the shared system, so that expansions and other summaries could be completed for a location and date range at the press of a button. This yielded far easier and more rapid analysis statewide, while ensuring that all biologists utilised the same analytical approach.

Creel data were collected all over the state using this standardised system during 2014–2022 without integrating with the official Iowa Department of Natural Resources' server, advancing creel standardisation but with some technological limitations and shortcomings. However, the most difficult challenge of aligning all staff to a standardised survey design and questions was achieved. Thus, when the opportunity arose to integrate the standardised creel system to the Iowa Department of Natural Resources' server, which provided vastly improved data acquisition forms and quality checking tools, the shift was rapid and relatively easy. A revised data architecture, online interface, and data acquisition app were designed in a matter of months and launched in May 2022. Historical data from 2014 to 2022 were rectified to the new architecture and prepared for import in a matter of days. This would not have been possible without having gone through the first process of consultation and collaboration with statewide stakeholders (i.e. biologists) to standardise the Iowa Department of Natural Resources' creel surveys.

3.3 *Two Examples of Regional Standardisation*

The management and study of recreational fisheries often extend beyond political boundaries. Some inland fish stocks such as salmonids, paddlefish, sturgeons, and a variety of shad and herring species, commonly migrate across inland fisheries managed by multiple agencies (Hooker et al. 2009; Temby et al. 2015; Phelps et al. 2017; Tripp et al. 2019). Many recreational fisheries are also part of networks, including rivers and streams, chains of lakes, and large lakes and reservoirs, that span more than one jurisdiction (Lynch and Taylor 2000; Tripp et al. 2019). When fisheries are viewed as social–ecological systems, in many cases, ecological dynamics and social dimensions extend across multiple jurisdictions (Nieman et al. 2021). Fishers are often not constrained to a single jurisdiction and may shift effort across boundaries based on the fisheries characteristics, differences in regulations, presence of desirable infrastructure or amenities, and for a variety of other reasons (Ward et al. 2016). To regulate a shared fishery, meet desired management outcomes, or better understand the factors that are influencing a fishery, it may be useful to utilise data from multiple jurisdictions. To do this effectively, it may be beneficial to standardise survey methodology and data management.

3.3.1 Pacific Northwest Salmon Fisheries

In the Pacific Northwest (United States of America and Canada), some salmon fisheries are co-managed by tribal, state, and territory agencies under the terms of an international treaty (Pacific Salmon Commission 2022). The Pacific Salmon Commission is an international decision-making organisation that leads efforts to cooperatively manage Pacific salmon fisheries. Effective and equitable management of the fishery, because of the anadromous life history of Pacific salmon, depends on managing harvest in both marine and inland systems that often span multiple jurisdictions (Cederholm et al. 2001; Ruckelshaus et al. 2002). Due to the sensitive nature, these fisheries are sometimes managed with strict harvest allocations across multiple jurisdictions to manage the entire stock in an equitable manner (Joint Columbia River Management Staff 2022). In the case of the Columbia River and its tributaries, harvest is co-managed by four treaty tribes as well as the states of Washington, Oregon, and Idaho, as outlined in US vs Oregon, 1969. The text of the treaty identifies the need for the development of a "stock assessment and management data system, including catch, effort, escapement, and coded-wire tag data that will yield reliable management information in a timely manner and develop analytical models along with standardised methods for monitoring fishing effort" (Government of the United States of America and Government of Canada 2014, pp. 159). Fisher surveys serve as a valuable tool to adequately monitor harvest and ensure quotas are not exceeded. The cooperative efforts of the parties to collect standardised information that can be shared to manage interjurisdictional fisheries serve as an example of the benefits of developing standardised methods and data management systems.

3.3.2 Upper Mississippi River Fisheries Management

The Upper Mississippi River is an inter-jurisdictional waterway located in the northern United States of America, which is managed by the states of Missouri, Illinois, Iowa, Wisconsin, and Minnesota. Cooperation among state natural resource agencies, through the Upper Mississippi River Conservation Committee, aids the management of this important fisheries resource. Addressing management challenges, such as the impacts of habitat modifications and invasive species, in this large river system requires cooperation across boundaries due to the interconnected, multi-jurisdictional nature of fish populations (Garvey et al. 2010). The ability to leverage fisher survey data collected from multiple agencies is likely to be a crucial factor in ensuring this interjurisdictional fishery can be effectively managed. The collection of consistent creel and recreational use data from the Upper Mississippi River was identified as being valuable for future management efforts (Schlesser 2020). Efforts are currently underway by the Upper Mississippi River Conservation Committee creel survey committee to develop a program of standardised creel surveys with interjurisdictional coverage of the system to enhance the ability of those agencies charged with managing the resource to do so with a more complete understanding of this complex system (UMRCC 2021). This cooperative effort highlights the value of standardisation and planning to achieve management objectives across political boundaries.

3.4 Approach to Standardisation

Standardisation of creel data requires recognition of the data management lifecycle in general: a continual process of planning, data acquisition and cleansing, data architecture design, data analysis and reporting, and data sharing and archiving (Krogman et al. 2023). Throughout this cycle, key elements are repeatedly emphasised: documentation and metadata, quality management, and security and backup (Fig. 2). Whether standardising previously collected data or collecting new

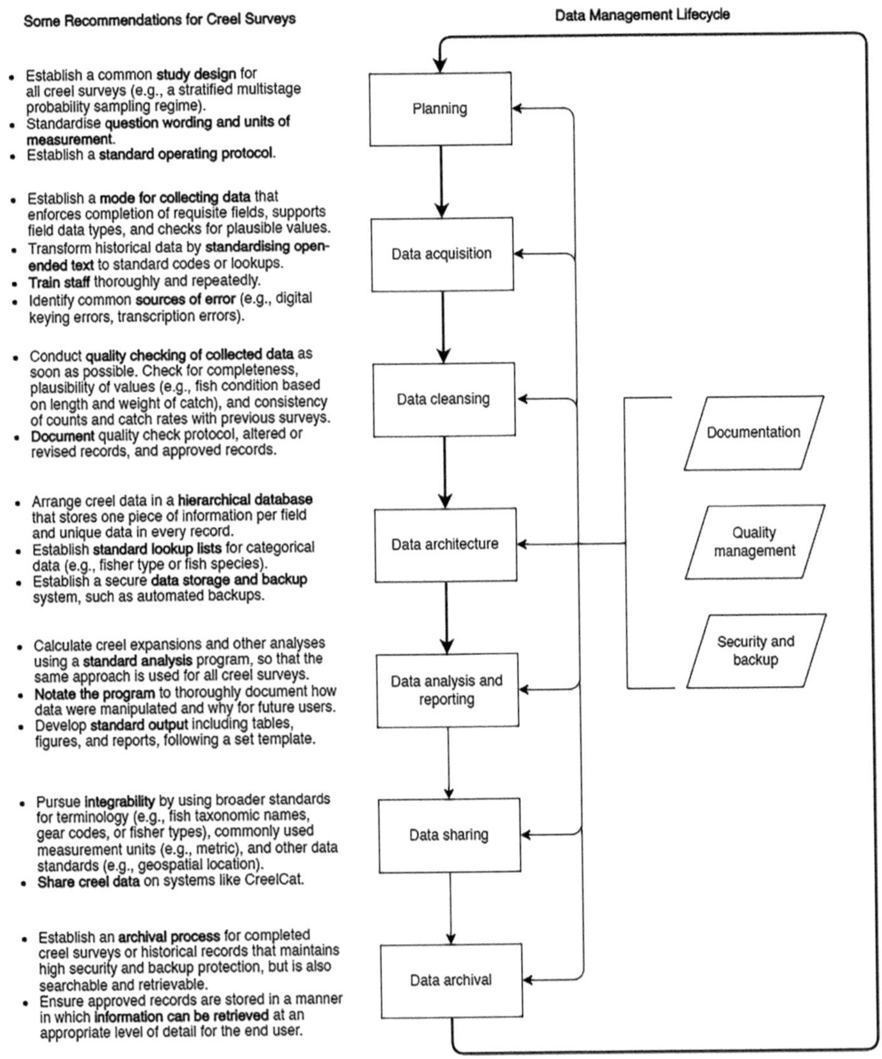

Fig. 2 The data management lifecycle with reference to creel surveys associated with each step

data, standardisation requires extensive planning before staff are deployed for data acquisition. Much of the documentation, quality assurance and checking, and data storage depends on planning early to avoid later data wrangling or data loss. Therefore, researchers must not wait to study each stage of the data management lifecycle as they come to it; instead, consideration of each step and the foundational elements should occur during planning. Note also that Krogman et al. (2023) considered quality management to be a continuous improvement process, and that both quality assurance and checking may evolve over time as common missteps or error sources are identified.

Although creel surveys can be incredibly diverse, consistent data and metadata standards can be used in surveys to achieve greater interoperability in the future (see Sect. 4 below). It is helpful to organise data in a hierarchical database that stores at least the following information: location, date, beginning and end times of work shift, pressure counts of recreational fishers in standardised categories, and interviews including fisher category, party size, fishing trip duration, and catch (Fig. 3). Catch should include fish species, number of fish, and fish size (length, weight, or both). When these data are collected, they can often be aligned to existing naming standards (Table 1). This hierarchical organisation differs from a flat file (e.g. a spreadsheet) in that individual pieces of information are never stored redundantly. For instance, the date of the survey is not repeated on every data line; instead, it is stored in a higher-level Survey_Day table that contains all the day's metadata, one data line for each day. This can help reduce keying and copying errors and ensures that a correction of a value only needs to be performed once. Databases can also support data types, standard responses, and quality assurance more effectively. For example, fish species may be a standardised code, which is best supported through a lookup dropdown menu; this limits the data entered to a set list and disallows deviant codes. A more detailed example with real-world creel data from the United States of America is provided by Lynch et al. (2021).

Components beyond catch, effort, and satisfaction may also behove fisheries science as a whole, including aspects that are useful in social–ecological systems modelling (Nieman et al. 2021). Specifically, more information is needed regarding fisher heterogeneity, catch-and-release and fish welfare orientations, and fisher attitudes (Arlinghaus et al. 2021; Nieman et al. 2021). Nieman et al. (2021) specifically suggested postal codes (as an index of distance travelled and source of demographic information) and questions regarding avidity and preferences (helping to understand site choice). Contextual information is also important for interpreting and analysing data derived from creel surveys, and standardised methods for incorporating relevant contextual information would be helpful. This includes information such as regulations at time of survey (because regulations may change over time), waterbody characteristics (i.e. waterbody area at time of survey, particularly for reservoirs that have pools that can change dramatically), and environmental conditions (i.e. ice duration can vary greatly from one year to the next and may influence the amount and types of fishing that occur). Standard methods of reporting survey characteristics and performance metrics, such as survey stratifications, interview and count methodology types, or standards such as percent of counts

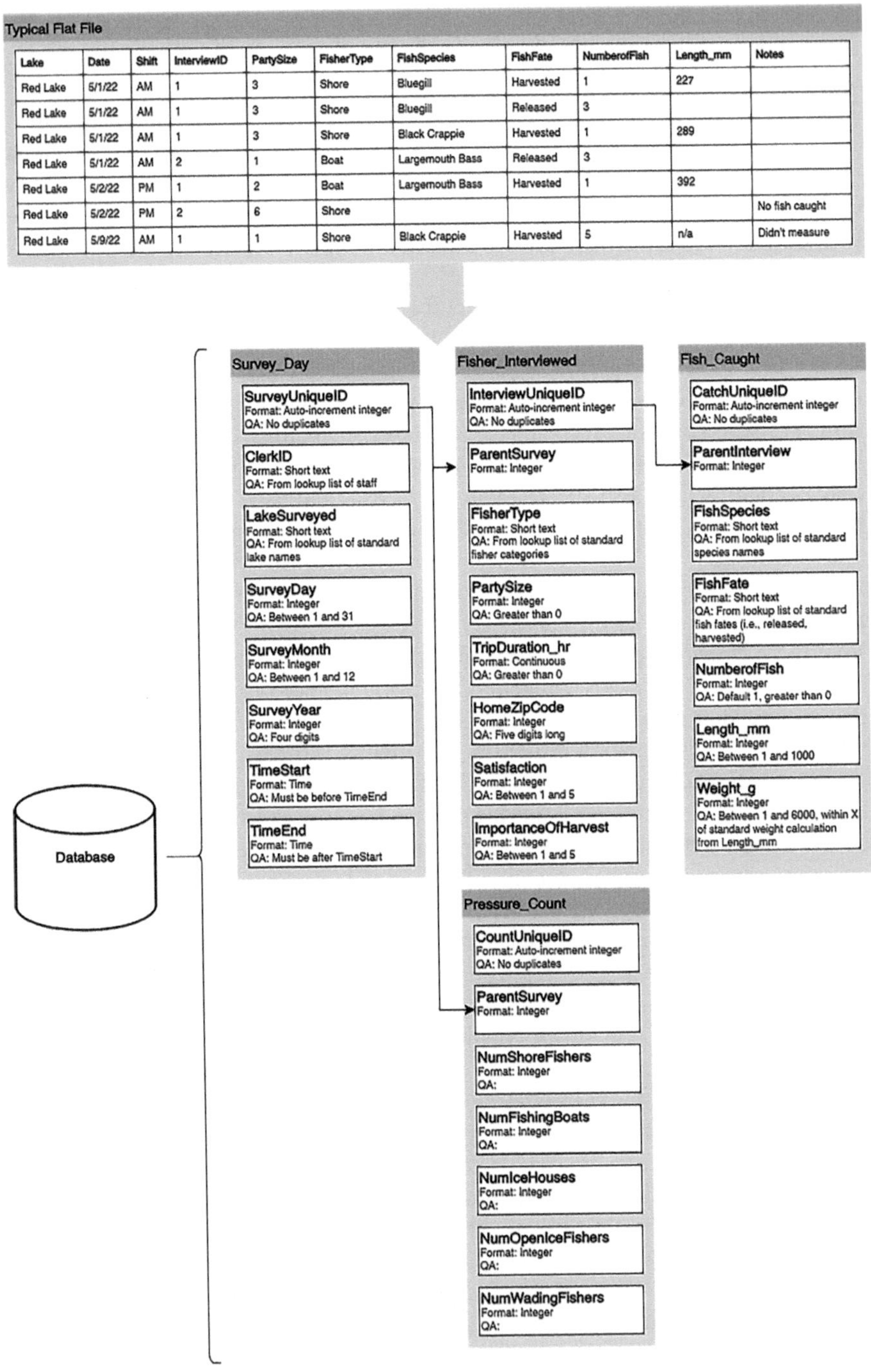

Typical Flat File

Lake	Date	Shift	InterviewID	PartySize	FisherType	FishSpecies	FishFate	NumberofFish	Length_mm	Notes
Red Lake	5/1/22	AM	1	3	Shore	Bluegill	Harvested	1	227	
Red Lake	5/1/22	AM	1	3	Shore	Bluegill	Released	3		
Red Lake	5/1/22	AM	1	3	Shore	Black Crappie	Harvested	1	289	
Red Lake	5/1/22	AM	2	1	Boat	Largemouth Bass	Released	3		
Red Lake	5/2/22	PM	1	2	Boat	Largemouth Bass	Harvested	1	392	
Red Lake	5/2/22	PM	2	6	Shore					No fish caught
Red Lake	5/9/22	AM	1	1	Shore	Black Crappie	Harvested	5	n/a	Didn't measure

Fig. 3 Example of converting storage of creel data from a single table, also known as a flat file, to a standard hierarchical structure with field formatting and quality assurance rules in place

Table 1 Data standards for creel surveys

Metric	Scope	Standard	Source
Geographic names, especially waterbody names	United States of America	Geographic names information system	U.S. Geological Survey, national hydrography dataset
Geospatial information	Global	ISO 19115:2003	International organisation for standardisation
Date and time	Global	ISO 8601	International organisation for standardisation
Fish species identification	North America	Common and scientific names of fishes in North America	Page et al. (2013)
Fish species abbreviated code	North America	Abbreviations for fishes of North America	Krogman (2016)
Fish species identification	Global	Integrated taxonomic information system	ITIS (2022)

completed, may be included to aid in the interpretation and use of data. We have included some examples of how these fields may be integrated with a standard creel data architecture (Fig. 3).

3.5 Constraints and Challenges for Standardisation Efforts

Despite their importance and benefits (de Kerckhove et al. 2026), creel surveys tend to be limited in terms of funding and staff. Survey programs must balance the demands of stakeholders who desire resources to be expended at popular fisheries and the need for broader spatial targeting of fishery monitoring efforts (Lester et al. 2003). Perhaps more importantly in the context of this chapter, limitations to data collection hinder the inclusion of all aspects that allow for social–ecological system analyses, such as ensuring interview length is acceptable and that survey exhaustion does not occur because of adding human dimensions questions (Nieman et al. 2021).

Not all aspects of a creel survey must be standardised. For instance, questions specific to a particular fishery may be asked as part of a creel survey. Opinions regarding restrictive regulations have been included in specific surveys for muskellunge (*Esox masquinongy*) in Virginia (Brenden et al. 2007), crappie (*Pomoxis* spp.) in Ohio (Petering et al. 1995), walleye (*Sander vitreus*) in Iowa (Krogman et al. 2022), and many other places. That said, it may be possible to standardise the way a question is asked, facilitating analysis. If asking about how a fisher feels about a proposed fish length restriction, the wording and measurement scale may be standardised to "Would you oppose or support a minimum length limit of 18 inches for walleye in Lake X?", with response options of "strongly oppose", "oppose", "neither oppose nor support", "support", "strongly support". If asking about preference between two possible regulations, the wording might be standardised to

"Which would you prefer: [Scenario A] a minimum length limit of 18" with a bag limit of 5 fish, or [Scenario B] a protected slot limit of 18–21 "with a bag limit of 1 fish above the slot"? Within a single management entity, this type of question standardisation is feasible, although scenarios may differ from fishery to fishery.

Though standardisation of methodology is a worthwhile goal, rarely is there a one-size-fits-all solution for survey protocols, especially at broader spatial scales. The timing and duration of a creel survey may be influenced by factors such as when seasons are open for frequently targeted species within a waterbody, length of day, or safety of creel technicians. The selection of appropriate survey methodology may be influenced by a variety of factors, including waterbody size, accessibility, patterns in fisher use, and more. Thus, it may be advisable to develop standards for different classes of fisheries based on the characteristics of the waterbody, taxa composition, or fisher population. For instance, having one set of standardised survey protocols for lakes and reservoirs and a separate set of protocols and metrics (i.e. harvest per acre or harvest per mile) for trout streams will allow comparisons and analyses to easily apply across multiple waterbodies within a class, while practically acknowledging differences among systems and associated logistics.

In some cases, the continuity of a long-term sampling program may be deemed more important than adjusting to a shared standard approach. Maintaining consistent estimation of fisher numbers, effort, or satisfaction may be necessary to compare to historical values, thereby limiting how much the protocol may be changed. It may be possible to expand such a protocol to help it fit with a broader standardised approach, or it may be possible to calculate an adjustment to align historical data with standardised data, as has been done with fish sampling gears (Peterson and Paukert 2009).

Finally, standardisation should be strategic, requiring not only careful establishment of standards but honest and ongoing evaluation. Its cost may entail the procurement of new equipment, the implementation of new processes, and the dedication of new staff. These costs balance with the value of high-quality data and the alternative cost of lost or poor-quality data (Sakpal 2021), as well as the risk of standards changing or becoming obsolete over time (Farrell and Saloner 1985). Technological advancement, environmental changes, and societal shifts are all likely to result in changes to the methods employed in recreational fisher research, resulting in the need to alter methodology over time. Consideration of a reasonable lifespan for a set of standards and anticipating future needs is an important part of the standardisation process and should always be considered. Although any set of standards is likely to have a finite lifespan, the ability to modify a relatively uniform set of standards may increase adaptability compared to the demands associated with altering numerous methodologies. Consideration of the future needs can reduce the burden of adapting standards, and identifying opportunities to maintain consistency can increase the long-term usability of a dataset. The benefits of standardisation can outweigh the costs and invite the reader to challenge and improve upon outdated standards, as necessary. This will only improve recreational fisher data quality and availability over time, contributing to increased trust and professionalism in recreational fishery science.

4 Data Integrability

4.1 Management for Integrable Data

Ensuring data can be shared beyond its initial purpose is critical to harnessing its potential for informing future management and research needs. Project planning (Sect. 4.2), data architecture (Sect. 4.3), completeness and accuracy (Sect. 4.4), metadata (Sect. 4.5), and storage and access (Sect. 4.6) each play critical roles in the ability to integrate disparate data sources (Box 1). When data are well managed, they are compliant with FAIR principles (Wilkinson et al. 2016). Issues with any of these qualities can limit the ability of the data to be integrated into new efforts within or outside of an organisation. Thoughtful data management plans that consider these characteristics will help ensure the long-term usefulness of a dataset. The following guidance documents some best practices both for managing directly collected datasets (i.e. information from a fisher survey) as well as efforts to compile data from multiple sources into a composite dataset (i.e. national or regional database of creel and fisher surveys). When making a data management plan, it is important to determine how data is to be formatted, what information is necessary to interpret the dataset, and how it can be stored, found, and accessed.

Box 1 Considerations for Developing Integrable Datasets (for More Detailed Information Please Consult the Associated Chapter Sections)

Project planning (Sects. 4.2, 4.3, 4.4, 4.5, 4.6, and 4.7)	
Stand-alone and long-term efforts Review related projects for standardisation of study design and data architecture. Identify information needs and how those data are best structured. Consider future uses of the data to inform study design and data architecture.	*Composite datasets from varied sources* Determine data needs for the project. Identify and acquire potential data sources. Review source data and determine which datasets are usable and how that data can best be combined. Consider how the data may be used by yourselves and others to inform dataset content and architecture.

Data architecture (Sect. 4.8) *Format*	
Avoid the use of reports or documents to store data Often incomplete or summarised limiting access to the full set of underlying data. Not machine-readable and often requires manual transcription.	*Tabular data and relational databases are appropriate methods for storing data* Use non-proprietary formats. Capable of providing machine-readable data. Non-hierarchical data as a tabular dataset. Hierarchical data as a relational database.

(continued)

Box 1 (continued)

Fields		
Select appropriate type Numeric. Class or factor. Character.	*Define attributes* Name. Format. Definition and units.	*Constrain and validate* Correct type. Within range.

Completeness and accuracy (Sect. 4.9)	
Ensure data are recorded in a complete manner All relevant data for both the current and any future efforts is retained. Retain directly assessed data (counts, measures) even if data being used take the form of calculations, summaries, rates.	*Validate accuracy of data* Ensure classes match a discrete set of options. Review text or character strings. Set range of acceptable values and verify values fall within range and check outliers. Confirm ID fields are unique. Compare across related columns to identify within range errors.

Metadata (Sect. 4.10)	
Provide thorough documentation for the dataset Study scope. Methodology. Each variables name, format, definition, and units. Variable format should be unambiguous. Methodology for calculated values should be described or a reference provided.	*Metadata files should meet the following criteria* Saved in an open, non-proprietary format (i.e. XML). Meet relevant standards (i.e. FGDC-CSDGM, ISO). A variety of tools are available for creating metadata.

Storage and access (Sect. 4.11)		
Avoid storage of datasets on local systems Often lost or forgotten. Challenging to find and access.	*Use network storage for internal datasets* Access may be limited via permissions. Existence or location of data may be unknown.	*Use web storage for internally or externally accessible datasets* Access may be limited via permissions. Often easier to find. Provide tools for versioning and archival.

4.2 Project Planning

Whether developing a stand-alone survey or study, a long-term program, or a composite dataset from a variety of outside sources, it is critical to thoroughly plan the effort to ensure that all data are collected and managed in a way that supports both the initial intended use and potential future use. For all efforts, the goals for the use of the data should be identified, and both required and beneficial supplementary data elements should be identified. When planning a stand-alone survey, previous related efforts should be reviewed, and opportunities for consistency and standardisation with those efforts should be explored. Additionally, it is also worth exploring if there may be related projects taking place in the future to plan a study design that will potentially allow standardisation and integration with those efforts. Planning a long-term program can benefit from including a thorough review of past efforts to identify both commonalities and unique elements of existing datasets. Future needs and opportunities can be explored, and efforts to standardise the program can be emphasised. When data are being compiled from a variety of sources, it is beneficial to determine the scope and requirements of the effort, determine methods for data acquisition, and develop a strategy for long-term data management.

The collection of data for use beyond the initial purposes of a dataset can be daunting and often faces limitations based on both data management issues with the source data and the project design of the new effort. It is important to employ the principles for integrable data described in the following sections whenever possible; however, given that the datasets being brought together are often variable in nature there are unique challenges that often exist in these types of efforts. Before beginning a data compilation effort, it is important to determine the scope and desired outcomes of the effort. A critical first step is to determine the desired spatial and temporal scope of the effort to identify data contributors, make clear and appropriate data requests, and develop the architecture for the dataset.

There are a variety of ways to identify potential source datasets. If source data have been published and are available online, it may be as simple as downloading the data from a repository and adhering to any licensing or other guidance about the use of the data. When the existence of source data is known but open access is not provided, it is necessary to reach out to the data holder with a data request. In other cases, the existence of datasets may be unknown, and requests for data may be made to groups of people who may have the needed data. Approaches for acquiring data in these situations may involve requesting data via an email listserv with relevant membership, via a presentation at a conference, or systematically reaching out to potential data sources, such as management agencies. Requests should clearly state the scope of data being sought, purpose of the request, a reasonable deadline for completing the request, required characteristics of the data (format, needed fields), and discussion of opportunities for acknowledgement of contributions (Midway et al. 2022). For large datasets and spatial data, it may also be necessary to set up file transfer accommodations in the cloud. In some cases, highlighting incentives for participation may increase the willingness of data providers to contribute. Incentives can include access to potentially useful data or tools, funding opportunities, or opportunities to participate in collaborative networks. Willingness to discuss a request and answer any questions the data provider has can aid in both identifying any potentially useful data and getting data providers to agree to fulfil the request.

Once data have been received, the challenging process of integrating data from multiple sources can begin. Source datasets should be reviewed to provide a preliminary assessment of what data are available and what issues may need to be addressed during the compilation process. Setting standards regarding the contents and quality of the data being incorporated should be completed before data integration begins. This should include identifying acceptable study scopes and methodologies and determining required data elements. Datasets meeting the requirements of the user should be separated from those that do not. In some cases, it may be necessary to follow up with data providers to address any issues, such as missing, unclear, or likely erroneous values and metadata. Once all datasets have been reviewed and acceptable datasets have been identified, the architecture of the compiled dataset can be developed. In many cases, combining these datasets will require modifications such as unit conversions, additional calculations, and the use of contextual fields that provide additional information necessary to interpret associated records.

4.3 *Considerations for Data Architecture*

The format of a dataset is critical to ensuring it is integrable and can be utilised for various future efforts. Datasets are commonly stored in reports or manuscripts, tabular formats, or in relational databases. These various approaches to data architecture offer different benefits and drawbacks for efforts to utilise the data beyond the scope of the initial effort (Box 1).

4.4 *Documents for Data Storage*

In many cases, datasets may not be retained once projects are complete, leaving only reports and manuscripts as representations of the data that were collected. This often creates a substantial loss of data, and efforts should be made to save underlying datasets, as well as products such as reports. Oftentimes, reports, manuscripts, and other documents are saved as portable document format (".pdf") or Word document (".docx") files. Utilising data directly from documents can present several challenges for future use. One issue with using data from documents is that often only items of interest may be included, leaving out information that might be valuable for future efforts. In some cases, the data contained in a report may be truncated or summarised and only include the most common elements (i.e. a creel survey recorded data related to a dozen taxa; however, only the five most common taxa were included in the report). In other cases, a report may be focused on specific taxa or metrics and potentially valuable information related to other aspects of the fishery may be omitted (i.e. a creel survey was completed to evaluate Centrarchid regulations, and the data included in the report may not include information that was collected for Ictalurid or Salmonid taxa). Efforts should be made to retain the complete dataset. When data are presented within the text of a report, they can present an enormous burden for data compilation efforts as it is often necessary to transcribe data value by value into the new, compiled dataset, rather than being able to copy and paste large volumes of data at once. In many cases, data contained within report files (.doc and . pdf formats) are limited in its ability to be read by machines. Ensuring data exists in a machine-readable format is critical in ensuring data can be accessed and extracted in an efficient and accurate manner. Even when data are presented within tables in a document, it is often difficult to transfer the information from the document into a tabular electronic format, as many times, correct columns and rows may not be recognised when extracting data from a document format into a tabular format for compilation and additional analyses.

4.5 *Tabular Datasets*

Storage of data in tabular formats has several advantages over data stored in reports, including ease of reformatting, ability to add new data, and accessibility to analytical software. Tabular data are often stored as text files with comma-separated values (". csv") or spreadsheets (".xlsx"). One caution in the use of spreadsheets is to avoid the use of formulas in the final saved version of a dataset as these can serve as a potential source of error if future users modify the file structure in ways that result in calculation errors. Other important considerations include the potential for truncated fields, field type conversions (i.e. numeric to character when values are stored in scientific notation), or other unintended modifications to the dataset when it is being read or converted. It is critical to verify that the values and field names have not been changed when changing formats or reading software packages. Typically, data stored in electronic tabular datasets are machine-readable, which allows data to be quickly and easily be accessed and extracted. Whenever possible, it is best to store data in non-proprietary formats (i.e. ".csv" or ".txt" files) so they can be accessed freely. In some cases, potential users of the dataset may not have access to the software needed to read data in proprietary formats. Additionally, the use of proprietary file formats can also cause issues if the developers of the software required for reading a given file type stop supporting it in the future, which could cause data to become inaccessible.

4.6 *Relational Databases*

Storage of data in relational databases is often the most effective and efficient way of storing a dataset. Relational databases allow data to be stored hierarchically so associations can be made across levels using a key field. In the case of fisher survey data, taxa-specific estimates may be recorded as separate records (rows) for which many taxa can be linked back to a single record related to survey or waterbody characteristics based on a key field such as a unique identifier for a survey or waterbody. By linking data across hierarchical levels, redundancy in data entry and storage can be minimised. Relational databases typically allow users to perform queries of a dataset to extract data meeting desired criteria. When compiling data from a relational database, users may select which fields they wish to include in the output dataset, which can be helpful when many fields are not of interest to a particular research or management question, and they may also filter records (rows) based on custom selection criteria. Relational databases can provide access to data in a machine-readable format, allowing easy access to the data and accurate extraction (no need for manual transcription). Though relational databases represent an efficient and functional way of storing data, constructing them typically requires additional effort and expertise relative to tabular data.

4.7 Fields

Datasets that are designed with formatting that aids in interpretation and use, and evaluated for accuracy, support integrable data management. Constraining fields to specific formats can be a useful tool for ensuring data are entered and stored correctly. Many types of software used for storing electronic tabular data include the ability to restrict the types of data stored in each field (e.g. numeric fields only allow numbers, factor or class fields can be limited to a discrete set of classes). Numeric data should be stored as such and formatted appropriately (e.g. counts as integers, measurements with an appropriate number of decimal places). For numeric fields, this includes checking whether values are within a reasonable range (negative values when all values should be positive, values that exceed a maximum threshold, such as percentages over 100 when not allowed, outliers that exceed reasonable expectations). Numeric fields should also be checked to identify possible non-numeric values (non-numeric characters included in any values, any records containing text). For descriptive fields, it can be helpful to create discrete factors for those fields that can be treated as a class, such as taxonomy, methodology, and habitat. Restricting data entry to discrete classes can make it easier to perform quality control, filter, or analyse the data later. For class fields, the list of unique classifications should be checked against the expected complete set to identify any potential erroneous classes. When data are stored as text or character strings, it is often helpful to place reasonable limits on the length of text which can be entered for a given field. When using a relational database, all key fields, which are used to link tables through defined relationships, should be evaluated for uniqueness to avoid conflicting relationships, and key fields in a related lower-level table should have a match in the main, higher-level table. Making sure all values contained within a dataset are accurate and formatted correctly is critical for ensuring a dataset meets the requirements of future users.

4.8 Data Architecture Considerations

Determination of an appropriate data architecture should be based on both the needs of the current effort and consideration of potential future efforts. For composite datasets, data architecture should also consider the content and structure of the source data. If data are easily represented at a single level, a simple table may be sufficient, whereas if data are hierarchical, then a relational database likely offers a more appropriate structure. When creating a relational database, it is important to consider how data are best linked from one level to another. If the dataset has a spatial component, consideration needs to be given as to both the format of the spatial data (points, lines, polygons, and rasters) and how spatial data will be linked with tabular data.

Tables require thoughtful consideration of their architecture (often with observations being represented with rows and variables, measurements, and calculations as fields), and all fields need to be given a name, format, definition, and units (where appropriate), which can be documented in the metadata. Fields representing measurements should be based on consistent units (e.g. length in centimetres, harvest estimates as counts). For any fields representing names (e.g. taxa, locations) a consistent nomenclature should be used (Table 1). Any sources used for names should be clearly documented (e.g. Integrated Taxonomic Information System, Geographic Names Information System). Any fields that involve a calculated value should be adequately defined so the user can determine what observed or measured values went into the calculation and how the calculation was performed. For example, though a field representing fisher preference called "Taxa_Target_Percent" and defined as the percentage of targeted effort allocated to a given taxa may seem sufficient, understanding whether targeted effort was based on fisher hours, trips, or days can provide useful context. Oftentimes, when compiling data from multiple sources, fields intended to capture the same information may be measured or calculated using slightly different techniques. In these cases data can be either stored in separate columns, or placed in a shared column with a contextual field (e.g. a "Target_Percent" field may contain values from 0 to 100 representative of the percent of angling effort, and a related column "Target_Percent_Calc" may document how the "Target_Percent" field was calculated for each record with class values such as "Percent of Fisher Hours Taxa Targeted" and "Percent of Fishers Indicating Taxa as Primary Target").

4.9 Completeness and Accuracy

The completeness and accuracy of a dataset are critical for its use beyond initial reporting and analysis. Data completeness refers to whether all relevant information has been stored within a dataset. Often, more data are collected than required to answer a specific management or research question. For example, a creel survey may collect information on all caught taxa, but the report may focus on the characteristics of a subset of those taxa. Though those select taxa may have been all that were required for the initial report and associated decision-making or management task, the more comprehensive dataset may have utility down the road, and a failure to adequately report the full dataset may result in it having limited utility. Another important related consideration is that it is advisable to store directly assessed data (measures, counts) wherever possible, rather than summarised data or outputs of calculations. For example, a survey may collect harvest and effort information to calculate harvest rates (harvest per hour of effort). If the stored data contains the directly measured data (harvest and effort), then future efforts can leverage those measured values while still having the ability to calculate harvest rate; however, if only the calculated harvest rate is reported (without measures of harvest or effort) neither of those values can be inferred. If future use of data requires harvest or effort information, the lack of the original measured values will render the dataset unusable. The same issues can arise when data are pooled. Pooling data often results in situations where the underlying values cannot be attributed. If metrics related to several taxonomically related species are pooled and a future effort is interested in assessing the taxa independently from the larger group, it will not be possible. For example, if values associated with largemouth bass (*Micropterus salmoides*), smallmouth bass (*Micropterus dolomieu*), and spotted bass (*Micropterus punctulatus*) are pooled into a single value for the genus black bass (*Micropterus*), then values cannot later be attributed back to the individual species. However, if values are reported for the individual species, they can be pooled to generate a genus-level value. Information representing confidence intervals, variability, or other measures of associated error is also helpful to include in a dataset as it allows users to associate the precision of a given value. Including sample size is also a particularly useful piece of contextual information and can often be used to convert among common measures of error.

It is important to review data to ensure accuracy. This includes verifying all values fall within acceptable bounds and checking for outliers. At times, it can be useful to evaluate multiple related fields in conjunction with each other. By evaluating outliers for fields that were calculated based on multiple related fields, it is possible to identify errors in reported values that otherwise fall within the reasonable range of values for the input fields. For example, a harvest value may fall within the range of reported values but appear as an outlier when converted to harvest per hour. Likewise, a fish weight may seem reasonable until fish condition is calculated using length; an outlier in the condition metric could indicate a measurement or recording error.

4.10 Metadata

The ability to interpret a dataset, via complete and accurate metadata, is an often-overlooked, yet critical aspect of data management. Metadata are typically stored as separate files that contain the information necessary to interpret a dataset. This includes documentation of the scope of the study, study methodology, analytical methods, definitions of fields, and other pieces of information necessary to interpret the dataset. Documentation of the scope of a study includes providing information that allows a user to identify when and where a study took place. These two factors are critical to determining the level of inference that can be drawn from a dataset. When providing spatial information, locations should not only be described (i.e. "Cedar Lake") but also georeferenced. In many cases, a descriptive name alone is not enough to conclusively determine where a study took place, as multiple locations often share names. In addition to providing a georeferenced location, it is important to provide the extent to which the study was representative. In many cases, results are representative of an entire, singular waterbody, however occasionally results may only be representative of a portion of a location or waterbody (particularly in larger systems), or in other cases a survey may encompass multiple locations with combined results (series of ponds in a park, chain of lakes). Defining survey extent is particularly important in stream networks, as the upstream and downstream bounds for which a study is representative should be noted. Although providing spatial information via coordinates and descriptions of extent can often be sufficient, creating shapefiles that document the extent of study or survey areas can be extremely helpful. Providing context related to the timing and duration of surveys is also important. Understanding whether results are representative of a snapshot of a moment in time or over a period of days, weeks, or months shapes how results should be interpreted. Similarly, the timing of when a survey occurred can also influence how results should be interpreted. Seasonal differences are well documented in recreational fisheries (Taylor and Reiling 1981; Margenau et al. 2003), and making sure that data users can account for when a survey occurred is often important. By accurately documenting the spatial and temporal scope of a dataset, data providers allow future data users to adequately determine the level of inference that can be drawn from the dataset. Diverse methodology exists in the assessment of recreational fishers and fisheries. Documentation of how a study is conducted can influence the interpretation of a dataset. Clear documentation of study protocols should exist, including how study sites were selected, what protocols were used to guide survey and sampling efforts, and what methods were used to analyse the data. Studies lacking clear documentation are liable to be misinterpreted or unused for future efforts. Clear documentation allows future data users to determine whether a study is suitable for a given future use and may even provide opportunities to account for potential issues by selecting appropriate analytical techniques. For complex methodologies (i.e. standard operating procedure documents with dozens of pages) it may be appropriate to include classifications or brief descriptions in the metadata (i.e. reference documentation of methodology and provide a short description), while also providing a link to a permanent location (such as a Digital Object Identifier) where additional documentation of methodology can be obtained.

Appropriate metadata are critical to accurately interpret tabular datasets. Each field should have a clear definition and units should be defined where applicable. Metadata for calculated values should define calculations or provide a reference to analytical methodology when a concise definition is not feasible. If a lack of information regarding the formatting of a field could lead to ambiguity (such as with dates "03/07/2021" or times "07:21"), then the format of a given field should be defined (i.e. DD/MM/YYYY or 24 hr). Fields expressed as discrete sets of factors should have the source of the classifications or range of possible values defined in the metadata. When data needs to be stored as text or character strings, typically for descriptive data or comments, information should still be recorded within the metadata that describes the information being characterised by a field. Spatial data requires metadata that captures both the tabular aspects of the data as well as the spatial components. Spatial metadata should contain indirect spatial references when used (descriptions of locations without coordinates), direct spatial reference method (point, vector, raster), and information about the objects, coordinate system details, and projection information.

Metadata that is created in an open, non-proprietary format supports FAIR data principles. Although metadata can be captured and reported in a variety of ways, the use of Extensible Markup Language (.XML) file format is convenient and common. A variety of tools exist that allow XML data to be easily created and read. A wide variety of standards have been developed to ensure the quality and completeness of metadata. In the United States of America, federal agencies have adopted the metadata standards endorsed by the Federal Geographic Data Committee (FGDC), which include both "Content Standard for Digital Geospatial Metadata" (CSDGM) and "International Organisation for Standardisation" (ISO) standards. Style sheets are available that can be applied to the XML file to make metadata documents easier to read and interpret. A wide variety of tools have been developed for generating metadata, including mdEditor (https://www.mdeditor.org/) and the US Geological Survey's Metadata Wizard (https://usgs.github.io/fort-pymdwizard/index.html). These tools allow users to efficiently create metadata files, including biological data profiles and other extensions. The Metadata Wizard can also be added as a toolbox within some geospatial software systems to easily create geospatial metadata.

4.11 Storage and Access

The ability to find and access data can be a major barrier to the ability to compile data for additional uses. Data can be stored in a variety of locations, from a filing cabinet in an office to a publicly accessible web portal or data repository. Oftentimes, there is a trade-off in up-front versus long-term resources needed to manage information, depending on where it is stored. Information that is stored in locations where a limited number of staff have access, such as in someone's desk drawer or on a local hard drive, can be difficult to find and time-consuming to obtain, and data stored in this manner are often lost over time and are not compliant with FAIR standards. Though there is little upfront investment in storing data this way, there is substantial opportunity cost when data becomes difficult to retrieve or lost entirely. The storage of data in shared locations such as a network drive can facilitate the transfer of data by making it easier to locate and access, but does require additional information technology resources relative to local storage, although free scientific data repositories minimise the need for internal information technology resources. Though storage in a network location can enhance accessibility to the data, there is still often legwork involved in identifying what data are available and determining the location of those data. The storage of data in online repositories or web portals can increase data availability by making it possible for those wanting to use the data to access it directly. Although the upfront resources needed to create a web portal can be greater, the ability for users to directly access the data may result in substantial time savings for staff who may otherwise be responsible for fulfilling data requests, although additional resources are often needed to maintain these outlets. Rather than developing an online data repository from scratch, it is also possible to use existing repositories, many of which are available independently or associated with journal publication outlets. Some governmental agencies have created their own repositories for governmental datasets, such as ScienceBase (U.S. Geological Survey), EarthData (National Aeronautics and Space Administration), and others. Uploading data to an outside repository for scientific data requires some initial effort, but requires minimal long-term resources for storing or archiving static datasets.

Without adequate documentation of what data exists and where they are located, data can easily be lost or forgotten, and retrieving them for future use can require substantial time and effort. Ensuring that it is possible to locate data is critical to the long-term value of those data. Staff turnover often results in situations in which data may be lost if those data have not been adequately stored and documented. Collaborative efforts within or among agencies can be hampered when the existence and location of datasets cannot easily be established. Careful consideration of where data are stored should be given to maximise the potential return on investment of a dataset. The creation of a globally unique and persistent identifier for a dataset is an effective way for making data findable. For data that exist online, the creation of a digital object identifier allows for unique and persistent identifiers for a dataset. A digital object identifier will persistently direct a user to a URL where a dataset (and associated metadata) exists online, regardless of changes in domain names and web spaces. Registering, indexing, or storing data within a searchable repository or resource can make it easier to find and access a dataset. Journals focused on documenting and sharing open datasets represent another way of making datasets findable and interpretable.

When making decisions on where to store data, it is also important to consider questions such as who needs access and who may benefit from access. At a minimum, data should be stored in a way that all staff who need access can do so simply and directly, and the number of potential users or perceived value of potential efforts employing the data should guide whether additional efforts to increase accessibility should be made. When determining where data are to be stored, it is also important to consider whether access needs to be limited based on the sensitivity of the information they contain. Any sensitive data stored in a shared location should have measures in place that manage who has access to the dataset. Additionally, all data stored in shared locations should have measures in place to manage which users can modify a dataset, and backups should be made to protect against data loss, corruption, or undesired modifications. Determination of who has the authority to share data is also a critical consideration, and having that role defined can streamline the process when requests for data sharing are made. It is also important to consider the use of versioning with datasets that are updated or modified to ensure that datasets can be referenced and retrieved according to how they appeared at a given point in time to ensure reproducibility of results. Data that are no longer being actively used should still be archived and preserved to allow for validation or follow-up on any analyses that used the dataset.

Open science, which emphasises methodological transparency and data sharing, is gaining traction across many scientific disciplines and across a range of governmental, academic, and non-governmental organisations (Foster and Deardorff 2017; Powers and Hampton 2019; Shrout and Rodgers 2018, European Commission 2021; UNESCO 2021; The White House 2023). Conducting and sharing science in an open manner can increase transparency, reproducibility, and accelerate the accumulation of knowledge (Hampton et al. 2015). Some who practice open science in their work now view their relationship with the data they generate as one of stewardship rather than ownership (Hampton et al. 2015). Data stewardship promotes responsible use of the data and emphasises transparency throughout the data management life cycle (Hampton et al. 2015). This transparency includes ensuring that the limitations and assumptions associated with the dataset are accurately documented and available to those who may use the data (Hampton et al. 2015). Though numerous benefits to conducting science in an open manner have been documented, barriers to its practice exist. In some cases, data may be sensitive in nature due to the presence of personal information, at-risk taxa, or other elements which may cause harm if widely available, and decisions about how or if these types of data can be shared needs to be carefully considered (Fox et al. 2019; Campbell et al. 2023). Fear of misuse is also a common cause for concern among those considering sharing data; however, in many cases, providing guidance and documentation related to proper use of the data can ameliorate these concerns (Sayogo and Pardo 2013). The incentive structures within the research community reward those who publish novel findings, and the fear of scooping, which is others publishing findings based on the data first, is also a barrier for many (Laine 2017). Alternatively, sharing data freely not only promotes replicable and robust science but also provides additional opportunities for collaboration to the benefit of data-providers (Hampton et al. 2015, Allen and Mehler 2019). As the emphasis on open science grows, research organisations, funding sources, and publishers are increasingly mandating or encouraging open science practices, including the management of data in ways that are consistent with the tenets of open science (Kozlov 2022; McKiernan et al. 2016).

4.12 *CreelCat as an Example of a Data Compilation Effort*

The compilation of data from a variety of sources into a composite database is an important approach for addressing large-scale issues in recreational fisheries. We present a recent example, CreelCat, where creel and fisher survey data from inland waters were compiled from across the United States of America (Lynch et al. 2021). This effort was undertaken to help inform regional and national-scale inquiries related to inland recreational fisheries across the United States of America.

The first step in this process was to determine the desired scope of the project (see Sect. 4.2 for a general description). The CreelCat effort focused on fisher intercept surveys that produced expansions based on fisher count data; however, when those surveys were not available, alternative survey types such as pressure estimates, unexpanded interview counts, mail or phone or internet surveys, and fisher diaries were included. The scope of the CreelCat effort is inland waters within the United States of America, with emphasis on data from 2010 to present. The request was limited to data more recent than 2010 to help make our request manageable for data providers. Data of interest included information on fisher effort, fish catch and harvest metrics, survey and waterbody characteristics, fisher demographics, and fisher preferences.

State natural resource agencies were identified as potential data providers for this effort. Emails requesting data were sent to contacts in each state, as well as Washington D.C. and Puerto Rico. These messages succinctly stated the purpose of our request, the scope of the data being requested, types of information for which we were interested, any restrictions or permissions required related to the use of the data, and acknowledgement of our appreciation for their consideration of our request. Most requests required follow-up correspondence or calls related to required permissions, logistics for large dataset transfers, or discussion of unique aspects of the data being provided. Over 80% of requests led to data being provided for use in this effort.

Once a dataset was received, it was reviewed to determine whether it met the criteria for inclusion based on whether it aligned with the desired scope and objectives of the effort. This review also included documenting which variables of interest were consistently available, sometimes available, rarely available, or never available. Units were reviewed to identify those that were commonly selected for various metrics to select standard units for the database that most consistently matched those used by the data providers (inches, pounds, acres, hectares).

Determination of the most appropriate data structure for CreelCat was determined based on both a review of the data received as well as the objectives for its use (see Sects. 4.3, 4.4, 4.5, 4.6, 4.7, and 4.8 for a general description). In addition to providing the dataset to researchers for their own research applications, the CreelCat effort also aimed to create a web application that would allow users the ability to interactively determine data availability, customise data tables based on both surveys and fields of interest with interactive tools for filtering, sorting, and searching the dataset, develop data summarisations, and create a variety of custom visualisations including maps, scatterplots, pie charts, time-series line plots, and bar charts.

Data were hierarchical with records at the level of survey for data related to waterbody and survey characteristics, fisher effort, and demographics, and the level of individual taxa for fish catch and harvest metrics and fisher preference metrics. Data were structured as a relational database with a unique identifier for each survey being used as a key field. Each record was associated with a single survey for tables containing information on survey and waterbody characteristics, fisher demographics, and effort metrics, whereas each record was associated with each taxon attributable to a survey for tables containing information on catch and harvest metrics and fisher preference metrics.

CreelCat contains over 125 fields. Identifier fields include the unique survey identifier used as the key field within the database as well as a unique waterbody identifier used to uniquely identify surveyed waterbodies and link them with associated spatial data. These identifier fields are created as character strings to both convey basic information about the associated survey or waterbody, such as the state, as well as additional characters to ensure the identifiers are uniquely attributed to a single survey or waterbody. Fields containing discrete classifications, such as methodology types, taxa names, and metric reference fields, were represented as factors. Numeric fields were formatted to appropriately convey associated values with the appropriate numbers of decimal places relative to the information being conveyed. Units were standardised to U.S. customary units (inches, pounds, acres, miles), which were how most data providers reported their data. Though the International System of Units is most used for scientific research globally, U.S. customary units were used to allow the data to take the format that was most consistent with the source data to ensure easier integration into provider efforts. Values that were not reported in the standardised units were converted as needed. Waterbody names were always reported to match the source datasets; however, taxonomic names were standardised to match the Integrated Taxonomic Information System. Taxa that were not represented within the Integrated Taxonomic Information System database, such as certain hybrids (i.e. *Salvelinus fontinalis* × *namaycush* commonly known as splake), variants (i.e. blacknose crappie), and non-taxonomic groups (i.e. panfish) were attributed based on the most widely used nomenclature that was as consistent with the Integrated Taxonomic Information System naming as possible. To make taxonomic attributions as accessible to the public as possible taxa names included both the associated scientific and common names (i.e. "*Lepomis macrochirus* (bluegill)"). Many of the fields contained in CreelCat were calculated based on other fields (i.e. "Harvest_Per_Acre" was calculated by dividing the harvest estimate by survey area). For all calculated fields, the methods used for performing the calculations were documented in the metadata. In many cases, data for a single field could not be completely standardised; in these cases, the inclusion of a separate field to provide context was required. For example, data providers for CreelCat calculated the number of outings as a metric of fisher effort. In some cases, this was reported as "days" spent fishing, whereas in other cases it was reported as "trips", the difference being that a fisher could make multiple trips in a single day or have a single trip which lasts multiple days, depending on how defined. A contextual field was used to identify which values in the "Outings" field were measured in days and which values were measured in trips, because these values are not interchangeable and cannot be directly converted.

Data were reviewed for completeness and accuracy, both as data were being entered and as part of a final review once data entry was complete (see Sect. 4.9 for general description). In some cases, the data provided to CreelCat were incomplete (only data on focal species were provided, or information needed to attribute survey characteristics, such as location or timing, were not provided). In these cases, warning fields (e.g. "Focal_Species", "Location_Warn", "Timing_Warn") were used to flag records that were incomplete due to missing information. In other cases, the dataset lacked completeness for taxonomic attribution based on the taxonomic attribution level. For example, if a survey attributed harvest at the genus level (i.e. *Pomoxis* [crappies]), then species level harvest (i.e. *Pomoxis annularis* [white crappie] or *Pomoxis nigromaculatus* [black crappie]) could not be attributed even though one species was or both species were harvested. To address this issue, a field flagging any potential unattributed taxa was created to alert users. This allows situations in which taxa metrics are potentially missing from the dataset due to the attribution level of taxa to be identified. Instances of incomplete data attribution are common and often unavoidable in data compilation efforts; however, it is important to either remove or document incomplete records because of incomplete data elements on analysis and interpretation.

The uniqueness of all identifier fields was evaluated, and values that included duplicates were rectified. For categorical fields, such as taxonomic names and survey methodologies, all values were confirmed to be contained within the list of discrete, acceptable values. For numeric fields, all values were evaluated to confirm they existed within a reasonable range (e.g. percentages between 0 and 100, no unreasonable outliers). For example, catch estimates may be very small for taxa that are not often targeted, have low population densities, or for surveys that are short, occur in small waterbodies, or have limited fisher effort, but may also be very high for taxa that are highly targeted, have high population densities, or in surveys with long durations, occur in large waterbodies, or have high levels of fisher effort. Because of this, a metric with a value in the middle of the reasonable range for a field may be identifiable as a substantial over- or underestimate when other factors such as survey duration, waterbody size, and fisher effort are considered. For example, a reasonable range of harvest estimates may be 10 to 10,000 and angling hours may be 100 to 100,000. A record with a harvest estimate of 5000 and an angling hours estimate of 500 both fall within reasonable ranges, but after reviewing the related fields of harvest and fisher hours together it may be apparent that this record has a harvest per hour (10.0) value substantially higher than other records (0.001 to 2.0). To resolve this potential error, it is helpful to evaluate the source data for potential errors either in the reported data or the transcription of those data (i.e. the harvest estimate of 5000 was supposed to be 50.00, but the decimal point was missing).

Metadata for CreelCat were created using the US Geological Survey's Metadata Wizard, which creates FGDC-CSDGM compliant metadata as a non-proprietary . XML file (see Sect. 4.10 for general description). The associated metadata file documents dataset characteristics, including citation information, associated works, data access constraints, thematic information, data quality and processing details, geospatial references and details, individual field details (column names, definitions, format, and units), distribution and access information, and reference information for the metadata. Complete definitions of the calculations that were performed or references to the calculation methodology were provided for all calculated fields, and each class was defined (i.e. survey types, warning flags) or the source of the class was referenced (e.g. state abbreviations from the U.S. Postal Service, taxa names from Integrated Taxonomic Information System) for fields that utilised categorical data.

The CreelCat dataset is accessible in multiple locations online (see Sect. 4.11 for general description). The primary home for the dataset is on ScienceBase (https://doi.org/10.5066/P9DSOPHD), an online repository for scientific data developed or maintained by U.S. Geological Survey employees or collaborators. CreelCat data can be directly downloaded from ScienceBase or accessed via an application programming interface (API) to make the content fully machine-readable. Additionally, the complete dataset can be directly downloaded from within the interactive CreelCat application (https://rconnect.usgs.gov/CreelCat/). The application also allows users to create custom datasets through the selection of surveys of interest based on survey characteristic filters or geospatial selections, and selection of which fields to include. This allows users to quickly and easily create a custom dataset that provides only the data they are interested in, rather than having to process a dataset with over a million values, many of which may not be of use for their efforts. In addition to providing access to the data, the application provides a variety of tools that facilitate the use of the dataset, including the ability to assess data availability, create custom summarisations, and construct a variety of visualisations, including maps, scatter plots, bar charts, time-series plots, and pie charts. CreelCat data can easily be found via a web search (currently returned on the first page of results of a Google search for "Creel Data"; 12/20/2022) or encountered in various locations around the web where it has been documented (https://www.usgs.gov/programs/climate-adaptation-science-centers/science/us-inland-creel-and-angler-survey-catalog).

5 Conclusion and Future Efforts

Increasing the level of standardisation and integrability of datasets containing data on recreational fisheries is likely to increase our understanding of fisher behaviour, needs, and fulfilment. Though recreational fisheries data are often collected across limited spatial and temporal scales, the ability to leverage data collected across many efforts may provide valuable insight for resource management. The ability to compare surveys across time and space can aid in identifying important patterns in resource use and condition. As the study and management of recreational fisheries continues to emphasise the complex social–ecological nature of these systems, there will be an increasing need for large-scale datasets. When planning a study or survey, it is important to review past efforts to promote potential integration with historical data, identify current goals to ensure information is collected in a way that allows project objectives to be met, and consider potential future efforts to provide information in a way that may support future data needs.

Standardisation needs to involve developing consistent sampling protocols, data management procedures, and analytical techniques. Efforts to standardise data help achieve the levels of consistency among datasets necessary to support robust analyses with adequate levels of inference. Future efforts to develop standardised frameworks and methodologies for fisher surveys could offer substantial benefits. These frameworks could streamline the process of conducting fisher surveys as many of the aspects of survey design and associated analytical techniques would already be developed. Additionally, surveys conducted using these standardised frameworks would help facilitate analysis and comparison. The ability to share data to support additional research and management efforts benefits from data management that considers principles of data integrability, adhering to the FAIR principles of being findable, accessible, interoperable, and reusable (Wilkinson et al. 2016). Integrable data requires project planning with consideration of other potential applications, use of an appropriate data architecture, verification of data completeness and accuracy, thorough metadata, and data storage that is both discoverable and accessible.

We acknowledge that the process of creating standardised data frameworks requires large commitments, both in terms of capacity and time. Further, advances in technology and societal changes may invalidate some approaches in the future without continued support and evaluation. However, many of the components to achieve standardised and integrable data are adaptable and customisable to shifting needs and resource availability. Future efforts to facilitate integrable data management may include the development of accessible repositories for fisher survey data, standardised templates for data formatting and storage, and additional guidelines for metadata creation. We believe managers and researchers will be better able to tackle current and future challenges in recreational fisheries management by working towards standardisation and data integrability in fisher surveys.

Acknowledgements We thank Zach Feiner and two anonymous reviewers whose feedback improved this chapter. Any use of trade, firm, or product names is for descriptive purposes only and does not imply endorsement by the U.S. Government. This research was supported in part by an appointment to the U.S. Geological Survey (USGS) Research Participation Program administered by the Oak Ridge Institute for Science and Education (ORISE) through an interagency agreement between the U.S. Department of Energy (DOE) and the U.S. Department of the Interior (DOI). ORISE is managed by the Oak Ridge Associated Universities (ORAU) under DOE contract number DE-SC0014664. All opinions expressed in this paper are the authors, supported by the U.-S. Geological Survey, and do not necessarily reflect the policies and views of DOE or ORAU/ORISE.

References

Allen C, Mehler DM (2019) Open science challenges, benefits and tips in early career and beyond. PLoS Biol 17:e3000246

Arlinghaus R, Aas Ø, Alós J, Arismendi I, Bower S, Carle S, Czarkowski T, Freire KM, Hu J, Hunt LM, Lyach R, Kapusta A, Salmi P, Schwab A, Tsuboi J-i, Trella M, McPhee D, Potts W, Wołos A, Yang Z-J (2021) Global participation in and public attitudes toward recreational fishing: international perspectives and developments. Rev Fish Sci Aquac 29:58–95

Bartley DM, De Graaf GJ, Valbo-Jørgensen J, Marmulla G (2015) Inland capture fisheries: status and data issues. Fisheries Manage Ecol 22:71–77

Braynen M, Perez VB, Callwood K, Carroll G, Castañeda A, Dahlgren C, Garbutt D, Gittens L, Glazer B, Green F, Haukebo S, Moore R, Poon S, Requena N, Sherman K, Sho V, Tate S (2022) A climate vulnerability assessment for Caribbean recreational fisheries. Fishery Solution Center, Environmental Defense Fund, San Francisco. Available at: https://fisherysolutionscenter.edf.org/resources/climate-vulnerability-assessment-caribbean-recreational-fisheries

Brenden TO, Hallerman EM, Murphy BR, Copeland JR, Williams JA (2007) The New River, Virginia, muskellunge fishery: population dynamics, harvest regulation modeling, and angler attitudes. Environ Biol Fish 79:11–25

Campbell R, Goodman-Williams R, Engleton J, Javorka M, Gregory K (2023) Open science and data sharing in trauma research: developing a trauma-informed protocol for archiving sensitive qualitative data. Psychol Trauma 15:819–819

Cass-Calay SL, Schmidt TW (2009) Monitoring changes in the catch rates and abundance of juvenile goliath grouper using the ENP creel survey, 1973–2006. Endanger Species Res 7: 183–193

Cederholm CJ, Johnson DH, Bilby RE, Dominguez LG, Garrett AM, Graeber WH, Greda EL, Kunze MD, Marcot BG, Palmisano JF, Plotnikoff RW, Pearcy WG, Simenstad CA, Trotter PC (2001) Pacific salmon and wildlife-ecological contexts, relationships, and implications for management. In: Johnson DH, O'Neil TA (eds) Wildlife-habitat relationships in Oregon and Washington. Oregon State University Press, Corvallis, pp 628–685

Chizinski CJ, Martin DR, Huber CG, Pope KL (2014) The influence of a rapid drawdown and prolonged dewatering on fishing effort, catch, and harvest in a Nebraska reservoir. Great Plains Res 24:145–152

Cook MF, Younk JA (1998) A historical examination of creel surveys from Minnesota's lakes and streams. Investigational Report 464. Minnesota Department of Natural Resources, St. Paul

Cowie ED (2021) Understanding angler dynamics in a recreational lake trout (*Salvelinus namaycush*) fishery in Algonquin Provincial Park using long-term access creel data. Master's thesis. Trent University, Ontario

de Kerckhove DT, Tucker CM, Gutowsky L, Morgan G, Guzzo M, Barajas VL (2026) Optimizing creel surveys. In: Pope KL, Arlinghaus R, Hunt LM, Lynch AJ, van Poorten BT (eds) Understanding recreational fishers: disciplinary and interdisciplinary approaches for fisheries management. Springer Nature, Cham, p 387-409

Embke HS, Beard TD Jr, Lynch AJ, Vander Zanden MJ (2020) Fishing for food: quantifying recreational fisheries harvest in Wisconsin lakes. Fisheries 45:647–655

European Commission (2021) Monitoring the open access policy of Horizon 2020: final report. https://data.europa.eu/doi/10.2777/268348. Accessed 14 July 2023

Farrell J, Saloner G (1985) Standardization, compatibility, and innovation. RAND J Econ 16:70–83

Fayram AH, Hansen MJ, Ehlinger TJ (2006) Influence of walleye stocking on angler effort in Wisconsin. Hum Dimens Wildl 11:129–141

Foster ED, Deardorff A (2017) Open science framework (OSF). J Med Libr Assoc 105:203

Fox R, Bourn NA, Dennis EB, Heafield RT, Maclean IM, Wilson RJ (2019) Opinions of citizen scientists on open access to UK butterfly and moth occurrence data. Biodivers Conserv 28: 3321–3341

Garvey J, Ickes B, Zigler S (2010) Challenges in merging fisheries research and management: the Upper Mississippi River experience. Hydrobiologia 640:125–144

Golebie E, van Riper CJ, Kochalski S, Nunan F (2026) Mixed methods research: applications and considerations for the human dimensions of recreational fisheries. In: Pope KL, Arlinghaus R, Hunt LM, Lynch AJ, van Poorten BT (eds) Understanding recreational fishers: disciplinary and interdisciplinary approaches for fisheries management. Springer Nature, Cham, p 481–506

Government of the United States of America and Government of Canada (2014) Treaty between the government of Canada and the government of the United States of America concerning Pacific Salmon

Hall K, Douglas J (2002) The use of creel surveys to answer fisheries management questions in Victoria, Australia. Proceedings of the 3rd world recreational fishing conference, Northern Territory Australia, pp 55–58

Hall-Arber M, Pomeroy C, Conway F (2009) Figuring out the human dimensions of fisheries: illuminating models. Mar Coast Fish 1:300–314

Hooker B, Meyers S, Olsen S, Donnelly K, Culver T, Ford K, Carey S, Bloom S (2009) Federal assistance for interjurisdictional and anadromous fisheries program report 2007–2008. United States, National Marine Fisheries Service, Washington, DC. Available at https://repository.library.noaa.gov/view/noaa/15505

Hampton SE, Anderson SS, BagbyS C, Gries C, Han X, Hart EM, & Zimmerman, N (2015) The Tao of open science for ecology. Ecosphere 6(7):1–13

Hunt LM, Sutton SG, Arlinghaus R (2013) Illustrating the critical role of human dimensions research for understanding and managing recreational fisheries with social-ecological system framework. Fish Manag Ecol 20:111–124

Hunt LM, Camp E, van Poorten B, Arlinghaus R (2019) Catch and non-catch-related determinants of where anglers fish: a review of three decades of site choice research in recreational fisheries. Rev Fish Sci Aquac 27:261–286

Hurley KL, Jackson JJ (2002) Evaluation of a 254-mm minimum length limit for crappies in two Southeast Nebraska reservoirs. N Am J Fish Manage 22:1369–1375

Hyder K, Weltersbach MS, Armstrong M, Ferter K, Townhill B, Ahvonen A, Arlinghaus R, Baikov A, Bellanger M, Birzaks J, Borch T, Cambie G, de Graaf M, Diogo HMC, Dziemian Ł, Gordoa A, Grzebielec R, Hartill B, Kagervall A, Kapiris K, Karlsson M, Kleiven AR, Lejk AM, Levrel H, Lovell S, Lyle J, Moilanen P, Monkman G, Morales-Nin B, Mugerza E, Martinez R, O'Reilly P, Olesen HJ, Papadopoulos A, Pita P, Radford Z, Radtke K, Roche W, Rocklin D, Ruiz J, Scougal C, Silvestri R, Skov C, Steinback S, Sundelöf A, Svagzdys A, Turnbull D, van der Hammen T, van Voorhees D, van Winsen F, Verleye T, Veiga P, Vølstad J-H, Zarauz L, Zolubas T, Strehlow HV (2017) Recreational Sea fishing in Europe in a global context – participation rates, fishing effort, expenditure, and implications for monitoring and assessment. Fish Fish 19:225–243

Integrated Taxonomic Information System (2022) U.S. Geological Survey and Smithsonian Institution. www.itis.gov. Accessed 15 Jan 2023

Joint Columbia River Management Staff (2022) 2022 Joint staff report stock status and fisheries for spring Chinook, sockeye, steelhead, and other species. Oregon Department of Fish and Wildlife and Washington Department of Fish and Wildlife

Kaemingk MA, Bender CN, Chizinski CJ, Bunch AJ, Pope KL (2021) Temporal invariance of social-ecological catchments. Ecol Appl 31:e02272

Kochalski S, Barz F, Pita P, Harrison HL (2026) Quantitative and qualitative content analysis of text and images. In: Pope KL, Arlinghaus R, Hunt LM, Lynch AJ, van Poorten BT (eds) Understanding recreational fishers: disciplinary and interdisciplinary approaches for fisheries management. Springer Nature, Cham, p 507–532

Kolb TL, Blukacz-Richards EA, Muir AM, Claramunt RM, Koops MA, Taylor WW, Sutton TM, Arts MT, Bissel E (2013) How to manage data to enhance their potential for synthesis, preservation, sharing and reuse – a Great Lakes case study. Fisheries 38:52–64

Kozlov M (2022) NIH issues a seismic mandate: share data publicly. Nature 602:558–559

Krogman RM (2016) A standard method for abbreviating the names of fishes of the United States, Canada, and Mexico. Fisheries 41:74–82

Krogman RM (2021) Fisheries and habitat response to lake restoration efforts. Federal Aid to Sport Fish Restoration Study 7045: 2021 Fisheries research section annual report for project F-160-R. Iowa Department of Natural Resources, Des Moines

Krogman RM, Stubbs TJ (2021) Strategic planning for catfish angling in the future: an example from urban Iowa. N Am J Fish Manage 41:S298–S313

Krogman RM, Dodd BJ, Otting A, Weber MJ, Weber RE (2022) Modeled walleye regulation outcomes in a Midwestern reservoir with varying forage and walleye density. N Am J Fish Manage 42:597–611

Krogman RM, Bayer JM, Cooper AR, Kopaska J, Leonard NJ, Pritt JJ, Roe C, Tracy EE, Venturelli P, Wieferich D, Infante DM (2023) Standard data management practices. In: Bonar SA, Mercado-Silva N, Pope KL (eds) Standard methods for sampling North American freshwater fishes, 2nd edn. American Fisheries Society, Bethesda

Krogman RM, Taylor SM, Adams AJ, Boucek RE, Lombardo SM, Crandall CA, Steffen SF, Acosta A, Boné Morón E, Haukebo S, Hyder K, Radford ZM, Potts WM (2026) Role of recreational fisher information in fisheries management. In: Pope KL, Arlinghaus R, Hunt LM, Lynch AJ, van Poorten BT (eds) Understanding recreational fishers: disciplinary and interdisciplinary approaches for fisheries management. Springer Nature, Cham, p 21–54

Laine H (2017) Afraid of scooping–case study on researcher strategies against fear of scooping in the context of open science. Data Sci J 16:29–29

Lester NP, Marshall TR, Armstrong K, Dunlop WI, Ritchie B (2003) A broad-scale approach to management of Ontario's recreational fisheries. N Am J Fish Manage 23:1312–1328

Lynch KD, Taylor WW (2000) Interjurisdictional cooperation and the management of Laurentian Great Lakes ecosystems. Int Ver The:Verhandlungen 27:353–356

Lynch AJ, Sievert NA, Embke HS, Robertson AM, Myers BJ, Allen MS, Feiner ZS, Hoogakker F, Knoche S, Krogman RM, Midway SR, Nieman CL, Paukert CP, Pope KL, Rogers MW, Wszola LS, Beard TD Jr (2021) The U.S. inland creel and angler survey catalog (CreelCat): development, applications, and opportunities. Fisheries 46:574–583

Margenau TL, Gilbert SJ, Hatzenbeler GR (2003) Angler catch and harvest of northern pike in northern Wisconsin lakes. N Am J Fish Manage 23:307–312

Martin DR, Chizinski CJ, Pope KL (2015) Reservoir area of influence and implications for fisheries management. N Am J Fish Manage 35:185–190

McKiernan EC, Bourne PE, Brown CT, Buck S, Kenall A, Lin J, McDougall D, Nosek BA, Ram K, Soderberg CK, Spies JR, Thaney K, Updegrove A, Woo KH, Yarkoni T (2016) How open science helps researchers succeed. elife 5:e16800

McWilliams RH (1984) Natural lakes investigations. Annual performance report, Federal aid to fish restoration, Project F-95-R-3. Iowa Conservation Commission, Fisheries Section, Des Moines

Midway SR, Lynch AJ, Peoples BK, Dance M, Caffey R (2021) COVID-19 influences on US recreational angler behavior. PLoS One 16(8):e0254652

Midway SR, Sievert NA, Lynch AJ, Whittier JB, Pope KL (2022) Asking nicely: best practices for requesting data. Ecol Inform 70:101729

National Academies of Sciences, Engineering, and Medicine (2016) Statistical challenges in assessing and fostering the reproducibility of scientific results: summary of a workshop. Committee on Applied and Theoretical Statistics, Board on Mathematical Sciences and Their Applications, Division on Engineering and Physical Sciences. National Academies Press, Washington, DC

Nieman CL, Iwicki C, Lynch AJ, Sass GG, Solomon CT, Trudeau A, van Poorten B (2021) Creel surveys for social-ecological-systems focused fisheries management. Rev Fish Sci Aquac 29: 739–752

Pacific Salmon Commission (2022) Treaty between the Government of Canada and the Government of the United States of America concerning Pacific salmon

Page LM, Espinosa-Pérez H, Findley LT, Gilbert CR, Lea RN, Mandrak NE, Mayden RL, Nelson JS (2013) Common and scientific names of fishes from the United States, Canada, and Mexico, 7th edn. American Fisheries Society, Special Publication 34, Bethesda

Petering RW, Isbell GL, Miller RL (1995) A survey method for determining angler preference for catches of various fish length and number combinations. N Am J Fish Manage 15:732–735

Peterson JT, Paukert CP (2009) Converting nonstandard fish sampling data to standardised data. In: Bonar SA, Hubert WA, Willis DW (eds) Standard methods for sampling north American freshwater fishes. American Fisheries Society, Bethesda, Maryland

Phelps QE, Hupfeld RN, Whitledge GW (2017) Lake sturgeon *Acipenser fulvescens* and shovelnose sturgeon *Scaphirhynchus platorynchus* environmental life history revealed using pectoral fin-ray microchemistry: implications for interjurisdictional conservation through fishery closure zones. J Fish Biol 90:626–639

Pollock KH, Jones CM, Brown TL (1994) Angler survey methods and their applications in fisheries management. American Fisheries Society, Bethesda

Powers SM, Hampton SE (2019) Open science, reproducibility, and transparency in ecology. Ecol Appl 29:e01822

Rose ET (1956) The quantitative creel census methods at Spirit Lake. Iowa Conservation Commission, Fisheries Section, Des Moines

Ruckelshaus MH, Levin P, Johnson JB, Kareiva PM (2002) The Pacific salmon wars: what science brings to the challenge of recovering species. Annu Rev Ecol S 33:665–706

Sakpal M (2021) How to improve your data quality. https://www.gartner.com/smarterwithgartner/how-to-improve-your-data-quality. Accessed 10 Jul 2023

Sass GG, Shaw SL (2020) Catch-and-release influences on inland recreational fisheries. Fish Sci Aquac 28:211–227

Sayogo DS, Pardo TA (2013) Exploring the determinants of scientific data sharing: understanding the motivation to publish research data. Gov Inf Q 30:S19–S31

Schlesser N (2020) UMRCC fisheries compendium, 4th edn. Upper Mississippi River Conservation Committee

Shrout PE, Rodgers JL (2018) Psychology, science, and knowledge construction: broadening perspectives from the replication crisis. Annu Rev Psychol 69:487–510

Smith CL (1994) Connecting cultural and biological diversity in restoring Northwest salmon. Fisheries 19:20–26

Sutinen JG (1989) Cost-effective information for fisheries management. In: Pollnac RB, Morrissey MT (eds) Aspects of small-scale fisheries development. International Center for Marine Resource Development, University of Rhode Island, Kingston, Rhode Island, pp 133–152

Taylor JL, Reiling SD (1981) A comparison of Maine open water and ice fishing activities and participants. University of Maine, Department of Agricultural and Resource Economics, Life Sciences and Agriculture Experiment Station

Temby O, Rastogi A, Sandall J, Cooksey R, Hickey GM (2015) Interagency trust and communication in the transboundary governance of Pacific salmon fisheries. Rev Policy Res 32:79–99

The Whitehouse (2023) FACT SHEET: Biden-Harris Administration Announces New Actions to Advance Open and Equitable Research. https://www.whitehouse.gov/ostp/news-updates/2023/01/11/fact-sheet-biden-harris-administration-announces-new-actions-to-advance-open-and-equitable-research. Accessed 14 July 2023

Thunberg EM, Fulcher CM (2006) Testing the stability of recreational fishing participation probabilities. N Am J Fish Manage 26:636–644

Tripp SJ, Phelps QE, Hupfeld RN, Herzog DP, Ostendorf DE, Moore TL, Brooks RC, Garvey JE (2019) Sturgeon and paddlefish migration: evidence to support the need for interjurisdictional management. Fisheries 44:183–193

UMRCC (2021) Upper Mississippi River Conservation Committee fish technical session fall meeting minutes. https://umrcc.org/wp-content/uploads/2022/04/6.-Fisheries-minutes-fall-2021.pdf. Accessed 22 Dec 2022

UNESCO (2021) UNESCO Recommendation on open science. https://unesdoc.unesco.org/ark:/48223/pf0000379949. Accessed 14 July 2023

van der Hammen T, Chen C (2020) Participation rate and demographic profile in recreational angling in The Netherlands between 2009 and 2017. Fish Res 229:105592

van Putten EI, Trudeau A, Mackay M, Obregón C, Post JR, Richter A, Schuch E, Solomon CT (2026) Economic experiments in recreational fishing studies. In: Pope KL, Arlinghaus R, Hunt LM, Lynch AJ, van Poorten BT (eds) Understanding recreational fishers: disciplinary and interdisciplinary approaches for fisheries management. Springer Nature, Cham, p 411–442

Venturelli P, Skov C, Audzijonyte A, Belmaker J, Ana Cardoso AS, Gordon L, Gundelund C, Hyder K, Jarić I, Lennox RJ, Potts WM, Vacura K, Vaz AS, Sbragaglia V (2026) Digital data mining. In: Pope KL, Arlinghaus R, Hunt LM, Lynch AJ, van Poorten BT (eds) Understanding recreational fishers: disciplinary and interdisciplinary approaches for fisheries management. Springer Nature, Cham, p 443–479

Ward HG, Allen MS, Camp EV, Cole N, Hunt LM, Matthias B, Post JR, Wilson K, Arlinghaus R (2016) Understanding and managing social-ecological feedbacks in spatially structured recreational fisheries: the overlooked behavioral dimension. Fisheries 41:524–535

Wilkinson MD, Dumontier M, Aalbersberg IJ, Appleton G, Axton M, Baak A, Blomberg N, Boiten JW, da Silva Santos LB, Bourne PE, Bouwman J (2016) The FAIR guiding principles for scientific data management and stewardship. Sci Data 3:1–9

Best Practices for Understanding Recreational Fishers

Brett T. van Poorten, Len M. Hunt, Robert Arlinghaus, Abigail J. Lynch, and Kevin L. Pope

Abstract In this closing chapter of our edited book, we summarize what we believe are best practices for understanding recreational fishers. Fishers are an integral part of the recreational fishery social-ecological system, and we emphasize the importance of placing them in that context. We begin with an overview of the process of developing a project and conclude with some broad suggestions for standardising approaches to gather data from and about fishers, mainly focusing on social science methods. Throughout, we emphasize tactics to promote the development of inter- and transdisciplinary tools and processes, as a means of more fully understanding the full social-ecological system. This chapter draws from methodological details and suggestions developed throughout this book; we describe how they work together but point the reader to these specific chapters to further understand how to build interdisciplinarity into research and management.

Keywords Best practices · Disciplinary teams · Experimental approaches · Interdisciplinary teams · Social-ecological systems

---

B. T. van Poorten (✉)
School of Resource and Environmental Management, Simon Fraser University, Burnaby, BC, Canada
e-mail: brett_van_poorten@sfu.ca

L. M. Hunt
Ontario Ministry of Natural Resources, Thunder Bay, ON, Canada

R. Arlinghaus
Department of Fish Biology, Fisheries and Aquaculture, Leibniz Institute of Freshwater Ecology and Inland Fisheries, Berlin, Germany

Division of Integrative Fisheries Management, Humboldt-Universität zu Berlin, Berlin, Germany

A. J. Lynch
National Climate Adaptation Science Center, U.S. Geological Survey, Reston, VA, USA

K. L. Pope
Cooperative Research Units Program, United States Geological Survey, Lincoln, NE, USA

K. L. Pope et al. (eds.), *Understanding Recreational Fishers*, Fish & Fisheries Series 45, https://doi.org/10.1007/978-3-031-99739-6_25

1 Introduction

This book (Pope et al. 2026) focuses on understanding recreational fishers for the sake of improving their management. Recreational fisheries are complex adaptive social-ecological systems (Arlinghaus et al. 2017). Such systems include multiple feedbacks and interactions between the ecological subsystem (the fish, the habitats, and the broader ecosystem) and the social subsystem (the fishers, organisations of actors and broader society), as well as interactions between each of these subsystems and the management and governance subsystems (managers, user communities, agencies and policy makers) at various scales (Ward et al. 2016; Arlinghaus et al. 2017). Recreational fishers are a key component of the social-ecological system of recreational fisheries because of their multiple connections between all subsystems. In fact, fishers play at least four important roles within the social-ecological system (Brown 1987; Ward et al. 2016): (1) they affect and are affected by the current state of the fishery and by potential future changes to the fishery; (2) their participation and well-being (e.g. satisfaction, utility) are typically key objectives of fishery management and thereby part of the objectives and targets of the governance subsystem; (3) they have information on the state of the fishery and ecosystem; and (4) they fundamentally shape the state of the fishery or the chosen management actions. These four overlapping roles underscore the importance of understanding and working with recreational fishers in both research and management if we are to understand recreational fisheries as complex adaptive systems (Hunt et al. 2013). Other important behavioural dimensions of recreational fisheries are the behaviours of individual decision-makers, for example, those in fish and wildlife agencies or in angler communities (Fujitani et al. 2020) and the broader policy setting, which affects decision-maker structures and agency (Boucquey et al. 2026).

In this closing chapter of our edited book (Pope et al. 2026), we summarise what we believe are best practices for understanding recreational fishers. We begin with an overview of the process of developing a project and conclude with some broad suggestions for standardising approaches to gather data from and about fishers, mainly focusing on social science methods. That said, the methodological details of the suggestions we provide are to be found in the detailed chapters that we reference from our book.

Any change to a recreational fishery will have a reciprocal impact on recreational fishers that may affect the level of well-being derived from their experiences and, in turn, impact future fishers' behaviours and intentions to return to the fishery (Carpenter and Brock 2004, Ward et al. 2016, Post et al. 2026). Recreational fishers derive non-monetary benefits from multiple aspects of fishing, including catch-related attributes such as species available, fish sizes, and catch rates, and non-catch related attributes such as access, scenic beauty, amenities available, the social aspects of fishing and certain regulations (i.e. restrictions on free harvest choice) (Hunt et al. 2019). Their reactions to local changes are thus different from those expected from commercial fisheries or natural predators (Johnson et al. 2009; Matsumura et al. 2019; Post et al. 2026). Recreational fishers assess the expected quality of a local experience within the social context of their trip (e.g. with whom they might fish) to decide where and when to fish among a series of alternatives (Hunt et al. 2019). The outcomes of these experiences compared to their initial expectations contribute to their levels of satisfaction (TenHarmsel et al. 2019; Birdsong et al. 2021; Landon et al. 2026) and to fisher utility more generally (Hunt et al. 2019; Melstrom et al. 2026), which in turn contributes to their decisions to return, to discuss their fishing experiences in social networks, or to set or reset expectations and norms about management (Arlinghaus and Mehner 2005, van Poorten et al. 2011; Matsumura et al. 2019). Measures of well-being from fishing and fisheries are often quantified in resource economics as utility (Melstrom et al. 2026) or in social psychology as satisfaction (Birdsong et al. 2021). These measures may serve as integrative management objectives because they integrate ecological changes through the impact on catch rates and angler well-being (Johnson et al. 2009). That said, the fact that there are non-catch-related pleasure components may result in fishing effort remaining stable or changing in less predictable ways, which can contribute to collapse (Post et al. 2002; Golden et al. 2022).

In many recreational fisheries, there is the objective of providing satisfactory experiences to fishers (Krogman et al. 2026), which is often complemented by another, more ecological objective of maintaining local populations (Cowx et al. 2010). Whereas fishers satisfy many of their expectations for non-catch-related attributes through their choices of where and when to fish (e.g. environmental quality, scenery), catch-related outcomes (e.g. catch numbers, harvest opportunities and size of fish in the catch) are much more variable and under less control by the person (Arlinghaus et al. 2007), and more closely linked to fish stock-directed management actions and the collective mortality exerted by the pool of recreational fishers as a whole (van Poorten and Camp 2019). Properly understanding the dynamics of recreational fisher satisfaction and well-being more generally is challenging and requires proper survey tools (Vaske et al. 2026a, b; Venturelli et al. 2026; Kyle et al. 2026; De Kerckhove et al. 2026). Methods to study individual-level well-being outcomes are available in both the social-psychological approach to recreational fisheries (Birdsong et al. 2021; Landon et al. 2026) and in microeconomic study approaches (Melstrom et al. 2026). However, more qualitative research approaches are also suited to study well-being aspects of communities and of individuals that are surveyed (Boucquey et al. 2026; Golebie et al. 2026; Kochalski et al. 2026).

Recreational fishers have a unique understanding and perspective of fisheries, both individually and in aggregate (Aminpour et al. 2020; Löki et al. 2023). Fishers often have first-hand knowledge of catch rate and size structure; they observe the states of water quality and habitat, and they comprise a vast network of information gatherers and sharers that help to determine where and when to fish (Eden and Bear 2011). Their perspectives can provide information across the landscape, which can be tremendously valuable for management agencies that may have dozens to hundreds of individual fisheries to monitor (Venturelli et al. 2026). However, these fishery-dependent data are not without bias, and they require a careful understanding of the cognition of fishers to recognize their motivations and constraints for reporting accurately, or even reporting at all (Venturelli et al. 2016). How to collect these data, while being respectful of fishers' time and helping them feel invested in the fisheries and the outcomes, remains an elusive challenge (Vaske et al. 2026a, b; Venturelli et al. 2026; de Kerckhove et al. 2026). Fortunately, these so-called non-probabilistic survey types (Howarth et al. 2024) co-exist together with a large family of probabilistic survey types (Vaske et al. 2026a, b; Kyle et al. 2026), qualitative (Kochalski et al. 2026; Thurstan et al. 2026) and mixed-mode approaches (Golebie et al. 2026), as well as experimental approaches (van Putten et al. 2026; Mackay et al. 2026). Each method comes with its own issues, and many lessons are available in this book (Pope et al. 2026) to improve fisher engagement and information quality (Hunt 2023).

Recreational fishers also impact the state of fisheries by affecting habitats, wildlife, fish abundance, size structure, and behaviours of exploited fishes (Post et al. 2002; Arlinghaus et al. 2007; Lewin et al. 2006). Understanding recreational fisher behaviour is crucially important to understanding these impacts (Johnson et al. 2009; Hunt et al. 2013; Ward et al. 2016; Matsumura et al. 2019). These impacts may be direct, through harvest, or incidental ("cryptic;" Coggins et al. 2007) mortality as well as behaviourally as fish adjust their behaviour to the effort of recreational fishers or to catch-and-release (Arlinghaus et al. 2016b). Incidental impacts include mortality of released fish, whether the release is mandated (e.g. size-based or possession limits) or voluntary (Arlinghaus et al. 2007), along with impacts resulting from the release of non-native fishes from bait buckets (Johnson et al. 2009). Other indirect impacts include habitat degradation and litter (Watson et al. 2022; O'Toole et al. 2009), crowding that may affect the well-being of fishers (Olaussen 2009), nutrient inputs (Niesar et al. 2004), and possible impacts to aquatic and riparian biodiversity (Schafft et al. 2021). The ways in which fishers interact with the resource also vary over time, with changes in social and personal norms and technology affecting how people fish (Cooke et al. 2021). For example, changes in the propensity for fishers to voluntarily release fish have increased dramatically for some species and fisheries (e.g. Myers et al. 2008), but not for others (Cooke et al. 2017). The practice of voluntarily releasing fish can have direct benefits to size structure, with larger fish increasingly being released (Rypel et al. 2016), but it may also cause stunting if populations suffer from density-dependent population control and populations become essentially unexploited release rates approach 100% (Sass and Shaw 2019). Novel sonar technology is altering the way recreational fishers are able to locate fish, which can increase fishing pressure and lead to conflicts among people over limited resources (Cooke et al. 2021). Understanding and managing the impacts of fishers on the resource can be addressed through fisheries ecological science (Post et al. 2026), including looking back towards past developments through the lens of historical ecology (TenHarmsel et al. 2021; Thurstan et al. 2026).

Whereas resource economists focus almost exclusively on measuring how changes within a fishery influence the benefits that fishers obtain from fishing (Melstrom et al. 2026), social psychologists (Landon et al. 2026) and behavioural economists (Mackay et al. 2026) examine more closely the mechanisms by which fishers behave, and political ecologists (Boucquey et al. 2026) focus on the power structures and more fundamental overarching relationships that fishers have with other people and non-human actors within a social-ecological system. Each of these different disciplinary perspectives approaches the understanding of fishers with different beliefs about realities (ontology) and ways to generate legitimate knowledge (epistemologies, Hunt 2023) that in turn lead researchers to produce knowledge using different quantitative, qualitative, or mixed methods approaches (Kochalski et al. 2026; Golebie et al. 2026). Social science approaches to understanding recreational fishers have historically developed in isolation of each other, creating academic silos and limited cross-referencing (Fenichel et al. 2012; Hunt et al. 2013)—an issue that the present book wants help overcome (Pope et al. 2026). The opportunities for integration across disciplines are ample. For example, most choice models developed by resource economists to understand economic values and preferences of recreational fishers constitute behavioural models of how fishers respond to attributes that characterize a fishery, which can be linked to sub-models of fish population dynamics via effort dynamics (Johnson et al. 2009; Hunt et al. 2013; Matsumura et al. 2019; Jensen et al. 2026). Similarly, qualitative and quantitative information that describe power structures and incentives to behaviour that developed from political ecology, microsociology or social-psychology are critical for a more holistic understanding of the entire system and can also be used to inform the behaviour of agents (managers or fishers) in coupled models (Jensen et al. 2026).

Numerous ways exist to collect data about fishers; however, up to the present day, many collection efforts continue to follow disciplinary lines. Within fisheries management, an ecological perspective dominates, with researchers primarily gathering data to inform fish ecological models and to test ecological theory. Therefore, most of the information about recreational fisher behaviour, at least in North America, is still gathered by management agencies through creel surveys (i.e. in-person interviews at fishing sites; de Kerckhove et al. 2026) and other surveys primarily aimed at estimating catch, effort, and harvest (Post et al. 2026). This neglects that through a slight modification of the survey approach, a much richer variety of information can be gleaned about the human dimensions of recreational fisheries related to values, beliefs, attitudes, or norms (Ditton and Hunt 2001; Nieman et al. 2021). Some researchers are starting to experiment with new means of gathering information from fishers, such as working directly with smartphone apps designed for recreational fishers, mining data across a variety of online platforms (Venturelli et al. 2026), and even creating large databases of creel data to leverage cross-jurisdictional learning (Sievert et al. 2026). In conjunction with these new and exciting opportunities, there are opportunities for developing new tools and techniques—rooted in both disciplinary and interdisciplinary efforts and increasingly relying on experiments (van Putten et al. 2026)—to elicit information from recreational fishers to better understand fishers' behaviours and outcomes, ultimately for the betterment of fisheries management.

2 Interdisciplinary Integration in Practice

Some fisheries questions or problems can and will always be addressed from a single disciplinary perspective (Arlinghaus et al. 2014). For example, if a researcher or management agency is interested in learning about the diversity of fisher types present in a local fisher population, a probabilistic fisher survey informed by social psychology is reasonable. Other times, however, it becomes obvious that researchers and managers may want to consider addressing certain fisheries problems or research questions from multiple disciplinary perspectives. For example, information needed to plan fisheries management across multiple lakes linked through mobile recreational fisheries in a landscape typically encompasses both ecological and social dimensions jointly, and importantly the interaction of the two, as one needs to understand how different policies might affect the landscape patterns of movements across scales (Carpenter and Brock 2004, Post 2012; Mee et al. 2016). This approach has been dominant since the 1970s, after "human dimensions" developed as a field in recreational fisheries. Yet, knowledge generated through multidisciplinary approaches do not necessarily integrate well with each other, may be overlooking procedural aspects of decision-making where stakeholders are marginalized and may in fact reinforce academic silos (Arlinghaus et al. 2017). Such approaches can also be insufficient to solve specific fisheries problems that have a greater complexity and are characterized by strong feedbacks between the ecological and the social system (Arlinghaus et al. 2017). In such situations, we encourage scientists and managers to consider the more integrative approach of interdisciplinary research that is outlined in this book. Putting an interdisciplinary approach in place requires a shared research framework to encourage joint learning, the construction of research questions in a way that at least two disciplines have to work together to answer the question or questions, a clear and agreed-upon vision about how to synthesize knowledge streams and generally a willingness for co-creation of interdisciplinary knowledge by a team of researchers from different disciplines (Arlinghaus et al. 2014). Perhaps even stakeholders, such as local recreational fishers, and rights holders (e.g. Indigenous communities, but may include recreational fisher communities in some countries), need to be part of the research team, leading to transdisciplinary co-production settings (Arlinghaus et al. 2026), which further increases the demands on the research team. Such a complex setting requires a complex adaptive social-ecological system lens (Arlinghaus et al. 2017; Lynch et al. 2017) that encourages researchers from ecological science to work with social scientists, but it can also help to illustrate the potential benefits of looking at a problem from different social science disciplines and fields (Hunt 2023). What design to choose and what level of cooperation among scientists, managers, and stakeholders to aim at crucially depends on the research problem. Neither disciplinarity nor inter- or even trans-disciplinarity is panacea—what design and approach to choose depends on the fisheries problem or research question to solve and needs careful thought by those initiating the project.

Of course, understanding recreational fishers is a necessary component of fisheries management and decision-making. In many contexts, recreational fisheries decision-making is commonly top-down, with government agencies determining appropriate fishing practices, but this need not be the case. There are other cases where decision-making benefits from the integration of suggestions, knowledge, and the inclusion of stakeholders and rights-holders. Working with stakeholders can involve fishers along a continuum that may include no participation in the actual science or management, soliciting fisher feedback through surveys, or full co-management (Camp et al. 2026), similar to the ladder of participation that (Arnstein 1969) describes. Although there are cases where no stakeholder participation is warranted, such as where legal frameworks mandate top-down action, increasing participation from the project onset will lead to improved trust, acceptance, and a broader system perspective of all involved (Camp et al. 2026; Arlinghaus et al. 2026). It should be noted, however, that with greater engagement with fishers and other stakeholders comes the risk of negotiating different values, norms, and interests. Gaining stakeholder trust, building relationships, and integrating different knowledge sources through the process are difficult and time and resource-intensive. Further, some groups (e.g. Indigenous groups in North America) feel integration is insulting and prefer acceptance of multiple worldviews (Reid et al. 2020). Also, in joint processes, researchers and managers may need to accept that stakeholders will often have divergent perspectives and that these perspectives are equally valued as the more "academic" views if trust is to develop. Generally, non-academic knowledge must be treated as similarly valuable as academic knowledge, and stakeholder knowledge should be formally represented in knowledge gap identification or at the end of the project synthesis. Proper ways of integration and synthesizing are needed (Cooke et al. 2026), which can involve formal models or using principles of collective intelligence if we are dealing with large numbers of people (Aminpour et al. 2020).

Whereas there are obvious differences in the steps needed to initiate and achieve any of the above-mentioned example problems and questions, there are certain similarities, particularly when, as is often the case in practice, recreational-fisheries management is considered from an adaptive management perspective (Robinson et al. 2026). In its simplest form, passive adaptive management includes understanding the system, predicting changes to the system after an intervention such as a regulation change, implementing those changes while monitoring outcomes, and using monitoring data to learn and adapt the management intervention in the future (Robinson et al. 2026). In the active counterpart, management interventions are seen as system-level experiments and are planned a priori (Robinson et al. 2026). Both approaches involve gaining an understanding of the system from multiple perspectives, building interdisciplinary models to derive hypotheses a priori (Walters 1986; Jensen et al. 2026), ideally with the shared perspective and participation of all involved (Arlinghaus et al. 2026; Jensen et al. 2026), generating hypotheses or predictions of how to intervene in the system, intervening in the system, setting up a monitoring program, and collecting ecological and social data before and after, and perhaps with treated and controlled replicates across the landscape. Most of these

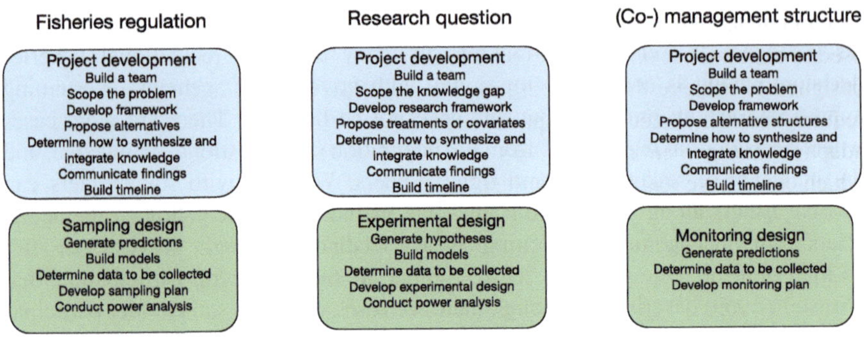

Fig. 1 Steps in implementing and evaluating different changes to the system (including making minor changes such as changing regulations, addressing a research question, or making a structural change to management, including implementing co-management; many of the steps involved are repeated across these scenarios, suggesting common strategies)

same steps are necessary regardless of whether one is imposing a regulation change, conducting a field experiment (Mee et al. 2016), or even considering restructuring the management responsibility of the fishery to include rights-holders or stakeholders (Fig. 1). In most cases, one will probably end up with the need to be interdisciplinary in such complex scenarios at the level of entire fisheries.

There are multiple overlapping stages of any interdisciplinary, or even transdisciplinary, stakeholder-inclusive endeavour. In the next section, we will describe them and their role in developing interdisciplinarity to illustrate how to achieve cooperation or integration of multiple disciplines to gain more holistic insights and understanding of the role recreational fishers play within a coupled social-ecological system. Although not all these stages will necessarily apply to all research or management projects, most are salient for any activity that may benefit from a diversity of viewpoints, approaches, and methods, including setting fishing regulations, developing a research question, or changing a (co)-management structure (Fig. 1). We consider the initial stages of any project to include project development, research framework design and designing the sampling or experimental program. But the reality of the specifics of a given project might deviate from the steps we outline, as it is extremely difficult to generalize to all situations.

2.1 *Project Development*

The initial task of developing almost any fisheries project involves the following steps:

1. Build a team as a collaboration among researchers from multiple disciplines, managers, stakeholders, and rights-holders.

2. Scope the problem being addressed by the team, including the concrete research questions.
3. Build a common research or managerial framework that is agreed upon by all.
4. Identify methods to address the various problems and questions.
5. Decide how to integrate and synthesize the various knowledge dimensions at the end of the project, again to be agreed by all.
6. Think and decide about a strategy of knowledge dissemination.
7. Create a project timeline and determine roles and responsibilities among the team.

When addressing a fisheries problem, the first step is to build a team involving multiple partners from across academic disciplines, and different perspectives of the system (fisheries managers, policymakers, stakeholders, and rightsholders). The diversity of perspectives provided by such a team can yield a variety of objectives, knowledge, and disciplinary approaches to developing, refining, and answering the questions to be addressed (Arlinghaus et al. 2026; Camp et al. 2026). To determine who should be part of the team, it is important to consider who interacts with the system and who would benefit or be impacted by the outcome of any research or system changes. If addressing management structure, it is crucial to proactively include proposed co-management partners immediately so trust and a collective understanding can be established. It is critical to think about balancing the team members such that it is not dominated by individuals from one disciplinary perspective or group, and to clarify the roles of each team member. If projects involve stakeholders, it may be necessary to encourage people to move beyond their "comfort zone" and be willing to listen and interact with people outside of their disciplines or with practitioners (Kelly et al. 2019, Arlinghaus et al. 2026). This is not for everybody.

The next step is to scope the problem or knowledge gap to be addressed with all team members as a group. There are several questions that one can ask to help understand the problem. *What are the goals and objectives of the system? What is the applied fisheries problem we seek answers about? Do we all see the same problem and if not, why not? What led us to this process? What is the primary research question or questions for which there is no accurate or recent, place-based information? Which group already has knowledge about these questions, what is the availability and robustness of scientific and non-scientific knowledge? If this information was available, can it be synthesized before new data are collected and used? Is information context-specific, suggesting there are broader questions that will colour results? Importantly, what are the logistical constraints, such as available budget and time, for this work?* These types of questions help the team to fully understand what approaches can and cannot be used to move forward, and importantly, nail down the knowledge gap the new project wants to fill.

Once a team is assembled and the knowledge gap identified, it is important that everyone can work well together throughout the project against a common procedural or operational framework that everybody agrees upon and wants to contribute to. It is especially important to take time to think about the system, gaps in understanding, and research questions that can best address these gaps through collaboration among diverse participants. These discussions are essential, often need to be personal, and they are key to building trust, respect, and mutual understanding among participants. The resulting project and research framework should be visualised, and the roles of each team member should be made clear and transparent within the framework (e.g. who contributes to aspect A of the shared questions, who to aspect B). It is important to be explicit and consider the values held by different people in the team. For interdisciplinarity, it is also instrumental to avoid disciplinary "arrogance" among team members, and similarly, scientists and stakeholders must feel respect so lasting relationships develop based on trust and mutual understanding. It is not uncommon that people trained in different disciplines are conditioned to reject knowledge that is generated using a different epistemology (Hunt 2023), and different disciplines respond to different reward systems (e.g. in which journal to publish, as sole author or as a team), which have to be negotiated and addressed. It is important to have at least one person on the team who is conversant in multiple disciplines to help each member understand the perspective of others (i.e. a "knowledge broker;" Arlinghaus et al. 2014). This knowledge or information broker can help to formulate a research framework that all participants buy into, guide discussions, and outline the synergies for different disciplines. Relations matter. It is thus particularly important to take the time and energy to have team members get to know each other both professionally and personally to break down barriers and enhance and importantly, secure collaboration in a process that can take many years to complete.

Once the team has a mutual understanding of the operating process, the next step is creating a research framework or system model that represents how the project team understands the managed ecological system (e.g. recruitment, angler harvest). Frameworks are a description of the most important presumed causal mechanisms, interactions, and feedbacks of a system (McGinnis and Ostrom 2014; Arlinghaus et al. 2016a, b) and the gaps that need to be addressed through the present research. The research framework should be related to the most important questions being addressed, and the team needs to devise appropriate methods to address different components of the fishery problem. Initially, all participants can suggest methods for addressing the problem, then the group can come together to discuss each and make final decisions. This is especially fruitful if there are strong differences of opinion because it allows each member to be heard and for the team to constructively debate each suggestion. Discussion within the team could focus on different perspectives of the problem or different sub-questions, which should help suggest numerous ways that team members can approach it. Getting a variety of options allows for creative solutions and potential for creating new tools that bridge disciplinary boundaries. Discussions about methods crucially depend on budgets and other constraints (e.g. project timelines, availability of sampling sites, availability of previous data sets, discussion on whether experiments are planned, or only observational data collected). If students are involved, feasibility becomes particularly important to complete student timelines, considering project demands.

Crucially, at this point of identifying input sources of information, it is very important to agree on an approach to synthesis of the various project components, such as how will disciplinary knowledge be integrated with stakeholder knowledge (Arlinghaus et al. 2026), or how will knowledge of discipline A (say resource economics) be integrated with discipline B (say fisheries ecology) to contribute to the project goals. Most collaborative multidisciplinary projects fail if they do not have a clear a priori strategy on how to synthesize across the various knowledge streams (Arlinghaus et al. 2014).

Planning how and where to communicate the findings of the project is also an important part of project development. This plan can include roles and responsibilities for team members with respect to communication and the process that will be used to recognize the contributions of team members on different communication products. Consider different outlets for communication, including news outlets, academic journals, social media, and communication back to broader stakeholder groups and the local public. By developing such a plan, potential issues related to who is privileged to communicate with media requests or to submit manuscripts have already been agreed upon by the team. All team members can contribute, review, and revise the plan. The plan may also provide some formal or informal way to assess the effectiveness of the efforts to communicate the project findings and be respectful of sensitive information, especially when stakeholders are involved. Effectiveness may be better measured through outcomes (e.g. measures of awareness of the project among a population of interest, new networks established among stakeholders and scientists) than through outputs (e.g. number of published papers). It is imperative to determine who leads each piece of communication and how all partners and stakeholders will contribute, and how to deal with the disclosure of possible sensitive information. An important conversation relates to recognition. For multi-authored communications, having an explicit discussion about who should be included as an author and a process to determine authorship order will prevent competitive behaviours or misunderstandings. These conversations may be more important to some researchers and partners than others, but being explicit will protect against problems later.

Finally, identifying roles and responsibilities for the team members and building a timeline will help to manage expectations for the team and build key deliverables to help keep the project on time. This timeline should include all components of the project, such as hypothesis generation, data collection, data analysis, writing, and communication, and who will lead each of these components. Some processes, such as redesigning a management structure, understandably will take months or years, and it is important that all members understand this reality. All team members should have appropriate time dedicated to addressing their component of the work and time to review other components to ensure there are no lags in the project that might impair the entire team. Importantly, the timeline should also include multiple dedicated opportunities for reflexive interactions. These communications help recommit all partners and stakeholders to the broader project questions and goals. These interactions also provide flexibility in the project in case certain tasks have problems or unexpected results. These can be seen as opportunities for partners and stakeholders to consider how to pivot in a way that further strengthens interdisciplinary collaboration and knowledge generation.

2.2 Sampling Design and Monitoring

Once the team establishes a framework for making decisions, addressing a knowledge gap, or restructuring management, it is important to generate hypotheses and predictions of how the system might respond to different changes to the system or the way it operates. These hypotheses should build on the interdisciplinary strengths and backgrounds of the project team. Each hypothesis should be brought to the team to see how it might support or contradict expectations from others based on their experience or disciplinary background. This might suggest a reconsideration of hypotheses, or creation of new hypotheses that consider multiple disciplinary perspectives.

Some researchers may choose to build mental or simulation models of the system and use these to generate plausible hypotheses based on prevailing (or competing) discipline-specific paradigms. Alternatively, models may be built to incorporate multiple disciplines, which helps to uncover uncertainties at the interface of these disciplines (Jensen et al. 2026). Models can also be built working with stakeholders and rights-holders to help understand their worldviews and use this experience to build additional predictions (Wilberg et al. 2026). Models may or may not be quantitative; they can be fully qualitative and conceptual. Models should be co-developed among all and shared among the research team for critique and discussion, which will enrich hypotheses or predictions of the model based on multiple perspectives and lived experiences. These predictions should open suggestions for how to monitor the system before, during, and after the proposed change or changes to the system occur, if experiments are designed. Predictions are equally relevant for designing fisher surveys or statistically analysing observational data. Fully integrative models may also be a synthesis tool and be used for experiments to study fish-fisher interactions (e.g. Hunt et al. 2013; Matsumura et al. 2019).

After the hypotheses are set, a sampling and analysis plan must be devised and implemented, which can involve experiments, surveys, analysis of past databases, and other approaches. Then, data need to be collected or experiments completed to determine if the proposed change or changes to the system are working in the ways predicted and to examine any hypotheses and predictions derived by the research team. If the system does not respond as expected, or if other deviations from expectations emerge, it will be easier to diagnose and adapt if data exist with which to infer where the information gap(s) exist. Identifying databases that are free and curated is critical here. Again, even if a new management structure is being proposed, thinking of this as an adaptive management experiment will help restructure thinking about the importance of data and monitoring (Robinson et al. 2026). Partners with different disciplinary backgrounds will have different suggestions on how to monitor the fishers and the fishery, and giving all consideration may broaden the perspective and encourage further engagement. This is an opportunity to identify efficiencies so that different perspectives of the fishery may be gained through the same data. In the end, although not all data collection strategies may be acceptable because of limits to time and resources or for ethical reasons, they should all be considered at this stage.

The models developed above will be useful in comparing different data collection strategies in a safe environment before any manipulation of the actual system occurs. Importantly, these models can be used to simulate all the ways the system might respond, and "data" can be gathered in the simulation model, which can then be analysed using different proposed approaches. Data analysis is not often considered until after the data are in hand, but the analytical approach should be established as part of the experimental design and will be a key factor in how data are collected. Models developed for quantitative approaches by the team provide an opportunity to evaluate and compare experimental designs and proposed analyses. Using a combination of simulation models and statistical tests or estimation models (where appropriate), the team can then evaluate how data collection scenarios directly influence the ability to detect trends in the population (Parkinson et al. 1988). This can be useful to determine if any necessary data were missed, or if some data are unnecessary to address the research questions. Qualitative projects necessarily require more flexibility; however, prior consideration is beneficial to ensure that the design is sufficiently flexible to adapt to themes that emerge from data, that the proper respondents will be included, and that the context will be set appropriately.

Power analysis can be performed on observational or experimental data, where analyses are repeatedly run on simulated data to determine the statistical power of the data collection process, or accuracy of parameter estimates (Kelley et al. 2003). Power analyses are ideally conducted using pilot data, which informs sampling variance (e.g. Parkinson et al. 1988). However, even without data, stakeholder knowledge is often invaluable for suggesting appropriate levels of sampling variation, which can provide a useful basis for comparing different data collection options. Of course, power analysis will only inform certain types of quantitative analyses, but, when possible, it is a powerful tool (Peterman 1990; Wagner et al. 2013). In qualitative studies, focus groups can serve a similar role, which can generate a wealth of prior information that can be used to develop coding schemes that work before the actual project and data generation start.

To better understand data collection processes using different methods, the reader is referred to the various chapters in this book.

3 Best Practices

This book covers many distinct aspects of working with and understanding recreational fishers. This understanding can come from traditional disciplinary perspectives and from interdisciplinary approaches. Here, we suggest some best practices to ensure that the knowledge gained is defensible, repeatable (as much as possible), and capable of being integrated into future work. We organize these suggestions into sections related to engaging fishers and knowledge integration.

3.1 Engaging Fishers

A lot of fisher engagement will come in the form of data collection, where the thoughts, perspectives, and actions of the broader fisher community will be sampled through, for example, a survey (Vaske et al. 2026a, b) or qualitative research (Kochalski et al. 2026). This is an important aspect of understanding fishers in general and will lay the groundwork for engaging fishers in higher-level decision-making or research. However, recreational fishers today are not only "informants," but engage with research and management through a variety of other collaborative formats and may be part of formal co-production settings where the boundaries among scientists and non-scientists blur (Arlinghaus et al. 2026). Engagement and trust are particularly relevant in such situations.

One may collect data from recreational fishers and their behaviours through a variety of means, including personal interviews, focus groups, semi-structured in-person interviews, on-site creel surveys, off-site surveys, text corpus analysis, mining digital sources or written text, public online databases or private databases (e.g. spreadsheets stored on hard drives), and field or laboratory experiments (Cooke et al. 2026; de Kerckhove et al. 2026; Golebie et al. 2026; Kochalski et al. 2026; Kyle et al. 2026; Post et al. 2026; Sievert et al. 2026; van Putten et al. 2026; Vaske et al. 2026a, b; Venturelli et al. 2026). Members of the project team will have varying thoughts on how to gather data, but careful discussion will help reduce the list of options to what the team can do to address the questions within the constraints of time and money, and respecting ethical issues of data collection. An especially important and increasingly rising consideration is about the personal boundaries that social science may cross. For example, respondents in surveys devote their time to answering what the researchers are interested in, and the researcher, in turn, has the sole power in interpreting the results. Difficult ethical questions are relevant here that one has to be aware of as a researcher because relationships to stakeholders may quickly break if stakeholders perceive that the social science results are breaching local custom or misrepresenting or disrespecting local culture and norms.

Choosing how to gather data from fishers is partly driven by the desired sample frame (i.e. the population of fishers from which to collect information) and the degree of individual- or population-level information collected. When, for example, the sample frame is fishing activity for particular waterbodies, an on-site (creel) survey is appropriate. When fishers across a region are of interest, interviewing current fishers will produce biased results, because active fishers will be over-represented, and occasional fishers may never be encountered. In this case, off-site surveys administered via telephone, mail, internet, or mixed modes are better suited.

Questionnaires can be administered using in-person, telephone, (e)mail, or other electronic modes. However, the means of contacting potential respondents and their representativeness of the population of interest (leading to coverage bias and possibly nonresponse error; Dillman et al. 2014; Vaske et al. 2026b) can be a real concern. If information about populations is gathered, researchers should ideally sample following some random process (i.e. a probabilistic sample), and participants should ideally be selected from complete sampling frames (i.e. no coverage error exists). If these sampling frames do not exist, quota sampling may be a solution, or researchers will need to invest effort to develop a sampling frame (e.g. from addresses from on-site surveys, if the population of interest is active fishers). Researchers are increasingly adopting mixed-mode surveys (where respondents can either choose to answer by mail or online) in part to try and rectify declining response rates to surveys of hunters and recreational fishers (Stedman et al. 2019). By allowing individuals to select their preferred mode for completing a questionnaire, it is believed that response rates should increase along with the quality of survey data (Hunt 2023). However, mixed-mode surveys are not a panacea (Millar and Dillman 2011), and they do not replace the need to follow standard practices and advice of experts when designing surveys (Dillman et al. 2014; Vaske et al. 2026a), constructing questionnaires (Dillman et al. 2014, Vaske et al. 2026b), and measuring key concepts through standard questions (Kyle et al. 2026). Furthermore, complementary to the standard off-site recreational fisher survey is to increasingly include experiments into survey design using principles from behavioural economics, such as nudges (see Mackay et al. 2026; van Putten et al. 2026). These experiments can, for example, help teams and other researchers identify cost-effective approaches to survey recreational fishers that trade off efforts to increase response rates against additional costs of implementing surveys (see Hunt 2023).

Another increasingly common approach for surveying is to place a survey online and encourage anyone to complete it (i.e. a non-probabilistic or convenience survey) or to rely on digital data only (Venturelli et al. 2026). Similarly, snowballing of respondents is common in qualitative research (Kochalski et al. 2026). Whereas this non-probabilistic sampling approach for surveying is useful to address specific questions or to contact hard to reach populations (Griffiths et al. 2010), this approach typically produces biased estimates (e.g. related to fishing avidity) for a population of interest due to selection biases (e.g. more avid fishers are most likely to complete the survey) and very low response rates (Howarth et al. 2024). In short, unless one has particularly effective ways of informed data weighing, non-probabilistic surveys cannot generalize to the population of interest, though there can be a strong temptation to do so. For example, a well-executed local snowball-based qualitative sample of recreational fishers will provide rich and detailed information on local discourse and even cause-and-effect reasoning by people, but it is impossible to generalize from this by designing a non-random sample to all recreational fisheries in a region. The same issue holds for quantitative convenience online surveys that are not weighted or otherwise bias-controlled. The researcher will need to properly frame their research results and clearly avoid population-level jargon in reports and papers. Generally, researchers may carefully think about how they will use the data from convenience surveys before using this "low cost" approach to collect data. There is potential for mixed-mode approaches where qualitative studies are used to inform a larger quantitative survey using probabilistic samples that may then provide more generalizable insights that hold across large populations of recreational fishers (Golebie et al. 2026).

The advent of the internet and the cost or absence of sampling frames of recreational fishers has increased the temptation of convenience sampling using internet surveys. Many such surveys are regularly released, to which people are invited through email lists or on social media. Many of these poorly designed surveys are created by people untrained in survey design, and these studies regularly enter natural science journals because reviewers may be similarly untrained in social sampling design. This issue has been raised long ago by human dimensions experts in recreational fisheries studies (Ditton 1996), but it continues to be a problem. In response, some leading biology-oriented journals have altered their policies and published papers calling for rigour in survey-based research (Kyle et al. 2020). Clearly, there is a large diversity of social science approaches to conservation and fisheries (Bennett et al. 2017), and thus sampling approaches differ. Independent of the tradition, carefully designing sampling, clearly understanding sampling units, and generalizing only to the population from which the sample originates is critically important for the robustness of our science.

A less common, but scientifically strong, method for understanding recreational fishers and their behaviours is to conduct experiments on recreational fishers and fisheries. Experiments have stronger inference than observational studies because variables of importance are actively manipulated and outcomes are compared against counterfactuals, which create stronger estimates for establishing causal pathways and impacts (Poteete et al. 2010; Janssen et al. 2010; see van Putten et al. 2026). This is especially valuable in social-ecological systems with many feedbacks. Fisher behaviour may be experimentally evaluated using revealed or stated preference approaches in surveys (see Melstrom et al. 2026) or using behaviours in direct response to treatments (van Putten et al. 2026). Revealed preference studies are used to explore how fishers make decisions across a range of natural or purposefully altered gradients in variables of interest. However, the lack of experimental methods used in the vast majority of revealed preference studies leads to the production of only associations that might not reflect causation between behaviours and variables of interest. Stated preference methods avoid this problem by using experimental methods to design hypothetical scenarios from which respondents choose a preferred alternative. Of course, stated preference outcomes may be biased as hypothetical or intended choices might differ from actual choices (see Melstrom et al. 2026; Landon et al. 2026). Many drawbacks of each method may be overcome by using combined revealed and stated preference models (Ben-Akiva and Morikawa 1990), multi-model inference, or Bayesian methods. Importantly, however, experiments can also be completed in the field at the system level, for which there are excellent examples in the recreational fisheries science (Stoop et al. 2012; Fujitani et al. 2017). For observational data, statistical methods of the field of causal inference should increasingly be used, which include clever ways of distilling cause-and-effect from natural experiments and other statistical approaches (Gangl 2010; Fujitani et al. 2012).

Finally, though it is often best to interact directly with the fishing public to understand their actions and motives, it is possible to instead gather the required information using existing sources of data. These may include archived information such as magazines, peer-reviewed publications, government documents, oral history, or broad-scale contemporary data that one can mine for useful information (Kochalski et al. 2026; Thurstan et al. 2026; Cooke et al. 2026). Historical information allows current social norms to be put in the context of their origins. New methods in digital data analysis from a field now called culturomics allow changes in attitudes and behaviours to be assessed and documented (Lennox et al. 2022; Thurstan et al. 2026). Alternatively, contemporary information such as digital data from fishing apps or websites is often focused on behaviours and outcomes (e.g. locations fished, species caught, gear used) rather than values or attitudes (Venturelli et al. 2026). Peer-reviewed and non-reviewed documents may be collected and analysed to gain broad insights into increasingly detailed processes related to angler motives and actions; specific guidelines for distinct levels of evidence synthesis and their analysis can help ensure appropriate rigour (see Cooke et al. 2026). Each of these options provides a wealth of information from which to better understand fisher values, behaviours, and motivations.

The wide variety of approaches available also suggests that many opportunities exist to combine approaches. Taking several approaches to understand the system provides a more complete understanding of the nuances within the system, while also providing opportunities for efficiencies in data collection. For example, can experiments be combined with outcomes from probabilistic surveys? Can a more complete understanding be gained by combining quantitative and qualitative data and data collection? There are multiple ways to engage in this mixed-modes approach to data collection and interpretation, each of which has different benefits and constraints (Golebie et al. 2026). This interweaving of approaches represents a crucial step forward for interdisciplinary collaboration and exploration of systems.

An alternative to collecting data is to actively co-produce knowledge among stakeholders and scientists. In the most elaborated variant of co-production, stakeholders become research partners (Arlinghaus et al. 2026). For co-production processes to work, scientists and stakeholders need to develop a trustworthy working relationship and design multiple phases to reflect on intermediate results so as to be adaptive to change routes of actions as new knowledge emerges. Transparency is key, as is a commitment to respecting each other and to not misusing information without consent. This also holds for publications. Importantly, for co-production to work, scientists must rethink their role, accept that stakeholders and their norms are equally important as the researcher's interest and values, and consider the research process as a truly two-way interaction where learning is pursued together. Whereas this is a delicate process when working with any stakeholders in this way, demands are particularly crucial when working with rights-holders. Although not all projects will demand or are suitable for co-production, if co-production is chosen, the research process can benefit from four phases: (1) listening to each other, (2) identifying common grounds and disagreement, which is often assisted by conducting probabilistic surveys, (3) collecting, analysing, and interpreting new data together, and (4) evaluating the entire process of co-production at multiple stages (Arlinghaus et al. 2026). Like many processes mentioned in this book, although co-production requires a commitment from all participants, the potential benefit to better understand recreational fishers is often well worth the effort and can result in better joint understanding of a topic through principles of collective intelligence (Arlinghaus and Krause 2013).

Today, especially in North America, researchers and managers often use creel interviews to gather on-site data on fisher catch and effort. Some surveys have continued for many years, either on a particular lake or in a particular region, providing a rich and extensive time series of data (de Kerckhove et al. 2026). Most surveys are conducted on a small scale, yet fisheries extend over larger geographic areas as fishers may travel extensive distances across regions and borders to fish (Martin et al. 2015). By standardizing these data, they can help examine new, broader questions (Sievert et al. 2026) and contribute to evidence-based syntheses such as meta-analyses (Cooke et al. 2026). This is particularly helpful given the expense of conducting creel surveys—giving them a second opportunity to contribute to broader inference that generates added value.

Whatever mode of data collection is used, one needs to follow norms of standardization and data reporting (Sievert et al. 2026). Similar to how the study of lake fish ecology benefitted from the use of standardized gill net sampling (Hansen et al. 2023; Giacomini et al. 2020), the field of human dimensions will benefit from standardization in survey formats and measurements scales, which includes data reporting and making datasets publicly available in an anonymous fashion. For example, survey researchers should agree on a standardized way of measuring various concepts of relevance in recreational fisheries studies, such as specialization, catch-orientation, motives and other, should apply the same scales whenever possible (e.g. see Kyle et al. 2026) in various surveys using the same question wording and answer scales. Data made publicly accessible facilitates evidence synthesis techniques that examine common patterns across recreational fisher populations (Cooke et al. 2026).

We end with something obvious, but often forgotten: within any given study, researchers must take due diligence to report basic research design (e.g. year of study, population of interest, sample statistics including sample sizes, sampling units, variances, and means as well as providing transparent information on survey formats, items lists, and construct reliability; Hunt et al. 2019). Standard reporting facilitates future integration into larger databases (Sievert et al. 2026) and potential meta-analysis (Cooke et al. 2026). Researchers and managers would be well served to consult with human dimensions researchers and adhere to survey standards and measurement scales as documented in this book and elsewhere (e.g. Dillman et al. 2014). We also do not talk further about the specific statistical and other methods that need to be followed, given that the literature is replete with directions and the methods are constantly developing. It is in researchers' best interests to follow the discipline-specific standards, but also to get inspired when applying solutions developed in a specific field (e.g. econometrics or psychometrics) and apply them to their own data sets. This demands open eyes and reading in different disciplines to learn about the newest technologies available to analyse complex recreational fisheries data sets. Many methods and tools that are mentioned in textbooks to analyse social science data with proprietary software (e.g. Vaske 2019) are today also possible in the open-access computing language.

3.2 Knowledge Integration

Especially in interdisciplinary projects, but also in co-production activities, knowledge harmonization and integration are key. This integration can happen in a number of ways: integrating diverse views and perspectives, integrating different disciplinary epistemologies, or integrating knowledge generated through a number of approaches or over time.

Knowledge integration and harmonization, both among stakeholders and scientists, or among scientific disciplines, can be facilitated through participatory modelling in workshop settings (Ehrlich et al. 2023; Wilberg et al. 2026) or other ways of devising models that integrate across disciplinary boundaries. This has been traditionally challenging as integrated models (bioeconomic or social-ecological systems models) are rarely true collaborations where two or more components are equally detailed and grounded in theory (Jensen et al. 2026). However, models can be an excellent synthesis tool, by providing a boundary object in which different knowledge streams inform different areas of the coupled model. Ideally, the development of all components, which should at least include fish and fisher dynamics, should each be developed by experts who each engage in careful, respectful, and continuous communication with one another. In a perfect world, the general model structure is devised at the start of the project, and parameters are developed or informed as the project unfolds and empirical data are collected. In that way, models are excellent synthesis opportunities for interdisciplinary cooperation, where each discipline contributes critical pieces. Model parameters can also integrate stakeholder knowledge (Ehrlich et al. 2023). Continued communication about model performance and uncertainties is the key to developing multiple parts that can easily and seamlessly work together. As with all models, minimizing complexity where possible may yield clearer results to general questions, although the level of detail should be based on the question being addressed (Jensen et al. 2026).

Participatory modelling is a special case of integrated model building and allows models to be built through discussions and collaboration with fishers. Here, problem-solving is achieved through the act of building models collaboratively with a team of fishers, managers, modellers, and decision-makers (Wilberg et al. 2026). This process helps to organize knowledge into that which is known and unknown, which may further lead to defined experiments, management strategy evaluation (Punt et al. 2014), or even adaptive management of the system (Walters 1986; Robinson et al. 2026). These models may involve building an understanding of different components of the social-ecological system or include the full integration of these different sub-models (Jensen et al. 2026). Building models with recreational fishers is challenging because of the need to keep fishers motivated, engaged, and valued, but it is possible (Ehrlich et al. 2023). This includes general suggestions such as using accessible language, developing visualizations, communicating model limitations, and fully evaluating participant information whenever possible (Wilberg et al. 2026). However, it also involves careful discussions regarding uncertainty, which stakeholders can find uncomfortable. Wilberg et al. (2026)

suggest instead framing discussions as the risk of exceeding unwanted thresholds. Finally, validating models is important and necessary to demonstrate the plausibility of models. In areas where models do not match expectations, it is important to learn why this occurs and consider revising the model. This process, like many involving knowledge integration and stakeholders, can take many meetings over months or years. It is important to remain realistic about expectations and timelines and maintain strong and respectful communications throughout the process.

The initial purpose of integrating knowledge is often to better understand fishers and the fishery overall, which enables better problem framing and ultimately improves management decision-making. However, some issues and decisions are hindered by large uncertainty in one or more processes, where "if you knew more, you might make a different decision" (Runge et al. 2011). This is an ideal situation for starting a co-production process where different disciplines, or science and society, collaborate to solve the critical uncertainties, for example, via adaptive management (Robinson et al. 2026). Adaptive management processes should necessarily involve fishers, who are impacted by, and may benefit from, the outcome of these processes. When fishers are included as partners in the adaptive management process, learning is enhanced by all parties (Fujitani et al. 2017). It should be noted that active adaptive experiments to learn about fisher behaviour or their intentions are rare (but see Mee et al. 2016, Fujitani et al. 2017), yet are an important source of reducing uncertainty in any decision process or to get answers to difficult fisheries questions that are controversially discussed. We emphasize that adaptive management is challenging to fully implement because of the long-term horizons and resources required; for this and other reasons, such experiments are often stopped prematurely or fail to deliver on the initial questions (Walters 2007). Therefore, it is important to fully partner with decision-makers, fishers, stakeholders, and rights-holders over a lengthy period and see adaptive management as a knowledge generation and integration process. Further, strong leadership and long-term resources are essential to success.

The final axis of knowledge integration is to relate knowledge currently being developed to knowledge in other areas or over time. This argues for data standardization, where the ways in which data are collected and stored are consistent. Although data standardization is valuable on its own, it is often seen as a critical component to data management that involves collecting and storing data, so that they are available for others to use. Data collected and documented, ideally in publicly and openly available databases, using the FAIR principles of data curation (Findable, Accessible, Interoperable, and Reusable; Wilkinson et al. 2016), are more useful in future research by others. Sievert et al. (2026) provide examples of how data standardization works for creel surveys, but these same principles apply to other collected data. This also demands a need for careful database creation, which helps structure data collection, but also helps reduce errors and redundancy across tables and researchers. Critically, knowledge integration will be massively improved if data sets generated locally and regionally become widely available, to allow large data synthesis projects in recreational fisheries to the same degree they exist in other fields of environmental science. This is currently a major shortcoming and generates incentives to waste a lot of time for no reason and not be able to generalize. For example, publicly publishing all data sets of angler surveys would be a major stepping stone for comparative analysis. The same holds for catch-and-release or management data, which would allow comparative analysis of key patterns that exist across regions or even globally.

4 Charting a Path Forward

Although it is not always necessary to address a fisheries problem with an interdisciplinary or transdisciplinary approach, this book provides ample evidence of the benefits these approaches will bring. Here we briefly summarize some of the key lessons to learn to move towards a more integrated approach in understanding recreational fishers to improve fisheries management.

The first is that interdisciplinarity demands a common framework for how to structure our beliefs on the fishery and the role that fishers play in it. We suggest viewing recreational fisheries as social ecological systems, which allows for explicit consideration of one or more groups of fishers, one or more populations or species of targeted species (fish and invertebrates), and potentially one or more management authorities (Lynch et al. 2017). Building this framework for any particular fisheries problem requires integration of disciplines that specialize in understanding dynamic interactions within aquatic ecosystems and one or more disciplines that specialize in understanding cognition and behaviour of humans and governance institutions, and greatly benefits from non-scientists who have lived experience in the system. The different disciplines involved often have different epistemologies (Hunt 2023), and there may be an "us versus them" attitude between scientists and non-scientists. Breaking down barriers between individuals and mental models requires a certain humility on the part of team members. The best approach to humility is to have an open mind to how different people view legitimate knowledge, which will help ease tensions and create opportunities for innovative ideas and approaches.

The second lesson is that although new interdisciplinary methods will develop and improve our understanding of recreational fishers from multiple perspectives, they should not necessarily replace traditional methods and data. New methods will only supplement and expand the quality of these metrics and data. Traditional methods and approaches, described in the second section of this book, have a rich history and will continue to inform our understanding of fisher values and behaviours. The desire to create new interdisciplinary approaches should not supplant continued research in each of the component disciplines.

Finally, it is important to view the knowledge and data gained from any research on recreational fishers as contributing to a broader understanding. Therefore, the value generated from any study or intervention should be seen as a long-term investment. That value is enhanced by using standardized methods and storing (and disseminating) data in publicly available databases (Sievert et al. 2026), which can then contribute to future evidence synthesis (Cooke et al. 2026). Further, investing in continually developing social capital among researchers and those invested in the system—stakeholders and rights-holders—will lead to continued research opportunities from interdisciplinary and co-management perspectives.

The study of recreational fishers and recreational fisheries social-ecological systems more generally has been limited by a distance between disciplines that should be working together (Fenichel et al. 2012; Solomon et al. 2020). The tools and suggestions provided in this book provide opportunities to break down barriers between disciplines and between researchers and stakeholders, which should lead to the development of novel interdisciplinary or transdisciplinary tools and projects. It is our sincere hope that this book helps take the first steps in this journey.

References

Aminpour P, Gray SA, Jetter AJ, Introne JE, Singer A, Arlinghaus R (2020) Wisdom of stakeholder crowds in complex social–ecological systems. Nat Sustain 3:191–199

Arlinghaus R, Krause J (2013) Wisdom of the crowd and natural resource management. Trends Ecol Evol 28(1):8–11

Arlinghaus, Robert, Thomas Mehner (2005) Determinants of management preferences of recreational anglers in Germany: habitat management versus fish stocking. Limnologica 35(1-2):2–17

Arlinghaus R, Cooke SJ, Lyman J, Policansky D, Schwab A, Suski C, Sutton SG, Thorstad EB (2007) Understanding the complexity of catch-and-release in recreational fishing: an integrative synthesis of global knowledge from historical, ethical, social, and biological perspectives. Rev Fish Sci 15:75–167

Arlinghaus R, Hunt LM, Post JR, Allen MS (2014) Not fish not meat: some guidance on how to study fisheries from an interdisciplinary perspective. In: Taylor WW, Lynch AJ, Léonard NJ (eds) Future of fisheries: perspectives for emerging professionals. American Fisheries Society, Bethesda, pp 223–230

Arlinghaus RJ, Beardmore AB, Daedlow K, Dorow M, Fujitani M, Hühn D, Haider W, Hunt L, Johnson BM, Johnston FD, Klefoth T, Matsumura S, Monk CT, Pagel T, Post JR, Rapp T, Riepe C, Ward H, Wolter C (2016a) Understanding and managing freshwater recreational fisheries as complex adaptive social-ecological systems. Rev Fish Sci Aquac 25(1):1–41

Arlinghaus R, Laskowski KL, Alós J, Klefoth T, Monk CT, Nakayama S, Schröder A (2016b) Passive gear-induced timidity syndrome in wild fish populations and its potential ecological and managerial implications. Fish Fish 18(2):360–373

Arlinghaus R, Alos J, Beardmore B, Daedlow K, Dorow M, Fujitani M, Hühn D, Haider W, Hunt LM, Johnson BM, Johnston F (2017) Understanding and managing freshwater recreational fisheries as complex adaptive social-ecological systems. Rev Fish Sci Aquac 25(1):1–41

Arlinghaus R, Fujitani M, Ehrlich E, Engel M, Cooke SJ (2026) Knowledge co-production in recreational fisheries science and management. In: Pope KL, Arlinghaus R, Hunt LM, Lynch AJ, van Poorten BT (eds) Understanding recreational fishers: disciplinary and interdisciplinary approaches for fisheries management. Springer Nature, Cham, p 533–572

Arnstein, S.R., 1969. A ladder of citizen participation. Journal of the American Institute of planners, 35(4):216–224

Ben-Akiva M, Morikawa T (1990) Estimation of switching models from revealed preferences and stated intentions. Transport Res A Gen 24(6):485–495

Bennett NJ, Roth R, Klain SC, Chan K, Christie P, Clark DA, Cullman G, Curran D, Durbin TJ, Epstein G, Greenberg A, Nelson MP, Sandlos J, Stedman R, Teel TL, Thomas R, Veríssimo D, Wyborn C (2017) Conservation social science: understanding and integrating human dimensions to improve conservation. Biol Conserv 205:93–108

Birdsong M, Hunt LM, Arlinghaus R (2021) Recreational angler satisfaction: what drives it? Fish Fish 22(4):682–706

Boucquey N, Fly J, Boonstra WJ, Drakopulos L, Leong KM, Marjadi MN, Wise S, Fairbanks L (2026) Political ecology for understanding recreational fishers and fisheries. In: Pope KL, Arlinghaus R, Hunt LM, Lynch AJ, van Poorten BT (eds) Understanding recreational fishers: disciplinary and interdisciplinary approaches for fisheries management. Springer Nature, Cham, p 197–232

Carpenter SR, Brock WA (2004) Spatial complexity, resilience, and policy diversity: fishing on lake-rich landscapes. Ecol Soc 9(1):8

Brown TL (1987) Typology of human dimensions information needed for Great Lakes Sport-Fisheries Management. T Am Fish Soc 116(3):320–324

Camp EV, van Poorten BT, Robinson KF, Harrison HL, NguyenV CCA, Lorenzen K, Casola W, Dedual M, Klizentyte K (2026) Integrating diverse stakeholder knowledge in fisheries decision making. In: Pope KL, Arlinghaus R, Hunt LM, Lynch AJ, van Poorten BT (eds) Understanding recreational fishers: disciplinary and interdisciplinary approaches for fisheries management. Springer Nature, Cham, p 607–660

Coggins LG Jr, Catalano MJ, Allen MS, Pine WE III, Walters CJ (2007) Effects of cryptic mortality and the hidden costs of using length limits in fishery management. Fish Fish 8(3):196–210

Cooke SJ, Twardek WM, Lennox RJ, Zolderdo AJ, Bower SD, Gutowsky LF, Danylchuk AJ, Arlinghaus R, Beard D (2017) The nexus of fun and nutrition: recreational fishing is also about food. Fish Fish 19:201–224

Cooke SJ, Venturelli P, Twardek WM, Lennox RJ, Brownscombe JW, Skov C, Hyder K, Suski CD, Diggles BK, Arlinghaus R, Danylchuk AJ (2021) Technological innovations in the recreational fishing sector: implications for fisheries management and policy. Rev Fish Biol Fish 31:253–288

Cooke SJ, Rytwinski T, Harper M, Howarth A, Nakagawa S, Hunt LM (2026) Evidence synthesis in recreational fisheries science and management. In: Pope KL, Arlinghaus R, Hunt LM, Lynch AJ, van Poorten BT (eds) Understanding recreational fishers: disciplinary and interdisciplinary approaches for fisheries management. Springer Nature, Cham, p 573–604

Cowx IG, Arlinghaus R, Cooke SJ (2010) Harmonizing recreational fisheries and conservation objectives for aquatic biodiversity in inland waters. J Fish Biol 76:2194–2215

de Kerckhove DT, Tucker CM, Gutowsky L, Morgan G, Guzzo M, Barajas VL (2026) Optimizing creel surveys. In: Pope KL, Arlinghaus R, Hunt LM, Lynch AJ, van Poorten BT (eds) Understanding recreational fishers: disciplinary and interdisciplinary approaches for fisheries management. Springer Nature, Cham, p 387–409

Dillman DA, Smyth JD, Christian LM (2014) Internet, phone, mail, and mixed-mode surveys: the tailored design method. Wiley, Hoboken

Ditton RB (1996) Human dimensions in fisheries. In: Ewert AW (ed) Natural resource management: the human dimension. Westview Press, Colorado, pp 73–90

Ditton RB, Hunt KM (2001) Combining creel intercept and mail survey methods to understand the human dimensions of local freshwater fisheries. Fisheries Manag Ecol 8:295–301

Eden S, Bear C (2011) Reading the river through 'watercraft': environmental engagement through knowledge and practice in freshwater angling. Cult Geogr 18:297–314

Ehrlich E, Bath AJ, Niessner D, Koemle D, Cyrus EM, Arlinghaus R (2023) Co-designing management recommendations with stakeholders: a case study about northern pike (*Esox lucius*) in the southern Baltic Sea. Fish Res 263:106687

Fenichel EP, Abbott JK, Huang B (2012) Modelling angler behaviour as a part of the management system: synthesizing a multi-disciplinary literature. Fish Fish 14:137–157

Fujitani ML, Fenichel EP, Torre J, Gerber LR (2012) Implementation of a marine reserve has a rapid but short-lived effect on recreational angler use. Ecol Appl 22:597–605

Fujitani M, McFall A, Randler C, Arlinghaus R (2017) Participatory adaptive management leads to environmental learning outcomes extending beyond the sphere of science. Sci Adv 3:e1602516

Fujitani ML, Riepe C, Pagel T, Buoro M, Santoul F, Lassus R, Cucherousset J, Arlinghaus R (2020) Ecological and social constraints are key for voluntary investments into renewable natural resources. Global Environ Chang 63:102125

Gangl M (2010) Causal inference in sociological research. Annu Rev Sociol 36:21–47

Giacomini HC, Lester N, Addison P, Sandstrom S, Nadeau D, Chu C, de Kerckhove D (2020) Gillnet catchability of walleye (*Sander vitreus*): comparison of North American and provincial standards. Fish Res 224:105433

Golden AS, van Poorten B, Jensen OP (2022) Focusing on what matters most: evaluating multiple challenges to stability in recreational fisheries. Fish Fish 23:1418–1438

Golebie E, van Riper CJ, Kochalski S, Nunan F (2026) Mixed methods research: applications and considerations for the human dimensions of recreational fisheries. In: Pope KL, Arlinghaus R, Hunt LM, Lynch AJ, van Poorten BT (eds) Understanding recreational fishers: disciplinary and interdisciplinary approaches for fisheries management. Springer Nature, Cham, p 481–506

Griffiths SP, Pollock KH, Lyle JM, Pepperell JG, Tonks ML, Sawynok W (2010) Following the chain to elusive anglers. Fish Fish 11:220–228

Hansen AG, Miller MW, Cristan ET, Farrell CJ, Winkle P, Brandt MM, Battige KD, Lepak JM (2023) Gill net catchability of walleye (*Sander vitreus*): are provincial standards suitable for estimating adult density outside the region? Fish Res 266:106800

Howarth A, Cooke SJ, Nguyen VM, Hunt LM (2024) Non-probabilistic surveys and sampling in the human dimensions of fisheries. Rev Fish Biol Fish 34. https://doi.org/10.1007/s11160-023-09831-1

Hunt LM (2023) Actively encouraging online responses to a mixed-mode mail and web survey: a case of nudging anglers in Ontario, Canada. Hum Dimens Wildl 29:121–128

Hunt LM, Sutton SG, Arlinghaus R (2013) Illustrating the critical role of human dimensions research for understanding and managing recreational fisheries within a social-ecological system framework. Fisheries Manag Ecol 20:111–124

Hunt LM, Camp E, van Poorten B, Arlinghaus R (2019) Catch and non-catch-related determinants of where anglers fish: a review of three decades of site choice research in recreational fisheries. Rev Fish Sci Aquac 27:261–286

Janssen MA, Holahan R, Lee A, Ostrom E (2010) Lab experiments for the study of social-ecological systems. Science 328:613–617

Jensen OP, Arlinghaus R, Golden AS, Janssen MA, Solomon CT, van Poorten BT (2026) Integrated models of the social-ecological dynamics of recreational fisheries. In: Pope KL, Arlinghaus R, Hunt LM, Lynch AJ, van Poorten BT (eds) Understanding recreational fishers: disciplinary and interdisciplinary approaches for fisheries management. Springer Nature, Cham, p 661–692

Johnson BM, Arlinghaus R, Martinez PJ (2009) Are we doing all we can to stem the tide of illegal fish stocking? Fisheries 34:389–394

Kelley K, Maxwell SE, Rausch JR (2003) Obtaining power or obtaining precision. Eval Health Prof 26:258–287

Kelly R, Mackay M, Nash KL, Cvitanovic C, Allison EH, Armitage D, Bonn A, Cooke SJ, Frusher S, Fulton EA, Halpern BS, Lopes PFM, Milner-Gulland EJ, Peck MA, Pecl GT, Stephenson RL, Werner F (2019) Ten tips for developing interdisciplinary socio-ecological researchers. Socio-Ecol Pract Res 1:149–161

Kochalski S, Barz F, Pita P, Harrison HL (2026) Quantitative and qualitative content analysis of text and images. In: Pope KL, Arlinghaus R, Hunt LM, Lynch AJ, van Poorten BT (eds) Understanding recreational fishers: disciplinary and interdisciplinary approaches for fisheries management. Springer Nature, Cham, p 507–532

Krogman RM, Taylor SM, Adams AJ, Boucek RE, Lombardo SM, Crandall CA, Steffen SF, Acosta A, Boné Morón E, Haukebo S, Hyder K, Radford ZM, Potts WM (2026) Role of recreational fisher information in fisheries management. In: Pope KL, Arlinghaus R, Hunt LM, Lynch AJ, van Poorten BT (eds) Understanding recreational fishers: disciplinary and interdisciplinary approaches for fisheries management. Springer Nature, Cham, p 21–54

Kyle G, Landon A, Vaske J, Wallen K (2020) Tools for assessing the psychometric adequacy of latent variables in conservation research. Conserv Biol 34:1353–1363

Kyle G, Landon AC, Pilgreen D, Golebie E (2026) Measurement instruments. In: Pope KL, Arlinghaus R, Hunt LM, Lynch AJ, van Poorten BT (eds) Understanding recreational fishers: disciplinary and interdisciplinary approaches for fisheries management. Springer Nature, Cham, p 355–385

Landon AC, Schroeder SA, Stefen SF, Arlinghaus R (2026) Social psychology for understanding recreational fishers and fisheries. In: Pope KL, Arlinghaus R, Hunt LM, Lynch AJ, van Poorten BT (eds) Understanding recreational fishers: disciplinary and interdisciplinary approaches for fisheries management. Springer Nature, Cham, p 157–196

Lennox, Robert J., Valerio Sbragaglia, Knut Wiik Vollset, Lene K. Sortland, Loren McClenachan, Ivan Jarić, Meaghan L, Guckian et al (2022) Digital fisheries data in the Internet age: Emerging tools for research and monitoring using online data in recreational fisheries. Fish and Fisheries 23(4):926–940

Lewin W-C, Arlinghaus R, Mehner T (2006) Documented and potential biological impacts of recreational fishing: insights for management and conservation. Rev Fish Sci 14:305–367

Löki V, Nagy J, Neményi Z, Hagyó A, Nagy A, Vitál Z, Mozsár A, Lukács BA (2023) Exploring ecological knowledge in recreational fishing for conservation purposes: a literature review. Glob Ecol Conserv 48:e02697

Lynch AJ, Cooke SJ, Beard TD Jr, Kao Y-C, Lorenzen K, Song AM, Allen MS, Basher Z, Bunnell DB, Camp EV, Cowx IG, Freedman JA, Nguyen JK, Nohner JK, Rogers MW, Siders ZA, Taylor WW, Youn S-J (2017) Grand challenges in the management and conservation of North American inland fishes and fisheries. Fisheries 42:115–124

Mackay M, van Putten EI, Richter A, Abatayo AL, Chuah S-H, Dankel DJ, Schadeberg A, Snyder H (2026) Behavioural economics for understanding recreational fishers and fisheries. In: Pope KL, Arlinghaus R, Hunt LM, Lynch AJ, van Poorten BT (eds) Understanding recreational fishers: disciplinary and interdisciplinary approaches for fisheries management. Springer Nature, Cham, p 261–284

Martin DR, Chizinski CJ, Pope KL (2015) Reservoir area of influence and implications for fisheries management. N Am J Fish Manage 35:185–190

Matsumura S, Beardmore B, Haider W, Dieckmann U, Arlinghaus R (2019) Ecological, angler, and spatial heterogeneity drive social and ecological outcomes in an integrated landscape model of freshwater recreational fisheries. Rev Fish Sci Aquac 27:170–197

McGinnis MD, Ostrom E (2014) Social-ecological system framework: initial changes and continuing challenges. Ecol Soc 19(2):30

Mee JA, Post JR, Ward H, Wilson KL, Newton E, Cantin A (2016) Interaction of ecological and angler processes: experimental stocking in an open access, spatially structured fishery. Ecol Appl 26:1693–1707

Melstrom RT, Jungers B, Bi X, Abbott JK (2026) Resource economics for understanding recreational fishers and fisheries. In: Pope KL, Arlinghaus R, Hunt LM, Lynch AJ, van Poorten BT (eds) Understanding recreational fishers: disciplinary and interdisciplinary approaches for fisheries management. Springer Nature, Cham, p 123–156

Millar MM, Dillman DA (2011) Improving response to web and mixed-mode surveys. Public Opin Q 75:249–269

Myers R, Taylor J, Allen M, Bonvechio TF (2008) Temporal trends in voluntary release of largemouth bass. N Am J Fish Manage 28:428–433

Nieman CL, Iwicki C, Lynch AJ, Sass GG, Solomon CT, Trudeau A, van Poorten B (2021) Creel surveys for social-ecological-systems focused fisheries management. Rev Fish Sci Aquac 29: 739–752

Niesar M, Arlinghaus R, Rennert B, Mehner T (2004) Coupling insights from a carp, *Cyprinus carpio*, angler survey with feeding experiments to evaluate composition, quality and phosphorus input of groundbait in coarse fishing. Fisheries Manag Ecol 11:225–235

O'Toole AC, Hanson KC, Cooke SJ (2009) The effect of shoreline recreational angling activities on aquatic and riparian habitat within an urban environment: implications for conservation and management. Environ Manag 44:324–334

Olaussen JO (2009) Bandwagon or snob anglers? Evidence from Atlantic salmon recreational fishing. Mar Resour Econ 24:387–403

Parkinson EA, Berkowitz J, Bull CJ (1988) Sample size requirements for detecting changes in some fisheries statistics from small trout lakes. N Am J Fish Manage 8:181–190

Peterman RM (1990) Statistical power analysis can improve fisheries research and management. Can J Fish Aquat Sci 47:2–15

Pope KL, Arlinghaus R, Hunt LM, Lynch AJ, van Poorten BT (eds) (2026) Understanding recreational fishers: disciplinary and interdisciplinary approaches for fisheries management. Springer Nature, Cham

Post JR (2012) Resilient recreational fisheries or prone to collapse? A decade of research on the science and management of recreational fisheries. Fisheries Manag Ecol 20:99–110

Post JR, Sullivan M, Cox S, Lester NP, Walters CJ, Parkinson EA, Paul AJ, Jackson L, Shuter BJ (2002) Canada's recreational fisheries: the invisible collapse? Fisheries 27:6–17

Post JR, Wilson KL, Johnston FD, Sass GG, Askey PJ, Ward HGM, Allen MS (2026) Ecology for understanding recreational fishers and fisheries. In: Pope KL, Arlinghaus R, Hunt LM, Lynch AJ, van Poorten BT (eds) Understanding recreational fishers: disciplinary and interdisciplinary approaches for fisheries management. Springer Nature, Cham, p 95–121

Poteete AR, Janssen MA, Ostrom E (2010) Working together: collective action, the commons, and multiple methods in practice. Princeton University Press, New Jersey

Punt AE, Butterworth DS, de Moor CL, De Oliveira JA, Haddon M (2014) Management strategy evaluation: best practices. Fish Fish 17:303–334

Reid AJ, Eckert LE, Lane JF, Young N, Hinch SG, Darimont CT, Cooke SJ, Ban NC, Marshall A (2020) "Two-eyed seeing": an indigenous framework to transform fisheries research and management. Fish Fish 22:243–261

Robinson KF, Arlinghaus R, Camp EV, Post JR, Stedman R, Sutton SG (2026) Adaptive management for recreational fisheries decisions in the face of uncertainty. In: Pope KL, Arlinghaus R, Hunt LM, Lynch AJ, van Poorten BT (eds) Understanding recreational fishers: disciplinary and interdisciplinary approaches for fisheries management. Springer Nature, Cham, p 719–746

Runge MC, Converse SJ, Lyons JE (2011) Which uncertainty? Using expert elicitation and expected value of information to design an adaptive program. Biol Conserv 144:1214–1223

Rypel AL, Lyons J, Griffin JD, Simonson TD (2016) Seventy-year retrospective on size-structure changes in the recreational fisheries of Wisconsin. Fisheries 41:230–243

Sass GG, Shaw SL (2019) Catch-and-release influences on inland recreational fisheries. Rev Fish Sci Aquac 28:211–227

Schafft M, Wegner B, Meyer N, Wolter C, Arlinghaus R (2021) Ecological impacts of water-based recreational activities on freshwater ecosystems: a global meta-analysis. P Roy Soc Lond B Bio 288:20211623

Sievert N, Krogman RM, Embke H (2026) Data standardisation and management to facilitate large-scale and interdisciplinary approaches. In: Pope KL, Arlinghaus R, Hunt LM, Lynch AJ, van Poorten BT (eds) Understanding recreational fishers: disciplinary and interdisciplinary approaches for fisheries management. Springer Nature, Cham, p 747–782

Solomon CT, Dassow CJ, Iwicki CM, Jensen OP, Jones SE, Sass GG, Trudeau A, van Poorten BT, Whittaker D (2020) Frontiers in modelling social-ecological dynamics of recreational fisheries: a review and synthesis. Fish Fish 21:973–991

Stedman RC, Connelly NA, Heberlein TA, Decker DJ, Allred SB (2019) The end of the (research) world as we know it? Understanding and coping with declining response rates to mail surveys. Soc Nat Resour 32:1139–1154

Stoop J, Noussair CN, van Soest D (2012) From the lab to the field: cooperation among fishermen. J Polit Econ 120:1027–1056

TenHarmsel HJ, Boley BB, Irwin BJ, Jennings CA (2019) An importance-satisfaction analysis of trout license holders in Georgia. N Am J Fish Manage 39:1227–1241

TenHarmsel HJ, Boley BB, Irwin BJ, Jennings CA (2021) Perceived constraints and negotiations to trout fishing in Georgia based on angler specialization level. N Am J Fish Manage 41:115–129

Thurstan RH, Chong-Montenegro C, Christensen J, Clark A, Gaynor A (2026) Environmental history and historical ecology for understanding recreational fishers and fisheries. In: Pope KL, Arlinghaus R, Hunt LM, Lynch AJ, van Poorten BT (eds) Understanding recreational fishers: disciplinary and interdisciplinary approaches for fisheries management. Springer Nature, Cham, p 233–259

van Poorten BT, Camp EV (2019) Addressing challenges common to modern recreational fisheries with a buffet-style landscape management approach. Rev Fish Sci Aquac 27:393–416

van Poorten BT, Arlinghaus R, Daedlow K, Haertel-Borer SS (2011) Social-ecological interactions, management panaceas, and the future of wild fish populations. P Natl Acad Sci USA 108: 12554–12559

van Putten EI, Trudeau A, Mackay M, Obregón C, Post JR, Richter A, Schuch E, Solomon CT (2026) Economic experiments in recreational fishing studies. In: Pope KL, Arlinghaus R, Hunt LM, Lynch AJ, van Poorten BT (eds) Understanding recreational fishers: disciplinary and interdisciplinary approaches for fisheries management. Springer Nature, Cham, p 411–442

Vaske JJ (2019) Survey research and analysis, 2nd edn. Sagamore-Venture Publishing, Champaign

Vaske JJ, Needham MD, Miller CA (2026a) Fundamentals of sampling and implementing human dimensions of fishing surveys. In: Pope KL, Arlinghaus R, Hunt LM, Lynch AJ, van Poorten BT (eds) Understanding recreational fishers: disciplinary and interdisciplinary approaches for fisheries management. Springer Nature, Cham, p 327–353

Vaske JJ, Needham MD, Miller CM, Sponarski C (2026b) Guidelines for writing and constructing recreational fisher surveys. In: Pope KL, Arlinghaus R, Hunt LM, Lynch AJ, van Poorten BT (eds) Understanding recreational fishers: disciplinary and interdisciplinary approaches for fisheries management. Springer Nature, Cham, p 287-325

Venturelli PA, Hyder K, Skov C (2016) Angler apps as a source of recreational fisheries data: opportunities, challenges and proposed standards. Fish Fish 18:578–595

Venturelli P, Skov C, Audzijonyte A, Belmaker J, Ana Cardoso AS, Gordon L, Gundelund C, Hyder K, Jarić I, Lennox RJ, Potts WM, Vacura K, Vaz AS, Sbragaglia V (2026) Digital data mining. In: Pope KL, Arlinghaus R, Hunt LM, Lynch AJ, van Poorten BT (eds) Understanding recreational fishers: disciplinary and interdisciplinary approaches for fisheries management. Springer Nature, Cham, p 443–479

Wagner T, Irwin BJ, Bence JR, Hayes DB (2013) Detecting temporal trends in freshwater fisheries surveys: statistical power and the important linkages between management questions and monitoring objectives. Fisheries 38:309–319

Walters CJ (1986) Adaptive management of renewable resources. MacMillan Publishing Company, New York

Walters CJ (2007) Is adaptive management helping to solve fisheries problems? Ambio 36:304–307

Ward Hillary GM, Micheal S, Allen, Edward V, Camp, Nick Cole, Len M, Hunt, Bryan Matthias, John R, Post, Kyle Wilson, Robert Arlinghaus (2016) Understanding and managing social–ecological feedbacks in spatially structured recreational fisheries: the overlooked behavioral dimension. Fisheries 41(9):524–535

Watson AR, Blount C, McPhee DP, Zhang D, Smith ML, Reeds K, Williamson JE (2022) Source, fate and management of recreational fishing marine debris. Mar Pollut Bull 178:113500

Wilberg MJ, Irwin BJ, Jones ML (2026) Participatory modelling to support evaluation of management actions for recreational fisheries. In: Pope KL, Arlinghaus R, Hunt LM, Lynch AJ, van Poorten BT (eds) Understanding recreational fishers: disciplinary and interdisciplinary approaches for fisheries management. Springer Nature, Cham, p 693–718

Wilkinson MD, Dumontier M, Aalbersberg IJ, Appleton G, Axton M, Baak A, Blomberg N, Boiten JW, da Silva Santos LB, Bourne PE, Bouwman J, Brookes AJ, Clark T, Crosas M, Dillo I, Dumon O, Edmunds S, Evelo CT, Finkers R, Gonzalez-Beltran A, Gray AJG, Groth P, Goble C, Grethe JS, Heringa J, 't Hoen AC, Hooft R, Kuhn T, Kok R, Kok J, Lusher SJ, Martone ME, Mons A, Packer AL, Persson B, Rocca-Serra P, Roos M, van Schaik R, Sansone S-A, Schultes E, Sengstag T, Slater T, Strawn G, Swertz MA, Thompson M, van der Lei J, van Mulligen E, Velterop J, Waagmeester A, Wittenburg P, Wolstencroft K, Zhao J, Mons B (2016) The FAIR guiding principles for scientific data management and stewardship. Sci Data 3: 160018

Glossary

Actor in fisheries, any person of civil society—including academics, managers, fishers, and policymakers—engaging with or being affected by recreational fisheries and aquatic ecosystems.

Actor, rational an assumed way that people behave such that they make rational choices based on calculations and the information that is available to them.

Adaptive management a process for iteratively making management decisions and monitoring system responses in such as a way as to systematically reduce uncertainty and increase learning in fisheries systems. Inherently and explicitly views iterative management decisions as experimental treatments.

Adaptive management, active a type of adaptive management in which intentional experiments are used to learn about a suite of alternative hypotheses of the system to inform management decisions. In contrast to passive adaptive management, management actions are used as experimental treatments to resolve alternative hypotheses about the system.

Adaptive management, passive a type of adaptive management in which implementation of adaptive management that seeks to understand the effect of a management action on the system, for example by implementing the model-predicted best option and monitoring to evaluate outcomes or by less rigorously monitoring the effects of past management actions and changing future procedures based on new knowledge. Unlike active adaptive management, hypotheses and treatments are not defined a priori.

Affect the feeling states experienced by humans including physiological reactions, expressive reactions, behavioural tendencies, and emotional experiences.

Analysis, confirmatory factor (CFA) a theory-driven approach that specifies (hypothesizes) the number of factors and the pattern of relationships in advance (Brown 2015).

Analysis, content a research method that, through a systematic classification process of coding, reduces complexity, handles large volumes of data, and uncovers meaningful themes and patterns, offering researchers a structured approach to

K. L. Pope et al. (eds.), *Understanding Recreational Fishers*, Fish & Fisheries Series 45, https://doi.org/10.1007/978-3-031-99739-6

organize, analyse, and interpret images, photos, videos, and textual data (Hsieh and Shannon 2005; Patton 2002).

Analysis, exploratory factor (EFA) a data-driven approach that does not specify the number of factors or the pattern of relationships between the common factors and the indicator (Brown 2015).

Analysis, meta- the quantitative re-analysis of previously published independent findings and studies on the same subject to reveal overall trends.

Analysis, multi-criteria decision a formal structure for making decisions that must account for multiple objectives or criteria and may be measured on different scales.

Analysis, sentiment a type of natural language processing that identifies, extracts, and quantifies the opinions, attitudes, and emotions that are present in a sample of text.

Anonymous personal data (including contact information) that cannot be associated with other data collected from people.

Archive (a) a collection of primary sources selected for preservation.(b) a physical place (e.g. a building or organization) where primary sources are kept.

Archive, embodied the recognition that historical knowledge and sources also exist in current human practices, requiring interrogation of the subjective self and individual or group experiences and practices for the purpose of historical enquiry.

Artefact (a) a material expression of communication, such as recordings, documents, images, and objects (Finnegan 2014).(b) a human-created object of cultural or historical significance.

Attitude a subjective evaluation of an object, person, group, issue, or concept on a dimension ranging from negative to positive. Attitudes provide summary evaluations of target objects and are often assumed to be derived from specific beliefs, emotions, and past behaviours associated with those objects.

Avidity a measurement based on the amount of fishing activity expended by recreational fishers, often measured for each individual fisher as days per year.

Baseline a reference or starting point that one can use to evaluate changes in people's behaviour or the state and condition of ecosystems across extended timescales (i.e. multiple years to centuries).

Behaviour an organism's (including people's) activities in response to external or internal stimuli, including objectively observable activities, introspectively observable activities, and nonconscious processes.

Behavioural control the ability or perceived ability that a person has for engaging or not engaging in a behaviour such as going fishing or following a regulation.

Behavioural interventions strategies and approaches designed to influence and modify the behaviour of individuals.

Belief a person's thoughts or cognitions about what is factual or true whether it is demonstrably factual or true in objective reality. It constitutes the personal acceptance of the truth, reality, or validity of something.

Bias occurs when parameter estimates derived from a sample are not representative of the population.

Bias, avidity the potential to oversample the most active users of the fishery because they are the most likely to encounter in the field and are often most willing to engage and provide responses with a variety of survey methodologies.

Bias, cognitive a systematic pattern of deviation from norm or rationality in judgement.

Bias, nonresponse checks the process of contacting a sample (often at least 10%) of nonrespondents and asking them questions from the questionnaire to see if respondents and non-respondents differ from each other.

Bias, prestige the potential for recreational fishers to overestimate their activity and catches to represent themselves as a preconceived desire to be a particular type of fisher.

Bias, recall the unintentional misrepresentation of past events through imperfect recollection and the tendency to homogenize similar memories, or round numbers.

Bias, social desirability a bias that occurs when respondents give answers to questions that they believe will make them look good to others or give answers they believe the interviewer wants to hear.

Bounded rationality an assumption that humans make decisions that depart from perfect economic rationality due to limits in our mental capacity, the information available to us, and time.

Catchability the proportion of a fish population captured per unit of fishing activity (effort).

Catch-per-unit-effort (CPUE) the number of fish caught in a fishery divided by the amount of time spent fishing (i.e. fishing effort). Often expressed in units of number of fish per fisher-hour.

CATI a survey data collection approach that uses computer-assisted telephone interviewing.

Co-creation an approach whereby scientists, policymakers, managers, and stakeholders co-design or co-develop projects, co-create knowledge, and co-implement results.

Co-delivery an approach whereby stakeholders and actors work together with scientists and managers to disseminate and apply the results of science-based knowledge.

Co-design an approach whereby scientists, stakeholders, and others work together to define research questions, approaches, and solutions to management.

Co-learning a joint learning process of scientists, managers, and stakeholders.

Co-management a process of management in which government or other decision makers and rights holders share power with resource users, with each given specific rights and responsibilities relating to information and decision making. Describes approaches for governing fisheries spanning from complete government control to self-governance (Sen and Neilsen 1996).

Co-production a process in which scientists, policymakers, managers, and stakeholders work together in knowledge generation and where stakeholders are directly involved in data collection, analysis, and interpretation.

Co-production, empowering creating partnerships that lead to better management and conservation.

Co-production, outcome-oriented outcome-oriented (improved understanding of an issue)

Co-production, practical and pragmatic increase of relevance, scope, and usability of science through building of adaptive capacity.

Co-production, transformative better and more democratic management of fish resources through synergies across knowledge systems.

Coding the systematic process of assigning categories, labels, or codes to units of data, such as words, phrases, sentences, or paragraphs (Gibbs 2007).

Cognition mental processes, such as ones' thoughts, intentions, values, beliefs, attitudes, feelings, memory, and learning.

Collapse a substantial reduction in the abundance of a fish population (less than 10–20% of unfished abundance) or other units including regulatory, economic, or other human systems.

Colonialism a set of practices or policies instigated to establish control over groups of people within a geographical area.

Compensation where due to one or more processes, the per capita fish production increases as populations decline in abundance.

Conditional logit a variant of the multinomial logit statistical model whereby the attributes of each choice alternative are allowed to vary (e.g. expected catch can vary among fishing sites).

Confidentiality in a survey context, where survey responses remain private by removing identification numbers and destroying address lists after the study ends. See *anonymous*.

Conflict a disagreement or argument, typically a protracted one, that takes one of two forms: (i) goal interferences whereby the actions of an individual (or group) prevent another individual from achieving their desired outcome from an experience, and (ii) social values whereby the actions of an individual or group violate fundamental norms of another group.

Congestion the number of people in a given area at a given time. The density for the measure is not agreed upon.

Constraint factors that interfere with individuals' satisfaction with or participation in outdoor recreation activities like fishing. Research has classified leisure constraints as: intrapersonal, interpersonal, and structural.

Constructivism one of a range of paradigms (worldviews) in modern science and social science that share key sets of ontological, epistemological, and methodological beliefs. Constructivists posit that multiple realities may exist simultaneously. For constructivists, knowledge emerges in a dialectical (back-and-forth) relationship between individuals and their world and is always tied to specific places and contexts (Lincoln et al. 2017; Obermeister 2017; Slevitch 2011).

Consumer surplus the benefit that accrues to fishers (and society) that is measured as the difference between the willingness to pay for an experience and the actual costs that are paid.

Corpus traditionally, a collection of a large amount of naturally occurring language (spoken or written) formatted to allow observation of patterns. Now, it is also used to describe collections of other forms of communication such as images, videos, and objects (Hunston 2022), Plural form is *corpora.*

Cover letter introductory text to a questionnaire that describes the study and attempts to persuade a person to answer a questionnaire and take the project seriously.

Creel optimization changes to protocol or analysis of an ongoing creel survey program to leverage a desired outcome that is not currently being well addressed.

Critical realism one of a range of paradigms (worldviews) in modern science and social science that share key sets of ontological, epistemological, and methodological beliefs. Critical realists maintain that an independent material reality exists. However, they argue that our knowledge of this reality must always be partial because it is filtered through human thought, which is inevitably situated in historical, cultural, and social contexts (Lincoln et al. 2017; Proctor 1998).

Crowding perception of an individual of the total number and density of people in a given area.

Data, cross-sectional observations made at one point in time (Babbie 2016).

Data, fisheries dependent data collected from ongoing fisheries operations and typically supplied by active fishers.

Data collection, active a form of fisher-generated digital data collection in which fishers are aware that their data are of interest to researchers or managers (e.g. use of fishing apps or websites within a citizen science framework, and surveys and questionnaires that can be shared *en masse* through general and specialized social media).

Data collection, passive a type of fisher-generated digital data collection in which fishers are unaware (or do not care) that their data are of interest to researchers or managers (e.g. aggregate data from general, online activity, and fisher-specific data from social media platforms and connected devices).

Data management, integrable an approach to conducting data management using methods that facilitates the integration of the dataset with others.

Data management lifecycle a strategy for managing data that incorporates the process of planning, data acquisition and cleaning, data architecture design, data analysis and reporting, and data sharing and archival.

Data mining, digital the process by which digital data are gathered, managed, and analysed to learn about the world or make predictions.

Database, noSQL or *newSQL* a modern database management system for reliably organizing and querying large volumes of growing and changing data in a manner that is both flexible and scalable.

Decision making, participatory a sub-category of participatory approaches that focuses on giving some decision-making power to stakeholders.

Decision making, structured a specific process in which decision makers and stakeholders can work together to define the problem, clearly articulate the objectives to be achieved, identify a set of alternatives or actions that could be implemented to achieve the objectives, predict the consequences of each alternative on each objective, and make trade-offs among objectives.

Denaturalize to deconstruct the taken-for-granted nature of something—a ubiquitous idea, discourse, or ideology, for example. For instance, political ecologists have been instrumental in denaturalizing tragedy-of-the-commons narratives.

Depensation where, due to one or more processes, the per capita production decreases as animal populations decline in abundance.

Design, complementary a mixed-methods research design that integrates qualitative and quantitative data to triangulate results and address the same research question (Creswell et al. 2003).

Design, embedded a mixed-methods research design that includes quantitative or qualitative data embedded within a larger design. Considered a variation of the complementary, exploratory, or explanatory design (Creswell et al. 2003).

Design, explanatory a mixed-methods research design in which qualitative research is used to provide a deeper understanding of data collected during a quantitative phase (Creswell et al. 2003).

Design, exploratory a mixed-methods research design in which quantitative research follows a qualitative phase, and is either informed by the qualitative phase, or seeks to understand the generalizability of qualitative results (Creswell et al. 2003).

Design, three-phase a mixed-methods research design including three phases of inquiry that build on each other, in either a qualitative → quantitative → qualitative form, or a quantitative → qualitative → quantitative form (Creswell et al. 2003).

Device, connected an electronic tool that is connected to the internet or other electronic tools via wireless networks and protocols (e.g. smartphones, boat-based electronics such as navigational devices and fish finders, drones, cameras, and smart rods).

Device, mobile a portable electronic tool that is compact, battery-powered, and capable of wireless communication (e.g. smartphones, tablets, and wearable devices).

Dialectical pluralism a research paradigm that accounts for multiple underpinning philosophies, attempts to give equal respect and consideration to these philosophies, emphasizes dialogue between qualitative and quantitative perspectives, and seeks to create a research style based on this dialogue (Johnson 2017).

Digital humanities the use of digital sources, methods, or technologies across humanities fields.

Disciplinary specialization and in-depth content knowledge of a certain type of curriculum or subject.

Discourses collections of texts, images, or other media that collectively advance narratives, ideologies, or themes (Adger et al. 2001; Neumann 2004).

Ecology a field, for the context of understanding recreational fishers, that is grounded in well-established theories of predator-prey dynamics through which the behaviours of both predator and prey jointly govern the outcomes.

Ecology, historical a research framework or community of practice that employs historical approaches to understand past human influences on ecosystems, and their social and ecological outcomes through time.

Ecology, political most succinctly, a field of study that examines how power and politics function in the relationship between socioeconomic and environmental changes in particular places.

Economic impact a macroeconomic measure relying on expenditures associated with recreational fishing and how these direct expenditures cycle through local, regional, or national economies through resulting purchases by service and product suppliers (indirect expenditure) and induced expenditures by individuals employed by the service or product supplier.

Economics, behavioural a field that can help understand how recreational fishers are likely to respond and behave (Wieczorek et al. 2021) by combining economic principles with knowledge from behavioural sciences, especially psychology, which broadly examines human behaviour and mental processes, and social psychology, which focuses on how behaviour is affected by the social and physical environment.

Economics, neoclassical a broad approach that attempts to explain the production, pricing, consumption of goods and services, and income distribution through supply and demand.

Economics, resource a field that examines the economic value and socioeconomic consequences of decisions that impact natural resources. For recreational fishing, these decisions can range from fishing at a favourite spot, to constructing public boats ramps, to putting restrictions on harvest.

Economy, moral the understanding that particular resource users have of the "right" or "just" way to use those resources.

Effect, order a measurement issue whereby the order of questions can influence how people answer.

Effect, primacy selecting from the first few answers in a "check all that apply" question.

Effect, recency a measurement issue whereby survey respondents select from the last few answers in a "check all that apply" question.

Elite capture a condition in which a cooperative stakeholder process falls under the control of one or more powerful stakeholders or organization and therefore, ceases to become more representative of the stakeholders at large.

End-arounds a condition in which a process for integrating diverse stakeholder knowledge into decision making is subverted near the actual decision, often by disgruntled stakeholders applying pressure on decision-makers to discount or discard whatever approach has been followed.

Epistemology theory of knowledge which is concerned with all aspects of the validity, scope, and methods of acquiring knowledge, such as what constitutes a knowledge claim and how knowledge can be produced or acquired (Moon and Blackman 2014). Epistemologies are influenced by the ontological beliefs of a discipline (its theories on the nature of reality), and in turn affect a discipline's methodology (i.e. beliefs about how researchers can best obtain knowledge in the world) (Lincoln et al. 2017; American Psychological Association 2023b).

Evidence the available body of facts, knowledge, or information indicating whether a belief or proposition is true or valid. Often referred to as the *evidence base*.

Expansion a process by which estimates for discrete periods are generated based on calculations using counts, samples, and interview data.

Experiment a quantitative research method to test hypotheses in a controlled setting.

Experiment, choice a survey-based research method to understand how people make decisions and trade-offs when choosing between goods or services that are described by different levels of attributes (e.g. expected catch and travel costs).

Experiment, field an experiment that is conducted in a real-world setting with some elements being controlled.

Experiment, lab-in-the-field an experiment conducted in real world environment but using a lab highly controlled experimental approach.

Experiment, laboratory (lab) an experiment conducted under highly controlled conditions.

Experiment, natural observational studies where there is no control over the independent variable as it occurs naturally in real life.

Facilitators members of a human group whose primary responsibilities are process and progress.

Factor (construct) an unobservable variable that statistically influences one or more observed measures and accounts for correlations among observed measures (indicators). This term is used within the context of confirmatory factor analysis but is generally synonymous with latent variable (Kline 2016; Brown 2015).

Factor loading a measure that quantifies the magnitude of the relationship between observed variable and its latent factor (Yang 2010).

Fair principles principles of data management that emphasize the findability (discoverability), accessibility, interoperability, and reusability of a dataset to facilitate future use and integration.

Fisheries management, ecosystem-based a holistic approach to fisheries management that recognizes all the interactions within a system—including economic, cultural, and social benefits—rather than considering a single species or issue in isolation.

Fisheries management action an activity, typically an intervention, conducted to achieve a fisheries management objective (e.g. instituting a protective harvest regulation or quota, stocking fish, restoring habitat, or adding a fishing access point).

Fisheries management goal a broad, typically qualitative, statement of benefits desired from a fishery resource.

Fisheries management objective a statement of what must be achieved to attain a fisheries management goal, ideally in terms of SMART (i.e. specific, measurable, achievable, realistic, and time bound).

Fishery, recreational a type of social-ecological system in which recreational fishing occurs, for which the primary motivation is for leisure and enjoyment and where the catch is usually not of substantial relevance for personal survival.

Fishing, artisanal traditional fisheries involving fishing households (as opposed to commercial companies), using relatively small amount of capital and energy, relatively small fishing vessels (if any), making short fishing trips, close to shore, and mainly for local consumption (FAO 2022).

Fishing, catch-and-release intentional release, whether voluntary or regulated, of caught fish after capture.

Fishing, commercial fishing undertaken primarily for sale of fish on the market, though it can include trade (OECD 2022a).

Fishing, recreational the act of attempting to catch fish or other aquatic animals with leisure as the primary motivation as opposed to meeting basic nutritional needs or sustaining livelihoods (OECD 2022b).

Fishing, subsistence generally understood to mean fishing for personal or household consumption, often in the context of maintaining cultural or historical traditions (OECD 2022c).

Fishing effort the amount of fishing activity expended by recreational fishers, typically measured for a population of fishers, and often expressed in hours or days fished.

Flat file a two-dimensional database that stores data in an easy-to-access tabular format (e.g. spreadsheets and comma-separated values [CSV] files).

Food web composition the relative abundance of various species within a species community.

Goodness of fit the extent to which observed data are predicted (reproduced) by the hypothesized model (Anderson and Gerbing 1988; Hu and Bentler 1998; Kline 2016).

Governance, self- the case in which stakeholders holding actual or effective complete power for governing their own fisheries and management agencies playing a minimal role if any.

Governance, top-down a non-participatory approach in which government decision makers (e.g. management agencies) essentially hold all decision-making power.

Governmentality the tactics and strategies of government via stated or unstated rules and norms.

Hermeneutics a philosophical approach that belongs to interpretivism. Its key assumption being that language has a non-verbalized meaning that the researcher can reconstruct (Moon and Blackman 2014).

Heuristic mental shortcuts that allow people to solve problems and make judgements quickly and efficiently.

Historical record sources of information about the past.

History, environmental the study of how humans have interacted with the natural world in the past and the outcomes for people and nature in the past and the present.

History, more-than-human an account of the past that does not place humans at the centre but instead emphasizes multiple species and processes.

History, oral the recording and preservation of oral testimony produced with the intention of creating a record of a person or group's experience of the past.

Homo economicus a hypothetical person who behaves in exact accordance with their rational self-interest.

Hyperdepletion rapid decreases in catch rate early in the development of a fishery followed by low but stable catch rates.

Hyperstability maintenance of relatively high catch rates despite declines in the abundance of fish within a fishery until collapse at low density.

Ideal free distribution a theory positing that individual predators will choose prey patches that minimize competition and maximize their fitness (i.e. patches that provide the highest consumption rates). It assumes that predators have perfect knowledge of the quality of all patches, that predators are ideal (e.g. choosing optimally), selecting patches that maximize their fitness, and free, implying no costs or constraints on patch choice.

Image the visual representation of concepts, people, or objects, as opposed to other meanings of the word image, such as mental images or brand images (Van Leeuwen and Jewitt 2001; Merriam-Webster n.d.).

Indicator (measure) (a) an observed variable used as an indirect statistical measure of a construct.(b) term used within the context of confirmatory factor analysis but is generally synonymous with manifest variable (Kline 2016).

Industrial revolution the shift from an agrarian economy dominated by production by hand tools, to an economy dominated by the use of machines in industrial manufacturing.

Institution (a) a social arrangement that is stable and predictable, created and maintained to serve the needs of society.(b) all rules in use by humans and their societies, whether formal or informal.

Instrumental drivers or motives behaviour initiated by a desire to avoid punishment (e.g. fines) or to receive positive rewards.

Interdisciplinary a research approach that combines more than one discipline throughout all phases of a research process to develop new information using new perspectives.

Internet panels self-selected individuals who willingly join survey panels, provide their contact information, and complete questionnaires on various topics several times per year.

Interpretivism a philosophical perspective that observes that human and social reality cannot be understood objectively but depend on the context and people's perspective (Moon and Blackman 2014).

Involvement a state of arousal, motivation, or interest towards an activity.

Justice, contributive or recognitive the recognition of contributions to the collective good.

Justice, distributive the fairness that people experience in relation to their context, in particular concerning the distribution of material or immaterial benefits.

Justice, environmental the distribution of benefits and harms that people experience due to their use and dependence on natural environments.

Justice, procedural the inclusiveness of collective decision making.

Knowledge a familiarity, awareness, or understanding of objects, processes, skills, or facts.

Knowledge, local in the context of fisheries, it describes multiple ways of knowing about fisheries systems that generally do not derive from formal or academic education, but come from sociocultural heritage, tradition, and lived experience. Includes terms such as *local traditional knowledge* (LTK), *local ecological knowledge* (LEK), *Indigenous knowledge* (IK), *Indigenous and local knowledge* (ILK), *community knowledge* (CK), and *translational ecology* (TE).

Knowledge co-production iterative and collaborative processes involving diverse types of expertise, knowledge, and actors (commonly involving scientists and practitioners) to produce context-specific knowledge and pathways towards a sustainable future (adapted from Norström et al. 2020).

Leaders members who prioritize issues, determine major directions, and help connect others.

Learning, double-loop, or institutional in the adaptive management framework, revisiting the elements and structure of the decision itself to evaluate changes that may have arisen, including in the problem framing, suite of objectives, actions that could be taken, and the predictive models.

Learning, two-way an attribute of participatory approaches in which both professional researchers or decision makers and stakeholders or rights holders learn about the fishery system and each other.

Lemmatization grouping of inflected word forms (e.g. "fished", "fishing,", and "will fish") so that they can be analysed as a single item (Ghavimi 2019).

Libertarian paternalism the idea that it is both possible and legitimate for private and public organizations to affect behaviour while also respecting freedom of choice.

Machine learning a type of artificial intelligence in which a computer uses statistical models to independently recognize patterns and make inferences.

Measurement instrument in surveying, it is another name for a questionnaire (Blalock 1979).

Method, birthday a sampling method that involves selecting an individual for a sample based on who had the most recently passed birthday or has the most recent upcoming birthday.

Method, contingent value a research method that values recreational site access and quality changes by measuring the hypothetical payments fishers are willing to pay to realize a good or service not currently provided, or are willing to pay to prevent loss of a good or service that is currently provided.

Method, Kish a sampling method that involves listing all members of a household (primary sampling units), by gender and age (or sometimes name and age), and then selecting one member (secondary sampling unit) from the list using previously developed random selection tables.

Method, travel cost a research method that values recreational site access and quality changes by measuring the relationship between travel cost, potentially other site attributes, and the sites individuals choose to visit.

Method, Waksberg a two-stage cluster design or sequential process used to reduce unproductive calls in telephone surveys. A listing of all area codes and prefix numbers in the area of interest are obtained in the first stage.

Methodology assumptions about the best means to obtain the most accurate knowledge about the world. Closely related to epistemology (Lincoln et al. 2017).

Methods, mixed research underpinned by both qualitative and quantitative paradigms that typically involves integrating multiple forms or phases of data collection, analysis, and interpretation.

Methods, revealed preference the use of actual, observed behaviour and expenditures to measure economic value.

Methods, stated preference using hypothetical or intended behaviour to predict choices of people including their willingness to pay.

Model, common factors a formal proposition that assumes each indicator in a set of observed measures is a linear function of one or more common factors and one unique factor (Thurstone 1947; Brown 2015).

Model, measurement a statistical model that defines the relationships between observed indicators and unobserved factors and evaluates how well those factors are measured by the indicators (i.e. provides empirical estimates of psychometric properties) (Kline 2016).

Model, mental cognitions of what a person or a group of people believe about an issue as well as how components of a system are related.

Model, operating a numerical model that describes the population, fishery, and management.

Model, random utility maximization a modelling framework that uses a utility function to relate behaviours and decisions to individual preferences, characteristics, and choice attributes, with individuals choosing the choice alternative with the highest utility. Observed characteristics and attributes are represented in the utility function by explanatory variables, whereas unobserved factors are represented as random variables.

Model, structural equation a family of statistical analyses that assess hypothesized causal connections among a set of latent variables and causal connections linking the latent variables to observed measures (Kline 2005).

Modellers members of a research group whose primary responsibilities are the construction and use of an operating or estimation model, commonly in mathematical terms but can also be conceptual.

Modelling, collaborative an approach to encourage stakeholders to contribute their beliefs and knowledge of the fishery system to the shared knowledge of the system.

Modelling, participatory a collaborative problem-solving approach that involves shared learning with a group of participants and a modelling team.

More-than-human a term used to de-centre humans in studies of social-ecological systems and to describe assemblages of humans and the living and non-living entities with which we share our world (Whatmore 2006).

Motivations the desired or expected outcomes that drive individuals' interest in activities.

Multidisciplinary a research approach that involves different disciplines working in parallel to accomplish a collaborative goal defined by the perspective of individual disciplines or by the desire to solve a collective problem (Stock and Burton 2011; Tress et al. 2004).

Narrative the practice of writing history as a story-based form.

Natural language processing a type of artificial intelligence in which computers use algorithms and models to understand and interpret human language (e.g. speech from online videos or text from fisher forums).

Noise ecological, environmental, or social stochasticity in a system that is irreducible.

Nomological network hypothesized pattern of relationships among variables that is inclusive of antecedents, mediators, and outcomes (Cronbach and Meehl 1955; Hagger et al. 2017).

Non-governmental organization (NGO) an organization, often non-profit, that works outside of official government channels to further its cause.

Normative motives or drivers behaviour initiated from the internalization of an attitude because it is coherent with the individual's value and belief system (e.g. social norms).

Norms shared standards of acceptable behaviour that imply an obligation. More simply, norms reflect what is considered "normal" in each culture or situation.

Norms, descriptive perceptions of which behaviours are typically performed by others.

Norms, injunctive people's perceptions of what behaviours are approved or disapproved by others.

Norms, personal own perceptions of acceptable behaviour.

Norms, social expectations of an individual what is believed to be expected from others about one owns behaviour.

Objectives specific statements about what managers want to achieve, often grounded in perceptions of what participants want to be able to experience.

Ontology deals with questions about the nature of what exists (Lincoln et al. 2017), and how different aspects of being are related to each other (American Psychological Association 2023a). Perspectives range from objectivity to a subjective reality that exists only in the minds of individuals (Moon and Blackman 2014).

Options actions that can be taken to solve a problem (e.g. regulations or stocking in management as a problem).

Orientation, catch a multidimensional attitudinal concept incorporating activity-specific attitudes related to catching fish, including: (i) catching something, (ii) catching many fish, (iii) catching big fish, and (iv) retaining (or releasing) fish.

Orientation, value patterns of beliefs about a topic, and the directionality and intensity of those beliefs. They are more specific and directional than values.

Parsing breaking up a sentence or paragraph into its separate components that are valid according to a given grammar (Ghavimi 2019).

Participant in fisheries, any person taking part in recreational fisheries or its management in some way. These individuals can offer insights on values, risks, and system form and functions.

Participant observation the ethnographic research technique involving the researcher spending (usually extensive) time engaged with research subjects (e.g. spending time recreationally fishing with a group of recreational fishers being studied), then documenting their observations and experiences (Bernard 2017).

Participatory approaches a category of approaches for integrating stakeholder knowledge into decision making that involves non-scientists in research or decision making itself.

Participatory knowledge creation a sub-category of participatory approaches that focuses on involving stakeholders in the development of knowledge.

Performance measures operating model outcomes that allow the comparison of options and the determination of whether an option meets an objective. Usually, there is at least one performance measure per option.

Place attachment a bond between people and their environments (Moore and Graefe 1994) that develops through meanings that individuals attach to places (Stedman 2002).

Positivism one of a range of paradigms (worldviews) in modern science and social science that share key sets of ontological, epistemological, and methodological beliefs. Positivists believe in a single true reality, which humans can come to know via objective empirical study. The positivist methodology is that of experimentation and hypothesis-testing (Lincoln et al. 2017).

Post-positivism a research paradigm that assumes that knowledge can only be generated from objective empirical observations obtained through the scientific method (Moon and Blackman 2014).

Power informs three different social scientific traditions that study power. A conflictual view on power focuses on power-over and domination. A consensual view focuses on power-to and power-with. A constitutive view focuses on the power of social structures (i.e. how these constitute people as social subjects with certain dispositions that define their power) (Haugaard 2014).

Power-over an asymmetrical relation between two or more people or groups (Pansardi and Bindi 2021).

Power-to the ability of a person to achieve certain outcomes (Pansardi and Bindi 2021).

Power-with the ability of a group of people that act together to achieve collective outcomes (Pansardi and Bindi 2021).

Pragmatism a research paradigm that can be underpinned by multiple philosophies and that advocates for an approach that "works best" and is most practical for addressing the research question. Such approaches are often *mixed methods* (Shan 2022).

Preference an evaluative judgement about liking an object or outcome over another. Methodologically, inferences can be made about what a person prefers from observing behaviours or asking about behavioural intentions.

Preference heterogeneity variation in the aspects of an experience as desired by people. This heterogeneity can be divided into observed (e.g. related to characteristics of a fisher) and unobserved (must be accommodated during modelling).

Prenotification letter a letter that alerts potential respondents to expect a mailed questionnaire in the future.

Psychometrics a field concerned with the quantification and measurement of mental attributes, behaviour, and design, analysis, and improvement of the tests, questionnaires, and other instruments used in such measurement (American Psychological Association 2018).

Qualitative a research paradigm that uses words or images as data, seeks to understand and interpret meaning, and tends to be theory generating and inductive (Braun and Clarke 2013).

Quantitative a research paradigm that uses numbers as data, seeks to identify relationships between variables, and tends to be theory testing and deductive (Braun and Clarke 2013).

Question, close-ended (i.e. fixed-scale) question that may include: (i) ordered response choices in a particular order or gradation, (ii) discrete unordered response choices, or (iii) partially close-ended response choices such as those with an extra option to write in such as "other, please specify".

Question, double-barrelled contains two different questions or components but only requires one answer (e.g. the question "Do you agree or disagree that fish and other marine species [e.g., crabs, sea stars, oysters] will benefit from the new fishing restrictions?" is double-barrelled because a respondent may agree that fish will benefit, but disagree that other marine species will benefit, or vice versa).

Question, open-ended a question that does not provide response alternatives from which to select a particular answer. Rather, respondents write or verbally express their answers however they choose.

Question, partially close-ended a question that offers a compromise between open-ended and close-ended questions. These questions give respondents flexibility to choose from a set of researcher-determined responses or create their own answers.

Question stem the most important part of a question in a questionnaire. The question stem provides explicit information about what is being asked, how people are to respond (e.g. "please check one"), and definitions, examples, or additional instructions (Roe 2008).

Questionnaire packet a survey mail package that typically contains the: (i) envelope where documents are contained, (ii) personalized and signed cover letter on letterhead, (iii) questionnaire, (iv) addressed and personally stamped postage-paid return envelope, and (v) sometimes a token incentive.

Random digit dialling (RDD) a sampling approach that uses internet or other directories to determine area codes and prefixes, and then randomly selects the next four digits of a telephone number to call.

Recreational fisher information information derived from or about recreational fishers, directly or indirectly, including but not limited to their fishing preferences, behaviours, opinions, and experiences as it pertains to a fishery resource.

Reliability the extent to which an instrument yields consistently repeatable scores (Cronbach 1951; Raykov 1997).

Research, archival the search for and extraction of information from historical primary sources including written documents, images, and material objects.

Research, collaborative an approach whereby researchers involve stakeholders in the scientific research, especially the collection and analysis of data. Includes terms such as *citizen science*, *co-creation*, and *community-based monitoring*.

Research, integrated research that involves direct collaboration among researchers from different disciplines. Includes both interdisciplinary and transdisciplinary research (Stock and Burton 2011; Tress et al. 2004).

Respondent burden a goal of surveying to ensure that questionnaires minimize the effort for individuals to answer a questionnaire. The goal is often operationalized by: (i) keeping completion time to a minimum, (ii) decreasing time that respondents need to think about questions, and (iii) respecting respondents by ensuring they will not be embarrassed by not understanding what is expected.

Response, fisher effort analogous to an ecological numerical response where fishers shift effort from fisheries providing lower benefit to fisheries providing greater benefit. From the perspective of a single fishery, this would be perceived as more fishers are active when the expected benefits are greater.

Response, functional a relationship between prey abundance (or density) and resulting prey consumption rates by predators.

Response, numerical an aggregation of predators in prey patches with greater abundance or production.

Response rate the proportion of completed questionnaires to the total number of eligible respondents, which indicates success in gaining cooperation of people.

Responses, mutually exclusive responses that cannot occur at the same time (e.g. a person cannot be a fisher and a non-fisher at the same time).

Responses, mutually exhaustive responses that comprise the entire range of possible outcomes.

Retrospective account an account about an event or happening that occurs sometime after the event in question.

Review, literature the synthesis of evidence using a variety of subjective or objective methods depending on the rigour or methods used.

Review, rapid a form of systematic review where decisions are made to exclude certain steps to increase the speed and reduce the cost of a traditional systematic review.

Review, systematic a synthesis of the evidence on a specific topic using critical methods to identify, define, and assess research on the topic and identify general trends. Methods are repeatable and transparent based on expert-determined best practices. Can be done using a narrative approach or may include an embedded meta-analysis. Considered to be the "gold standard" of evidence synthesis which is the norm in health care.

Rights holders indigenous peoples or recreational fishing communities with legally binding fishing rights.

Sample, non-probabilistic a non-representative subset of a population that is generated without ensuring that each member of the population has an equal or known chance of being included in the subset.

Sample, probabilistic (probability) a representative subset of a population that is generated by ensuring that each member of the population has an equal or known chance of being included in the subset. Creel and mail surveys are expected to generate probabilistic samples if they are properly designed and executed, which allows results to be inferred to the population from which the sample was selected.

Sampling the process of selecting population units (e.g. people) and collecting data that describe the larger population based on a selection of that population.

Sampling, cluster a sampling approach that involves conducting a random sample of groups or clusters (e.g. households) and then sampling people within the selected clusters.

Sampling, disproportionate deliberately sampling subgroups at rates unequal to their size in the population (i.e. over- or under-sample groups).

Sampling, multistage a sampling approach that uses both primary (e.g. households) and secondary (e.g. individuals within households) sampling units and a combination of sampling techniques (e.g. cluster sample combined with stratified random sample).

Sampling, representativeness in the sample selection an outcome of sampling that occurs when all constituents in the population have an equal chance (or known chance that is accounted during analyses) of being selected.

Sampling, simple random a sampling approach whereby people are selected so that each member of the population has an equal chance of being selected.

Sampling, single-stage sampling approach (can be simple random, systematic, or stratified random sampling) for which there is a single set of primary strata or sampling units (i.e. non-nested strata).

Sampling, stratified random a random sampling approach with two steps: (i) dividing the population into nonoverlapping groups (i.e. strata) that are of interest because of project objectives or hypotheses, and then (ii) selecting a random sample from each stratum or group.

Sampling, systematic a sampling approach that randomly selects the first person in a population, choosing an appropriate fixed interval, and systematically selecting subsequent individuals.

Sampling interval the distance between people selected in the sample (e.g. every fifth fisher encountered along a stream).

Sampling ratio the proportion of units in the population that are selected (e.g. ⅕ of fishers encountered along a stream).

Satisfaction a measure of the net benefit that arises when outcomes from an experience meet or exceed the expectations from that experience (*dissatisfaction* occurs when expectations are not met).

Scalar politics the social and political struggles that occur either between institutional scales (such as local, regional, and national), or in the development of framings of scale itself (e.g. small-scale fisheries).

Scale (a) subsets of questions within the questionnaire that measure a particular concept (Blalock 1979).(b) the temporal or spatial extent of a problem.

Scale, bipolar a response scale with options in two directions (e.g. strongly disagree, disagree, neither disagree nor agree, agree, strongly agree).

Scale, unipolar a survey measurement approach whereby a scale has responses in one direction (e.g. not at all important, slightly important, moderately important, and very important).

Scenarios alternative versions of the operating model that represent different hypotheses about how processes in the operating model work, or alternative outcomes based on potential decision choices. These are usually developed to address critical uncertainties.

Semi-structured interviews a qualitative research method whereby the researcher develops a set of structured (usually open-ended) questions to ask participants, but conversations are allowed to evolve and grow beyond the set of prepared questions (Bernard 2017).

Shifting baseline syndrome a systematic change in accepted norms (the baseline) over time against which a system or component of a system is measured (Pauly 1995).

Size structure the relative abundance of fishes of various sizes within a population.

Skip patterns within questionnaires, the instructions for respondents who to skip over and not answer one or more questions that do not apply to them given their responses to previous questions.

Social construction a term used to describe how the often taken-for-granted concepts around us (e.g. "clean" or "wild") do not reflect an objective reality, but rather socially created meanings that are very particular to specific times, places, and cultures (Proctor 1998).

Social psychology a field that is broadly cited as the study of how an individual's thoughts, feelings, and actions are affected by the actual, imagined, or symbolically represented presence of other people (Mitchell 2009).

Source, primary sources written at the time of or shortly after an event, either by direct witnesses or those who present views expressed at the time.

Source, secondary a data source where the accounts written after a historical event, and which are based on the collation and interpretation of primary sources.

Source comparison the cross-comparison of multiple sources with the aim to determine their consistency or incompatibility.

Source criticism the critical examination of the content and context of historical sources, with the aim to determine their value and limitations as historical evidence.

Sources, material artefacts, human-made or natural structures, and archaeological finds that are used in historical research.

Specialization multidimensional measure of commitment to recreation activities that incorporates psychological, cognitive, and behavioural components that can help summarize the heterogeneity within a fisher population.

Stakeholders in fisheries, individual, or group that has an interest in any decision or activity regarding fisheries resources. This term is offensive to some groups (e.g. Indigenous peoples are rights holders not stakeholders).

Standardization the process by which various elements of a study or survey are made to be consistent with each other.

State dependence in resource economics, fidelity to selecting a site (e.g. a fishing site) that is unrelated to the modelled attributes of the site.

Stationarity the assumption that natural systems and processes fluctuate around an average. Researchers are finding that stationarity is often not present, as global climate change and other system variables are causing the underlying mechanisms that drive these systems to evolve, leading to *non-stationarity*. Non-stationarity in social processes is possible as well if objectives, and potentially values, change over time.

Stemming the process of reducing a group of related words to their stem, base, or root form so that they can be analysed as a single item (Ghavimi 2019).

Survey, creel a survey conducted to primarily gather information about recreational fishers and the catch and harvest of fish.

Survey, electronic the use of electronic correspondence to obtain web-based survey responses.

Survey, fisheries independent typically, a biological survey conducted outside the normal operations of fishers using standardized sampling gears and methods appropriate to estimate the abundance and size-structure of a fish population.

Survey, mail survey delivered by mail to an individual from the sample, and thus, no direct interaction between the researcher and the respondent exists.

Survey, mixed mode the use of more than one type of survey method (e.g. on-site and mailed or mailed and electronic).

Survey, off-site creel surveys involving a remote interaction with recreational fishers through web, telephone, mobile app, or mail-in protocols, which typically rely on recreational fishers recalling fishing activities or self-reporting activity, catch, and harvest.

Survey, on-site creel surveys involving direct interaction with recreational fishers through counting, interviewing, or catch sampling that typically involve the presence of a survey crew on the waterbody containing the fishery.

Survey, on-site or intercept survey conducted at the research site (e.g. a river). Typically one page in length and can be self-administered or completed by an interviewer who asks the respondent the questions.

Survey, telephone survey administered over the phone by a trained interviewer.

Survey, voluntary a type of survey approach that allows people to choose whether to participate without any consequence.

Survey fatigue loss of respondents' interest in a survey caused by numerous survey requests or lengthy questionnaires that often leads to rushed completion and low response rates.

System, complex adaptive a system characterized by multiple nonlinear feedbacks acting across heterogeneous individuals functioning over multiple spatial- and temporal-scales.

System, social-ecological a complex system that includes interconnections and feedbacks between one or more ecological subsystems and one or more mutually dependent social, social-economic, or socio-technical subsystems (and governance organizations that affect both) (Ostrom 2009; Arlinghaus et al. 2017).

Systematic map a transparent, robust, and repeatable method to identify and collect relevant literature to a research question in policy or management. A useful method to show the quantity and quality of evidence available rather than yielding the answer to a specific question. Often used as a pre-cursor to systematic review.

Theoretical saturation the empirical confidence in conceptual categories when the same research hypotheses appear again and again in the advanced stages of data analysis and no new hypotheses emerge (Glaser and Strauss 1967).

Timeframes the period over which respondents should aggregate their experiences.

Tokenization the process of dividing sections of text into substrings, words, or other meaningful units that can be used for parsing (Ghavimi 2019).

Transdisciplinary (a) combining approaches from different disciplines and fields with the goal to develop entirely new theoretical or practical frameworks (Austin et al. 2008).(b) process of knowledge production, which includes both scientists and non-scientists, with the goal of generating socially relevant outcomes for sustainability, through consideration of different forms of knowledge and different scientific disciplines, often in co-creation by scientists and society.

Travel cost the monetary costs of travelling to and from a site that include vehicle operating costs and often a cost for the travel time.

Triangulation a process of comparing different types of data, observers, methods, or theories relevant to the same research question to identify similarities and differences, reduce bias, and ultimately increase the credibility and validity of research results.

Uncertainty, critical uncertainty identified by the participatory modelling group that is considered important enough to explicitly model.

Uncertainty, parametric uncertainty in parameter estimates used in a model of a system.

Uncertainty, structural uncertainty about biological, ecological, or social processes of a system, represented as a set of hypotheses about the state of knowledge.

Utility a latent (unobserved) measure related to well-being that arises to people such as to a recreational fisher from a fishing experience.

Validity the extent to which the instrument measures what it claims to measure (Cronbach and Meehl 1955; Campbell and Fiske 1959; Brown 2015).

Validity, construct an overarching principle of validity that refers to the extent to which a psychological measure, test, or instrument measures the concept it purports to measure (Brown 2015).

Validity, convergent measures of theoretically similar or overlapping constructs being strongly interrelated (Brown 2015).

Validity, discriminant measures of theoretically distinct constructs not highly intercorrelated (Brown 2015).

Validity, external the validity of applying the conclusions of a scientific study outside the context of that study.

Validity, internal the extent to which a piece of evidence supports a claim about cause and effect, within the context of a particular study.

Valuation, contingent a survey-based research method where a researcher asks whether an individual would be willing to pay a certain amount of money for an item or a quality change.

Valuation, nonmarket valuing environmental goods and services that are not traded in a market.

Value positive or negative evaluations of abstract ideals that transcend specific contexts that are held by individuals or present within a society. Values are also seen as underlying thoughts that people use to examine themselves, other people, and occurrences for which actions are upheld. In contrast to *attitudes*, values are few, strongly held, and enduring.

Value, economic (a) monetary value of, for example, a day of recreational fishing that includes the costs of fishing plus the willingness to pay more than those costs to enjoy the good.(b) the trade-off that individuals are willing to make for a particular good or service given the current distribution of resources.

Value, expected of perfect information a method to quantify the gains to be made by reducing uncertainty by comparing the expected value of implementing an action after the uncertainty has been resolved with the expected value of implementation of an action under complete uncertainty.

Value, expenditure the actual amount paid to use or interact with an environmental good or service.

Value, marginal the change in value from an incremental change in the amount or quality of a good or service.

Value, non-use the value that individuals passively experience from a natural resource's existence, which can be for reasons such as bioethics, cultural heritage, or altruism towards others.

Value, surplus the remaining amount of economic value that an individual has from consuming a good or service after expenditure.

Value, total economic an overarching measure the value of an item that includes the aggregated benefits all individuals obtain through use of the environmental good (use values) as well as the benefits from the good they do not actually use or encounter (passive-use or non-use values).

Value, use the value that individuals derive from actively using or interacting with a good.

Variable, dependent a variable whose value depends on the value of other variables.

Variable, independent a variable whose value does not depend on the value of other variables.

Variable, latent (unobserved variable) a variable with a theoretical basis presumed to reflect (explain) a concept that cannot be directly observed or measured (Kline 2016).

Variable, manifest (observed variable) an observed measure that underlies a latent variable it is presumed to represent (i.e. the data collected) (Kline 2016).

Willingness to pay the maximum amount of income an individual is willing to pay for a product or service.

References

Adger WN, Benjaminsen TA, Brown K, Svarstad H (2001) Advancing a political ecology of global environmental discourses. Dev Chang 32:681–715

American Psychological Association (2018) Psychometrics. APA dictionary of psychology. APA. https://dictionary.apa.org/psychometrics. Accessed 6 Apr 2023

American Psychological Association (2023a) Ontology. APA dictionary of psychology. APA. https://dictionary.apa.org/onology. Accessed 6 Apr 2023

American Psychological Association (2023b) Epistemology. APA dictionary of psychology. APA. https://dictionary.apa.org/epistemology. Accessed 6 Apr 2023

Anderson JC, Gerbing DW (1988) Structural equation modeling in practice: a review and recommended two-step approach. Psychol Bull 103:411–423

Arlinghaus R, Alós J, Beardmore B, Daedlow K, Dorow M, Fujitani M, Hühn D, Haider W, Hunt LM, Johnson BM, Johnston F (2017) Understanding and managing freshwater recreational fisheries as complex adaptive social-ecological systems. Rev Fish Sci Aquac 25(1):1–41

Austin W, Park C, Goble E (2008) From interdisciplinary to transdisciplinary research: a case study. Qual Health Res 18:557–564

Bernard HR (2017) Research methods in anthropology: qualitative and quantitative approaches. Rowman & Littlefield, New York

Blalock H (1979) Social statistics, 2nd edn. McGraw Hill, New York

Brown TA (2015) Confirmatory factor analysis for applied research, 2nd edn. Guildford Press, New York

Campbell DT, Fiske DW (1959) Convergent and discriminant validation by the multitrait-multimethod matrix. Psychol Bull 56(2):81

Cronbach LJ (1951) Coefficient alphas and the internal structure of tests. Psychometrika 16(3): 297–334

Cronbach LJ, Meehl PE (1955) Construct validity in psychological tests. Psychol Bull 52:281–302

FAO (2022) Artisanal fisheries. Family farming knowledge platform. FAO. https://www.fao.org/family-farming/detail/en/c/335263/. Accessed 28 Oct 2022

Finnegan R (2014) Communicating: the multiple modes of human communication. Routledge

Ghavimi P (2019) Big data analytics methods: analytics techniques in data mining, deep learning and natural language processing. De Gruyter, Boston/Berlin

Gibbs GR (2007) Thematic coding and categorization. In: Analyzing qualitative data. Sage Publications, Ltd, London, pp 38–55

Glaser B, Strauss A (1967) Grounded theory: the discovery of grounded theory. Sociology 12(1): 27–49

Hagger MS, Gucciardi DF, Chatzisarantis NLD (2017) On nomological validity and auxiliary assumption: the importance of simultaneously testing effects in social cognitive theories applied to health behavior and some guidelines. Front Psychol 8:1933

Haugaard M (2014) Power. In: Gibbons MT, Coole D, Ellis E, Ferguson K (eds) The encyclopedia of political thought. Wiley, Oxford, pp 2965–2978

Hsieh HF, Shannon SE (2005) Three approaches to qualitative content analysis. Qual Health Res 15(9):1277–1288

Hu LT, Bentler PM (1998) Fit indices in covariance structure modeling: sensitivity to underparameterized model misspecification. Psychol Methods 3(4):424

Hunston S (2022) Corpora in applied linguistics. Cambridge University Press

Kline TJB (2005) Psychological testing: a practical approach to design and evaluation. Sage Publications, Thousand Oaks

Kline RB (2016) Principles and practice of structural equation modeling, 5th edn. Guilford Press, New York

Lincoln YS, Lynham SA, Guba EG (2017) Paradigmatic controversies, contradictions, and emerging confluences, revisited. In: Denzin NK, Lincoln YS (eds) The SAGE handbook of qualitative research. Sage Publications, Los Angeles, pp 97–128

Merriam-Webster (n.d.) Image. https://www.merriam-webster.com/dictionary/image. Accessed 17 Sept 2023

Mitchell JP (2009) Social psychology as a natural kind. Trends Cogn Sci 13(6):246–251

Moon K, Blackman D (2014) A guide to understanding social science research for natural scientists. Conserv Biol 28(5):1167–1177

Neumann RP (2004) Moral and discursive geographies in the war for biodiversity in Africa. Polit Geogr 23:813–837

Obermeister N (2017) From dichotomy to duality: addressing interdisciplinary epistemological barriers to inclusive knowledge governance in global environmental assessments. Environ Sci Pol 68:80–86

OECD (2022a) Commercial fishing. Glossary of Statistical Terms. https://stats.oecd.org/glossary/detail.asp?ID=2990. Accessed 28 Oct 2022

OECD (2022b) Recreational fishers. Glossary of Statistical Terms. https://stats.oecd.org/glossary/detail.asp?ID=2255. Accessed 28 Oct 2022

OECD (2022c) Subsistence fishing. Glossary of Statistical Terms. https://stats.oecd.org/glossary/detail.asp?ID=2600. Accessed 28 Oct 2022

Ostrom E (2009) A general framework for analyzing sustainability of social-ecological systems. Science 325:419–422

Pansardi P, Bindi M (2021) The new concepts of power? Power-over, power-to and power-with. J Political Power 14(1):51–71

Patton MQ (2002) Two decades of developments in qualitative inquiry: a personal, experiential perspective. Qual Soc Work 1(3):261–283

Pauly D (1995) Anecdotes and the shifting baseline syndrome of fisheries. Trends Ecol Evol 10(10): 430

Proctor JD (1998) The social construction of nature: relativist accusations, pragmatist and critical realist responses. Ann Assoc Am Geogr 88(3):352–376

Raykov T (1997) Scale reliability, Cronbach's coefficient alpha, and violations of essential tau equivalence with fixed congeneric components. Multivar Behav Res 32:329–353

Roe D (2008) Question stem. In: Lavrakas PJ (ed) Encyclopedia of survey research methods. Sage Publications, Thousand Oaks

Slevitch L (2011) Qualitative and quantitative methodologies compared: ontological and epistemological perspectives. J Qual Assur Hosp Tour 12:73–81

Thurstone LL (1947) Multiple factor analysis. University of Chicago Press, Chicago

Van Leeuwen T, Jewitt C (eds) (2001) The handbook of visual analysis. Sage, Los Angeles

Whatmore S (2006) Materialist returns: practising cultural geography in and for a more-than-human world. Cult Geogr 13:600–609

Wieczorek AM, Schadeberg A, Krogh Hallin J, van Putten I, Kraak SBM, Richter A, Clay PM, Goti Aralucea L, Pedreschi D, Hamon KG, Dankel DJ, Mackay M (2021) Behavioural economics in fisheries: a systematic review protocol. PLoS One 16(8):e0255333

Yang H (2010) Factor loadings. In: Salkind NJ (ed) Encyclopedia of research design. Sage Publications, Thousand Oaks, pp 481–483

Index

K. L. Pope et al. (eds.), *Understanding Recreational Fishers*, Fish & Fisheries Series 45, https://doi.org/10.1007/978-3-031-99739-6

D

GPSR Compliance

The European Union's (EU) General Product Safety Regulation (GPSR) is a set of rules that requires consumer products to be safe and our obligations to ensure this.

If you have any concerns about our products, you can contact us on ProductSafety@springernature.com

In case Publisher is established outside the EU, the EU authorized representative is:

Springer Nature Customer Service Center GmbH
Europaplatz 3
69115 Heidelberg, Germany

Batch number: 09423678

Printed by Printforce, the Netherlands